GALACTIC RADIO ASTRONOMY

INTERNATIONAL ASTRONOMICAL UNION
UNION ASTRONOMIQUE INTERNATIONALE

SYMPOSIUM No. 60
HELD AT MAROOCHYDORE, QUEENSLAND, AUSTRALIA,
3–7 SEPTEMBER, 1973

GALACTIC RADIO ASTRONOMY

EDITED BY

F. J. KERR AND S. C. SIMONSON, III
University of Maryland, College Park, Md., U.S.A.

D. REIDEL PUBLISHING COMPANY
DORDRECHT-HOLLAND/BOSTON-U.S.A.

1974

Published on behalf of
the International Astronomical Union
by
D. Reidel Publishing Company, P.O. Box 17, Dordrecht, Holland

Sold and distributed in the U.S.A., Canada, and Mexico
by D. Reidel Publishing Company, Inc.
306 Dartmouth Street, Boston,
Mass. 02116, U.S.A.

Library of Congress Catalog Card Number 74–81939

ISBN-13:978-90-277-0502-0   e-ISBN-13:978-94-010-2263-7
DOI: 10.1007/978-94-010-2263-7

TABLE OF CONTENTS

PART 2 / GALACTIC H II REGIONS

PART 3 / SUPERNOVA REMNANTS

PART 4 / STELLAR AND CIRCUMSTELLAR SOURCES

PART 5 / THE GALACTIC CENTER

PART 6 / LARGE-SCALE GALACTIC STRUCTURE

PREFACE

'Galactic Radio Astronomy' was chosen as the subject of this Symposium, which was held in conjunction with the IAU General Assembly that took place in Sydney in August 1973, largely because it is a very suitable Southern Hemisphere topic. This results in part from the advantages of a southern location in studying the Galaxy and in part from the long association of Australia with radio astronomy.

Following the General Assembly, the Symposium was held at the Surfair International Hotel in Maroochydore, Queensland, from 3 to 7 September, 1973. The conference participants were effectively isolated from the rest of the world during the Symposium, and the excellent spring weather and geographical situation led to the development of an unusually good rapport.

The Symposium was sponsored by Commissions 40, 33, and 34. The Organizing Committee was composed of A. H. Barrett (chairman), J. E. Baldwin, D. S. Heeschen, F. J. Kerr, J. Lequeux, S. W. McCuskey, P. G. Mezger, B. Y. Mills, Yu. N. Parijskij, B. J. Robinson, H. van der Laan, and H. F. Weaver. The Local Committee, consisting of B. J. Robinson, N. G. Seddon, and P. J. Kelly, looked after the arrangements in very fine style. The Symposium was supported financially by the IAU, the Australian Academy of Science, the CSIRO Division of Radiophysics, Union Carbide Australia Limited, and the Science Foundation for Physics within the University of Sydney.

Seventy-three papers were presented, with discussion periods following nearly every one. Each session contained one, and in some cases several, review papers. We have changed the order of presentation of the sessions in the volume from that of the Symposium itself, to bring related sessions more closely together. The session on supernova remnants was actually the last one held. The Symposium program arrangement was subject to some constraints which do not apply to the printed version.

For several papers where abstracts only were submitted, we used the tape recordings and our own notes to prepare a summary of the author's remarks. For review papers this was extended over several pages. For other papers only one page was written, but we felt this useful, especially to provide some basis for the discussion. Sometimes we summarized already published papers. All papers prepared by us from the tapes are so indicated.

As part of the philosophy governing the organization of the Symposium, the Organizing Committee encouraged wide discussion based on the review papers and the small number of short contributed papers. Therefore we have attempted to include almost all of the discussion, using two procedures. First we took the discussion slips that were filled out by the commentator and the speaker on the spot. These were also checked by the original authors after typing. Second, we used transcriptions from the tape recordings and our notes to summarize any significant parts of the discussion that had not been written down. Essentially every relevant comment appears in the printed discussion.

We were greatly assisted in the preparation of the manuscript by Robin Farber, Ann Frantz and Nancy Rosander. The discussion was included only by dint of the exertions of R. A. Batchelor, N. Fourikis, J. W. Brooks, and J. N. Clarke, who organized the discussion slips at the sessions, and Joyce McIntosh and Jenny Stainer, who typed many of them on the spot. The technical expertise of D. C. Dunn in recording the papers and discussion was indispensable in making this volume complete. The participants* in the Advanced Galactic Research Seminar at the University of Maryland assisted the Editors by critically reading the manuscripts.

A note on the jansky. We have used the jansky (Jy) for the unit of flux density; $1 \text{ Jy} = 10^{-26} \text{ W m}^{-2} \text{ Hz}^{-1}$. This unit was adopted by the IAU at the General Assembly in Sydney, and it has been accepted by the editors of several major journals.

F. J. KERR

S. C. SIMONSON, III

* D. L. Ball, R. H. Becker, P. F. Bowers, J. Carmody, K.-H. Fogg, E. J. Grayzeck, L. K. Hutton, P. D. Jackson, G. R. Knapp, R. A. Perley, R. P. Sinha, and N. R. Vandenberg.

LIST OF PARTICIPANTS

J. G. Ables, CSIRO Division of Radiophysics, Sydney, Australia

J. W. Baars, Netherlands Foundation for Radio Astronomy, Dwingeloo, The Netherlands

J. E. Baldwin, Mullard Radio Astronomy Observatory, Cambridge, U.K.

A. H. Barrett, Massachusetts Institute of Technology, Cambridge, Mass., U.S.A.

R. A. Batchelor, CSIRO Division of Radiophysics, Sydney, Australia

M. J. Batty, University of Sydney, Sydney, Australia

J. Borgman, Kapteyn Observatory, Roden, The Netherlands

L. L. E. Braes, Sterrewacht, Leiden, The Netherlands

J. W. Brooks, CSIRO Division of Radiophysics, Sydney, Australia

R. D. Brown, Monash University, Clayton, Victoria, Australia

B. F. Burke, Massachusetts Institute of Technology, Cambridge, Mass., U.S.A.

B. J. Burns, CSIRO Division of Radiophysics, Sydney, Australia

W. B. Burton, National Radio Astronomy Observatory, Green Bank, W. Va., U.S.A.

J. L. Caswell, CSIRO Division of Radiophysics, Sydney, Australia

W. N. Christiansen, University of Sydney, Sydney, Australia

E. B. Churchwell, Max-Planck-Institut für Radioastronomie, Bonn, F.R.G.

D. H. Clark, University of Sydney, Sydney, Australia

J. N. Clarke, CSIRO Division of Radiophysics, Sydney, Australia

C. H. Costain, Dominion Radio Astrophysical Observatory, Penticton, B. C., Canada

D. F. Crawford, University of Sydney, Sydney, Australia

D. D. Cudaback, University of California, Berkeley, Calif., U.S.A.

R. D. Davies, Nuffield Radio Astronomy Laboratories, Jodrell Bank, U.K.

G. A. Day, CSIRO Division of Radiophysics, Sydney, Australia

H. R. Dickel, University of Illinois, Urbana, Ill., U.S.A.

J. R. Dickel, University of Illinois, Urbana, Ill., U.S.A.

B. Donn, NASA Goddard Space Flight Center, Greenbelt, Md., U.S.A.

R. D. Ekers, Kapteyn Astronomical Institute, Groningen, The Netherlands

N. Fourikis, CSIRO Division of Radiophysics, Sydney, Australia

F. F. Gardner, CSIRO Division of Radiophysics, Sydney, Australia

P. D. Godfrey, Monash University, Clayton, Victoria, Australia

D. W. Goldsmith, State University of New York, Stony Brook, N.Y., U.S.A.

M. A. Gordon, National Radio Astronomy Observatory, Green Bank, W. Va., U.S.A.

J. M. Greenberg, State University of New York, Albany, N.Y., U.S.A.

M. Grewing, Institut für Astrophysik und Extraterr. Forschung, Bonn, F.R.G.

M. Guélin, Observatoire de Paris, Meudon, France

H. J. Habing, Sterrewacht, Leiden, The Netherlands

P. A. Hamilton, University of Tasmania, Hobart, Tasmania

C. G. Haslam, Max-Planck-Institut für Radioastronomie, Bonn, F.R.G.

R. F. Haynes, CSIRO Division of Radiophysics, Sydney, Australia

C. Heiles, University of California, Berkeley, Calif., U.S.A.

H. Hirabayashi, Tokyo Astronomical Observatory, Tokyo, Japan

J. A. Högbom, Stockholm Observatory, Saltsjöbaden, Sweden

B. Höglund, Onsala Space Observatory, Onsala, Sweden

W. K.-H. Huchtmeier, CSIRO Division of Radiophysics, Sydney, Australia

V. A. Hughes, Queen's University, Kingston, Ont., Canada

A. R. Hyland, Mt. Stromlo and Siding Spring Observatory, Canberra, Australia

E. B. Jenkins, Princeton University Observatory, Princeton, N.J., U.S.A.

K. J. Johnston, Naval Research Laboratory, Washington, D. C., U.S.A.

F. D. Kahn, University of Manchester, Manchester, U.K.

P. Kaufmann, Universidade Mackenzie, São Paulo, Brazil

F. J. Kerr, University of Maryland, College Park, Md., U.S.A.

J. Lequeux, Observatoire de Paris, Meudon, France

A. G. Little, University of Sydney, Sydney, Australia

A. G. Lyne, Nuffield Radio Astronomy Laboratories, Jodrell Bank, U.K.

D. A. MacRae, David Dunlap Observatory, Richmond Hill, Ont., Canada

D. S. Mathewson, Mt. Stromlo and Siding Spring Observatory, Canberra, Australia

A. Maxwell*, Harvard Radio Astronomy Station, Fort Davis, Texas, U.S.A.

T. K. Menon, Tata Institute of Fundamental Research, Bombay, India

P. G. Mezger, Max-Planck-Institut für Radioastronomie, Bonn, F.R.G.

G. K. Miley, Sterrewacht, Leiden, The Netherlands

B. Y. Mills, University of Sydney, Sydney, Australia

D. K. Milne, CSIRO Division of Radiophysics, Sydney, Australia

G. Monnet, Observatoire de Marseille, Marseille, France

M. Morimoto, Tokyo Astronomical Observatory, Tokyo, Japan

J. D. Murray, CSIRO Division of Radiophysics, Sydney, Australia

J. H. Oort, Sterrewacht, Leiden, The Netherlands

P. Palmer, University of Chicago, Chicago, Ill., U.S.A.

Yu. N. Parijskij, Main Astronomical Observatory, Pulkovo, U.S.S.R.

P. Pishmish de Recil, Las Universidad Nacional de Mexico, Ciudad Universitaria,
    Mexico

R. M. Price, Massachusetts Institute of Technology, Cambridge, Mass., U.S.A.

V. Radhakrishnan, Raman Research Institute, Bangalore, India

J.-C. Ribes, CSIRO Division of Radiophysics, Sydney, Australia

B. J. Robinson, CSIRO Division of Radiophysics, Sydney, Australia

R. Sancisi, Kapteyn Astronomical Institute, Groningen, The Netherlands

* Attended only one session.

F. Sato, Chiba Prefecture Education Center, Chiba, Japan

E. Scalise, Universidade Mackenzie, São Paulo, Brazil

J. R. Shakeshaft, Mullard Radio Astronomy Observatory, Cambridge, U.K.

W. L. H. Shuter, University of British Columbia, Vancouver, B.C., Canada

S. C. Simonson, University of Maryland, College Park, Md., U.S.A.

M. W. Sinclair, CSIRO Division of Radiophysics, Sydney, Australia

O. B. Slee, CSIRO Division of Radiophysics, Sydney, Australia

F. G. Smith, Nuffield Radio Astronomy Laboratories, Jodrell Bank, U.K.

L. F. Smith, Max-Planck-Institut für Radioastronomie, Bonn, F.R.G.

J. M. Sutton, University of Sydney, Sydney, Australia

G. Swarup, Tata Institute for Fundamental Research, Bombay, India

K. Takakubo, Tohoku University, Sendai, Japan

H. D. Tananbaum, American Science and Engineering, Cambridge, Mass., U.S.A.

Y. Terzian, Cornell University, Ithaca, N.Y., U.S.A.

C. H. Townes, University of California, Berkeley, Calif., U.S.A.

B. E. Turner, National Radio Astronomy Observatory, Green Bank, W. Va., U.S.A.

A. J. Turtle, University of Sydney, Sydney, Australia

H. van Woerden, Kapteyn Astronomical Institute, Groningen, The Netherlands

A. Watkinson, CSIRO Division of Radiophysics, Sydney, Australia

H. F. Weaver, University of California, Berkeley, Calif., U.S.A.

R. M. West, European Southern Observatory, Geneva, Switzerland

G. Westerhout, University of Maryland, College Park, Md., U.S.A.

J. B. Whiteoak, CSIRO Division of Radiophysics, Sydney, Australia

R. Wielebinski, Max-Planck-Institut für Radioastronomie, Bonn, F.R.G.

R. W. Wilson, Bell Telephone Laboratories, Holmdel, N.J., U.S.A.

C. G. Wynn-Williams, California Institute of Technology, Pasadena, Calif., U.S.A.

C. Yuan, City College of New York, New York, N.Y., U.S.A.

B. M. Zuckerman, University of Maryland, College Park, Md., U.S.A., and University of California, Berkeley, Calif., U.S.A.

PART 1

THE INTERSTELLAR MEDIUM

"It seems to me these structures are not understood."
C. Heiles, in the discussion following his paper

THE OBSERVATIONAL
EVIDENCE FOR AN INTERCLOUD MEDIUM

V. RADHAKRISHNAN

Raman Research Institute, Bangalore, India

Abstract. The 21-cm evidence for an intercloud medium is reviewed. The observations include the sky distribution and velocity structure of 21-cm emission profiles, self-absorption features in emission profiles, and absorption profiles in the directions of discrete sources. It is concluded that the intercloud medium in the solar neighbourhood has a temperature between 10^3 and 10^4 K, a column density of $\sim 1.4 \times 10^{20}$ cosec $|b|$ atoms cm^{-2}, and a brightness temperature of ~ 4.4 cosec $|b|$ K wherever the line of sight does not intersect optically thick cold concentrations.

The notion that interstellar matter occurs in clouds or concentrations had its beginnings in the patchy appearance of both bright and dark nebulae, and in the grouping of lines noted by Adams (1949) in optical interstellar absorption spectra. In over two decades since the detection of the 21-cm line, more and more evidence has accumulated to indicate that dust, molecules and the heavier atoms occur in concentrations wherein the density is vastly greater than in the in-between spaces separating these concentrations.

Well, what of interstellar neutral hydrogen, the most abundant constituent and one which is clearly detectable in emission in whatever direction of the sky one looks and whatever the angular resolution of the instrument used? Does it also occur only in discrete concentrations or is there a substantial amount of intercloud neutral hydrogen as well? It is the observational evidence relating to this question which I shall attempt to review briefly here.

The column density of hydrogen along any line of sight is usually calculated from the integral of the emission profile. Unfortunately, this calculation is valid only in the situation where the hydrogen is optically thin to a first approximation. Because of the flattened distribution of hydrogen and the enormous column lengths available at low galactic latitudes, measurements in these directions tend to refer to high optical depths. Such low latitude measurements therefore, while telling us almost all we know about the kinematics of hydrogen on a galactic scale, have made very little sense in terms of its temperature and density structure on a smaller scale.

Observations at intermediate and higher latitudes, on the other hand, have invariably demonstrated variations which had less to do with galactic rotation and more with the gas distribution. The many such studies made away from the galactic plane refer perforce to the solar neighbourhood only; however, they have all contributed towards providing a picture that is probably valid for most parts of the Galaxy.

Because 21-cm emission measurements can be made in any direction in the sky, whereas absorption measurements can only be made in the directions of sufficiently intense continuum sources, the number of available measurements of the emission exceeds by a few orders of magnitude the number of absorption measurements. In

F. J. Kerr and S. C. Simonson, III (eds.), Galactic Radio Astronomy, 3–12. All Rights Reserved.
Copyright © 1974 by the IAU.

spite of this, the small number of absorption measurements have contributed much to our understanding of the structure of neutral hydrogen because they enable us to get spin temperatures, true column densities and a measure of the turbulence when combined with emission measurements in the same directions.

Interferometers are particularly suited for 21-cm absorption measurements and the early work done at Caltech culminated in Clark (1965) proposing a 'raisin-pudding' model for the neutral hydrogen distribution of cold clouds ($T_s < 100$ K) embedded in a hot medium with $T_s > 1000$ K.

Among the many reasons which led to this suggestion were:

(1) the consistent difference in appearance between absorption and emission profiles, the absorption components invariably appearing narrower;

(2) the presence of self-absorption effects seen occasionally in emission profiles showing that there were significant differences in the spin temperatures characterising different regions in the neutral hydrogen distribution; and

(3) the difficulty of holding the clouds together; the calculated masses in several cases where one could measure or estimate both size and density indicated that these clouds were gravitationally unbound and should disperse in a very short time in the absence of pressure balance from a hot medium or other such restraining influence.

The important point to note about Clark's proposal is that it invoked the existence of large temperature differences in the hydrogen distribution. Most previous models assumed an effectively uniform temperature of around 125 K and relied on density variations alone to explain the various observed phenomena.

Although this important step resulted from absorption studies, it is not as if emission measurements had provided no clues as to the existence of intercloud gas. One of the earliest studies not confined to the plane was that of McGee and Murray (1961). From a sky survey of neutral hydrogen emission they were led to the conclusion that the local distribution was substantially horizontally stratified and that embedded in it were concentrations of gas. Carl Heiles (1967) published an account of high-resolution observations of a small region at a galactic latitude of 15° undertaken with a view to studying the small-scale spatial structure. He found that his observations did not correspond at all with the predictions of the 'standard cloud model' and in particular that about $\frac{3}{4}$ of the integrated 21-cm emission from the observed region originated in a diffuse smooth background.

In both the above studies it was the spatial distribution of the integrated column density that led these authors to their conclusions. Information is also available in the frequency or velocity structure of the profiles and can be extracted by analysing the profile shapes. Grahl et al. (1968) and later Mebold (1972) have analysed some 1300 emission profiles obtained at a galactic latitude of 30° and with longitude values between 0° and 360°. It was found that at this latitude profiles had the typical shape illustrated in Figure 1 of Mebold's paper (1972), and that they could be very well approximated by a narrow Gaussian component with a dispersion ≈ 3 km s^{-1} superposed on a wide shallow component with dispersion ≈ 10 km s^{-1}.

Figure 2 in the same paper is a normalized histogram of component areas versus velocity dispersion and the double peaked structure of this histogram clearly demonstrates that there are two types of components contributing to the emission profiles. The narrow components are interpreted as arising in conventional 'clouds' and the wide components as caused by widely dispersed gas. The latter are not necessarily Gaussian and their form has a dependence on the galactic longitude as would be expected from an extended medium acted upon by differential galactic rotation. For this extended gas distribution Mebold (1972) derives a density,

$$n(z) \approx 0.2 \exp(-|z|/210) \text{ cm}^{-3}$$

and a dispersion ≈ 9 km s^{-1} after removing the effects of galactic rotation. The Doppler temperature corresponding to this dispersion is 9400 K, suggesting that for any reasonable amount of turbulence the kinetic temperature must be at least in the hundreds of degrees.

An earlier survey of emission over a range of intermediate latitudes was that of Takakubo and van Woerden (1966). Enormous effort was put into resolving their profiles into Gaussian components and the procedures adopted are discussed in great detail. As in other studies of the emission only, it was not possible to extract any information on the spin temperature from the observations, and in fact one has to put in an assumed spin temperature to convert brightness temperatures to optical depths before any resolution into Gaussian components can be attempted. Components of all the different values of dispersion were assumed to have arisen in clouds, but it was noted (Takakubo, 1968) that for the wider components with dispersion > 7 km s^{-1} the velocities were correlated over much larger angles, suggesting that if they were due to clouds then these clouds would have to be large ones.

Mebold (1972) has made a histogram (Figure 6 of his paper) of the Gaussian components obtained by Takakubo and van Woerden (1966) similar to the one using his own data mentioned earlier. Here again the double-peaked nature is evident as also the increased line-broadening effect of differential galactic rotation at the lower latitudes of their observations. As these observations span a range of latitudes, they provide the possibility of demonstrating that the wide shallow components arise in a stratified medium rather than in large individual clouds. In Figure 1, I have plotted the mean column density at each latitude of the widest Gaussian component observed at each of the 191 positions listed by Takakubo and van Woerden. The good fit to the cosecant curve shown in Figure 1 clearly demonstrates that the widest Gaussian component in each profile originates not in individual clouds but in a fairly smoothly stratified distribution of hydrogen in the solar neighbourhood. Very recent work by Schwarz and van Woerden reported at this symposium has also led to the conclusion that the wide components arise in a stratified approximately plane parallel medium affected by differential galactic rotation.

Turning now to absorption studies, Hughes *et al.* (1971) have observed over 90 extragalactic sources for 21-cm absorption with the Caltech interferometer and found measurable amounts in 64 cases. Although they did not make any emission *measurements themselves*, they have included emission profiles taken from other observers at points close by, and the brightness temperature values taken from these emission profiles have been used in a statistical analysis of the absorption data obtained by them. Their analysis takes into account only the total optical depth of each prominent absorption feature as no Gaussian analysis was performed on the absorption profiles. In their analysis it was assumed that a two-temperature structure ex-

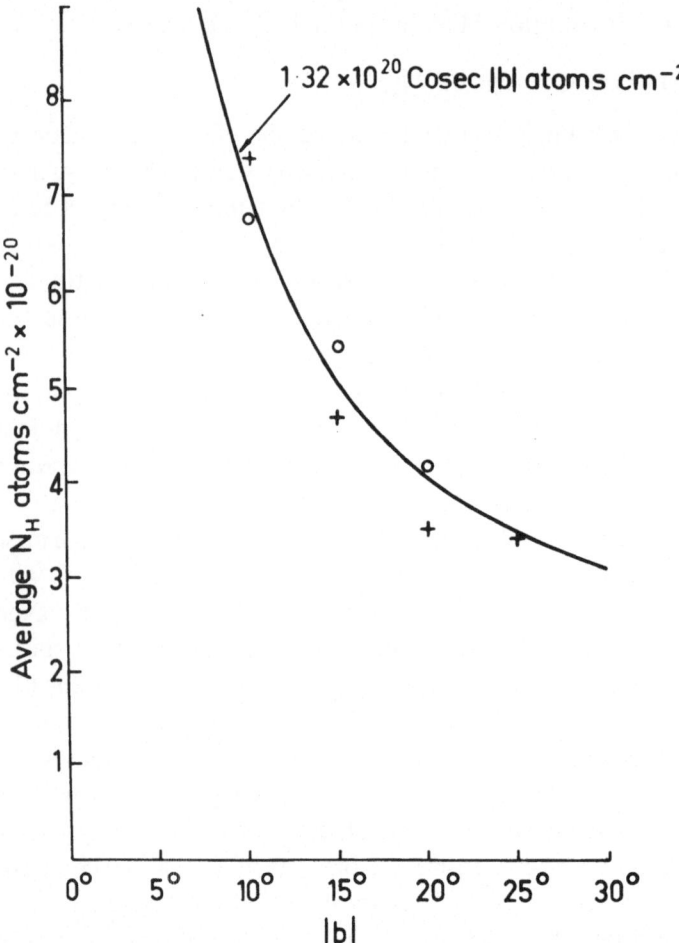

Fig. 1. The mean column density of the widest Gaussian component at each latitude observed by Takakubo and van Woerden (1966) is shown plotted as a function of the absolute value of the latitude. The crosses and circles refer to observations at north and south latitudes. The cosecant dependence on the latitude illustrates the stratified nature of the hydrogen distribution giving rise to the wide components in the profiles.

isted for the interstellar gas and the data were used to calculate the parameters for the two components with different temperatures. The figures arrived at by them for the mean temperature of the cool absorbing gas was 71 ± 9 K and the highest lower limit for the temperature of the hotter gas was approximately 600 K. They also concluded that the fraction of the neutral atomic hydrogen in the cool state was in the range between 40% and 75%.

Concurrently with the investigations referred to above, an extensive survey of 21-cm absorption in discrete source spectra was carried out at the Parkes Observatory. One section (Paper II) of this survey (Radhakrishnan *et al.*, 1972) was specifically designed to throw light on the question of intercloud gas. A detailed comparison was made of emission and absorption spectra obtained in the direction of extragalactic sources situated at intermediate or higher latitudes. As the sources were extragalactic, all the hydrogen seen in emission in their directions is in the path of the continuum radiation from the sources, and would have been expected to produce an absorption profile similar to the corresponding emission profile. However, as shown in Figure 2 there is a consistent dissimilarity in the two types of profiles in

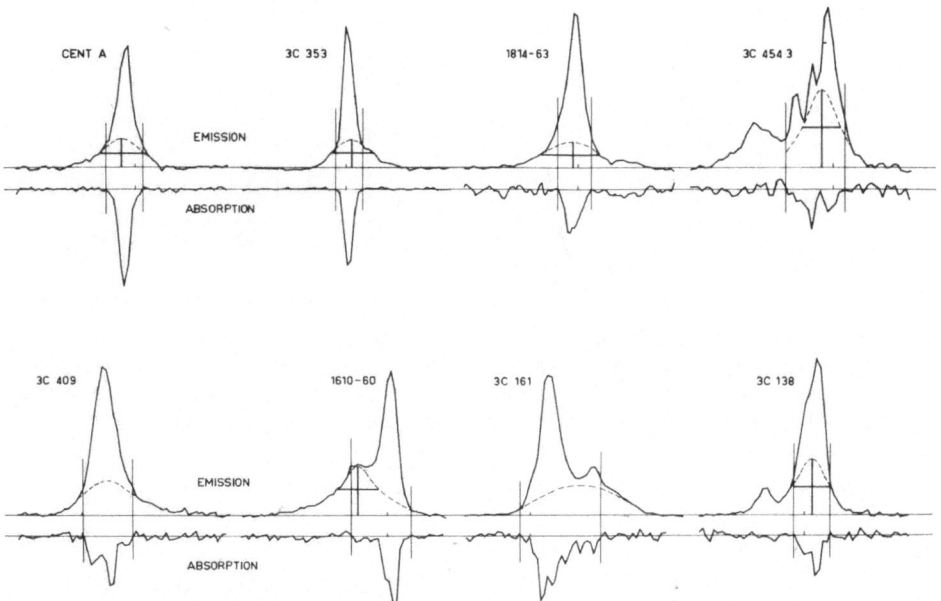

Fig. 2. Comparison of eight emission and absorption spectra obtained at intermediate latitudes and selected on the basis of high signal-to-noise ratio. The velocity limits of the absorption spectra are demarcated, showing clearly the presence of an optically thin component in every emission spectrum. These low, wide components are shown by dashed lines, and in some cases (crosses) their parameters can be determined by a computer analysis into Gaussians.

that there is always a low wide component in the emission spectrum which is absent in the absorption spectrum to the limits of instrumental sensitivity. As seen in Paper II of the Parkes survey, there are continuum sources in the spectra of which 21-cm absorption could not be detected. In the corresponding emission spectra, the narrow components were missing but the wide shallow component was always present. That this optically thin component belongs in a separate category from those with measurable optical depths is seen in Figure 3 where the number of both types of components is plotted as a function of their full widths. In Figure 4, the integrated column density in the wide optically thin component is shown plotted against the absolute

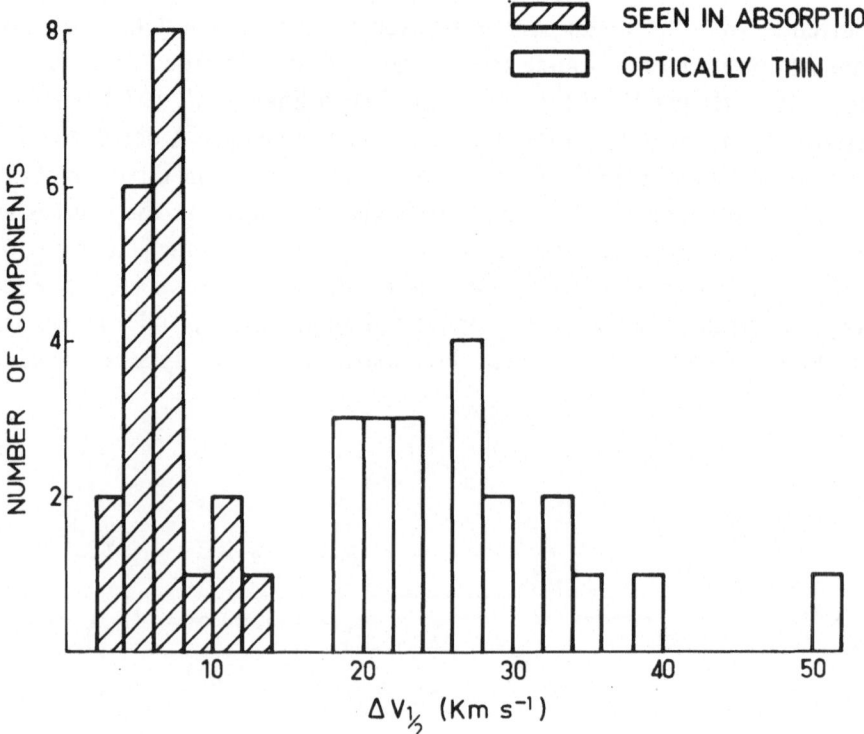

Fig. 3. A histogram of the number of Gaussian components as a function of their full widths observed in the direction of extragalactic sources at intermediate and high latitudes. A clear separation is evident between the narrow components with measurable optical depths and the wide optically thin components. The former originate in low-temperature concentrations and the latter in a high-temperature, widely dispersed medium.

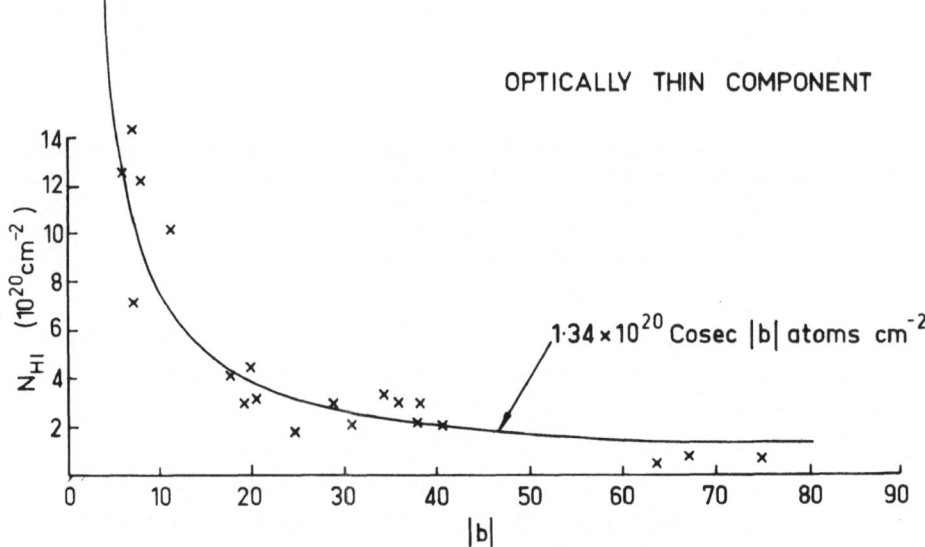

Fig. 4. The integrated column density in the wide optically thin component is shown plotted against the absolute value of the galactic latitude. The stratified nature of the hydrogen distribution giving rise to this component of the emission is shown by the fit to the cosecant curve.

value of the galactic latitude. The reasonable fit to a cosecant curve indicates again the stratified nature of the hydrogen distribution giving rise to this component of the emission.

If the narrow Gaussian components seen in both emission and absorption profiles arise in clouds, the spin temperatures characterising these clouds can be calculated and these are shown plotted in Figure 5. It is seen that there is a concentration around

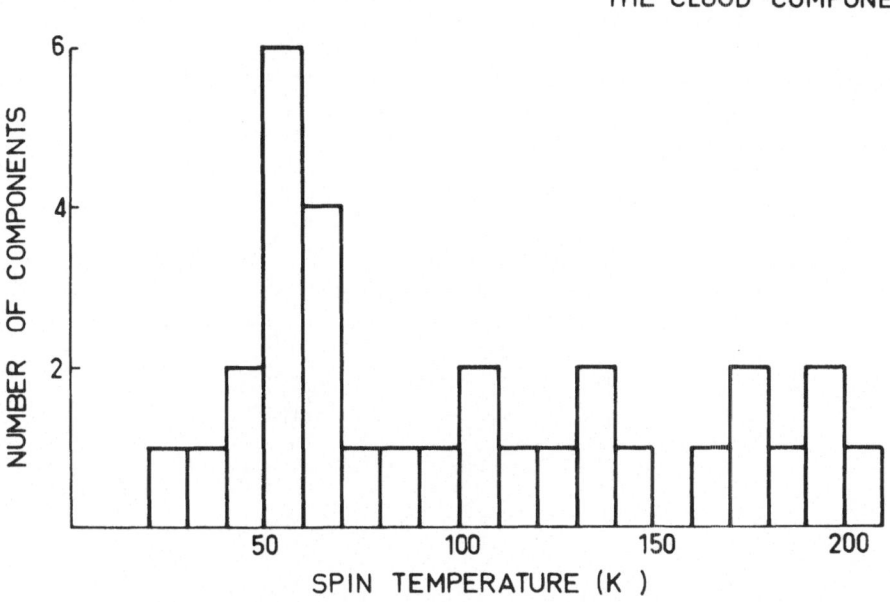

Fig. 5. A histogram of the calculated spin temperatures for the narrow Gaussian components observed in both emission and absorption. The mean for the distribution is around 80 K.

60 K with a shallow distribution extending up to 200 K or so. The corresponding Doppler temperatures for these clouds obtained from their line widths is invariably several times the kinetic temperature showing that there is considerable turbulence within most clouds.

For the optically thin gas, lower limits only can be obtained for the spin temperature, and the highest such lower limit presently available is only of the order of 1000 K. When measurements of adequate sensitivity provide us with values for the spin temperature of the optically thin gas, it is almost certain that a histogram of column density versus temperature for hydrogen of all optical depths will show a clear double-peaked structure like some of the other histograms discussed above. The Doppler temperatures corresponding to the observed line widths of this widely dispersed hot gas is of the order of 10^4 K, but it has not yet been possible to separate the turbulent and thermal contributions making up this apparent temperature.

All of the observations in the Parkes survey put together lead us to the following picture of the neutral hydrogen distribution in the solar neighbourhood. Concentrations with a mean column density of 3×10^{20} atoms cm^{-2}, a mean spin temperature of 80 K and a typical line of sight separation of one full scale thickness of the galactic disk are immersed in a hot medium containing an equal mass of gas at a temperature of somewhere between 10^3 and 10^4 K. The column density of this hot gas is $\approx 1.4 \times 10^{20}$ cosec $|b|$ atoms cm^{-2} and its brightness temperature is ≈ 4.4 cosec $|b|$ kelvins whenever the line of sight through it does not intersect optically thick cold concentrations. When the line of sight does intersect one or more cold concentrations, the profile appearance gets modified in a way depending, among other factors, on the latitude. Extreme examples are shown in Figure 6 where the narrow component is seen as

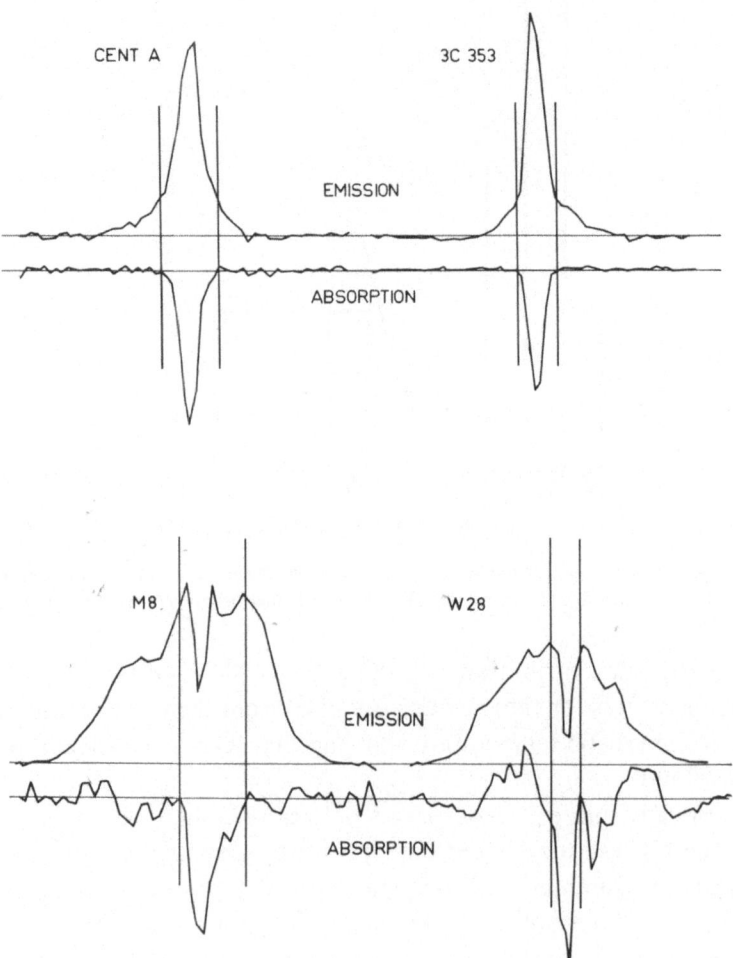

Fig. 6.   (Upper) 21-cm emission and absorption spectra obtained in the directions of two intermediate-latitude sources ($b \approx 20°$). The absorption in both cases corresponds to the peaks in the emission profiles caused by low-temperature, high-opacity concentrations of hydrogen. The wings of the emission profiles are from the optically thin high-temperature medium and are not seen in absorption.

(Lower) The absorption spectra of two sources in the plane (M8 and W28) showing correspondence not to peaks but to dips in the associated emission spectra. The cold absorbing concentrations in the foreground are seen in the emission profiles in self-absorption against the integrated contributions of the background diffuse medium.

a peak at intermediate (or higher) latitudes and as a self-absorption dip at latitudes very close to the galactic plane.

In conclusion, I would like to draw attention to the major areas of ignorance concerning the intercloud neutral hydrogen. We have no actual values for its spin temperature and consequently no knowledge of possible temperature variations from region to region. We do not know how much turbulence there is in the medium, and we are not likely to find out until very sensitive measurements are made of the optical depths in various directions leading to determinations of the spin temperatures and thence to a separation of the thermal and turbulent motions. The degree of ionization of the intercloud medium is another unknown of importance to both theorists and those concerned with the relationship of pulsar distances to their dispersion measures. And lastly, it would be of great interest to know how the medium is disturbed by the formation of a condensation bearing in mind that the mean column density of observed concentrations is equal to that through one full-scale thickness of the intercloud medium.

References

Adams, W. W.: 1949, *Astrophys J.* **109**, 354.
Clark, B. G.: 1965, *Astrophys. J.* **142**, 1398.
Grahl, B. H., Hachenberg, D., and Mebold, U.: 1968, *Beitr. Radioastronomie* 1, 1.
Heiles, C.: 1967, *Astrophys. J. Suppl.* **15**, 136.
Hughes, M. P., Thompson, A. R., and Colvin, R. S.: 1971, *Astrophys. J. Suppl.* **23**, 323.
McGee, R. X. and Murray, J. D.: 1961, *Australian J. Phys.* **14**, 260.
Mebold, U.: 1972, *Astron. Astrophys.* **19**, 13.
Radhakrishnan, V., Murray, J. D., Lockhart, P., and Whittle, R. P. J.: 1972, *Astrophys. J. Suppl.* **24**, 15.
Takakubo, K.: 1968, *Bull. Astron. Inst. Neth.* **20**, 107.
Takakubo, K. and van Woerden, H.: 1966, *Bull. Astron. Inst. Neth.* **10**, 488.

V. Radhakrishnan
Raman Research Institute,
Bangalore, India

DISCUSSION

Zuckerman: What are the lower limits to the spin temperature of the broad component?

Radhakrishnan: They range in the hundreds of kelvins, 400 or 600 to about 1000. The highest lower limit – I think it was done in the direction of M87 by Hughes, Thompson and Colvin – reached 750 or 1000 K. Actually, you can get very low values too, because it simply depends on the sensitivity of your measurement. There are no measurements yet that actually give you the temperature for any one of these. This is one of the most important things which will have to be done, by pushing the sensitivity up.

Habing: Could you give us an idea of the scale size over which the intercloud medium is homogenous?

Radhakrishnan: From the observations available so far, it would seem that within a few hundred parsecs of the Sun there is no evidence for density variations of over a factor of two from the mean density.

Van Woerden: Schwarz and I find, in a detailed study of a region in Camelopardalis (see paper later in this session), for the intercloud medium a column density $N_{\rm H} = 1.5 \times 10^{20}$ cosec b cm^{-2}, in excellent agreement with Radhakrishnan's figure. The run of this quantity with latitude shows fluctuations of $\pm 10\%$; part of this must be due to error.

Burton: You remarked at the beginning of your talk that hydrogen profiles observed at low latitudes are saturated. I think that this is not the case except near certain directions ($l \approx 0°$, 75°, 180°) where the radial velocity varies slowly with distance. This opinion is suggested by the intensity cut-off observed near zero velocity in profiles observed in the galactic plane at, say, $20° < l < 60°$. Intensities at positive velocities are typically twice those at negative velocities. This cut-off is probably due to the double-value of the velocity-distance relation at positive velocities, whereby two regions contribute to each positive velocity; on the other hand, only one region contributes to negative velocities. Both positive velocity regions would contribute to the profile, producing the observed intensity cut-off at zero velocity, only if the nearer region is transparent. This comment applies only to the gross appearance of the profiles; the small-scale self-absorption features apparent in the low-latitude profiles are undoubtedly optically thick.

Radhakrishnan: Absorption measurements on sources lying in the longitude range in question reveal considerable optical depth in almost every case. Since some of these sources lie within the solar circle, the true optical depth along the full line of sight must be even greater. Any satisfactory explanation for the observed cut-off in the emission profiles cannot therefore be based on the assumption that the nearer region is transparent.

Davies: At Jodrell Bank we have used the Mark IA radio telescope (beamwidth 12') to study the H I absorption spectra of the strong sources Cas A, Cyg A and Vir A. By using the adjacent emission spectra a direct measurement is made of the spin temperature of the hot component of the neutral hydrogen seen in these directions. Vir A, an extragalactic source, lies near the NGP and has only 'hot' H I in its line of sight. The spin temperature of the 0 km s^{-1} component is ~ 950 K and that of the -50 km s^{-1} component ~ 1500 K. No 'normal' cool clouds with $T_s \sim 100$ K are seen. In the case of Cas A and Cyg A, which both lie near the galactic plane, there are regions of the velocity profile which show gas with $T_s = 1000$–2000 K. This gas is most readily identified at velocities which correspond to interarm regions. It is clear from this work that the velocity spread found in these 'hot' clouds is much larger than can be explained by thermal broadening of the profiles.

Baldwin: If there are essentially empty holes in the intercloud medium, what fraction of space could they occupy and yet have escaped detection?

Radhakrishnan: A large number of empty holes distributed homogeneously would appear the same as a lower intercloud density of hydrogen with no holes.

A MODERN LOOK AT 'INTERSTELLAR CLOUDS'

CARL HEILES

University of California, Berkeley, Calif., U.S.A.

Abstract. We compare past and present modes of investigation of the structure of the interstellar gas. Many aspects of the interstellar cloud model are invalid.

Interstellar optical absorption lines and H I 21-cm emission lines show a number of very large aggregates with properties similar to those of 'cloud complexes'. At nonzero velocities especially for $b < 0°$, exist optical lines which have no H I counterparts. These are almost certainly produced in low-density gas clouds; perhaps the intercloud medium is itself cloudy.

Maps of H I column density taken over large velocity ranges do not reveal much small-scale structure. This fact cannot easily be reconciled with the statistical analyses of interstellar reddening. The maps do reveal large, coherent gas structures which are often filamentary in shape and at least sometimes aligned parallel to the interstellar magnetic field.

Maps of H I column density over small velocity ranges show much small-scale structure, often filamentary in shape. The filaments are almost universally oriented parallel to the interstellar magnetic field and have Doppler velocity gradients along their lengths. In one area the geometry of the field and gas almost exclusively suggests Alfvén-type motions.

I. Introduction

The 'cloud model' of the interstellar medium (see, for example, Spitzer, 1968a, b) pictures the gas as distributed in two components: regions of high density (the clouds), and the intercloud region where the density is much smaller. The clouds are pictured to be of random sizes (within certain limits), shapes, distribution in space, and velocity. For purposes of simplification, the whole range of cloud sizes is often replaced by a single one, the 'standard cloud.' Much of the observational work concerning the interstellar medium has been directed toward the determination of the spectrum of cloud sizes, and much of the theoretical work has used the cloud model.

The origin of the cloud model appears to lie mainly in old statistical studies. The average reddening per unit length in the Galactic plane must arise from both the uniformly-distributed dust and that portion which is concentrated into clouds. The latter will, in addition, provide a spatially fluctuating component. The classic studies of this by Chandrasekhar and Münch (1952) and Münch (1952) indicate two types of cloud, with reddenings of about 0.07 and 0.4 mag and line-of-sight intersection frequencies of 6.5 and 0.6 per kpc, respectively. These frequencies, combined with the statistical estimates of number of clouds per kpc^3 by Ambartsumian and Gordeladse (1938), imply cloud radii of 5 and 50 pc, respectively. The reddenings imply corresponding H I column densities of 3.5×10^{20} and 20×10^{20} cm^{-2} (cf. Savage and Jenkins, 1972), volume densities of about 10 cm^{-3}, and thus masses of 90 and 80000 $M_\odot$ respectively. The larger of these are often called 'cloud complexes' after Oort (1953), with the implication that they contain smaller clouds within. Similar results (but with somewhat smaller radii and larger volume densities) are derived from more modern data by Scheffler (1967a). The larger of these two types of clouds is not in-

F. J. Kerr and S. C. Simonson, III (eds.), Galactic Radio Astronomy, 13–44. All Rights Reserved.
Copyright © 1974 by the IAU.

consistent with results derived from completely independent data (see Section II).

A more modern analysis by Scheffler (1967b) shows that the picture of two types of clouds must be replaced by one of a continuous distribution of cloud parameters. He derives a mass spectrum for clouds which varies as $(mass)^{-\beta}$ with $\beta = 1$ to 2 for different mass ranges. The upper mass limit is about the same as for a cloud complex, discussed above. His spectrum fits quite well with the existing, but very sparse, H I data on cloud masses (Field and Hutchins, 1968).

These data have, until recently, seemed fully consistent with existing interpretations of H I data and optical line data, although the exact quantitative results were never certain. However, the cloud model does not fully describe the situation. In particular, how good are the fundamental tenets concerning randomness? Where do the dense and massive dust clouds, which contain nearly half of the total mass of the interstellar gas, fit in? Authors, when carefully describing the observational situation in review articles such as this one, have generally been careful to point out many of these uncertainties. But many other authors, both theoretical and observational, go right ahead and ignore them. After enough repetition the standard assumptions have come to be regarded as being observational fact. The purpose of the present paper is to critically discuss these uncertainties.

Until recently, insufficient data have been available to make such a discussion. We will find that some aspects of the cloud model remain valid. Other aspects, especially the assumptions concerning randomness, are incorrect. Much of the observable gas is affected by the interstellar magnetic field and/or huge explosions. Many large aggregates contain hierarchical structure with non-random shapes and velocities. Outside these aggregates, the gas is often distributed in long, delicate, interconnected filaments rather than clouds.

II. Comparison of Optical (Na I and Ca II) and Radio (H I) Studies

Although mechanisms governing the strengths of Na I and Ca II interstellar lines are ill-understood (see, for example, Pottasch, 1972), much can be learned from the study of these lines. Most of the following discussion, which emphasizes the comparison of optical and radio results, is concerned only with stars having $|b| > 10°$. The line observations in these directions delineate individual interstellar gas structures and avoid confusion with H I beyond the star which occurs to such a large degree near $b = 0°$.

(a) THE GENERAL PICTURE: GOOD CORRELATION AT LOW VELOCITIES

(i) *Velocities*

The early definitive study by Howard *et al.* (1963) compared Adams' (1949) and Münch and Zirin's (1961) interstellar Ca II lines with 21-cm H I lines and showed that there is in general a good correlation for both velocity and intensity. Disagreements were always ascribable to distance effects, with some nearby stars showing no optical absorption lines.

This single general and important conclusion – that interstellar optical lines are similar to the 21-cm line of H I – has remained unchanged, with one exception. These authors restricted themselves to stars showing single interstellar lines, which nearly always appear at the same velocity as the 21-cm line peak. For $|b| > 10°$, this is always near 0 km s^{-1} (LSR). New studies by other workers, however, have included spectra which show more than one velocity component. These include the relatively small studies by Takakubo (1967) and van Woerden (1967) and the large studies by Goldstein and Macdonald (1969) and Habing (1969), all in the Northern Hemisphere; and the large study by Goniadzki (1972) in the South. It has become apparent from these studies that interstellar optical lines may occur at nonzero velocities which have no obvious corresponding H I lines. Here and below the word 'nonzero' means $|v_{LSR}| > 10$ to 20 km s^{-1}. We discuss these 'noncoincident' components in the next section (IIb).

The more recent optical interstellar Na I line data of Hobbs (1969a, 1971) and Ca II data of Marschall and Hobbs (1972) and to some extent Rickard (1972), obtained by scanning with a series of Fabry-Pérot interferometers, carries the comparison of Ca II, Na I, and H I to new heights in terms of precision and detail because of the much higher velocity resolution. Many of the older optical data were almost marginal in this regard. At present the comparisons with Hobbs' newer, higher quality data have yielded more precise quantitative information and increased confidence in the abovementioned correlations.

The line widths in the interstellar medium are almost *never* determined purely by thermal broadening. The 21-cm line usually has a full halfwidth of more than 5 km s^{-1}, which would imply temperatures of many hundreds of degrees in the absence of turbulent broadening. H I absorption studies (Radhakrishnan *et al.*, 1972; Hughes *et al.*, 1971) show that temperatures are more often lower. Furthermore, if thermal broadening were predominant the Na I and Ca II linewidths would be smaller than the H I values by factors of 5 to 6. Instead the optical lines are usually more than half as broad and often nearly as broad as the radio lines. As perhaps extreme examples, the three stars, ζ Oph, ζ Per, and 32 Peg have optical lines which essentially mirror the 21-cm lines (Hobbs, 1971; Marschall and Hobbs, 1972) as shown in Figure 1. The solid angle sampled by radio telescopes looking at H I in emission is usually a factor of some 10^{14} larger than that sampled by optical telescopes looking at absorption against a star, which by itself can only make the radio lines appear broader. Hence, line width differences cannot be regarded as significant and the widths must be determined almost exclusively by turbulence, as concluded by Hobbs (1969a).

(ii) *H I in Absorption*

Comparison of optical lines with H I seen in *absorption* against bright radio sources would reduce effects caused by solid angle differences, and provide unique information on the temperature and ionization as well. Both the optical and radio absorption measurements are biased towards cold gas, while the optical measurements are in addition biased toward regions of high recombination rate of the prevalent ions (Na II and Ca III) and thus also to regions of high electron density (see Habing (1969)

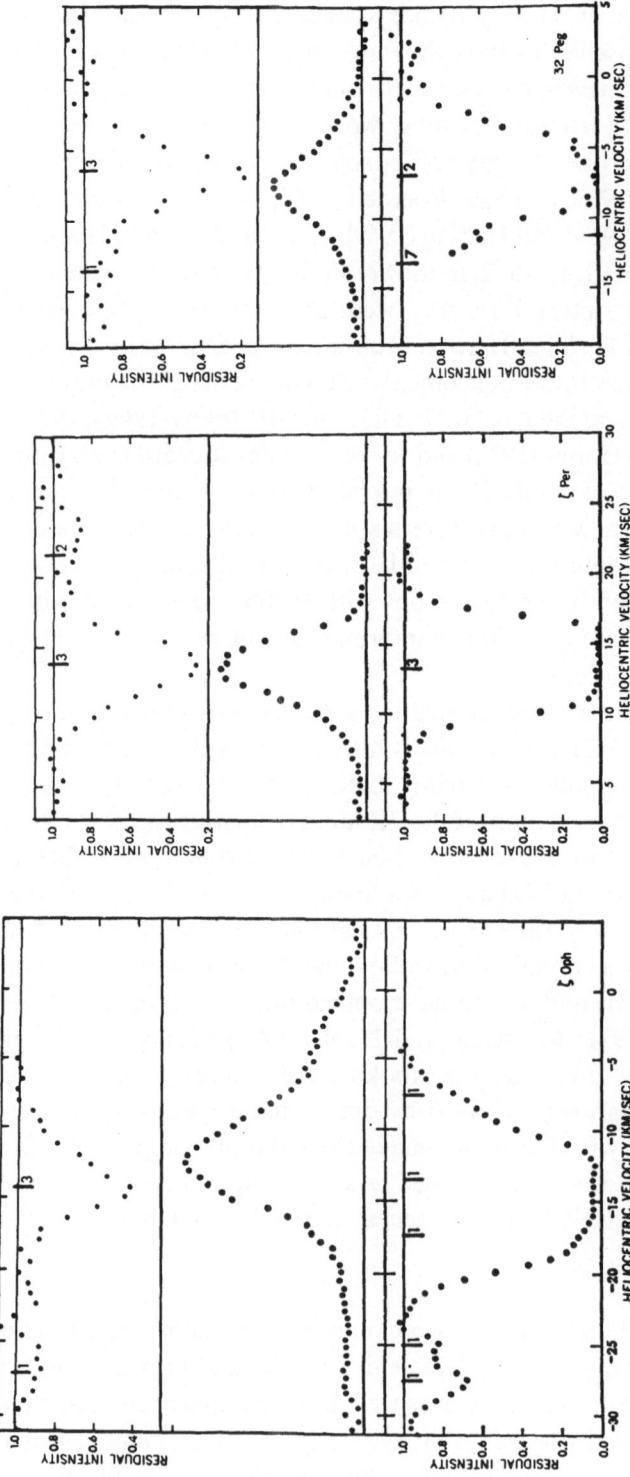

Fig. 1. Comparison of Na I (top), H I (middle), and Ca II (bottom) lines for 3 stars for which the lines have components which are nearly identical in shape. For ζ Oph, much of the difference in shape undoubtedly occurs because the star is in front of most of the H I. From Hobbs (1971) and Marschall and Hobbs (1972).

for a recent discussion of the ionization balance of Na and Ca). If the optical spectra are of high enough quality to provide line shapes, comparison of the velocity dispersions of H I, Na I, and Ca II can be used to separate thermal and turbulent broadening due to the differences in molecular weight. The resultant temperature could then be used in determining the recombination rate, and hence the electron density in the cloud. Both the Na I and Ca II lines are more easily detected, for a given column density of H I, than the corresponding H I absorption lines. Even without good velocity resolution in the optical spectra comparisons can provide information on electron density, although less complete. The only objects for which such data exist are extragalactic radio sources. Such data must certainly exist in profusion; unfortunately, it all remains unanalyzed in the plate files of the great optical observatories.

Although the comparison of optical and radio lines in absorption is potentially very interesting, it is most useful if in fact the lines are all produced in the same region. This might not be the case. Histograms of $(1/e)$ line widths in absorption (Clark, 1965; Radhakrishnan and Goss, 1972) show peaks at 1.5 to 1.8 km s^{-1}, usually somewhat smaller than H I widths in emission. Since H I widths in emission seem to be close to the optical line widths, the optical lines might not compare well with H I in absorption. Another indication that they might not be produced in the same region is the comparison of the number of line components seen per kpc in the line of sight for the two types of observation. Blaauw (1952) finds 8 to 12 optical components per kpc for Ca II. This is much larger than the 2.5 per kpc derived by Radhakrishnan and Goss (1972) for H I in absorption. One contributory cause to the large number seen by Blaauw is the association of high-velocity optical lines with low-density ionized regions (see Section IIb). Another is his correction for instrumental blending which is based on the assumption of complete randomness of the clouds in space and velocity. This assumption is unjustified (see Section IV), and he has therefore overestimated the correction for instrumental broadening. Nevertheless, the discrepancy between the absorption results obtained in the radio and optical probably remains real.

It is therefore our impression that H I absorption probably correlates with optical absorption less well than does H I emission. If so, this would seem contrary to theoretical expectation. Comparative studies of optical and H I absorption against the same background source is a potentially rich and rewarding field, presently unexploited.

(iii) *Line Strengths*

In his comparison of H I and Na I, Hobbs (1971) found that the average apparent ratio $Q = [\text{Na I}]/[\text{H I}] = 3.2 \times 10^{-9}$. However, velocity components in five stars were found to have ratios which are larger by factors ranging from 4 to 20. One of these is in κ Ori and is a nonzero velocity component; given the large number of early-type stars in the vicinity of the Orion complex, ionization of H I would not be unexpected and would account for the large value of Q. The remaining three cases are more *interesting.*

One occurs in ζ Oph; in a large region in this direction, Heiles and Jenkins (1974) show that the apparent H I/dust ratio is lower than usual. This is probably due to saturation of the 21-cm line. Although this would give an above-average value of Q, the value of $Q = 20$ for this star is even higher than expected from saturation effects alone. Perhaps this O9.5V star, distance 170 pc, is ionizing enough gas to enhance the Na I line (which will always appear stronger in H II than H I regions, see Hobbs, 1969). Can this hypothesized H II region be detected by other observational techniques?

Two other stars, 35 Ari and ζ Per, are located in the general region of Perseus. ζ Per $(l = 162°, b = -17°)$ is very close to the small region (about $10°$ diameter) in Perseus where the extinction is so large that it merits classification as a separate 'island' in the zone of avoidance (Shane and Wirtanen, 1967). Heiles and Jenkins (1973) have shown that an H I deficiency exists in part of this region, which is most straightforwardly interpreted as a result of conversion of H I to H_2. The large value of Q for the star then simply shows that this chemical transformation has not significantly affected the Na I abundance. Comparison by eye of the Ca II line profile (Marschall and Hobbs, 1972) with the Na I profile indicates the usual low-velocity abundance ratio; however, this estimate must be done again with much care to reach a definite conclusion due to the saturation in the Na I profile. If the ratio is in fact the same as for other low-velocity interstellar line components, it argues against depletion of Ca by dust as being responsible for the usually observed underabundance of Ca II relative to Na I. This is because the region is very dense and very dusty. 35 Ari $(l = 151°, b = -29°)$ is about $10°$ away from the apparent edge of this dark region in Perseus. A combination of 21-cm line saturation and H I conversion to H_2 may be responsible for its large value of Q. Due to its large value of Q we can surmise that all the H I seen in this direction is closer than the star, whose distance is 170 pc.

There are, of course, stars which show absolutely no indication of H I at velocities seen in Na I and/or Ca II. At least two of these appear in Habing's (1969) list discussed in Section IIb. It seems most sensible to ascribe these to circumstellar matter which is invisible with the large telescope beamwidths employed for studies of the 21-cm line. However, proof of this hypothesis would be comforting. If it is instead H II, we are missing a most interesting, though undoubtedly relatively insignificant in mass, portion of the interstellar medium!

(b) DEVIATIONS FROM THE GENERAL PICTURE: H I→H II, A CLOUDY INTERCLOUD MEDIUM?

It was mentioned above that optical components at nonzero velocity, i.e., $|v_{LSR}| > 10$ to 20 km s^{-1}, often have no associated H I line component. Habing (1969) makes this velocity distinction particularly clear by presenting a list of 17 stars (14 of which have $|b| > 10°$) showing this 'noncoincidence.' All but one of these 14 stars are located at negative galactic latitudes! This discovery caused the author to look at the interstellar line velocities relative to the star velocities, given by Adams (1949); again, those components showing relative velocities greater than 20 km s^{-1} preponderate for negative latitudes if stars having $|b| \leqslant 10°$ are excluded. It has been thought that the high velocity components are a result of expansion of matter around the star, since negative velocities predominate (see Spitzer, 1969b). However, if further study shows the bias towards negative latitude to be real, this thought must obviously be abandoned. Similar comments cannot yet be made for the southern sky work by Goniadzki (1972) because she was, of course, restricted to stars for which optical material exists (Buscombe and Kennedy, 1968). Most of these are located close to the galactic plane. The general lack of optical line work outside of the galactic plane in the southern sky is understandable, but regrettable.

These nonzero velocity components, which show no H I peak corresponding to the Ca II peak, are just those in which Ca II appears stronger than Na I. This is the reverse from the usual low-velocity situation (Toutly and Spitzer, 1952). These authors postulated that Na I lines weaken in these nonzero velocity clouds due to selective collisional ionization of Na I relative to Ca II, which can occur if the gas is hot. Most authors today instead believe that Ca is depleted in normal low velocity clouds by adhering to dust grains (see Pottasch, 1972). However, there seems to be little in the way of direct evidence favoring either hypothesis. Below we will see that these nonzero velocity components almost certainly come from ionized gas; this greatly strengthens the original interpretation of Routly and Spitzer (1952).

Comparison of Habing's (1969) abovementioned list of 14 stars with the Hat Creek Survey (Heiles and Habing, 1973) of 21-cm line radiation for $b \geqslant 10°$ reveals that some of the non-coincidences are probably a result of ionization of H I. Many of these stars have weak, broad H I features at or close to the same velocity as the optical lines which appear *nearby* in the sky, but not necessarily *at* the position of the star. Typical angular distances between the H I features and stars are 5°. The H I clouds show diameters of up to 10°, of course not centered near the star; they have total velocity widths of 5 to 10 km s^{-1} and brightness temperatures of 1 to 2 K. This kind of association occurs definitely for six stars, probably for four, and does not occur for three; the optical line for one star lies outside the velocity range of the H I survey. Further definite associations might be revealed with higher sensitivity in the H I line, since the usual brightness temperature of 1 to 2 K is close to the sensitivity limit. This kind of uncertain association can, of course, be criticized as being due to chance, and I have not attempted to assess the probability that the associations are significant.

Two definite associations are shown in Figure 3. HD 25558 has a relatively narrow (4 km s^{-1}) and intense (3 K) feature nearby in position ($l = 170$ to 180°) which weakens somewhat at the position of the star ($l = 185°$). For a single 21-cm line profile taken at the position of the star this feature is not distinguishable as a distinct peak and would only appear as a weak, extended line wing. HD 34816 ($l = 215°$) has a weak, large diameter feature nearby ($l = 206$ to 214°); in Figure 3a the star seems to be located in the boundary region of this feature. Again, for a single 21-cm line profile this feature would appear only as an extended line wing, even at the most intense position near $l = 210°$.

It is tempting to speculate that these 21-cm features are clouds which are essenti-

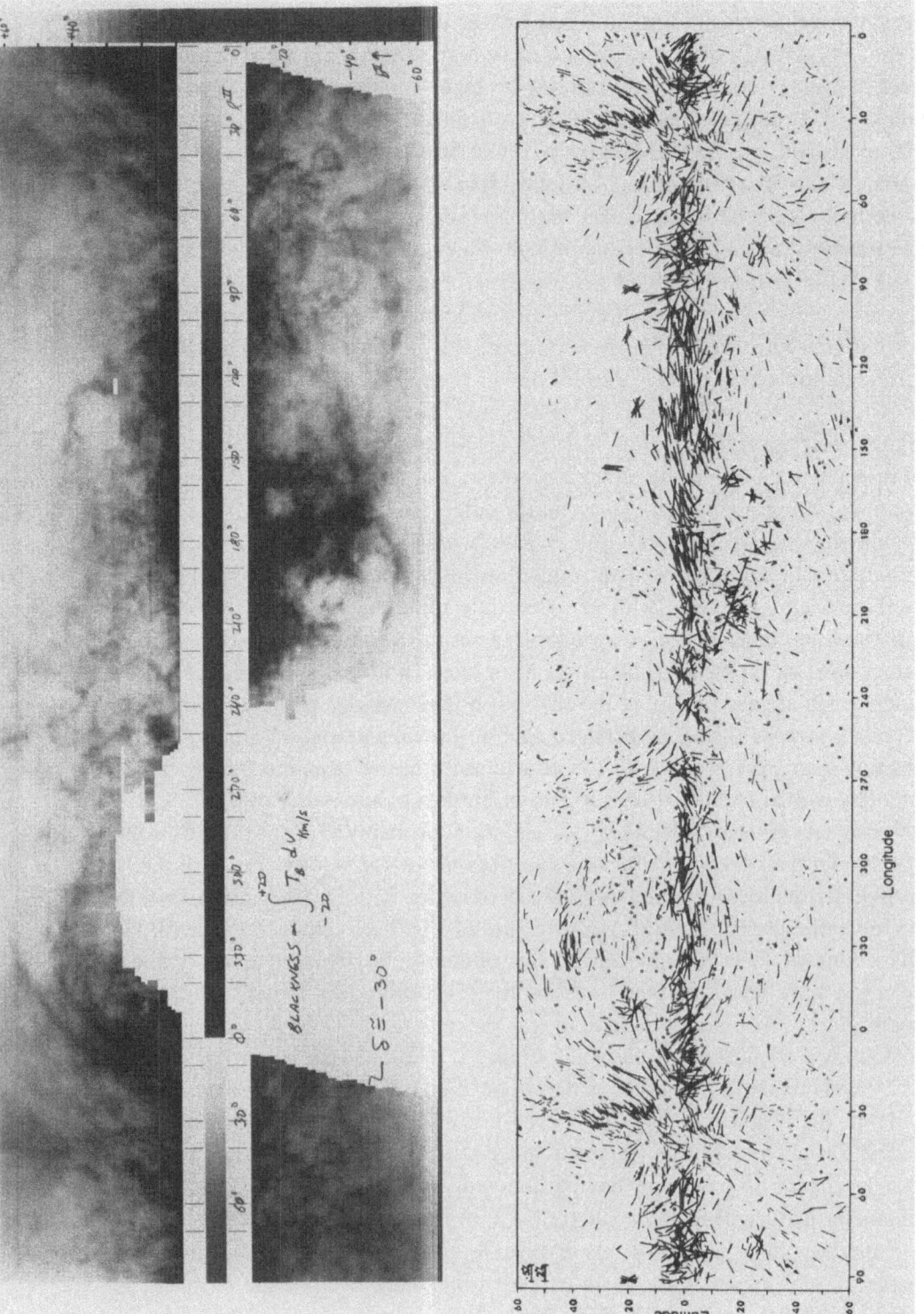

Figs. 2a–b.   (a) A photographic presentation of the distribution of H I column density in the sky over the velocity range −20 to +20 km s^{-1} (LSR). From Heiles and Jenkins (1974). (b) The directions of polarization of optical starlight, and hence presumably the direction of the interstellar magnetic field. From Mathewson and Ford (1970).

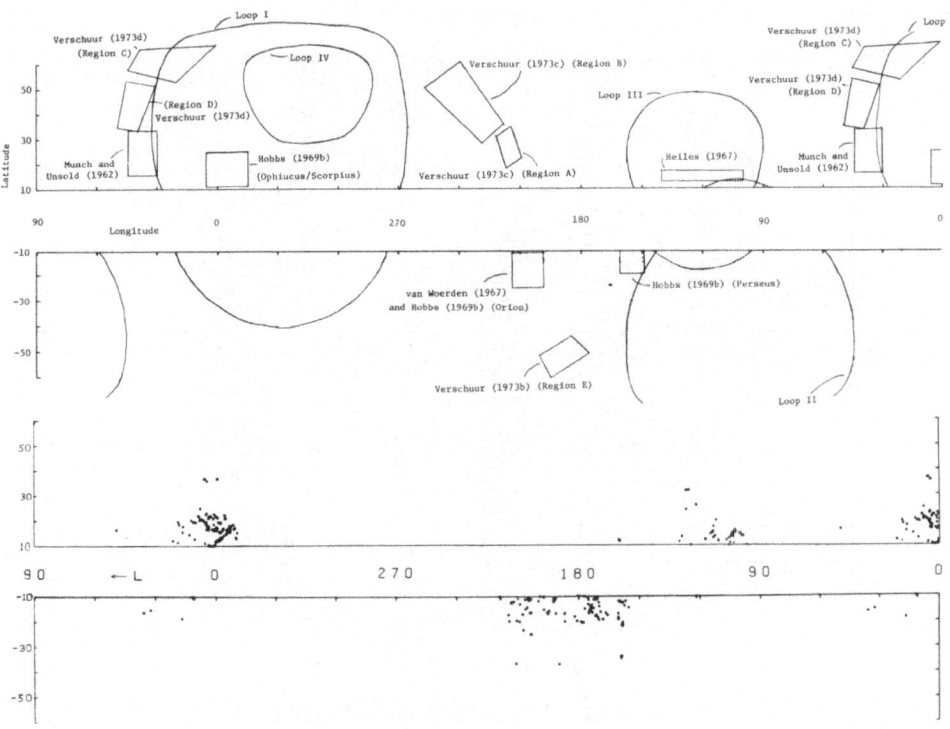

Figs. 2c–d.   (c) Regions referred to in the text. (d) The locations of dust clouds catalogued by Lynds (1962). Made from computer cards recently generated by Lynds.

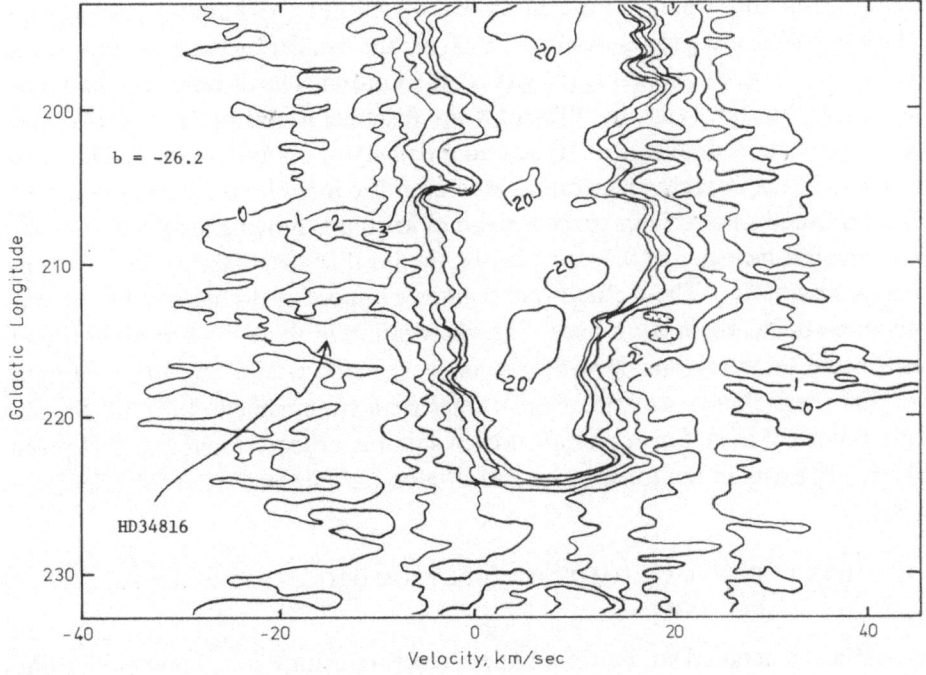

Figs. 3a–b.   Contour maps of antenna temperature vs. galactic longitude and velocity, for fixed galactic latitude. *Each covers a star showing an interstellar* Ca II *line with no corresponding* H I *line. The position of the star and the velocity of the line are located at the tip of the arrow. Note that* H I *emission occurs nearby in angle. From Heiles and Habing (1973).*

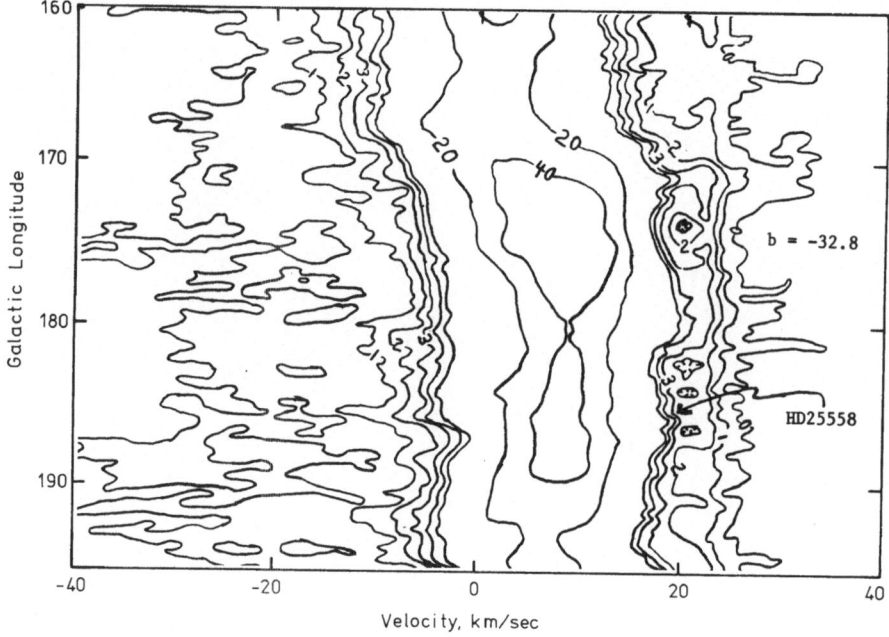

Fig. 3b.

ally completely ionized at the position of the star, either by random chance or perhaps even by the star itself. Values for the physical parameters of these clouds can be estimated. Typical equivalent widths for the Ca II components are only a few hundredths of an ångstrom unit. Using the ionization equilibria and Ca/H abundance ratio given by Habing (1969) and a temperature of 3000 K, one can derive the emission measure, about 0.1 cm^{-6} pc, by assuming the gas is fully ionized. The distance can be no more than that of the star, typically 300 pc; if the distance is 100 pc, the angular size of about 10° implies a diameter of 10 pc and a density of about 0.1 cm^{-3}. This density reminds us of the intercloud medium. However, the intercloud medium is supposed to be distributed smoothly in space instead of in clouds moving at specific velocities.

An emission measure of 0.1 is far below the limit of detectability for any type of emission observation. These clouds are therefore impossible to observe by any means other than optical absorption lines. The observation of these lines in more than one object in the same region, showing the same velocity, would constitute irrefutable proof that these clouds are interstellar rather than circumstellar and that the above picture is correct in its fundamentals. Much suitable observational material presumably already exists in the form of spectra of globular clusters, external galaxies, and quasars.

(c) ATTEMPTS TO DERIVE SPATIAL STRUCTURE FROM OPTICAL LINES ALONE:
   USUALLY UNSUCCESSFUL

The number of detailed studies of angular structure using optical interstellar lines is small. All deal with low-velocity lines, so it is no surprise that conclusions drawn from optical and radio data are consistent. One of the earliest studies (Schluter et al.,

1953) grouped Adams' (1949) stars by position and showed that groups of stars contained in small areas show low-velocity lines which all have about the same velocity. Thus interstellar clouds are larger than these areas. More recent and accurate data for these regions, and others, has been obtained by Hobbs (1969b); in the discussion below, the regions are shown on Figure 2c, and the physical parameters of the associated clouds are summarized in Table I.

TABLE I

Gross properties of Hobbs' (1969b) clouds
For other cloud parameters, see Section IVa

Region	Size[a]		$10^{-11} N_{\mathrm{Na\,I}}$ [b]	$10^{-20} N_{\mathrm{H\,I}}$ [c]	$n_{\mathrm{H\,I}}$	Mass
	(deg)	(pc)	(cm^{-2})	(cm^{-2})	(cm^{-3})	$(M_\odot)$
Perseus	17	$>45 \atop <100$ (72)	53	4^{d}	1.8	2200
Pleiades	15	<30 (30)	3 to 10	1	1.1	390
Scorpius	34	$>46 \atop <80$ (63)	8	8	4.0	13600
Orion	21	$>47?^{\mathrm{e}} \atop <160$ (103)	–	14	4.3	65000

[a] Geometrical mean of orthogonal diameters.
[b] From Hobbs (1971).
[c] From H I alone (Heiles and Habing, 1973).
[d] See discussion of ζ Per in Section IIa (iii).
[e] See text.

In *Perseus* Hobbs (1969b) finds coherence over the whole area sampled, which is 15° in angular extent. The distance of the gas is approximately 150 pc, implying clouds of size at least 40 pc. Examination of the 21-cm line contour maps (Heiles and Habing, 1973) show that the true size of this feature is larger. At $b = -18°$ the high-longitude end is cut off by what is probably a large, dense cloud of H_2 (Heiles and Jenkins, 1974) which appears as a white area in Figure 2a. Closer to the galactic plane, e.g., at $b = -12°$, there is no obvious boundary until $l \cong 235°$. A boundary may be rendered indistinct by blending of H I at different distances; alternatively, there really might be no boundary! Further from the galactic plane, e.g., $b = -30°$, a distinct velocity change appears near $l = 170°$ which should perhaps be taken as the high longitude boundary at all latitudes. The velocity structure of the feature appears to change character between $b = -30°$ and $-40°$; thus, the extent of the features in latitude is probably about 25°. Gross properties of this cloud are summarized in Table I under these assumptions, using the H I line data.

Six stars were observed in the *Pleiades* cluster, all in an area of about 1 deg². Here the 21 cm data show two peaks, one at -4 km s^{-1} and another at $+10$ km s^{-1} (both LSR). It is this latter peak which is seen in Na I. The former, which is the more intense, must therefore be more distant than the cluster. (This situation, in which the more-negative-velocity gas is the more distant, also occurs 30° to 50° lower in longitude, see Ames and Heiles, 1970.) In contrast to the situation in Perseus, the Na I seems to show considerable spatial structure. This implies angular structure on the scale of 1°, or about 2 pc for this cloud. H I data shows that this cloud extends about 12° in longitude at $b = 23°$, and about 20° in latitude. There is apparent rotation with $dv/dl = 0.5$ to 1 (km s^{-1}) per degree. However, there are a number of sub-condensations within the cloud. The Pleiades cluster lies near an internal boundary where the H I changes particularly rapidly with position. Thus the optical data show the structure within the cloud rather than the angular extent of the cloud as a whole, and an interpretation based on the latter assumption is misleading. Gross physical properties of this cloud are given in Table I.

In *Scorpius* Hobbs (1969b) finds an intense line near 0 km s^{-1} covering all stars sampled within an area bounded by $l \cong 346°$ to 6°, $b \cong 15°$ to 24°. The lines are all well-correlated with H I (Hobbs, 1971). Again, the sample of stars does not cover enough area to fully cover the cloud. Examination of the H I 21-cm emission line data shows the profile looks similar for the 50° interval $l \cong 342°$ to 30° at $b = 24°$. This large concentration is visible in Figure 2a and is at positive galactic latitudes where Gould's belt rises highest above the galactic plane. This distance lies between 100 pc (from star counts; Bok, 1956) and 170 pc, the distance of the stars (which of course, may well be imbedded in the gas rather than located behind all of it). The angular extent of nearly 50° in longitude corresponds to a linear size of some 80 to 140 parsec. Gross properties of this cloud are summarized in Table I.

In *Orion* the situation is more complex, with three velocity peaks in Na I located at -10, -7, and $+7$ km s^{-1} (LSR). Although Hobbs suspects the -10 km s^{-1} component to be circumstellar since it is seen in only one of his 10 stars, it is in fact interstellar and is associated with a spectacularly large and intense H I feature. This feature has relatively sharp boundaries and extends from $l \cong 203°$ to 210° and $b \cong -19°$ to $-26°$, reappearing again below $b \cong -29°$ at somewhat different longitudes. The -7 km s^{-1} line appears obvious only in four stars near NGC 2024 (Orion B) where there is a corresponding H I feature with a small angular size. If this small feature has been accelerated by the H II region, the 22 km s^{-1} velocity of Orion B (Mezger and Höglund, 1967) implies an ejection velocity of about 15 km s^{-1}. Finally, the $+7$ km s^{-1} Na I line coincides with the main H I peak and appears in the spectra of all the stars. Again, the feature is so large that its boundaries fall outside the regions sampled by the stars. It is the large concentration surrounding the Orion region shown in Figure 2a. This concentration has large linear velocity gradients in parts amounting to $|dv/dl| = 1$ (km s^{-1}) per degree, but restricted in angular extent so that the total velocity change is no more than 10 km s^{-1} or so. One of these gradients is responsible for the shift in velocity of the Na I line for the star κ Ori as compared to the other stars. All of the stars observed are located in the Orion association, 450 parsec distant. There is therefore no lower bound to the distance to the interstellar gas. One star, λ Ori, is closer (130 pc) and was observed by Hobbs to have no Na I for velocities less than -3 km s^{-1} (LSR), but unfortunately he did not obtain complete data for this star. For the purposes of expediency we assume in Table I that this star also lies in front of the $+7$ km s^{-1} (LSR) component feature; this assumption is not unreasonable since this gas is probably physically associated with that in the nearby Perseus region.

Münch and Unsöld (1962) examined 13 stars in a region in which the gas is probably affected by the North Polar Spur, as revealed by both continuum and H I radio astronomical studies (see Section IVb). They found the usual low-velocity association of Ca II and H I at -2 km s^{-1} (LSR) for seven of the stars. More interesting was a second component at -8 to -11 km s^{-1} (LSR) in three of the six stars not showing the -2 km s^{-1} component; all of these stars are at small distances. One of these, α Oph (distance $= 18$ pc), is surrounded by a number of other more distant stars which did not show the -8 to -11 km s^{-1} component. The angular extent of this component is thereby strictly limited to a few degrees. For one of these stars there is no corresponding H I component. For two of these stars there exist very weak H I components which are discernable only as very small bumps on a contour map of antenna temperature versus position and velocity; on a single profile they are too weak to be easily distinguished as separate components and would normally be considered part of the 21-cm line wing. They are small in angle and the strength varies rapidly with position. (The contour maps of Heiles and Habing (1973) are currently at the printers; in examining these stars I was forced to use my personal set of copies of maps, which is incomplete. The above statements are made on the basis of the map at $b = 23.2°$, which is 0.6° to 1.2° away from the stars of interest. I am therefore unable to state with absolute confidence that my statements about H I apply at the particular positions of these stars; they may instead apply only at nearby positions). The distance of these clouds is limited by the distance of α Oph, 18 parsec. On this basis Münch and Unsöld assign a diameter of 1 pc. However, as mentioned above, there must be considerable angular structure in this cloud on much smaller length scales. This cloud is strange in showing a large Ca II/Na I line intensity ratio which normally occurs only for nonzero velocity components. The velocity of this component is definitely different from that of the H I main peak, however, so the high ratio is perhaps not surprising. It is tempting to speculate that the North Polar Spur is somehow responsible for this feature.

We summarize Habing's (1969) work here even though strictly speaking it fits in a slightly different category. He found high-velocity components in Ca II data for three stars of Adams (1949) and Münch and Zirin (1961) which he then mapped in the 21-cm line. One of these was associated with a negative intermediate-velocity cloud which was so large in angular extent that he abandoned the job of mapping it. The second had a velocity dispersion of 5 km s^{-1}, central velocity from 20 to 30 km s^{-1} (LSR) depending on position, $\langle N_{H I} \rangle = 0.6 \times 10^{20}$ cm^{-2}, and an angular diameter of 15°. Its distance lies between 140 and 3400 pc. Assuming a distance of 1000 pc, its diameter is 260 pc; its density, 0.08 cm^{-3}; its mass, 24000 $M_\odot$. The density is not inconsistent with that of the noncoincident, probably ionized components discussed above in Section IIb; since both are at intermediate nonzero velocities, they may be similar. The third cloud often showed two velocity peaks with similar angular structure. The H I velocities vary from -40 to -60 km s^{-1} and the feature measures $12°$ by $3°$ in angular extent. If its distance is 1000 pc, its mass is 8000 $M_\odot$ and its density about 0.5 cm^{-3}. As Habing points out, the double-peaked velocity structure may indicate the presence of a strong shock front.

To summarize, thirty years of effort in studying optical absorption lines has not been particularly profitable for the purpose of mapping the angular extents of inter-stellar gas structures. Boundaries have been located only for one small cloud, (Münch and Zirin, 1961), which is itself much smaller than clouds which are normally re-sponsible for optical lines. However, due to the coincidence of the radio and optical data, at least for the main body of gas, the radio results are expected to give the same results as would be obtained from a closely-sampled field of stars in the optical lines. The optical results which do exist are usually in full accord with the known radio results.

We note, however, that those statements refer to comparisons only of equivalent widths and central velocities of the optical and H I lines. If, for example, both sets of lines were subject to the scrutiny of van Woerden's (1967) Gaussian procedure, they might compare less closely. Such comparisons will always be subject to addi-tional uncertainties due to the different beamwidths; however, given the close com-parison existing for some objects (see Figure 1), it is most reasonable to assume that the structural details shown would in fact be identical.

(d) SUMMARY

In the main, the optical interstellar absorption lines arise mainly from gas which is *inter*stellar rather than *circum*stellar. The line width of low velocity gas is always much too large to result from thermal broadening and is therefore determined by macroscopic gas motions. The 8 to 12 optical components per kpc is an overestimate, but the true number is probably larger than the 1 or 2 found from H I absorption, even when accounting for the various velocity components which are seen in optical absorption but not in H I emission.

These noncoincident components occur at nonzero velocities and are almost cer-tainly formed in H II regions. The high temperature may be responsible for the en-hanced Ca II/Na I line intensity ratios in these objects, instead of the usually assumed depletion of Ca in low velocity clouds. These nonzero velocity features have low emission measures and are detectable only with the Na I and Ca II lines themselves. A large amount of suitable observational material presumably already exists in the form of spectra of external galaxies, quasars, and even globular clusters. Optical spectra of extragalactic radio sources would also permit the comparison of optical and radio absorption, which should be fruitful.

III. The Magnetic Field

Parker (1969) has pointed out that the correlation lengths of the gas and magnetic field in interstellar space should be the same if the magnetic energy density is smaller than that of the gas; the field which is frozen into the gas is randomized by the gas motions. Some of this random field component lies in the z direction. This z component is necessary, in the picture of Jokipii and Parker (1969), to enable cosmic rays to travel to high z distances and, ultimately, escape the Galaxy altogether by inflation of the magnetic tubes by cosmic ray pressure (Parker, 1966). (Cosmic ray particles are known to traverse less than a few g cm^{-2} of interstellar matter because of the absence of substantial quantities of nuclei heavier than H[1]). This general picture has generated a series of papers attempting to derive the correlation length for the magnetic field (as opposed to the interstellar gas).

(a) THE CORRELATION LENGTH OF THE FIELD

Jokipii and Parker (1969) used the rms deflections of direction of starlight polarization in the galactic plane as determined by Hiltner (1956). The result was a correlation length of 100 to 300 pc.

Jokipii et al. (1969) have analyzed more recent starlight polarization data (Behr, 1959) in more detail and found similar results. However, the functional form used to match the data was such that equally good fits to the observational data could have been obtained with much different values of the correlation length simply by changing the required mean square value of the magnetic field. We therefore do not regard their quantitative results as being very reliable and hope that they will repeat the analysis with a more extensive set of data, for example the compilation of 7000 stars by Mathewson and Ford (1970). Jokipii and Lerche (1969) analyzed the dispersions of Faraday rotation measure for polarized extragalactic radio sources within a number of bins of galactic latitude and again obtained similar results. We regard this procedure with suspicion because they used a statistical approach which by its very nature requires a large number of independent samples. However, the derived correlation length, about 250 pc (equal to the total thickness of the galactic disk), is so large that the observational data in fact refer to only two or three samples (in this context, a sample is an individual magnetic bulge) over the whole area of sky ($|b| > 10°$) used in their analysis. Their 'dispersions' therefore simply represent differences between the small total number of independent samples visible from the Earth, and the dependence of dispersion on galactic latitude which they found simply reflects the fact that the average rotation measures become bigger in absolute value near the galactic plane because the average magnetic field is directed parallel to the galactic plane. All of this is self-evident in global plots of rotation measure, e.g., Wright (1973), or starlight polarization (Mathewson and Ford, 1970; see Figure 2b). That their derived correlation length is comparable to the thickness of the galactic disk is almost inevitable, given the nature of these global plots and their analysis technique. While it therefore follows that their numerical result is roughly correct, it does not necessarily follow that the result applies anywhere except immediately locally, or that the field really is randomly distributed as they assumed.

(b) THE CORRELATION LENGTHS OF THE FIELD AND GAS: DIFFERENT?

We regard the first determination by Jokipii and Parker (1969) as reasonably reliable.

It has the additional advantage that it refers to a substantial volume of space and is not restricted to the volume within a few hundred parsecs of the Sun. The correlation length for the magnetic field derived by Jokipii and Parker (1969) is larger than the 70-parsec correlation length for the gas derived by Kaplan (1966). Although one would hesitate to take the quantitative difference seriously, there are other data which indicate a scale difference between field and gas. These are the starlight polarization maps of Mathewson and Ford (1970) and the H I photographs of Heiles and Jenkins (1974) (see Figure 2). Although H I filaments arch out from the tops of some of the large cloud complexes, it is clear that at least in some regions (e.g., $l \cong 0°$, $b > 0°$) the large clouds (which would make the largest contribution to the correlation length) appear to be physically smaller than the loops of magnetic field. This difference in angular scale is almost certainly equivalent to a difference in linear scale, since the similarity of the directions of H I filaments and magnetic field strongly suggests a close physical association.

We should not forget that there certainly exist structural features in both the gas and the field on much smaller length scales than those discussed in the above paragraph. This is evident from looking at Figure 2. The large correlation lengths discussed above are those that are, or would be, derived by smoothing out all the small-scale irregularities. If these are not smoothed one will derive a whole spectrum instead of a single correlation length. At some stage it will probably be of theoretical interest whether this spectrum is more intense at small or large length scales. At present neither theory nor observation has really reached that stage of sophistication.

(c) SUMMARY

We believe that the difference in angular scale is real, in which case the z component of the field cannot be produced exclusively by gas motions as pictured by Parker (1969). The general picture of a field being produced exclusively by the gas motions would then also be incorrect. It is not clear to us that the cosmic ray problem actually requires a random field of the sort explicitly envisioned by Jokipii and Parker (1969); one can easily imagine that the cosmic ray particles could still reach substantial z distances and leave the Galaxy by magnetic tube inflation in a more organized (helical? – see Mathewson and Nichols, 1968) field which is not necessarily produced exclusively by random gas motions. It is less easy to imagine that the cosmic ray velocities will be made isotropic by such a more organized field.

The 21-cm data sheds more light on this relationship between gas and field, and strongly suggests the field is controlling the motion of the gas. These data are discussed in Section IV.

IV. H I 21-cm Line Observations

H I 21-cm line studies have the unique advantage over all others that the interstellar gas can be sampled in emission. Hence, the selection of positions to be observed is not limited by the availability of background sources. This very fact, which is of definite advantage in mapping individual features, has one disadvantage: individual features too near the galactic plane cannot be easily observed because of blending with hydrogen at other distances in the line of sight. The majority of work done in H I has been in the galactic plane, related to problems of galactic structure. In this paper we usually restrict our attention to studies at higher latitudes, since we are concerned with the spatial and velocity distribution of interstellar matter on a scale small compared to that of the Galaxy. Furthermore, in the main we consider only low-velocity gas. Much discussion has been aired on the subject of intermediate and high velocity gas in recent years, and the picture is still not clear. We doubt it will become so without high quality surveys of the southern sky. Thus, we avoid this material – which is itself sufficient for another whole symposium, anyway.

A number of H I studies have derived structural properties using statistical arguments applied to profiles taken at positions widely separated compared to the telescope beamwidth. These studies discard the fundamental advantage of emission studies – the ability to directly map structural features – and, in addition, always make some assumptions about the nature of interstellar gas structures in order to do the statistical analyses. This seems to us improper, since the goal is to derive the nature of the structural features in an unbiased manner. Accordingly, we omit such studies from consideration and restrict ourselves in the main to studies in which the gas has been directly mapped by 21-cm line emission, small-scale features noted, and their properties discussed.

The number of extensive mapping programs away from the galactic plane has burgeoned in recent years. This is because, by their very nature, emission studies require the acquisition of many profiles because structures exist on angular scales ranging from less than $10'$ to about $60°$ or so, a range of about 10^5 in solid angle. Much smaller features exist, down to the smallest scale resolved by aperture synthesis techniques in the absorption spectra of some bright radio continuum sources (Greisen, 1973; Elliot *et al.*, 1973); we exclude these from present consideration. Studying the small structures to the exclusion of the large ones not only provides a biased view of the size scales in the interstellar medium, but also fails to show how the various size scales are related. Receivers and automatic data-taking and data-handling equipment of the quality required for really large survey programs became available only in the mid 1960's.

The first definitive emission study showing structural features of the interstellar H I was performed by Helfer and Tatel (1959) in the regions of Perseus and the Pleiades. This early study is remarkable in several respects. It used the first multichannel line spectrometer, of the same type as the one now being used at the IAR in Argentina, but without the aid of automatic digital data recording and analysis equipment. The authors presented data in the now standard formats of contour maps of antenna temperature vs one position coordinate and velocity, and contour maps of column density in certain velocity ranges vs both position coordinates. Their conclusions concerning the structure of the interstellar gas were prophetic. Six gas aggregates were found, none of which fell completely inside the area surveyed, a total of 225 deg². Some of the aggregates were definitely noncircular in shape. Unfortunately, the direction of polarization of interstellar starlight (hereafter referred to simply as 'optical polarization') is confused in this region so no comparisons can be made with the directions of elongation. They noticed that the map of column density for a velocity range of about 20 km s^{-1} showed much less structure than would be seen for a smaller velocity range. Thus they suggested that "many of the fine details in the line profile result from local turbulence rather than from the existence of sizable discrete gas clouds". Although the use of the word 'turbulence' may be incorrect and misleading, this suggestion is fully substantiated by the more extensive and recent studies by Verschuur (see below). If gas and dust are well mixed, as we suspect, there would then be no angular structure of interstellar reddening on small length scales, as has apparently been found (see Section I). This is a discrepancy which is not easily resolved. The reddening data used in those studies are usually confined to the galactic plane which, from our rather uninformed radio astronomical vantage point, makes the results suspect.

Since this early study, structural features have been seen in several forms which are easily classified. Some are definite *clouds*: structures which appear roughly circular on the sky, and are thus most likely roughly spherical in shape. They have diameters ranging from one to perhaps 100 pc with a corresponding range in mass. A surprising number of authors find structures which are distinctly different from clouds: *filaments and sheets*. Within these are sometimes imbedded cloud-like structures. The clouds are sub-condensations within the filaments or sheets, which themselves are single, coherent structures. Sometimes the 'clouds' are instead filamentary and are associated with the interstellar magnetic field. Finally, some of the gas is in a temporary state, having been affected by some (presumably) external process such as a supernova explosion: below we refer to this as *active* interstellar gas, because it definitely shows significant kinematic effects whereas the clouds, sheets, and filaments often show little obvious kinematic structure apart from relative motions of up to about 10 km s^{-1}. Discussion of this active gas violates, to some extent, our avowed intention to avoid the subject of intermediate velocity gas; however, we will restrict ourselves exclusively to studies of gas whose velocities are obviously not (even to Verschuur, 1973a) a result of galactic structure.

(a) CLOUDS

Raimond (1966) found 10 clouds with random shapes in the direction of the stellar associations I Mon and II Mon. Half of these seem to be related to the stellar associations themselves, and the others to dark clouds. Those associated with dark clouds are not dissimilar to that observed in the 21-cm line by Simonson (1973), who derives $M = 3400\ M_\odot$ and $n_H = 2.2\ \text{cm}^{-3}$ for the dark cloud Kh 713 (Khavtassi, 1960). The clouds associated with I Mon have masses ranging up to 90000 $M_\odot$ and diameters to 115 pc; those associated with II Mon are less massive by a factor of 10 or so. The densities are the usual 5 to 15 cm^{-3} found for most interstellar gas aggregates when seen in the 21-cm line. In addition to clouds, Raimond found an expanding shell around the Rosette nebula which is discussed below as 'active' gas.

Heiles (1967) studied the region $l = 100°$ to 140°, $b = 13°$ to 17°, and found a host of cloud-like structures. He divided these into two classes, large clouds and 'cloudlets'. Large clouds typically have mass $\cong 3000\ M_\odot$, density $\cong 5\ cm^{-3}$, and diameter $\cong 30$ pc. These values contrast with those of the cloudlets, which typically have mass $\cong 4\ M_\odot$, density $\cong 2\ cm^{-3}$, and diameter $\cong 5$ pc. Field and Hutchins (1968) point out that the large clouds and the cloudlets fit roughly onto the mass spectrum of interstellar clouds derived from the interstellar reddening data by Scheffler (1967a, b) but that there is an absence of objects in the mass range 24 to 280 $M_\odot$. This absence may simply reflect Heiles' bias in selecting objects for study, but it also may be real. All of the large clouds and cloudlets were found to exist in two huge sheet-like structures extending tens of degrees in angle and thus having characteristic dimensions of perhaps 100 pc; the material within each sheet moves coherently, apart from small random motions whose dispersion amounts to about 2.1 and 3.4 km s^{-1} for the two sheets.

Another huge aggregate was studied in detail by Riegel and Crutcher (1972). This object is distinguished by its self-absorption dip, seen in all profiles over an area 30° in longitude by 12° in latitude centered near $l = 15°$, $b = 3°$. The object is seen at three distinct velocities and has $N_H \leqslant 1.2 \times 10^{20}$. It contains much small-scale structure. Mass estimates are rendered very uncertain due to its completely uncertain distance, which is probably in the range of a few dozen to 1000 pc. The authors derive densities of 0.1 to 0.5 by assuming the gas is spread out uniformly over the line of sight to the upper distance limit of 900 pc. In fact we argue that the structure is much more compressed along the line of sight simply because it has a characteristic angular size of about 20° and contains a profusion of small-scale structure. If the distance is what we consider more reasonable, 100 pc, the density would probably be about 30 times larger, i.e., 3 to 15 atoms cm^{-3}, comparable with most other H I concentrations seen in the 21-cm line. Given its low temperature, one would a priori expect a density at least this large, but probably not excessively larger due to the absence of dust and molecules which always seem to appear in detectable amounts in dense concentrations.

A similar feature, but at a somewhat higher temperature, has been found by Baker (1973a) near $l = 180°$, $b = -15°$ to $-40°$. Such objects cannot appear in simple self-absorption at intermediate galactic latitudes because of the absence of a bright background temperature in the 21-cm line. Baker has invented a new analysis technique, the 'deviation defect method', to detect such objects; use of this technique should provide a new look at the colder portions of the interstellar gas in the coming years.

Van Woerden (1967) mapped 140 deg^2 of the Orion region, distinguishing 31 clouds by careful Gaussian analysis of all the profiles. He mapped the clouds by mapping the Gaussian parameters and found that the clouds are not dissimilar from the 'standard' interstellar cloud (Spitzer, 1968b). Both the velocities and shapes appear randomly distributed, in accord with the standard cloud model. Assuming a distance of 280 pc as used in Table I rather than the 480 pc used by van Woerden, cloud diameters are typically 25 pc, densities a few atoms cm^{-3}, and masses about 1000 $M_\odot$. (These values were estimated only very crudely from information given in van Woerden's paper.) The character of his results seems different from all the others quoted above. There are probably two reasons for this. First of all, the Orion region is a very special region in the sky. Examination of the contour maps of Heiles and Habing (1973) shows that the 21-cm line is stronger and contains more structure with larger, more extensive velocity gradients than for the typical region of interstellar space. The Orion nebula itself could not be responsible for this kinematical behavior because it is so young. Although older star associations exist in the same region, any old H II regions produced by them could not be responsible because the total extent of the disturbed region is huge, extending almost 30° in angle. We believe instead that the conditions which enable stars to form – presumably the preponderance of self-gravitation over disruptive forces – are themselves generated by, or are themselves responsible for, the kinematical structure. That is, we feel that the cause of the kinematical structure lies in gas-dynamical processes residing within the gas itself, rather than in any external influence such as hot stars. This feeling is based on little evidence, however. We state the above partially to justify the distance estimates for the gas used in Table I, which (for lack of definite evidence one way or the other) differ from the distance to the Orion nebula and the older stellar associations in its vicinity.

The second reason why van Woerden's results differ from those of others might be his extensive and thorough use of the technique of Gaussian analysis of individual profiles. Other observers usually discern structural features directly from contour maps. It would be extremely useful to see what van Woerden's data would look like if presented in the more usual form of contour maps of antenna temperature vs one position coordinate and velocity, or of contour maps of column density vs. both position coordinates for specific velocity ranges. We hope that the conclusions drawn are independent of the details of the analysis process; however, only a thorough comparative investigation will provide a definitive answer to this urgent question.

Finally, we come to the extensive high-resolution studies of Verschuur and his collaborators, made with the NRAO 300-ft telescope, the largest used for this type of work. Knapp and Verschuur (1972) analyze the properties of cold (temperature $\cong 25$ K) clouds near $l \cong 230°$, $b \cong 45°$. Physical properties are derived by assuming pressure equilibrium with the intercloud medium ($n_{\mathrm{H}}T$ taken as 2000 K cm^{-3}); distance $\cong 19$ pc, mass $\cong 0.2\,M_{\odot}$, $n_{\mathrm{H}} \cong 50$ cm^{-3}, and diameter $\cong 0.5$ pc for one and similar values for the other cloud.

Verschuur (1973c, d) has surveyed more than 500 deg^2 in two large regions, designated A and B, centered near $l = 235°$, $b = 44°$, and $l = 215°$, $b = 27°$, respectively. He (1973b) also observed the large filament, and its internal cloudy structure, which arches out of the Perseus-Orion gas aggregate near $l = 180°$, $b = -55°$, visible clearly in Figure 2a and here called region E (this nomenclature is not used by Verschuur).

In all of these regions he finds many small clouds of mass $\cong 10\,M_{\odot}$, diameter 3 pc, and density 40 cm^{-3}. Many of these clouds are really elongated filaments, parallel to the magnetic field; they are discussed more fully in Section IVb. Heiles's cloudlets, which are smaller and less dense, seem to be similar to these in qualitative aspects (except perhaps with regard to orientation of the magnetic field). Verschuur finds higher velocity dispersion, however: 4 to 5 km s^{-1} in regions A and B and the major

H I filament, with somewhat higher values in two minor filaments associated with the major filament. The velocity dispersions within Heiles's sheets are smaller, only about 3 km s^{-1}.

(b) SHEETS AND FILAMENTS

(i) *Structure*

We have recently learned that sheets and filaments – especially the latter – are very common in the interstellar medium. Sheets appear to have been seen first by Heiles (1967) in his detailed study of the region $l = 100°$ to $140°$, $b = 13°$ to $17°$. These sheets are coherent over a large fraction of the whole area surveyed, or perhaps even a larger area. Motions within a sheet are small, with dispersions of 2.1 and 3.4 km s^{-1} for two sheets. The velocity of a sheet varies slowly, if at all, with position. Another sheet has been found in the Taurus region by Baker (1973b), who suspects the sheets may form a ring-like structure around the solar neighborhood. Herbig (1968), using optical techniques, finds a large sheet in front of ζ Oph from interstellar absorption lines.

Sancisi and van Woerden (1970) saw a filamentary-shaped feature near $l = 350°$, $b = 20°$. This object is barely visible on Figure 2. It is 4° wide and 14° long, perhaps longer because it may extend beyond their southern declination limit of $-30°$. Its linear dimensions are probably about 15×45 pc, with a density about 2 cm^{-3}. This object is elongated perpendicular to the magnetic field. However, it is bounded on one side by dust (which would tend to convert the H I into H$_2$, see Heiles, 1971) and on the other by bright nebulae (converting H I to H II). Thus its apparent elongation as seen in the 21-cm line of H I is probably simply a result of conversion of H I to other forms. In this case its orientation with respect to optical polarization is irrelevant, and it is a filament of a completely different class than those discussed immediately below.

A large number of filaments have recently been seen in the interstellar gas. Many of these are apparently aligned with optical polarization, hence presumably the magnetic field. Of course, this result is not without its precedents; dust filaments on widely different length scales have long been known to be aligned with the magnetic field. Shajn (1955) finds large dust filaments, degrees in angular size, oriented parallel to the field in large regions in Perseus, Taurus, Cygnus, Ophiuchus, and near IC 1396. He and others also find the very small filaments in the Pleiades oriented parallel to the magnetic field (see review by Hall and Serkowski, 1963). Furthermore, dust clouds and intermediate-velocity clouds (Verschuur, 1970a) and some low-velocity clouds (Verschuur, 1970b and private communication; Heiles and Jenkins, 1974) are aligned with the field.

Figure 2 (a) shows the large filaments seen in H I especially well. Near $l = 0°$ at high positive latitudes a number of filaments appear to arch out of the large Ophiuchus cloud where Gould's belt achieves its maximum displacement to positive latitudes (see Table I); these are also parallel to optical polarization. An apparently similar filament arches out of the Orion-Perseus gas complex near $l = 180°$ at negative latitudes; the scant optical data which exist in this region also suggests alignment with the interstellar magnetic field. This filament, in contrast to the one just mentioned, contains appreciable velocity structure along its length. A portion of this filament has been studied in detail, at high resolution, by Verschuur (1973b). Yet another large filament rises perpendicular to the galactic plane near the north celestial pole ($l \cong 120°$, $b \cong 30°$); again, optical polarization data is scanty here, but suggests alignment. This filament has appreciable velocity structure both along and perpendicular to its axis. Some of the filamentary structures are extremely long, extending up to 80° or so in projection on the sky. A typical large filament has density $\cong 30$ cm^{-3}, diameter $\cong 2$ pc, and mass $\cong 6\, M_{\odot}$ per parsec of length.

Long filamentary structures similar to those seen in Figure 2 have been observed in regions other than the solar neighborhood. (All those seen in Figure 2a are located at intermediate to high galactic latitudes, and are therefore nearby, probably all within a few hundred parsecs and some within 100 pc). Vieira (1971) has examined a portion of the southern sky near $l = 306°$, $b = 6°$, and finds two prominent filamentary structures and an elongated cloud (which might in fact be another filament). The two

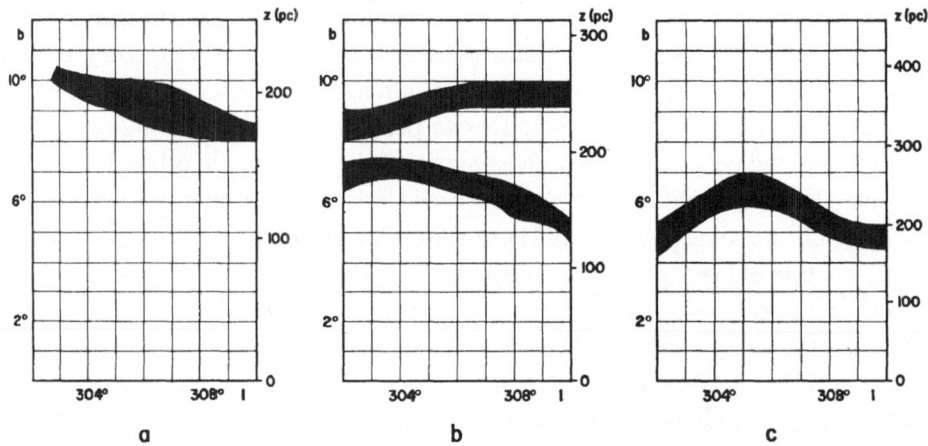

Fig. 4.   Filamentary structure at a distance of 1 to 2 kpc. From Vieira (1971).

filaments were seen at two velocities, one at -20 and one at -30 km s^{-1}. The -20 km s^{-1} feature is really a double filament with $z \cong 160$ and 240 pc, both of which run more or less in a straight line across the whole extent of the observed region. The -30 km s^{-1} feature has $z \cong 190$ to 250 pc, variable because this filament meanders crookedly. Both the elongated cloud and the double filament at -20 km s^{-1} are aligned parallel to the optical polarization; we cannot specify the polarization direction for the other filament because it is too distant (2.2 kpc). Physical properties must be derived using kinematical distances, which are rather uncertain since the radial velocities are not large. All have projected densities $\cong 10^{21}$ cm^{-2}, diameters of almost 1° (about 30 pc), and volume densities $\cong 10$ cm^{-3}. The latter two parameters may be affected to a significant extent by the angular resolution of the telescope, about 0.5°.

All the filaments contain about 30 $M_\odot$ per parsec of length, not inconsistent with that for the nearby filaments discussed above, especially considering the uncertainties in distance.

Weaver's (1974) map of the z extensions of the outer arm of the Galaxy also shows the presence of filamentary structure. Apparently, then, the existence of the large filaments, at least some of which seem to arch up out of the galactic plane, is common.

There are other intermediate and high galactic latitude regions which contain filamentary structure with a smaller characteristic angular size, and therefore presumably linear size as well. Dieter (1964, 1965) surveyed the galactic poles, in particular the North Galactic Pole. She found definite cloudlike structures, some at normal low velocities and some at the characteristic intermediate negative velocities. Here, of course, we concentrate on the low velocity gas. She found several condensations with properties very close to those of the 'standard cloud' (Spitzer, 1968b): diameter 7 pc, $n_H = 10$ cm^{-3}, and mass $= 30$ $M_\odot$. However, these clouds are elongated, in a direction parallel to the magnetic field as inferred from both optical (Mathewson and Ford, 1970) and radio (Bingham, 1966; Spoelstra, 1972) polarization. These filaments lie just outside of the North Polar Spur in the region where it reaches its maximum extent in galactic latitude.

Near $l = 40°$ for positive latitudes nearer to the galactic plane there is also filamentary structure, here in profusion (see Figure 2a). These also lie just outside the NPS. They are aligned perpendicular to the galactic plane, and parallel to the magnetic field as inferred from both optical (Mathewson and Ford, 1970) and radio (Spoelstra, 1971, 1972) polarization. They are partially contained in Verschuur's (1973d) regions C and D, studied with the high angular resolution afforded by the NRAO 300-ft telescope. Verschuur (1973c) also found filamentary structure aligned with the magnetic field in regions A and B. He confirms the general impression gained from Figure 2a.

Other regions contain filamentary structure whose orientation with respect to the magnetic field is presently uncertain, mainly because of insufficient optical polarization data. These occur at positive latitudes for $l = 180°$ to 270° and at negative latitudes for $l = 30°$ to 140°. Many of these are extremely delicate and beautifully shaped and can be fully appreciated only by staring at the relevant areas of Figure 2a for a minute or so.

(ii) *Internal Motions*

What are the velocity fields within these filaments? First we consider the very long filaments visible in Figure 2a. Little in the way of detailed analysis has been done, but a qualitative feeling can be obtained from the color photographs in which color indicates velocity (Heiles and Jenkins, 1974). These show that some filaments have essentially no relative Doppler motions along their length while some show considerable velocity differences but with small velocity gradients, as indicated above. The one near $l = 120°$, $b = 30°$, has rather chaotic velocity structure along both its length and width. The kinematics therefore depends on which filament we examine. However, we must remember that we can only measure the line-of-sight component of velocity, and unfortunately the line-of-sight geometry (i.e., distance as a function of position along a filament) is unknown. Even for those filaments showing relative motion with position along the length, we cannot say whether the gas is expanding, contracting, or moving perpendicular to the axis of the filament.

The smaller scale filaments discussed by Verschuur (1973b, c, d) in regions A, B, C, D, and E show kinematical structure whose magnitude depends on the regions. The sense of the structure is always the same, with velocity differences occurring along the length of the filament. Region A shows only small velocity gradients, regions B and E intermediate gradients, and regions C and D very large gradients. Only in one case do we have definitive information on the line-of-sight field geometry, in regions C and D. This comes from optical and radio polarization studies. These show that the component of the field projected on the plane of the sky runs nearly perpendicular to the galactic plane, and of course parallel to the filaments; the lack of Faraday rotation in the radio shows that there can be very little component of the field parallel to the line of sight.

Therefore, since the Doppler gradients run along the length of the filaments which are themselves aligned with the magnetic field and have no line-of-sight component, we have the following possibilities:

(i) The gas is actually moving perpendicular to the magnetic field. This possibility violates the fundamental theoretical precepts of magnetogasdynamics.

(ii) The filaments are not really aligned parallel to the magnetic field, but only appear so in projection on the sky. This possibility requires a favored location of the Earth, since the apparent alignment of field and filaments appear in several regions located in completely different parts of the sky.

(iii) There are no electrons in these regions, so that a line-of-sight field component would remain undetected by Faraday rotation studies. This violates the generally accepted theories and observations of the ionization balance in the interstellar gas.

(iv) The gas is moving parallel to the field and the field has a very small inclination to the line of sight. This would mean that the observed velocities represent only a tiny fraction of the true gas velocities because of projection effects.

(v) Both the gas and field are moving perpendicular to the magnetic field together as in an Alfvén wave. The waves would be propagating either towards or away from the galactic plane, in a direction along the magnetic field but perpendicular to the line of sight. Alfvén waves, being transverse waves, would allow us to see the mass motion involved in the transverse vibration.

The last possibility seems the only acceptable one. Verschuur's maps show significant velocity changes in angles of a few degrees; given the distance to these filaments (about 60 pc, from optical polarization work) this implies a wavelength scale of a few pc. If the Alfvén velocity is 10 km s^{-1}, the relevant time scale is about 10^5 yr. This is comparable to the cooling time scale. The Alfvén wave is strong enough so that compressibility of the gas cannot be neglected. A theoretical treatment of these waves would not be simple.

Although the above argument is based on only a cursory look at Verschuur's data, the fundamental conclusion that we see motion of gas and field together in a direction perpendicular to that of the average field direction seems almost inescapable.

(c) ACTIVE REGIONS

As a preliminary, we emphasize that all masses quoted below are HI only as derived from the 21-cm line. The velocities of these objects are large enough so that one would naively expect most of the gas to be HII. Thus, total masses may be seriously under-estimated.

A number of authors have found HI in a kinematical state which is apparently produced by an external event, such as a supernova explosion. The most clear-cut case of this specific type of association, i.e., with the supernova HB 21, has been made by Assousa and Erkes (1973). They find a shell of 3000 $M_{\odot}$, density 2.5 cm^{-3}, radius 25 pc, expansion velocity 25 km s^{-1}; the dependence of velocity on position mimics that of an ideal uniform expanding shell to a surprising degree, given the expected degree of density fluctuation in the interstellar medium on that length scale.

An apparently expanding, very large (diameter 30°) shell of HI with an apparent center of $l = 310°$, $b = 45°$, was found by Fejes (1971). It is characterized by triple-peaked profiles and has an expansion velocity of 30 to 40 km s^{-1}. Its total mass is perhaps 2000 $M_{\odot}$. One edge of this object is close to the southern declination limit of the Dwingeloo telescope, and it would be desirable for the southern observers to survey the nearby portion of sky to ensure that the object is really self-contained within a diameter of 30°.

Another expanding shell around the Rosette nebula was found by Raimond (1966). Its expansion velocity is 25 km s^{-1}, diameter 8 pc, mass 4000 $M_{\odot}$. This is an example of an expanding shell presumably produced by an object less energetic than a super-nova, although the amount of kinetic energy in the shell is nevertheless very large.

A number of observers have found correlations of HI and the radio continuum loops. The correlation of HI and the North Polar Spur (NPS), or Loop I, is quite apparent (Berkhuijsen et al., 1971). This correlation takes two forms. One is the correlation of column density HI with the outer (?) *gradients* of the continuum loops. This is evident in Figure 2a for the NPS; a comparison with the radio continuum map of Berkhuijsen (1971), particularly when presented in photographic form (Heiles and Jenkins, 1974), shows very clearly that the hydrogen is lined up just like the continuum, but about 5° higher in galactic longitude. The clouds of Dieter (1965) near the North Galactic Pole continue this association and are also elongated in the direction of the optical polarization, which is in turn aligned with the NPS (Loop I).

The second form taken by the HI continuum correlation concerns the velocity structure. Within five degrees longitude of the points where Loops I and III cross $b = 30°$ there are relatively small-diameter features with very high velocity dispersions, greater than 40 km s^{-1}. Correlations with other loops discussed by Berkhijsen et al. (1971) are not convincing to this author. Fejes and Verschuur (1973), however, find what appears to be a significant correlation of HI density and velocity structure with Loop III. At $l=90°$, where loop III crosses $b=+17.8°$, there is a deficiency of H I relative to that at surrounding longitudes; at $l=155°$, where Loop III crosses the same latitude again, the H I profile splits into two components separated by about 15 km s^{-1}. Maps of H I column density within narrow velocity intervals centered at -32, $+5$, and $+12$ km s^{-1} show that the hydrogen structure is oriented parallel to the loops in this region. The positive velocity components are the same cloud seen in optical Na I by Hobbs (1969) in front of the Pleiades, discussed in Section IIc of the present paper. Whether the correlations found by Fejes and Verschuur (1973) are truly significant or simply accidental can best be determined by studying the H I-continuum relation around the whole perimeter of Loop III. Although examination of Figure 2a does not suggest that the correlation will necessarily continue around the whole perimeter, Figure 2a is made by integrating over a 40 km s^{-1} interval in velocity. Fejes and Verschuur show that one must consider smaller velocity intervals to make many of these elongated H I features stand out.

Finally, there may exist other somewhat less active regions in which H I is influenced by stars. Baker (1973a) has found a number of these, which are easily visible on the contour maps of Heiles and Habing (1973). Verschuur (1969) has found what appears to be a tunnel in H I produced by a B2 star moving through the gas. The tunnel is about 10 pc in diameter, 31 pc long, and contains 500 $M_\odot$ at density 9 cm^{-3} of what is now H II. It shows as a relative deficiency of H I emission. This seems vaguely reminiscent of the ideas of Hills (1972) concerning the effects of stars on the structure of H I and H II in the interstellar medium. Such interactions between hot stars and gas in relative motion might represent a mechanism to provide significant kinetic and perhaps thermal energy input to the gas.

(d) SUMMARY

In his latest studies Verschuur reaches the same conclusion regarding the spatial velocity structure of the gas as did Helfer and Tatel (1959) in the earliest study. Much more structure is visible in column density maps made from small velocity ranges than in the corresponding map from a large velocity range. Thus, to repeat the words of Helfer and Tatel (1959), "many of the fine details in the line profile result from local turbulence rather than from the existence of sizable discrete clouds." When the "turbulence" is integrated over velocity, regions look quite smooth in the distribution of column density. It is then hard to understand how the analyses of the angular distribution of interstellar reddening arrive at small length scales (see Section I), and we instinctively suspect (without having seriously investigated the question) these analyses to be incorrect in this regard. The clouds found with today's modern instrumentation are very small, typically with mass $\cong 20$ $M_\odot$, density $\cong 50$ cm^{-3}, and diameter $\cong 3$ pc. This phenomenon of small-scale structure in velocity and position, having been noticed in all regions observed with the narrow beam of the NRAO 300-ft, may be ubiquitous, at least within large gas aggregates.

However, the turbulence referred to by Helfer and Tatel, and by many others more recently, is not turbulence in the rigorous sense. The word 'turbulence' connotes the existence of momentum transfer by turbulent viscosity (cloud collisions, in astronomical terminology). These do not necessarily occur. Both the shapes and velocities of Verschuur's 'turbulent' elements are highly organized with respect to the magnetic field. One gains the impression that the gas is moving along the magnetic field, and in one case we see something resembling an Alfvén wave. The field, in turn, has uniformity on an angular scale which is much larger than the 'turbulent' elements. Like cars travelling on a highway, they might well never collide – they are all moving in the same direction together. The qualitative difference between turbulence and the observed situation in these regions is immense, and we must be less capricious in our use of this word.

The randomness in position implied by the interstellar cloud model does not usually exist. The delicate filaments seen in regions containing no large gas aggregates in Figure 2a can hardly fit the spirit of the cloud model. Neither do the large filaments, nor the gas associated with the North Polar Spur in which we apparently see Alfvén type motions, nor do the active regions.

The internal structural details of the large gas aggregates themselves have not yet been adequately studied. One of the few studies of these is van Woerden's (1967); if his results are representative, and there is reason to believe they are not, the standard cloud model might indeed apply *within* the aggregates.

V. Dust Clouds

Dust clouds are usually invisible in the H I line (Heiles, 1969; Mahoney, 1972) because of conversion to H_2. This conversion was predicted several years ago (Hollenbach *et al.*, 1971), and recent UV data have completely corroborated the theoretical predictions (Carruthers, 1970; Spitzer *et al.*, 1973). The properties of dust clouds have been discussed in several recent review papers (Heiles, 1971; Rank *et al.*, 1971; Gordon and Snyder, 1973; Solomon, 1973) and accordingly we shall treat this subject very briefly and emphasize some points which we feel to be particularly important.

The amount of dust contained in a cloud is easily estimated if the distance and extinction are known. The distance can readily be obtained, at least approximately. Often the extinction is so high that essentially nothing can be seen through the cloud; in such cases only lower limits can be obtained. Grasdalen (1973) has analyzed the appearance of small reflection nebulae within some dark clouds and finds that these lower limits may be smaller than the true extinctions by a factor of 10 or so.

Once the amount of dust is known the total mass can be derived simply by assuming that hydrogen and helium are not contained in the dust grains. A lower limit to the total mass is then obtained simply by multiplying the mass of dust by the mass ratio of (hydrogen + helium)/heavy elements, a factor of about 100. The derived masses are lower limits, often in two senses, then: one, the derived mass of dust is itself a lower limit; two, not all the heavy elements exist in the form of solid dust particles (the observation of molecular lines assures us that this is the case).

There is an independent and much more direct way to obtain a lower limit to the mass of gas in the clouds. Many molecular lines require high collision rates to bring their excitation temperature above the 3 K blackbody temperature so that they can be seen in emission. Collisions with neutrals, particularly H_2, is the generally accepted mechanism for this excitation. In many cases the excitation temperatures are known quite well, and this provides lower limits on the required H_2 volume density. This, together with the angular extent of the molecular emission and a distance estimate for the cloud obtained optically, provide a lower limit for the H_2 mass directly. Radiation trapping for optically thick molecular lines can reduce the required density to a degree which is currently uncertain, but these effects are probably important only for the very large clouds near the galactic center and H II regions, rather than the usual interstellar dust clouds.

Lower limits for mass derived using the two methods agree quite well. Derived densities range up to 2400 cm^{-3}, masses to 2000 $M_\odot$, and radii to 6 pc (see Hollenbach *et al.*, 1971). Self-gravity may overcome internal pressure in many clouds; numerical estimates come close to the borderline. If the true values of mass are much larger than the lower limits – for example, if Grasdalen's extinction estimates are indeed both correct and representative – then gravity certainly dominates in at least some clouds. This is perhaps surprising in view of the tendency for elongated dust clouds to be aligned parallel to the interstellar magnetic field (Shajn, 1955; Hall and Serkowski, 1963; Verschuur, 1970a). Perhaps the magnetic field pressure is stronger than the internal gas pressure. There is some observational support for this view, since grains in the Ophiuchus cloud show a high degree of alignment (Carrasco *et al.*, 1973).

Clouds located near H II regions or the galactic center dwarf the usual interstellar dust clouds discussed above. The distinction between these may not be too sharp, however. An intermediate case has been found by Kutner *et al.* (1973), who have detected a 10^5 $M_\odot$ cloud in the vicinity of Orion B, far enough away so that it is not really associated directly with this H II region. The clarification of this situation requires a large scale CO survey.

Hollenbach *et al.* (1971) have performed a statistical analysis of the dust clouds in Lynds' (1962) catalog with the aim of deriving the total mass resident in these clouds. They find the astounding result that *fully 40% of all the interstellar matter is located in dust clouds, and is invisible* to radio astronomers because the H I has turned to H_2. The distribution of these clouds in the sky, shown in Figure 2d from a catalog on computer cards kindly supplied recently by Lynds, shows that the distribution of these clouds is not random. Most are imbedded within the large gas complexes visible in the H I map of Figure 2a. Although this may affect the statistical results of Hollenbach *et al.* to some degree, the basic conclusion remains. A large fraction of the gas in these complexes is located in very dense clouds where the hydrogen is mainly H_2.

Do these clouds move randomly, with the resultant possibilities of cloud-cloud collisions? Probably not. The gas they now contain must have flowed along the magnetic field when the cloud was forming; it seems unlikely, then, that two clouds could exist close together on the same tube of magnetic flux. Thus they can collide only by moving across the field. But the dust clouds are elongated in the direction of the field (Verschuur, 1970a), so the field is important in their dynamics and the cloud cannot move across the field. Therefore the clouds can collide only rarely, if ever. They should evolve completely independently.

VI. Summary

A comprehensive summary of this paper would simply repeat the points emphasized in the summaries of individual sections. Comments showing our disapproval of the standard cloud model have been liberally interspersed among the material of this paper. Accordingly, a summary seems superfluous.

For those readers who cannot afford the time to read the entire article, the sections having the broadest range of interest are probably Sections IVb, V, and the summaries of Sections II and IV.

Acknowledgements

I thank a number of people, and particularly G. Verschuur, for preprints. Dr B. Lynds kindly provided a copy of her dark and bright nebula catalogs on computer cards. This work was supported by the National Science Foundation. In addition, I wish to thank the U.S. National Academy of Sciences, the University of California, and the American Astronomical Society for travel support.

References

Adams, W. S.: 1949, *Astrophys. J.* **109**, 354.
Ambartsumian, V. A. and Gordeladse, S. G.: 1938, *Abastumansk. Obs. Bull.* **2**, 37.
Ames, S. and Heiles, C.: 1970, *Astrophys. J.* **160**, 59.
Assoua, G. E. and Erkes, J. W.: 1973, *Astron. J.* **78**, 885.
Baker, P. L.: 1973a, submitted to *Astron. Astrophys.*
Baker, P. L.: 1973b, submitted to *Astron. Astrophys.*
Behr, A.: 1959, *Nach. Akad. Wiss. Göttingen* **126**, 201.
Berkhuijsen, E. M.: 1971, *Astron. Astrophys.* **14**, 359.
Berkhuijsen, E. M., Haslam, C. G. T., and Salter, C. J.: 1971, *Astron. Astrophys.* **14**, 252.
Bingham, R. G.: 1966, *Monthly Notices Roy. Astron. Soc.* **134**, 327.
Blaauw, A.: 1952, *Bull. Astron. Inst. Neth.* **11**, 459.
Bok, B. J.: 1956, *Astron. J.* **61**, 309.
Buscombe, W. and Kennedy, P. M.: 1968, *Monthly Notices Roy. Astron. Soc.* **139**, 417.
Carrasco, L., Strom, S. E., and Strom, K. M.: 1973, *Astrophys. J.* **182**, 95.
Carruthers, G.: 1970, *Astrophys. J. Letters* **161**, L81.
Chandrasekhar, S. and Münch, G.: 1952, *Astrophys. J.* **115**, 103.
Clark, B. G.: 1965, *Astrophys. J.* **142**, 1398.
Dieter, N. H.: 1964, *Astron. J.* **69**, 288.
Dieter, N. H.: 1965, *Astron. J.* **70**, 552.
Elliot, D., Wright, M., Kiegel, K., and Heiles, C.: 1973, in preparation.
Field, G. B. and Hutchins, J.: 1968, *Astrophys. J.* **153**, 737.
Fejes, I.: 1971, *Astron. Astrophys.* **11**, 163.
Fejes, I. and Verschuur, G. L.: 1973, *Astron. Astrophys.* **25**, 85.

Goldstein, S. J. and MacDonald, D. D.: 1969, *Astrophys. J.* **157**, 1101.

Goniadzki, D.: 1972, *Astron. Astrophys.* **17**, 378.

Gordon, M. A. and Snyder, L. E.: 1973, *Molecules in the Galactic Environment*, Wiley, New York.

Grasdalen, G.: 1973, private communication.

Greisen, E. W.: 1973, *Astrophys. J.* **184**, 363.

Habing, H. J.: 1969, *Bull. Astron. Inst. Neth.* **20**, 171.

Hall, J. S. and Serkowski, K.: 1963, *Stars and Stellar Systems* **3**, 293.

Heiles, C.: 1967, *Astrophys. J. Suppl.* **15**, 97.

Heiles, C. and Habing, H. J.: 1973, *Astron. Astrophys. Suppl.* **14**, 1.

Heiles, C. and Jenkins, E.: 1974, this volume, p. 65 and p. 625, and in preparation.

Helfer, H. L. and Tatel, H. E.: 1959, *Astrophys. J.* **129**, 565.

Herbig, G. H.: 1968, *Z. Astrophys.* **68**, 243.

Hills, J. G.: 1972, *Astron. Astrophys.* **17**, 155.

Hiltner, W. A.: 1956, *Astrophys. J. Suppl.* **2**, 389.

Hobbs, L. M.: 1969a, *Astrophys. J.* **157**, 135.

Hobbs, L. M.: 1969b, *Astrophys. J.* **158**, 461.

Hobbs, L. M.: 1971, *Astrophys. J.* **166**, 333.

Hollenbach, D. J., Werner, M. W., and Salpeter, E. E.: 1971, *Astrophys. J.* **163**, 165.

Howard, W. E., III, Wentzel, D. G., and McGee, R. X.: 1963, *Astrophys. J.* **138**, 988.

Hughes, M. P., Thompson, A. R., and Colvin, R. S.: 1971, *Astrophys. J. Suppl.* **23**, 323.

Jokipii, J. R. and Lerche, I.: 1969, *Astrophys. J.* **157**, 1137.

Jokipii, J. R. and Parker, E. N.: 1969, *Astrophys. J.* **155**, 799.

Jokipii, J. R., Lerche, I., and Schommer, R. A.: 1969, *Astrophys. J. Letters* **157**, L119.

Kaplan, S. A.: 1966, *Interstellar Gas Dynamics*, Pergamon Press, New York, p. 109.

Khavtassi, J.: 1960, *Atlas of Galactic Dark Nebulae*, Abastumani Astrophysical Observatory.

Knapp, G. R. and Verschuur, G. L.: 1972, *Astron. J.* **77**, 717.

Kutner, M., Thaddeus, P., and Tucker, K. D.: 1973, *Astrophys. J. Letters* **186**, L13.

Lynds, B.: 1962, *Astrophys. J. Suppl.* **7**, 1.

Mahoney, M. J.: *Astrophys. J. Letters* **12**, 43.

Marschall, L. A. and Hobbs, L. M.: 1972, *Astrophys. J.* **173**, 43.

Mathewson, D. S. and Ford, V. L.: 1970, *Mem. Roy. Astron. Soc.* **74**, 139.

Mathewson, D. S. and Nicholls, D. C.: 1968, *Astrophys. J. Letters* **154**, L11.

Mezger, P. G. and Höglund, B.: 1967, *Astrophys. J.* **147**, 490.

Münch, G.: 1952, *Astrophys. J.* **116**, 575.

Münch, G. and Unsöld, A.: 1962, *Astrophys. J.* **135**, 711.

Münch, G. and Zirin, H.: 1961, *Astrophys. J.* **133**, 11.

Oort, J. H.: 1953, *Bull. Astron. Inst. Neth.* **12**, 177.

Parker, E. N.: 1966, *Astrophys. J.* **145**, 811.

Parker, E. N.: 1969, *Astrophys. J.* **157**, 1129.

Pottasch, S. R.: 1972, *Astron. Astrophys.* **20**, 245.

Radhakrishnan, V. and Goss, W. M.: 1972, *Astrophys. J. Suppl.* **24**, 161.

Raimond, E.: 1966, *Bull. Astron. Inst. Neth.* **18**, 191.

Rank, D. M., Townes, C. H., and Welch, W. J.: 1971, *Science* **174**, 1083.

Rickard, J. J.: 1972, *Astron. Astrophys.* **17**, 425.

Riegel, K. W. and Crutcher, R. M.: 1972, *Astron. Astrophys.* **18**, 55.

Routly, P. M. and Spitzer, L., Jr.: 1952, *Astrophys. J.* **115**, 227.

Sancisi, R. and van Woerden, H.: 1970, *Astron. Astrophys.* **5**, 135.

Savage, B. and Jenkins, E.: 1972, *Astrophys. J.* **172**, 491.

Scheffler, H.: 1967a, *Z. Astrophys.* **65**, 60.

Scheffler, H.: 1967b, *Z. Astrophys.* **66**, 33.

Schluter, A., Schmidt, H., and Stumpff, P.: 1953, *Z. Astrophys.* **33**, 194.

Shajn, G. A.: 1955, *Astron. Zh.* **32**, 381.

Shane, C. D. and Wirtanen, C. A.: 1967, *Pub. Lick. Obs.* **22**, 1.

Simonson, S. C., III: 1973, *Astron. Astrophys.* **23**, 19.

Solomon, P. M.: 1973, *Phys. Today* **26**, 32.

Spitzer, L.: 1968a, *Diffuse Matter in Space*, Interscience Publishers, p. 82.

Spitzer, L.: 1968b, *Stars and Stellar Systems* **7**, 1.

Spitzer, L., Drake, J. E., Jenkins, E. B., Morton, D. C., Rogerson, J. B., and York, D. G.: 1973, *Astrophys. J. Letters* **181**, L116.

Spoelstra, T. A. Th.: 1971, *Astron. Astrophys.* **13**, 237.

Spoelstra, T. A. Th.: 1972, *Astron. Astrophys.* **21**, 61.

Takakubo, K.: 1967, *Bull. Astron. Inst. Neth.* **19**, 125.

Verschuur, G. L.: 1969, *Astron. J.* **74**, 597.

Verschuur, G. L.: 1970a, in H. J. Habing (ed.), 'Interstellar Gas Dynamics', *IAU Symp.* **39**, 150.

Verschuur, G. L.: 1970b, *Astrophys. J. Letters* **6**, 215.

Verschuur, G. L.: 1973a, *Astron. Astrophys.* **22**, 139.

Verschuur, G. L.: 1973b, *Astron. J.* **78**, 573.

Verschuur, G. L.: 1973c, *Astrophys. J. Suppl.* **27**, 65.

Verschuur, G. L.: 1973d, in preparation.

Vieira, E. R.: 1971, *Astrophys. J. Suppl.* **22**, 369.

Weaver, H. F.: 1974, this volume, p. 573.

Williams, D. R. W.: 1965, in I. Robinson, A. Schild, and E. L. Schucking (eds.), *Quasi-Stellar Sources and Gravitational Collapse*, p. 212.

Woerden, H. van: 1967, in H. van Woerden (ed.), 'Radio Astronomy and the Galactic System', *IAU Symp.* **31**, 3.

Wright, W. E.: 1973, Ph.D. Thesis, California Institute of Technology.

Carl Heiles
Astronomy Department,
University of California,
Berkeley, Calif. 94720, U.S.A.

DISCUSSION

Heiles: This distribution of shapes seems completely different from what you found in Orion, which I think is very significant, because I think it says that the shape depends on where you look. This particular area, on the Hat Creek survey photograph produced by Jenkins and myself, shows a filamentary-type structure with a great big hole in the middle. The hole in the middle is right on top of a hole in the radio continuum. So I think this region is an unusual one where there has been some sort of event, probably an explosion, which has blown both the radio continuum and the neutral hydrogen out of this hole, and it is all concentrated around the edge, with a diameter of some 20° or 30°. So maybe some of these particular shapes are not representative of the interstellar medium in general but only this funny kind of region in particular. I hope you continue your studies in other regions.

Van Woerden: It is quite a big job. We had selected this region because it had no optical peculiarities about it. In Orion, though, I would say that there are quite irregular shapes, at least in some of the clouds. But maybe the resolution into clouds is not as good there as in this region.

Greenberg: What are the scales of the various filaments we have seen? Are the scales of the dust clouds the same as the scales of these filaments?

Heiles: In the Pleiades you see filaments that are much less than 1 pc – 0.01 pc or something like that. And then you see a continuous distribution of sizes all the way up to things on the hydrogen photograph which are 40° to 70° long and an equivalent number of parsecs long. On the galactic scale, in Weaver's picture those things occupy 10° in length but they are much further away. In the middle of Andromeda things sort of look filamentary, but on a size scale of kiloparsec. So it's all the way from 0.01 pc or less up to kiloparsec. I suppose one ought to try and derive the spectrum.

Oort: In connection with the problem of the stability or the collapse of dust clouds which Heiles has discussed, I want to draw attention to the fact that even the denser dust clouds often have shapes which show clearly that they are not equilibrium structures. Furthermore, it should be remembered that every cloud condensed from the interstellar medium must have considerable angular momentum. This will prevent rapid collapse. In fact, it is more of a problem how stars can be formed at all in the presence of this angular momentum than how their formation would rapidly use up the gas in these dense clouds.

Heiles: Theoreticians tell us that clouds will be unstable to either collapse or expansion. However, the existence of so much matter in dust clouds implies that they are stable. So it seems to me these structures are not understood.

Mathewson: This filamentary structure which is evident at high latitude seems to me to be more of a general characteristic of the interstellar medium. My unified magnetic field model has a helical field, which may also be a general characteristic of this region of a spiral arm. There is also another model of lots of supernova remnants, lots of loops, which has been put forward. It seems to me that with your data you could distinguish between these two. Looking at your velocities, I would consider it favors the characteristic of the general magnetic field of the spiral arm rather than lots of explosions. How do you feel about that?

Heiles: I am of two minds. I think it depends on where you look. The one filament around the north celestial pole, near $l \sim 30°$, which we were just talking about, I think, is lined up precisely on a hole in the continuum. Much of the intermediate velocity gas is lined up on the continuum loops. Both of these look like explosive events. On the other hand, some of the major filaments coming out of Ophiuchus follow the field very obviously. You can trace them all the way around to the southern declination limit of $-30°$. Then starting over in Orion, you can trace a filament to the southern declination limit of $-30°$. Those filaments may be actually one and the same. If so, it would be in favor of the helical field with the gas aligned right along it. So I think it may depend on just which filaments one talks about.

Mathewson: Would you expect the continuum to also coincide with your filaments? It's quite natural to think that if these things are explained by an explosion then the magnetic field is concentrated in the filaments, and also the dust. It would be a natural consequence even if it were just a general characteristic of the interstellar medium. I don't think you would need any explosion.

Heiles: The only reason I say that some of those are probably explosions is because of the velocity fields in the hydrogen which occur there – a very large velocity difference of some sort. In other ones, only 5 or 10 km s^{-1}, so no problem. That's why I distinguish between those two.

Burke: Although the interstellar magnetic field certainly exists, it is not so clear that it has an important effect on the H I motions, as Heiles implies. For example, the mere existence of filaments does not require a magnetic field. The known properties of the hot and cold components allow them to be in approximate pressure equilibrium. Filamentary structures in a gas in approximate pressure equilibrium are then entirely natural, with no intervention of a magnetic field. This can be seen easily in the case of cigar smoke curling through the air in a room even when the air is quite actively stirred. The production of thin sheets by expanding shock fronts is also quite natural. The magnetic field, therefore, may well be a passive participant, dragged along by more energetic processes, having minor effects on the gas dynamics.

Heiles: It's only an impression from looking at things that the field controls the gas rather than vice-versa. We are going to try to pin this down by Zeeman measurements.

PROPERTIES OF NEUTRAL HYDROGEN CLOUDS

ULRICH J. SCHWARZ and HUGO VAN WOERDEN

Kapteyn Astronomical Institute, University of Groningen, Groningen, The Netherlands

Abstract. In the region $+19° < b < +30°$, $122° < l < 143°$, 21-cm line profiles were analyzed into Gaussian components, and these components combined into clouds, by means of computerized procedures. Maps of some of the clouds so defined are presented, and their properties discussed. Many clouds are elongated and/or irregular in shape. Histograms of cloud parameters are presented and subjected to statistical analysis. There is evidence for tenuous clouds of large angular extent but very low column density $(N_H < 10^{19} \text{ cm}^{-2})$.

I. Introduction

This paper is a preliminary report on the results of a determination of interstellar cloud properties in a region at intermediate galactic latitudes. The region $(122° < l < 143°, +19° < b < +30°)$ was selected because it is relatively free from optical peculiarities, and the structure of 21-cm H I line profiles found in earlier investigations of the region (van Woerden *et al.*, 1962; Takakubo, 1968) appeared well-defined.

The analysis of hydrogen-line profiles in this region represents an attempt to define interstellar clouds, and determine their properties, by quantitative, computerized methods, consistent over the whole region. The need for information of this kind has been emphasized at a recent IAU Symposium (van Woerden, 1970).

II. Method of Analysis

The observations (beamwidth $0°6$, bandwidth 2 km s^{-1}, rms noise in T_b 0.18 K) cover the $20° \times 10°$ region with a $1° \times 1°$ grid. The 21-cm emission line profiles were analyzed into Gaussian components; in determining the 'best' among several possible solutions of one profile (always independently for each position), an improved form of the criterion defined by Kaper *et al.* (1966, Equation (5.6)) was applied.

Next the profile components were compared with those at surrounding positions $(|\Delta l|, |\Delta b| \leqslant 2°)$; components indistinguishable by Gaussian analysis if placed in the same profile were considered as related, and combined into one 'cloud'; separable components were considered unrelated. The comparisons assess both ΔV and $\Delta \sigma$ in terms of the velocity dispersion, σ; the component strength only influences the accuracy of the comparisons. A 'cloud' thus defined is essentially a concentration in (l, b, V, σ)-space.

The process described furnished a total of 88 clouds in the region. Of these, 10 are believed to belong to large-scale structures, including the intercloud medium; we shall not discuss these in detail in the present context.

F. J. Kerr and S. C. Simonson, III (eds.), Galactic Radio Astronomy, 45–50. All Rights Reserved.
Copyright © 1974 by the IAU.

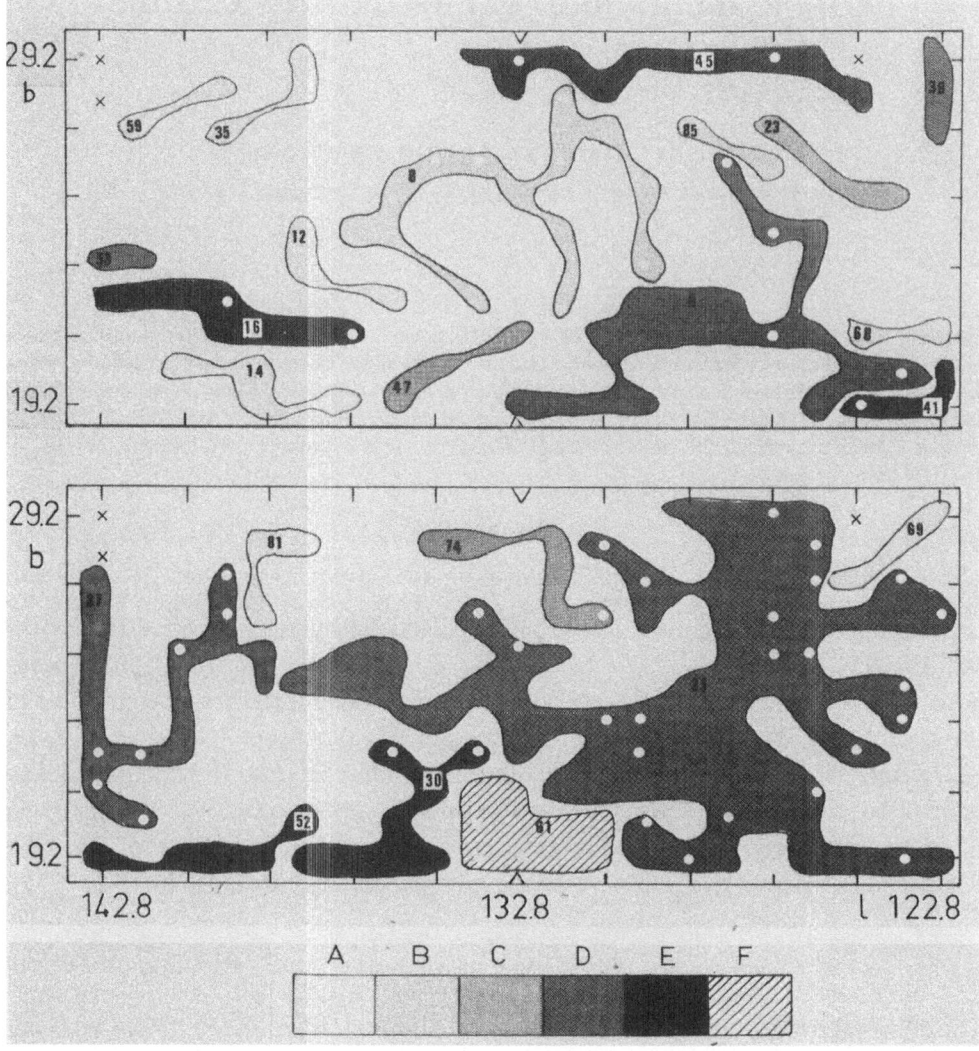

Fig. 1. Positions and shapes of 23 hydrogen clouds. Each cloud is labeled with a number. The grey shades indicate column densities $\bar{N}_H$, as follows:

Shade	A	B	C	D	E
$\bar{N}_H$ limits (10^{20} atom cm^{-2})	0	0.1	0.2	0.4	1	∞

Shade F indicates clouds with velocity dispersions, $\bar{\sigma}_0$, exceeding 10 km s^{-1}.

The outline of a cloud encloses all grid positions (and *only* those) where a profile component is found belonging to the cloud. Dots indicate positions where the component's relationship to the cloud is uncertain. Examples: cloud 61 is certainly present at 7 positions ($b = +19°2$, $l = 130°8$ and $131°8$; $b = +20°2$, $l = 130°8$, $131°8$ and $132°8$; $b = +21°2$, $l = 132°8$ and $133°8$) and uncertain at 3; it has $\bar{V}_0 = -17.8$ km s^{-1} and $\bar{\sigma}_0 = 29$ km s^{-1}. Cloud 68 ($\bar{V}_0 = -39.3$ km s^{-1}, $\bar{\sigma}_0 = 1.6$ km s^{-1}, $\bar{N}_H = 0.035 \times 10^{20}$ atom cm^{-2}) has components at ($122°8$, $+21°2$) and ($124°8$, $+21°2$), but not at $l = 123°8$; this is a 'tenuous cloud' (Section 4). Note the irregular and filamentary shapes of many clouds.
Crosses indicate 3 unobserved grid positions.

III. Cloud Shapes, Parameters and Properties

(a) SHAPES

Figure 1 shows the *shapes* for 23 clouds, selected at random from the total. The outlines sketched are determined by the positions where components belonging to a cloud are seen, but they remain arbitrary to some extent, since the beamwidth ($0°6$) is significantly smaller than the grid spacing ($1°0$). The most striking characteristics of Figure 1 are the irregular or filamentary shape of many clouds, and the loose, incoherent structure of some. The sample of 23 is not atypical in this respect. Note that some of these structures are not unlike those of parts of supernova remnants or reflection nebulae.

(b) OBSERVED PARAMETERS

For each of the 88 clouds, a number of *observed parameters* were calculated. For most of these parameters, the statistics are displayed in the histograms of Figure 2. We make the following notes.

(i) The distributions of optical depth $\bar{\tau}_0$, column density $\bar{N}_H$ and total hydrogen content S (related to mass, see below) increase steeply towards the lowest values. This is in agreement with earlier findings (e.g., Clark, 1965; Radhakrishnan and Goss, 1972; Heiles, 1967; Field and Hutchins, 1968), but extends them to much smaller values of the parameters.

(ii) The distribution of radial velocity $\bar{V}_0$ is surprisingly wide, much wider than is compatible with external velocity dispersions usually assumed (cf. van Woerden, 1967, Table 4). The cause of this discrepancy is under further investigation.

(iii) The distribution of velocity dispersion $\bar{\sigma}_0$ peaks between 1 and 3 km s^{-1}. This result is closely similar to those obtained in *absorption* by Muller (1959), Clark (1965) and (also in emission) Radhakrishnan *et al.* (1972; they give $\Delta V_{1/2} = 2.35\ \sigma$ in Tables 2 and 4). In our earlier study of *emission* profiles in Orion (van Woerden, 1967, Figure 10 and Table 5), where we followed much the same methods as in the present work, somewhat larger values of $\bar{\sigma}_0$ were derived.

(iv) For most clouds, the minor angular diameter β (for definition see the caption of Figure 2) is much smaller than the major diameter α, in agreement with the filamentary structure noted above. Some clouds may be unresolved across their major diameter.

(c) INTRINSIC PROPERTIES

The observed parameters α, β, S and $\bar{N}_H/\beta$ are related to *intrinsic properties* via relations involving various powers of the unknown distance, r:

major linear diameter, $a \sim \alpha r$

minor linear diameter, $b \sim \beta r$

hydrogen mass, $M_H \sim S r^2$

hydrogen volume density, $n_H \sim \bar{N}_H \beta^{-1} r^{-1}$

Assuming a specific distribution of distances, one may derive from the distributions

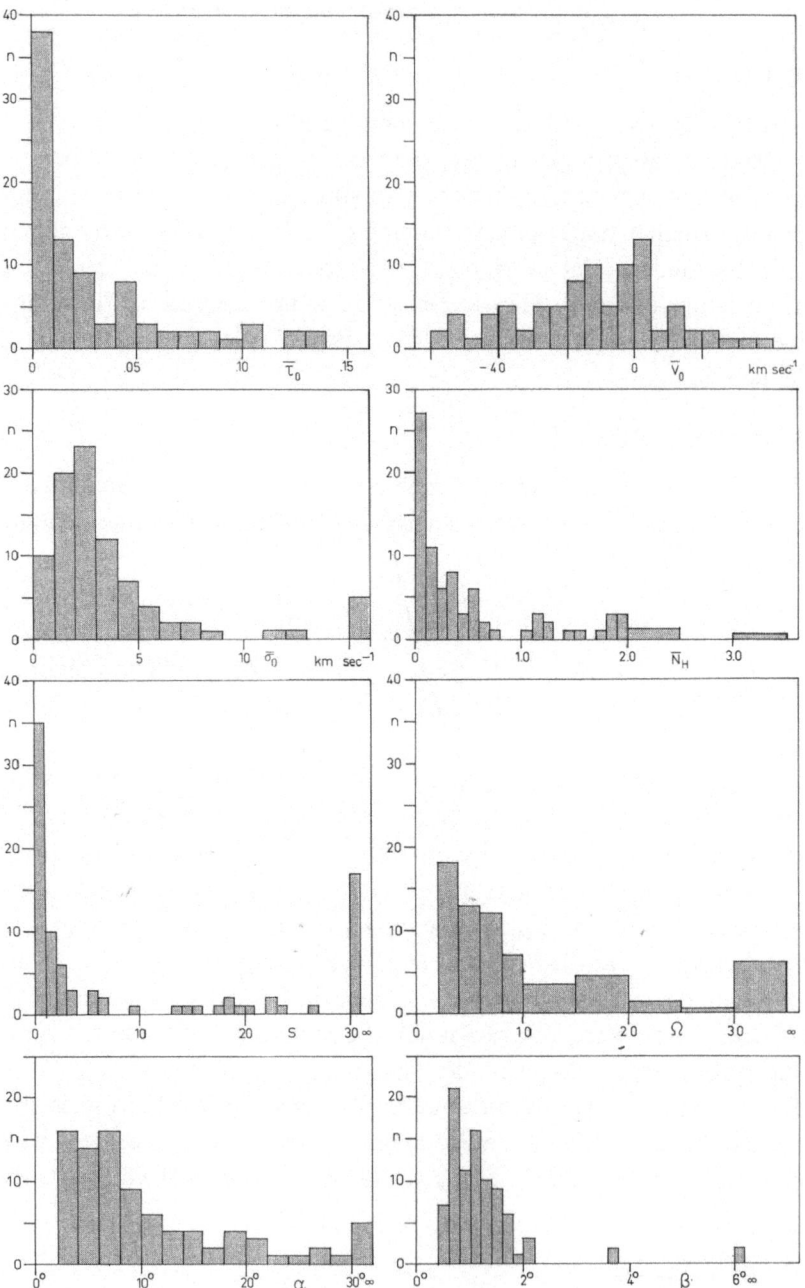

Fig. 2.   Histograms of observed cloud parameters:
$\bar{\tau}_0$, average peak optical depth (assuming a spin temperature of 125 K); $\bar{V}_0$, average radial velocity (with respect to LSR); $\bar{\sigma}_0$, average velocity dispersion (corrected for instrumental broadening); $\bar{N}_H$, average column density (unit: 10^{20} atom cm^{-2}); $S \equiv \Sigma N_H \equiv \bar{N}_H \times \Omega$ (unit: 10^{20} atom cm^{-2} degree^2); Ω, area covered $\equiv$ number of grid positions where cloud is present (unit: degree^2); α, major angular diameter; $\beta \equiv \Omega/\alpha$, minor angular diameter.

For clouds reaching the edge of the observed region, the quantities S, α and Ω are multiplied by 2.

of observed parameters those of intrinsic properties. Preliminary results suggest the following power laws:

$$\psi(a) = Ca^{-1 \text{ to } -2}, \quad \psi(M) = CM^{-1}, \quad \psi(n) = Cn^{-1.5}.$$

IV. Tenuous Clouds

A large number of clouds have very small column densities, $\bar{N}_H < 0.1 \times 10^{20}$ atom cm^{-2}. This finding is due to the high sensitivity of observations and analysis. For the 7 'tenuous clouds' with $\bar{N}_H < 0.05 \times 10^{20}$ and $\Omega \geqslant 3$, Table I compares the average parameters with those of Heiles' (1967) cloudlets. The differences are obvious.

TABLE I

Tenuous clouds and cloudlets

	Average tenuous cloud (this paper)	Average cloudet (Heiles, 1967)
Peak brightness temperature, T_{b0} (K)	0.6	7
Column density, $\bar{N}_H$ (10^{20} atom cm^{-2})	0.044	0.3
Hydrogen content, S (10^{20} atom cm^{-2} degree^2)	0.28	0.6
Velocity dispersion, $\bar{\sigma}_0$ (km/s)	1.5	0.7
Major diameter, α (degrees)	7	0.5
Minor diameter, β (degrees)	1	
Upper limit to kinetic temperature: T_k (K)	$\leqslant 280$	$\leqslant 70$

The considerable angular size of these objects results in a very low density – unless they are extremely nearby. A distance of 1 kpc (the expected distance of most of the hydrogen in this direction) would make the sizes 100×20 pc, mass 100 $M_\odot$, density 0.1 cm^{-3}; at 10 pc distance, the sizes are 1×0.2 pc, mass 0.01 $M_\odot$, density 10 cm^{-3}. Equilibrium with an intercloud medium of kinetic temperature, $T_k \approx 10000$ K, $n_H \approx$ ≈ 0.17 cm^{-3} would require for the 'tenuous clouds' $n_H \approx 6$ cm^{-3}, thus a distance close to 10 pc. Possible alternatives include the following:

(i) The clouds are at about 1 kpc, but their minor diameters are overestimated (unresolved) and densities underestimated; the clouds are then extremely elongated.

(ii) The clouds are in a transient phase, not in equilibrium with their surroundings.

(iii) The kinetic pressure of the intercloud medium is offset by magnetic rather than kinetic pressure inside the clouds.

(iv) The intercloud medium is much cooler than 10000 K.

This work will be submitted in more detail to *Astronomy and Astrophysics* or its Supplements.

Acknowledgements

The observations were done at Dwingeloo, with financial support from the Netherlands Organization for the Advancement of Pure Research (ZWO).

References

Clark, B. G.: 1965, *Astrophys. J.* **142**, 1398.
Field, G. B. and Hutchins, J.: 1968, *Astrophys. J.* **153**, 737.
Heiles, C.: 1967, *Astrophys. J. Suppl.* **15**, 97.
Kaper, H. G., Smits, D. W., Schwarz, U. J., Takakubo, K., and Woerden, H. van: 1966, *Bull. Astron. Inst. Neth.* **18**, 465.
Muller, C. A.: 1959, in R. N. Bracewell (ed.), 'Paris Symposium on Radio Astronomy', *IAU Symp.* **9**, 360.
Radhakrishnan, V. and Goss, W. M.: 1972, *Astrophys. J. Suppl.* **24**, 161.
Radhakrishnan, V., Murray, J. D., Lockhart, P., and Whittle, R. P. J.: 1972, *Astrophys. J. Suppl.* **24**, 15.
Takakubo, K.: 1968, *Bull. Astron. Inst. Neth.* **20**, 107.
Woerden, H. van: 1967, in H. van Woerden (ed.), 'Radio Astronomy and the Galactic System', *IAU Symp.* **31**, 3.
Woerden, H. van: 1970, in H. J. Habing (ed.), 'Interstellar Gas Dynamics', *IAU Symp.* **39**, 369.
Woerden, H. van, Takakubo, K., and Braes, L. L. E.: 1962, *Bull. Astron. Inst. Neth.* **16**, 321.

Ulrich J. Schwarz
Hugo van Woerden
Kapteyn Astronomical Institute,
University of Groningen,
Postbus 800,
Groningen 8002, The Netherlands

FREE ELECTRONS OUTSIDE H II REGIONS

M. GUÉLIN

Observatoire de Paris, Meudon, France

Abstract. This paper reviews the recent observational results on the electron distribution outside the H II regions. The mean electron density in the solar neighbourhood is of the order of 0.03 cm^{-3}, and does not seem to vary very much between the H I clouds and the intercloud medium, or between the arms and the interarm region. The z thickness of the electron layer in the solar neighbourhood is found to be larger than or of the order of 0.8 kpc. No reliable figure of the electron distribution can be deduced for the moment in the inner parts of the Galaxy.

I. Introduction

In directions of the sky where no discrete H II region can be recognized from the radio continuum maps, a number of radio or optical observations show the presence of ionized interstellar matter.

The aim of this review is to try to determine which part of this matter is concentrated in the small H II regions which may be crossed by the line of sight, and which part refers to what we shall call the 'general' interstellar gas, that is to say the hot or cold gas only partly ionized, outside the Strömgren spheres of the hot stars. As will be seen below, this distinction is not always clear, in particular because the characteristics of possible H II regions surrounding the B stars (for instance the electron density) are badly known. Hence, while trying to avoid the large emission nebulae, we shall deal with data relative to these weak H II regions as well as to the general medium.

We shall review the recent important results inferred from:

(i) the pulsar dispersion measure and distance measurements; (ii) the interstellar absorption lines; (iii) the radio recombination lines and the free-free absorption, and (iv) the diffuse Hα and [N II] emissions.

The pulsar dispersion measurements give clear-cut information on the average electron density in interstellar space. The other observations yield quantities depending both on electron density and on temperature. These two parameters cannot be properly separated, and apart from the interstellar absorption lines and some recombination lines which clearly originate in cold gas, these observations are for the moment unreliable for giving either the density or the temperature of the emitting electrons.

II. Pulsar Dispersion Measure and Distance

A straightforward determination of the mean electron density n_e in interstellar space can be made in the direction of the pulsars for which we know the dispersion measure and the distance. The dispersion measure, $DM = \int n_e \, dl$, which represents the electron column density in the line of sight, can be generally derived with good accuracy (see,

F. J. Kerr and S. C. Simonson, III (eds.), Galactic Radio Astronomy, 51–63. All Rights Reserved.
Copyright © 1974 by the IAU.

e.g., McCulloch *et al.*, 1973). This has been done now for about a hundred pulsars. The dependence of their dispersion measure on galactic latitude shows clearly that, in the mean, most of the dispersion occurs in the interstellar medium and not in the immediate vicinity of the pulsars.

The determination of distance is much more difficult. It results from the analysis of the 21-cm line absorption of the pulsar signal, or, as for the Crab nebula pulsar, from the association with a supernova remnant of known distance. As most of the pulsars are very faint at 21-cm wavelength, the absorption method can be applied only to a few sources. As an example, Figure 1 shows the absorption spectrum of

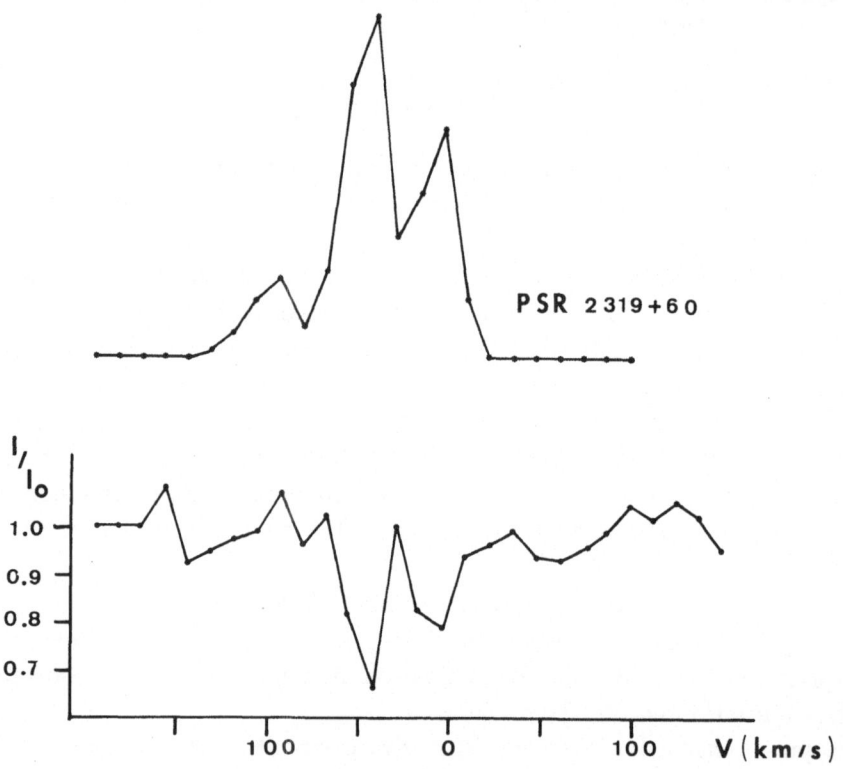

Fig. 1. Emission (upper) and absorption (lower) spectra in the direction of pulsar PSR 2319 + 60. These spectra were obtained with the Nançay radio telescope using a 12.7 km s^{-1} velocity resolution. The absorption features near 0 km s^{-1} and −40 km s^{-1} correspond to absorption in the local and Perseus arms respectively.

PSR 2319 + 60 which was obtained with the Nançay radio telescope after 20 h of observation. Absorption is clearly present at 0 km s^{-1} and −40 km s^{-1}, the velocities of the local and Perseus arms. This indicates that PSR 2319 + 60 is located farther than 3 kpc in the Schmidt galactic rotation model.

Table I lists the pulsars for which distances (or limits to distance) are known independently of the dispersion measure. The distances of PSR 0531 + 21, 0611 + 22 and 0833 − 45 are taken equal to the distances of the Crab nebula (Trimble, 1968), IC 443

TABLE I

Pulsars with known distances

Name	l	b	DM (cm^{-3} pc)	Distance (pc)	$\langle n_e \rangle$ (cm^{-3})
PSR 0329+54	145°	—1°	26.8	$1000 < D < 2000$	$0.015 < n_e < 0.03$
PSR 0525+21	184	−7	51.0	2000	0.025
PSR 0531+21	185	—6	56.8	2000	0.03
PSR 0611+22	189	2	99	2000	0.05
PSR 0740−28	244	—2	73.8	$1500 < D < 2500$	$0.03 < n_e < 0.06$
PSR 0833−45	264	−3	69.2	500	0.14 (Gum Nebula)
PSR 1749−49	1	−1	50.9	$D < 1000$	$0.05 < n_e$
PSR 1818−04	26	5	84.4	$D < 1500$	$0.06 < n_e$
PSR 1929+10	47	−4	3.2	$D < 1000$	$0.003 < n_e$
PSR 1933+16	52	−2	158.5	$6000 < D$	$n_e < 0.025$
PSR 2016+28	68	−4	14.2	$1000 < D$	$n_e < 0.015$
PSR 2020+28	69	− 5	44	$1000 < D$	$n_e < 0.04$
PSR 2021+51	88	8	22.6	$D < 1000$	$0.02 < n_e$
PSR 2319+60	112	—1	96	$3000 < D$	$n_e < 0.03$

(Minkowsky, 1959) and Vela X (Ilovaisky and Lequeux, 1972) respectively. The other pulsar distances result from 21-cm line absorption profiles observed at Nançay (Guélin, 1973; Guélin and Gomez-Gonzalez, 1974).

The range of variation of the line-of-sight-averaged electron density, $\langle n_e \rangle$, must be derived with care from Table I, as in most of the cases only limits to $\langle n_e \rangle$ are reported. These limits are very loose in the case of nearby pulsars such as PSR 1929 +10, because the relative errors in their distances are large. Nevertheless, the mean electron density appears to vary at least between 0.015 and 0.14 cm^{-3}. This scatter is reduced if we only consider the distant sources. Rough lower limits to $\langle n_e \rangle$ towards PSR 1933+16 and 2319+60 can be set by assuming that these pulsars do not lie farther than 15 kpc from the galactic centre. Using the upper limits of Table I, we then derive $0.01 < n_e < 0.025$ cm^{-3} for PSR 1933+16 and $0.015 < n_e < 0.03$ cm^{-3} for PSR 2319+60. These values, as well as those of Table I in the direction of the other distant pulsars, agree within a factor of two with the mean value of 0.03 cm^{-3} which may be derived from Table I. As these distant sources are scattered over a large range of galactic longitude, this suggests a roughly uniform large-scale electron density in the solar neighbourhood. The limits of 0.015 and 0.14 cm^{-3} show, on the other hand, relatively important local variations.

To derive the average electron density outside the H II regions, one has first to check whether the line of sight to a pulsar cuts across the Strömgren sphere of known very hot stars or clusters of hot stars. They are few in the solar vicinity and this check can be made relatively easily (Prentice and ter Haar, 1969). For instance, the line of sight to PSR 0833−45 very probably crosses some large H II regions in the Gum nebula. Thus, the average value of n_e is unusually high in this direction and has not to be taken into account in the estimate of n_e for the general medium.

This discussion becomes much more difficult when we go further away from the

Sun or when we consider the possible contribution of cooler stars like B stars (Grewing and Walmsley, 1971). Simple considerations, however, suggest that the contribution of the 'classical' H II regions is relatively small in the mean.

Let us consider the pulsars of known dispersion measure, and among them, those with $DM \cos |b| < 30$ cm^{-3} pc (b is the galactic latitude). Because of the mean value of $\langle n_e \rangle$ derived above, most of these pulsars lie in a region one kiloparsec in radius centred on the Sun. In this region, we know the distribution of the O and B type stars and we can recognize those pulsars for which the line of sight may cross the Strömgren sphere of one of these stars and those pulsars for which this is unlikely. This separation depends on the size of the Strömgren spheres and thus on the density inside the H II regions. Assuming that the latter is not smaller than 0.5 cm^{-3}, one remains with about twenty-five 'non-contaminated' cases.

If, on the average, the contribution of H II regions to the pulsar dispersion measures were important, $\langle n_e \rangle$ would be substantially smaller along the 'clean' lines of sight than along the 'contaminated' ones. Thus, the distances and heights above the galactic plane, z, derived from the dispersion measures by using $\langle n_e \rangle = 0.03$ cm^{-3} would be strongly underestimated for the 'clean' cases. In fact, the z distribution of the 'clean' pulsars is very similar to that of the other pulsars and the mean scale height $\langle |z| \rangle$ is very close to 200 pc for the 'clean' nearby pulsars as well as for the remaining nearby ones or for the distant ones.

The same type of considerations shows that there is probably no large difference between $\langle n_e \rangle$ in the arms and in the interarm region comprised between longitudes 30° and 50°.

We will thus tentatively conclude that in the solar neighbourhood, the mean electron density in the general interstellar medium is about 0.03 cm^{-3} and is, on a scale of a few kiloparsecs, constant within a factor two. The resulting z distribution of all the pulsars of known dispersion measure (Figure 2) shows moreover that the electron layer is thicker than 0.8 kpc. This is in good agreement with the determinations resulting from free-free absorption or Faraday rotation measurements (Bridle and Venugopal, 1969; Falgarone and Lequeux, 1973).

A more detailed electron density distribution cannot be derived in this way as pulsar distance estimates are inaccurate (typical uncertainties are of the order of 1 kpc).

III. Interstellar Absorption Lines

Observation of interstellar absorption lines in front of hot stars offers an opportunity to study the electron density on a smaller scale. Measurements of the relative abundance of different ions of the same atom lead to the electron density, under the usual assumption of ionization equilibrium, provided that the recombination coefficient and the ionization rate are known. If the absorption lines originate in concentrations similar to those detected in the 21-cm line absorption surveys (Radhakrishnan et al., 1972), the recombination coefficient and the ionization rate can be properly estimated, probably within a factor of 3. The electron densities derived in this way are typically

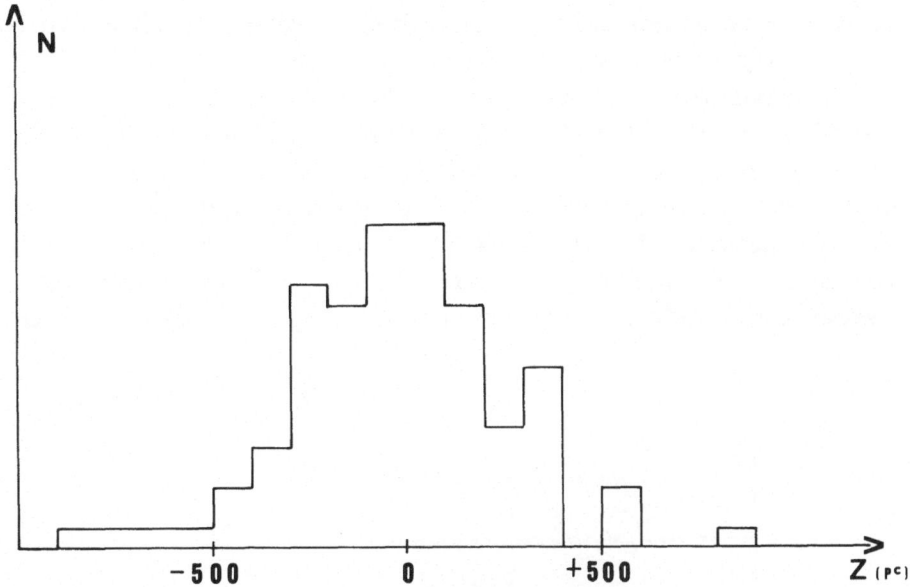

Fig. 2. Distribution of pulsars in galactic z-distance. The distances are derived from the dispersion measures by using $<n_e> = 0.03$ cm^{-3}.

a few times 10^{-2} cm^{-3} (White, 1973; Jenkins, 1974), thus close to the average value in the interstellar medium derived from pulsar measurements. Since the cold H I concentrations fill only a small fraction of the line of sight, this means that n_e does not vary very much between the clouds and the intercloud medium.

IV. Hydrogen and Carbon Radio Recombination Lines in Cold Clouds

Estimates of the electron density and temperature can also be derived from the radio recombination lines of hydrogen, carbon and heavier elements in some cool gas concentrations. The comparison of the width of the hydrogen lines with that of the carbon lines, assuming that the two emissions originate in the same region, allows us in principle to separate thermal broadening from turbulence. Chaisson and Lada (1974) have recently done this for four clouds and find typically $400 \pm ^{1000}_{400}$ K. The uncertainties are too large to show anything else than that these lines do not arise in H II regions. On the other hand, the ratio of intensities of the carbon and hydrogen lines may lead to the relative abundance of the electrons, n_e/n_H, provided that carbon ions and hydrogen ions share the same volume in the cool region. This assumption, of course, is very questionable. Assuming moreover that the abundances of carbon and heavier elements are equal to or smaller than the cosmic abundances, Chaisson and Lada find $n_e/n_H \leqslant 10^{-3}$.

V. Diffuse Radio Recombination Line Emission from the Galactic Plane

In opposition to these narrow hydrogen and carbon lines which are observed only

in front of some strong H II regions, broad hydrogen recombination lines have been
observed from regions of the galactic plane free of discrete radiosources.

Figure 3 summarizes the observations (Gordon and Cato, 1972; Gordon *et al.*,
1972; Matthews *et al.*, 1973; Jackson and Kerr, 1974). The observed directions are
plotted on a neutral hydrogen longitude-velocity map (Burton, 1971). The heavy lines
indicate the velocity interval explored and the dashed portions the velocity extent of
the lines when detected. Three points have to be made. First, the width of the lines
is large compared with what may be expected from the emission of a single region;
this suggests that the radiation originates in different regions widely spread over the

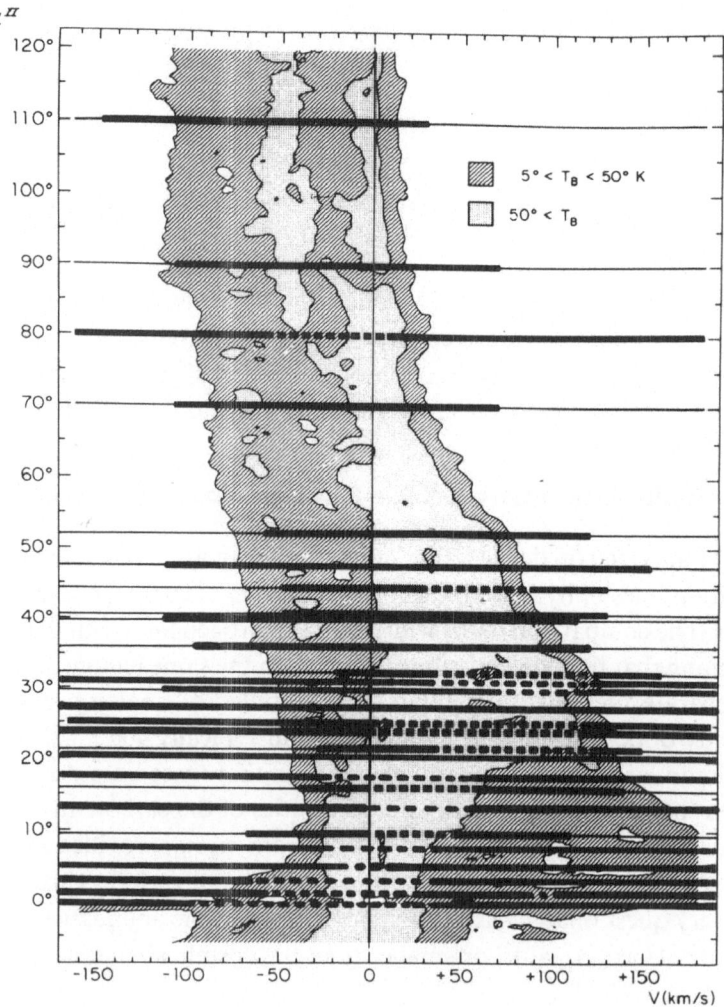

Fig. 3. Diffuse radio recombination emission from the galactic plane. The observed directions (Gordon
and Cato, 1972; Matthews *et al.*, 1973; Jackson and Kerr, 1974) are plotted on a neutral hydrogen longitude-
velocity map (Burton, 1971). The heavy lines indicate the velocity interval explored and the dashed portions
the velocity extent of the lines.

line of sight. Second, the diffuse line emission is detected in nearly all the observed directions with longitude smaller than $30°$ and nearly disappears for $l \geqslant 35°$. Third, the latitude extent of the line (at least near $l = 33°$) is narrower than $1°$. The diffuse recombination emission appears thus to be mainly concentrated in the zone of strong radio continuum emission from the galactic plane ($l < 35°$, $|b| < 1°$). In the same way, the broad band line emission observed at $l = 80°$ in Cygnus also coincides with a zone of strong continuum. More generally, Matthews *et al.* (1973) and Jackson and Kerr (1974) have found that, in fact, the line strength of the radio recombination profiles is correlated with the temperature of the continuum at 6-cm wavelength, thus probably with the thermal radiation.

The question now arises whether these lines are predominantly emitted by unresolved H II regions, or by the general interstellar medium, and what are the temperature and density associated with the ionized gas.

The line strength of the radio recombination emission yields directly the integral:

$$\int n_e^2 T_e^{-1.5} \, dl$$

in the case of local thermodynamic equilibrium and of low optical depth (Gordon and Cato, 1972). In order to separate n_e and T_e, one has to compare this integral with a different function of n_e and T_e, measured along the same line of sight and, as far as possible, with the same angular resolution. This is usually done in two ways: (i) with the help of free-free thermal emission measurements, or (ii) with the help of free-free absorption measurements.

In the case of low optical depth and LTE, the ratio ϱ of the line strength integral to the thermal continuum temperature may be expressed as:

$$\varrho \sim \frac{\int n_e^2 T_e^{-1.5} \, dl}{\int n_e^2 T_e^{-0.35} \, dl},$$

or, if the emitting gas is isothermal,

$$\varrho \sim T_e^{-1.15}.$$

Using this ratio, Matthews *et al.* (1973) and Jackson and Kerr (1974) find temperatures ranging from 4000 to 8000 K, which leads to emission measures $EM = \int n_e^2 \, dl$ of a few times 10^3 cm^{-6} pc in a 10 kpc path.

These estimates assume however: (i) that the lines are formed in LTE; (ii) that the thermal continuum is accurately evaluated; and (iii) that the emitting region is isothermal. These assumptions, particularly the last one, make the derived T_e and EM very tentative.

On the other hand, the comparison of the radio recombination line power to the free-free absorption in the same direction may also allow one to evaluate the temperature if the emitting gas is isothermal. This has been tried by Cesarsky and Cesarsky (1973a) who have observed the H92α emission in the direction of the supernova remnant 3C 391 and compared their results to the free-free absorption measurements of this source (Dulk and Slee, 1972). At the frequency of their observations (8.3 GHz), the ratio ϱ' of the recombination line power to the free-free optical depth becomes strongly dependent on the temperature. Cesarsky and Cesarsky obtain a very low value of ϱ', which means that the radio recombination line originates in a low temperature medium (20^{+300}_{-20} K), that is to say in the cold clouds. This conclusion is somewhat supported by the detection, in the same direction, of a line which is probably due to recombination of carbon (Cesarsky and Cesarsky, 1973b).

This temperature estimate, which contradicts the previous ones, seems however not very reliable since: (i) the temperature could be as high as 10^4 K if the actual error in ϱ' is only two times its quoted value; and (ii) the correlation of the free-free emission with the recombination line emission on the one hand, as well as the high resolution observations on the other hand (Altenhoff et al., 1973; Jackson and Kerr, 1974), suggest that part of the 'diffuse' recombination line emission arises in a hot medium.

Altenhoff et al. (1973) have mapped, at 11-cm wavelength and with a 4.5' resolution, one of the typical regions where radio recombination lines of large velocity extent have been reported by Gottesman and Gordon (1970) from 18' resolution observations. Within 36' of the Gottesman and Gordon position ($l=24°$), they have found ten discrete radiosources. At least three of these sources show radio recombination line emission with characteristics typical of H II regions (half-power line-width of 25 km s^{-1} and line-to-continuum ratio of about 3%); whereas between the sources a weak and broad line emission is probably still present (half-power line-width of 80 km s^{-1}). Thus, at least part of the emission reported originally by Gottesman and Gordon at this position arises in normal H II regions. But at the same time, it remains possible that emission from H I regions may be responsible at least partly for the weak and broad background line radiation seen between the patches of continuum and recombination line emission.

VI. Weak Hα and [N II] Emission from the Galactic Plane

Weak diffuse Hα and [N II] emission has been detected towards many directions between $\pm 30°$ of the galactic plane (Reynolds et al., 1973; Monnet, 1974). It has been investigated in detail by Reynolds, Scherb and Roesler between galactic longitudes $0°$ and $120°$.

A very good correlation exists in velocity between this emission and the 21-cm line emission from the H I gas nearer than 3 kpc (see Reynolds et al., 1973). This shows that the emitting regions are spread in the neutral hydrogen arms. A good correlation exists also between the distribution in the sky of the brightest of these Hα features and the distribution of the emission nebulae (Lynds, 1965). This is particularly clear for the Cygnus ($l=80°$) and the Ophiucus ($l=10°$) regions where the bulk of the Hα emission very probably is related to the H II regions.

Hα/[N II] intensity ratios have been observed in the direction of Orion. They range from 2 to 5 and are thus similar to those found in typical H II regions. In the same way, the width of the lines in the anticenter direction implies temperatures of the order of 6000 to 8000 K, provided that no turbulent motions larger than 7 km s^{-1} occur. Assuming a temperature of 6000 K and correcting the observed intensities for visual extinction, Reynolds *et al.* (1973) find emission measures in a 2 kpc path ranging from 10^2 to a few times 10^3 cm^{-6} pc.

A similar Hα emission has been detected in external galaxies (Comte, 1973). Figure 4a represents an Hα filter photograph of the southern spiral arm of M33, compared with a Pérot-Fabry Hα interferogram of the same region (Figure 4b). The stellar con-

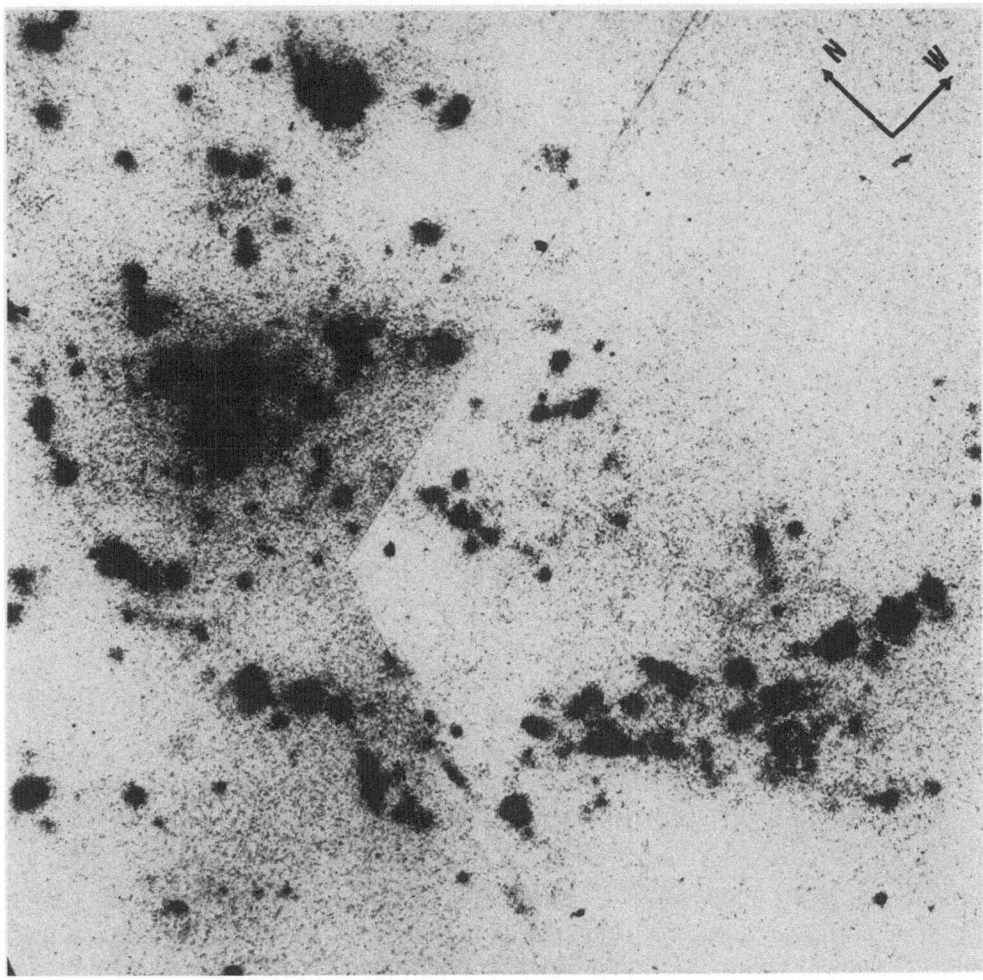

Fig. 4a.   Hα filter photograph of the southern arm of M33 (Boulesteix, Observatoire de Haute Provence). The field of the photograph is 15'. The large emission region in the left part of the photograph is the centre of the galaxy. Note the diffuse or filamentary structure which appears in the spiral arm between the bright H II regions. This diffuse emission looks very similar to the diffuse Hα emission detected in our Galaxy (Reynolds *et al.*, 1973).

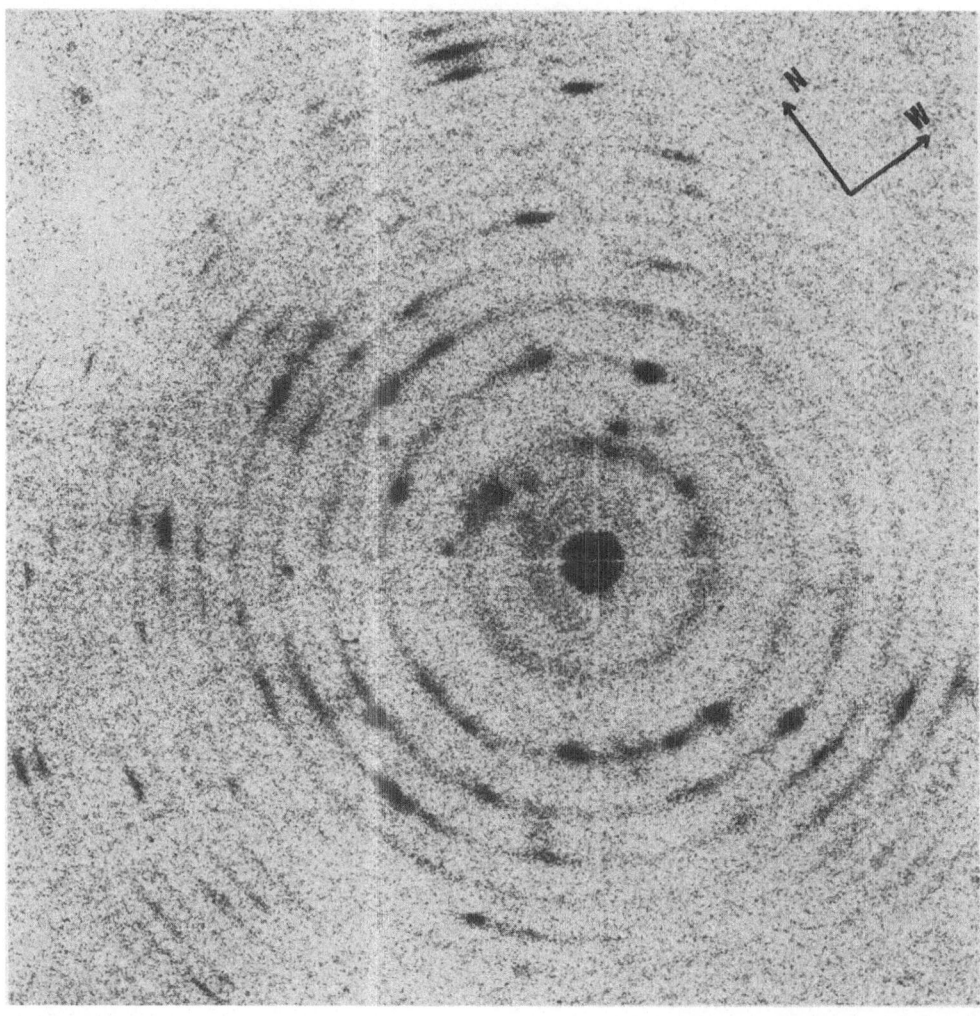

Fig. 4b.   Hα Pérot-Fabry interferogram of the same region to the same scale (Boulesteix, Observatoire de Haute Provence). Hα emission from the geocorona falls between two Pérot-Fabry rings and is eliminated. The stellar continuum radiation which may remain on the filter photograph is strongly reduced here. Note that the diffuse filaments are still present.

tinuum radiation which may remain on the 6 Å passband photograph is strongly reduced on the interferogram. By comparing the two pictures, one can recognize the ionized gas distribution. Three types of ionized regions are present in Figure 4 besides the nuclear region:

(i) The bright knots with rather sharp boundaries which are 'classical' H II regions with emission measures about 10^4 to 10^5 cm^{-6} pc;

(ii) between these knots, a relatively bright filamentary structure which appears only in the arms, with Hα/[N II] and Hα/[S II] ratios similiar to those in the knots and with emission measures typically about 10^2 cm^{-6} pc for a temperature of 6000 K (Comte, 1973);

(iii) a general background much weaker than the filamentary structure and with Hα to [N II] and [S II] ratios much smaller than in the knots.

The diffuse filamentary structure of M33 appears very like the weak galactic Hα structure studied by Reynolds, Scherb and Roesler: (i) both have comparable Hα intensities, if one allows for the difference in path length; (ii) both appear in spiral arms and in the vicinity of H II regions, and (iii) they present comparable Hα/[N II] ratios. Because of their similarity with the classical H II regions and of their location in the arms, Comte has identified the filaments of M33 with extended H II regions, excited by ultraviolet radiation escaping from bright nearby H II regions. An alternative explanation may be numerous small H II regions surrounding relatively cool stars. In both cases the filaments are hot, and the emission measure across one filament amounts to a few times 10^2 cm^{-6} pc in the mean. On the other hand, scattering of Hα and [N II] light from the bright H II regions by the interstellar grains may also contribute substantially to this emission. The same explanations may apply for the diffuse Hα emission in our Galaxy.

If actually this diffuse Hα emission arises in hot gas (its probable association with [N II] emission argues for temperatures larger than 3000 K), the emission measure along a 2 kpc path towards Sagittarius, in directions free of any strong H II region, is of the order of 10^3 cm^{-6} pc (Reynolds *et al.*, 1973). From this, one can predict a radio recombination integrated brightness temperature of a few K kHz at 18-cm wavelength. In fact, the values observed in the same directions (Gottesman and Gordon, 1970) appear very comparable to this figure if one takes into account that the radio intensities refer to a path length of about 10 kpc. This discussion, however, remains tentative as important extinction corrections have to be applied to the optical intensities in the Sagittarius region, and as the radio intensities, which may be predicted from the observed Hα intensities in directions of small visual extinction, are below the sensitivity of the present surveys.

VII. Conclusion

The only reliable data on the electron distribution outside the H II regions result thus from the pulsar and interstellar absorption line measurements. They refer for the moment to the solar neighbourhood and only to a small number of directions. Along these directions, it appears that the electron density does not vary very much between the H I clouds and the intercloud medium or between the arms and the interarm region. Its value, averaged over a few kiloparsecs, is constant within a factor of two in the direction of the six distant pulsars of Table I and equals 0.03 cm^{-3}.

Very little is known about the electron distribution in the other regions of the Galaxy. About twenty pulsars have a dispersion measure larger than 150 cm^{-3} pc and are thus very probably more distant than 4 or 5 kpc from the Sun. Their scale height above the galactic plane, $\langle |z| \rangle$, derived from their dispersion measures by assuming a constant electron density, is very close to that of the nearby pulsars. This suggests that even at distances of the order of 4 kpc, the mean electron density remains of the order of 0.03 cm^{-3}. The small number of high dispersion pulsars prevents however extending this argument to more distant regions. No one of the pulsars of Table I is located in the inner part of the Galaxy so that no direct determination of the electron density can be made in this region. Diffuse radio recombination lines are observed in the central region of the Galaxy but, as has been seen, they come in an unknown proportion from normal H II regions, and are thus unreliable for giving either the temperature or the density in the general medium.

References

Altenhoff, W., Churchwell, E. B., Mebold, U., and Walmsley, M.: 1973, private communication.
Bridle, A. H. and Venugopal, V. R.: 1969, *Nature* **224**, 545.
Burton, W. B.: 1971, *Astron. Astrophy.* **10**, 76.
Cesarsky, D. A. and Cesarsky, C. J.: 1973a, *Astrophys. J.* **184**, 83.
Cesarsky, D. A. and Cesarsky, C. J.: 1973b, *Astrophys. J. Letters* **183**, L143.
Chaisson, E. J. and Lada, C. J.: 1974, *Astrophys. J.* **189**, 227.
Comte, G.: 1973, Thèse de 3eme cycle Université de Provence.
Dulk, G. A. and Slee, O. B.: 1972, *Australian J. Phys.* **25**, 429.
Falgarone, E. and Lequeux, J.: 1973, *Astron. Astrophys.* **25**, 253.
Gordon, M. A. and Cato, T.: 1972, *Astrophys. J.* **176**, 587.
Gordon, M. A., Brown, R. L., and Gottesman, S. T.: 1972, *Astrophys. J.* **178**, 119.
Gottesman, S. T. and Gordon, M. A.: 1970, *Astrophys. J. Letters* **162**, L93.
Grewing, M. and Walmsley, M.: 1971, *Astron. Astrophys.* **11**, 65.
Guélin, M.: 1973, *Proc. Inst. Elec. Electron. Engrs.* **61**, 1298.
Guélin, M. and Gomez Gonzalez, J.: 1974, *Astron. Astrophys.* **32**, 441.
Ilovaisky, S. A. and Lequeux, J.: 1972, *Astron. Astrophys.* **20**, 347.
Jackson, P. D. and Kerr, F. J.: 1974, *Astrophys. J.*, in press.
Jenkins, E. B.: 1974, this volume, p. 65.
Lynds, B. T.: 1965, *Astrophys. J. Suppl.* **12**, 163.
McCulloch, P. M., Komesaroff, M. M., Ables, J. G., Hamilton, P. A., and Rankin, J. M.: 1973, *Astrophys. J. Letters* **14**, 169.
Matthews, H. E., Pedlar, A., and Davies, R. D.: 1973, *Monthly Notices Roy. Astron. Soc.* **165**, 149.
Minkowski, R.: 1959, in R. N. Bracewell (ed.), 'Paris Symposium on Radio Astronomy', *IAU Symp.* **9**, 315.
Monnet, G.: 1974, this volume, p. 249.
Prentice, A. J. R. and ter Haar, D.: 1969, *Monthly Notices Roy. Astron. Soc.* **146**, 423.
Radhakrishnan, V., Murray, J. D., Lockhart, P., and Whittle, R. P. J.: 1972, *Astrophys. J. Suppl.* **24**, 15.
Reynolds, R. J., Scherb, F., and Roester, F. L.: 1973, *Astrophys. J.* **185**, 869.
Trimble, V. L.: 1968, *Astron. J.* **73**, 535.
White, R. E.: 1973, *Astrophys. J.* **183**, 81.

M. Guélin
Observatoire de Paris,
Section d'Astrophysique de Meudon,
92190 Meudon, France

DISCUSSION

Gordon: It is important *not* to compare available $\langle n_e \rangle$ derived from pulsars and from radio recombination lines. While dispersions are seen in pulsars from a range of longitudes, radio recombination line emission is seen *only* from directions in which the line of sight lies inside the solar radius. No recombination lines have been seen in the direction of pulsars. The data imply that the dispersion measures and the lines arise from regions of very different characteristics.

Guélin: I agree. The dispersion measures depend on n_e whereas the radio recombination lines depend on n_e^2.

Zuckerman: You mentioned the observation of carbon and hydrogen recombination lines by Chaisson and his co-workers and indicated that these observations have been used to deduce the fraction of hydrogen that is ionized in H I regions. As I will discuss in my talk that follows, I do not believe that these re-combination lines can be used in this way, because the hydrogen and carbon lines both come from very dense regions and, thus, may not be coextensively ionized. Furthermore, as we have seen from the Copernicus results, one may not assume that a unique 'cosmic' C/H ratio generally obtains in interstellar clouds. Finally, the carbon lines originate in very dense clouds near H II regions, and these special clouds are hardly representative of the type of regions that you were otherwise discussing.

Terzian: I hope that we do not go away from here and try to use $n_e = 0.03$ cm^{-3} to derive distances for all the pulsars. In an analysis of pulsar dispersion measures which we have performed we find a range of n_e from 0.01 to 0.1 cm^{-3}, with a mean of ~ 0.04 cm^{-3}.

Guélin: Many people and you too have made some estimates of distances of pulsars by using the disper-sion measure. These were done before the 21-cm line absorption measurements. Now every time we get a measure by the 21-cm line, the dispersion measure distance appears to be wrong.

Greenberg: How do you take into account the possibility that the O and B stars do not sit in a cloud but rather in an intercloud region? The Strömgren sphere then extends far beyond its 'normal' distance from the star, i.e., extending to perhaps 500 pc from the star. It therefore may include material in the line of sight to the pulsar.

Guélin: In order to compute the radius of Strömgren spheres around the O and B type stars, I have assumed that the density is equal to 1 cm^{-3} inside the O star H II regions and 0.5 cm^{-3} inside the B star H II regions. These values are small compared to those in the known H II regions (Murdin and Sharpless). If a few stars sit in an intercloud region, the density inside the corresponding H II regions may be lower than 0.5 cm^{-3}, the H II regions will be larger, and some of the lines of sight assumed to be 'clean' may well cross them. However, because of this low density, the contribution to the dispersion measure will very probably be smaller, in the mean, than that of the more dense H II regions supposed to lie on the 'non-clean' lines of sight. The argument thus still holds. If all the stars sit in an intercloud medium with $n_e \ll 0.5$ cm^{-3}, of course this is no longer valid.

LINE ABSORPTION STUDIES
OF THE INTERSTELLAR GAS NEAR 10^{15} Hz

EDWARD B. JENKINS

Princeton University Observatory, Princeton, N.J., U.S.A.

Abstract. The *Copernicus* satellite now opens up the ultraviolet region for inspection, and the number of resonance lines which may be studied has increased from 6, seen in the visible, to more than 30. The distribution and properties of an important constituent of the interstellar gas, molecular hydrogen, can be studied in detail using this instrument. A more comprehensive picture may now be developed for element depletion factors, electron densities, and sources of ionization (UV photons, low energy cosmic and X-rays).

I. General Perspective

In studying the interstellar medium, one of the fundamental questions we can explore is, "What is its composition?" Aside from the merits of answering this question for its own sake, a comprehension of the relative abundances of the primary constituents of the interstellar material plays an important role in our understanding of the different physical and chemical phenomena which may occur in space. As a starting point, we could maintain the premise that the distribution of elements conforms to cosmic abundances and then ascertain the apportionment of these elements into various ions, atoms, free radicals and molecules, or solid particles. Until some evidence compels us to think otherwise, this assumption seems reasonable since the cosmic abundances are consistent with the composition of the atmospheres of B type stars which have recently formed out of the interstellar material (Traving, 1966).

In its atomic form, the presence of the most abundant element, hydrogen, is easily detected by line emission and absorption at 21 cm. As many of the contributions in this symposium volume attest, the study of this radiation has provided us with extensive information on the distribution, kinematics and temperature of interstellar H I. Microwave emissions from a wide variety of molecules have been detected, although most of the molecules are in appreciable abundance only in very dense concentrations of gas. In many H II regions, information on the abundances of ionized H, He, N, O and Ne has come from observations of emission lines at radio and optical wavelengths, resulting from recombinations and collisional excitations (Churchwell, 1974; Osterbrock, 1970). For the measurement of interstellar atoms and ions in the general H I regions, however, we must rely upon absorption lines appearing in the visible and ultraviolet spectra of O and B type stars. (Although high-level recombination lines have been detected from H I regions, they are difficult to interpret quantitatively.) A survey of the conclusions from these optical absorption line studies will be covered in the discussion which follows.

F. J. Kerr and S. C. Simonson, III (eds.), Galactic Radio Astronomy, 65–77. *All Rights Reserved.*
Copyright © 1974 by the IAU.

II. Studies at Visible Wavelengths

Let us consider the principal atoms and ions we might expect to find in the interstellar gas. For all elements having a cosmic abundance greater than 10^{-8} that of hydrogen, the diagram in Figure 1 shows a vertical slot whose width is proportional to the logarithm of the abundance (deuterium has been shown explicitly). This width serves as a crude but reasonably realistic measure of the strengths of lines which may appear,

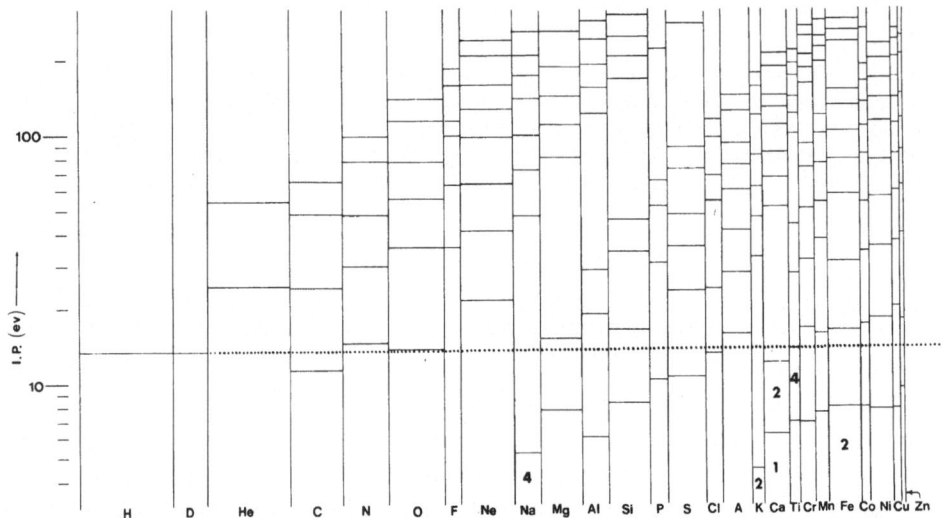

Fig. 1.   A representation of the more cosmically abundant ions and atoms. The width of each column is proportional to log (element abundance relative to hydrogen) -8; if this quantity is negative the element is omitted. The columns are partitioned vertically into segments which represent successively higher levels of ionization, starting with neutral atoms at the bottom. The upper boundaries for each ion are positioned according to their respective ionization potentials. The ionization potential of hydrogen is continued across the diagram as a dotted line because of its significance in determining the dominant stage of ionization of H I regions. The numbers in the boxes indicate the number of interstellar absorption lines detected in the visible spectra of stars.

as well as an indication of the importance of an element's contribution to the composition of the interstellar material. Each of the slots is divided vertically into boxes which represent the successive stages of ionization. The boundaries of the boxes are placed at heights which correspond to the appropriate ionization potentials in each case. These energies are relevant to the principal phenomena which are responsible for distributing an element into various levels of ionization.

Ionization by starlight is the dominant means for ionizing atoms in space. Photons at wavelengths between the Lyman limit and the soft X-ray region cannot penetrate H I regions, and hence starlight will not produce ions requiring more energy than 13.6 eV. For most elements (a notable exception is calcium) the recombination with free electrons is so much slower than the ionization that most of the atoms are found in the stage intersected by the horizontal dotted line in the diagram. For instance, carbon should be predominantly singly ionized in the interstellar medium; less than 1% of it would be neutral in the ordinary, low density H I regions.

Virtually none of the interstellar atoms and ions would have excited electronic levels (except for fine structure states), and thus all absorption lines must originate from the ground state. A recent compilation of such transitions, with their f-values if known, has been given by Morton and Smith (1973). As shown in Figure 1, relatively few of the more abundant elements have strong enough transitions of this sort to be seen as interstellar lines in the visible region of the spectrum. The numbers in the boxes show how many transitions have been detected by ground-based telescopes over many decades of observing. Not shown are a number of atoms or ions sought after with negative results; these have led to upper limits for the abundances, sometimes below the cosmic values. Except for Ti II all of the transitions observed are from levels of ionization which may be ionized by starlight (i.e., they are below the dotted line at 13.6 eV in Figure 1). Thus total element abundances may be derived only after solving for the ionization equilibrium. The most extensively studied lines in the visible are the Na I D lines and H and K lines of Ca II. The results have shown that, regardless of uncertainities in the corrections for ionization, the abundances of calcium and titanium are far below their cosmic values, while sodium is moderately below normal. The abundance of potassium is relatively close to the cosmic value. Habing (1969) has summarized and discussed many of the results on abundances from various studies of visual interstellar lines.

In addition to investigating abundances, the early surveys established that the distribution of gas was not uniform, and distinct concentrations of material, often referred to as clouds, moved about at their own peculiar velocities. A statistical description of the dispersion of cloud velocities, as well as a specification for the number of clouds per unit volume in our region of the Galaxy evolved from these data. Conclusions of this sort have been summarized by Spitzer (1968) and Heiles (1974).

III. Ultraviolet Research

Soon after the launch of the *Copernicus* satellite in August 1972, we were able to realize a manyfold increase in the number of lines which could be analyzed. This instrument was designed to record absorptions from just above the Lyman limit (912 Å) to 3100 Å where our atmosphere becomes transparent. There is a lack of coverage from about 1450 to 1600 Å, and longward of 1600 Å the effective sensitivity of the spectrometer is somewhat reduced by the high noise level resulting from cosmic rays and trapped radiation interacting with the near UV detectors. However, the greatest wealth of lines occurs at the shorter wavelengths where instrument performance has been excellent. Figure 2 summarizes, in a format identical to that of Figure 1, the number of transitions and the ions detected during the first year in orbit.

Not only has *Copernicus* allowed us to measure the most abundant ion states of 13 elements (and also deuterium, see Section VII), but in many cases we have had the opportunity to examine the relative distribution over a range of different ionization

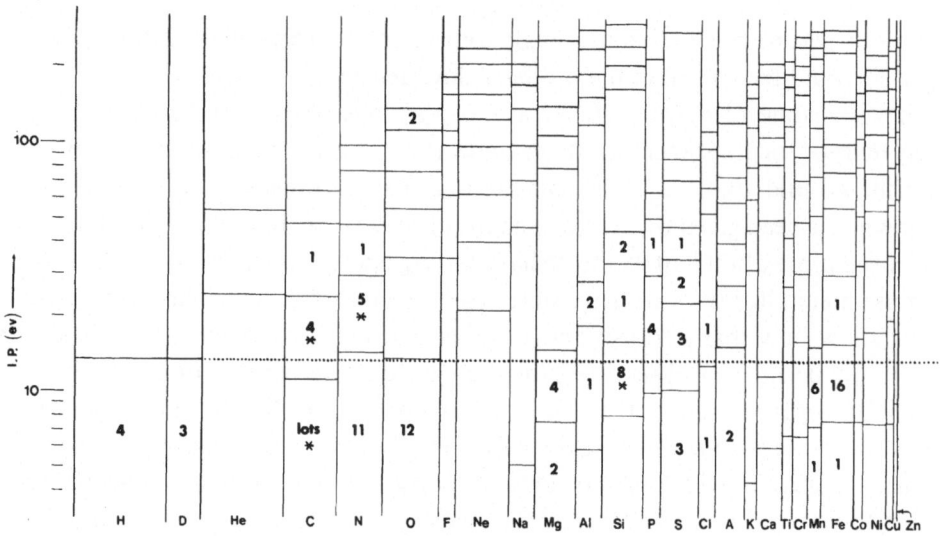

Fig. 2.   Same as Figure 1 except the numbers denote how many transitions for a given atom or ion have
          been observed in the ultraviolet spectra of stars recorded by the *Copernicus* instrument.

stages. The populations of ions above those intersected by the 13.6 eV line allow us to assess the importance of ionizations by X-rays and cosmic rays. When an element is predominantly ionized (by starlight), the amount in neutral form will provide an indication of the electron density if we know the ambient starlight flux and the cross sections for ionization and recombination, provided the ions and atoms reside in the same regions of space. An analysis of this sort has already been successfully derived from the ground-based observations of the Ca I to Ca II ratio; for instance, White (1973) derived electron densities in clouds which are consistent with the general results $n_e \approx 0.03$ cm^{-3} from pulsar dispersion measures (Guélin, 1974).

The asterisks within some of the boxes in Figure 2 identify those atoms or ions for which interstellar absorptions from fine-structure excited levels have been observed. These levels, which arise from the J splitting of the ground state, can be populated under interstellar conditions by collisions at typical H I temperatures (~ 60 K) when the densities are sufficiently high (generally 10 to 100 cm^{-3}). The creation of these excitations is balanced with collisional and radiative de-excitation. One can arrive at an estimate for various combinations of temperature and density of the gas by evaluating this equilibrium and relating it to the observed relative populations (Bahcall and Wolf, 1968; Dalgarno and McCray, 1972). The infrared radiation from these excited levels is a principal source of heat loss for the interstellar gas; a measurement of the number density of excited atoms or ions, when multiplied by the infrared transition's energy and Einstein A coefficient, provides a direct indication of the cooling rate by this mode.

IV. Element Abundances

Although it is evident from Figure 2 that a good coverage of the more important elements has been obtained by studying ultraviolet spectra, there are some significant gaps. There are no lines available from the ground state of helium which have a wavelength longward of the Lyman limit. This deficiency is not too serious since helium abundances are available from both optical and radio observations of recombinations occurring in H II regions. Many lines of neutral oxygen have been observed, but all of the oxygen ions except O VI are undetectable. If a reasonable fraction of the oxygen is singly ionized, as somewhat to our surprise we find for nitrogen, we would underestimate the oxygen abundances. Another significant omission is Ne; once again the optical studies of emissions for H II regions can fill this need.

For most stars, a substantial fraction of the observed lines are saturated, and we must rely upon a curve-of-growth analysis to determine a column density. We are fortunate enough in some cases to have at our disposal a large collection of lines with appreciably different oscillator strengths. N I, O I, Si II and Fe II are particularly favorable in this respect. On the other hand there are several astrophysically important species, such as C III, N II, N III, Al II, and Si III for which only one line has been available for measurement. (Although Figure 2 lists five lines for N II, all but one are from the excited levels for a single multiplet.) When one is confronted with a single line which may be saturated, one may derive a column density either by using the curves of growth for other elements or ionization stages, or one may derive a curve of growth from high resolution data taken in the visible, such as the Na I and Ca II line scans of Hobbs (1969) and Marschall and Hobbs (1972). It must be remembered, however, that the velocity profiles of various species along a line of sight to a star may not all be the same. This may be especially true for different levels of ionization of a single element, which may have markedly different distributions in space.

The initial results from *Copernicus* were reported in a series of six articles (which are given in the references and will be referred to as Papers I through VI). In addition to information on element abundances and relative ionizations, these articles discussed the observations of interstellar H_2, HD and CO and also the far UV extinction properties of dust grains. Paper I gave a basic description of the instrument and its performance. Papers II and III indicated that elements are depleted from the interstellar gas phase in both interstellar clouds and in the intercloud medium (the distinction between the two cases was based purely on the amount of reddening toward the stars). Figure 3, taken from Paper II, exemplifies this depletion for five stars which have a moderate amount of reddening. Although there is a significant variability from one star to the next for the depletion of a given element, the overall indication is that the deficiencies are a general phenomenon.

It has often been suggested that the interstellar dust grains are responsible for the depletion of various elements. It is not at all unreasonable that interstellar dust represents a significant fraction of the heavy element material available: Jenkins and Savage (1974) estimate that the ratio of the mass of dust to that of hydrogen in the interstellar medium is 0.006 while the cosmic ratio of heavy elements (excluding He) to hydrogen is 0.022 (from the results of Withbroe, 1971). Field (1973) has estimated the characteristic (maximum) temperature at which various elements would form compounds and could condense onto, or actually form, dust grains. He then examined

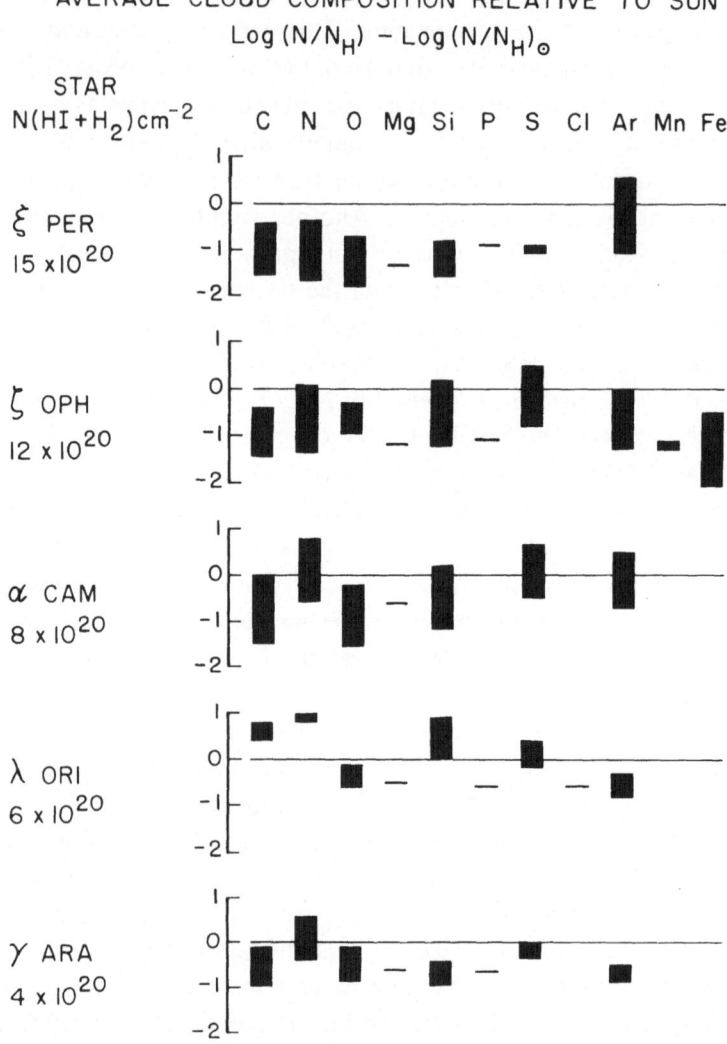

Fig. 3.   Relative interstellar gas abundances of elements, in all observable levels of ionization, compared with solar abundances. The tops and bottoms of the bars are the upper and lower error limits, respectively. The less certain values (depicted with tall bars) are for elements whose lines are on the flat portion of the curve of growth; the velocity dispersions of the material are not known accurately. This figure is adopted from Paper II of the original *Copernicus* series, except for the correction of mistakes in the drawing of the bars for Si toward ζ Oph and C toward λ Ori (Reproduced by courtesy of *The Astrophysical Journal*, University of Chicago Press, publisher. © 1973. American Astronomical Society. All Rights Reserved).

the general pattern for the depletion of elements in the direction of ζ Orph, based on the results from visual lines and the ultraviolet lines observed by *Copernicus*, and showed that the depletion was strongest for elements having the higher condensation temperatures. Grain growth, he concluded, could begin in the outer layers of stellar atmospheres where densities are large, and result in a strong depletion of the more refractory elements. Then, as the gas is ejected and spreads into space where the densities and temperatures become progressively lower, the other elements condense but the depletion is not as complete.

Future observations by the *Copernicus* instrument may help to clarify the involvement of element depletion with the formation of dust. For instance, it would be interesting to ascertain whether or not excesses and deficiencies in the ratio of extinction to the amount of hydrogen present (in the form of both molecules and atoms) correlates with any elemental deficiencies. Alternatively, it might be profitable to see whether or not differences in the character of the extinction correlate with any changes in the overall pattern of depletion. A few stars have markedly atypical extinction curves in the far ultraviolet (Bless and Savage, 1972). A discussion by Greenberg (1974) emphasizes the possibility that the extinction shortward of 1400 Å may be caused by very small silicate grains, and when mantles of ice and other volatile compounds form on these grains the extinction in the visible increases. One could check the compatibility of this hypothesis with the comparison of depletion patterns and the ratios of far ultraviolet to visual extinctions. Strom (1973) has emphasized the possible importance of the value for λ_{max}, the wavelength where maximum interstellar polarization occurs, as a determinant of the character of dust grains.

V. Degrees of Ionization

An immediate conclusion from the results of Paper III was that relatively few of the atoms were to be found in very high stages of ionization. For instance, upper limits for the intercloud abundances of N v, Si iv, and S iv generally ranged from 10^{-2} to 10^{-3} of the respective total element abundances. Meszaros (1973) concluded the observed amounts of C iii and N iii were too small to be consistent with the presence of a large enough flux of low energy galactic X-rays or cosmic rays to explain the high apparent ionization rate in H i regions, as deduced from various measures of interstellar electron densities. Such hypothetical large fluxes have also been suggested as a means for supplying enough heat gain to sustain the observed temperature of the H i gas (Goldsmith *et al.*, 1969). On the other hand, the *Copernicus* data on the ionization structure seem consistent with the presence of the *observed* diffuse X-ray flux down to 100 keV, which may be sufficient to explain the heating of the intercloud medium (with $n_H = 0.2$ cm^{-3}) but not the observed electron densities (Grewing and Walmsley, 1974).

A fundamental difficulty in measuring the abundances of highly ionized atoms in H i regions is separating out the contributions from each (observed) star's own H ii region. In reducing the *Copernicus* data, we are just now achieving enough accuracy and confidence in our wavelength scales to identify lines of differing radial velocity. In some favorable cases the velocity of the gas within the star's Strömgren sphere will differ enough from the intervening H i material to allow a distinction to be made. The relative importance of the H ii regions can also be reduced by choosing stars having both *large distances* and *low effective temperatures*.

VI. Molecules

Over the past decade there has been a dramatic series of discoveries of new molecules in space which has followed from line-radiation observations at centimeter and milli-meter wavelengths (Rank *et al.*, 1971). These radio studies have probed very dense clouds where a variety of moderately complex compounds could form rapidly and be protected from destruction by ultraviolet starlight. The large extinctions through these clouds, even at visible wavelengths, precludes our studying these rich molecule regions optically. In less dense areas, which are more representative of the 'general' interstellar medium, lines from the ground states of CH, CH^+ and CN have been observed in the visible spectra of some early-type stars. The observed abundances of these diatomic molecules at times exceed 10^{-9} that of hydrogen, a pattern which seems consistent with recent theoretical discussions by Watson and Salpeter (1972) and Solomon and Klemperer (1972). Measurements of absorption (or lack thereof) from the lowest rotationally excited levels of these molecules have helped to confirm that the cosmic background radiation indeed conforms to a 3 K black-body curve at millimeter wavelengths (Bortolot *et al.*, 1969).

The most abundant molecule in space, molecular hydrogen, has eluded detection until only recently. Except for pressure induced dipole transitions and some very weak quadrupole lines, this molecule does not have any resonance lines from the ground state at wavelengths longward of about 1100 Å (Field *et al.*, 1966). Hence visual and radio astronomical techniques have been unable to answer any questions concerning the distribution of H_2 in space, and even many of the early rocket and satellite ultraviolet experiments were not capable of observing the strong absorption by the Lyman bands because of insufficient sensitivity below about 1150 Å. Carruthers (1970) accomplished the first detection of interstellar H_2 features in his rocket obser-vation of the spectrum of ξ Per. The ability to measure the transitions of H_2 was an important objective for the *Copernicus* telescope and was a strong justification for having a system sensitivity which extended well below 1100 Å. Shortly after the launch of *Copernicus* a survey of many stars was carried out to determine not only the abun-dances of H_2 but also the distribution of the molecules in various stages of rotational excitation.

The results of the initial H_2 survey were given in Paper IV of the original series of *Copernicus* articles. The most striking property of the abundances was that for most cases either a significant fraction of the atoms appeared in molecular form ($> 10\%$) or else less than 10^{-7} of the atoms were bound in H_2. Figure 4 illustrates this bistable situation for the stars reported in Paper IV, where the H_2 column den-sities are plotted against the $B-V$ color excesses of the stars. The $E(B-V)$ values are a good representation of the total amount of interstellar material present. The data clearly show that concentrations of H_2 are found preferentially in regions of space where the gas (and dust) density is above a certain amount.

The qualitative features of the contrasting H_2 abundances are in accord with theo-retical expectations. Stecher and Williams (1967) pointed out that when H_2 is excited into the $B_1\Sigma_u^+$ state by starlight photons in the Lyman bands, a significant fraction of the subsequent decays ends up in the vibrational continuum of the ground state, and the molecule is destroyed. Hollenbach *et al.* (1971) investigated the importance of self-shielding by H_2 in the Lyman bands and absorption by dust. They concluded that while destruction by unattenuated starlight can keep the equilibrium fractional

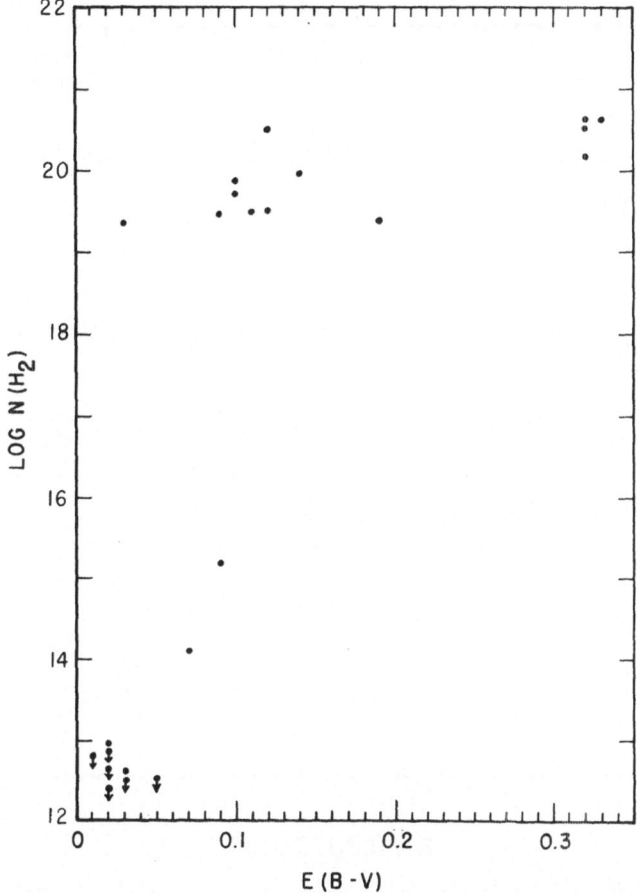

Fig. 4.   A plot of measured H_2 column densities (or their upper limits) against interstellar reddening for the 23 stars reported in Paper IV of the original series of articles on the *Copernicus* results.

abundance of H_2 at around 2×10^{-7}, when dense enough concentrations of material are present there is a fairly rapid shift toward much of the hydrogen being in molecular form.

Paper IV also discussed the observation of lines from rotationally excited H_2, with absorptions appearing for molecules up to $J = 6$. In a followup on this finding, Spitzer and Cochran (1973) measured the various column densities $N(J)$ for a number of stars. They found that $N(1)/N(0)$ was generally consistent with a temperature of 80 K if $N(0)$ exceeded 10^{17} molecules cm^{-2}. Collisions with protons can be effective in transferring H_2 *between these two levels* (Dalgarno *et al.*, 1973), and hence the level populations are strongly coupled to, and a good measure of, the kinetic temperature of the gas. For the higher J levels $(J > 3)$, the temperatures abruptly increase to values ranging from 180 to 390 K. Figure 5 illustrates the two temperatures for observations toward ξ Per. This pattern is rather typical of cases where $N(0) > 10^{17}$ molecules cm^{-2}. The high J molecules have a rapid enough radiative decay to preclude their

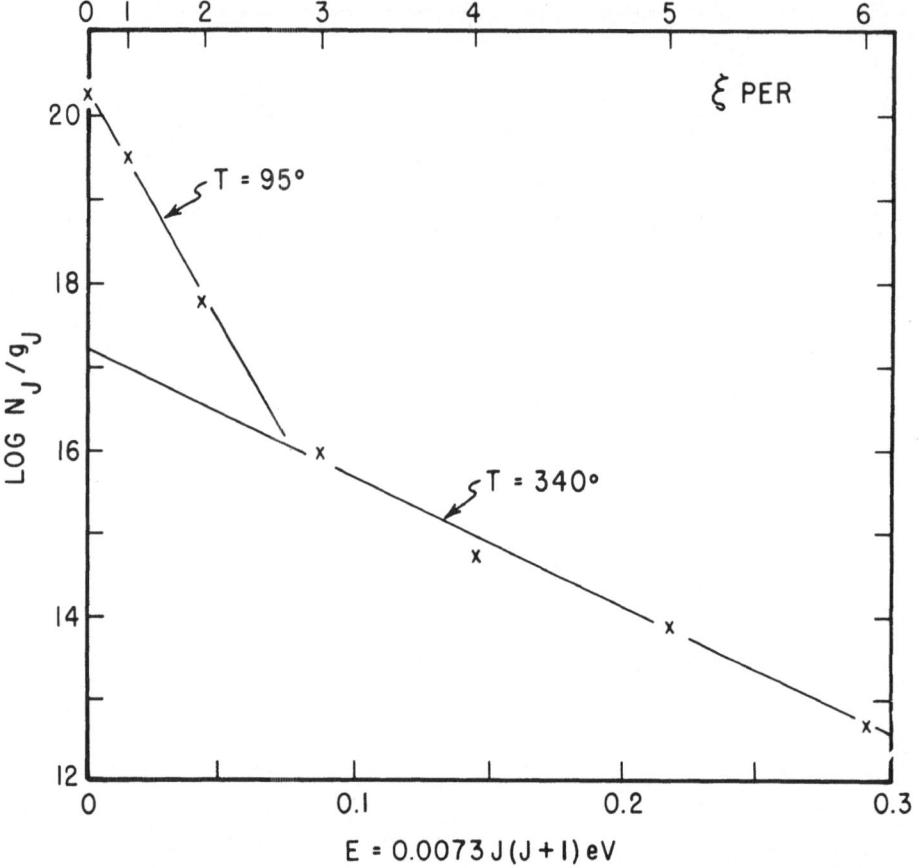

Fig. 5.   An illustration of the two rotation temperatures of H_2 observed toward ξ Per, a star which shows the typical pattern found for large H_2 column densities. The ordinate of each point is the log column density of molecules in a particular rotation level J divided by the level's statistical weight, $2J + 1$ or $3(2J + 1)$ for even and odd J, respectively. The abscissa represents the energy of each level, with the J values identified at the top.

coming into equilibrium with the kinetic temperature, unless the total densities are somewhat higher than normal. Spitzer and Cochran also noted that the velocity dispersion (as deduced from the line widths) increased as the higher J levels were reached. They suggested the large rotational and translational excitation could be attributed to newly formed molecules. In addition, molecules which were not dissociated during the decay following an absorption of a Lyman-band starlight photon would usually cascade to a high J level. Aanestad and Field (1973) have explored the possibility that shocks in the interstellar gas could be responsible for the observed high rotation temperatures.

VII. Deuterium

The abundance of deuterium is of special interest because of the hypothesis that it may all be produced in the early stages of the big bang of the Universe. Unlike some other elements which are also produced in the primordial fireball, deuterium has a cosmic abundance which depends critically upon the present average density of the Universe, and hence a measurement of D/H is a useful discriminant of this quantity. A summary of various determinations of the cosmic D/H ratio has been prepared by Reeves *et al.* (1973), but the methods generally suffer from either giving rather high upper limits or indirect values based upon measurements with uncertain corrections. Observations of deuterium in the interstellar medium would aid in narrowing the uncertainty in the cosmic D/H ratio.

Paper IV listed a number of reasonably accurate determinations of the column densities of HD. Unfortunately, the ratios of HD to H_2 (generally around 10^{-6}) give us little insight on the actual abundance of deuterium. Two effects significantly alter the HD/H_2 ratio from that of the total D/H. First, by virtue of its lower abundance, HD is much less effective in protecting itself against dissociation in the Lyman lines. To compensate for this difference, corrected HD/H_2 ratios were derived in Paper IV which were higher than the observed values by factors of about 10^4. The second important effect, which works in the opposite sense to the first (and was not treated in Paper IV), arises from the importance of the exchange reaction, $H_2 + D^+ \rightarrow HD + + H^+$. This reaction is exothermic owing to the difference in the zero-point vibrational energies of H_2 and HD. Hence, with other exchange reactions and the fractional ionization rate for H and D being equal, the one-sidedness of this reaction at interstellar temperatures favors a relative enrichment of HD, to an extent which is not too certain (Watson, 1973; Black and Dalgarno, 1973). Recent radio measurements of the DCN/HCN ratio (Jefferts *et al.*, 1973) also suffer from the difficulty of having to correct for the chemical fractionation effect (Solomon and Woolf, 1973).

The ideal measurement for a deuterium abundance in the interstellar medium is the ratio of atomic deuterium to atomic hydrogen in regions where no molecules are present. Although the separation of the H and D Lyman lines is only about $\frac{1}{4}$Å, it is possible to distinguish the two in Ly-β and higher order lines when the hydrogen column densities are not too large. From an analysis of the interstellar Ly-β, -γ and -δ lines in the spectrum of β Cen, a star which is bright at short wavelengths (due to the negligible reddening) and which has no H_2 in the line of sight, Rogerson and York (1973) found a D/H number ratio of 1.4×10^{-5}.

After a correction for the deuterium consumption by stars during the age of our Galaxy, the aforementioned result implies a primordial abundance which is consistent with a present density of 1.5×10^{-31} g cm^{-3} for the Universe. This density is somewhat above the estimate by Shapiro (1971) of 5×10^{-32} g cm^{-3} (with H revised to 50 km s^{-1} Mpc^{-1}) for the amount of *visible* material present, but is considerably smaller than the critical density of 4×10^{-30} needed to close the Universe. There is some uncertainty in the magnitude of the correction for consumption by stars, but even if the effect is neglected altogether – a most conservative stance – the deuterium measure would still not be consistent with a closed Universe. Colgate (1973) has suggested that significant deuterium production may occur in supernova shocks. Some indication of whether or not this possibility may be an important source of deuterium may be accomplished by looking for a variability in the abundance ratio over various lines of sight or, better yet, by looking for an enhancement in the amount of deuterium along a path through a young supernova remnant.

References

Bahcall, J. N. and Wolf, R. A.: 1968, *Astrophys. J.* **152**, 701.
Black, J. H. and Dalgarno, A.: 1973, *Astrophys. J. Letters* **184**, L101.
Bless, R. C. and Savage, B. D.: 1972, *Astrophys. J.* **171**, 293.
Bortolot, V. J., Clauser, J. F., and Thaddeus, P.: 1969, *Phys. Rev. Letters* **22**, 307.
Carruthers, G.: 1970, *Astrophys. J. Letters* **161**, L81.
Churchwell, E. B.: 1974, this volume, p. 195.
Colgate, S. A.: 1973, *Astrophys. J. Letters* **181**, L53.
Dalgarno, A. and McCray, R. A.: 1972, *Ann. Rev. Astron. Astrophys.* **10**, 375.
Dalgarno, A., Black, J. R., and Weisheit, J. C.: 1973, *Astrophys. Letters* **14**, 77.
Field, G. B.: 1973, preprint.
Field, G. B., Somerville, W. B., and Dressler, K.: 1966, *Ann. Rev. Astron. Astrophys.* **4**, 207.
Goldsmith, D. W., Habing, H. J., and Field, G. B.: 1969, *Astrophys. J.* **158**, 173.
Greenberg, J. M. and Hong, S.-S.: 1974, this volume, p. 155.
Grewing, M. and Walmsley, C. M.: 1974, *Astron. Astrophys.* **30**, 281.
Guélin, M.: 1974, this volume, p. 51.
Habing, H. J.: 1969, *Bull. Astron. Inst. Neth.* **20**, 176.
Heiles, C. E.: 1974, this volume, p. 13.
Hobbs, L. M.: 1969, *Astrophys. J.* **157**, 135.
Hollenbach, D. J., Werner, M. W., and Salpeter, E. E.: 1971, *Astrophys. J.* **163**, 165.
Jefferts, K. R., Penzias, A. A., and Wilson, R. W.: 1973, *Astrophys. J. Letters* **179**, L57.
Jenkins, E. B. and Savage, B. D.: 1974, *Astrophys. J.* **187**, 243.
Jenkins, E. B., Drake, J. F., Morton, D. C., Rogerson, J. B., Spitzer, L., and York, D. G.: 1973, *Astrophys. J. Letters* **181**, L122. (Paper V).
Marschall, L. A. and Hobbs, L. M.: 1972, *Astrophys. J.* **173**, 43.
Mészáros, P.: 1973, *Astrophys. J. Letters* **185**, L41.
Morton, D. C. and Smith, W. H.: 1973, *Astrophys. J. Suppl.* **26**, 333.
Morton, D. C., Drake, J. F., Jenkins, E. B., Rogerson, J. B., Spitzer, L., and York, D. G.: 1973, *Astrophys. J. Letters* **181**, L103. (Paper II)
Osterbrock, D. E.: 1970, *Quart. J. Roy. Astron. Soc.* **11**, 199.
Rank, D. M., Townes, C. H., and Welch, W. J.: 1971, *Science* **174**, 1083.
Reeves, H., Audouze, J., Fowler, W. A., and Schramm, D. N.: 1973, *Astrophys. J.* **179**, 909.
Rogerson, J. B. and York, D. G.: 1973, *Astrophys. J. Letters* **186**, L 95.
Rogerson, J. B., Spitzer, L., Drake, J. F., Dressler, K., Jenkins, E. B., Morton, D. C., and York, D. G.: 1973, *Astrophys. J. Letters* **181**, L97. (Paper I)
Rogerson, J. B., York, D. G., Drake, J. F., Jenkins, E. B., Morton, D. C., and Spitzer, L.: 1973, *Astrophys. J. Letters* **181**, L110 (Paper III).
Shapiro, S. L.: 1971, *Astron. J.* **76**, 291.
Solomon, P. M. and Klemperer, W.: 1972, *Astrophys. J.* **178**, 389.
Solomon, P. M. and Woolf, N. J.: 1973, *Astrophys. J. Letters* **180**, L89.
Spitzer, L.: 1968, *Diffuse Matter in Space*, Interscience, New York.
Spitzer, L., and Cochran, W. D.: 1973, *Astrophys. J. Letters* **186**, L13.

Spitzer, L., Drake, J. F., Jenkins, E. B., Morton, D. C., Rogerson, J. B., and York, D. G.: 1973, *Astrophys. J. Letters* **181**, L116. (Paper IV)

Stecher, T. P. and Williams, D. A.: 1967, *Astrophys. J. Letters* **149**, L29.

Strom, S. E.: 1973, private communication.

Traving, G.: 1966, in H: Hubenet (ed.) 'Abundance Determinations in Stellar Spectra', *IAU Symp.* **26**,213.

Watson, W. D.: 1973, *Astrophys. J. Letters* **182**, L73.

Watson, W. D. and Salpeter, E. E.: 1972, *Astrophys. J.* **175**, 659.

White, R. E.: 1973, *Astrophys. J.* **183**, 81.

Withbroe, G. L.: 1971 in K. B. Gebbie (ed.), *The Menzel Symposium* (NBS Special Pub. 353, U.S. Government Printing Office, Washington).

York, D. G., Drake, J. F., Jenkins, E. B., Morton, D. C., Rogerson, J. B., and Spitzer, L.: 1973, *Astrophys. J. Letters* **182**, L1. (Paper VI)

Edward B. Jenkins
Princeton University Observatory,
Peyton Hall,
Princeton, N.J. 08540, U.S.A.

DISCUSSION

Zuckerman: The Copernicus results appear reasonably consistent with ionization by starlight in H I clouds plus a contribution from the H II regions that surround the observed stars. There is no apparent reason to invoke a non thermal (i.e., X-ray or cosmic ray) source of ionization for the intercloud medium which might, therefore, be an extended very low density H II region. Interpretation of the diffuse radio recombination line radiation also suggests that much of space may be filled with very low density H II regions. Thus, except possibly for the 21-cm emission results, there appears to be no observational evidence for a hot partially ionized (H I) intercloud medium. Perhaps the broad 21-cm lines observed at high latitudes are due to a blend of relatively hot H I clouds having a fairly wide dispersion of velocities rather than the 'intercloud' medium.

Jenkins: I would tentatively agree with your remarks. We are improving our wavelength scale to a point where we can separate with some confidence gas clouds of different velocities. For some of the data now being analyzed we can differentiate between absorption lines produced by the star's H II region and lines arising from the intercloud regions.

Heiles: If you look at stars in some high-latitude regions you may expect not to find any intercloud medium because there is a great big region up in positive latitudes where all the intermediate and high velocity gas is concentrated and where all the local-velocity gas disappears. I think this is shown most beautifully in the slides van Woerden showed in Sydney, and also I see it in my profiles. So if you ever look at a star and see only 0.02 cm^{-3} along the line of sight to the star, you should check if this star is in that particular region, which occupies a very large solid angle in the sky, and, if it does, then all it tells you is that the intercloud medium in that particular region has been disturbed by the energetic events which have produced the intermediate and high velocity gas.

van Woerden: The low-velocity hole at high northern latitudes is not *empty* of hydrogen; the column density there, for $-20 < V < +15$ km s^{-1}, is still about 0.9×10^{20} H cm^{-2} (Wesselius and Fejes, *Astron. Astrophys.* **24** (1973), 25, Figs. 6 and 7). Takakubo (in preparation) has found that the hole is prominent for narrow components ($\sigma < 7$ km s^{-1}), but absent for the wide components ($\sigma > 7$ km s^{-1}), which may be identified with the intercloud medium. Thus, here as almost everywhere, observations are consistent with a smooth intercloud medium.

Heiles: I agree.

ON MICROWAVE RECOMBINATION LINES
FROM H I REGIONS

B. ZUCKERMAN*

University of Maryland, College Park, Md., U.S.A., and
University of California, Berkeley, Calif., U.S.A.

and

JOHN A. BALL

Center for Astrophysics, Harvard College Observatory and Smithsonian
Astrophysical Observatory, Cambridge, Mass., U.S.A.

Abstract. There has been considerable uncertainty in the nature of the regions from which carbon and narrow hydrogen lines have been observed. Specifically there are two different models. In model I the lines originate in low-density H I regions that are basically atomic, and stimulated emission enhances the line intensity by an order of magnitude. In model II the lines originate in the outer layers of very dense molecular regions and are due primarily to spontaneous emission. The first model is generally accepted and used by the astronomical community. However, we believe that the second is correct.

We observed recombination lines attributed to carbon at 21-, 30-, 36-, and 43-cm wavelengths toward IC 1795 (W3), Orion A, and NGC 2024; and a narrow hydrogen line toward NGC 2024.

A comparison of a considerable body of carbon recombination-line and radio molecular-line data indicates that these two types of lines are probably formed in dense contiguous regions – the carbon lines in a thin layer facing a hot star and the molecular lines in the rest of the cloud shielded from the stellar radiation by the carbon-emitting slab.

The high-frequency carbon lines toward Orion A are probably due to spontaneous rather than stimulated emission, which might not play an important role in any carbon-line emission region. In NGC 2024 the H I regions responsible for the observed hydrogen and carbon lines are probably dense ($n_H > 10^3$ cm^{-3}). Previous estimates of the fractional ionization of hydrogen may be inaccurate because different volumes of gas may be observed in the hydrogen and carbon recombination lines. Without further calculations and observations it is not possible to decide whether the narrow hydrogen line originates in a dense 'cloud' or in an ionization front.

We give a fuller account in the *Astrophysical Journal* **190** (1974), 35.

B. Zuckerman
University of Maryland Astronomy Program,
College Park, Md. 20742, U.S.A.

* Alfred P. Sloan Foundation Fellow.

John A. Ball
Center for Astrophysics,
Harvard College Observatory and Smithsonian Astrophysical Observatory,
60 Garden Street,
Cambridge, Mass. 02138, U.S.A.

(Discussion follows the paper by F. J. Kerr et al., p. 84)

TWO NEW RECOMBINATION-LINE RESULTS

F. J. KERR, P. D. JACKSON and G. R. KNAPP

University of Maryland, College Park, Md., U.S.A.

and

R. L. BROWN

National Radio Astronomy Observatory, Green Bank, W. Va., U.S.A.

Abstract. This paper reports two new recombination-line results. The first is the detection of carbon line emission from the dark cloud near ϱ Ophiuchi, and the second discusses the origin of hydrogen recombination line emission associated with ionized gas outside known discrete continuum sources.

Current theories of heating and ionization processes in the interstellar medium suggest that many dark dust clouds should be sufficiently dense and cold, and have low enough velocity dispersions, that radio recombination-line emission should be detectable. Previous searches have proved negative (see Gordon, 1973), but a more recent search at 21 cm (the 166α and 167α lines) of four clouds by Brown and Knapp (1974) has shown the existence of a carbon emission line from one of them, the ϱ Ophiuchi cloud. The fact that only carbon, and not hydrogen, emission lines are seen suggests that the ionization is due to UV and not cosmic rays. Such UV could originate either in newly-formed stars in the cloud itself or in the early-type stars (such as ϱ Ophiuchi) near the cloud. The ϱ Ophiuchi cloud is the only dust cloud so far observed which is associated with such early-type stars.

The four clouds observed by Brown and Knapp contain large amounts of atomic hydrogen (Knapp, 1972), and the fact that ionized gas was seen in only one of the four clouds suggests that the processes responsible for ionization in dust clouds, and for the presence of atomic hydrogen, are not related.

The second recombination-line result comes from a new study by Jackson and Kerr of H110α recombination-line emission associated with ionized hydrogen outside of known discrete continuum sources. This type of distributed ionization has been studied by several workers at various wavelengths, mostly near 18–21 cm. Both the present observations, and earlier work of ours, were carried out near 6-cm wavelength. The use of a shorter wavelength gives the advantage of greater directional resolution and also ensures that the thermal component of the continuum emission is relatively more important. Unfortunately, the lines are more difficult to detect at the shorter wavelengths since the line brightness temperatures are roughly proportional to wavelength.

Our new observations were largely concerned with point-to-point variability of the emission. Figure 1 shows the results of integrating for 2 to $4\frac{1}{2}$ h on each of a series of positions approximately along the galactic equator from $l = 23°.92$ to $l = 32°.32$. Although the profiles are noisy, there are clear variations from position to position. Nine closely spaced positions were observed, the central one being $l = 25°.07$, $b = 0°.01$,

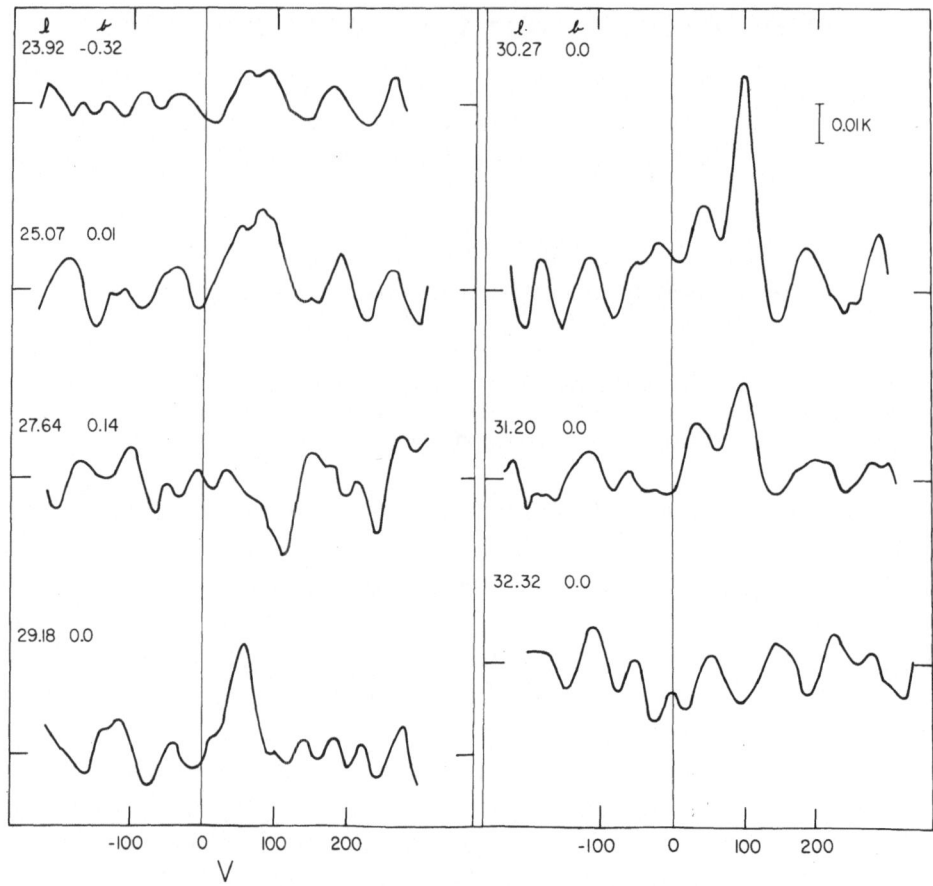

Fig. 1. Profiles of H110α recombination-line emission from a series of positions near the galactic equator between $l=23°92$ and $l=32°32$, The vertical axis is brightness temperature, and the horizontal axis is radial velocity in km s^{-1} relative to the local standard of rest after correction for 'standard' solar motion. The galactic coordinates for each profile are also indicated.

a position observed near 18-cm wavelength by Gottesman and Gordon (1970) and by Gordon and Gottesman (1971). We observed the nine closely spaced positions in order to build up a 6-cm profile corresponding to the beam size at 18 cm so that we could look at the spectral characteristics of the line radiation.

In Figure 2 is plotted the integrated line brightness L in K km s^{-1} against the concurrently observed 6-cm continuum temperature, T_b, for the seven positions of Figure 1. Despite the noise in the profiles, the correlation between L and T_b is quite good. A fit to the data is given by the line $L=2.3\,(T_b-0.7)$. If we identify the portion of T_b which correlates with L as the thermal component of the continuum emission, then we can infer that the residual portion of T_b, 0.7 K, represents the general non-thermal background over the whole region.

The slope $L/T_{th}=2.3$ allows us to derive an electron temperature $T_e=4400\pm600$ K as an average electron temperature for the distributed ionized gas in the region $l=24°-33°$. This computation assumes LTE, which seems to be a reasonable approximation, since the ratio of our value of L for $l=25°07$. $b=0°01$, with that obtained near 18 cm by Gordon and Gottesman (1971) agrees well with that predicted by LTE.

The fact that the LTE equation seems to be suitable for the emission, combined with the rapid point-to-point variability, the correlation with the continuum emission, and the relatively high associated electron temperatures all lead to the con-

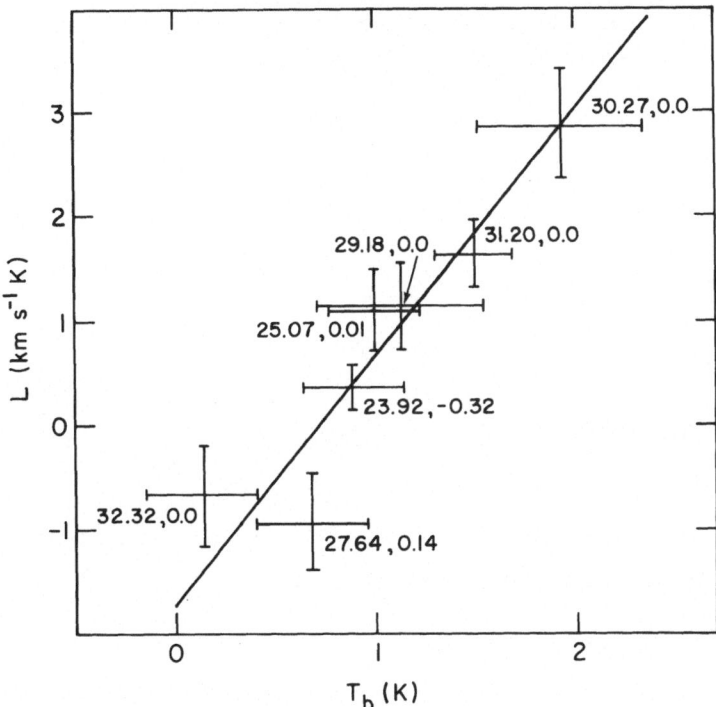

Fig. 2.   Integrated H110α recombination-line emission, L, as a function of the continuum brightness temperature, T_b, at the same wavelength of observation, for the 7 positions whose profiles are given in Figure 1. The galactic coordinates for each position are indicated next to its plotted symbol. The estimated uncertainties in L and T_b are also indicated by error bars for each position. The straight line fit has the form
$$L = 2.3(T_b - 0.7).$$

clusion that most of the line radiation originates in small H II regions. Some calculations of the amount of line emission based on likely ionization around early-type stars come out a few times too low to explain our observations, using data for the solar neighborhood. However, we can reasonably assume that the small H II regions become more important in the inner regions of the Galaxy studied here; we know that a corresponding distribution is true for the giant H II regions.

Acknowledgements

The observations described herein were carried out with the 43-m telescope of the National Radio Astronomy Observatory. We thank the NRAO for the observing time to carry out these projects. The NRAO is operated by Associated Universities

Inc., under contract with the National Science Foundation. This research is supported by the NSF.

References

Brown, R. L. and Knapp, G. R.: 1974, *Astrophys. J.* **189**, 253.
Gordon, M. A.: 1973, *Astrophys. J.* **184**, 77.
Gordon, M. A. and Gottesman, S. T.: 1971, *Astrophys. J.* **168**, 361.
Gottesman, S. T. and Gordon, M. A.: 1970, *Astrophys. J. Letters* **162**, L93.
Knapp, G. R.: 1972, Ph.D. thesis, University of Maryland.

F. J. Kerr
P. D. Jackson
Astronomy Program,
University of Maryland,
College Park, Md. 20742, U.S.A.

G. R. Knapp
Owens Valley Radio Observatory,
California Institute of Technology,
Pasadena, Calif. 91109, U.S.A.

R. L. Brown
National Radio Astronomy Observatory,
Edgemont Road,
Charlottesville, Va 22901, U.S.A.

DISCUSSION

Guélin: The proportionality of the integral of the radio recombination line strength to continuum temperature does not mean that the emitting electron gas is isothermal. Your temperature estimate may thus be not very meaningful.

Menon: Have you high enough angular resolution in the Orion nebula carbon observations to tell how the carbon is distributed? There is an H I emission cloud superimposed on the Orion nebula, corresponding to the dark bay, with the same velocity as the carbon line.

Zuckerman: At 2' or 3' resolution the carbon emission comes from all around the nebula. Most of it probably comes from the background molecular cloud, but some could be coming from the dark bay also.

Gordon: I have searched for recombination lines from five other dust clouds without success. Combining these results with those of Brown and Knapp, I feel that the detection in one of ten clouds searched means that the ϱ Oph dust cloud is indeed unusual and not characteristic of dust clouds in general.

Churchwell: It seems to me that the high-density model for the carbon recombination line does not answer two problems:

(1) If the carbon emission cloud is behind the H II gas in Orion A, how does one explain the increased line intensity of the carbon line at lower frequencies, particularly with LTE emission as you suggest?

(2) In W3 why does one not observe the carbon line in absorption at lower frequencies?

Zuckerman: In response to your question about the increased carbon line intensity in Orion A at low frequencies one must be careful to distinguish between non-LTE emission and stimulated emission. That is, although the latter may be unimportant this does not necessarily imply that I believe that the emission region is in LTE. Theoretical calculations by Dupree of the carbon level populations (the b_n's) indicate that they increase rapidly towards higher n. This would result in increased low frequency emission in the situation where only spontaneous emission contributes to the line intensities. As for W3 one would not necessarily expect to observe the carbon line in absorption at low frequencies for two reasons. Firstly, the central (bright) regions of the nebula fill only a small fraction of the beam of a single antenna at $\lambda \gtrsim 21$ cm, and, in any case, Dupree's calculations still suggest a small inversion of the levels even for dense clouds near $n \sim 166$.

J. R. Dickel: All the evidence we have seen today has pointed to the recombination lines arising in hot regions, but Gordon says he still favors cool regions. Would he comment as to why he favors low temperatures for the component emitting the diffuse recombination lines?

Gordon: I believe the source of the 'diffuse' recombination lines to be uncertain, both in $\langle T_e \rangle$ and $\langle n_e \rangle$. However, the beautiful analytical work for 3C 391 by Cesarsky and Cesarsky, leading to $\langle T_e \rangle \approx 20 \, \mathrm{K}$, is difficult to refute. However, in other directions, different physical conditions may prevail; there the lines might be generated by weak H II regions.

Zuckerman: I don't favor Chaisson's interpretation of some of these re-combinations results. He has a preprint out where he's analyzed that same supernova and he finds that he can reproduce the observations with the hot gas, essentially the H II regions.

THE GALACTIC ELECTRON
DISTRIBUTION FROM A PULSAR SURVEY

A. G. LYNE

*University of Manchester, Nuffield Radio Astronomy Laboratories,
Jodrell Bank, Cheshire, United Kingdom*

Abstract. The recent high-sensitivity pulsar survey at Jodrell Bank has allowed a statistical study of more distant objects. The longitude distribution suggests that many of the pulsars observed have distances greater than 5 kpc, leading to an upper limit of about 0.03 cm^{-3} for the mean electron density. The electron density averaged over distances of a few hundred parsecs seems to be very constant. The width of the electron distribution in the z-direction appears to be greater than about 600 pc.

One year ago about a total of 65 pulsars had been discovered by eight observatories, all observing with different instruments, frequencies and sensitivities. The search at Molonglo was the only survey which had a high sensitivity, a thorough coverage and well-defined selection effects. This enabled Large (1971) to study the distribution and statistical properties of pulsars on a sample of 29 objects, scattered over a large area of sky.

From this work it was possible to predict that with a greater sensitivity many, more distant pulsars would be observable at low galactic latitudes and it would be possible to study pulsar distribution on a much larger galactic scale. A survey of the galactic plane was therefore carried out at Jodrell Bank at 408 MHz using the Mk I telescope. (Davies *et al.*, 1972, 1973). The experiment used a 4-MHz bandwidth and consisted of on-line period analysis on 11-min sections of data taken at points in the search area.

During the survey, 51 pulsars were detected. An indication of the improvement in sensitivity over previous searches was that only 11 of these had been discovered at other observatories. I shall describe only a few of the results of the survey, namely those which have a bearing upon the general distribution of electrons in the Galaxy.

Figure 1 shows the area covered by the survey and the positions of the pulsars detected, superimposed on a contour map of the background brightness distribution at 400 MHz (Seeger *et al.*, 1966). There are a number of small selection effects which are present in our data, but in general these are well defined and can be taken into account in our analyses. The most important of these are the area and thoroughness of the search over the area, and the system sensitivity which varies over the sky because of the galactic continuum radio background. The main result of the latter is that a number of weak pulsars at low longitudes and latitudes could have been missed.

Figure 2 shows the longitude distribution of those pulsars within 5° of latitude of the plane. The top diagram is the distribution corrected only for the thoroughness of coverage and the ordinate is the deduced number of pulsars observable per square degree. The lower diagram indicates the estimate of the true distribution after allowing

F. J. Kerr and S. C. Simonson, III (eds.), Galactic Radio Astronomy, 87–95. All Rights Reserved.
Copyright © 1974 by the IAU.

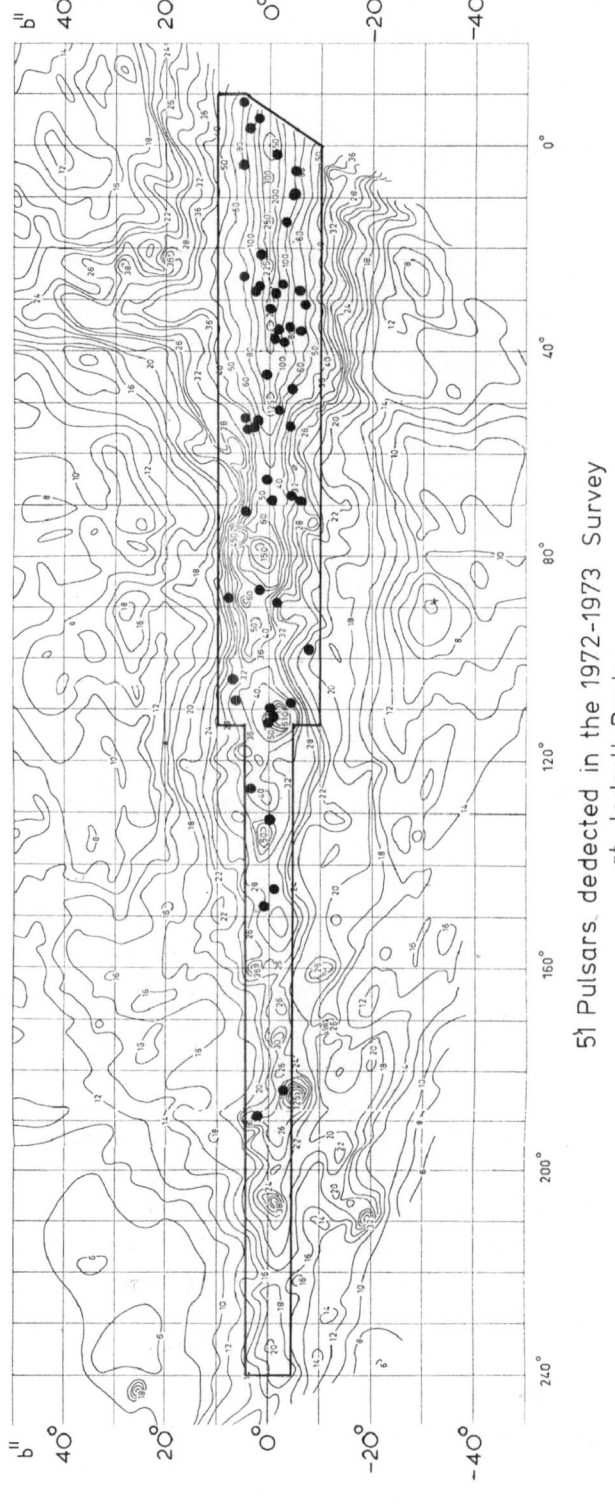

51 Pulsars detected in the 1972-1973 Survey at Jodrell Bank

Fig. 1.   The distribution of the pulsars detected, superimposed upon a contour map of the galactic background radiation. The heavy lines enclose the area covered by the survey.

for system temperature variations over the sky under the assumption that the pulsars are a disc population. The justification for this will be seen shortly.

We see that there is a strong longitude concentration of pulsars, the bulk of them lying at longitudes of less than 100°. There is about a three to one ratio in densities

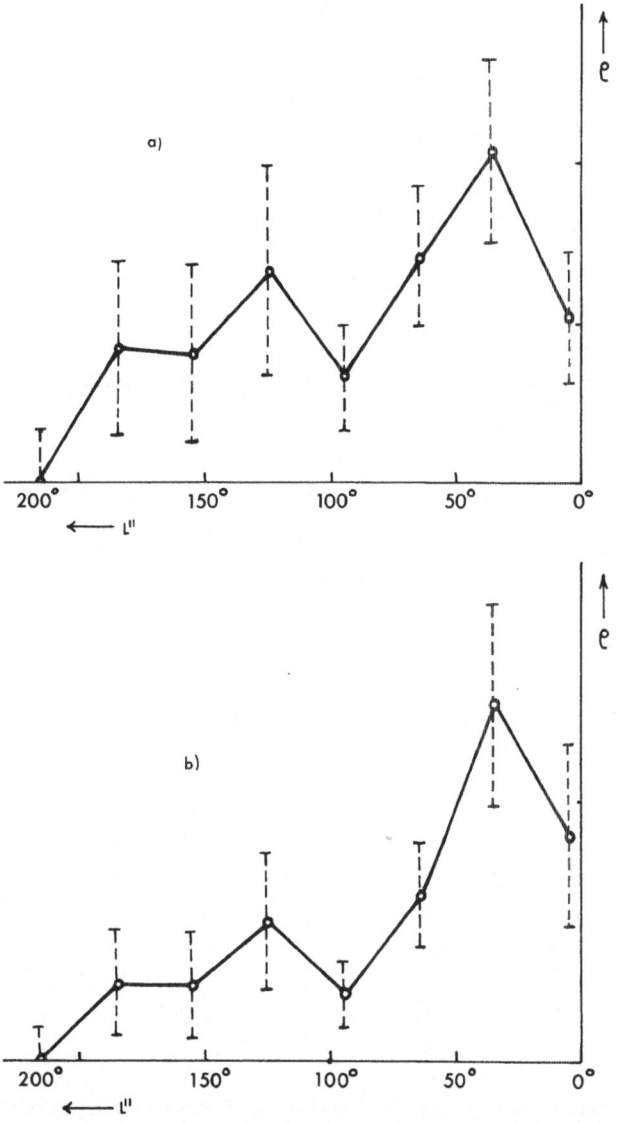

Fig. 2.   (a) The observed longitude density of pulsars having $|b| \leqslant 5°$ after correction for thoroughness of coverage. (b) The same as Figure 2a but after correction for variation of sky background temperature.

between the galactic centre and anticentre regions. This is clearly not just a local clustering due to the Sun's position with respect to the local spiral structure because of the observed narrowness of the latitude distribution indicated in Figure 3. Again, the top diagram is the observed density, while the lower one is that corrected for the background temperature.

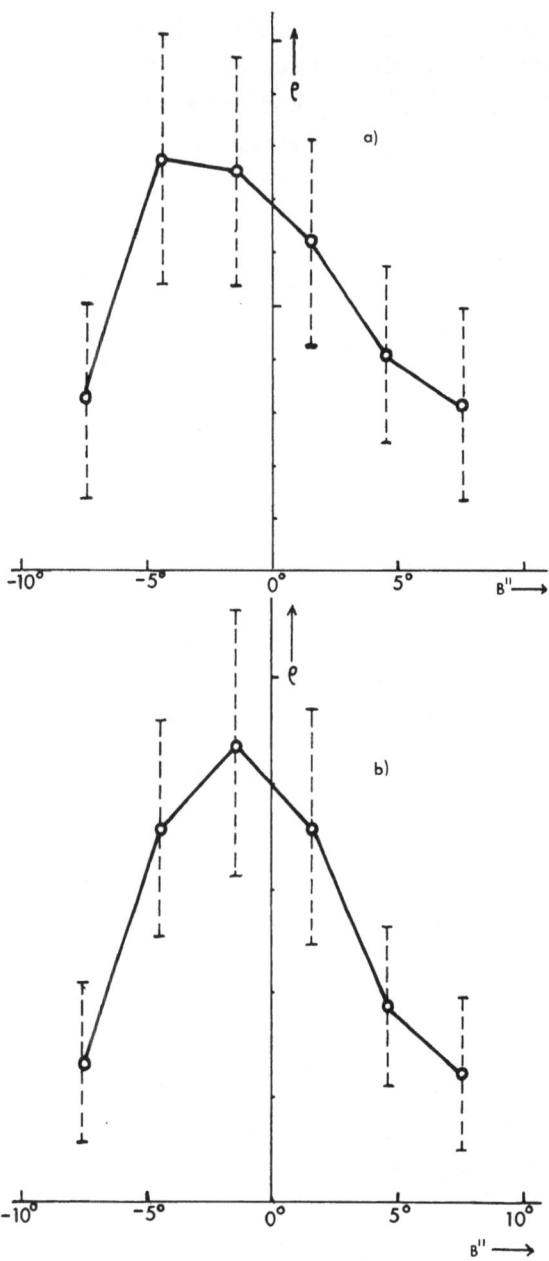

Fig. 3.   (a) The observed latitude density of pulsars having $350 < l < 114°$ after correction for thoroughness of coverage. (b) The same as Figure 3a but after correction for variation of sky background temperature.

We are therefore seeing pulsars on a galactic scale and to obtain the strong longitude dependence of their distribution, without putting any formal assumptions upon the way we might expect pulsars to be distributed through the Galaxy, I would estimate that on average they are at least half the distance of the galactic centre from the Sun, about 5 kpc away. This conclusion is reached by considering the nature of the radial dependence of the densities of various galactic populations, in both our own Galaxy and in M31. (See for example Arp, 1964). This dependence is usually smooth and gradual. In the case of the OB associations observed in M31 (van den Bergh, 1964), there does seem to be quite a sharp fall-off in density near the outer regions of the Galaxy and if the Sun were situated near such a fall-off in the pulsar population, the pulsars might be generally a little closer than 5 kpc. However, there is no evidence for such an effect in any populations at the Sun's distance from the galactic centre. The mean dispersion measure of the pulsars is about 150 pc cm^{-3}, leading to a rough estimate of 0.03 cm^{-3} for an upper limit for the mean electron density.

This is nicely compatible with the arguments presented by Guélin earlier. The value of the electron density might be refined further if we could identify the distribution with the spiral structure of the Galaxy. The clusters at longitudes 50° and 30° correspond quite closely to the tangential points of the Sagittarius and Norma-Scutum arms, suggesting that we are seeing some spiral structure. In fact, if a mean electron density of about 0.025 is assumed, there does seem to be quite a good correspondence between the pulsar population and the spiral structure, but with only about 50 pulsars, I do not consider the case to be very strong.

Having argued about the distribution of pulsars in the plane of the Galaxy, I would like to give a brief discussion of the distribution of pulsars above and below the plane in z distance. As we will see, this leads to conclusions concerning the width of the ionised layer and its uniformity.

We can deduce a pulsar's z distance from its dispersion measure under the assumption of a uniform extended ionized layer wider than the pulsar distribution. In Figure 4 we see how the z distance so deduced and quoted in units of dispersion measure varies with distance away from the Sun. Of the 105 known pulsars, I have not included in this diagram 5 pulsars with poorly determined dispersion measures and the 14 pulsars found in low latitude surveys, namely the high-longitude Jodrell Bank pulsars (6) and the pulsars recently discovered at Parkes (8) (Komesaroff et al., 1973a, b), because these have a strong selection effect upon this diagram.

The Parkes pulsars, for instance, would mostly lie within the two lines marked $b = \pm 2°$ and hence on this diagram would all be necessarily low z distance objects. The latitude limitation of the main Jodrell Bank survey ($b < 10°$) is not very serious, although it may give rise to a small excess of low z distance local pulsars. In fact many of these pulsars were discovered in other, wider-latitude surveys suggesting that many of the higher z distance local pulsars missed by the Jodrell Bank survey would have been discovered and are included in the diagram.

I find the most remarkable thing about this diagram is the constancy of the mean distance of pulsars off the plane as a function of distance away from the Sun at least out to dispersion measures of about 200. Table I shows how the mean distance off the plane, $\langle |z| \rangle$, varies with the distance, D, along the plane, both quoted in units of dispersion measure. The errors are the standard errors.

This presentation of the data simply demonstrates quantitively how well the pulsar *dispersion measures obey the cosec b law.*

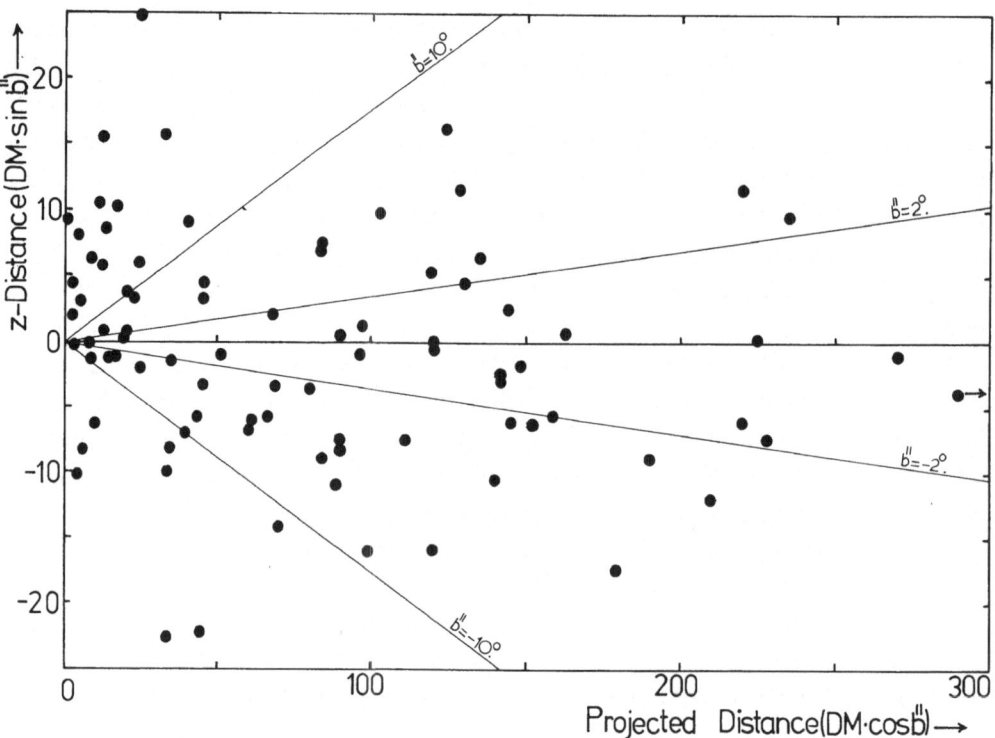

Fig. 4.   The derived z-distances of pulsars plotted against the distances along the galactic plane from the
Sun. The units of both quantities are in units of dispersion measure, pc cm^{-3}.

TABLE I

Mean distances of pulsars from the plane

D (pc cm^{-3})	$\langle |Z| \rangle$ (pc cm^{-3})
0– 50	6.8 ± 0.8
50–100	6.2 ± 0.8
100–150	6.5 ± 1.0
> 150	7.0 ± 1.1

If the line of sight to any pulsar intercepted a large H II region, then the apparent
z distance could be increased considerably. The data is thus consistent with a very
simple model of the distribution of electrons and pulsars, in which the electrons have
a uniform density out to a distance of about 6 kpc from the Sun, and large H II regions
contribute only a minor amount to the dispersion measures of pulsars in a statistical
sense. In particular, the mean electron density over distances of this order is very
similar to that immediately in the Sun's neighbourhood. Prentice and ter Haar (1969)
and others from a more local selection of pulsars have come to similar conclusions
for distances out to about 1 or 2 kpc.

Finally, Figure 5 shows the pulsar z distribution. This shows a substantially monotonic fall in numbers with z distance. If the ionised layer is substantially thinner than the pulsar layer, this would show a peak at a non-zero z distance. The absence of this effect suggests therefore that the ionised layer has a full width of greater than about 600 pc in the Sun's neighbourhood.

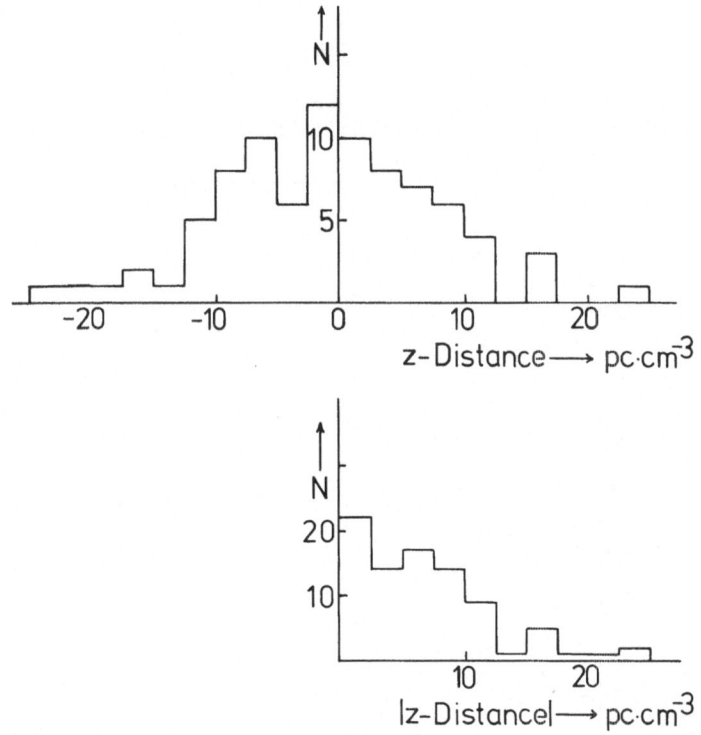

Fig. 5.   Histograms of the numbers of pulsars as a function of the z-distance and modulus of the z-distance in units of pc cm^{-3}.

The deduced width of the pulsar distribution depends upon the precise form of the electron distribution. Whatever this is, the observed width of the z distribution in Figure 5 and an assumed mean electron density of 0.03 cm^{-3} lead to a full-width to half-power points of at least 460 pc for the pulsar distribution.

References

Arp, H.: 1964, *Astrophys. J.* **139**, 1045.

Bergh, S. van den: 1964, *Astrophys. J. Suppl.* **9**, 65.

Davies, J. G., Lyne, A. G., and Seiradakis, J. H.: 1972, *Nature* **240**, 229.

Davies, J. G., Lyne, A. G., and Seiradakis, J. H.: 1973, *Nature* **244**, 84.

Komesaroff, M. M., Hamilton, P. A., McCulloch, P. M., Ables, J. G., and Cooke, D. J.: 1973a, *IAU Circ.* No. 2505.

Komesaroff, M. M., Hamilton, P. A., McCulloch, P. M., Ables, J. G., and Cooke, D. J.: 1973b, *IAU Circ.* No. 2563.

Large, M. I.: 1971, in R. D. Davies and F. G. Smith (ed.), 'The Crab Nebula', *IAU Symp.* **46**, 165.

Prentice, A. J. R. and ter Haar, D.: 1969, *Monthly Notices Roy. Astron. Soc.* **146**, 423.
Seeger, Ch. L., Westerhout, G., Conway, R. G., and Hoekama, T.: 1965, *Bull. Astron. Inst. Neth.* **18**, 11.

A. G. Lyne
University of Manchester,
Nuffield Radio Astronomy Laboratory,
Jodrell Bank, Cheshire SK11 9DL, United Kingdom

DISCUSSION

Guélin: You have said that you can see correlations, taking $n_e = 0.025$ cm^{-3} for the pulsars, with the spiral structure. I have made that calculation, and I don't find it. I think I have taken all the pulsars. I mean that the distribution is just chance along these directions.

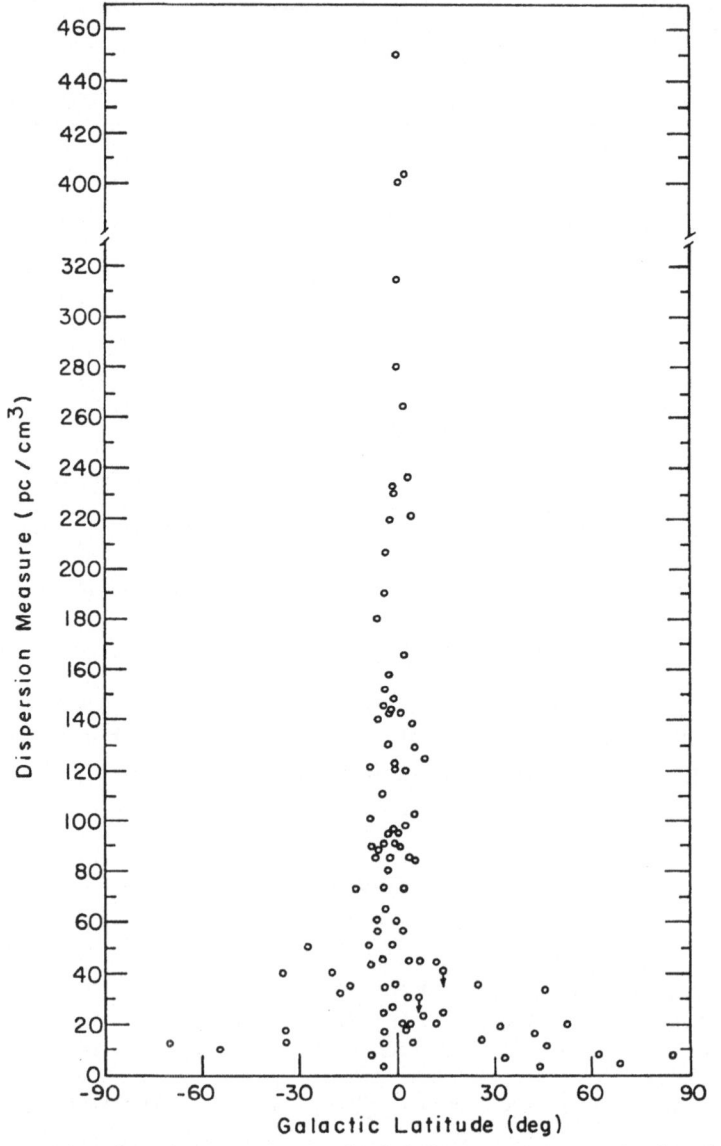

Fig. 1.   Pulsar dispersion measures vs galactic latitude as of August 1973 (Terzian).

Lyne: Yes, I do not consider the case to be very strong. There is an indication of it; I don't believe it is more than that.

Van Woerden: As a matter of definition: one may use $DM \sin b$ and $DM \cos b$ for *estimates* of vertical and horizontal distances, but I think it inadvisable to call these quantities 'distances.'

Terzian: In Figure 1 I would like to show the up-to-date data (August 1973) on pulsar dispersion measures shown as a function of galactic latitude; 105 pulsars are included.

F. G. Smith: Many of the 105 pulsars have been found in surveys with strong selection effects. To establish a sample of 51 pulsars it was necessary to extend the survey into relatively non-productive areas of the sky, in contrast to surveys which looked only in the best places.

HEATING AND IONIZATION OF THE
INTERSTELLAR MEDIUM

MICHAEL GREWING

Institut für Astrophysik und extraterrestrische Forschung der Universität Bonn und
Max-Planck-Institute für Radioastronomie, Bonn, F.R.G.

Abstract. Ionization and heating mechanisms are reviewed; some of them apply to the interstellar medium as a whole, others only to localized regions. Cooling processes are briefly summarized. Results are given from a recent model calculation for an intercloud medium heated by X-rays.

I. Introduction

Over the past few years a large number of both observational and theoretical studies have been undertaken to reveal the physical state of the interstellar medium (ISM) and the processes that cause its ionization and heating. Observationally, new results have come from radioastronomical measurements, from optical studies, and – in the most spectacular form – from the OAO-C Copernicus satellite in the ultraviolet range. Theoretically, new processes that could cause an energy input into the interstellar gas have been studied, and further details of the so-called 'two-component model' of the ISM have been worked out. Also, the first successful attempts to study time-dependent phenomena in interstellar space have been made.

As far as the theory is concerned, an excellent and very extensive review of the present state of the art has recently been given by Dalgarno and McCray (1972) so that in our discussion of the heating and ionization of the interstellar medium we can restrict ourselves to a broad outline of the relevant physical processes the details of which may be found in their paper. We shall start out by a discussion of various heating mechanisms. This is followed by a brief summary of cooling processes, and finally, we shall present results from a new model calculation for an X-ray heated intercloud medium in which an attempt was made to account for the observations from the Copernicus satellite.

II. Heating Mechanisms

The heating processes so far discussed in the literature may be divided into three categories depending on whether they are assumed to operate (a) continuously in space and time, (b) continuously in time but only in localized regions in space, or (c) discontinuously in both space and time.

(a) PROCESSES OPERATING CONTINUOUSLY IN SPACE AND TIME

A large variety of heating mechanisms have been suggested that fall in this category. Among these are the heating due to the light of the stars at $\lambda > 912$ Å, low energy

F. J. Kerr and S. C. Simonson, III (eds.), Galactic Radio Astronomy, 97–109. All Rights Reserved.

cosmic rays (1–10 MeV) and soft X-rays (~ 100–300 eV), as well as the heating due to cloud collisions, hydromagnetic waves and the dissipation of the turbulent energy of the ISM. In our further discussion we shall concentrate upon the first three of these processes.

Basic to these heating mechanisms is the conversion of energy originally stored in some background radiation field (including the cosmic rays) into kinetic energy which, carried away e.g. by electrons, will then be thermalized in elastic collisions. The primary interaction process may be denoted as

$$A + \begin{Bmatrix} h\nu \\ p \end{Bmatrix} \rightarrow A^+ + e^- + \begin{Bmatrix} - \\ p' \end{Bmatrix},$$

where A is an atom or ion that is transferred into its next higher stage of ionisation. Energy conservation determines the amount of kinetic energy carried away by the liberated electron E_e. Heating occurs if

$$E_e > \langle E_e \rangle,$$

where $\langle E_e \rangle$ is the mean kinetic energy of the gas, determined by its temperature. For $10 \leqslant T \leqslant 10^4$ K, $8.6 \times 10^{-4} \leqslant \langle E_e \rangle \leqslant 0.86$ eV.

(i) Starlight at $\lambda > 912$ Å

If all Lyman-continuum photons are indeed trapped in H II regions around the stars producing them, the stellar radiation field terminates at the Lyman limit. It can therefore ionize only trace elements like C, Mg, Si, S, and Fe which are transferred to their first stage of ionization. In this process the liberated electrons obtain an average energy of roughly 2 eV which is rapidly thermalized. However, due to the fact that this interaction process relies on the abundance of the trace elements the total heat input into the gas is small compared to what is required.

(ii) Low-Energy Cosmic Rays

The inadequacy of the stellar radiation field to explain the observed interstellar temperatures led Hayakawa (1960) to propose that low-energy cosmic rays might be a more favourable heating source. The primary ionization rate of neutral hydrogen in a collision with a fast proton is given by

$$\zeta_H = \int_{E_0}^{\infty} J(E)\,\sigma(E)\,dE$$

where $J(E)$ denotes the flux of cosmic ray protons and $\sigma(E)$ their interaction cross section. Since $J(E) \propto E^{-2.6}$ and $\sigma(E) \propto E^{-1}$, roughly, the value of the above integral is determined by the flux of the least energetic particles with energy E_0 that are still available in sufficient numbers. This energy E_0 depends on the mean free path of cosmic rays in the Galaxy which for a gas of unit density is approximately 530, 57, and 0.7 pc for 3, 1, and 0.1 MeV particles. These numbers suggest the choice $E_0 = 2$ MeV, the value most often quoted in the literature.

In an ionization of hydrogen by a 2-MeV proton, the liberated electron will carry away an average energy of 32 eV. Only part of this energy is available for heating, typically 5 eV, whereas the rest goes into secondary ionizations and excitations. The exact amount that is thermalized depends on the degree of ionization of the medium as shown e.g. by Dalgarno and McCray (*loc. cit.*, Figure 7).

Despite the fact that theories assuming the heating of the ISM to be due to low-energy cosmic rays have been worked out to various degrees of sophistication (see, e.g., Spitzer and Scott, 1969; Habing and Goldsmith, 1971; Bergeron and Souffrin, 1971; Jura and Dalgarno, 1972; Vogel, 1972 and references quoted therein), they are all subject to a major uncertainty which results from the difficulty of determining the flux of 2-MeV particles applicable to the ISM. These particles strongly interact with the solar wind, being scattered back and forth between solar wind magnetic field inhomogeneities, so that by the time they reach the Earth their spectral energy distribution is completely modified, even during periods of minimum solar activity. Earlier attempts to demodulate the observations have e.g. been made by Gloeckler and Jokipii (1967) and Webber (1968) (see also Silk and Steigman, 1970). However, as demonstrated by Urch and Gleeson (1972) measurements made near the Earth are indeed very badly suited for reconstructing the interstellar flux of low energy cosmic rays. Their results are shown in Figure 1.

As shown in this figure, Urch and Gleeson calculated the energy distribution resulting at the Earth from three vastly different input spectra, taking into account adiabatic expansion effects which had previously been overlooked. These results demonstrate that reliable estimates of the 2-MeV cosmic-ray flux in interstellar space will have to await measurements made at much larger distances from the Sun which can be carried out from space probes flying to the outer planets.

With actual measurements still to come, indirect arguments have been used to determine the hydrogen ionization rate due to 2-MeV protons as $\zeta_H = 4 \times 10^{-16}$ s^{-1} (e.g., Field *et al.*, 1969). This requires an energy density of 2-MeV particles of 6×10^{-14} erg cm^{-3}, corresponding to a flux of roughly 0.2 particles cm^{-2} s^{-1} per one MeV energy range, which is very large on the basis of present observations.

(iii) *Soft X-Rays*

In view of the doubts that remain whether low-energy cosmic rays are sufficiently numerous to keep the interstellar gas heated and ionized, Silk and Werner (1969) suggested the heating and ionization by soft X-rays as an alternative process. The primary ionization rate is given by

$$\zeta = \int_{E_0}^{\infty} J(E)\, \sigma(E)\, dE,$$

where $J(E)$ denotes the flux of X-ray photons, $\sigma(E)$ the photoionization cross-section

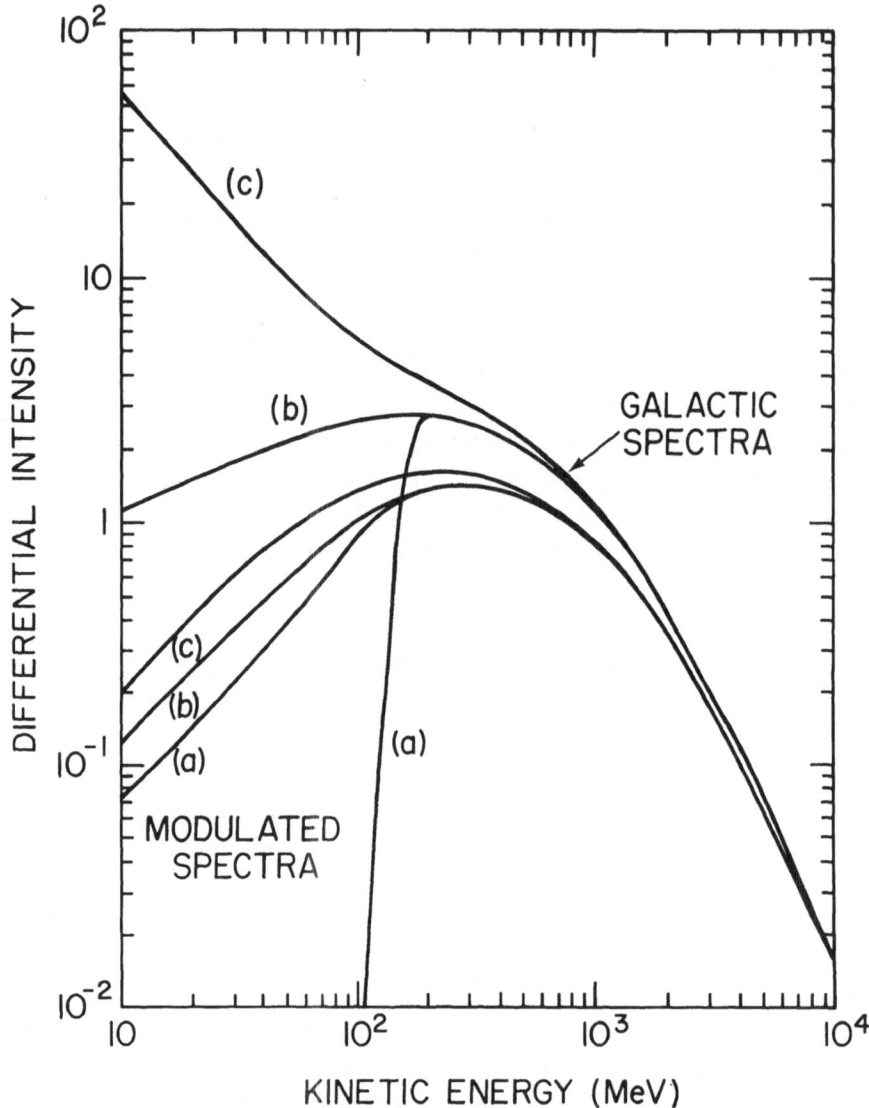

Fig. 1.   Urch and Gleeson (1972) calculated the modulation effects that occur as low-energy cosmic-ray protons penetrate into the solar system where they are scattered back and forth between solar wind magnetic field inhomogeneities. Due to adiabatic expansion effects the cosmic-ray protons are degraded in energy and their spectral energy distribution is completely modified when the particles finally reach the Earth. This is shown here for three different (assumed) galactic spectra (a, b, c) which would hardly be distinguishable by measurements from near the Earth.

(including inner shell contributions, Auger- and Coster-Kronig reactions), and E_0 the ionization threshold energy. Since $J(E)$ falls off with increasing energy, and since $\sigma(E) \sim E^{-3}$, roughly, the softest X-ray photons still available in sufficient numbers determine the ionization rate. The best information on the low-energy photon flux comes probably from the recent experiment by Yentis *et al.* (1972), which covers the 0.1–0.28-keV range. Results based on their data which are shown in Figure 2 will be presented in Section IV.

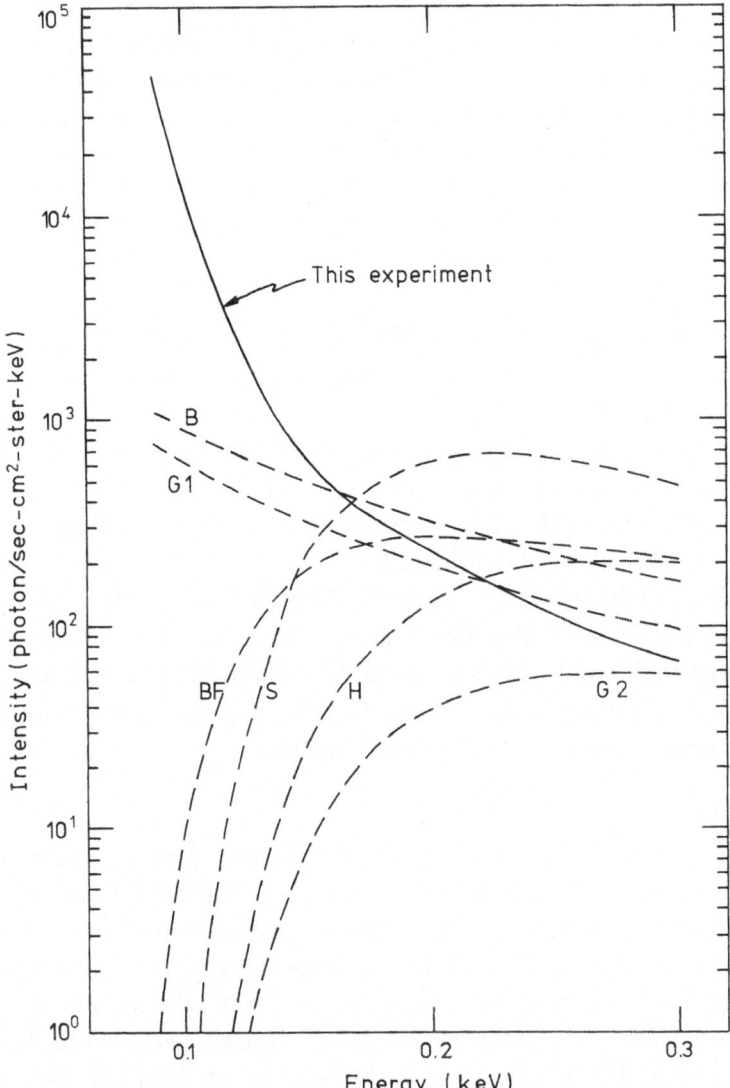

Fig. 2.   This figure is taken from the paper of Yentis *et al.* (1972) who succeeded in extending the measurements of the soft X-ray background to the 100–200 eV range in a rocket experiment.

As shown in Figure 2, the observed flux of soft X-rays increases strongly towards lower energies, making the results rather crucially dependent on the choice of E_0. The mean free path of soft X-rays in the Galaxy is approximately 130, 9, and 1 pc for 250, 100, and 50-eV photons, assuming a gas density of 1 particle cm^{-3}. These numbers suggest that for $n < 1$, i.e. in the intercloud gas, $50 \leqslant E_0 \leqslant 100$ eV, whereas for $n > 1$, i.e. for interstellar clouds, $E_0 > 100$ eV seems more appropriate.

(b) PROCESSES OPERATING CONTINUOUSLY IN TIME, BUT ONLY IN LOCALIZED
     REGIONS IN SPACE

Among the processes that fall into this category are the ionization and heating due to photons with $\lambda < 912$ Å, which originate (i) from O- and early-type B stars, (ii) from UV-stars, and (iii) from He-recombination processes in the case of an X-ray heated ISM. These mechanisms are particularly relevant for the intercloud regions, whereas the ones listed next pertain to interstellar clouds: (iv) heating due to photo-electron emission from irradiated grains, (v) collisional de-excitation of vibrational levels of H_2 after photo-excitation, and (vi) photodestruction of H_2-molecules.

(i) *O- and Early-Type B-Stars*

Most obvious of course is the severe effect that Lyman-continuum photons generated by O- and early-type B-stars have on the nearby interstellar space in which they produce an H II region.

The enhanced matter density in regions where new stars form would normally guarantee that these H II regions are ionization-bounded. However, in particular B-stars may be sufficiently old to have moved into more normal regions of interstellar space where the densities are considerably lower so that much larger volumes can be ionized. This possibility was considered by Prentice and ter Haar (1969), Grewing and Walmsley (1971), Walmsley and Grewing (1971), and Richstone and Davidson (1972). Though their results are somewhat discrepant due to differences in the assumptions made, these authors show that O- and early-type B-stars play a substantial role for the ionization of the intercloud medium.

(ii) *UV-Stars*

A potentially very powerful further source of Lyman-continuum quanta are the so-called UV-stars (Hills, 1972). Model calculations predict that on their way from the red giant to the white dwarf stage stars will have surface temperatures of $T_{eff} \gtrsim 10^5$ K for times $t \sim 10^6$ yr. Their luminosity would be $L \sim 10^2 L_\odot$. The source of this energy is the gravitational contraction of the star, and the numbers given depend rather crucially on what amount of the energy liberated is lost, e.g. in neutrinos. This question is still open, and no UV-star has – to my knowledge – as yet been seen, contrary to what one would expect on the basis of the number density estimate given by Hills (*loc. cit.*). However, *if* one uses the numbers adopted by Hills, all the intercloud matter in the Galaxy could be ionized by these stars.

(iii) *EUV-Photons from Helium-Recombinations*

In those regions of interstellar space which are heated and ionized by soft X-rays, recombinations of He II and He III lead to a further production of photons in the Lyman-continuum band. As shown in Section IV, these photons play an essential role in the ionization of hydrogen in such regions and add a little to the heating.

(iv) *Photoelectron Emission from Interstellar Grains*

Watson (1972) first noted that for likely interstellar grain materials except ice, photo-emission of electrons is expected from laboratory data to be a relatively efficient process for photon energies of ~ 10–13.6 eV. The average energy of the ejected photoelectron depends somewhat on the grain material and on the photon energy, but will typically be 2 eV, well in excess of the mean kinetic energy in interstellar clouds. Using recent determinations of the galactic ultraviolet flux, Watson arrives at the conclusion that the energy input into H I clouds due to this mechanism is comparable to that from proposed fluxes of low energy cosmic rays.

(v) Collisional De-Excitation of Vibrational Levels of H_2

After photo-excitation of molecular hydrogen in the Lyman and Werner bands, in the majority of cases the molecules return to the ground electronic state in a vibrationally excited level. As pointed out by Stecher and Williams (1973) collisional de-excitation of these levels will provide a heating mechanism, which generally however is small compared to the one to be discussed next.

(vi) Radiative Dissociation of H_2

As stated above in the majority of all cases ($\sim 73\%$) the excitation of molecular hydrogen by UV-photons will lead to vibrationally excited molecules in their electronic ground state. In the remaining cases ($\sim 27\%$), however, the molecules cascade into the vibrational continuum of the ground state whereby they dissociate. The pair of H-atoms that comes out of this photo-destruction mechanism will carry an average kinetic energy of ~ 0.5 eV according to an estimate by Dalgarno, which exceeds the mean kinetic energy in interstellar clouds. Milgrom et al. (1973) pointed out the importance of this heating mechanism and compared it to the heating due to the photo-effect on grains.

We should emphasize that the processes (iv), (v), and (vi) depend critically on the penetration of photons with energies 10–13.6 eV into the clouds. Along with the molecules and the grains, carbon will be competing for these photons. For clouds of sufficient optical thickness one will have to distinguish between a C I-core and an C II-envelope in which photons with energies 10–11.2 eV will all be absorbed. A first attempt to take these effects into account has been made by Walmsley (1973).

(c) TIME-DEPENDENT PROCESSES IN LOCALIZED REGIONS IN SPACE

As a third category we shall briefly consider time-dependent heating mechanisms that apply to localized regions in space. It was only fairly recently that the potential importance of short-term large-scale energy releases connected with gravitational collapse as it occurs e.g. during a supernova explosion for the heating and ionization of the ISM was noted. In their analysis of data pertaining to the Gum nebula, Brandt et al. (1971) were led to the hypothesis that this region in space is a fossil H II region generated by a SN-flash some 10^4 yr ago when the Vela pulsar was formed. Though the model they suggest for the Gum nebula is still open to discussion, it served to trigger a rapid series of theoretical studies of time-dependent models for the ISM (e.g., Schwarz, 1973; Kafatos, 1973; Gerola et al., 1973).

The most recent discussion of the possible role of supernovae for the ionization and heating of the ISM was given by Cox (1973). From it one may conclude that

SN provide so much energy (in the form of UV quanta, cosmic rays and shockwaves) that their influence on the ISM will indeed not be limited to localized regions in space but that instead SN rather than the X-ray background or the cosmic ray background provide the fundamental source of energy at least for the intercloud medium in our Galaxy.

III. Cooling Mechanisms

An excellent and very detailed review of the cooling processes operating in interstellar space has been given in the paper by Dalgarno and McCray (1972). We shall therefore be very brief here.

Basic to all cooling mechanisms is the conversion of kinetic energy by inelastic collisions into line radiation. Except in dense interstellar clouds where one has to solve a radiation transfer problem, the ISM may generally be assumed to be optically thin at the relevant wavelengths. The photons produced can thus freely escape, causing a loss of energy to the system.

The cooling processes so far considered in the literature may be grouped into three main classes depending on whether they are based on collisions involving free electrons, H-atoms, or H_2-molecules. Most of the mechanisms belonging to the second class have already been considered in the classical papers by Spitzer (1948, 1949). He concentrated on these processes since at that time free electrons were thought to be extremely rare in interstellar space ($\sim 10^{-4}$ cm^{-3}). The mechanisms that fall into this group are

$$L_{Hi}, L_{Ha}, L_{Hm}, L_{Hg},$$

denoting the cooling due to hydrogen-ion and hydrogen-atom collisions in which ground-state fine-structure transitions are excited, the cooling produced by the collisional excitation of rotational levels of molecules, and the cooling due to collisions of hydrogen atoms with grains. For L_{Hi} the cases $i =$ C II, Si II and Fe II are particularly important. For L_{Ha} the cases $a =$ O I and C I have been considered. For L_{Hm} the cases $m =$ H_2, HD, CO, CN, and CH have been discussed in the literature. A compilation of all these processes and detailed references to rate estimates etc. may be found in Dalgarno and McCray's paper (loc. cit.).

In their paper Dalgrano and McCray also discuss the processes

$$L_{H_2i}, L_{H_2a}, L_{H_2m},$$

which denote the excitation of ions, atoms, and molecules in collisions with H_2 molecules. These processes may be relevant in interstellar clouds where the H_2 abundance is known to be high from observations of the Copernicus satellite (Spitzer et al., 1973). The cases so far considered are $i =$ C II, $a =$ O I, C I, and $m =$ H_2, and H_2CO.

The processes that belong to the first class mentioned may be denoted as

$$L_{ei}, L_{ea}, L_{em}.$$

In electron-ion collisions the excitation of ground-state fine-structure levels dominates at lower temperatures (i=C II, Si II, and Fe II) whereas at intermediate temperatures (600–6000 K) the excitation of metastable levels is most important (i=C II, O II, Si II, S II, and Fe II). At higher temperatures the collisional excitation of allowed dipole transitions of atoms (a=H I, O I, and N I) must be taken into account. In their paper Dalgarno and McCray give a very detailed summary of all these processes as well as of the L_{em} collisions where the cases m=CO, CN, CH, and CH^+ are considered.

IV. Results from Recent Model Calculations for an X-Ray Heated Intercloud Medium

Recent advances in measuring the soft X-ray background in the 100–280 eV range (Yentis *et al.*, 1972), improved photo-ionization cross-sections (Barfield *et al.*, 1972), and a wealth of detailed observational data from the Copernicus satellite (Rogerson

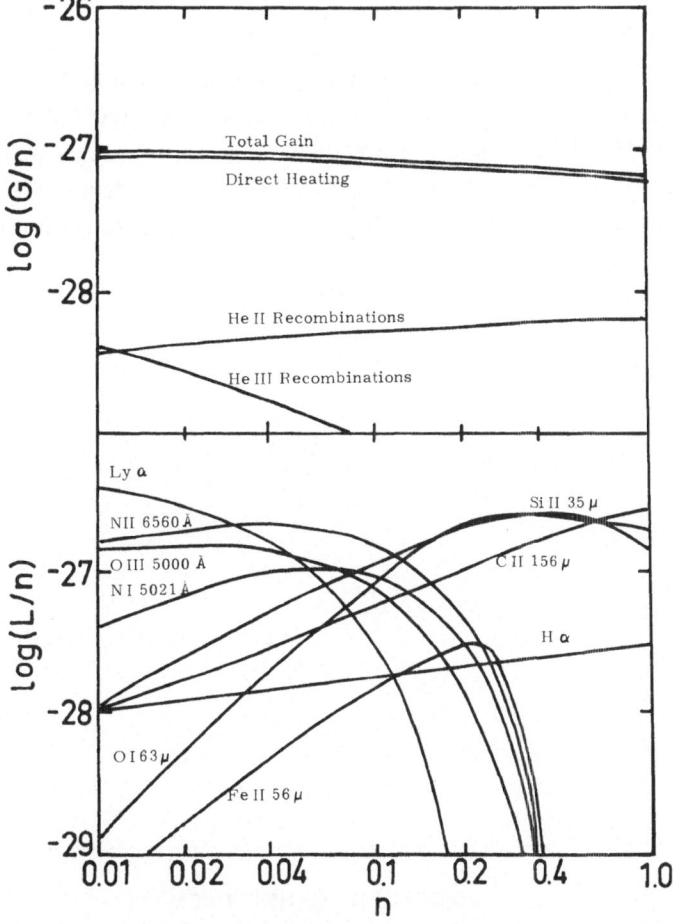

Fig. 3.   This as well as the following two figures are taken from the paper by Grewing and Walmsley (1974) in which results are given from model calculations for an X-ray heated intercloud medium. In the top part of this figure the energy gain of the gas is shown that results from the observed X-ray background shown in Figure 2. In the bottom part of the figure the energy loss of the gas due to various line transitions is plotted.

et al., 1973; Morton *et al.*, 1973) made it seem worthwhile to re-compute the ionization and thermal balance of an X-ray heated intercloud medium. A full description of this new model is given by Grewing and Walmsley (1974) so that we can restrict ourselves to a few main points.

Adopting the X-ray spectrum as measured by Yentis *et al.* (*loc. cit.*) the primary ionization rate for hydrogen is found to be 3.4×10^{-17} s^{-1}. Primary ionization rates have also been calculated for 47 further atoms or ions, using the photo-ionization cross-sections of Barfield *et al.* (*loc. cit.*). Auger- and Coster-Kronig contributions have been considered but appear to be negligible except for the ions Al II, Si II, and Fe II. Large effects, however, are found to come from secondary ionization processes. For hydrogen, where they are largest, we obtain a ratio of secondary to primary ionization rates of 3.1 and 5.1 for a density of 0.1 and 1 cm^{-3}, respectively. Indeed, even the rate of hydrogen ionizations following He-recombinations is found to exceed the primary rate by factors 1.45 and 1.53 at these same densities.

The total heat input into the gas from the observed X-ray background is shown in the top part of Figure 3. The energy gain per particle is rather constant with density, and is for all densities essentially determined by the direct heating whereas the heating that results from He recombinations is only of secondary importance.

In calculating the thermal equilibrium of the gas, cooling rates were used which differ from that of previous authors (e.g. Bergeron and Souffrin, 1971; Habing and Goldsmith, 1971) in two respects. Firstly, the effects of the ionization balance were

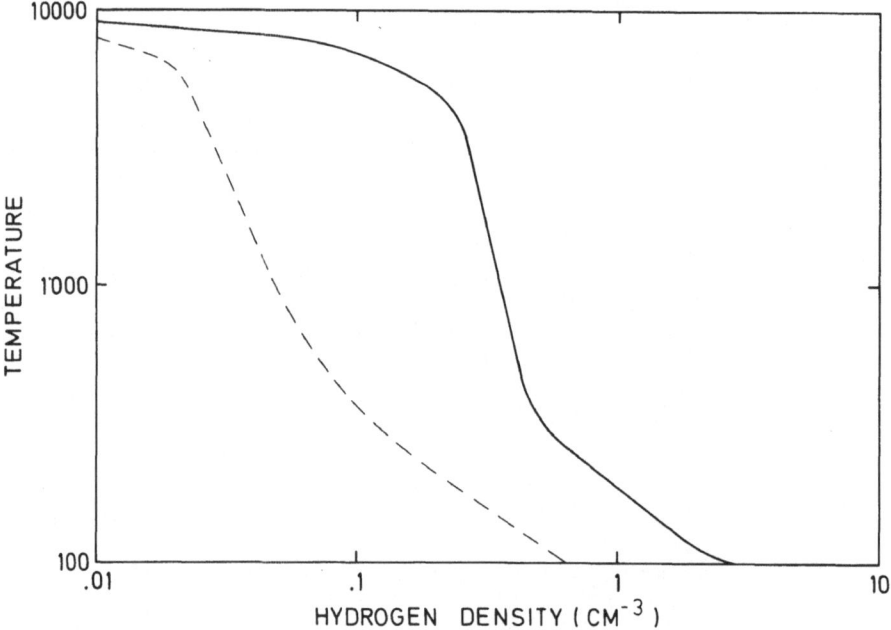

Fig. 4.   By solving the thermal balance equation the equilibrium temperature of the gas was determined as a function of the gas density. This result depends strongly on the assumed abundances of heavy elements. The full line shows the temperature that is obtained using abundances determined by the OAO-C satellite (see text), whereas the dashed line is based on solar abundances.

taken into account, and secondly, heavy element abundances as determined by the OAO-C satellite for the line of sight towards the star α Leo were used (Rogerson *et al.*, *loc. cit.*). The cooling losses that then result are shown in the bottom part of Figure 3.

In Figure 4 the equilibrium temperature is plotted as a function of the hydrogen density. The full line corresponds to the OAO-C abundances. It differs significantly from the temperature derived for normal solar abundances shown as a dashed curve. It is particularly interesting that the X-ray flux determined by Yentis *et al.* (*loc. cit.*) combined with the abundance determinations from the Copernicus satellite allows the maintenance of a gas density of 0.2 cm^{-3} at a temperature of 5000 K. These values

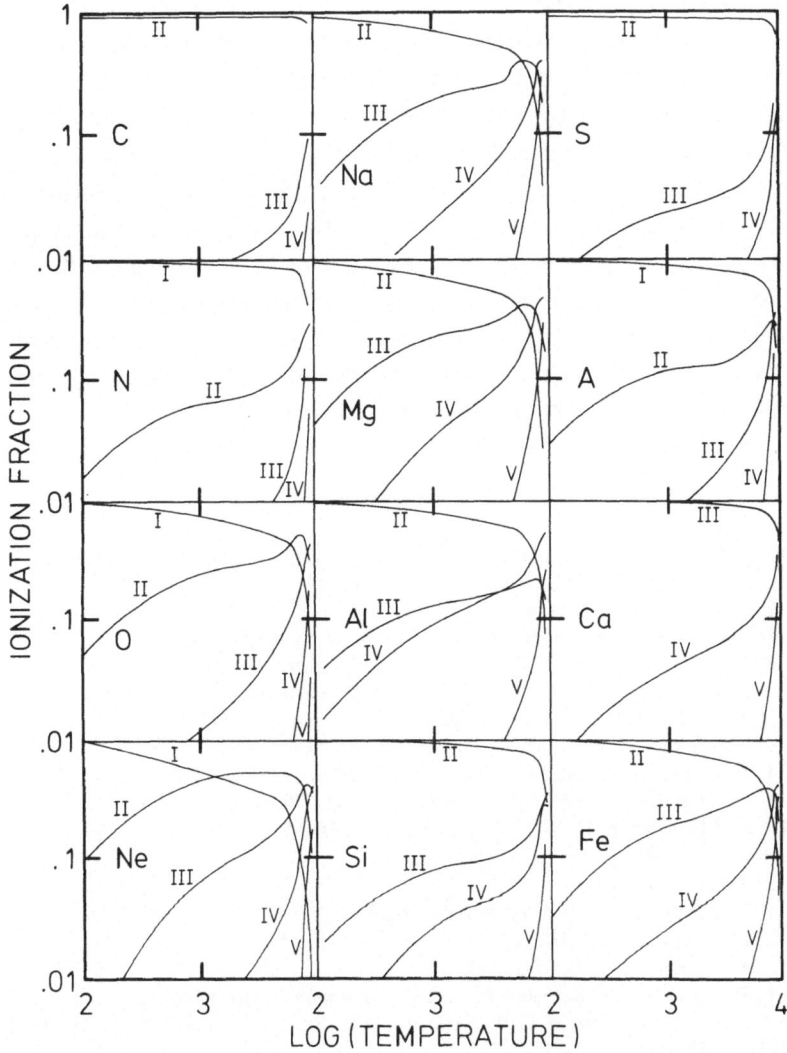

Fig. 5. The fractional ionization of the elements C, N, O, Ne, Na, Mg, Al, Si, S, A, Ca, and Fe is shown as a function of the equilibrium temperature (see text for further details of the calculations).

agree well with those derived from an analysis of 21-cm hydrogen line profiles (e.g., Mebold, 1972).

In Figure 5 the fractional ionization of the elements C, N, O, Ne, Na, Mg, Al, Si, S, A, Ca, and Fe is shown as a function of the equilibrium temperature T for the OAO-C abundances. These results should serve as a test whether the intercloud H I gas is indeed ionized by X-rays.

V. Conclusions

Over the past few years our understanding of many individual physical processes that may be relevant in interstellar space has greatly increased. In particular, the 'two-component model' of the ISM has been worked out in great detail, and considerable progress has also been made on the front of time-dependent models. It remains doubtful, however, whether this has really helped us to derive from the observations a quantitatively more reliable picture of the actual conditions existing in interstellar space. By looking at a distant star our line of sight might intersect interstellar clouds, intercloud matter heated by cosmic rays, X-rays or UV quanta, a tenuous H II region around an early-type B-star, a normal H II region, a fossil Strömgren sphere and even the supernova-produced tunnel-network proposed recently by Cox (1973). In our measurement we are integrating over physically very different regimes. In order to interpret such observations in a quantitative manner we would therefore firstly have to really understand what is going on in each of the different regions that can be located along a given line of sight, and secondly, we would have to find methods to recognize their actual contributions to what we see.

References

Barfield, W. D., Koontz, G. D., and Huebner, W. F.: 1972, *J. Quant. Spectr. Radiative Transfer* **12**, 1409.
Bergeron, J. and Souffrin, S.: 1971, *Astron. Astrophys.* **11**, 40.
Brandt, J. C., Stecher, T. P., Crawford, D. I., and Maran, S. P.: 1971, *Astrophys. J. Letters* **163**, L99.
Cox, D.: 1973, Report given at the IAU General Assembly, Sydney, Australia.
Dalgarno, A. and McCray, R. A.: 1972, *Ann. Rev. Astron. Astrophys.* **10**, 375.
Field, G. B., Goldsmith, D. W., and Habing, H. J.: 1969, *Astrophys. J. Letters* **155**, L149.
Gerola, H., Iglesias, E., and Gamba, Z.: 1973, *Astron. Astrophys.* **24**, 369.
Gloeckler, G. and Jokipii, J. R.: 1967, *Astrophys. J. Letters* **148**, L41.
Grewing, M. and Walmsley, C. M.: 1971, *Astron. Astrophys.* **11**, 65.
Grewing, M. and Walmsley, C. M.: 1974, *Astron. Astrophys.* **30**, 281.
Habing, H. J. and Goldsmith, D. W.: 1971, *Astrophys. J.* **166**, 525.
Hayakawa, S.: 1960, *Publ. Astron. Soc. Japan* **12**, 110.
Hills, J. G.: 1972, *Astron. Astrophys.* **17**, 155.
Jura, M. and Dalgarno, A.: 1972, *Astrophys. J.* **174**, 365.
Kafatos, M.: 1973, *Astrophys. J.* **182**, 433.
Mebold, U.: 1972, *Astron. Astrophys.* **19**, 13.
Milgrom, M., Panagia, N., and Salpeter, E. E.: 1973, *Astrophys. Letters* **14**, 73.
Morton, D. C., Drake, J. F., Jenkins, E. B., Rogerson, J. B., Spitzer, L., and York, D. G.: 1973, *Astrophys. J. Letters* **181**, L103.
Prentice, A. J. R. and ter Haar, D.: 1969, *Monthly Notices Roy. Astron. Soc.* **146**, 423.
Richstone, D. O. and Davidson, K.: 1972, *Astron. J.* **77**, 298.

Rogerson, J. B., York, D. G., Drake, J. F., Jenkins, E. B., and Spitzer, L.: 1973, *Astrophys. J. Letters* **181**, L110.

Schwarz, J.: 1973, *Astrophys. J.* **182**, 449.

Silk, J. and Steigman, G.: 1970, *Astrophys. Space Sci.* **9**, 304.

Silk, J. and Werner, M.: 1969, *Astrophys. J.* **158**, 185.

Spitzer, L.: 1948, *Astrophys. J.* **107**, 6.

Spitzer, L.: 1949, *Astrophys. J.* **109**, 337.

Spitzer, L. and Scott, E. M.: 1969, *Astrophys. J.* **158**, 161.

Spitzer, L., Drake, J. F., Jenkins, E. B., Morton, D. C., Rogerson, J. B., and York, D. G.: 1973, *Astrophys. J.* **181**, L116.

Stecher, T. P. and Williams, D. A.: 1973, *Monthly Notices Roy. Astron. Soc.* **161**, 305.

Urch, I. H. and Gleeson, L. J.: 1972, *Astrophys. Space Sci.* **16**, 55.

Vogel, U.: 1971, Thesis, University of Bonn.

Walmsley, C. M.: 1973, *Astron. Astrophys.* **25**, 129.

Walmsley, C. M. and Grewing, M.: 1971, *Astrophys. Letters* **9**, 185.

Watson, W. D.: 1972, *Astrophys. J.* **176**, 103.

Webber, W. R.: 1968, *Australian J. Phys.* **21**, 845.

Yentis, D. J., Novick, R., and van den Bout, P.: 1972, *Astrophys. J.* **177**, 365.

Michael Grewing

Institut für Astrophysik und extraterr. Forschung der Universität Bonn,
Auf dem Hügel 71,
5300 Bonn, F.R.G.

THEORETICAL CALCULATIONS
OF INTERSTELLAR CLOUD FORMATION

DONALD W. GOLDSMITH*

State University of New York, Stony Brook, N.Y., U.S.A.

Abstract. Numerical calculations of the growth of thermal instabilities show that condensations tend to form with the proper sort of density contrast but the wrong shapes (and, to some extent, sizes) to qualify as the observed interstellar clouds analyzed by Heiles.

During the last few years, several theoretical models for the heating and ionization of the interstellar medium have been proposed (cf. Grewing (1974)). Each of these models for the overall behavior of the medium allows for a calculation (by numerical hydrodynamics) of how small perturbations in the local temperature and density might grow into sizable contrasts in these parameters that could pass for observed interstellar 'clouds'. Such calculations have been carried out in some detail by Goldsmith (1970) for the steady-state model based on heating by low-energy cosmic rays or X-rays, by Schwarz *et al.* (1972) for the time-dependent model of heating by localized supernova outbursts, and by Ames (1973) for the steady-state model with the inclusion of the Rayleigh-Taylor instabilities proposed by Parker (1966) that are important for perturbation wavelengths of ~ 1 kpc or more.** All of these calculations (cf. the review by Goldsmith, 1973) show that if the assumed parameters of the model do indeed characterize the interstellar medium, then condensations with a density increase of $\sim 10^2$ over the original density are likely to form. This is, of course, good news from the theoretical front.

The bad news is that the condensations which are formed by the numerical hydrodynamics tend to have dimensions less than those of 'typical' interstellar clouds, and, worse, that the condensation process occurs along the magnetic lines of force, producing flat pancakes for clouds. Unfortunately, as Heiles (1974) has shown in his review of interstellar H I in this volume, cloud material appears in filaments elongated *along* the magnetic field lines (as estimated from optical polarization measurements). Before we therefore abandon the numerical calculations, let us recall that they are the only calculations we have, and look briefly at the fundamental mechanism of thermal instability.

As outlined in Field's classic paper (1965), the process of thermal instability hinges on the possibility that gas somewhat cooler than its surroundings may quickly become *much* cooler. Because of the perfect gas law, such gas will most likely become much denser as it cools by radiating optical and infrared photons. For the interstellar medium, the approximation that the gas is optically thin to its own radiation ranks

---

* Present address: 1655 12th Ave., San Francisco, Calif. 94122.
** Because of the complexity of the problem studied by Ames, her calculations concern only the linearized *hydrodynamical equations* and do not enter the non-linear region of sizable density contrasts.

F. J. Kerr and S. C. Simonson, III (eds.), Galactic Radio Astronomy, 111–114. All Rights Reserved.
Copyright © 1974 by the IAU.

among the safest of the many that enter the calculations. In the steady-state model of the interstellar medium (Field *et al.*, 1969), a characteristic kink in the (p, ϱ) curve for thermal and ionization equilibrium that occurs for temperatures near 6000 K leads naturally to a situation where a slight density increase (or temperature decrease) produces a phase transition from hot, rarefied gas ($T \simeq 6000$ K, $n \simeq 0.2$ cm^{-3}) to a cold, dense gas ($T \simeq 20$ K, $n \simeq 40$ cm^{-3}) in a time of $2–3 \times 10^7$ yr. On the other hand, the calculations by Schwarz *et al.* (1972) of thermal instability within a medium that is cooling and recombining after a supernova outburst show that a similar density increase will arise from an initially small (10%) perturbation, but in a time of 10^6 yr.

The maximum wavelength affected by the thermal instability is in both cases approximately ct, where c is the speed of sound in the initial medium, about 8 km s^{-1}, and t is the time mentioned above. Thus the steady-state and time-dependent models have maximum wavelengths for the initial perturbation of $\sim 10^{20.5}$ or 10^{19} cm respectively. These perturbations could yield spherical clouds with acceptable sizes of 10^{20} or $10^{18.5}$ cm for the largest clouds formed, *if* the condensation could proceed in three dimensions. However, if the instability occurs in regions with initial magnetic fields of 1 μG or more, then the contraction perpendicular to the field lines cannot keep up with the contraction along the field lines after the first *e*-folding of the density. The result would be a cloud with a thickness of $\sim 10^{18.5}$ or 10^{17} cm along the field lines, but a hundred or more times wider in directions perpendicular to the field. Although accurate data on the sizes and shapes of clouds are lacking, such observations as are now available (Heiles, 1974) fail to reproduce the calculated models. The discrepancy between the elongation predicted for theoretical clouds and the actual (reverse) elongation that has been observed powerfully reminds us that we have no reliable understanding of how clouds form in the presence of a magnetic field. A realistic (more than one-dimensional) treatment of the numerical magnetohydrodynamics of condensations may someday provide the key to resolving this problem.

Meanwhile, in 1973, two conclusions emerge from the theoretical studies. First, condensations will form through thermal instabilities in a gas that is cooling from $T \simeq 10^4$ K and recombining after a supernova outburst. The size, and hence the overall importance, of the regions subject to such condensation processes depends on the flux of ionizing photons from the supernova. A small ionizing flux ($\ll 10^{51}$ erg) would make the Schwarz *et al.* (1972) condensations a relatively local accompaniment of supernova outbursts. Second, if the interstellar medium is heated and ionized primarily by a steady flux (steady on time scales greater than 10^7 yr) of low-energy cosmic rays or X-rays, then thermal instabilities will produce condensations wherever the local density exceeds a certain critical value (near $n = 0.2$ cm^{-3} with the assumed model parameters). The significance of this mechanism depends on the importance of the steady flux as compared to the time-dependent sources of heating and ionization. If the ionization rate from the unmeasured portions of the X-ray and cosmic-ray spectra does not exceed that from the measured X-ray background, then steady-state theories can not explain the observed temperatures in interstellar clouds, and are unlikely to explain the formation of these clouds.

Calculations of the thermal instabilities that could produce interstellar clouds suffer from gross uncertainties concerning the initial conditions that could lead to condensations. Such uncertainties can be invoked in partial explanation of why clouds formed in computers differ from clouds we find in our Galaxy, but apparently some key element of the problem still eludes us. Cox (1973) has suggested that supernova blast waves play the dominant role in ordering the interstellar medium, and perhaps this favorite theoreticians' retreat, the supernova event, can thus save us once again.

References

Ames, S.: 1973, *Astrophys. J.* **182**, 387.

Cox, D.: 1973, paper presented at the meeting of IAU Commission 34, Sydney, Australia.

Field, G.: 1965, *Astrophys. J.* **142**, 531.

Field, G., Goldsmith, D., and Habing, H.: 1969, *Astrophys. J. Letters* **155**, L149.

Goldsmith, D.: 1970, *Astrophys. J.* **161**, 41.

Goldsmith, D.: 1973, review paper presented at the meeting of IAU Commission 34, Sydney, Australia (available as report no. 560 of CRSR, Cornell University).

Grewing, M.: 1974, this volume, p. 97.

Heiles, C.: 1974, this volume, p. 13.

Parker, E.: 1966, *Astrophys. J.* **145**, 811.

Schwarz, J., McCray, R., and Stein, R.: 1972, *Astrophys. J.* **175**, 673.

D. W. Goldsmith
1655 12th Ave.,
San Francisco, Calif. 94122, U.S.A.

DISCUSSION

Habing: It is possible to obtain n_e in any specified cloud from interstellar absorption lines of an element in two ionization stages. Dunham tried this already in 1939. This probably requires higher spectral resolution than is available even with the Copernicus satellite.

Goldsmith: This is probably the best way to determine the electron density accurately, and in the few cases where Ca II/Ca I line strengths have been measured (by White), it has already given valuable results. Also, the Mg II doublet can be used in a similar way if we can observe at the appropriate near-ultraviolet wavelength.

Jenkins: I'd like to ask Grewing about how dense a cloud in your calculations of the ionization equilibrium complies with this. You must have a problem when it gets so dense that it is opaque to X-rays in the cloud bed. Is that in a regime of observability with optical ultraviolet lines?

Greenberg: And I would like to ask, in your calculation of the reduced cooling rate due to the evidence of reduced abundance of the heavier elements based on Copernicus, did you then take into account the additional cooling due to the additional dust and consequent molecules?

Grewing: Our model calculations were not really aimed at treating interstellar clouds but rather the intercloud gas. To treat clouds adequately, we would have to incorporate further heating and cooling processes into our computer program. We are in the process of doing this.

Goldsmith: The cloud and intercloud medium could be heated in entirely different ways. Soft X-rays may heat the intercloud medium, but almost certainly they don't penetrate the clouds. You need another explanation. It may be some unknown kind. Then again it would not be surprising that we don't know much about how to form clouds.

Gordon: As Burke mentioned, perhaps the association of magnetic fields with filaments means *only* that the matter controls the fields and not the other way around. In this case, we should ignore the fields in modeling the formation of clouds.

Goldsmith: We still seem to face the problem either of moving material across field lines on time scales $\sim 10^7$ yr (or less), or of explaining the compression of material into clouds without raising the field strength by two orders of magnitude.

Oort: How about a supernova explosion? You can form clouds in a supernova explosion.

Goldsmith: Perhaps everything else is ruled out and you do need a huge energy input. You still need an extra power of linkage because otherwise the magnetic field will dominate.

Kahn: Do you not think that one ought to compare the magnetic energy density with the turbulent energy density in interstellar space, in order to decide whether the cloud motions can distort the magnetic field pattern?

Goldsmith: It seems that the greatest problem is not to have clouds distort the magnetic field pattern, but rather to explain how an initial perturbation in the intercloud medium can become a cloud elongated along the magnetic field.

Townes: In the case of dense molecular clouds, many shape irregularities have been found, but none so far can be described as filamentary. In this case also, the kinetic energy density is probably much greater than the magnetic field energy density, so that one has a simpler, or at least a different, theoretical problem.

While observations of dense clouds characteristically indicate blob-like shapes, with great irregularities rather than filaments, it is also clear that in some cases there is finer structure within the clouds that is not yet resolved. This finer structure might possibly be filamentary, but there is no such indication at present.

Goldsmith: Dense (molecular) clouds may indeed not suffer from the same problems of magnetic field restrictions as those which arise in the formation of 'ordinary' clouds. Unfortunately we probably still must form 'ordinary' clouds on the way to forming dense clouds.

Davies: In reply to Goldsmith's query whether the magnetic field penetrates H I clouds, I would point out that Zeeman measurements show that the denser ($n_H \simeq 100$ cm^{-3}) clouds are permeated by magnetic fields. Further, the OH maser sources ($n_H \sim 10^6$ to 10^7 cm^{-3}) contain fields whose magnitude is that expected from the compression of these dense H I clouds on the assumption of flux conservation. Magnetic fields of the observed strengths play a significant role in the pressure equilibrium of all these clouds.

Goldsmith: However, the key question that remains is how to form clouds from intercloud material without being prevented (by the magnetic field) from contracting in directions perpendicular to the field.

Burke: The Crab nebula is the best-observed example of a magnetic, hot gas which exhibits condensations. Have you considered this example?

Goldsmith: The formation of condensations in material ionized by a supernova explosion has been considered by Schwarz, McCray and Stein, and by Kafatos and Morrison. I think it more difficult to form interstellar clouds in this way than to form such localized filaments.

EXPANDING SHELLS OF
NEUTRAL HYDROGEN AS BIRTHPLACES OF
STELLAR ASSOCIATIONS

R. SANCISI

Kapteyn Astronomical Institute, University of Groningen,
Groningen, The Netherlands

Abstract. Expanding shells of neutral hydrogen have been found connected with the associations Per OB2 and Sco OB2. The shells also contain dust and molecules. A model of the spatial configuration and the kinematics of gas, dust and stars is discussed. The stars of the associations lie outside, close to the shells; gas and stars move outward from a common centre of expansion. If the shells are interpreted as old supernova remnants which have been strongly decelerated by the interstellar medium, ages of the order of 1×10^6 yr are obtained. A genetic relationship between stars and shells is suggested. The stars formed in the densest gas shell; afterwards they separated from the gas, as the shell was further slowed down by intervening interstellar matter. Limits are set on the time scale of star formation. The geometry and the motion of the shell at the moment of star formation account also for the kinematics of Per OB2 and Sco OB2.

I. Introduction

The study of neutral hydrogen in regions of young stellar associations provides information on the physical conditions of interstellar matter which favour star formation, and on the interaction between young stars and interstellar gas. Previous 21-cm line studies (cf. Raimond, 1966) have established the existence of massive and dense concentrations of neutral hydrogen connected with some OB associations; however, they have failed to reveal the nature of the connection between stars and gas. In particular, no conclusive evidence has ever been found for expanding motions of the gas in the regions of the associations.

In this paper we report the results of a new, detailed study of neutral hydrogen in the regions of the associations Per OB2 and Sco OB2. These objects are known to be very young and in rapid expansion (Blaauw, 1964). Some of their characteristics are listed in Table I. Both lie in the solar neighbourhood, at intermediate galactic latitudes; their membership, state of motion and age are fairly well known. For all these reasons they represent ideal candidates for a detailed investigation of the interstellar neutral hydrogen in regions of recent star formation.

The 21-cm observations were made with the 25-m radiotelescope at Dwingeloo (beamwidth $0°6$; bandwidth 2 km s^{-1}). The linear resolution is 3.5 pc at the distance of Per OB2 and 1.8 pc at that of Sco OB2. For this study the observational data have been displayed in three different sets of contour maps, which show the distribution of brightness temperature in longitude and velocity at a constant galactic latitude, in latitude and velocity at a constant longitude, and in longitude and latitude at a constant radial velocity.

In the following section we show that expanding shells of neutral hydrogen do exist

F. J. Kerr and S. C. Simonson, III (eds.), Galactic Radio Astronomy, 115–128. All Rights Reserved.
Copyright © 1974 by the IAU.

TABLE I

General properties of the associations Per OB2 and Sco OB2

Property		Per OB2	Sco OB2
Position l		156° to 165°	341° to 2°
b		−11° to −21°	+10° to +30°
α (1950)		3^h25^m to 4^h4^m	15^h30^m to 16^h35^m
δ (1950)		+30° to +36°	−30° to −15°
Distance from the Sun (pc)		330	170
Distance from galactic plane (pc)		−85	+55
Mean radial velocity (LSR) (km s^{-1})		+15	+6
Total mass in stars ($10^3\ M_\odot$)		1.5	1.9
Estimated average	(from cm diagrams)	4	10
age (10^6 yr)	(from kinematic data)	1.5	20

Notes to Table I

All the data in the table are quoted from Blaauw (1960, 1964) with the exception of the mean radial velocities; these have been derived in this study from the available, best determined radial velocities of the individual stars.

Slightly lower values for the age of Per OB2 have been obtained recently by Lesh (1969) and by Maeder (1972). The age of Sco OB2 is probably much smaller than in the table; Maeder finds 4.5×10^6 yr.

in the regions of the associations Per OB2 and Sco OB2; from the optical and H I data we derive a model of the spatial and kinematical arrangement of stars and gas. In the last section we discuss a possible interpretation of the observed H I shells as old supernova remnants and derive some conclusions on the process and the time scale of star formation.

II. Results of the Observations

A few representative maps showing the observational data are reproduced in Figures 1 and 2. All radial velocities in this paper are with respect to the local standard of rest (LSR). A detailed discussion of the observations will be given elsewhere; we here restrict ourselves to a discussion of the main results.

(a) EXPANDING SHELLS OF NEUTRAL HYDROGEN

Large concentrations of neutral hydrogen, associated with complexes of dark nebulosity, exist in the regions of the associations Per OB2 and Sco OB2. They stand out for their small velocity dispersion and their large intensities. Their velocity and spatial structure indicate that they are shells forming part of expanding systems of gas.

Most of the hydrogen forming the expanding system is concentrated in a dense shell of roughly semispherical shape. The shells are visible in Figure 1. Figure 1a shows the distribution of brightness temperature, T_b, in the longitude-velocity plane at constant galactic latitude, $b = -18°$, through the association Per OB2. The semispherical shell is the high-intensity feature around longitude $l = 160°$, with the peak-

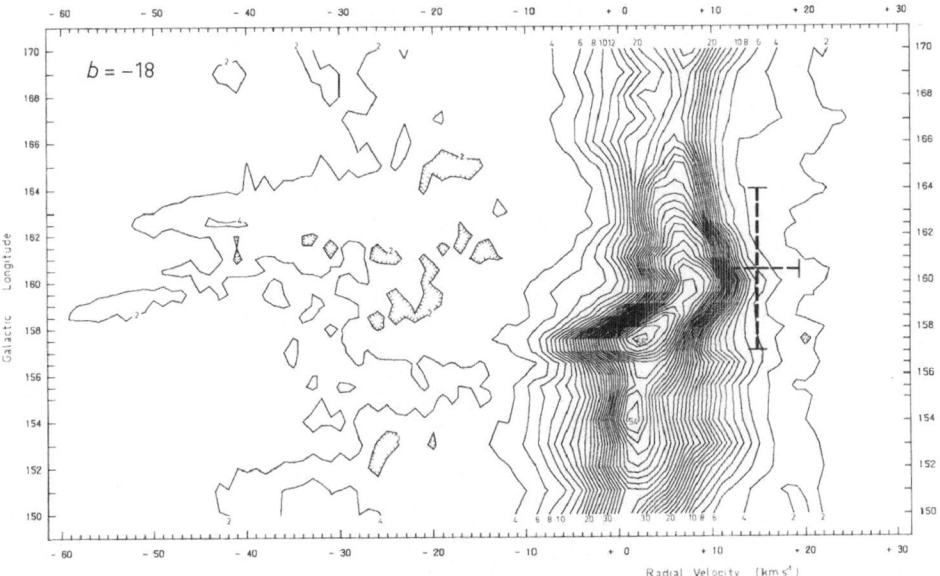

Fig. 1a. Map of contours of equal brightness temperature in a constant latitude cut ($b = -18°$) through the region of the association Per OB2. The contour interval is 2 K. The high-intensity feature around longitude $l = 160°$, with the peak velocity gradient from 0 to $+8$ km s^{-1}, forms the semispherical shell discussed in the text and sketched in Figure 3. The vertical, broken line shows the longitude extent of the association at the mean radial velocity ($+15$ km s^{-1}, LSR); the horizontal line gives the average velocity residual.

velocity gradient from 0 to $+8$ km s^{-1}. The velocity field of this feature has been already described in detail by Sancisi (1970; Figure 3). Figure 1b shows the distribution of T_b in the latitude-velocity plane at constant galactic longitude, $l = 353°$, across the area of the association Sco OB2. The H I feature discussed here is the ridge of high intensity and low velocity dispersion seen between $b = +17°$ and $b = +27°$ with peak velocity ranging from -1 to $+3$ km s^{-1}.

The three-dimensional structure of the shell in Perseus can also be seen in Figures 2a and b, which give the distribution of H I column density in two different velocity intervals. Map (a) (velocity range from -7 to 0 km s^{-1}) shows in the direction of Per OB2 an empty region, from which all the gas has been swept away; around it one can see some of the gas which is presumably moving away in directions perpendicular to the line of sight. The large H I concentration in the centre of map (b) (velocity range $+8$ to $+15$ km s^{-1}) is the receding part of the dense semispherical shell.

The parameters found for these dense shells are given in Table II. The actual masses and densities may be larger, perhaps by a factor of two, as undetected amounts of cool H I and molecules may also be present. For example, the H_2 molecules observed by Carruthers (1970) in the direction of the star ξ Per may well be associated with the H I shell in Perseus. The values of the densities are probably further underestimated, because the adopted values for the shell thickness may be too large. The clouds of dust seen in these regions are probably associated with the hydrogen shells; in fact position

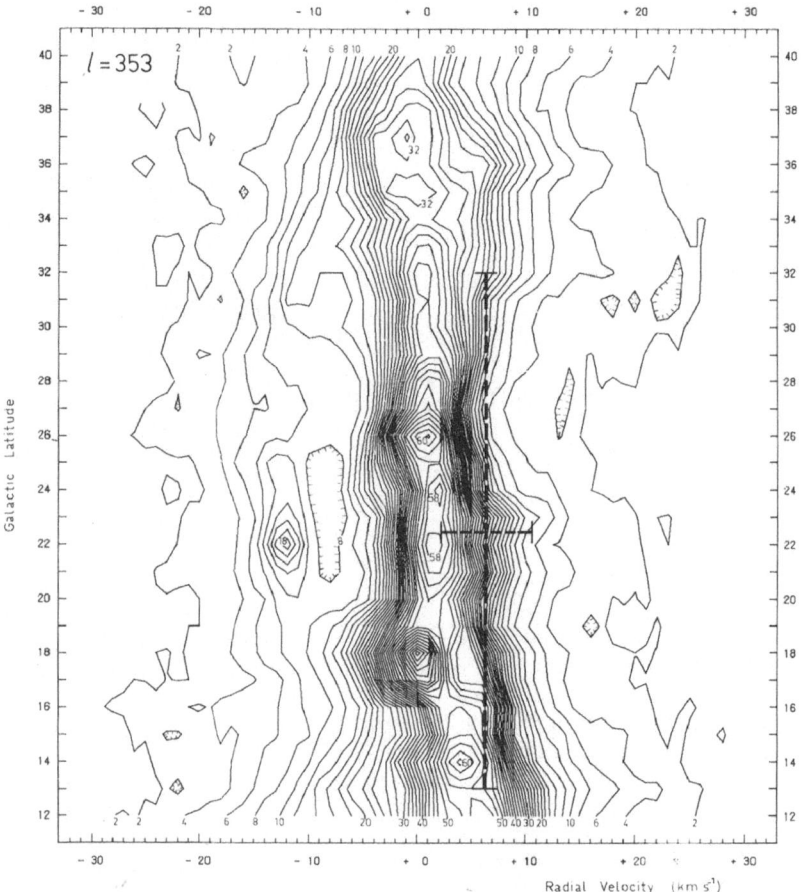

Fig. 1b.   Latitude-velocity diagram at constant galactic longitude ($l = 353°$) through the association Sco OB2. The dense shell of semispherical shape discussed in the text is the feature of high intensity and low velocity dispersion seen between $b = +17°$ and $b = +27°$, with peak velocity from -1 to $+3$ km s^{-1}. The concentration at $V = -12$ km s^{-1} (Sancisi and van Woerden, 1970) is the part of the low-density shell, which moves towards the Sun. The mean radial velocity and latitude extent of Sco OB2 are given by the vertical broken line; the horizontal line gives the average velocity residual.

and extent on the sky of H I and dust agree closely. In Perseus this agreement is particularly striking, as a comparison of Figures 2b and c shows. In this region the spatial and physical correlation of dust and H I is also supported by the excellent agreement between the radial velocity of the H I shell and that ($+8$, $+9$ km s^{-1}) of the 'normal' OH emission, which has been detected recently in the direction of some of these dust concentrations (Crutcher, 1972; Knapp and Kerr, 1973). In Scorpius the correlation of dust and H I is particularly well established in the 'Nebula' region (close to ϱ Oph), which forms the low latitude edge of the shell. We note also that a number of small concentrations of dust, which are similar to Bok's globules, are present in these regions.

Other, smaller shell pieces have also been tentatively identified as belonging to the same system of gas. Their brightness temperatures are from 4 to 20 K; the masses

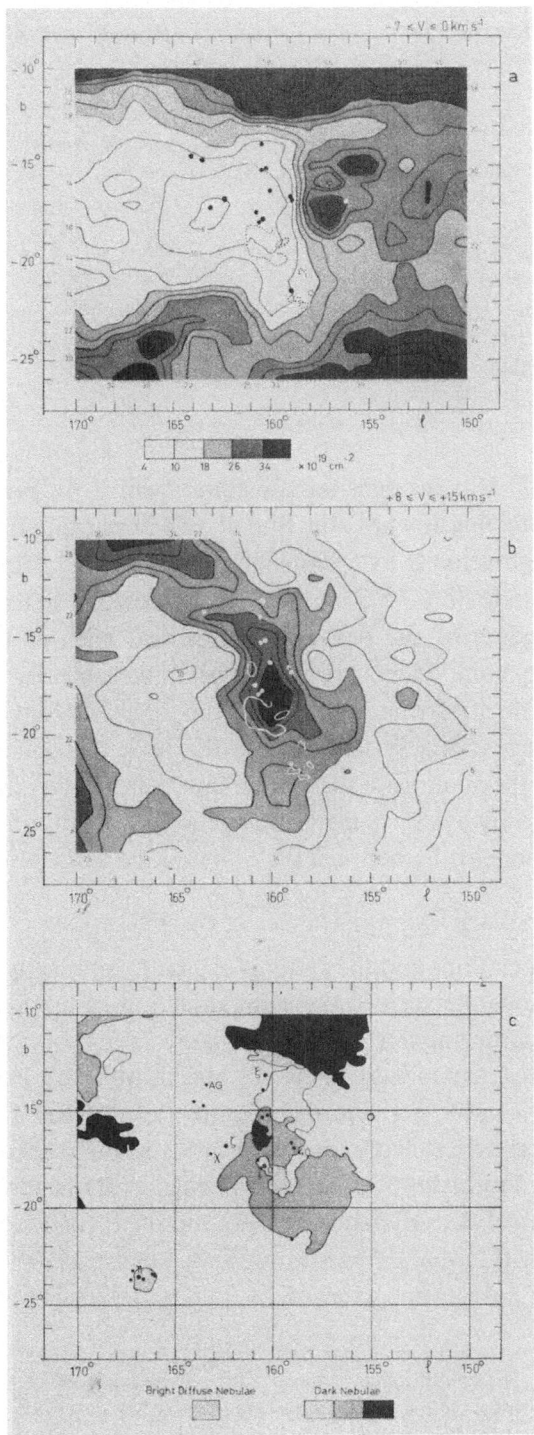

Fig. 2.   The maps (a) and (b) show the distribution of the column density of neutral hydrogen in Perseus in two different velocity intervals, respectively -7 to 0 km s^{-1} and $+8$ to $+15$ km s^{-1}; the gas was

TABLE II

Parameters for the dense semispherical H I shells in Perseus and in Scorpius

Property	Perseus	Scorpius
Centre	$l=160°\ b=-17°$	$l=354°\ b=+23°$
Systemic radial velocity (km s^{-1}) (LSR)	+3	−1
Radius	4°	5°
Radius (pc)	20	13
Thickness (pc)	5	5
Expansion velocity (km s^{-1})	5	3
Total mass ($M_\odot$)	1.4×10^4	3×10^3
Volume density (cm^{-3})	30	30
Peak brightness temperature (K)	40	40
Velocity dispersion (km s^{-1})	3	2

range from 10 to 10^3 $M_\odot$, the densities are about 5 cm^{-3}. In Perseus there are traces of gas (Figure 1a) moving towards the Sun at velocities around -45, -50 km s^{-1}; these seem to be high-velocity extensions of more intense H I components which are visible at lower latitudes ($b=-16°$) in the velocity range -30 to 0 km s^{-1}; the latter connect with the edges of the denser semispherical shell of positive velocity. In Scorpius (Figure 1b) components are seen at negative velocities, up to -12 km s^{-1}.

The space and velocity structure of these components strongly suggests that they belong to the same expanding system of gas as the dense semispherical shells discussed above; in Scorpius they complete the system on the negative-velocity side. This view is further supported by evidence from interstellar absorption lines, which indicates that these H I components lie between the Sun and the associations.

(b) KINEMATICAL AND SPATIAL ARRANGEMENT OF SHELLS AND ASSOCIATIONS

The stars of the associations and the systems of gas are physically related. The densest shells described above are seen in the same direction as the associations and have approximately the same angular extent (see Figure 2a and b); there is good evidence that the stars lie close behind the H I shells and move away from them with mean velocities of about 7 km s^{-1} in Perseus (Figure 1) and 5 km s^{-1} in Scorpius. This evidence is based on interstellar absorption line and interstellar extinction data. A close agreement is found between the interstellar sodium absorption lines in the spectra of some stars of Per OB2 and Sco OB2 and the H I emission components from the expanding shells (Hobbs, 1969a and b, 1971; Sancisi, 1970). The optical data for the Perseus region have been reviewed by Lynds (1969). From analysis of colour

---

assumed optically thin. The contour step is 4×10^{19} cm^{-2}. The positions of the stars of Per OB2 and an outline of the densest dust clouds, as given by Delhaye and Blaauw (1953), are also marked by dots and dashed lines respectively. The large H I concentration in the centre of map (b) is the receding part of the dense semispherical shell (sketched in Figure 3). The stars of Per OB2 lie behind this hydrogen and move away from it with a mean recession velocity of about 7 km s^{-1}. Map (c) is a reproduction from Khavtassi (1960). The dark nebulosity seen in the centre is probably associated with the H I shell: it has approximately the same position and extent as the H I feature shown in (b).

excesses as well as star and galaxy counts, she concluded that the association Per OB2 lies adjacent to or just behind a region of variable obscuration (absorption 1 to 3 mag). On the other hand this is undoubtedly the same layer of dust which was found to be associated with the neutral hydrogen in the shells.

Also consistent with this picture are the results of recent observations of the Ly-α line in absorption (Savage and Jenkins, 1972; Morton *et al.*, 1972a and b); the surface densities of neutral hydrogen derived for the stars of Per OB2 and Sco OB2 are equal to or even larger than those obtained from 21-cm measurements, contrary to the general result of lower N_H values from Lα than from 21-cm observations. This strongly suggests that all the hydrogen observed in the 21-cm emission line lies in front of each of the associations. The hydrogen in the shell is an important fraction of this total column density.

From these results and the distances to Per OB2 and Sco OB2 (Table I) we can infer that the distances to the shells are approximately 300 pc for Perseus and 150 pc for Scorpius.

The total momenta and kinetic energies estimated for the two systems of gas and stars are given in Table III. The reported values are underestimates, however, since

TABLE III

Total momentum and kinetic energy of the systems
of gas and stars in Scorpius and Perseus

	Perseus	Scorpius
Momentum ($M_\odot$ km s^{-1})	1×10^5	4×10^4
Kinetic energy (erg)	1×10^{49}	3×10^{48}

the masses of the identified shells are probably lower limits (see above), and certainly not all pieces of the shells, particularly those of high space velocity, have been identified.

III. Discussion

We propose the picture in Figure 3 to represent the geometry and the kinematical behaviour of gas and stars in Perseus. This is a model constructed from the available optical and radio data, as partly reviewed in the previous section; moreover, it has been assumed that stars and gas move outward from a common centre of expansion. The expansion speed of the gas is about 5 km s^{-1} and that of the stars 12 km s^{-1}. The latter value is simply the difference between the mean LSR radial velocity of the stars of Per OB2 (Table I), and the systemic radial velocity of the shell (Table II); therefore it is probably an underestimate of the actual expansion velocity of the association. In Scorpius the picture is quite similar; the expansion speeds are 3 and 8 km s^{-1} for gas and stars, respectively. Of course this simple semispherical picture represents only a rough approximation of the actual spatial and kinematic configuration of stars and gas; in particular, the H I shell is quite irregular in shape and morphology.

The picture suggests a few main problems for discussion: (a) the source and input mechanism of kinetic energy, (b) the ages of the shells, and (c) the process of formation of the associations.

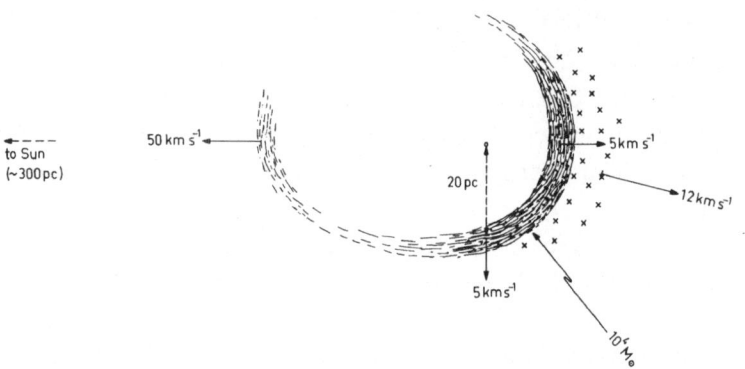

Fig. 3.   Sketch representing, in a plane containing the line of sight to the association, the proposed spatial arrangement and state of motion of the gas, dust and stars (X's) in the region of Per OB2. Stars, gas and dust move out from a common centre; the stars have larger velocities than the gas, and move ahead of it.

(a) SOURCE OF KINETIC ENERGY

Almost twenty years ago it was suggested for the first time that supernova shells (Öpik, 1953) or newly born O stars (Oort, 1954) could account for the observed motion of interstellar gas clouds and for the formation and acceleration of OB associations. But conclusive observational evidence for gas shells related to stellar associations was never found. Gas shells connected with the associations Per OB2 and Sco OB2 have been observed now, but no direct evidence has been found to support either of those two mechanisms. For instance, no radio -continuum or optical emission are found showing any obvious large-scale connection with the expanding shells.*

In the case of the O star ('old H II region') mechanism various difficulties arise; in particular it is doubtful (Kahn, 1967) whether, in the presence of high densities and efficient cooling, H II regions can release sufficient amounts of 'organized' kinetic energy. Also, known H II regions show no clear evidence of a coherent pattern of expanding H I motions of the kind observed in the present two cases; in general in the direction of H II regions one observes only an increased internal motion of H I and a broadening of the 21-cm line profile.

An alternative, more likely interpretation is that the H I shells observed in Perseus and Scorpius are 'very old' supernova remnants, which have been strongly decelerated by a dense, large interstellar cloud. This hypothesis may account for the observed

* Hα photographs shown by G. Monnet (Observatoire de Marseille) during this Symposium indicate that a weak Hα ring is probably associated with the H I shell in Scorpius.

total kinetic energies and radial momenta of the expanding systems (Table III) and also for their shape and morphology.

(b) AGES OF THE SHELLS

For the determination of the ages of the shells from the observed radii and expansion velocities (Table II), it is necessary to make assumptions about the mechanism of expansion of the gas.

If the expansion speed is assumed constant in time, ages of 4×10^6 yr are derived for the Perseus and Scorpius shells. On the other hand, if the shells are supernova remnants strongly decelerated by interstellar gas, as is most likely in the present case, a 'snow-plow' approximation gives ages of the order of 1×10^6 yr.

It should be realized that these values of the ages are obtained from radii and expansion velocities which are not determined precisely, but are based on the rough approximation of shells having semispherical shape. Therefore every age estimate has an intrinsic uncertainty of perhaps a factor of two.

The estimated ages and the small expansion velocities suggest that these shells may be at the last stage of evolution of a supernova remnant before it loses its identity and merges into the interstellar medium. Therefore any optical, radio or extended X-ray emission from these objects is expected to be totally absent or very weak. Also, none of the pulsars discovered to date lies in the direction of the shells; of course, a pulsar might well have moved quite far from the shell in such a long time since its birth. The only objects in the regions of sky of Per OB2 and Sco OB2 that, as far as we know, might be related with supernova events, are the X-ray sources Sco X-1 and 2U 0352+30. Sco X-1 is in the direction of one edge of the shell in Scorpius, and 2U 0352+30 lies in the direction of the centre of the shell in Perseus.

The dimensions of these shells (Table II) are rather small if compared to those of younger supernova remnants, like, for instance, the Cygnus Loop. But the ambient densities of interstellar gas in the present cases are at least 10 to 20 cm^{-3}, i.e., approximately 10 to 100 times larger than the ambient densities found for the Cygnus Loop.

It should be noted that the whole expanding systems of gas described in the first section include also the smaller, higher-velocity shell pieces like the ones at -12 km s^{-1} in Scorpius and at -50 km s^{-1} in Perseus; obviously they have larger dimensions than the dense, semispherical shells of Table II.

Finally we note that the supernova outbursts, suggested here to explain the existence of the H I shells, may be the same events which were advocated by Blaauw (1961) in these two regions in order to explain the existence of the runaway stars ξ Per and ζ Oph. These stars are connected respectively with Per OB2 and Sco OB2. The hypothesis of a common origin of shells and runaway stars is consistent with the kinematical ages of ξ Per and ζ Oph (1.6 and 1.1×10^6 yr, respectively), which are similar to those of the shells. Alternatively, the supernova outbursts which possibly released these runaway stars may have occurred after the associations were formed; in that case they would have no connection at all with the formation of the observed shells, which in our view must be older than the associations.

(c) FORMATION OF THE ASSOCIATIONS

The geometry and the motion of stars and gas described above and represented in Figure 3 strongly suggest that the formation of the associations Per OB2 and Sco OB2 is closely related to the expanding, dense shells. The following sequence of events may explain the observations: (1) the supernova remnant sweeps up dense interstellar gas and is decelerated; (2) star formation occurs in the decelerated, condensed shell; (3a) the shell of gas sweeps up more interstellar material and is further slowed down; (3b) the stars move on with constant velocity and therefore are separated from the gas.

From the observational data we can set restrictions on the time scale of star formation. In our view the period of time elapsed since the moment of star formation should be the same as that taken by the gas shell to slow down from the stellar velocities (12 km s^{-1}) to the present gas velocity (5 km s^{-1}). In the snow-plow approximation, assuming a constant density of the ambient gas, this further deceleration of the shell should have taken about 0.7×10^6 yr for both Per and Sco. Consequently, the ages of the stars should also be 0.7×10^6 yr. Of course this determination is not very precise; the actual ages of Per OB2 and Sco OB2 may be somewhat larger, but not larger than those derived above for the shells, which are in the range 1 to 4×10^6 yr.

Since estimates of the ages of these associations from kinematical data and from color-magnitude diagrams are available (Table I), a comparison with the present results is interesting. The kinematic ages of the associations should be equal to the ages of the shells; but it must be noted that the former were derived from the simple assumption of constant expansion speed of the stars, while in the model proposed here the present expansion of the association must be regarded as the result of two distinct phases: one of decelerated motion of the shell before the stars are formed, and a second in which the stars move with constant speed equal to that observed now. Therefore, according to this model, the reported values (Table I) of the kinematic ages of Per OB2 and Sco OB2 would be too large and should be reduced by about a factor of 1.5 or 2 before they are compared with the ages of the decelerated shells. These corrections, however, are not very critical for the present comparison. Considering the uncertainties, the agreement between the kinematical ages of the associations and the ages of the shells is quite good, especially for Per OB2. On the other hand, contrary to expectation, the ages derived from the c-m diagrams are larger than the 'time since formation' derived here; they are even larger than the kinematic ages, as already pointed out by Lesh (1969) and discussed by Maeder (1972), and larger than the ages of the shells.

For how long has star formation been going on? So far, we have assumed that all stars have formed and separated from the gas almost at the same time in the stage of the expansion, when the velocity of the shell was around 12 km s^{-1} in Perseus and 8 km s^{-1} in Scorpius. The large spread of stellar velocities suggests that the actual situation is more complicated. In fact, the total spread of the line-of-sight velocities of the stars, from about $+6$ to $+19 \text{ km s}^{-1}$ in Per and from -1 to $+19 \text{ km s}^{-1}$ in Sco (with respect to the centre of expansion), does not seem to be entirely accounted for by intrinsic errors, random motions and expansion, particularly in the case of the association Per OB2. It is possible, therefore, that this large velocity spread is partly the result of a formation process extending in time, and perhaps still going on. This would imply a corresponding spread in the ages of the stars and a correlation between the ages and the space velocities of the stars. In this connection we point out also that the runaway star ξ Per which is moving in the same direction as the other association stars with a line of sight velocity of $+64$ km s^{-1}, could represent a case of star formation at an earlier stage of expansion of the shell, when the latter still had a high velocity (about 64 km s^{-1}). This would obviously conflict with the 'proto-binary' hypothesis (Blaauw, 1961), which is the generally accepted explanation for the 'runaway' phenomenon.

We further note that groups of T-Tauri stars are found in the regions of Per OB2 (close to IC348, Herbig, 1954) and Sco OB2 (Herbig, 1962). They are associated with dust concentrations, which are probably local condensations of the expanding shells; if they had formed before the process of expansion started, presumably their parent interstellar material would have been blown away by the event which caused the expansion. Therefore, it is likely that these stars have been forming quite recently, after the OB stars, in the dense and cool regions of the shells. Unfortunately, proper motions and radial velocities of these low-luminosity stars are not available, and their possible connection with the shell cannot be tested.

Finally we note that the braking inferred from the velocity decrease of the gas shells with respect to the stars, from 12 to 5 km s^{-1} in Perseus and from 8 to 3 km s^{-1} in Scorpius, implies densities, 10 to 20 cm^{-3}, which are quite normal for interstellar clouds. The present mean separation in space between stars and shell, resulting from the deceleration of the latter, is inferred to be only a few parsec. Such a closeness in space of stars and shells is consistent with the presence in the expanding shells of H II regions excited by the stars of the associations, such as δ Sco and π Sco (Sharpless, 1959), and of bright reflection nebulosity (around o Per).

IV. Concluding Remarks

The results of this work bring some new elements for a discussion of: (a) the kinematics of Per OB2 and Sco OB2, (b) the connection of associations with interstellar matter, and (c) the expansion of the Gould Belt system.

(a) The geometry and the motion of the shell at the time of star formation account for: (i) the presence and the rate of expansion of Per OB2 and Sco OB2, and (ii) their mean recession velocities, which are, as pointed out by Blaauw (1970), anomalously high compared with the general interstellar medium and the average B-star population in the solar neighborhood. In our view these recession velocities represent the component of the expansion of the associations in the direction of the line of sight, away from the Sun (Figure 3). Star formation occurred only in the densest shell of the expanding system, the one receding from the Sun; therefore the resulting picture is largely different from the conventional one of an approximately spherical system of stars expanding in all directions.

(b) Per OB2 and Sco OB2 probably are not the only associations having such a physical connection with interstellar matter. In fact, similar expanding shells are also found in the associations Lac OB1 (part of our 21-cm survey in associations) and Cep OB3 (van Someren Greve, private communication), but the nature of their connection with the stars is more difficult to establish. In some cases the identification of such shells may be very difficult. For instance, successive events of the type discussed above, which are quite likely to occur in the lifetime of an association, would produce a very chaotic situation in the interstellar medium; this may have happened in the Orion region, where star formation has been going on for a long time, and stars of quite different ages are present. On the other hand, in such successive events the history of Per OB2 and Sco OB2 may well repeat itself and lead to the formation of subgroups of different ages and to the shifting of the place of star formation, which were discussed by Blaauw (1962, 1964).

(c) The recognition of the peculiar origin and nature of the large space motions of Per OB2 and Sco OB2 also has some direct consequence on studies of the kinematics of the Gould Belt system. The inclusion (see for instance Lesh, 1968) of these two – and possibly also of the other – associations, together with all Gould Belt stars, in solutions for the general expansion of the system may not be physically justified and may seriously affect the results.

To conclude we point out that probably these shells contain fairly large amounts of molecules; detailed surveys of OH, H_2CO and CO in these regions would be of great interest and may help unravel the kinematic structure of the shells.

References

Blaauw, A.: 1960, Nuffic Intern. Summer Course on Present Problems Concerning the Structure and Evolution of the Galactic System (The Hague, The Netherlands).

Blaauw, A.: 1961, *Bull. Astron. Inst. Neth.* **15**, 265.

Blaauw, A.: 1962, in L. Woltjer (ed.), *Interstellar Matter in Galaxies*, W. A. Benjamin, Inc., New York, p. 32.

Blaauw, A.: 1964, *Ann. Rev. Astron. Astrophys.* **2**, 213.

Blaauw, A.: 1970, in W. Becker and G. Contopoulos (eds.), 'The Spiral Structure of our Galaxy', *IAU Symp.* **38**, 199.

Carruthers, G. R.: 1970, *Astrophys. J. Letters* **161**, L81.

Crutcher, R. M.: 1972, Ph.D. Thesis, University of California, Los Angeles.

Delhaye, J. and Blaauw, A.: 1953, *Bull. Astron. Inst. Neth.* **12**, 72.

Herbig, G. H.: 1954, *Publ. Astron. Soc. Pacific* **66**, 19.

Herbig, G. H.: 1962, *Adv. Astron. Astrophys.* **1**, 47.

Hobbs, L. M.: 1969a, *Astrophys. J.* **157**, 135.

Hobbs, L. M.: 1969b, *Astrophys. J.* **158**, 461.

Hobbs, L. M.: 1971, *Astrophys. J.* **166**, 333.

Kahn, F. D.: 1967, in H. van Woerden (ed.), 'Radio Astronomy and the Galactic System', *IAU Symp.* **31**, paper 15.

Khavtassi, J.: 1960, *Atlas of Galactic Nebulae*, Abastumani Astrophysical Observatory.

Knapp, G. R. and Kerr, F. J.: 1973, *Astron. J.* **78**, 453.

Lesh, J. R.: 1968, *Astrophys. J. Suppl.* **16**, 371.

Lesh, J. R.: 1969, *Astron. J.* **74**, 891.

Lynds, B. T.: 1969, *Publ. Astron. Soc. Pacific* **81**, 496.

Maeder, A.: 1972, in G. Cayrel de Strobel and A. M. Delplace (eds.), 'Age des Étoiles', *IAU Colloq.* **17** XXIV-1, Observatoire de Paris-Meudon.

Morton, D. C., Jenkins, E. B., Matilsky, T. A., and York, D. G.: 1972a, *Astrophys. J.* **177**, 219.

Morton, D. C., Jenkins, E. B., and Macy, W. W.: 1972b, *Astrophys. J.* **177**, 235.

Oort, J. H.: 1954, *Bull. Astron. Inst. Neth.* **12**, 177.

Öpik, E. J.: 1953, *Irish Astron. J.* **2**, 219.

Raimond, E.: 1966, *Bull. Astron. Inst. Neth.* **18**, 191.

Sancisi, R.: 1970, *Astron. Astrophys.* **4**, 387.

Sancisi, R. and van Woerden, H.: 1970, *Astron. Astrophys.* **5**, 135.

Savage, B. D. and Jenkins, E. B.: 1972, *Astrophys. J.* **172**, 491.

Sharpless, S.: 1959, *Astrophys. J. Suppl.* **4**, 257.

R. Sancisi
Kapteyn Astronomical Institute,
University of Groningen, Groningen, The Netherlands

DISCUSSION

J. R. Dickel: A younger example of such an expanding 21-cm cloud is that around the northeast part of the SNR HB21. (Assousa and Erkes, *Astron. J.*, in press). It is also of interest that the position of the densest part of the H I cloud is also the region of HB21 where there is an apparent turnover in the low-frequency spectrum of the source. If the latter is interpreted as free-free absorption and if the ionized and neutral regions are coincident the gas must be cold and about 10% ionized at the current time.

Hughes: There appears to be further evidence for expanding rings of H I with associated star formation. Rickard in 1968 detected one in the Cas-Per arm from 21-cm observations and showed that the motion of early-type stars agreed with the H I velocities. Some of our more recent work (V. A. Hughes and D. Routledge, 1972, *Astron. J.* **77**, (1972), 210) showed evidence for a similar expanding ring of H I surrounding the Sun, which also appeared to form an envelope for the dust in the solar neighborhood and which may also be associated with Gould's belt.

Monnet: There is considerable evidence for shells in external galaxies – when there are early-type stars to ionize them. They have been observed in Hα by Baade in NGC 6822 and K 342, and I have found six of them in M33. A curious point is that they are observed only in the borders of the galaxy – as well as the shells you have observed in Perseus.

Sancisi: Expanding shells of neutral hydrogen have been reported several times in the past, with or without any connection with stars and stellar associations. However, I want to emphasize that in order to establish the reality of expanding shells a careful study of the spatial and velocity structure of the neutral hydrogen is necessary. In the present case also the physical relationship between the expanding shell and the stellar association has been studied.

Greenberg: I would like to suggest that this shell region you have begun to study would be an interesting place to study the way in which the dust appears or is modified in a condensation. The time scale you mentioned ($\sim 2 \times 10^6$ yr) would be sufficient to have permitted a very substantial accretion onto grains. It would therefore appear useful to examine this specific region for the wavelength dependence of the extinction to see if it is modified from 'normal' or not. What is the hydrogen density there?

Sancisi: This would be normal, because the dust has been gathered up so that we get an increase in surface density of the dust particles. With column densities $\sim 5 \times 10^{20}$ cm^{-2} and a thickness ~ 5 pc the density is only ~ 30 cm^{-3}. This probably is a low value because it has been obtained assuming the gas to be optically thin, and probably the thickness of the shell is larger.

Jenkins: Ly-α absorption data taken by OAO-2 for stars in the Persei OB2 aggregate indicate H I column densities generally in excess of 10^{21} cm^{-2}. These results suggest that significant saturation is probably occurring with the 21-cm radiation since your maximum column densities seem to be around 5×10^{20} cm^{-2}.

Sancisi: The column density of 5×10^{20} cm^{-2} refers only to the H I shell. The total H I column density from 21-cm observations in the direction of the Per OB2 stars is about the same as derived from Ly-α absorption. This probably means that all the H I gas in that direction is located between the stars of Per OB2 and the Sun. Some saturation may certainly be occurring with the 21-cm radiation from the shell, but our 21-cm data do not give any clear indication in that sense.

Mills: Can you give any information about the energetics of the events which produce these expanding shells?

Sancisi: The only relevant information I have is the observed present total kinetic energy of the expanding system, about 1×10^{49} erg, and the total radial momentum of the order of $10^5 \, M_\odot$ km s^{-1}. It is likely, however, that these values are underestimates. They seem to be of the same order as expected for very old ($\sim 1 \times 10^6$ yr) type II supernova remnants.

INTERSTELLAR MOLECULES – THEORY

R. D. BROWN

Monash University, Clayton, Victoria, Australia

Abstract. A number of possible mechanisms for the formation of molecules in the interstellar medium are discussed. Plausible suggestions have been made for the production of most of the known diatomics, but there is little understanding of polyatomics so far.

Over the last two decades a number of possible mechanisms for formation of molecules in the interstellar medium have been proposed. Until recently attention was focussed entirely on diatomic molecules because until December 1968 no serious consideration was given to polyatomic molecules.

The classical study by Bates and Spitzer (1951) was concerned with radiative recombination of atoms, in particular

$$C + H \rightarrow CH + hv,$$
$$C^+ + H \rightarrow CH^+ + hv.$$

These processes were dismissed as too slow under interstellar conditions to account for the observed amounts of interstellar molecules. More recently Solomon and Klemperer (1972) have reconsidered these and other radiative recombinations. By including the effect of 'atomic-multiplet trapping' from the higher components of ^3P and ^2P states of C and C^+, the effect of quantum-mechanical tunnelling through the centrifugal barrier and by using revised values for oscillator strengths of the molecular transitions they obtained rates that make it plausible that in H I regions with H atom densities $\geqslant 10 \text{ cm}^{-3}$ a rich mixture of diatomics will be produced, including CH, CH^+, CO, CN, C_2 and C_2^+. Reactions such as

$$CH^+ + O \rightarrow CO + H^+,$$
$$CH^+ + N \rightarrow CN + H^+,$$

are important in yielding interstellar CO and CN. Their work was primarily aimed at clouds of moderate density such as those between us and ζ Ophiuchi.

In the meantime various people (Kahn, 1955; McCrea and McNally, 1960; Gould and Salpeter, 1963; Hollenbach and Salpeter, 1971; Knaap *et al.*, 1966) studied in increasing detail the feasibility of formation of H_2 from H on the interstellar dust grains.

Aannestad (1973) has presented elaborate calculations combining radiative recombination, dust grains and shock waves in an apparently all-embracing theoretical treatment of interstellar molecule formation.

Other mechanisms of molecule formation that have been proposed include formation in the dense atmospheres of stars with subsequent expulsion into the interstellar medium (Herbig, 1970), or by photolysis of molecules within the mantles of dust grains to produce free radicals and, subsequently, more complex molecules (Greenberg *et al.*, 1972). However the former proposal has to include regular production of molecules to balance the steady removal of molecules by condensation on grains (Rank *et al.*, 1971) and the molecules require constant protection by dust grains from the rapid photolysis that would be produced by the average interstellar UV radiation field (Stief *et al.*, 1972). The mantle photolysis mechanism has not yet been developed to the quantitative stage for comparison with observed molecular abundances. Most recently the formation of interstellar molecules from negative ions has been proposed by Dalgarno and McCray (1973).

The overall picture that emerges is the following:

(i) The only plausible mechanism proposed so far for the formation of H_2 in cool interstellar clouds (e.g., $T = 50$ K, $n_H = 100$ cm^{-3}, $\tau_v = 1$) is on interstellar grain surfaces. However at higher temperatures the alternative mechanism (Dalgarno and McCray):

$$e + H \rightarrow H^- + h\nu,$$
$$H^- + H \rightarrow H_2 + e,$$

becomes increasingly important, both because catalysis by grains becomes less effective and also because the rate of formation of H^- becomes greater.

(ii) In 'young' clouds ($\sim 10^7$ yr) CO, CH^+, CH and CN are next most readily formed (essentially as analysed by Solomon and Klemperer) but, in older clouds (say $> 5 \times 10^7$ yr) OH and later NH become more abundant, being produced by recombination on grain surfaces that have grown considerably in size in the ageing cloud. The calculations of OH abundance by Aannestad for a cloud of 500 $M_\odot$ and a depletion factor $d(C^+) = 0.1$ are compatible with observation. The observations on the clouds in front of ζ Ophiuchi can also be accounted for reasonably if the clouds are assumed to be young.

Aannestad (1973) has also attempted to evaluate the chemical effects of shock waves that occur when clouds collide or when they encounter expanding H II regions; he published typical results as a function of time after passage of the shock front. The most abundant molecules surviving for some time ($>$ a few hundred years) after the shock are (after H_2) CO, CN and CH in that order.

Thus we are left with a picture that the formation and rough relative abundances of H_2, CO, CN, CH, CH^+ and probably OH in interstellar clouds can be accounted for by invoking the intervention of interstellar grains as catalysts for a variety of reactions. However, a formidable array of input data, most of it soft, is involved in the calculations, making it hard to assess the confidence limits for the final results.

The situation for the other observed polyatomic molecules is more obscure. Aannestad noticed that his model did not easily account for observed abundances of H_2CO. Dalgarno and McCray have suggested that a number of polyatomic molecules could plausibly be formed by negative ion radiative attachment (Table I). No specific calculations on these mechanisms were offered and they must be regarded at present as interesting armchair galactochemistry. We might especially note the presence of a mechanism for formation of NO with a rate very similar to that of the preceding reaction for water formation. Searches for NO have so far been unsuccessful (1970).

TABLE I

Negative ion radiative
attachments
(Dalgarno and McCray,
1973)

$H^- + CO \rightarrow HCO + e$
$H^- + HCO \rightarrow H_2CO + e$
$O^- + CH_2 \rightarrow H_2CO + e$
$CH_2^- + O \rightarrow H_2CO + e$
$O^- + H_2 \rightarrow H_2O + e$
$O^- + N \rightarrow NO + e$
$O^- + CS \rightarrow OCS + e$
$S^- + CO \rightarrow OCS + e$
$S^- + H_2 \rightarrow H_2S + e$
$CH^- + N \rightarrow HCN + e$
$CN^- + H \rightarrow HCN + e$
$CH_2 + S^- \rightarrow H_2CS + e$
$CH_2^- + HCN \rightarrow CH_3CN + e$

While no persuasive quantitative theory of formation of polyatomic molecules has been published, there are a number of factors that have to be incorporated in a successful theory. We note that the studies of Stief *et al.* (1972) imply that the molecules must be formed within the photolytic protection of dust clouds of at least moderate opacity.

Firstly no cyclic molecules have been found (Table II), although in some cases integration to very low noise levels has been performed. While all of these molecules contain more atoms than any molecule yet found (the smallest is aziridine C_2H_5N) one wonders whether the mechanism of formation makes ring formation highly unlikely. Chemically there is a better prospect of producing ring structures by reactions on surfaces of catalysts than by reactions in the gas phase and so there may be a

TABLE II

Some cyclic molecules for
which searches have been
negative (Sgr B2)

Molecule	Limiting noise (peak to peak) (K)
Aziridine	0.01[a]
Pyrrole	0.02[a]
Imidazole	0.02[a] 0.08[b]
Pyrazole	0.04[a]
Pyridine	0.05[a]
Furan	0.08[b] 0.03[c]

[a] Parkes 64-m
[b] Green Bank 43-m
[c] Hat Creek 26-m

hint here that the larger molecules are formed in gas phase reactions. We might also note in passing that pyridine is a constituent of the bis pyridyl magnesium tetrabenz porphine that Johnson (1970) has proposed as the absorber of the diffuse interstellar lines.

Secondly molecules with N bound to O have not been detected, yet two molecules containing both N and O have been found: HNCO and $H-C{\overset{\displaystyle\nearrow O}{\searrow NH_2}}$. Solomon and Klemperer (1972) have shown that NO does not form by radiative recombination but it is hard to see why NO bonds should be avoided on, for example, grain surfaces. Incidentally we have recently had a search for HNO at Parkes without success. One would have thought $O + NH$, or perhaps $O^- + NH \rightarrow HNO + e$, are plausible.

Finally some of the polyatomic molecules that have been detected are known to chemists as highly reactive. I especially instance thioformaldehyde H_2CS and methanimine H_2CNH. The formation of such species in low temperature gas reactions is understandable but one is a little uneasy about catalytic formation of such species. One would expect them to undergo further rapid reactions on the grain surface to yield more chemically stable species.

This then seems to me to be the present state of understanding of molecule formation – plausible suggestions for production of most of the known diatomics but no real understanding of polyatomics. Hopefully, further observations, positive and negative, will provide a stronger fabric from which galactochemists will fashion a more wearable account of the origin of interstellar molecules.

References

Aannestad, P. A.: 1973, *Astrophys. J. Suppl.* **25**, 205 and 223.
Bates, D. R. and Spitzer, L.: 1951, *Astrophys. J.* **113**, 441; see also Kramers, H. A. and ter Haar, D.: 1946, *Bull. Astron. Inst. Neth.* **10**, 137.
Dalgarno, A. and McCray, R. A.: 1973, *Astrophys. J.* **181**, 95.
Gould, R. J. and Salpeter, E. E.: 1963, *Astrophys. J.* **138**, 393.
Greenberg, J. M., Yencha, A. J., Corbett, J. W., and Frisch, H. L.: 1972, *Mem. Soc. Roy. Liège* **3**, 425.
Herbig, G. H.: 1970, *Mem. Soc. Roy. Liège* **19**, 13.
Hollenbach, D. J. and Salpeter, E. E.: 1971, *Astrophys. J.* **163**, 155.
Johnson, F. M.: 1970, *Bull. Amer. Astron. Soc.* **2**, 323.
Kahn, F. D.: 1955, *Mem. Soc. Roy. Liège* **15**, 578.
Knaap, H. F. P., van den Meijdenberg, C. J. N., Beenakker, J. J. M., and van de Hulst, H. C.: 1966, *Bull. Astron. Inst. Neth.* **18**, 256.
McCrea, W. H. and McNally, D.: 1960, *Monthly Notices Roy. Astron. Soc.* **121**, 238.
Rank, D. M., Townes, C. H., and Welch, W. J.: 1971, *Science* **174**, 1083.
Solomon, P. M. and Klemperer, W.: 1972, *Astrophys. J.* **178**, 389.
Stief, L. J., Donn, B., Glicker, S., Gentieu, E. P., and Mentall, J. E.: 1972, *Astrophys. J.* **171**, 21.
Turner, B. E., Heiles, C. E., and Scharlemann, E.: 1970, *Astrophys. Letters* **5**, 197.

R. D. Brown
Monash University, Department of Chemistry,
Clayton, Victoria 3168, Australia

DISCUSSION

Robinson: Brown has drawn attention to the absence of molecules with the NO combination. There is also no known case of a carbon-carbon double bond, as pointed out by Winnewisser. The absence of NO and $C=C$ may provide useful clues to the mechanisms of molecule formation.

Turner: The calculations by Aannestad, based on the Solomon-Klemperer theory, have been shown to be irrelevant by recent work by Smith, Lizst and Lutz (*Astrophys. J.* **183**, (1973), 69) and by Yoshimine, Green and Thaddeus (*Astrophys. J.* **183**, (1973), 899), which shows that the direct radiative recombination rates of CH^+ and CH are much lower than previously assumed. Further, Julienne and Kraus (in M. A. Gordon and L. E. Snyder (eds.), *Molecules in the Galactic Environment*, John Wiley and Sons, New York, 1973) have shown that *indirect* radiative recombination (inverse predissociation) proceeds at comparable or larger rates under interstellar conditions, and since this process favours a different set of molecules (CN, CO, OH) than the direct process, relative abundances would be predicted quite differently to those of Aannestad.

The probable formation mechanism is *positive* ion-molecule reactions in gas phase, as studied by Herbst and Klemperer (*Astrophys. J.* **185**, (1973), 505) and by Watson (*Astrophys. J. Letters* **181** (1973), L129, and **183** (1973), L17). Such reactions proceed with much larger rates than those involving negative ions (Dalgarno and McCray, *Astrophys. J.* **181** (1973), 95). Such reactions will operate over a wide range of cloud densities, including the tenuous clouds ($A_v \sim 2$–3 mag) where molecules are seen optically, and the 'black' clouds ($A_v \sim 200$ mag) where complex molecules are seen at radio wavelengths. In these latter clouds, surface reactions cannot operate (cf. Watson and Salpeter, *Astrophys. J.* **174** (1972), 321 and **175** (1972), 659), and positive ion-molecule reactions (initiated by cosmic ray ionization) appear to be the only formation processes whose rates exceed or at least are comparable with the rate of destruction of molecules by permanent adsorption onto grain surfaces.

Certain points of comparison, of a qualitative nature, may be made between predicted abundances, via positive ion-molecule reactions, and observed abundances. One is that these processes favor formation of molecules with unsaturated carbon bonds (cf. Turner, *J. Roy. Astron. Soc. Canada* (1974), in press) and discriminate against saturated bonds. This agrees with observation. Another is that these processes predict that OH abundances should be independent of total density, if OH is destroyed primarily by absorption onto grains. This also agrees with observation (Turner, in *Galactic and Extragalactic Radio Astronomy*, Springer-Verlag, New York, 1973; Turner and Heiles, *Astrophys. J.* (1974), in press). Qualitative predictions are useful, because these do not depend on calculational difficulties associated with unknown reaction rates, or omission of reactions, difficulties which affect attempts at quantitative predictions in all theories of molecular formation.

In summary, arguments can be made against all interstellar formation mechanisms proposed to date except positive-ion molecule reactions, which predict correctly certain qualitative observational aspects, and which have yet to be thoroughly tested in a quantitative sense.

Brown: Mechanisms of formation based on homogeneous gas-phase positive ion-molecule reactions are particularly attractive since there is a better chance of developing a thorough quantitative theory than one could hope for in the case of heterogeneous reactions. However, until the articles by Herbst and Klemperer and Watson have been published and subjected to full scrutiny by scientists it is difficult to comment further.

Donn: The ζ Oph cloud is rather anomalous. The high dispersion work of Hobbs shows that four clouds are present. The CH^+ line differs in velocity from other species. It is therefore dangerous to use this region as standard for comparing theoretical and observed molecular abundance calculations.

The interstellar clouds are characterized by extreme deviations from thermodynamic equilibrium. The use of laboratory reaction rates requires a Boltzmann distribution for molecular vibrational and rotational levels. Chemical techniques now are beginning to yield rates for specific initial energy states and these may differ considerably among themselves and from equilibrium values. It is therefore also dangerous to use laboratory data for interstellar calculations without great care.

Brown: It is very dangerous to go to the literature and take an expression that chemists have derived for the rate of a reaction just by studies at room temperature and above. When activation energies and frequency factors obtained from experimental data at room temperature are used to predict reaction rates at low temperatures very sizeable errors can result – perhaps one or more orders of magnitude.

Morimoto: Nakagawa and others calculated relative production rate of molecules on the grain surfaces assuming that atoms chemically absorbed on the grain surface move on the surface until they form a molecule and leave the surface. This gives large production rates for molecules found astronomically (plus those without dipole moment) including polyatomic molecules. (Nakagawa, *Pub. Astron. Soc. Japan* (1974), in preparation).

Brown: The simple model used by Nakagawa *et al.* could point to some interesting possibilities for further study. However, rates of heterogeneously catalyzed reactions are often enormously affected by details of the surface structure of catalysts and so the assumptions employed by Nakagawa *et al.* must be regarded with reserve.

Parijskij: I would like to inform you that an experimental approach to molecular formation is in progress at the Institute of Physics and Technique in Leningrad. Conditions in cold and dense clouds were simulated, and a great amount of large molecules were visible in their mixture.

EXTENDED OH EMISSION AT 1720 MHZ*

W. M. GOSS

Kapteyn Astronomical Institute, University of Groningen, Groningen, The Netherlands

L. E. B. JOHANSSON, J. ELLDÉR, and B. HÖGLUND

Onsala Space Observatory, Onsala, Sweden

NGUYEN-Q-RIEU

Observatoire de Meudon, Meudon, France

and

A. WINNBERG

Max-Planck-Institut für Radioastronomie, Bonn, F.R.G.

Abstract. Low-velocity OH emission lines at 1720 MHz have been investigated in the direction of W41, W43, W51, and 3C 353. The emission lines in the directions of W43 and W51 show no detectable circular polarization and are about 27 and 32 arcmin in size, respectively. The inferred peak brightness temperatures are 0.6 K for W43 and 2.0 K for W51. This emission may be similar to the nonthermal 1720-MHz emission in dust clouds.

Low-velocity OH emission lines at 1720 MHz have been investigated in the direction of W41, W43, W51 and 3C 353. The work was prompted by the observations of Goss (1968) and Goss and Robinson (1968), who found low-velocity ($|V| < 10$ km s^{-1}) 1720-MHz emission in the directions of several sources. Since the gas giving rise to this emission is presumably local, no physical connection between the OH and the background discrete sources was postulated. (For W43 and W51 the background sources are H II regions with velocities $|V| > 10$ km s^{-1}). In most cases H I absorption lines are found at approximately the same velocities as the 1720-MHz OH in these directions. Later work by Hardebeck (1971) failed to detect the W51 line using an interferometer with a fringe separation of $\sim 5'$. This negative result suggested to us that the OH source might be extended with a size $> 5'$.

Observations were carried out with the 25-m telescope of the Onsala Space Observatory at 1720 and 1667 MHz, and with the higher resolution of the Nançay Radio Telescope at 1720 MHz. The emission lines in the directions of W43 and W51 show no detectable circular polarization and are about 27 and 32′ in size, respectively. The inferred peak brightness temperatures are 0.6 K for W43 and 2.0 K for W 51. This emission may be similar to the nonthermal 1720-MHz emission in dust clouds.

The sources OH 31-0 and OH 49-0 (in the vicinity of W43 and W51, respectively) exhibit OH emission at 1720 MHz and absorption at the other three 18-cm lines. The 1720-MHz emission is extended, with angular sizes of the order of 30′. The intensity ratio between the main lines is close to that expected for LTE conditions with small optical depth, whereas the 1612-MHz line is abnormally strong. This suggests a

* The full paper has been published in *Astron. Astrophys.* **28** (1973) 89.

smaller excitation temperature for this transition. The 1720-MHz emission line does not show any strong circular polarization, and the velocity width is broader (up to 10 times) than the emission lines from Class I OH sources. The emission and absorption probably take place in OH gas concentrations close to the Sun in which the ground state energy levels appear to be populated in a way similar to that found in some dust clouds. The pumping is probably caused by IR radiation. The observed properties of the 1720-MHz emission may be summarized as follows:

(i) the lines are broader than for Class I OH masers;

(ii) the intensity of the lines is not strongly correlated with the continuum brightness due to background sources;

(iii) the lines are not strongly circularly polarized as in the case of Class I sources; and

(iv) the apparent angular extent is many orders of magnitude greater than the Class I sources. The OH emission and absorption in the direction of 3C 123, 3C 353, Cas A, Sgr A and OH 23-0 probably have properties similar to OH 31-0 and OH 49-0. These properties are consistent with an enhanced excitation temperature for the 1720-MHz transition, although the presence of an actual population inversion in this transition cannot be ruled out.

References

Goss, W. M.: 1968, *Astrophys. J. Suppl.* **15**, 131.
Goss, W. M. and Robinson, B. J.: 1968, *Astrophys. Letters* **2**, 81.
Hardebeck, E.: 1971, *Astrophys. J.* **170**, 281.

W. M. Goss
Kapteyn Astronomical Institute,
University of Groningen,
Postbus 800, Groningen 8002, The Netherlands

L. E. B. Johansson
J. Elldér
B. Höglund
Onsala Space Observatory,
S-430 34 Onsala, Sweden

Nguyen-Q-Rieu
Observatoire de Meudon,
92 Meudon, France

A. Winnberg
Max-Planck-Institut für Radioastronomie,
Auf dem Hügel 69,
53 Bonn 1, F.R.G.

THE OBSERVED CHARACTERISTICS
OF THE LOCAL MAGNETIC FIELD

J. B. WHITEOAK

Division of Radiophysics, CSIRO, Sydney, Australia

Abstract. The main investigations of the local magnetic field are reviewed and are found to contain some conflict in interpretation. At radio wavelengths, studies have been made using both the Faraday rotation of the polarized radiation from extragalactic sources and pulsars, and the polarization of the galactic background radiation. With the former type of observation, although more data are available for extragalactic sources, any interpretation may be complicated by the influence of distant field structure. The results are consistent with a large-scale field parallel to the galactic plane, with a field strength of about 2 μG, and which is directed towards $l = 90°$. This field contains irregularities in direction and strength on a scale of about 100–200 pc. The polarization of galactic background radiation may yield the most detailed information about the local field structure – the results to date show loops of magnetic fields extending along the radio spurs.

The interpretation in terms of small-scale irregularities embedded in a large-scale field parallel to the galactic plane differs from that proposed to explain the optical polarization of starlight, in which a helical field configuration near the Sun was preferred to a more disordered pattern.

I. Introduction

A magnetic field is an important constituent of the Galaxy. On theoretical grounds it is believed by some (e.g. Piddington, 1972) to play a major role in the formation and evolution of galactic spiral structure. The existence of the field is manifested in many ways: in the Faraday rotation of extragalactic polarized radiation, the presence of galactic synchrotron radiation, the Zeeman splitting of spectral lines, and the optical polarization of starlight. However, the investigation of it is similar to that of spiral structure in its perversity, in that the more data are amassed, the more uncertain the solution becomes. Despite numerous studies, no satisfactory interpretation has been obtained for the structure of the local magnetic field. In this review I will attempt to describe the most widely accepted features by an appraisal of the various results, some of which appear to be in conflict. (The variation in point of view is demonstrated well in the differing conclusions of the reviews by Verschuur (1970) and Verschuur (1972).)

II. Results at Radio Wavelengths

(a) THE DISTRIBUTION OF FARADAY ROTATION

(i) *Polarization of Extragalactic Radio Sources*

One of the main methods used to investigate the local magnetic field is based on the Faraday rotation of the polarized radiation from extragalactic radio sources. The Faraday rotation is proportional to the product of electron density n_e and the line-of-sight component of magnetic field B_L, with the product integrated along the line of sight. Therefore, provided the electron density and magnetic field strength are reason-

F. J. Kerr and S. C. Simonson, III (eds.), Galactic Radio Astronomy, 137–150. All Rights Reserved.
Copyright © 1974 by the IAU.

ably constant, the distribution of rotation with galactic coordinates should provide information about the structure of the magnetic field. The method appeared promising in the initial observations of Gardner and Whiteoak (1963), where it was shown that the average rotation decreased as expected with increasing galactic latitude. However, later investigations of the variation with latitude and longitude were made by Morris and Berge (1964), Gardner and Davies (1966), Berge and Seielstad (1967) Gardner *et al.* (1969), Mitton (1972) and Wright (1973), but no simple field configuration was derived. Aside from the possibility that the field possesses complex structure, there are several factors that may contribute to this failure. The rotation intrinsic to the sources is assumed to be small, and although this is generally true, since few sources at high latitudes have large rotation measures, some exceptions are known, e.g., Cyg A (Hollinger *et al.*, 1964; Mitton, 1973). The rotation measure is derived on the assumption that the direction of polarization is proportional to wavelength squared, and this may not hold for objects with several differently polarized components with different spectra (such as quasars). The greatest limitation is probably that the line of sight for each source passes through the entire galaxy, and so the rotation may be affected not only by the local magnetic field but also by distant fields. This may even apply at high latitudes in the directions of the high-velocity clouds of neutral hydrogen which may indicate the presence of outer spiral arms (Davies, 1972; Verschuur, 1973a, b).

Figure 1 shows the distribution with galactic latitude of the rotation measures of about 530 sources observed at Parkes. Many of the values are based on unpublished observations. As found in 1963, the concentration of high rotation measures towards the equator is greater than expected from the distribution of electrons alone, but can be explained in terms of a large-scale magnetic field in the plane of the Galaxy. Figure 2 shows the distribution of rotation measures with respect to galactic coordinates. The broken lines show the loops of radio emission as depicted schematically by Haslam *et al.* (1971). The point-to-point variation in rotation measure is generally large, and Lerche (1970) and others have argued that there are large fluctuations in both magnetic field and electron density that dominate any large-scale field configuration that might exist. However, the distribution shows patterns in the sign of the rotation measures that persist over many square degrees, suggesting a large-scale order in the magnetic field. At present, the overall interpretation of the results that is most widely accepted is probably that advanced by Gardner *et al.* (1969). They argued that the data at negative galactic latitudes are consistent with a large-scale field which is parallel to the galactic plane and is directed towards $l = 80°$. At northern latitudes, the rotation measures between $l = 300°$ and $l = 30°$ are opposite in sign to that expected for this field alone (as was first pointed out by Morris and Berge (1964)), and Gardner *et al.* (1969) suggested that the reversal could be due to a looped field extending north of the galactic plane and possibly associated with Loop I (which contains the North Polar Spur). At $l = 180°$, $b = -30°$, and $l = 15°$, $b = +10°$, the rotation measures are too large in amplitude for directions nearly orthogonal to the field, and Gardner *et al.* (1967) concluded that the anomalies were caused by loops of magnetic field associated with spurs of gas taking part in the general expansion of Gould's Belt. The more

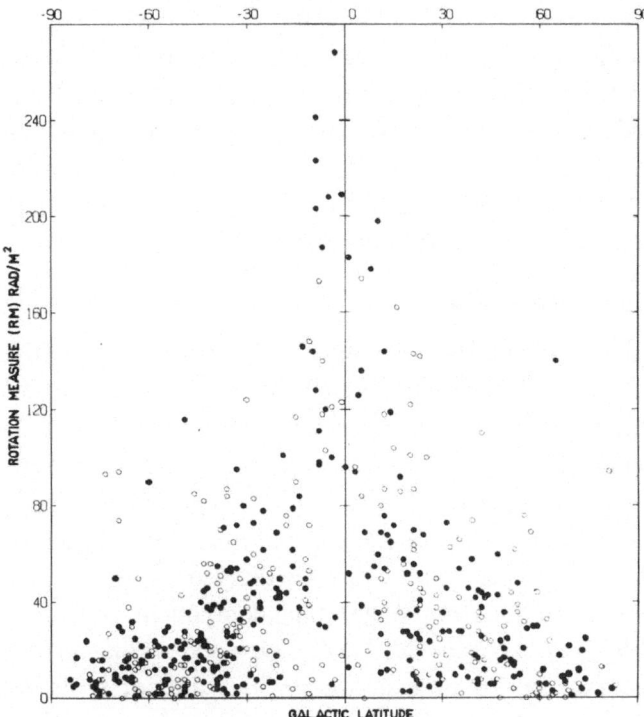

Fig. 1. The distribution with galactic latitude of the rotation measures for sources observed at Parkes. The open circles represent negative values, the closed circles show positive values. Positive values correspond to line-of-sight magnetic fields directed toward the observer.

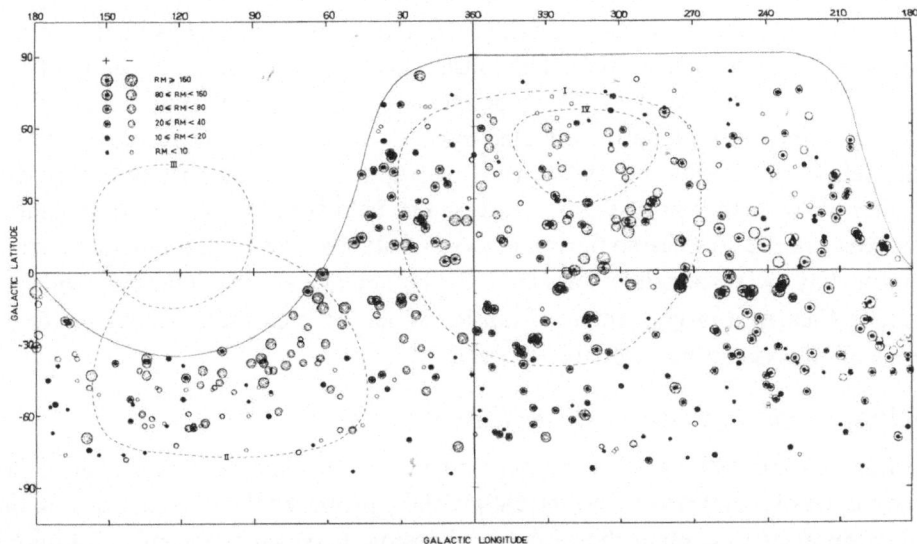

Fig. 2. The distribution with galactic coordinates of the rotation measures for sources observed at Parkes. Positive rotation measures correspond to line-of-sight magnetic fields directed toward the observer. The continuous line shows the limits of observation of the 64-m telescope. The broken lines show the loops of radio emission according to Haslam *et al.* (1971).

extensive data shown in Figure 2 suggest that such loops or field irregularities are more common than indicated by the results presented in 1969. In the most recent investigation of Faraday rotation (Wright, 1973), which included for the first time a substantial sample of sources at declinations inaccessible to Parkes, the large-scale field was found to be directed towards $l = 94° \pm 3°$, $b = -8° \pm 8°$, and has a strength of about $2\mu G$ for an assumed electron density of 0.06 cm^{-3}.

(ii) *Faraday Rotation of Pulsars*

The use of pulsars, many of which are highly polarized, in the investigation of Faraday rotation reduces some of the difficulties encountered with extragalactic sources. The dispersion of the pulses provides an approximate measure of the distance of the pulsar and therefore offers the possibility of studying the local field without the influence of more distant magnetic structure. Furthermore, by combining the dispersion measure (*DM*) with the rotation measure (*RM*), a weighted mean value for the lie-of-sight component of magnetic field can be directly derived: $\bar{B}_L = \int B_L n_e d_l / \int n_e d_l \propto RM/DM$. This value is similar to the mean longitudinal component, provided that the fluctuations in n_e and B_L are not correlated. There is no evidence that any significant fraction of the observed Faraday rotation occurs in the pulsar or in its immediate vicinity (Manchester, 1972).

Unfortunately, the pulsars for which both the dispersion measure and rotation measure have been determined are small in number and concentrated near the galactic equator. Therefore only the large-scale features of the magnetic field can be traced. Manchester (1973) has carried out the most comprehensive study using a total of 38 pulsars (Table I). The distribution of these results in galactic coordinates (Figure 3) indicates a field pattern that is surprisingly similar to that deduced using the Faraday rotation of extragalactic sources. The pulsars near the galactic plane, and with calculated distances of less than 2 kpc, delineate a large-scale field of 2.2 μG that is aligned parallel to the plane and directed towards $l = 94° \pm 11°$. The sense and strength of the field in the direction of the North Polar Spur show anomalies similar to those encountered with the extragalactic sources. The field is more uniform than deduced from extragalactic sources, so the influences of distant structure and intrinsic Faraday rotation may be responsible for many of the small-scale irregularities evident in Figure 2. Clearly, a larger sample of pulsars is needed before the results for the two groups of objects can be compared in detail.

(iii) *Irregularities in the Local Magnetic Field*

The scale of irregularities in the magnetic field is of considerable interest, particularly in the studies of cosmic rays. Fermi (1949) initially proposed the existence of a random field component to explain the isotropy of cosmic rays. More recently, Jokipii and Parker (1969) and others have argued that such a component would enable the escape of cosmic rays from the Galaxy in the relatively short period of time implied from observations of abundances of light cosmic-ray nuclei. Using observations of 20-cm polarization in extragalactic sources, Bologna *et al.* (1965) argued that a decrease in

TABLE I

Values of longitudinal magnetic fields derived
from pulsar observations (Manchester, 1973)

PSR	l	b	$\bar{B}_L$ [a] (μG)
0031 − 07	110°	− 70°	+ 1.1 ± 0.1
0105 + 65	125	+ 3	− 1.2 ± 0.2
0138 + 59	129	− 2	− 1.8 ± 0.2
0301 + 19	161	− 33	− 0.65 ± 0.02
0329 + 54	145	− 1	− 2.93 ± 0.02
0450 − 18	217	− 34	+ 0.46 ± 0.06
0525 + 21	184	− 7	− 0.960 ± 0.006
0531 + 21	185	− 6	− 0.92 ± 0.02
0611 + 22	188	+ 2	+ 0.88 ± 0.02
0628 − 28	237	− 17	+ 1.58 ± 0.02
0740 − 28	244	− 2	+ 2.50 ± 0.01
0809 + 74	140	+ 32	− 2.5 ± 0.3
0818 − 13	236	+ 13	− 0.08 ± 0.05
0823 + 26	197	+ 32	+ 0.37 ± 0.02
0833 − 45	264	− 3	+ 0.59 [b]
0834 + 06	220	+ 26	+ 2.3 ± 0.3
0950 + 08	229	+ 44	+ 0.7 ± 0.3
1112 + 50	155	+ 61	+ 0.43 ± 0.07
1133 + 16	242	+ 69	+ 0.99 ± 0.06
1237 + 25	252	+ 87	− 0.07 ± 0.05
1508 + 55	91	+ 52	+ 0.05 ± 0.04
1604 − 00	11	+ 35	+ 0.75 ± 0.12
1642 − 03	14	+ 26	+ 0.58 ± 0.09
1706 − 16	6	+ 14	− 0.12 ± 0.05
1749 − 28	1	− 1	+ 2.30 ± 0.05
1818 − 04	26	+ 5	+ 1.0 ± 0.1
1822 − 09	24	+ 1	+ 4.4 ± 0.2
1919 + 21	56	+ 3	− 1.80 ± 0.06
1929 + 10	47	− 4	− 3.3 ± 0.7
1933 + 16	52	− 2	− 0.015 ± 0.003
2016 + 28	68	− 4	− 3.0 ± 0.2
2020 + 28	69	− 5	− 3.74 ± 0.02
2021 + 51	88	+ 8	− 0.36 ± 0.05
2045 − 16	31	− 33	− 1.15 ± 0.04
2111 + 46	89	− 1	− 1.95 ± 0.03
2154 + 40	90	− 11	− 0.76 ± 0.04
2217 + 47	98	− 8	− 1.00 ± 0.05
2319 + 60	112	− 1	− 2.8 ± 0.3

[a] A positive field component is directed towards the observer.
[b] Komesaroff et al. (1971).

the mean polarization with decreasing latitude indicated an increase in depolarization
with increasing line of sight through the Galaxy and attributed this depolarization to
irregularities on a scale of about 1 pc. Their conclusions were rejected by Gardner
et al. (1969) who determined that the mean depolarization between 20 cm and 11
cm was essentially independent of latitude. However, despite the smooth variation

of Faraday rotation over ten square degrees of Cen A (Cooper *et al.*, 1965), an extended source at latitude $b=20°$, the rapid variation in rotation measure from source to source in some galactic directions (see Figure 2) and across some supernova remnants (Kundu and Velusamy, 1972) is evidence of considerable irregularities in the product $n_e B_L$. Any variation in sign is an indication that B_L is varying; otherwise the changes

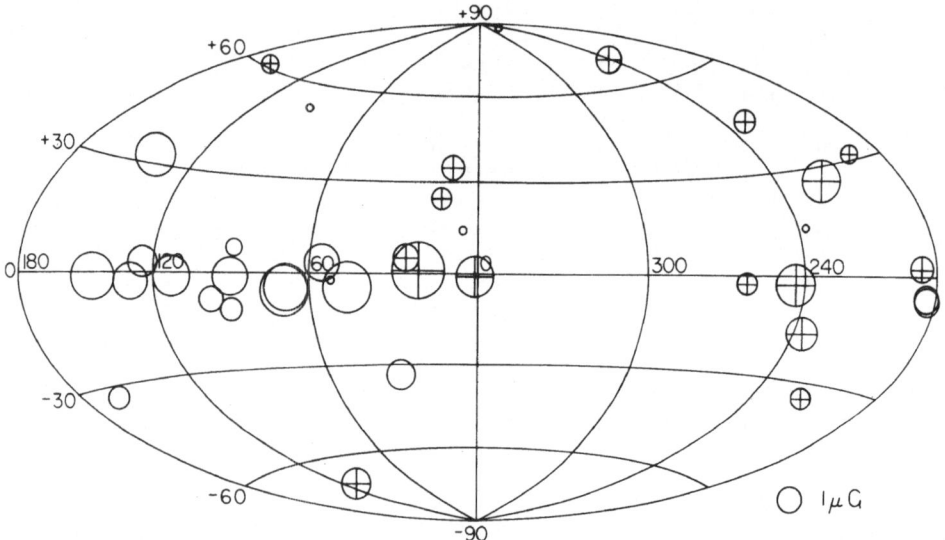

Fig. 3. The distribution with galactic coordinates of the line-of-sight components of magnetic field that were derived from observations of pulsars (Manchester, 1973). The diameters of the circles are proportional to the magnitudes of the derived field; circles containing a positive sign represent a field directed towards the observer.

could be due to a variation of n_e alone. From a statistical analysis of the rotation measures of extragalactic sources, Jokipii and Lerche (1969) deduced that the product $n_e B$ was constant only over distances of 250 pc. With more extensive data, Wright (1973) derived a scale of 100–200 pc. However, as mentioned earlier, the variations in rotation measure may not be due only to the local magnetic field, and therefore the scale may be larger than calculated.

In an investigation of the fluctuations of the magnetic field strength, Wright (1973) concluded that the variations could be as high as two to three times the average value. Since the rotation measures of extragalactic sources were used in the analysis, the values may be too high. From observations of pulsars, Manchester (1973) deduced that the strength of the local field fluctuates by only 50% of the average value. The pulsar sample is small, and not well-distributed with galactic coordinates, and the best estimate probably lies somewhere between the two values.

(b) THE POLARIZATION OF THE GALACTIC BACKGROUND RADIATION

The polarization of the synchrotron radio emission from the galactic background yields information about the field transverse to the line of sight, in contrast to Faraday rotation which depends on the line-of-sight component. As a tracer of the local magnetic field, it has considerable value, since the substantial polarization observed between the frequencies 400–1400 MHz is believed to have an origin limited by Faraday depolarization to distances within a few hundred parsec of the Sun. In one of the first investigations, Berkhuijsen *et al.* (1964) interpreted the directions of polarization and the Faraday rotation near $l = 140°$ in terms of a field parallel to the galactic plane and directed towards $l = 50°$. At about the same time Mathewson and Milne (1965) and Mathewson *et al.* (1966) deduced a similar result (but with a direction $l = 70°$) from the location of the great circle containing the most polarized emission, and from the intrinsic directions of polarization of this emission. A recent, more detailed study by Berkhuijsen (1971) added further support to these conclusions. She showed that the percentage polarization at 829 MHz varied with galactic latitude and longitude in a manner consistent with a field aligned parallel to the galactic plane and directed towards $l = 60°$.

The first attempts at interpreting the polarization results in terms of small-scale structure of the magnetic field were made by Bingham and Shakeshaft (1967) and Bingham (1967). The most detailed results, although limited to directions containing the spurs of radio emission, were obtained by Spoelstra (1972), who used observations at several frequencies to derive the intrinsic position angles of polarization. The spurs form parts of the loops of radiation that are shown schematically in Figure 2, and which have been fully discussed in articles by Haslam *et al.* (1971), and Berkhuijsen *et al.* (1971). For the North Polar Spur, the region of Loop I which extends north of the plane at longitude $l = 30°$, Spoelstra's results at 1415 MHz for latitudes north of $+50°$ are shown in Figure 4. Since the Faraday rotation in these directions is small, the directions of polarization are consistent with a magnetic field component extending along the ridge of radio emission. If the distance of the loop is 50–100 pc, and the electron density in the intervening space is 0.06 cm^{-3}, the low rotation measure of 4 rad m^{-2} yields a mean longitudinal field strength of 1–2 μG. At low latitudes, the field in the direction of the spur is aligned parallel to the galactic equator, and may represent the field associated with the radio emission in the plane of the Galaxy. Similar field configurations were also obtained for the Cetus Arc, a spur of radio emission defining loop II near $l = 150°$, and for loop III. For the former region, the rotation measures are lower than for background extragalactic sources, which is evidence that the spur is a local feature.

Thus it appears that a detailed study of the polarization of the galactic synchrotron radiation may provide the best means of investigating the fine structure of the local magnetic field. The results are reasonably consistent with those obtained from the distribution of Faraday rotation – the discrepancy in the direction of the large-scale field may indicate that within a few hundred parsec of the Sun the field orientation differs from that at larger distances. On the other hand, the spatial distribution of the thermal electrons responsible for the Faraday rotation may not be the same as that of the relativistic electrons producing the synchrotron radiation.

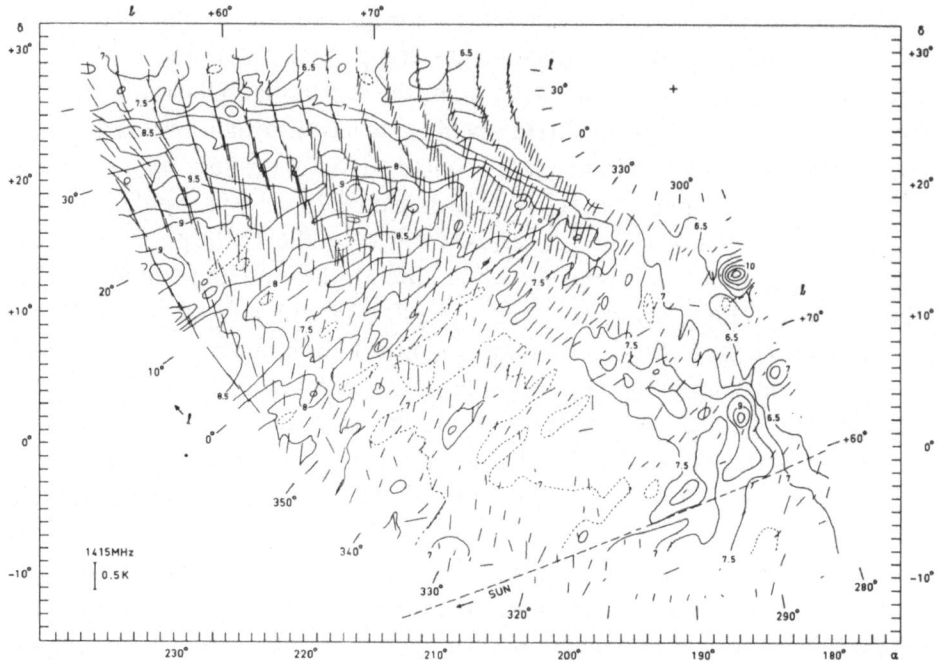

Fig. 4. Polarization measurements at 1415 MHz for the top of the North Polar Spur (Figure 4, Spoelstra, 1972). The vectors show the polarization beam brightness temperatures and directions of polarization; the curves are contours of total beam brightness temperature.

(c) THE ZEEMAN EFFECT

The Zeeman splitting of the circularly polarized components of spectral lines is proportional to the strength of the line-of-sight magnetic field in an emitting or absorbing region. To date the results of investigations have been disappointing because the uncertainties in the measurements have exceeded a few microgauss. The absorption-line studies of H I by Davies *et al.* (1968), Brooks *et al.* (1971), and Verschuur (a series of observations best summarized by Verschuur (1971)), have yielded only the few positive results listed in Table II. The values are considerably higher than for the

TABLE II

Estimated longitudinal magnetic
fields from H I Zeeman splitting

Continuum source	Line velocity (km s^{-1})	Observed field[a] (μG)
Tau A	+11	−6
Ori A	+7	−50
M17	+14	+30
Cas A	−38	+18
	−48	+11

[a] A negative field is towards observer.

fields observed by other techniques in other parts of the Galaxy, and probably represent compressed magnetic fields associated with high-density gas in, or in the immediate vicinity of, each source (Verschuur, 1970). The technique will be of little interest for the investigation of the general galactic field until it is possible to detect field strengths as low as 1 μG.

A similar conclusion applies to the investigations carried out at OH frequencies. The observations of Zeeman splitting in OH absorption (Turner and Verschuur, 1970) yielded upper limits in the range 30–130 μG. On the other hand, the presence of features which are similar in intensity, but displaced in velocity, in emission-line profiles measured with right and left circular polarization, has been interpreted in terms of fields with strengths of a few milligauss in high-density regions of W3 (Davies *et al.*, 1966; Yen *et al.*, 1969), W49 (Davies *et al.*, 1966) W75 and W3OH (Rydbeck *et al.*, 1970), and NGC 6334 (Gardner *et al.*, 1970).

(d) OTHER PHENOMENA CORRELATED WITH MAGNETIC FIELD DIRECTIONS

(i) *Interstellar Clouds*

Interstellar clouds can be elongated under the action of magnetic fields (Mansfield, 1973). The existence of dust clouds aligned parallel to magnetic fields has been noted in the Galaxy (Shain, 1955) and in other galaxies (Appenzeller, 1967), and Verschuur (1970) has presented evidence that H I clouds are similarly elongated. In addition, Berkhuijsen *et al.* (1971) have pointed out that neutral hydrogen features near the loops of radio emission are extended parallel to the loops. However, the subject has not been investigated in detail; nor has the possibility that the elongation might be due to causes not related to magnetic fields.

(ii) *Supernova Remnants*

According to the theoretical models developed for supernova remnants by van der Laan (1962), the remnant has an observed structure that is related to the magnetic field transverse to the line of sight. The structure of two remnants at a relatively high galactic latitude (Whiteoak and Gardner, 1968) and a statistical study by Shaver (1969) supported the models. However, a more extensive study by Milne (1970) failed to yield any correlation between structure and the orientation of the galactic plane. Therefore, the structure of remnants may not be of much use as field tracers, although it is worth noting that Spoelstra (1973) interpreted the galactic loops as supernova remnants, and suggested that the structure of these features is consistent with a magnetic field directed towards $l = 40°$, a longitude not much different from the values quoted earlier in this review.

III. Results at Optical Wavelengths

According to the generally accepted model of Davis and Greenstein (1951), starlight is polarized by foreground dust particles aligned along a magnetic field, with a plane of polarization parallel to the field direction. The original model required a relatively high field strength, but a modified version (Jones and Spitzer, 1967) is effective with a more realistic value of 3 μG. Means of alignment other than by the paramagnetic absorption of the Davis-Greenstein mechanism have been sought – e.g., Martin (1971) has suggested that the field influences the alignment of the particles by causing a precession of the angular momentum of charged grains. Other investigators (e.g. Harwit, 1970; Michel and Yahil, 1973) have questioned on theoretical grounds whether the optical polarization shows unambigously the magnetic field orientation. However, the use of the direction of polarization as a tracer of magnetic fields seems to be supported by studies of the Magellanic Clouds (Mathewson and Ford, 1970b) and other galaxies (Appenzeller, 1968) where, in general, the direction of polarization and galactic structure are correlated. The correlation fails, as expected, in regions of reflected radiation; it also fails in lanes of obscuration that cross the spiral arms, in which case the directions of polarization are parallel to the lanes and presumably delineate the field associated with the dust.

For the Galaxy, the first extensive study, involving distant early-type stars (Hiltner, 1956), suggested a magnetic field aligned along the local spiral arm. Observations of more than 1600 southern stars by Klare *et al.* (1971) indicated that the field was directed towards $l = 80°$. However, in the most extensive investigation, of 7000 stars including many nearby objects, Mathewson (1968) noted that the directions of polarization of stars within 500 pc of the Sun were not consistent with a magnetic field parallel to the galactic plane. He interpreted the distribution of these directions in latitude and longitude in terms of a tightly-wound helical field with an axis in the directions $l = 90°, 270°$.

The elegant concept of a local helical field, although in existence in one form or another for some years (Ireland, 1961), has not gained wide acceptance, even though Mathewson (1968) has claimed that it accounts for not only the polarization, but also the distribution of the spurs of galactic radio emission. For instance, an analysis by Seymour (1969), based on a smaller number of stars within 500 pc of the Sun, yielded a field parallel to the galactic plane and directed towards $l = 50°$, with irregularities on the scale of about 30 pc. In addition, the results showed a prominent field extension previously noted by Bingham (1967) along the North Polar Spur. The scale of the irregularities was less than the value (150 pc) derived by Jopikii *et al.* (1969) using a more distant sample of stars. In an independent investigation, Appenzeller (1968) claimed that the polarization vectors for stars near the south galactic pole were inconsistent with a helical field.

Because of this difference in interpretation, it is of interest to examine the basic observations. Figure 5 shows the distribution of polarization vectors (Mathewson and Ford, 1970a) that was analysed by Mathewson (1968). The vectors near the galactic equator are consistent with a major component of a magnetic field parallel to the plane and directed towards $l = 50°-80°$, at which longitudes there is a great confusion in vector direction consistent with a purely longitudinal field. In a recent review of these observations. Verschuur (1972) concluded that the most likely direction was

$l = 50°$, arguing that the directions near $l = 80°$ contained the Cygnus X complex which could contribute to the scatter in the polarization vectors. He also argued that directions near $l = 80°$ were not significant from the viewpoint of spiral structure, but there is a prominent spiral feature at $l = 80°$ in the distribution of neutral hydrogen (see review by Kerr (1969)). At higher galactic latitudes, the dominating feature is the

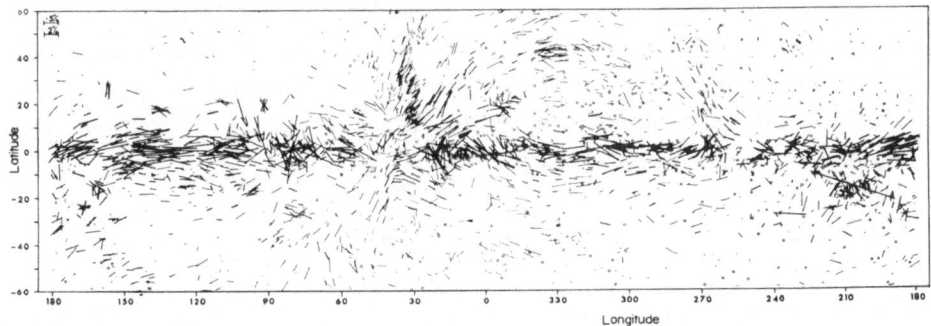

Fig. 5.   The distribution with galactic coordinates of the vectors of polarization of starlight (Mathewson and Ford, 1970a). The length of each vector is proportional to the percentage polarization P. Small circles are drawn about stars with $P < 0.08\%$.

magnetic loop that is probably associated with the North Polar Spur. The loop is local, since it can be traced in the distribution of vectors for stars in the distance range 50–100 pc (Figure 3a, Mathewson and Ford, 1970a). Another loop can be traced between the longitudes $l = 45°$ and $l = 160°$, the direction of the Cetus Arc. The conflict in interpretation depends largely on whether the loops are regarded as parts of an overall helical configuration or whether they are independent irregularities in the local field structure.

Although the observations can be qualitatively interpreted in terms of field structure that is also deduced from studies of radio emission, the interpretation of the optical results in terms of a local helical field is not convincingly disproved. However, a helical field is not consistent with the distribution of Faraday rotation for extragalactic sources (Wright, 1973) or pulsars (Manchester, 1973), or the distribution of galactic radio emission (Spoelstra, 1972). On the basis of the obvious irregularities that exist in the spiral structure of most galaxies, a large-scale field extending along a spiral arm, plus small-scale loops associated with spurs of radio emission, lanes of dust, or spiral arm extensions, seems the more realistic interpretation.

IV. Conclusion

The main points of this review are:

(i) Observations at radio and optical wavelengths are consistent with a large-scale magnetic field parallel to the galactic equator, which is directed towards longitude $l = 50°–90°$ and which has a field strength of about 2 μG. Within a few hundred parsecs of the Sun, the direction may be $l = 50°$, while at larger distances, the direction derived from the observations of Faraday rotation ($l = 90°$) appears more appropriate. On the other hand, the variation in direction could arise because the interstellar dust producing the optical polarization is differently distributed with respect to the synchrotron emission and the magnetoionic plasma.

(ii) Within a few hundred parsecs of the Sun, the field contains several loops associated with regions containing spurs of radio emission and interstellar dust.

(iii) Statistical studies suggest that there are field irregularities with a scale of about 100–200 pc. Within these regions, the field strength may be as much as twice the average value. In addition, compressed fields with strengths as high as several milligauss have been detected in several high-density regions in the Galaxy.

References

Appenzeller, I.: 1967, *Publ. Astron. Soc. Pacific* **79**, 600.
Appenzeller, I.: 1968, *Astrophys. J.* **151**, 907.
Berge, G. L. and Seielstad, G. A.: 1967, *Astrophys. J.* **148**, 367.
Berkhuijsen, E. M.: 1971, *Astron. Astrophys.* **14**, 359.
Berkhuijsen, E. M., Brouw, W. N., Muller, C. A., and Tinbergen, J. A.: 1964, *Bull. Astron. Inst. Neth.* **17**, 465.
Berkhuijsen, E. M., Haslam, C. G. T., and Salter, C. J.: 1971, *Astron. Astrophys.* **14**, 252.
Bingham, R. G.: 1967, *Monthly Notices Roy. Astron. Soc.* **137**, 157.
Bingham, R. G. and Shakeshaft, J. R.: 1967, *Monthly Notices Roy. Astron. Soc.* **136**, 347.
Bologna, J. M., McClain, E. F., Rose, W. K., and Sloanaker, R. M.: 1965, *Astrophys. J.* **142**, 106.
Brooks, J. W., Murray, J. D., and Radhakrishnan, V.: 1971, *Astrophys. Letters* **8**, 121.
Cooper, B. F., Price, R. M., and Cole, D. J.: 1965, *Australian J. Phys.* **18**, 589.
Davies, R. D.: 1972, *Monthly Notices Roy. Astron. Soc.* **160**, 381.
Davies, R. D., Booth, R. S., and Wilson, A. J.: 1968, *Nature* **220**, 1207.
Davies, R. D., de Jager, D., and Verschuur, G. L.: 1966, *Nature* **209**, 974.
Davis, L. and Greenstein, J. L.: 1951, *Astrophys. J.* **114**, 206.
Fermi, E.: 1949, *Phys. Rev.* **75**, 1169.
Gardner, F. F. and Davies, R. D.: 1966, *Australian J. Phys.* **19**, 129.
Gardner, F. F. and Whiteoak, J. B.: 1963, *Nature* **197**, 1162.
Gardner, F. F., Whiteoak, J. B., and Morris, D.: 1967, *Nature* **214**, 371.
Gardner, F. F., Morris, D., and Whiteoak, J. B.: 1969, *Australian J. Phys.* **22**, 813.
Gardner, F. F., Ribes, J. C., and Goss, W. M.: 1970, *Astrophys. Letters* **7**, 51.
Harwit, M.: 1970, *Nature* **226**, 61.
Haslam, C. G. T., Kahn, F. D., and Meaburn, J.: 1971, *Astron. Astrophys.* **12**, 388.
Hiltner, W. A.: 1956, *Astrophys. J. Suppl.* **2**, 389.
Hollinger, J. P., Mayer, C. H., and Menella, R. A.: 1964, *Astrophys. J.* **140**, 656.
Ireland, J. G.: 1961, *Monthly Notices Roy. Astron. Soc.* **122**, 461.
Jokipii, J. R. and Lerche, I.: 1969, *Astrophys. J.* **157**, 1137.
Jokipii, J. R. and Parker, E. N.: 1969, *Astrophys. J.* **155**, 777.
Jokipii, J. R., Lerche, I., and Schommer, R. A.: 1969, *Astrophys. J. Letters* **157**, L119.
Jones, R. V. and Spitzer, L.: 1967, *Astrophys. J.* **147**, 943.
Kerr, F. J.: 1969, *Ann. Rev. Astron. Astrophys.* **7**, 39.
Klare, G., Neckel, T., and Schnuur, G.: 1971, *Astron. Astrophys.* **11**, 155.
Komesaroff, M. M., Ables, J. G., and Hamilton, P. A.: 1971, *Astrophys. Letters* **9**, 101.
Kundu, M. R. and Velusamy, T.: 1972, *Astron. Astrophys.* **20**, 237.
Lerche, I.: 1970, *Astrophys. Space Sci.* **6**, 481.
Manchester, R. N.: 1972, *Astrophys. J.* **172**, 43.
Manchester, R. N.: 1973, *Astrophys. J.* **186**, 637.

Mansfield, V. N.: 1973, *Astrophys. J.* **179**, 815.

Martin, P. G.: 1971, *Monthly Notices Roy. Astron. Soc.* **153**, 279.

Mathewson, D. S.: 1968, *Astrophys. J. Letters* **153**, L47.

Mathewson, D. S. and Ford, V. L.: 1970a, *Mem. Roy. Astron. Soc.* **74**, 139.

Mathewson, D. S. and Ford, V. L.: 1970b, *Astrophys. J. Letters* **160**, L43.

Mathewson, D. S. and Milne, D. K.: 1965, *Australian J. Phys.* **18**, 635.

Mathewson, D. S., Broten, N. W., and Cole, D. J.: 1966, *Australian J. Phys.* **19**, 93.

Michel, F. C. and Yahil, A.: 1973, *Astrophys. J.* **179**, 771.

Milne, D. K.: 1970, *Australian J. Phys.* **23**, 425.

Mitton, S.: 1972, *Monthly Notices Roy. Astron. Soc.* **155**, 373.

Mitton, S.: 1973, *Astrophys. Letters* **13**, 19.

Morris, D. and Berge, G. L.: 1964, *Astrophys. J.* **139**, 1388.

Piddington, J. H.: 1972, *Cosmic Electrodyn.* **3**, 129.

Rydbeck, O. E. H., Kollberg, E., and Elldér, J.: 1970, *Astrophys. J. Letters* **161**, L25.

Seymour, P. A. H.: 1969, *Monthly Notices Roy. Astron. Soc.* **142**, 33.

Shain, G. A.: 1955, *Astron. Zh.* **32**, 381.

Shaver, P. A.: 1969, *Observatory* **89**, 227.

Spoelstra, T. A. T.: 1972, *Astron. Astrophys.* **21**, 61.

Spoelstra, T. A. T.: 1973, *Astron. Astrophys.* **24**, 149.

Turner, B. E. and Verschuur, G. L.: 1970, *Astrophys. J.* **162**, 341.

van der Laan, H.: 1962, *Monthly Notices Roy. Astron. Soc.* **124**, 125.

Verschuur, G. L.: 1970, in H. J. Habing (ed.), 'Interstellar Gas Dynamics', *IAU Symp.* **39**, 150.

Verschuur, G. L.: 1971, *Astrophys. J.* **165**, 651.

Verschuur, G. L.: 1972, Review Paper, *IAU Colloquium* **23**.

Verschuur, G. L.: 1973a, *Astron. Astrophys.* **22**, 139.

Verschuur, G. L.: 1973b, *Astron. Astrophys.* **27**, 407.

Whiteoak, J. B. and Gardner, F. F.: 1968, *Astrophys. J.* **154**, 807.

Wright, W. E.: 1973, 'Polarization Observations of 3C Radio Sources and Galactic Faraday Rotation', Ph.D. Thesis, California Institute of Technology.

Yen, J. L., Zuckerman, B., Palmer, P., and Penfield, H.: 1969, *Astrophys. J. Letters* **156**, L27.

J. B. Whiteoak
Division of Radiophysics, CSIRO,
P.O. Box 76, Epping, N.S.W. 2121, Australia

DISCUSSION

MacRae: Based on extensive new data on rotation measures, Kronberg and Vallee have developed a model of the local galactic magnetic field. Data on distances of stars and optical polarization are combined with the *RM*'s to demonstrate a local irregularity in the vicinity of the Sun, superimposed on a general spiral arm magnetic field. (*Nature* **246**, 49 (1973)).

Whiteoak: There are two interpretations on that. The polarization of starlight and the galactic radio emission suggests that the direction of the field within 15–200 pc of the Sun suggests a direction of $\sim 66°$, whereas the extragalactic polarization suggests $\sim 90°$. On the other hand, one might interpret this as a difference in the distribution of the nonthermal electrons responsible for synchrotron radiation and galactic radio emission and the dust which is causing the starlight polarization.

Mathewson: I split up the 7000 stars for which optical polarization measurements had been made into the distance intervals 0–50 pc, 50–100 pc and so on up to about 3 kpc. An analysis of the directions of the *E* vectors in each group enabled me to determine the spatial extent of the helical and longitudinal magnetic fields. (*Proc. Astron. Soc. Australia*, **1** (1969)). The helical field only occupies a volume within about 300 pc of the Sun whilst the longitudinal field which is in the direction $l = 90°$ and $270°$ is more widespread and is probably the general spiral arm magnetic field. I would like to emphasize that this analysis of the optical polarization data gives a longitudinal field as well as helical because in recent years authors of papers on galactic magnetic fields (as in your present review paper) only seem to refer to the helical field. I would also like to point out that only six out of the 38 pulsars investigated by Manchester lie within the helical field region while the other 32 lie in the longitudinal field region as defined by the optical polarization vectors.

To sum up there is good agreement between the model based on optical polarization work and the other methods of determining the structure of the galactic magnetic field.

Van Woerden: Does this mean that both sides of parties agree to the same models?

Whiteoak: Yes.

F. G. Smith: Polarization measurements at longer wavelengths show both depolarization and rotation. Away from the Spur regions there is relatively little rotation, indicating the presence of a random field component with rms value comparable to the organized field. In the Spur region there is more rotation, showing that there is a component of field in the line of sight, between the spur source and the Sun.

Whiteoak: The extragalactic source data at high latitude particularly in the direction of the North Polar Spur gives a rotation measure ~ 4 rad m^2. This is in accordance with what you were saying. This is not a very large rotation compared to what one can get close to the plane. This rotation doesn't affect the alignment of the fields to any great extent; it doesn't affect the interpretation.

A HIGH-RESOLUTION POLARIZATION SURVEY
OF THE NORTH POLAR SPUR

G. WESTERHOUT and D. BECHIS

University of Maryland, College Park, Md., U.S.A.

Abstract. Observations have been made at 21 cm with a resolution of 11' to look for fine structure in the polarization distribution. In the North Polar Spur, the angular scale of the polarization parameters varies with latitude. This is attributed to an increase in the irregularity of the magnetic field in the Spur with latitude.

I. Introduction

Following the extensive polarization observations by the Dutch group, notably Brouw and Spoelstra (1974) and Spoelstra (1971, 1972), it seemed appropriate to try and find out whether a resolution of 10' would reveal smaller structures in certain regions where the polarization distribution is still rather complex, even at 21 cm with a 36' beam. Setting a limit to the size of the irregularities in the magnetic field in a region such as the North Polar Spur might help in deciding what mechanism is causing depolarization of the radio radiation in this direction and what mechanism is generating the intense synchrotron emission observed in this presumably old supernova remnant. It might also provide a measure of the 'clumpiness' of the medium.

Observations were made over a total of about 75 days spread over several periods between June 1971 and March 1972, with the 300-ft transit telescope of the National Radio Astronomy Observatory* in Green Bank, West Virginia, at a wavelength of 21 cm where the beamwidth is approximately 11'. At this time, we have only reduced the region of the North-Polar Spur, i.e. between Dec. 6°–16°, R.A. 15:45–19:00 $(b=5°–55°, l=45°–15°)$. Points were observed every 10' in declination and every 6' in right ascension.

II. Observational Results

The following conclusions may be drawn from the observations:

(1) The scale over which the intensity of the polarized radiation and its position angle remain uniform varies with latitude:

b	
55°	$\sim 3°$ or more
35°	$\sim 1°$ to 2°
15°	$\sim 0.5°$ to 1°

If there are any polarization features smaller than the 11' beamwidth, they exist mostly

---

* The National Radio Astronomy Observatory is operated by Associated Universities Inc. under contract with the National Science Foundation.

F. J. Kerr and S. C. Simonson, III (eds.), Galactic Radio Astronomy, 151–154. All Rights Reserved.
Copyright © 1974 by the IAU.

at lower latitudes in the Spur. However, over the most intense regions of the Spur there do not seem to be any features with angular size less than $0°.5$. (2) Measurements over the brightest points of the Spur show that the degree of polarization (calculated by extrapolating Berkhuijsen's (1971) 820-MHz map to 1400 MHz using a spectral index of -2.7) decreases with latitude from about 70% for $b > 40°$ to about 35% around $b = 20°$. (3) The low polarization intensities measured in the Spur for $b < +50°$ by the Dutch surveys between 408 and 820 MHz may well be due partly to averaging small-scale polarization features over their 2° to 1° telescope beam. But this cannot explain the 'depolarization' of the 1400-MHz radiation at low latitudes as polarization structure much smaller than $0°.5$ does not seem to be present over the most intense regions of the Spur.

III. Discussion

Three possible explanations exist. The depolarization could be due to differential Faraday rotation caused by random fluctuations along the line of sight in the magnetic field and/or the electron density either (1) in the intervening medium, or (2) in the Spur itself. On the other hand, (3) random fluctuations in the Spur's magnetic field alone will tend to lower the intrinsic degree of polarization of the radiation emitted from the Spur. Berkhuijsen (1971) has considered these explanations in some detail. The present survey, because of its high angular resolution, adds one important piece of information: the size of the irregularities. We take the angular size of uniformity to be about 1° for $b < 50°$, corresponding to a linear size of a 'cell' of uniform magnetic field of about 1.6 pc, since the Spur's emission regions are at a distance of at most 100 pc. Using this figure we shall now discuss the three possible explanations; our conclusions are similar to those of Berkhuijsen.

Burn (1966) has shown that in the case that the depolarization occurs in the intervening medium, the degree of polarization we observe varies as

$$P(\lambda^2) = p_i \exp\{-2K^2 (n_e H_\mu)_f^2 \lambda^4 Rd\}$$

Here, p_i is the initial degree of polarization, d is the size of a cell in which the fluctuation occurs, $(n_e H_\parallel)_f^2$ is the variance of the product of the electron density and the line-of-sight magnetic field of a cell, R is the path length through the intervening medium, and $K = 2.62 \times 10^{-17}$, with all quantities in cgs units. For significant depolarization to occur, the quantity in brackets must be at least unity, which means that with $R = 100$ pc and $d = 1.6$ pc we have $(n_e H_\parallel)_f \gtrsim 2 \times 10^{-6}$ G cm^{-3}. Assuming a random component of $H_\parallel$ as high as the mean value observed in the interstellar medium, $\sim 3 \times 10^{-6}$ G, we find that a mean value of the electron density $n_e \gtrsim 1$ cm^{-3} is needed to explain the depolarization. Since this is far in excess of the average electron density in the interstellar medium, which is generally assumed to be < 0.1 cm^{-3}, we can rule out depolarization in the intervening medium.

Let us now see whether differential Faraday rotation within the Spur's emission regions can account for the depolarization. Since Burn's calculations are more complex in this case, we will deal with them in a more simplified way. Because the pathlength L through the Spur's emission regions is of the same order as their distance, 100 pc, each line of sight passes through at most $N \sim 60$ cells, each of length 1.6 pc. For the case where $N \gg (H_r/2H_0)^2$, with H_r and H_0 equal to the magnitudes of the random and the uniform components of the magnetic field, respectively, Burn calculates

$$P(\lambda^2) = p_i \frac{\sin \delta}{\delta}, \quad \text{where} \quad \delta = K n_e H_0 L \lambda^2$$

Significant depolarization will occur only when $\delta > 1$, corresponding to rotation measures in excess of 24 radians m^{-2} at 21 cm. Since the rotation measures calculated by Spoelstra (1971) are no greater than 15 radians m^{-2} for latitudes between $b = +25°$ and $b = +49°$ in the Spur, it is hard to explain the high degrees of polarization (around 40%) observed here. It is possible that if rotation measures could have been calculated on a closer angular spacing in this part of the sky, rotation measures in excess of 24 radians m^{-2} might have been found. However, Berkhuijsen (1971) finds in the Spur that the degree of polarization at 820 MHz is always about 0.4 to 0.6 times that at 1411 MHz over a wide range in degree of polarization. Such behavior would not be anticipated if differential Faraday rotation were an important factor in causing depolarization. Therefore, this model also fails to account for the latitude-dependent depolarization seen in the Spur.

If, however, the random fluctuations around the general direction of the magnetic field inside the Spur's emission regions increase toward lower latitudes, all observations are satisfied simply. Moreover, we can obtain an estimate of the strength of the random component of the magnetic field. For the case in which a random field H_r is superimposed on a uniform field H_0 and $H_0 \gg H_r$, Ginzburg and Syrovatskii (1965) give

$$\Pi = \left[1 - \frac{2}{3} \left(\frac{H_r}{H_0} \right)^2 \right] \frac{\gamma + 1}{\gamma + 7/3}$$

for the degree of polarization of radiation of spectral slope γ passing out of the source. Since the Spur has $\gamma \approx -2.7$ (Berkhuijsen, 1971), the gas giving rise to radiation with a a polarization of 35% has $H_r \sim 0.8 \, H_0$. Berkhuijsen (1971) found $H_r \approx 1.2 \, H_0$ from data at 829 MHz with a beamwidth of 1° at $b = +40°$ in the Spur. Of course for $H_0 \approx H_r$ this expression no longer holds, but the simple calculation gives at least a rough estimate. It is interesting that Jokipii and Lerche (1969), using Faraday rotations of extragalactic sources and dispersion measures of pulsar signals, found a fluctuating magnetic field of 1.4×10^{-6} G, about half the magnitude of the regular field, $\approx 3 \times 10^{-6}$ G.

We conclude then that Faraday rotation is unimportant in causing the decrease of the degree of polarization with galactic latitude in the Spur. It appears that instead the irregularity of the magnetic field in the Spur increases with decreasing latitude and that cells of constant magnetic field have a size of the order of 1.6 pc at intermediate latitudes. It seems somewhat unsatisfactory that in such a local phenomenon as the Spur there would be a latitude dependence. However, it should be noted that the intensity of the total radiation from the Spur also increases with decreasing latitude, at least down to $b = +15°$. Therefore, it might well be that the field irregularities are somehow related to the intensity of the radiation (or vice versa).

Acknowledgement

Part of this work was supported by the National Science Foundation.

References

Berkhuijsen, E. M.: 1971, *Astron. Astrophys.* **14**, 359.
Brouw, W. N. and Spoelstra, T. A. Th.: 1974, *Astron. Astrophys.*, submitted.
Burn, B. J.: 1966, *Monthly Notices Roy. Astron. Soc.* **133**, 67.
Ginzburg, V. L. and Syrovatskii, S. I.: 1965, *Ann Rev. Astron. Astrophys.* **3**, 297.
Jokipii, J. R. and Lerche, I.: 1969, *Astrophys. J.* **157**, 1137.
Spoelstra, T. A. Th.: 1971, *Astron. Astrophys.* **13**, 237.
Spoelstra, T. A. Th.: 1972, *Astron. Astrophys. Suppl.* **7**, 169.

G. Westerhout
D. Bechis
Astronomy Program, University of Maryland,
College Park, Md. 20742, U.S.A.

DISCUSSION

Baldwin: Are the percentage polarizations you quote relative to the total continuum emission or to a notional spur contribution obtained by some subtraction procedure? And do you feel secure about the variation with latitude?

Westerhout: The percentage polarization is derived by extrapolating Berkhuijsen's 820 MHz total intensity map using a spectral index of -2.7. Therefore, they are very uncertain (I estimate $\pm 30\%$) and we did not subtract a background. However, I feel quite sure about the decrease in degree of polarization with decreasing latitude (or, alternatively, with increasing intensity in the Spur). This has of course already been noted and discussed extensively by other authors, notably Berkhuijsen and Spoelstra.

Wielibinski: In the narrow neck of the Spur there is certainly structure on a smaller scale than described in this paper. On a continuum map made at 11 cm by Baker, Berkhuijsen and Haslam there is structure on the scale of some 5'.

Westerhout: Unfortunately, we only have limited coverage of the narrow neck; these data have not yet been reduced, and it will be very interesting indeed to see whether the polarization structure in the narrow neck will be smaller than the 0°5 at higher latitudes.

THE CHEMICAL COMPOSITION AND DISTRIBUTION
OF INTERSTELLAR GRAINS

J. MAYO GREENBERG and SEUNG-SOO HONG

State University of New York at Albany and Dudley Observatory, Albany, N.Y., U.S.A.

Abstract. The chemical composition of interstellar grains is derived here on the basis of (1) the cosmic abundance of the elements; (2) the wavelength dependence of extinction and polarization; (3) the average total extinction; (4) the ratio of polarization to extinction; (5) the predominantly dielectric character of grains in the visible spectral region; and (6) infrared spectral characteristics of grains. It is indicated that the major portion of the grains, by mass, consist of core-mantle particles in the 0.1-μm size range, whose cores consist largely of silicates and whose mantles are a solid mixture of O, C, and N with H in a heterogeneous combination of simple and complex molecules with frozen free radicals. A minor constituent of the solid particles exist in the form of very small uncoated particles generally less than 10^{-6} cm in size whose precise composition is not certain. Inferences of the core-mantle model with respect to spatial distribution are consistent with the proposition that growth of the mantles occurs in the galactic shock region predicted by the density-wave theory. Estimates of the total visual extinction toward the galactic center and the consequent estimates of the total amount of far infrared radiation are shown to depend critically on the grain model. Variations of the ratio of far ultraviolet to visual extinction are correlated with the conditions for growth of mantles on the bare small particles which are generally prevented from accreting mantles primarily because of their extreme temperature fluctuations produced by the ultraviolet photons in the radiation field.

I. Introduction

Since the early establishment of the solid particle nature of interstellar grains, a wide variety of possible candidates for their material constituents have been proposed. Initially, iron particles (Schalen, 1936; Greenstein, 1938) were considered largely because of the then current belief that meteorites were of cosmic origin. In a very significant way the circle is being closed and nowadays this basic concept is a most serious consideration as will be discussed later. Subsequently the suggestion by B. Lindblad (1935) that the grains could condense (or more properly, accrete) directly out of the interstellar medium led to the likelihood that ices of water, methane, and ammonia would be the major grain constituents (van de Hulst, 1943). This hypothesis appeared to be supported by the correlation between gas concentration and extinction. In the meanwhile the discovery of interstellar polarization by Hall (1949) and Hiltner (1949) led to some apparent difficulties in accounting for the observed amount of polarization relative to extinction. Even ideal alignment mechanisms seemed to preclude the ices (van de Hulst, 1957). Cayrel and Schatzman (1954) attempted to resolve this difficulty by the suggestion that a minor component of the interstellar grains in the form of graphite could form in stellar atmospheres and was sufficiently anisotropic in its optical properties as to provide adequate polarization. This suggestion was not pursued until somewhat later (Hoyle and Wickramasinghe, 1962). At about this time it was shown (Greenberg *et al.*, 1963a, 1963b) that given a sufficient degree of particle alignment, the dielectric ice grains of non-exotic shape would be quite adequate to provide the observed degrees of polarization. The concept of grain production in stellar atmospheres led others (Kamijo, 1963) to the conclusion that silicates could be injected into space and perhaps constitute a major constituent of the grains.

The reasons why so many different materials with such a wide range of optical properties could be used to explain the extinction and polarization were that the free parameters defining the size distribution were sufficient to explain the available observational results which spanned the somewhat limited wavelength region from the near infrared to the near ultraviolet.

As data has accumulated in the far infrared and the far ultraviolet, the situation would be assumed to become clearer. However, even now, it is only possible by combining a wide range of observational and theoretical calculations and inferences to draw a reasonably consistent picture of grains. In the next section we shall summarize what we consider to be the key observational and theoretical criteria for determining the principal chemical composition. The subsequent sections will use these data to give a self-consistent picture of the grains which is then used to infer a particular grain model. This is not expected to be unique, but it appears to contain the most essential chemical ingredients in configurations which seem to follow from entirely separate considerations such as the distribution of dust in spiral galaxies.

II. Basic Observations and Interpretations

(a) WAVELENGTH DEPENDENCE OF EXTINCTION AND POLARIZATION

The wavelength dependence of the extinction and polarization provide the means of determining the typical sizes of the interstellar grains as dependent on the particular optical material of the chosen model. Thus it is simply demonstrated from a comparison of the average wavelength of the maximum polarization and the wavelength dependence of the polarization produced by aligned infinite cylinders of index of refraction $m = 1.33$ that a characteristic particle with this optical property should have a radius of about 0.15 μm (Greenberg, 1973c). Detailed considerations confirm this. For the purposes of this paper we refer to the wavelength dependence of extinction primarily for inferences on particle sizes.

In Figure 1 is shown a schematic representation of the extinction curve from $\lambda^{-1} = 0$ to $\lambda^{-1} = 10$ μm^{-1}. Curves of this type have been obtained from OAO data (Bless and Savage, 1972) as well as from Copernicus (York et al., 1973). The general character of the extinction curve seems to divide itself into three regions. The region from 1 μm$^{-1} \leqslant$ $\leqslant \lambda^{-1} \leqslant 3$ μm^{-1} has been well established for many years and henceforth we call it the 'classical' region. The extinction over this wavelength range is characterized by the same-sized particles as those producing the polarization maximum at about $\lambda^{-1} = 1.8$ μm^{-1}. Excluding all other contributions to the extinction we would expect to see the extinction leveling off as shown by curve '2' in Figure 1. The continued rise in the extinction beyond $\lambda^{-1} \simeq 6$ μm^{-1} and, as indicated by the Copernicus results, even beyond $\lambda^{-1} = 10$ μm^{-1} can only be produced by solid particles whose sizes are characterized by radii of the order of or less than 10^{-6} cm (Greenberg, 1973a. 1973b).

These particles are associated with the curve '4'. The hump at $\lambda^{-1} = 4.6 \ \mu m^{-1}$ has been variously ascribed to graphite (Gilra, 1971) or silicates (Huffman and Stapp, 1971). The graphite particles needed to produce this feature are of sizes similar to those required to produce curve '4' and may indeed be the same particles. In the succeding

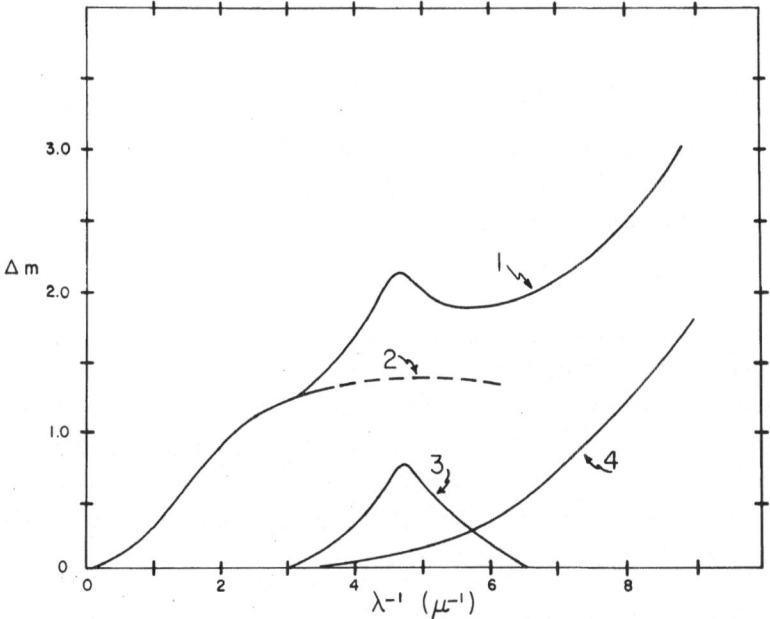

Fig. 1.   Representative extinction curve and schematic separation into several contributions. Curves are labeled as follows: (1) Typical OAO extinction curve, (2) Dashed extension showing contribution by classical sized particles (as discussed in text), (3) Contribution of absorption in the 0.22 μm band, (4) Contribution by very small particles.

discussions the possibility that curve '4' is produced by either silicates or carbon will be taken up along with several other materials.

(b) TOTAL EXTINCTION

Unless otherwise stated, by the term total extinction we will mean the total *visual* extinction. In combination with a knowledge of the particle sizes and composition, the total extinction provides estimates of the solid particle contribution to the mass distribution in space. The following calculation of grain mass uses a spherical grain model. It will be shown later that this idealized representation does not modify the conclusion qualitatively because the effect of shape on the mass estimation is not significant.

A canonical value for the average extinction per unit distance is generally stated as

$$\Delta m_v/D = 2 \text{ mag kpc}^{-1}. \tag{1}$$

The average mass density of interstellar particles of a characteristic size (radius) $\bar{a}$

and specific density s may be shown to be given by (Greenberg, 1968)

$$\varrho_d = \tfrac{4}{3}\bar{a}s\Delta m/D, \tag{2}$$

where we implicitly assume that the extinction is produced primarily by the particles of effective size $\bar{a}$. As already noted, contributions to the extinction in other spectral regions are typified by other particle sizes. For $\Delta m_v/D = 2$ mag kpc^{-1}, Equation (2) gives

$$\varrho_d = 0.80 \times 10^{-21}\, \bar{a}s\, \text{g cm}^{-3}, \tag{3}$$

Let $\bar{M}$ be the average molecular weight of the grain material and n_M be the number density of such molecules derived from Equation (3). From $n_M = \varrho_d/\bar{M}m_H$, where m_H is the atomic mass unit, we obtain

$$n_M = 0.048\, \bar{a}s/\bar{M}. \tag{4}$$

It should be noted that a homogeneous dirty-ice grain model ($s = 1$) with $\bar{a} = 0.2\ \mu$m gives a ratio of gas-to-dust density of 100 for an average number density of hydrogen atoms $n_H = 1$ cm^{-3}. This value of n_H will be used throughout as the basic reference for defining the ratio n_M/n_H from which relative atomic abundances are obtained.

In deriving Equations (2) and (3) we have used the fact that at the wavelength of evaluation of the total extinction the individual particles extinction efficiences are $Q = C_{ext}/\pi a^2 \simeq 1$, which gives (Greenberg, 1968)

$$\Delta m = 1.086\, n_d \pi \bar{a}^2 D, \tag{5}$$

where n_d is the number density of grains.

(c) EFFECT OF SHAPE ON THE MASS ESTIMATION

Suppose we had considered truncated circular cylinders of radius a and length $2\varepsilon a$ ($\varepsilon = $ length/diameter). Would this have affected the mass estimate significantly? The answer lies in evaluating the relative extinction per unit mass (or volume) for varying elongation.

It may be readily shown that the average projected area of an arbitrarily oriented convex particle is equal to one-fourth of the total surface area. Thus for an average extinction efficiency of $\bar{Q}$ we obtain the extinction averaged over orientation to be

$$C = \bar{Q}S/4,$$

where $S = $ particle surface area. For a sphere and $\bar{Q} = 1$ this reduces to $\bar{C} = \pi a^2$. The surface-to-volume ratio for the cylinders is

$$S/V = (1 + 2\varepsilon)/a\varepsilon, \tag{6}$$

from which we see that the change in the mass estimate using the *same* value of a is only about 17% in going from $\varepsilon = 1$ to $\varepsilon = 2$ and is only about 50% in going to $\varepsilon = \infty$! In view of the fact that the effective cylinder radius needed to produce the same wavelength dependence of the extinction as the spheres is less by perhaps 20% (Greenberg, 1968), we may infer that the net *overall* modification is largely cancelled; i.e., as ε increases we should decrease the value of a in Equation (6), thus leading to an almost invariant value of S/V. This general result is independent of the particular nonspherical shape chosen (Greenberg, 1973c, 1960). The detailed exact calculations will be presented in a later paper (Greenberg and Hong, 1974a).

(d) CIRCULAR POLARIZATION AND THE RATIO OF POLARIZATION TO EXTINCTION

The wavelength dependence of linear polarization gives no more definitive answer to the question of whether the grains are metallic or dielectric than does the extinction. However, recently the important discovery of the wavelength dependence of interstellar circular polarization (Kemp, 1973; Martin *et al.*, 1973) and its theoretical interpretation as being due to the birefringence of the interstellar medium created by aligned particles leads strongly to the conclusion that the grains which produce the linear polarization must be essentially dielectric (Martin, 1973). Metallic grains such as graphite are thus excluded as contributing substantially to the linear polarization.

The maximum ratio of polarization to extinction $\Delta m_p/\Delta m \simeq 0.06$ may be used to show that the major portion of the grains which produce the polarization must also be producing the extinction in the classical portion of the extinction curve unless we permit exceptional degrees of alignment. Let us assume that the visual extinction is produced partly by a non-polarizing component NP and partly by a polarizing component P. The ratio of polarization to extinction requirement is then (in optical depths rather than magnitudes) (Greenberg, 1969)

$$\frac{2(\tau_1^P - \tau_2^P)}{\tau^{NP} + (\tau_1^P + \tau_2^P)} = 0.06. \tag{7}$$

If we consider magnetic alignment, we find that even with field strengths as large as 10^{-5} G in clouds of density $n_H = 10$ cm^{-3} and temperature $T = 100$ K the best attainable value of $2(\tau_1^P - \tau_2^P)/(\tau_1^P + \tau_2^P)$ is *less* than about 0.08 even when *perfect* spinning alignment of the particles gives a ratio of polarization to extinction of 0.4 (the value for the core-mantle model discussed later). Thus Equation (7) gives $0.08/[(\tau^{NP}/\tau^P) + 1] = 0.06$, which allows a nonpolarizing contribution to the total extinction of $<1/4$. In general, we conclude that certainly most of the particles producing the visual extinction are dielectric, although we can not exclude on this argument alone a contribution by metallic non-polarizing particles of perhaps 10–20%.

(e) INFRARED SPECTRA

The only direct spectral information on the chemical composition of the grains occurs in the infrared. The 0.22-μm hump (curve '3' in Figure 1) is not considered to provide as direct information. In general, the implication of a silicate component in circumstellar regions and in interstellar space depends on a broad feature of emission and absorption, respectively, centered around 9.7 μm. The search for a 3.07-μm absorption band in H_2O ice is extremely difficult and there is only one case in which both the

9.7-μm and 3.07-μm have been found to be simultaneously present. Figure 2, taken from Gillett and Forrest (1973), gives the basic datum which we use as a starting point in inferring that at least some ice is present in the interstellar medium.

We may calculate the relative volumes of ice and silicate material implied from the relative absorptions at 3.07 μm and 9.7 μm shown as essentially equal in Figure 2. If we base our result on the relative absorptivities of small separate spheres of ice and

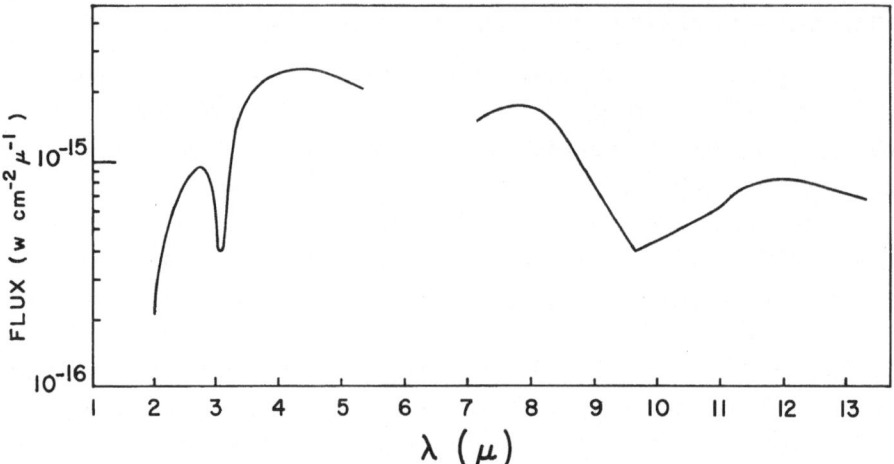

Fig. 2.   Infrared spectrum of the Becklin-Neugebauer object. Adapted from Gillett and Forrest (1973).

silicate and using the obvious fact that $a/\lambda \ll 1$ in both cases, we arrive at a ratio $V_{ice}/V_{sil} = 0.21$ (Greenberg, 1973c). The assumption of *independent* particles of silicates and ices is probably unrealistic and may in some conditions produce spurious results. The condition under which the independent particle model gives reliable or incorrect results will be discussed later in some detail. At this point all that we wish to demonstrate is that: (1) both silicates and ice may be presumed to exist, and (2) the relative proportion of ice to silicate is *at least* 0.21. This value may be taken to be a lower bound if we generalize our definition of ice to include besides frozen H_2O various combinations of O, C N, and H as molecules and frozen free radicals. This has been justified on the basis of theoretical and experimental studies of the effects of ultraviolet irradiation of the grains (Greenberg, 1973c).

(f) COSMIC ABUNDANCE

The principal argument in this paper is based on a comparison between the abundances of various elements relative to hydrogen which would be demanded by postulated grain models and the cosmic abundance ratios which are presumed. For reference, Table I is abstracted from the latest compilation (Cameron, 1973).

III. Chemical Composition

In this section we shall compare a variety of interstellar grain models with respect to their relative compatibility with the cosmic abundances of Table I. We consider first

TABLE I

Selected abundances of the elements

Element	Relative number of atoms
H	1
C	3.70×10^{-4}
N	1.17×10^{-4}
O	6.76×10^{-4}
Mg	0.34×10^{-4}
Si	0.32×10^{-4}
Fe	0.26×10^{-4}

models for grains which produce the classical portion of the extinction curve labeled '2' because, as will appear later, these particles contain most of the mass.

(a) HOMOGENEOUS MODELS

Even though the assumption of a single chemical component to represent the interstellar grains is not realistic, it serves as a very useful starting point in defining the limitations imposed by chemical abundances. The representative sizes selected for each material are based only on the broadest characteristics of curve '2'; i.e., we attempt only to match approximately the single particle position of extinction saturation with the value $\lambda^{-1} \simeq 4 \ \mu m^{-1}$ in curve '2'. Table II summarizes the results for a number of grain ingredients.

We see from Table II that the only grain ingredients which fit comfortably on cosmic

TABLE II

Atomic densities for homogeneous grain models
$\Delta m/D = 2$ mag kpc^{-1}, $n_H = 1$ cm^{-3}, $\bar{Q} = 1^d$

Material	$\bar{M}$	$\bar{a}$	$\bar{s}$	n_M	$\frac{[O]}{[H]}$	$\frac{[Si]}{[H]}$	$\frac{[Mg, Fe]}{[H]}$	$\frac{[Fe]}{[H]}$	$\frac{[C]}{[H]}$
		(μm)		(cm^{-3})			$\times 10^{-4}$		
'Ice'	17	0.2	1.0	5.61	3.28[a]				1.80
Orthopyroxene[b]	116	0.1	3.6	1.48	4.44	1.48	1.48		
Olivine[c]	172	0.1	3.8	1.05	4.20	1.05	2.10		
Magnetite	232	0.05	6.0	0.62	2.47			1.86	
Iron	56	0.05	7.9	3.36				3.36	
Silicon Carbide	38	0.05	3.2	2.02		2.02			2.02
Carbon	12	0.05	2.2	4.40					4.40
Cosmic Abundance					6.76	0.32	(Mg)0.34	0.26	3.70

[a] 3.28 is reduced from 5.61 because 'Ice' means [O, C, N] in relative cosmic abundances; i.e., O:C:N = 6.76:3.70:1.17 comprising a variety of complex organic molecules and frozen free radicals.
[b] Orthopyroxene = (Mg, Fe) SiO$_3$.
[c] Olivine = (Mg, Fe)$_2$ SiO$_4$.
[d] For metals (magnetite, iron, carbon) the appropriate $\bar{Q}$ value is about 1.3 and for dielectrics $\bar{Q} \approx 1.5$. The last six columns of the table should be divided by these factors.

abundance arguments are the dirty ice model and possibly the carbon which is discrepant only by a factor of two. All others produce discordant abundances by factors of at least 4. Thus, while orthopyroxene requires only about twice the CA amount of Mg and Fe it requires six times the CA of silicon and this is a fairly severe restriction. (These factors of rejection are reduced by using $\bar{Q}>1$ but not by enough to change the conclusions.) It would appear that SiC is precluded from providing a major portion of the visual extinction as would seem to be required by some grain models (Gilra, 1971). The carbon must be excluded as a major constituent primarily because of the presumed dielectric optical properties of the grains producing visual extinction and polarization. Note that we have yet to add a grain component to give the far ultraviolet extinction.

(b) CORE-MANTLE PARTICLES – CLASSICAL EXTINCTION

It seems highly unlikely that if silicates and ices are present in space they occur as disconnected objects. The most reasonable way for ice to form in interstellar space is by accretion on some nucleation points. We therefore assume that silicates and ices appear as cores and mantles respectively. Other materials may also provide cores but at this time we restrict ourselves to silicate cores alone for the classical particles. A summary picture of possible processes leading to this model is presented later.

In the following discussion the particles are treated as concentric spheres. The results for a more detailed model using concentric cylinders are given later and compared directly with observations.

Let a_c and a_m be the core and mantle radii respectively, and let $\alpha = a_m/a_c$. Again, letting $Q=1$, Equation (5) is replaced by

$$\Delta m = 1.086 \, n_d \pi a_m^2 D, \tag{8}$$

from which we obtain

$$\varrho_c = \frac{4}{3} \frac{a_m s}{\alpha^3} \frac{\Delta m}{1.086 D} = 0.80 \times 10^{-21} (a_m s/\alpha^3), \quad (\Delta m/D = 2 \text{ mag kpc}^{-1}), \tag{9}$$

where ϱ_c means space density of the core material only.

Based on the concentric cylinder model we choose $a_c = 0.08 \, \mu m$ and find that in order to bring the [Si]/[H] ratio down to its CA level we must have approximately $\alpha = 1.8$, neglecting the possible contribution of the very small particles which are considered in the next section. This means that the value of $V_{\text{ice}}/V_{\text{sil}}$ must be closer to 0.8 than the 0.2 derived directly from the Becklin-Neugebauer object for H_2O ice. It should be kept in mind that the choice of $a_c = 0.08 \, \mu m$ is probably not unique and may, indeed, not even be the best value. It was determined in an essentially empirical fashion based on 'cut and try' calculations on a few concentric cylinder models.

(c) VERY SMALL PARTICLES – FAR-ULTRAVIOLET EXTINCTION

As is well-known, the classical-sized particles can not produce the shape of the far ultraviolet extinction curve '4'. It is not at all unusual to find the extinction at 1000 Å

$(\lambda^{-1} = 10 \ \mu m^{-1})$ to be perhaps 3 or 4 times larger than the visual extinction. In the following calculations we let $\Delta m(10) = 4\Delta m_v$. Then the contribution of the small particles alone to the extinction at $\lambda^{-1} = 10 \ \mu m^{-1}$ is $4\Delta m_v - 2\Delta m_v = 2\Delta m_v$ where we have used the approximate result that the saturation extinction by the classical particles is twice the visual extinction. We thus have

$$\varrho_b = 0.40 \times 10^{-21} a_b s \ \frac{2\Delta m_v}{D}$$

$$= 1.6 \times 10^{-21} a_b s \quad (\Delta m_v/D = 2 \text{ mag kpc}^{-1}), \tag{10}$$

$$n_M = 0.095 \ a_b s / \bar{M},$$

where a_b is the radius of the very small particles assumed to be *bare* for reasons which will be demonstrated later and where ϱ_b = space density of the bare particles.

Using $a_b = 0.01 \ \mu m$ we have the results shown in Table III.

TABLE III

Abundances for bare particles

Material	n_M (cm^{-3})	$\dfrac{[Si]}{[H]}$	$\dfrac{[Fe, Mg]}{[H]}$	$\dfrac{[Fe]}{[H]}$	$\dfrac{[C]}{[H]}$
			$\times 10^{-4}$		
Orthopyroxene	0.295	0.295	0.295		
Olivine	0.210	0.210	0.420		
Magnetite	0.246			0.738	
Iron	1.34			1.34	
Carbon	1.74				1.74
Silicon Carbide	0.80	0.80			0.80
Cosmic Abundance		0.32 (Mg)	0.34	0.26	3.70

We note immediately that the contribution of the very small particles to the heavy element mass is comparable with that of the cores. However, we must keep in mind that our particle parameters are only chosen on a semi-quantitative basis so that factors of two may not be sufficient to exclude a model. On the other hand, factors of 4 or more must be taken seriously. In any case, there is abundant evidence from the above arguments that most of the heavy elements in interstellar space must be bound up in the form of solid particles. This is particularly true of the Si, Fe, Mg group, and to a lesser extent for O, N, C. These results appear to be compatible with the observed depletion (Morton *et al.*, 1973; Rogerson *et al.*, 1973). An additional depletion of C relative to O and N, which has been reported, could be attributed to the existence of very small graphite particles required for producing the 0.22-μm absorption hump.

It appears that iron particles may contribute no more than about 10% of the far ultraviolet extinction. Magnetite also is not an important constituent.

Incidentally, none of the above arguments exclude any heavy-element constituent from being part of the grain-core population but merely excludes them from providing the major optical manifestations of interstellar grains.

We conclude that cosmic-abundance arguments lead us to require the major optical effects of grains in the classical region to be produced by the mantles and the major additional effects in the far-ultraviolet to be produced by very small uncoated particles whose chemical composition is primarily silicate and/or carbon with the 0.22-μm hump leaning to support of the latter.

(d) RELATIVE NUMBERS OF CORE-MANTLES AND BARE PARTICLES

The number of grains is a rather artificial concept because it is usually based on the assumption of a single representative particle size. Nevertheless, it is a useful working model so long as one does not apply it without appropriate reservations.

The typical extinction values in the visible and far-ultraviolet, as defined in Sections IIIb and IIIc respectively, result in the number of core-mantle, n_{C-M}, and the number of bare particles n_b being given by

$$n_{C-M} = 1.3 \times 10^{-12} \text{ cm}^{-3}, \qquad \Delta m_v/D = 2 \text{ mag kpc}^{-1}$$
$$n_b = 290 \, n_{C-M}, \qquad \Delta m_{10} = 4\Delta m_v,$$

where the core radius is given by $a_c = 0.08$ μm, the ratio of mantle to core radius is $\alpha = 1.5$, and the bare particle radius is $a_b = 0.01$ μm.

IV. Small-Particle Temperatures

One of the properties of small grains which may at first be thought to provide a condition for distinguishing between the accretion possibilities of the core particles and the much smaller 'bare' particles is the fact that, in general, the temperatures of small particles increase with decreasing size. We see in the next section that this effect is not large enough to be particularly important. In the subsequent section, however, we show that the very small size of the bare particle leads to another phenomenon associated with radiant energy absorption; namely, energy fluctuations which are significantly larger than the average energy content of the grains and it is this effect which is presumed to be the agency for inhibiting accretion.

(a) EQUILIBRIUM TEMPERATURES

It has long been recognized that very small particles immersed in the interstellar radiation field arrive at equilibrium temperatures substantially higher than those of a black body (van de Hulst, 1946). The reason for this is apparent from elementary considerations of the equation relating the particle radiation absorption to emission.

$$\int_0^\infty \varepsilon_{ab}(a, \lambda) \, R(\lambda) \, d\lambda = \int_0^\infty \varepsilon_{em}(a, \lambda) \, B(\lambda, T_d) \, d\lambda$$

or

$$\bar{\varepsilon}_{ab} \int_0^\infty R(\lambda) \, d\lambda = \bar{\varepsilon}_{em} \int_0^\infty B(\lambda, T_d) \, d\lambda,$$

where ε_{ab} and ε_{em} are the wavelength-dependent absorption efficiencies of the particles of size a, $R(\lambda)$ is the interstellar radiation field, and $B(\lambda, T_d)$ is the Planck function for the grain temperature T_d. It is convenient to consider Equation (11) in terms of average efficiencies $\bar{\varepsilon}_{ab}$ and $\bar{\varepsilon}_{em}$. Further, to arrive at a size dependence it is useful to let $\bar{\varepsilon}_{em} = \bar{\varepsilon}_{eff} = 33.3 a T_d$, which will give a lower bound on the particle temperatures (Greenberg, 1971) because it is defined by maximum particle emissivity. The values of ε are strongly dependent on the value of a/λ, approaching unity as $a/\lambda \to \infty$ and zero as $a/\lambda \to 0$. Since the radiation is generally absorbed at values of λ for which a/λ is much larger than its value at the important emission regions, we find that the temperature T_d defined by Equation (11) is higher than the black-body temperature. For example we note that, for an 'ice' particle, $\bar{\varepsilon}_{ab} = 0.1$ for $a = 0.1$ μm and $\bar{\varepsilon}_{ab} = 1.0$ for $a = 1.0$ μm if the interstellar radiation field is represented by $R(\lambda) = WT^4$, with $W = 10^{-14}$ and $T = 10000$ K (Greenberg, 1968). Using the maximum long-wavelength emissivity, ε_{eff}, we get $T_d = 8$ K for the 0.1-μm particle which is only about 3 K less than a more correctly obtained value. We should not be surprised to find that a 0.01-μm (10^{-6} cm) particle does not become significantly hotter when we realize that while $\bar{\varepsilon}_{em}$ is decreasing with size, so also is $\bar{\varepsilon}_{ab}$. This may be shown by extrapolation of the curves showing the variation of grain temperature with size (Greenberg, 1968). Thus the very small particles referred to earlier as bare particles differ little in temperature from the classical-sized cores, and consequently this alone can not provide a reason for the bare particles remaining bare.

(b) TEMPERATURE FLUCTUATION OF BARE PARTICLES

It has been mentioned (Greenberg, 1968) that the concept of grain equilibrium temperature is meaningful only when the rate of energy input per second from the radiation field is generally very small compared to the average total energy content of the grain. This condition is fulfilled for the grains larger than 0.05 μm in radius. For sufficiently small grains it was shown that this condition is not met.

The energy content of a grain is given by (Kittel, 1956)

$$U = \frac{3\pi^4}{5} NkT_d(T_d/\theta)^3, \tag{12}$$

where $N =$ number of atoms per cm^3 and $\theta =$ Debye temperature. This, of course, is the result for a reasonably defined solid and therefore we must assume that the bare particles are of this character. The value of N in Equation (12) is obtained from

$$N = \frac{4}{3} \frac{\pi a^3 s}{\bar{M} m_H}; \tag{13}$$

665 K is used for the Debye temperature of olivine and 420 K for that of graphite. The value of 696 K for orthopyroxene is so close to that of olivine that the latter serves to give an adequate illustration of the temperature fluctuation for silicates. Graphite is a two dimensional lattice with a more complicated low temperature spe-

cific heat. Our use of a T^3 law with an effective Debye temperature is approximate and probably underestimates the temperature fluctuations.

In Figures 3 and 4 the ratio of the final temperature, after absorbing energy $h\nu$, to the equilibrium dust temperature $T_f/T_d = [(U + h\nu)/U]^{1/4}$ is plotted against the energy of the absorbed photon in olivine and graphite particles. With the Equations (12) and (13) this ratio is given by

$$\frac{T_f}{T_d} = \left(1 + \frac{5\bar{M}m_H\theta^3 h\nu}{4\pi^5 ska^3 T_d^4}\right)^{1/4}.$$ (14)

Hence we see that the final temperature is roughly independent of the initial temperature which was taken as the equilibrium dust temperature, and that the final temperature is proportional to $a^{-3/4}$ as a approaches zero. We see, for example, that when a 0.01-μm olivine grain whose equilibrium temperature is either 5 or 10 K absorbs a 5-eV photon ($\lambda = 240$ nm) its temperature jumps to 46.5 K or 48 K respec-

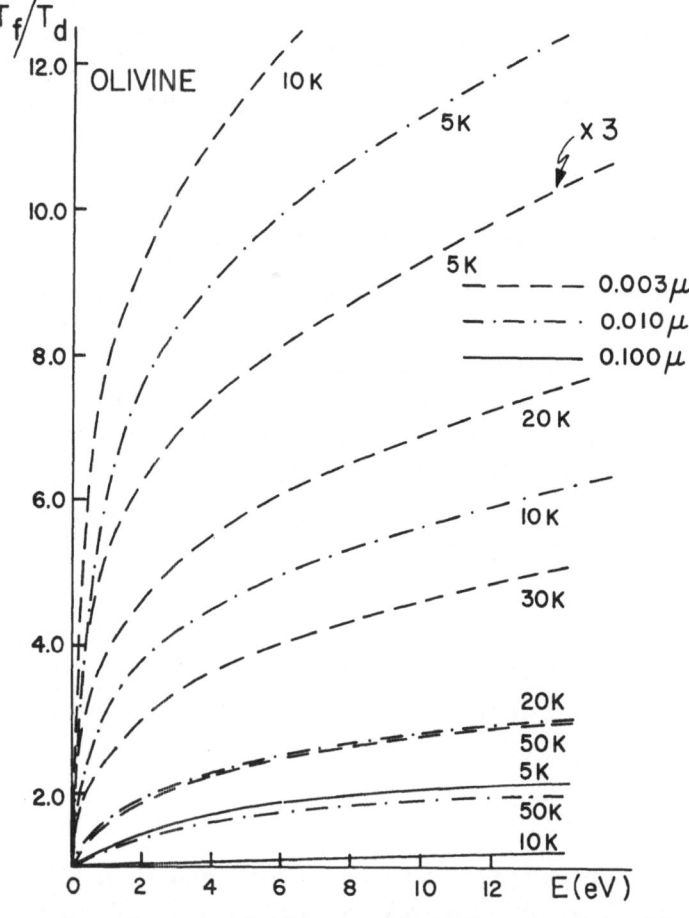

Fig. 3.   Temperature fluctuations of small olivine particles of three sizes induced by single photon absorption. Each curve is labeled with its appropriate equilibrium temperature. As indicated the curve for $a = 0.003$ μm and $T_d = 5$ K should be multiplied by a factor of three.

tively, and that going from $a=0.01$ μm to $a=0.03$ μm the final temperature rises to approximately 150 K.

This is reminiscent of a remark made earlier (Greenberg, 1968) with regard to the temperature of Platt (1956) particles.

It seems reasonable to conclude that such temperatures are adequate to prevent accretion on the bare silicate particles if they are sufficiently small and in the size

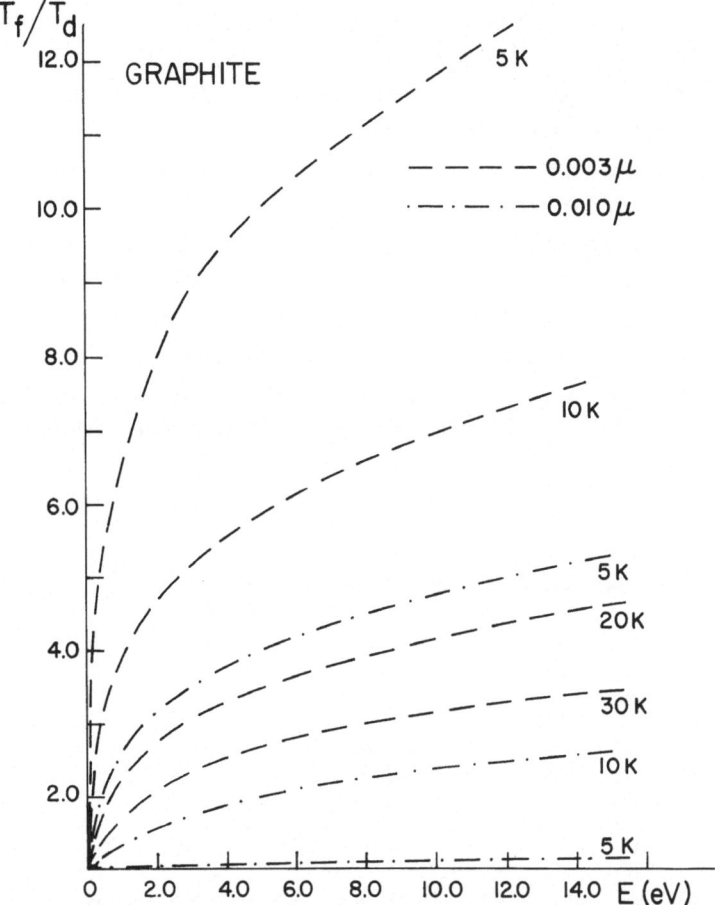

Fig. 4. Temperature fluctuations of small graphite particles of two sizes induced by single photon absorption. Each curve is labeled with its appropriate equilibrium temperature.

range required for the far ultraviolet extinction. The normal graphite temperatures are larger than the silicate temperatures; and their fluctuating values are somewhat more violent because graphite has a larger effective Debye temperature. Therefore this mechanism should play a similar role in inhibiting accretion in normal radiation fields.

V. Extinction to the Galactic Center

There are many reasons why the sizes of the mantles relative to the cores should be different from region to region in the Milky Way: different time scales for accretion due to the different physical conditions from place to place, possibilities for evaporation and sputtering, etc. It is by simultaneously measuring both the total extinction and the silicate band absorption that one may obtain such a measure most directly. As we shall see in the discussion of models, the variation of total-to-selective extinction is not a very sensitive discriminant of the mantle size over a wide range.

The silicate band absorption at 9.7 μm toward the galactic center is about $\Delta m_{9.7-\mu m} = 3$ (Woolf, 1973). Using a value of $m_{sil} = 1.7 - 0.71\,i$ leads to an extinction per unit volume of silicate core of $\Delta m_{9.7-\mu m}/V = 0.609$. Thus from Equation (5) we get $\Delta m_V = 4.06/\bar{a}$ for pure silicate grains of mean representative size $\bar{a}$ with no mantles. For $a = 0.1$ μm this leads to $\Delta m_V = 40.6$ or about 27, which seems generally in the range of the accepted value of 27 mag extinction to the center (Borgman, 1974). We note that if $\bar{a}$ is as much as 0.2, which has been occasionally used in the literature, the extinction would be only about 20 mag.

In order to evaluate the effect of mantles we borrow the core and mantle parameters from those which are used in our model calculations. We see in Table IV that consistent with the model parameters leading to essentially constant values of the ratio of total-to-selective extinction, there is a very wide range in possible values of the extinction to the galactic center. The equation for Δm_V is modified from the above to $\Delta m_V = 4.1\,\alpha^2/a_1$ where α is the ratio of the mantle-to-core radii and a_1 is the core radius.

TABLE IV

Mantle size effect on extinction to galactic center
$a_1 = 0.08$ μm

a_2	α	Δm_V
0.1	1.25	98
0.12	1.50	113 (92)
0.20	2.50	312

The values for Δm_V given in Table IV are correct if we assume that the particles producing the far-ultraviolet extinction are either non-existent or consist of some material (like graphite) not composed of silicates. If there are small bare silicate particles the values of Δm_V are reduced accordingly so that, for example, if we use the number of bare particles implied by our typical extinction curve we would reduce the value 113 to 92 as shown in parentheses.

According to our model, then, a value of $\Delta m_V = 41$ implies no mantles whatsoever on grains in the galactic center. It should be pointed out that more carefully calculated values of Q (the extinction efficiency) would modify our results somewhat and that such calculations are in progress. However, no significant changes in our conclusions are anticipated and if the value of $\Delta m_V = 27$ is independently confirmed, the mantles on dust grains near the galactic center must be very thin.

VI. Core-Mantle Cylinder Model

We have chosen as representative of a class of core-mantle cylinder models the case in which all cores are the same radius, $a_c = 0.08$ μm, and the mantles are distributed in sizes according to the form $n(a_m) = \exp\{-5[(a_m - a_c)/a_i]^3\}$, where $a_i = 0.12, 0.14, 0.16$ μm. The average or effective single value of a_m has been shown to be given by $a_m - a_c \cong 0.3$ a_i Greenberg (1968). Single equivalent mantle thicknesses are respectively 0.036, 0.042, 0.048 μm, which give $a_m = 0.116, 0.122, 0.128$ μm.

In order to obtain a realistic value of the polarization, we have performed the calculations for spinning cylinders corresponding to perfect Davis-Greenstein orientation. The computer program for arbitrarily oriented cylinders has been generalized from that for homogeneous cylinders (Lind and Greenberg, 1969) to concentric core and mantle cylinders (Shah, 1970). The numerical calculations are performed to obtain the extinction from the total cross sections for orthogonal polarizations of the incident radiation,

$$
C_E = \frac{2}{\pi} \int_{a_c}^{\infty} n(a_m)\, da_m \int_0^{\pi/2} C_E(a_c, a_m, \chi)\, d\chi,
$$

$$
\tag{15}
$$

$$
C_H = \frac{2}{\pi} \int_{a_c}^{\infty} n(a_m)\, da_m \int_0^{\pi/2} C_H(a_c, a_m, \chi)\, d\chi,
$$

where χ is the tilt angle of the cylinder axis with respect to the direction of incident radiation. The integration intervals used were $\Delta a_m = 0.01$ and $\Delta\chi = 9°$.

The normalized extinction values of $\Delta m = [(C_E + C_H)/2]$ $[\Delta m(1) = 0, \Delta m(3) = 1]$ for the three values of a_i are shown in Figure 5, where comparison in each case is made with the observations which have been obtained from appropriate renormalization of the data of Whiteoak (1966) and Bless and Savage (1972). Within the range $1 \leqslant \leqslant \lambda^{-1} \leqslant 3$ μm^{-1} all three calculated curves are almost indistinguishable from each other and give an excellent reproduction of the observational results. The total variation in mantle thickness in the three models is a factor of about 1.5 and the variation in total thickness is about 10%. A most significant consequence of the core-mantle model therefore is the essential invariance of the extinction curve to variation in mantle thickness. This is quite different from the result for homogeneous particles (see for example the numerous cases presented by Greenberg (1968) as well as elsewhere) where a 20% change in size produces a large change in curvature of the extinction around $\lambda^{-1} = 2$ μm^{-1}.

The *general* uniformity of the shape of the extinction in the visual region with its concomitant constancy in the ratio of A_v/E_{B-V} has long represented a puzzle to those who wished to represent the grains by particles which accrete matter in space and are therefore of variable size depending on their history. The statistical growth-destruction calculation of Oort and van de Hulst (1946) answered the question but

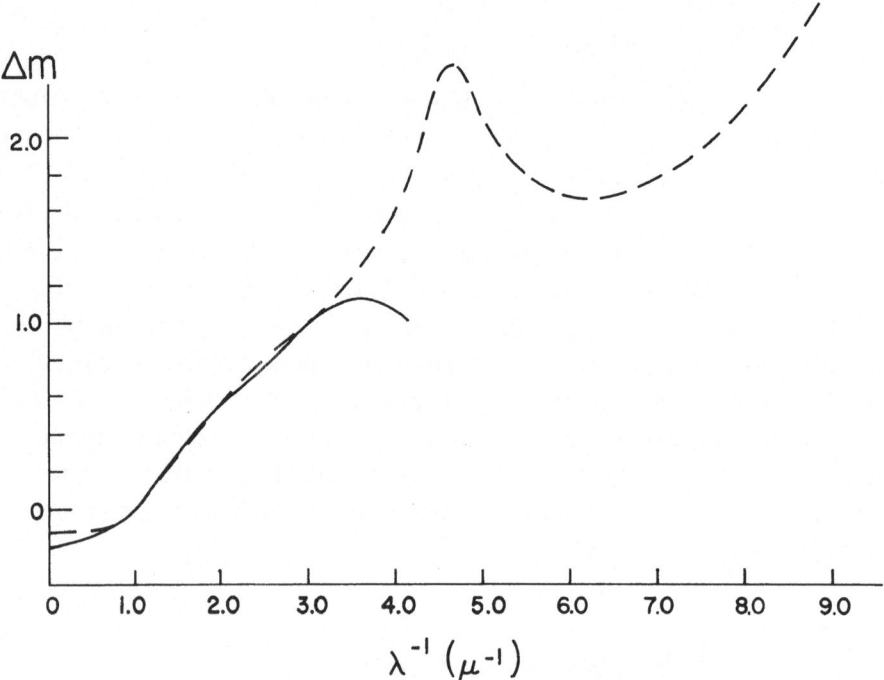

Figs. 5a–c. Comparison of calculated extinction by spinning core-mantle cylinders (solid curves) with observations (dashed curves). $R_{obs} = 3.20$

Fig. 5a. $a_c = 0.08 \ \mu m$, $n(a_m) = \exp\left[-5\left(\frac{a_m - 0.08}{0.12}\right)^3\right]$

$R_{calc} = 3.19$.

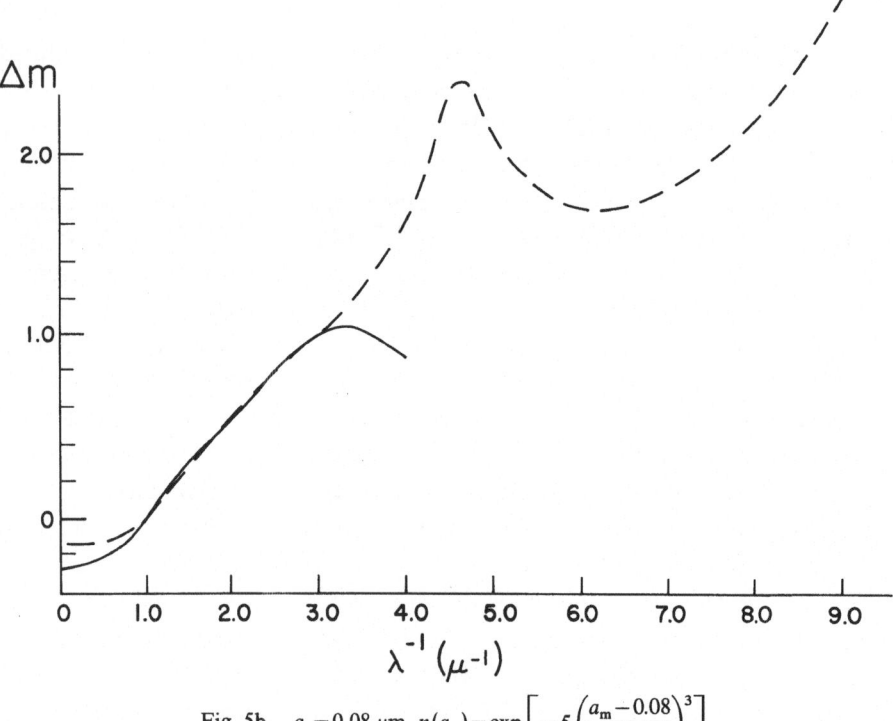

Fig. 5b. $a_c = 0.08 \ \mu m$, $n(a_m) = \exp\left[-5\left(\frac{a_m - 0.08}{0.14}\right)^3\right]$

$R_{calc} = 3.27$.

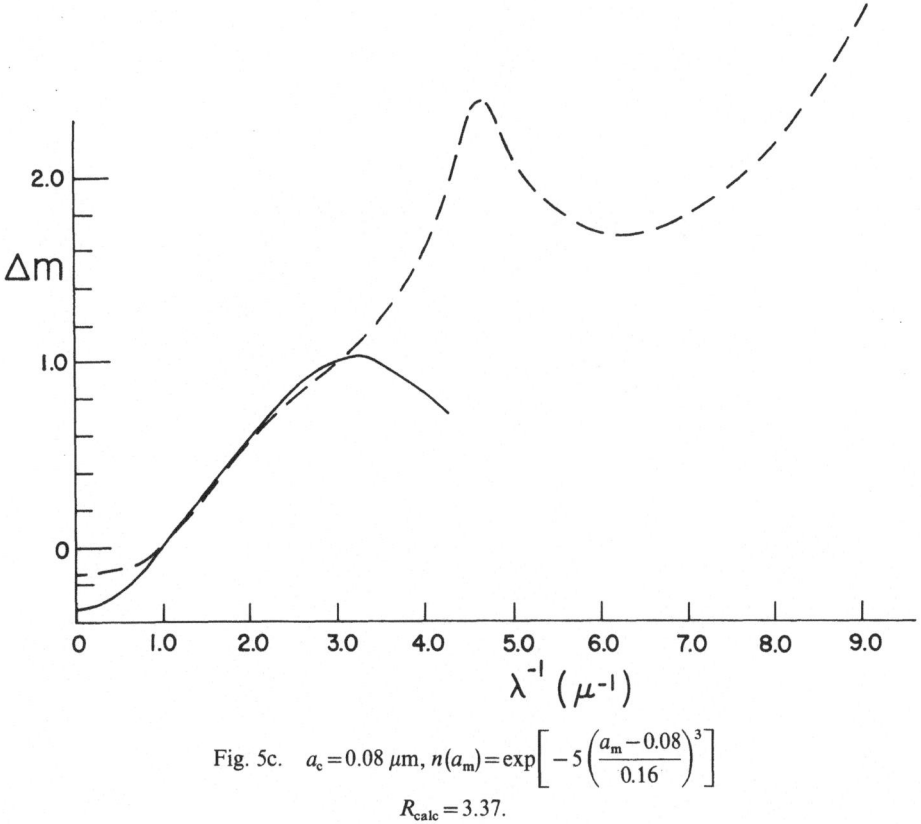

Fig. 5c. $a_c = 0.08\ \mu m$, $n(a_m) = \exp\left[-5\left(\dfrac{a_m - 0.08}{0.16}\right)^3\right]$

$R_{calc} = 3.37$.

only on the basis of an *average* over many dust clouds. Even disregarding the question of validity of random cloud-cloud collisions providing the grain destruction mechanism it was still difficult to answer the question of why the ratio A/E varied so little (with some clear exceptions for which good theoretical grounds existed). It is thus satisfying to have a grain model which has the character of preserving the invariability of A_v/E_{B-V} over such a wide range of physical conditions and history as evidenced by large differences in mantle thickness.

(b) POLARIZATION

The polarization is obtained from $\Delta m_p \sim C_E - C_H$. The wavelength dependence of polarization and ratio of polarization to extinction are shown in Figure 6. The position of P_{max} is seen to shift toward lower values of λ^{-1} as a_i increases in just about the same way as would be expected for homogeneous particles. Thus the wavelength dependence of polarization is still a good discriminant of particle size. We note also that the ratio of polarization to extinction is significantly larger for the core-mantle particles which satisfy the extinction than for homogeneous particles. They will thus more readily produce the observed ratios of polarization to extinction for a *given* magnetic field. In addition, because they are smaller in outer dimension, their inertial properties are better (lower moment of inertia) for orientation.

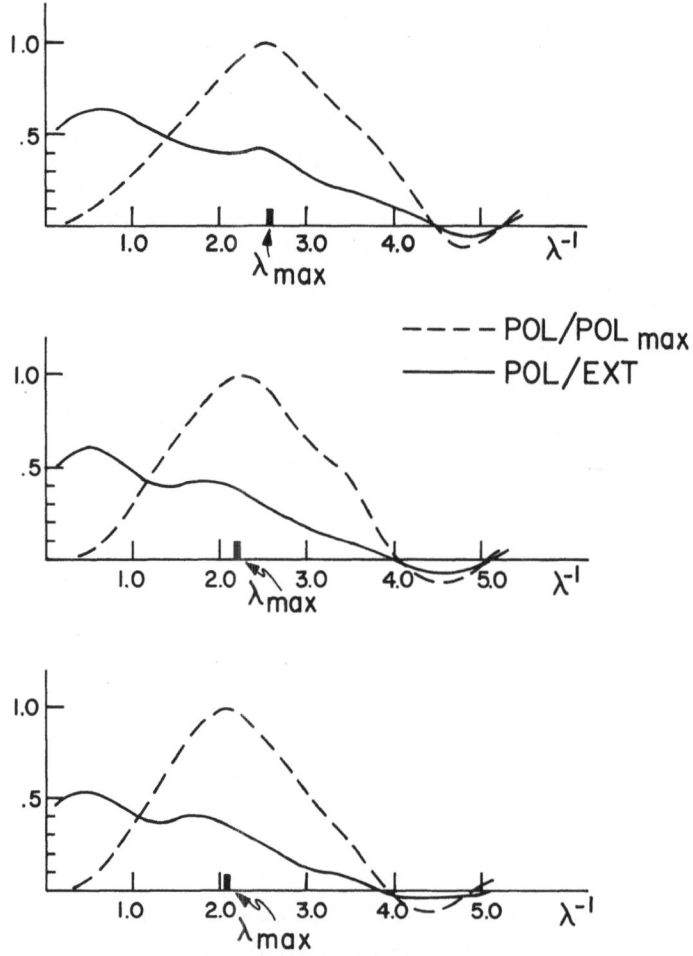

Fig. 6. Calculated wavelength dependence of polarization and ratio of polarization to extinction by spinning core-mantle cylinders. Upper figure corresponds to Figure 5a; middle corresponds to Figure 5b; bottom corresponds to Figure 5c.

(c) IR ABSORPTION ESTIMATES

One factor which has not hitherto been seriously considered in interpretation of infrared absorption bands is that these bands are produced in a combination of one material imbedded in the other rather than in isolated particles of different materials. A preliminary calculation has been made for several combinations of absorptive properties of core and mantle materials in order to determine how important this effect may be in modifying estimates of core and mantle masses based on isolated particle calculations.

Even when the mantle has no absorption at the position of the core absorption band there is some deviation – although not great; similarly when the roles are reversed. If, however, the mantle material has absorption in the vicinity of the core absorption, then the interpretation of the core mass (volume) based on the extinction will be significantly altered. Clearly an absorbing mantle hides the core. A proper calculation of this effect would have to be performed over the full extent of the absorption band in order to see not only the modification on the depth but also the shape of the band. Such calculations are in progress for selected core and mantle materials. However, as an indication of the magnitude of what to expect we present in Table V the volume of material inferred from the depth of an absorption band produced by a core-mantle particle compared with that which would be given by the *actual* volume of the core or mantle material if it occurred in isolated spherical

TABLE V

Ratios of estimated to actual volumes of core and mantle materials based on absorption band depths

Sphere α	I		II		III		IV	
	sil	ice	sil	ice	sil	ice	sil	ice
1.12	1.076	0.817	1.016	0.817	1.076	0.846	1.016	0.846
1.25	1.164	0.865	1.023	0.865	1.164	0.879	1.023	0.879
1.50	1.396	0.921	1.026	0.912	1.396	0.928	1.026	0.928
2.00	2.151	0.967	1.028	0.967	2.151	0.969	1.023	0.969
2.50	3.400	0.982	1.033	0.982	3.400	0.984	1.033	0.984
Cylinder								
1.12	1.057	0.998	1.013	0.998	1.057	1.051	1.013	1.051
1.25	1.121	0.992	1.026	0.992	1.121	1.017	1.025	1.017
1.50	1.271	0.996	1.054	0.996	1.271	1.008	1.054	1.008
2.00	1.679	0.997	1.140	0.997	1.679	1.002	1.140	1.002
2.50	2.264	0.990	1.296	0.990	2.264	0.992	1.296	0.992

	sil	ice		dark	clear
I	clear	dark	$m_{sil}(10)=1.52-$	$i\,0.68$	
II	clear	clear	$m_{ice}(3.1)=1.375-$	$i\,0.815$	
III	dark	dark	$m_{sil}(3.1)=1.6-$	$i\,0.01$ or	1.6
IV	dark	clear	$m_{ice}(10)=1.6-$	$i\,0.115$ or	1.6

Dark and clear refer to absorbing or non-absorbing representations of silicate and ice materials.

or cylindrical particles. It appears clear that if the mantle 'ice' were actual ice (solid H_2O) the silicate core 9.7-μm absorption would be almost entirely masked by the mantle absorption and any estimates of silicate mass based on it would be overestimated by 30% (cylinder) to 40% (sphere) even when the mantle is only half as thick as the core. Since we do not really know the absorption spectrum of the mantle composition, we should not overinterpret this result. It is to be remembered that it is so only as a possibility. Experiments are planned to measure the infrared absorption properties of ultraviolet irradiated dirty ice at low temperatures using experimental methods which have been reported earlier (Greenberg, 1973b).

VII. Variability of Grain Characteristics with Distribution

The core-mantle plus bare particle model for the grains leads to a number of iden-

tifiable changes which may be expected in the physical – and observational – characteristics of the grains as they go through various stages in their development. For
this paper we will introduce and briefly discuss the kinds of changes in the grains
which are particularly an outcome of this model. Detailed considerations are de-
ferred to a later paper.

(a) THE INTERARM REGION

Let us suppose, following the general scheme outlined by Greenberg (1970), that, as
the gas and accompanying grains pass through the outer edge of the density-wave
spiral arm, they are of 'normal' composition and size distribution, i.e., the relative
number of core-mantle and bare particles is as defined in Section IIId. We expect
sputtering of the mantles to begin to take place as a consequence of the He and H
atom bombardment (H atoms are less effective because of their mass) at temperatures greater than 1000 K.

If P_{sp} is the probability that the average colliding He atom causes a molecule of
molecular weight M_{mol} to be sputtered off, the rate of decrease of the radius of the
grain is

$$\dot{a} = -P_{sp} \frac{n_{He} v_{He} M_{mol} m_H}{4s}, \tag{16}$$

where n_{He} is the number density of He atoms, v_{He} is the average He atom velocity,
and s is the specific gravity of the grain mantle material. Unfortunately, we do not
know the sputtering rates with much reliability at the low energies corresponding to
temperatures of 1000–10000 K. Furthermore, the chemical composition of the mantle
is not precisely known except that it consists of *various* combinations of O, C, N,
and H. If these combinations are predominantly of large molecular weight, results
based on H_2O and CH_4 sputtering (Aannestad, 1973) would have to be revised.
Finally, surface characteristics undoubtedly play a very important role in the sputtering process.

Low-energy-sputtering yield rates for H and He on solid H_2O and CH_4 have been
given by Aannestad (1973). Greenberg (1973) used an average value of $\langle P_{sp} \rangle = 10^{-2}$
for H atom collisions in the 1000–10000 K range.

For $n_H = 0.2$ cm^{-3} ($n_{He} = 0.02$ cm^{-3}), $\bar{v}_{He} = (\bar{T}/100)^{1/2}(0.8) \times 10^5$ cm s^{-1}, $\bar{T} = 5000$ K,
$M_{mol} = 20$, $s = 1$, we get

$$\dot{a}/P_{sp} = -9.4 \times 10^{-20} \text{ cm s}^{-1} = 2.8 \times 10^{-6} \text{ cm } (10^6 \text{ yr})^{-1}.$$

Thus taking P_{sp} typically as 10^{-1} we find that a mantle thickness of 0.05 μm is sputtered off in 17×10^6 yr. In the time it takes the grain to arrive at the inner edge of
the next arm (10^8 yr) all mantles up to 0.28 μm thick (!) would be sputtered away.
This is substantially greater than the maximum in the mantle thickness of 0.13 μm
in the largest size distribution we have used to give a normal classical extinction for
the core-mantle model. Therefore, if our estimate of the sputtering rate is reasonable
– and this is yet to be firmly established – we expect that it is only silicate cores of
the large grains and the small bare particles which survive the hostile interarm region.

An obvious observational consequence is that the ratio of the far-ultraviolet extinction to the visual extinction should be much larger than normal because the visual extinction per large grain is everywhere (in wavelength) decreased while the far-ultraviolet extinction by the bare particles may be essentially constant. Thus estimates of the far-ultraviolet attenuation in the interarm region are greater than would be given by extrapolation from the E_{B-V} color excess.

(b) INNER EDGES OF ARMS

As the residual silicate cores arrive in the compression region of the inner edge of an arm they may be expected to accrete new mantles at a rate fast enough to provide a rather sharp discontinuity in the grain characteristics as seen in photographs of external galaxies (Lynds, 1970) and as discussed earlier (Greenberg, 1970). It should be pointed out here that a factor of $\frac{1}{4}$ was inadvertently omitted from Equation (6) of that paper, but the conclusions are the same.

(c) DARK CLOUDS

The dark clouds which are regions of very high dust and gas density contain substantial volumes of material which are well shielded from the interstellar radiation field. Consequently, we expect that the bare particles are not only at lower temperatures than normal (Greenberg, 1971) but more importantly their temperature fluctuations are greatly reduced in frequency. Under these conditions the bare particles become suitable nucleation points for accretion of the condensible gases. The total extinction increases but the relative amount of ultraviolet to visual extinction decreases. For example, the rate of accretion of mantles in a medium with gas density of $(O+C+N)$ of 10 – corresponding to a number density of H atoms (free or bound) of 10^4 – would be sufficient to produce 'ice' mantles of 0.05 μm in thickness in about 5×10^4 yr. This time is shorter than the collapse time of such a typical dark cloud. With such particles the saturation extinction would be at about $\lambda^{-1} = 5 \mu m^{-1}$ (where strong absorption sets in) rather than at $\lambda^{-1} = 20$ for the bare 0.01-μm particles. If there are substantial numbers of still smaller bare particles they might still have sufficient temperature fluctuations to prevent accretion. However, it is clear that even if we start with grains which normally give very high UV extinction, when they go to form dark clouds they must lose the ability to produce the sharp rise in extinction labeled '4' in Figure 1. It is important to realize here that if the 'bare' particles are adding mantles they start contributing to the classical extinction range '2' as additional small core-mantle particles while at the same time the already existing core-mantles are growing. Further investigation of this process (Greenberg and Hong, 1974b) leads to the conclusion that the bare particles must be inhibited from accreting even in very dense clouds.

(d) H II REGIONS

The process of star formation starting with condensation in dark clouds leads to effects on the grains inverse to those in dark clouds. When the star is turned on, and even before, in the not too distant regions of the solar nebula, the grains are elevated in temperature so that the mantles must evaporate. If the star is very hot then in the H II region the sputtering mechanism also becomes operative and the strong ultraviolet extinction component should reappear as the particles lose their mantles. However, as already mentioned the classical extinction should not be grossly modified. In general, dust associated with young H II regions will have depleted but still larger than normal mantles and relatively weak ultraviolet extinction.

VIII. Concluding Remarks

Cosmic abundance considerations lead very strongly to a silicate core-ice mantle model for the grains. The depletion of the atoms heavier than O, C, N observed by Copernicus is another indication of this type of grain composition. The far-ultraviolet extinction curves imply an additional population of dust grains which is much smaller than those which are generally considered as cores for the accretion of condensible atoms. A temperature-fluctuation mechanism is proposed to account for the distinction in accretion capabilities of these very small particles. A specific preliminary core-mantle cylinder model seems to offer a basis for establishing the connection with observations of grain variability. Detailed calculations are in progress in order to make more definitive predictions of the dependence of extinction and polarization through the entire observed spectral ranges on local conditions and the past history of the grains.

Acknowledgement

This work was supported in part by NASA grant #NGR-33-011-043.

References

Bless, R. C. and Savage, B. C.: 1972, *Astrophys. J.* **171**, 293.
Borgman, J.: 1973, *Space Sci. Rev.* **15**, 121.
Cameron, A. G. W.: 1973, preprint.
Cayrel, R. and Schatzman, E.: 1954, *Ann. Astrophys.* **17**, 555.
Gilra, D. P.: 1971, *Nature* **229**, 237.
Gillett, F. C. and Forrest, W. J.: 1973, *Astrophys. J.* **179**, 483.
Greenberg, J. M.: 1960, *J. Appl. Phys.* **31**, 82.
Greenberg, J. M.: 1968, *Stars and Stellar Systems* **7**, Ch. 6.
Greenberg, J. M.: 1969, *Physica* **41**, 67.
Greenberg, J. M.: 1971, *Astron. Astrophys.* **12**, 240.
Greenberg, J. M.: 1973a, *On the Origin of the Solar System* (Proc. of Symposium, Nice, France, April 1972, Centre National de la Recherche Scientifique, 1973), p. 135.
Greenberg, J. M.: 1973b, in M. A. Gordon and L. E. Snyder (eds.), *Molecules in the Galactic Environment*, John Wiley and Sons, p. 93.
Greenberg, J. M.: 1973c, 'Interstellar Dust and Related Topics', *IAU Symp.* **52**, 3.
Greenberg, J. M. and Hong, S. S.: 1974a, in A. G. W. Cameron and G. B. Field (eds.), *Proc. of the 'Dusty Universe Symp.'*, Cambridge, Mass. 1973, in press.
Greenberg, J. M. and Hong, S. S.: 1974b, *Proc. of 8th ESLAB Symp. on* H II *Regions and Galatic Center*, Frascati, Italy, 1974, in press.
Greenberg, J. M., Lind, A. C., Wang, R. T., and Libelo, L. F.: 1963a, *Electromagnetic Scattering* (Proc. of

Interdisciplinary Conf. held at Clerkson College of Technology, Potsdam, New York, August 1962), p. 123.

Greenberg, J. M., Libelo, L. F., Lind, A. C., and Wang, R. T.: 1963b, in E. C. Jordan (ed.), *Electromagnetic Theory and Antennas*, Macmillan Co., New York, Part. 1, p. 81.

Greenberg, J. M., Yencha, A. J., Corbett, J. W., and Frisch, H. L.: 1972, *Mem. Soc. Roy. Sci. Liège*, 6e series, 3.

Greenstein, J. L.: 1938, *Harvard Obs. Circ.*, No. 442.

Hall, J. S.: 1949, *Science* **109**, 166.

Hiltner, W. A.: 1949, *Science* **109**, 165.

Hoyle, F. and Wickramasinghe, N. C.: 1962, *Monthly Notices Roy. Astron. Soc.* **124**, 417.

Huffman, D. R. and Stapp, J. L.: 1971, *Nature Phys. Sci.* **229**, 45.

Hulst, H. C. van de: 1943, *Ned. Tijdschr. v. Natuurkunde* **10**, 251.

Hulst, H. C. van de: 1946, *Rech. Astron. Obs. Utrecht*, ii, Part 1.

Hulst, H. C. van de: 1957, *Light Scattering by Small Particles*, John Wiley and Sons, Inc., New York; Chapman and Hall, Ltd., London.

Kamijo, F.: 1963, *Publ. Astron. Soc. Japan* **15**, 440.

Kamijo, F. and Jong, T. de: 1973, *Astron. Astrophys.* **25**, 371.

Kemp, J. C.: 1973, in J. M. Greenberg and H. C. van de Hulst (eds.), 'Interstellar Dust and Related Topics', *IAU Symp.* **52**, 181.

Kittel, C.: 1956, *Solid State Physics*, John Wiley and Sons, Inc., New York, p. 130.

Lind, A. C. and Greenberg, J. M.: 1966, *J. Appl. Phys.* **37**, 3195.

Lindblad, B.: 1935, *Nature* **135**, 133.

Lynds, B. T.: 1970, in K. O. Kiepenheuer (ed.), 'The Spiral Structure of Our Galaxy', *IAU Symp.* **38**, 26.

Martin, P. G.: 1973, in J. M. Greenberg and H. C. van de Hulst (eds.), 'Interstellar Dust and Related Topics', *IAU Symp.* **52**, 161.

Martin, P. G., Illing, K., and Angel, J. R. P.: 1973, 'Interstellar Dust and Related Topics', *IAU Symp.* **52**, 169.

Morton, D. C., Drake, J. F., Jenkins, E. B., Rogerson, J. B., Spitzer, L., and York, D. G.: 1973, *Astrophys. J.* **181**, L103.

Oort, J. H. and Hulst, H. C. van de: 1946, *Bull. Astron. Inst. Neth.* **10**, 187.

Platt, J. R.: 1956, *Astrophys. J.* **193**, 486.

Rogerson, J. B., York, D. G., Drake, J. F., Jenkins, E. B., Morton, D. C., and Spitzer, L.: 1973, *Astrophys. J.* **181**, L110.

Schalén, C.: 1936, *Medd. Uppsala Astron. Obs.* **64**.

Shah, G. A.: 1970, *Monthly Notices Roy. Astron. Soc.* **148**, 93.

Whiteoak, J. B.: 1966, *Astrophys. J.* **144**, 305.

Woolf, N. J.: 1973, in J. M. Greenberg and H. C. van de Hulst (eds.), 'Interstellar Dust and Related Topics', *IAU Symp.* **52**, 485.

York, D. G., Drake, J. F., Jenkins, E. B., Morton, D. C., Rogerson, J. B., and Spitzer, L.: 1973, *Astrophys. J.* **182**, L1.

J. Mayo Greenberg
Seung-Soo Hong
State University of New York at Albany,
Astronomy and Space Science Department,
Albany, N.Y. 12222, U.S.A.

DISCUSSION

Wynn-Williams: The OAO data indicate a correlation between the slope of the extinction curve at about 1000 Å and the prominence of the 2200 Å peak. Does your model take any account of this?

Greenberg: Qualitatively the answer must be yes although we have not carried out any detailed calculations on this. Within the framework of our bi-modal model consisting of relatively large core-mantle particles and very small bare particles, it is the bare particles which would produce both the bump and the far UV extinction. Under those conditions in which the core mantle particles have larger than normal mantles the far UV extinction is obviously reduced along with the relative size of the 2200 Å bump.

THE GAS-TO-DUST RATIO IN THE GALAXY

F. J. KERR and G. R. KNAPP

University of Maryland, College Park, Md., U.S.A.

Abstract. We have investigated the gas-to-dust ratio in the Galaxy by comparing 21-cm H I column densities with the color excesses of globular clusters. We find a constant gas-to-reddening ratio in interstellar clouds and the intercloud medium. This ratio is also independent of galactic latitude.

The gas-to-dust ratio in our Galaxy can be investigated observationally by comparing the visual absorption and the neutral hydrogen column density along convenient lines of sight. This has been done for RR Lyrae color excesses and 21-cm H I column densities by Sturch (1969) and for OB star color excesses and H I Ly-α absorption by Savage and Jenkins (1972).

We have extended this method by measuring 21-cm profiles in the directions of 81 globular clusters for which color excesses are available [mainly from van den Bergh (1967)] with the NRAO* 140-ft telescope.

Globular clusters are halo objects and they can be considered to be outside the layer of interstellar matter, so that all the H I seen along the line of sight is in front of the cluster (except in a very few low-latitude cases). Also, the sampling geometries for the color excess and 21-cm column density data are almost identical, the globular clusters typically being comparable in angular size to the beamwidth of the telescope.

The H I profiles were corrected for any obvious self-absorption effects (although these only occurred for a few low-latitude clusters), and the integrated brightness $B_{H I}$ calculated from:

$$B_{H I} = 1.823 \times 10^{18} \int T_b(v) \, dv, \tag{1}$$

where $T_b(v)$ is the measured H I brightness temperature at velocity v. If the H I optical depth is small at all velocities, $B_{H I}$ is numerically equal to $N_{H I}$, the H I column density in atom cm^{-2}.

A plot of $B_{H I}$ vs E_{B-V} is shown in Figure 1. For low values of $B_{H I}$ and E_{B-V} ($B_{H I} \lesssim \lesssim 40 \times 10^{20}$, $E_{B-V} \lesssim 0^m.5$) the points lie close to the straight-line relationship:

$$B_{H I} = [(51.4 \pm 5.2) E_{B-V} - (0.1 \pm 1.9)] \times 10^{20} \text{ mag}^{-1}. \tag{2}$$

This relationship was calculated by least squares, with errors in both coordinates. The slope corresponds to a gas: dust ratio by mass of $\sim 100:1$ and a mean visual absorption coefficient for the interstellar medium of $\kappa_v \sim 300$ gm^{-1} cm^2 (see Knapp *et al.*, 1973).

Equation (1) shows that, to well within the uncertainties, the straight-line relation-

---

* The National Radio Astronomy Observatory is operated by Associated Universities Inc., under contract with the National Science Foundation.

F. J. Kerr and S. C. Simonson, III (eds.), Galactic Radio Astronomy, 179–182. All Rights Reserved.

ship goes through zero, i.e., the gas and dust are well mixed throughout the interstellar medium, and dust exists in the intercloud medium.

We examined our data for latitude effects, since the intrinsic luminosity of globular clusters allows us to observe at very low latitudes. Statistically, we found a very good correlation between $B_{HI} \sin b$ and $E_{B-V} \sin b$. The ratio E_{B-V}/B_{HI} averaged over each $10°$ of latitude is shown in Figure 2; it can be seen that any deviations from a con-

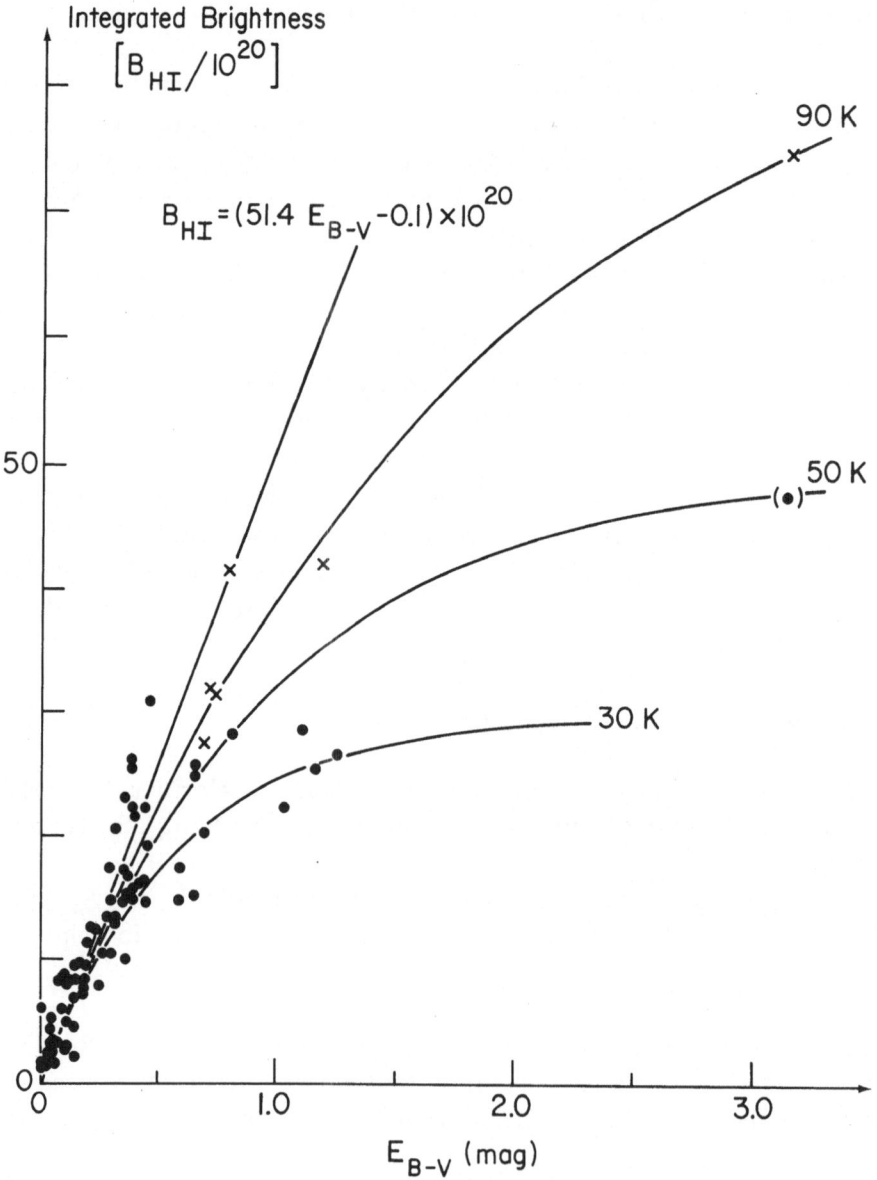

Fig. 1.   Plot of B_{HI} vs E_{B-V} for 81 globulars. The crosses correspond to points corrected for self-absorption. The uncorrected point for the highly reddened cluster Terzan 5 is included in parentheses. The straight line relationship is as defined in Equation (2); the curves represent the $B_{HI} - E_{B-V}$ relationship assuming a constant value for N_{HI}/E_{B-V} and various values of the harmonic mean spin temperature of the H I.

stant ratio are not significant. These results imply that the gas and dust have the same scale height.

The scatter of points with high values of E_{B-V} and B_{HI} to the high color excess side of the mean relationship shown in Figure 1 may be due mainly to a combination of saturation and molecule formation. In Figure 1 we also show curves corresponding

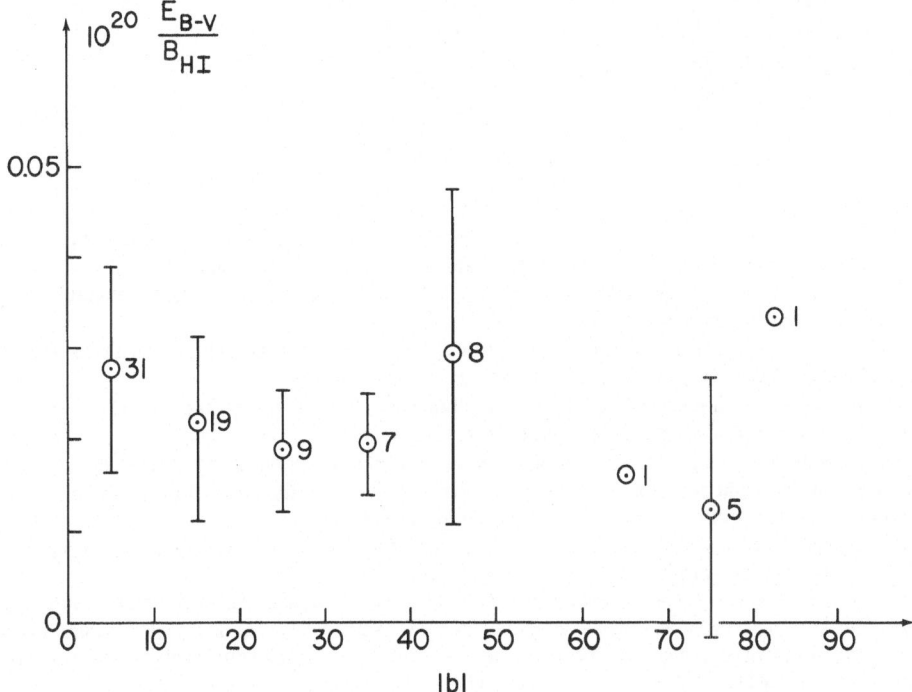

Fig. 2.   Plot of E_{B-V}/B_{HI} averaged over 10° latitude intervals. The number beside each point corresponds to the number of individual observations; the error bars represent the rms deviation of each sample.

to various HI spin temperatures and a gas-to-dust ratio as given by Equation (1), and it can be seen that saturation of the HI can easily explain much of the scatter. Points with particularly low values of B_{HI}/E_{B-V} were all found to correspond to clusters lying behind discrete dense dust clouds, so that some of the scatter is probably also due to the formation of hydrogen molecules.

In summary, we find that the gas and dust in the Galaxy are well mixed both in and out of clouds, and that the gas-to-dust ratio is independent of latitude. Apparent departures are due to a combination of saturation of the gas and the formation of H$_2$.

Acknowledgements

We thank the NRAO for the observing time for this project. This research is supported by the U.S. National Science Foundation.

References

Knapp, G. R., Kerr, F. J., and Rose, W. K.: 1973, *Astrophys. Letters* **14**, 187.
Savage, B. D. and Jenkins, E. B.: 1972, *Astrophys. J.* **172**, 491.
Sturch, C.: 1969, *Astron. J.* **74**, 82.
van den Bergh, S.: *Astron. J.* **72**, 70.

F. J. Kerr
G. R. Knapp
Astronomy Program, University of Maryland
College Park, Md. 20742, U.S.A.

DISCUSSION

Jenkins: One should not underestimate the importance of H_2 formation in causing a significant straggling of points below your correlation line when color excesses are large. In our most recent study of the gas and dust correlation from OAO-2 observations of Ly-α absorption, Savage and I found a significant improvement in the relation if we replaced $n(H_I)$ with $n(H_I) + 2n(H_2)$ for those stars for which H_2 column densities were available from Copernicus observations.

Kerr: The radio observations cannot easily distinguish between saturation and molecule formation in individual cases. Therefore, your ultraviolet observations are very important.

Oort: How were the color excesses of the clusters obtained? How did you get the normal colors?

Kerr: These were taken from van den Bergh's list, mainly.

Shuter: In your plot of E_{B-V} vs N_{H_I} is it not true that departures from linearity at large E_{B-V} occur for observations at low galactic latitudes or long galactic paths and are therefore more likely to be due to saturation in the H I profiles rather than by molecule formation.

Kerr: In general, yes, but there are cases at quite low galactic latitudes where E_{B-V} is low – for example, Baade's famous 'window' near the direction of the center.

Greenberg: How do you know for certain that H I and dust are in the same region along the line of sight?

Kerr: This is not necessary, as long as the globulars are outside the layer of interstellar material. In that case, it is only the integral that counts. There are however differences in solid angle between the radio beam and the cluster size, which may sometimes have an effect.

Radhakrishnan: You stated that you felt that gas and dust were well mixed in both cloud and intercloud regions. What is the evidence for dust in the intercloud medium?

Kerr: It is the fact that the straight line giving the mean relationship between the gas column density and the color excess passes through the zero point, implying that all gas has dust associated with it. Some 21-cm emission can be seen at all points in the sky, whereas the reported color excess is sometimes zero, but this is because the 21-cm measurement is more sensitive.

Baldwin: Is the variation in N_H from point to point mainly due to the influence of galactic latitude? If so, it is surely difficult to be sure that the scale height of the gas and dust are really the same.

Kerr: We examined the data carefully, and have shown that the ratio of gas column density to color excess is the same, irrespective of latitude.

ABSORPTION CHARACTERISTICS
OF DUST IN THE LYMAN CONTINUUM REGION

P. G. MEZGER, E. B. CHURCHWELL, and L. F. SMITH

Max-Planck-Institut für Radioastronomie, Bonn, F.R.G.

Abstract. The absorption cross-sections of dust grains in the Lyman continuum region are calculated based on integrated spectra of OB star clusters and observed values of the IR excess radiation from H II regions.

The IR excess radiation from H II regions, i.e., IR radiation in excess of that expected for heating by trapped Ly-α photons, is determined by both the absorption cross-sections of dust grains and the radiation spectrum of the associated star cluster shortward and longward of the Lyman continuum limit. We compute integral spectral parameters of OB-star clusters and apply these to the problem of dust heating in H II regions. Comparison of our results with observations indicates that the effective absorption cross-section of dust in the Lyman-continuum region is about 50% of the extinction cross-section in the visual. Absorption of Lyman-continuum photons by dust grains appears to be selective, with an effective absorption cross-section shortward of 504 Å about three times the effective absorption cross-section shortward of 912 Å.

Large values of IR excess radiation are to be expected for both dense and compact H II regions and for giant H II regions with large intrinsic stellar Lyman-continuum photon fluxes. At densities $n_e > 10^5$ to 10^6 cm^{-3}, H II regions will emit predominantly in the far IR as a result of opacity in the radio continuum and the fact that most of the Lyman-continuum photons are absorbed by dust. Values of $(IR)_{exc} \sim 4$ appear to be typical for galactic H II regions; this $(IR)_{exc}$ corresponds to optical absorption depths of < 1 shortward and longward of the Lyman-continuum limit.

We have given a fuller account in *Astronomy and Astrophysics* **32** (1974), p. 269.

P. G. Mezger
E. B. Churchwell
L. F. Smith
Max-Planck-Institut für Radioastronomie,
Auf dem Hügel 69,
53 Bonn 1, F.R.G.

DISCUSSION

Wynn-Williams: I think there are dangers in using Hoffman, Frederick and Emery's data to derive far infrared luminosities, since they used a comparatively narrow bandwidth with the result that their fluxes depend strongly on the dust temperature in the source. If you use Harper and Low's or Emerson, Jennings and Moorwood's results there is a very low dispersion in the ratio of radio to infrared fluxes among the more luminous H II regions.

F. J. Kerr and S. C. Simonson, III (eds.), Galactic Radio Astronomy, 183–184. All Rights Reserved.
Copyright © 1974 by the IAU.

Mezger: We multiplied the IR-flux densities given by Hoffman *et al.* by a factor of 4; this corresponds to a color temperature of the dust of about 70 to 80 K. To my knowledge there are no indications that the color temperatures of dust vary strongly from H II region to H II region. Most values of $(IR)_{exc} \simeq$ $\simeq (4\pi D^2 F_{IR}/N_c h v_a) - 1$ fall within the range 1 to 4; however, some H II regions, notably those in the galactic center, have very high values of $(IR)_{exc}$.

Zuckerman: I agree with Dr Mezger that when one looks carefully at H II regions one finds that the apparently good correlation between far infrared luminosity and radio continuum flux breaks down. For example the Kleinman-Low nebula, NGC 2264 and DR21 (OH), all strong far infrared emitters, produce essentially no radio continuum flux.

Lequex: Could Dr Greenberg comment about the ratio of color excess in this region?

Greenberg: I chose not to discuss the problem of the very high albedo in the ultraviolet. I think this is a very confusing question. The albedo for these particles should be of the order of 0.5, in the very, very far ultraviolet.

SPECTRUM OF GALACTIC BACKGROUND RADIATION

J. R. SHAKESHAFT

Mullard Radio Astronomy Observatory, Cambridge, United Kingdom

Abstract. New measurements of the galactic background radiation between 151 and 1407 MHz have been made by Sironi and by Webster. The radio spectrum shows a bend at ~ 315 MHz in Bridle's Region I and ~ 100 MHz in Region II, implying magnetic fields of 7.6 μG and 2.4 μG, respectively.

(The following summary was prepared from the tapes by the Editors.)

It is well known that studies of the galactic radio background spectrum provide evidence on the magnetic field and cosmic ray electrons in the Galaxy, and over the past 10–12 yr there have been various attempts to measure the radio spectrum more precisely. These have gone along with measurements of the cosmic-ray-electron spectrum to study the fit between the two spectra in terms of standard synchrotron theory.

The radio measurements must be made using scaled aerials, having the same beams at different wavelengths, to avoid side-lobe problems, for example the observations of Turtle, Bridle and Howell at various frequencies. The only exception is the work of Wielebinski and others using lunar occultations, the assumption being that the Moon is essentially the same size at all wavelengths.

This paper reports new measurements by Sironi (1974), using similar arrays at 151 and 408 MHz, and by Webster (1974) at 408, 610, and 1407 MHz with 15° horns. These beams are much larger than would be desirable, but the scaled-antenna method cannot give a very high resolution all the way from 17.5 to 1407 MHz.

The method of interpretation of measurements of this kind is usually in terms of 'a T–T plot,' in which the temperature of a given area at one frequency is plotted against that at another frequency, with the slope giving the temperature spectral index, β. In this way, any constant background components are eliminated, but it does mean that variations over an area are smoothed out to some extent.

In the time available, we will concentrate on the results for Bridle's 'Regions I and II,' which correspond essentially to the local spiral arm and an interarm region. Plotting temperatures at 408 and 610 MHz against each other for different right ascensions, we get reasonably good straight lines, but there are variations. One such distortion is associated with the North Polar Spur. The temperatures are higher for 408 than for 1407 MHz, implying that the Spur has a steeper spectral index between these two frequencies than does the average radiation.

Plotting the results at all available frequencies for the two regions, we obtain very smooth curves, but the two curves are not the same. We can now make use of the observed cosmic-ray electron spectrum for the differential number of electrons with given energy against the energy; this has a slope of -1.8 below an energy of about 3 GeV and -2.6 above 3 GeV, although there are some disagreements about details.

F. J. Kerr and S. C. Simonson, III (eds.), Galactic Radio Astronomy, 185–186. All Rights Reserved.

The observed radio spectrum shows a bend at around 315 MHz for Region I and 100 MHz for Region II. Applying the standard synchrotron formula, we can relate the frequency of the bend to the magnetic field and the critical energy, leading to a field of 7.6 μG for Region I and 2.4 μG for Region II.

References

Sironi, G.: 1974, *Monthly Notices Roy. Astron. Soc.* **166**, 345.
Webster, A. S.: 1974, *Monthly Notices Roy. Astron. Soc.* **166**, 355.

J. R. Shakeshaft
Mullard Radio Astronomy Observatory,
Cavendish Laboratory,
Free School Lane,
Cambridge CB2 3RQ, United Kingdom

DISCUSSION

Price: What is the latitude range of the regions considered in this study?
   Shakeshaft: They cover 40° or 50° in galactic latitude.

GALACTIC BACKGROUND RADIATION
AT 2 cm AND THERMAL ELECTRONS IN THE GALAXY

H. HIRABAYASHI

Tokyo Astronomical Observatory, Tokyo, Japan

Abstract. The galactic background radiation at 4.2 and 15.5 GHz was observed with a resolution of 11' at nine points on the galactic equator free of confusion from discrete sources. Relative amounts of thermal and nonthermal radiation were determined, and conditions in the interstellar gas are discussed.

Galactic background radiation was observed at 4.2 and 15.5 GHz with a resolution of 11' for 9 points on the galactic equator free of confusion from discrete sources. The brightness temperature obtained for the 'ridge' of the background radiation was typically 0.07 K at 15.5 GHz; the observations were made by on-off methods.

The results of these observations were compared with those at lower frequencies and the spectrum of the galactic background radiation between 1.4 and 15.5 GHz was determined. Though the spectrum in the Cygnus-X region was found to be mainly thermal, points in inter-arm and arm regions had positive curvature in the spectra showing a mixture of thermal and nonthermal emission. The turnover occurred near 3 GHz, and the spectral index for the nonthermal emission was found to be about -3.0 for these points.

The amount of thermal emission separated from nonthermal emission was compared with the strength of the recombination lines measured by Gordon and Gottesman (1971) and Jackson and Kerr (1971) to determine the kinetic temperature and emission measures of thermal electrons in the diffuse interstellar gas. They were about 3000 K and 4000 pc cm^{-6}, respectively. A value of about 0.5 cm^{-3} was derived for the rms electron density in the Galaxy, from the emission measure with an assumed pathlength.

A preliminary report has already appeared in *Nature* (Hirabayashi *et al.*, 1972), and full details will appear in *Publ. Astron. Soc. Japan*.

References

Gordon, M. A. and Gottesman, S. T.: 1971, *Astrophys. J.* **168**, 361.
Hirabayashi, H., Yokoi, H., and Morimoto, M.: 1972, *Nature Phys. Sci.* **237**, 54.
Jackson, P. D. and Kerr, F. J.: 1971, *Astrophys. J.* **168**, 29.

H. Hirabayashi
Tokyo Astronomical Observatory,
Mitaka, Tokyo, Japan

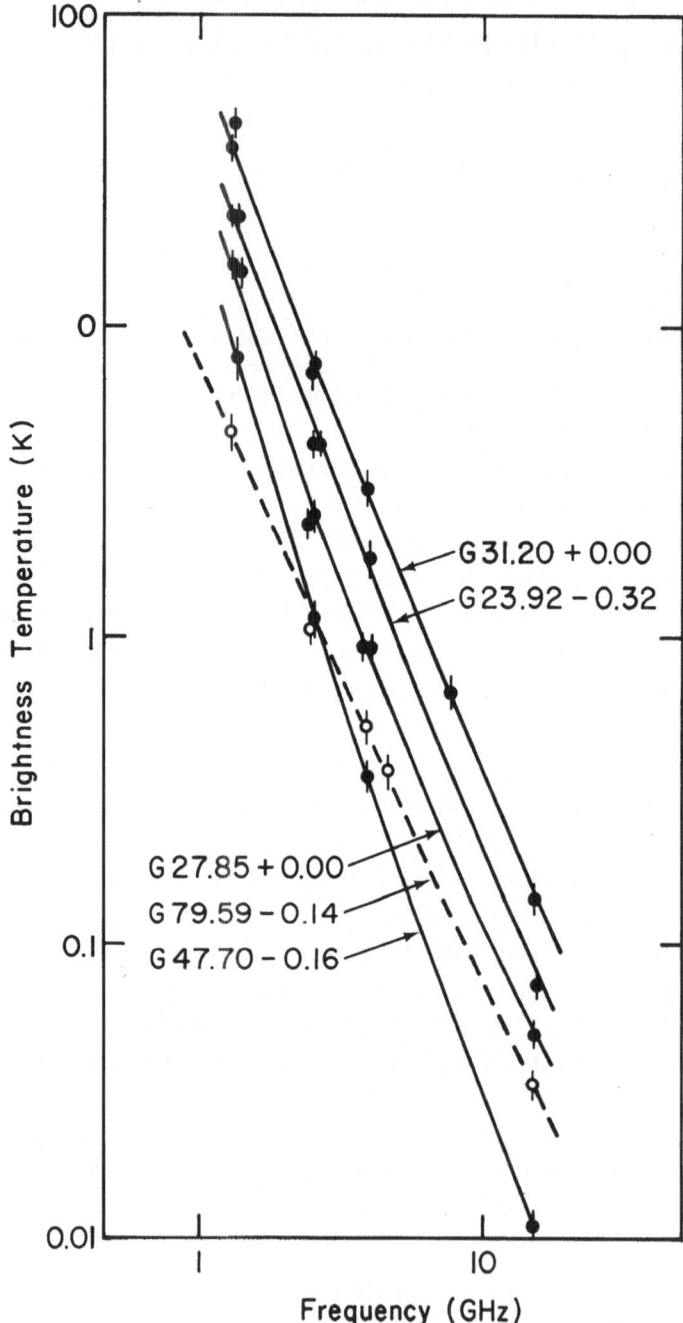

Fig. 1.   Spectra of the Galactic background in the frequency range of 1.4 to 15.5 GHz at 5 points on the galactic plane. The ordinate is for the brightness temperature and the abscissa is for the frequency. The points are designated by the new galactic longitude and lattitude. The bars represent the amount of uncertainties estimated by the author. For the points at 1.4, 2.7, 5.0 GHz, the results of NRAO and Parkes galactic plane survey were used. The broken line with open circles shows the spectrum of the point G79.59 − 0.14 and which is just the theoretical spectrum of the optically thin thermal radiation. For other points positive curvatures are seen.

DISCUSSION

Parijskij: We are preparing a radio map of the Milky Way visible from the Pulkovo Observatory at 4 cm with the large Pulkovo radio telescope. Usually we find the typical nonthermal spectrum of the galactic background. In some places the brightness temperature is extremely low, with an upper limit to the rms electron density of $\ll 0.1$ cm^{-3} (with $T_e \sim 10^4$ K). There is also a very deep minimum not far from the galactic center, symmetrical to that found in the Parkes survey at 6 cm.

Hirabayashi: I have examined the lower frequency results down to 19.7 MHz (Shain *et al.*, 1961) and derived the f-f optical depth from the absorption of the galactic background radiation by thermal electrons. The optical depth at 19.7 MHz was larger than that derived from my results by factor of 3. The difference may be due to different spatial resolution for the higher and lower frequency observation. Anyhow, these observational results both at the highest and lowest ends seem to prove the existence of this much thermal electrons in the galactic plane.

Reference

Shain, C. A., Komesaroff, M. M., and Higgins, C. S.: 1961, *Australian J. Phys.* **14**, 508.

THERMAL AND MASER PHENOMENA
IN ROTATIONALLY EXCITED OH

PATRICK PALMER

University of Chicago, Chicago, Ill., U.S.A.

Abstract. Both maser emission and thermal emission and absorption have been observed from the rotationally excited $^2\pi_{3/2}, J=5/2$ and $^2\pi_{1/2}, J=1/2$ states of OH. Conditions in the interstellar gas are discussed.

My colleagues (B. Zuckerman, Univ. of Maryland and Univ. of California at Berkeley, and L. J. Rickard, Univ. of Chicago) and I have been studying radiation from the rotationally excited $^2\pi_{3/2}$, $J=5/2$ and $^2\pi_{1/2}$, $J=1/2$ states of OH. Both maser emission and thermal emission and absorption have been observed. Maser emission has been found in several new sources, including NGC 7538, M17, and OH69-1, and time variations in the previously known sources have been monitored. In the later study, variations with as rapid an *e*-folding time as 8 h have been seen.

The thermal emission feature in the $^2\pi_{1/2}$, $J=1/2$ state in Sgr B2 (discovered by Gardner and Ribes) was re-observed and the corresponding features in the $^2\pi_{3/2}$, $J=5/2$ state were detected. In W3, however, absorption was found in the $^2\pi_{3/2}, J=5/2$ state. This is the shortest-lived state from which absorption is seen in the interstellar medium at any wavelength, radio or optical. The short lifetime of the state suggests very high densities for its excitation. Such densities in an extended source have previously been suggested only for the densest of the mm-wavelength like sources.

Full details of these studies will appear in the *Astrophysical Journal*.

Patrick Palmer
Department of Astronomy and Astrophysics,
Ryerson Laboratory 162,
1100-14 East 58th Street,
Chicago, Ill. 60637, U.S.A.

DISCUSSION

Wynn-Williams: New Caltech measurements by Werner, Wilson and myself of the position of the 1720 MHz OH maser in W3 (continuum) indicate that it does not coincide with either an H_2O, an IR or radio component, and therefore may not be so easily related to the 5-cm OH absorptions.

Townes: What is the linear size and the total mass of this dense region?

Palmer: Placing it at the edge of the W3 complex, its linear size is 0.3 pc. At a density of 10^6 cm^{-3}, the mass is a few hundred solar masses.

F. J. Kerr and S. C. Simonson, III (eds.), Galactic Radio Astronomy, 191. All Rights Reserved.
Copyright © 1974 by the IAU.

PART 2

GALACTIC H II REGIONS

"Those southern sources!"
"You'll have to take our word for it that they exist."

B. Zuckerman and B. J. Robinson, in the discussion
following the paper by B. J. Robinson

RECENT RADIO OBSERVATIONS OF GALACTIC H II REGIONS

E. CHURCHWELL

Max-Planck-Institut für Radioastronomie, Bonn, F.R.G.

Abstract. The structure and physics of H II regions are reviewed. The discussion is principally based on the radio mapping of intermediate-size H II regions to measure the energy balance, the mapping of small-scale structure by aperture-synthesis techniques to define the ionization conditions, and determination of helium abundances from radio recombination-line observations of H and He in H II regions.

I. Introduction

An H II region is an almost fully ionized cloud of gas (and dust) which is ionized, in most cases, by the ultraviolet radiation of a hot star(s) embedded in the cloud. H II regions are of astrophysical importance because: they offer a possibility to study the physics of low-density hot plasmas; they imply the radiation properties of O-stars; they provide conditions in the interstellar gas which make the determination of relative abundances possible; they are good tracers of spiral structure in galaxies; they are indicators of recent and in some cases still occurring star formation; and they are an important energy source for the interstellar gas.

Our understanding of the structure and physics of H II regions has improved significantly in the past five years. At radio wavelengths, this is the result of: (1) mapping of intermediate-size H II regions of low surface brightness with beamwidths 2–11′; (2) mapping by the technique of aperture synthesis of small-scale structure which is apparently embedded in an extended, lower-density plasma; and (3) spectroscopic observations of atoms and molecules toward H II regions. This paper briefly reviews the first two areas, as well as radio recombination-line observations of H and He in H II regions.

II. Radio Observations of Optically Visible H II Regions with Intermediate Resolution

Since 1970 several large-scale mapping programs of galactic sources have been published. The most extensive of these are: (1) the Altenhoff *et al.* (1970) survey of the galactic plane in the interval $l = 335°$ to $75°$ and $b = -4°$ to $+4°$ at 1.4, 2.7 and 5.0 GHz, with a resolution of 11 arcmin (356 sources were tabulated); (2) the Goss and Shaver (1970) and Shaver and Goss (1970) 5.0- and 0.4-GHz surveys of 206 southern galactic objects with a resolution of $\sim 4′$; (3) the survey by Wendker (1970) at 2.7 GHz of ~ 90 sources in the Cygnus-X region; the Parkes 2.7-GHz survey of the southern galactic plane in the interval $l = 190°$ through 360° to 61° and $b = \pm 2°$, which was carried out in part by Beard (1966), Thomas and Day (1969a, b), Day *et al.* (1969), Beard *et al.* (1969), Beard and Kerr (1969), Goss and Day (1970), Day *et al.* (1970), and Day *et al.* (1972); and (5) the survey of 137 optically visible H II regions by

F. J. Kerr and S. C. Simonson, III (eds.), Galactic Radio Astronomy, 195–218. All Rights Reserved.

Felli and Churchwell (1972) at 1.4 GHz with a resolution of 10'. The last survey has shown that most optically visible nebulae have associated radio continuum emission.

In this section I will review some results from radio observations of optically visible H II regions. The following statistics are based on a catalogue of exciting stars and radio data which is presently being prepared by P. Angerhofer, M. Walmsley and myself.

(a) STATISTICAL RESULTS

The nebulae listed by Sharpless (1959) comprise a wide variety of objects, few of which are typical of bright radio H II regions. Many are more extended and have much lower surface brightnesses than bright radio H II regions and therefore may represent a later stage of development. Others in this list may be representative of regions in which star formation is occurring under quite different conditions than that in giant radio H II regions. Hence conclusions based on this sample may not apply to bright radio H II regions.

We eliminate from consideration all nebulae where the identification of the optical with the radio emission is doubtful, and several large optical nebulae which contain numerous physically separate H II regions (e.g., S 108 and S 109 together encompass the whole Cygnus-X radio complex). This leaves a sample of 293 nebulae out of the 313 listed by Sharpless (1959).

TABLE I

Some statistical results for sources in the Sharpless (1959) catalog

1. Nebulae with one or more proposed exciting stars:	$\sim 34\%$	6. Nebulae searched for but not detected at radio frequencies:	$\sim 19\%$
2. Radio flux density measured at least at one frequency:	$\sim 56\%$	Of these,	
3. Nebulae with determinate radio spectral index:	$\sim 32\%$	a) exciting stars proposed for:	26%
Of these,		b) optical brightness:	
(a) thermal spectra:	$65 \pm 10\%$	bright:	10%
(b) nonthermal:	$18 \pm 2\%$	intermediate:	37%
(c) extragalactic:	$1.5 \pm .5\%$	faint:	53%
(d) proposed exciting stars for:	47%	c) Optical size:	
4. Nebulae in the region of sky surveyed by Hoffmann et al. (1971) with measured 100 μm emission	$16 \pm 2\%$	$\theta_s > 30'$	24%
		$10' < \theta_s < 30'$	21%
5. Nebulae observed at two or more frequencies but with uncertain spectral indices:	$\sim 11\%$	$\theta_s < 10'$	55%
		7. Nebulae, with proposed exciting stars, that have been *searched for* and *detected* at radio frequencies:	91%

From Table I several conclusions can be drawn. First there is a striking paucity of both radio and stellar data for these nebulae. Reliable stellar and radio data exists for only $\sim 30\%$ of all optically visible nebulae. Hα or Hβ surface brightnesses are available for even fewer nebulae. A large percentage of these sources have nonthermal radio spectra. At least three are apparently extragalactic objects: S 197 (= Maffei II), S 191 (= Maffei I), and S 172. From entry 6 we see that small optically faint H II regions are less likely to be detected at radio frequencies than other types. Finally we note that practically all nebulae with observable exciting stars, which have been searched for at radio frequencies, have been detected.

(b) SELECTED NEBULAE

We have selected from our catalogue all nebulae which satisfy the following criteria:

(i) the optical and radio positions agree to $\lesssim 10\%$ of the optical size or to $\lesssim 2'$ if the optical size is $< 20'$;

(ii) the optical and radio sizes and shapes basically agree.

These two criteria imply that the identification of the optical emission with the radio emission is correct and that the optical extinction is not large. Hence exciting stars in such nebulae should be observable, unless they are 'cocoon' stars.

Sixty-four nebulae, or 22% of our sample, satisfy these criteria. Twenty-six of these have fairly certain exciting star identifications and thermal radio spectra. Thirty-five have no known exciting star but have apparently thermal radio spectra. Three have nonthermal spectra.

An example of the set of nebulae with thermal spectra and an identified exciting star is S 125 (IC 5146). An 11-cm map with a resolution of $\sim 5'$ made with the MPIfR 100-m radio telescope is shown overlayed on the blue print of the Palomar Sky Survey in Figure 1. A B0 v star is located at the center of the nebula and the radio and optical emission agree in position and extent.

Assuming the distance derived from the ionizing stars, we have determined the excitation parameter (U_{obs}) from the radio flux densities of these 26 nebulae. The theoretical excitation parameter as a function of spectral type and luminosity class has been derived from stellar atmosphere models of Morton (1969) and Auer and Mihalas (1972) by Churchwell and Walmsley (1973). These are plotted as solid curves in Figure 2 and labeled as ZAMS, V, III and I for the zero-age main sequence, dwarfs, giants and supergiant stars respectively. The derived radio excitation parameter is plotted in Figure 2 with symbols indicating the spectral type and luminosity class of the proposed exciting star.

For this sample of nebulae, Figure 2 suggests that H II regions ionized by O8 or hotter stars are generally density-bounded, whereas those ionized by O9 or cooler stars are generally ionization-bounded. If all H II regions ionized by early O stars are density-bounded, the Lyman continuum (L_c) photon flux which escapes H II regions could make a substantial contribution to the ionization of the interstellar gas. Whether significant L_c-photon fluxes escape depends on what fraction of the stellar L_c-photon flux is absorbed by dust in H II regions. Table II lists three apparently density-bounded H II regions for which the ionizing stars are known and 100-μm and 5-GHz flux densities have been measured. The data in column 3 are from Schraml and Mezger (1969), those in columns 4, 6 and 7 were taken from Churchwell et al. (1974). The 100-μm fluxes were taken from Hoffmann et al. (1971). For these three

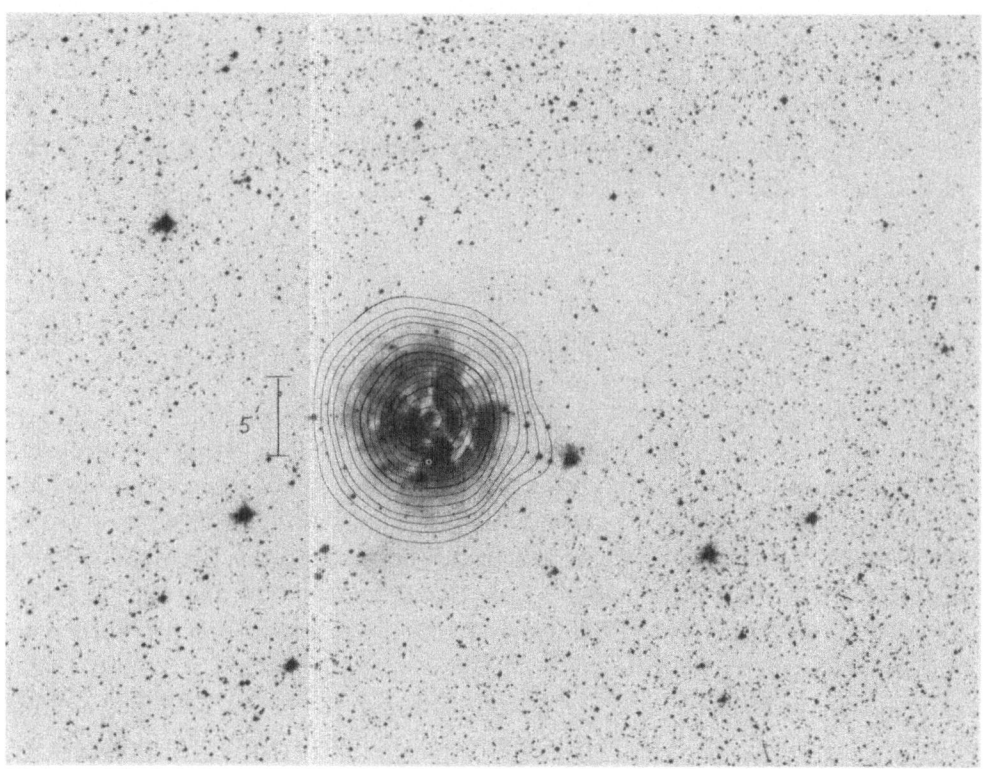

Fig. 1.   Overlay on the blue print of the Palomar sky survey of an 11-cm map of S 125 (IC 5146) (resolution ∼ 5′) by Walmsley, Schwartz and Churchwell. (Copyright National Geographic Society – Palomar
Observatory Sky Survey. Reproduced by permission from the Hale Observatories.

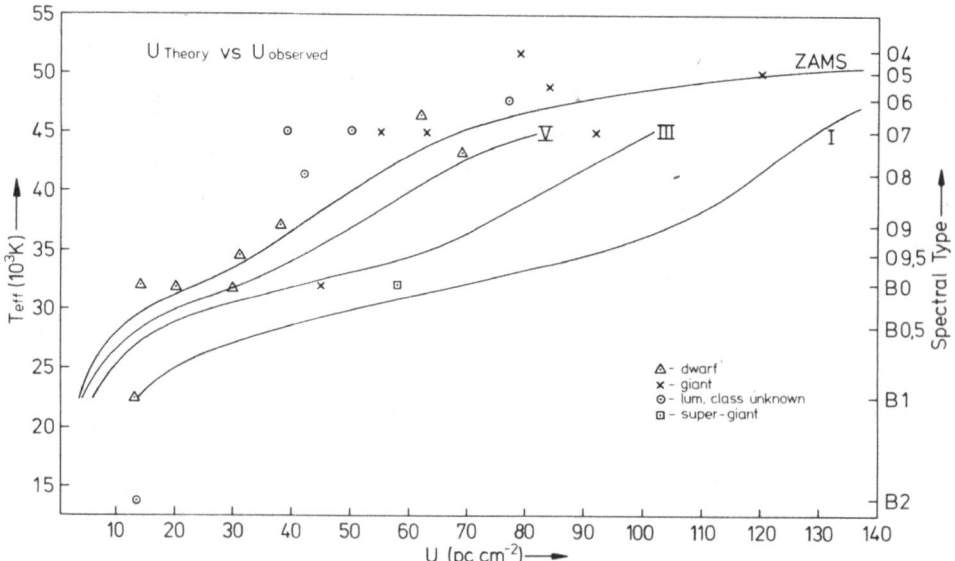

Fig. 2.   Comparison of the excitation parameters of several nebulae, as determined from radio-con-
tinuum observations, with that of the ionizing star as calculated by Churchwell and Walmsley (1973).

TABLE II

Energy balance in several H II regions

Source	Proposed ionizing star	n_e (cm^{-3})	N_c (star) $\times 10^{48}$ (s^{-1})	100 μm flux $\times 10^4$ (Jy)	$S_{5\ GHz}$ (Jy)	N_c (gas) $\times 10^{48}$ (s^{-1})	N_c (dust) $\times 10^{48}$ (s^{-1})	% L_c-photons escaping
Orion A	O6	2237	24	35	382	9.0	7.0	~33
M8	O7	391	9.7	3.4–3.7	85–192[a]	7.0	0.8	~20
NGC 6357 (353.2+0.9)	O7	774	9.7	5.4	87	7.8	1.9	~0

[a] Lower $S_{5\ GHz}$ value for M8 is the value taken from Reifenstein *et al.* (1970), and $S_{5\ GHz} = 192$ is from Goss and Shaver (1970).

nebulae it is possible to determine (with some reasonable assumptions) what fraction of the L_c-photon flux is absorbed by gas and dust, and what fraction escapes. N_c (dust) was calculated using the stellar properties given by Mezger *et al.* (1974) and the data in column 5.

In order to have N_c (star) = N_c (gas) + N_c (dust) in Orion A one would have to double $S_{5\ GHz}$ or triple the 100-μm flux. This is far outside the uncertainties of both the radio and IR data. It has been implicitly assumed that the 100-μm flux originates from heated dust that is well mixed with the ionized gas which gives rise to the radio source and that the 100-μm flux represents ~25% of the total IR flux, which would be the case if the dust grains have a temperature 80–100 K and radiate like black-bodies. Thus relatively large uncertainties enter the calculation of N_c (dust). Also N_c (star) values are probably not accurate to better than a factor of two; therefore an inequality in the above relation of ~30% is not meaningful. On the other hand the sense of the inequality in Table II (i.e., density-boundedness) appears to be supported by other evidence. The interstellar radio recombination lines observed by Gordon and Cato (1972) seem to be confined to regions in the Galaxy where OB stars are concentrated; particularly, in tangential directions to known spiral arms. Observations of Hα and Hβ emission by Carranza *et al.* (1968), Monnet (1971), and Benvenuto *et al.* (1973) have shown that much of the gas outside well defined discrete H II regions in the spiral arms of M33 is ionized. Courtès (1960) and Cruvellier (1967) have observed the same phenomenon in our own Galaxy. Finally, the Hα and/or Hβ observations of Dachler *et al.* (1968), Reay and Ring (1969), Johnson (1971), and Reynolds *et al.* (1973) have shown that in the solar neighborhood ionized gas not associated with known H II regions and with a low emission measure is seen over large solid angles.

To get a rough estimate of the effect on the interstellar radiation field by density-bounded H II regions let us suppose that 30% of the L_c-photon flux from all O6 stars escapes into the surrounding interstellar gas. The space density of O6 stars in the solar neighborhood is ~5×10^{-8} O6 stars pc^{-3} (Allen, 1963), which gives a mean distance between O6 stars of ~340 pc. The excitation parameter of O6 stars is ~90 pc cm^{-2} (Churchwell and Walsmley, 1973); therefore the effective excitation parameter for the interstellar gas is $(0.3)^{0.33} \times 90 = 60$ pc cm^{-2}. Hence the interstellar medium would be fully ionized for an electron density of $n_e = [60$ pc cm$^{-2}/170$ pc$]^{3/2} \simeq$ $\simeq 0.2$ cm^{-3} (neglecting absorption by dust). The mean photon density in the galactic plane from O6 stars alone would be $\sim 2 \times 10^{-4}$ cm^{-3}. The mean energy per L_c-photon from O6 stars is ~ 24 eV (Mezger et al., 1974 thus the average kinetic energy transmitted to the interstellar gas per ionization would be ~ 10 eV.

A further point related to this consideration is the question of the relative and cumulative contribution by each spectral type to the total L_c-photon flux of a statistically typical O-star cluster. This has been computed by Mezger et al. (1974) and is shown in Figure 3. The non-LTE atmosphere models of Auer and Mihalas (1972) and Salpeter's (1955) 'original' luminosity function were used for this calculation.

O6 stars apparently make the major contribution to the L_c-flux of an O-star cluster and early O stars dominate the integrated cluster luminosity.

A consequence of these considerations is that, if H II regions ionized by O6 stars or earlier are density-bounded, even by a small amount, they will make a major contribution to the interstellar radiation field and should be surrounded by a very extended, low density ionized region.

III. Small-Scale Structure in H II Regions

(a) OVERVIEW

More than 30 galactic H II regions have now been synthesized at Cal. Tech., Cambridge, NRAO and Westerbork with effective resolutions ranging from $\sim 90''$ (at 0.4 GHz) to $\sim 2''$ (at 5 GHz). Almost every H II region so far synthesized has small, dense components (exceptions are IC 1318b and c, G78.1 + 0.6 and G79.3 + 1.3 [Baars and Wendker, 1974]), which generally represent only a small fraction of the total radio emission from the H II region. The component sizes so far observed range from that of the whole Orion nebula ($\lesssim 0.5$ pc) in Sgr B2 (Martin and Downes, 1972) to that expected for single protostar clouds ($\lesssim 0.1$ pc) in Orion A itself. Similarly, the electron densities and emission measures range respectively from 10^3–10^5 cm^{-3} and 10^6–10^8 pc cm^{-6}.

Harper and Low (1971) have shown that H II regions emit strongly at IR wavelengths. It is generally believed that this is due to radiation by heated dust grains which are well mixed with the ionized gas. Maps of IR emission from several H II regions (e.g., W3 in the interval 1.65–20 μm by Wynn-Williams et al., 1972; M17 at 21 μm by Lemke and Low, 1972; and Orion A at 21 μm by Lemke, private communication) show close agreement with comparable resolution radio maps. Sources of OH and H_2O emission lie close to compact H II regions (Raimond and Eliasson, 1969; Robinson et al., 1970; Hardebeck, 1971; Hills et al., 1972).

It is not possible to discuss the peculiarities of each synthesized H II region in this paper. Instead, I have chosen to consider in some detail the W3 main source complex and to use it as an illustrative example of several points which may pertain to small-

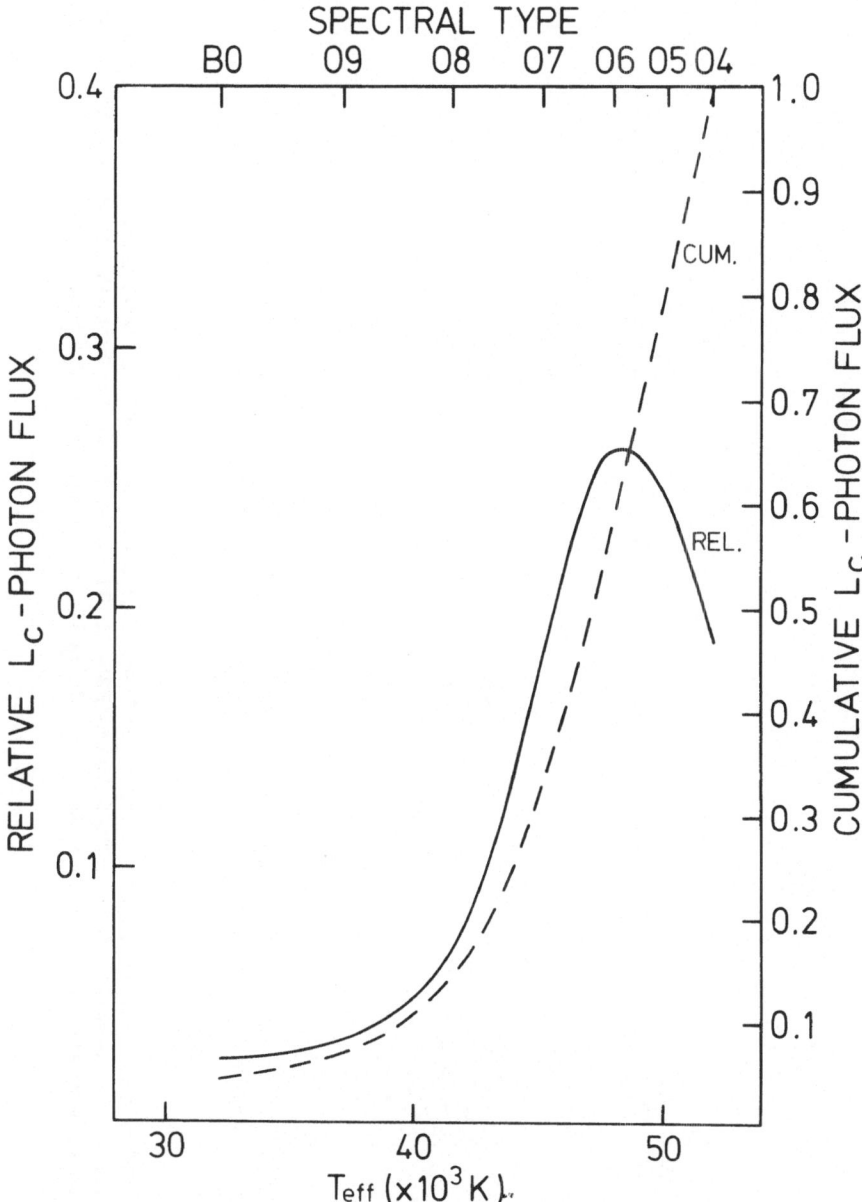

Fig. 3. The relative (solid curve) and cumulative (dashed curve) contributions to the integrated-Lyman-continuum photon flux of a statistically typical O star cluster as calculated by Mezger *et al.* (1974).

scale structure in H II regions in general. W3 is a good source for this purpose for several reasons: (1) its distance is ~ 3 kpc so that at the best resolution so far attained ($\sim 2''$) components of ~ 0.03 pc can be resolved; (2) its high declination permits uncomplicated synthesized beam shapes; (3) it has been observed in the IR, in radio recombination and molecular lines, and it has been synthesized at at least five frequencies; and (4) it has several kinds of compact components.

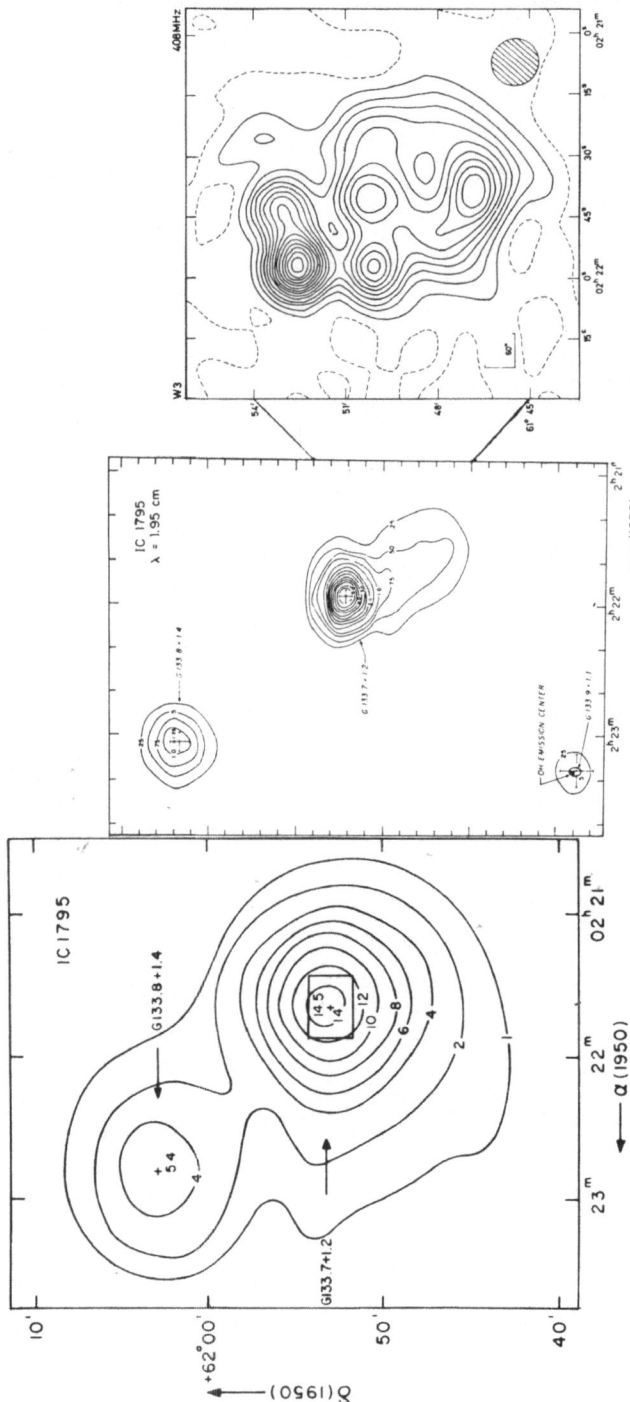

Fig. 4. *Left* – The 5.0-GHz (6-cm) map of W3 by Mezger and Henderson (1967) with a HPBW ~6'. The rectangle on G133.7 + 1.2 indicates the synthesized region in Figure 5. *Middle* – The 15 GHz (1.95-cm) map of W3 by Schraml and Mezger (1969) with a HPBW ~2'. The scale is the same as that on the left. *Right* – The 0.4-GHz (73.5-cm) synthesis map of the southern extension of G133.7 + 1.2 seen at 15 GHz by Wynn-Williams (1971) with a resolution of ~90".

(b) OBSERVATIONS OF W3

Figures 4 to 6 contain some of the radio data on W3. Figure 4 shows the Mezger and Henderson (1967) 5-GHz map (HPBW $\sim 6'$) and at the same scale the Schraml and Mezger (1969) 15-GHz map (HPBW $\sim 2'$).

The rectangle on the 5-GHz map indicates approximately the synthesized area at 2.7, 5.0 and 8.1 GHz. The 0.4-GHz map by Wynn-Williams (1971) resolves the southern extension seen at 15 GHz into at least 3 and perhaps 4 separate components. Figure 5 shows synthesis maps at 2.7 and 8.1 GHz by Wink (1973) with effective resolutions of $6'' \times 7''$ and $3'' \times 3''$ respectively, at 5 GHz by Wynn-Williams (1971) with a resolution of $6.5'' \times 7.4''$, and a synthesis map of H_2CO absorption at 4.83 GHz in front of W3 A and B by Fomalont and Weliachew (1972) with a resolution of $40'' \times 60''$.

The high-resolution continuum synthesis maps all show the four components designated A, B, C and D by Wynn-Williams (1971). The strongest component, W3 A, is apparently a ring or shell with some local density fluctuations. At the highest resolution shown ($3''$), component B appears to have even smaller structure; Wink (1973) had to use five gaussian components to fit the observed brightness distribution.

A recent synthesis map at 1.4 GHz has been made by Sullivan and Downes (1973) with a resolution of $25'' \times 28''$ and is shown in Figure 6.

The maps by Wynn-Williams *et al.* (1972) at 2.2 and 20 μm (diaphragm $\sim 10''$) are shown in Figure 7. The 1720-MHz OH emission source (Wynn-Williams *et al.*, 1973) has been added for comparison with the H_2O, IR and radio continuum positions.

Several points of interest are: (1) components A and B have IR counterparts at both 2.2 and 20 μm; (2) component C has only a 20 μm counterpart; (3) IRS2 coincides with the hole in component A and is thought to be the highly reddened radiation from the ionizing star of component A; (4) IRS5, IRS6, and IRS7 have no radio counterparts, although H_2O emission appears to be coincident with IRS5; and (5) neither radio continuum nor IR radiation comes from the 1720-MHz OH source.

(c) IONIZATION OF COMPACT COMPONENTS

Are the compact components in H II regions ionized from outside or from inside? Are they simply density fluctuations in the diffuse ionized gas or are they newly formed, hot stars surrounded by the ionized remnant of the cloud out of which they formed? Obviously the answer to this question is critical to our understanding of the nature of small-scale structure in H II regions.

Mezger and Henderson (1967) give for the flux density and size of G133.7 + 1.2 (the main component of W3 with spatial resolutions $\gtrsim 2'$) at 5 GHz, $S_{5\,GHz} = 76.5$ Jy and $\theta_s \simeq 4.'3$. This implies an excitation parameter of 124 pc cm^{-2} and a stellar L_c-photon flux of $N_c \gtrsim 7.3 \times 10^{49}$ s^{-1}, which corresponds to a star of spectral type O5 (ZAMS) or earlier (Churchwell and Walmsley, 1973). To see whether a single O5 star can account for the ionization of components A, B, C, and D in W3 three cases are considered:

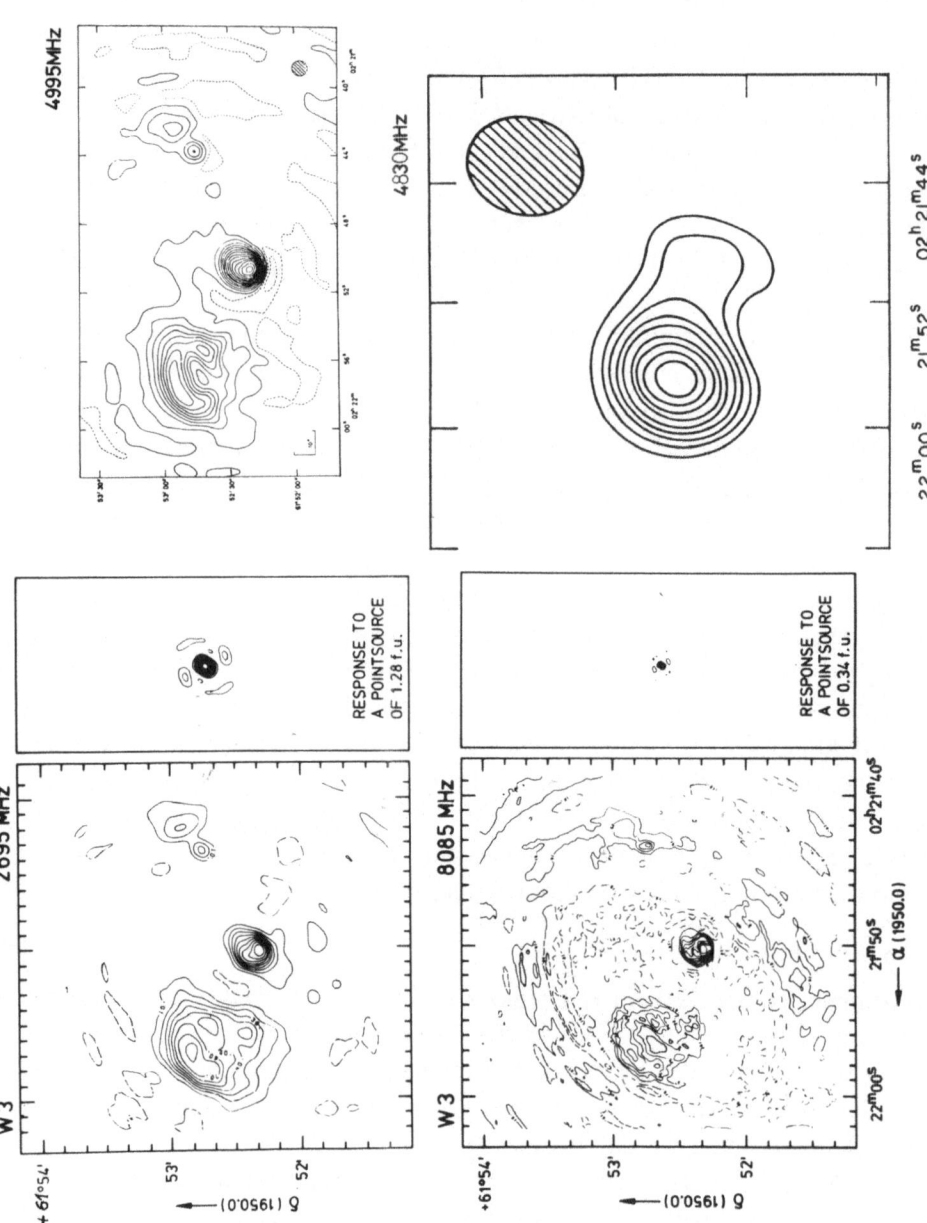

Fig. 5. Aperture synthesis maps of W3. *Upper left* – 2.7-GHz map by Wink (1973) with a resolution of 3" × 3". *Upper right* – 5.0-GHz map by Wynn-Williams (1971) with a resolution of 6.5" × 7.4". *Lower right* – map of H₂CO absorption at 4.83 GHz by Fomalont and Weliachew (1972) with a resolution of 40" × 60".

(1) an O5 (ZAMS) star at the center of W3 *A* (ring or shell structure);

(2) an O5 (ZAMS) star at the center of component *B*;

(3) an O5 (ZAMS) star at the geometrical center between components *A*, *B*, *C*, and *D*.

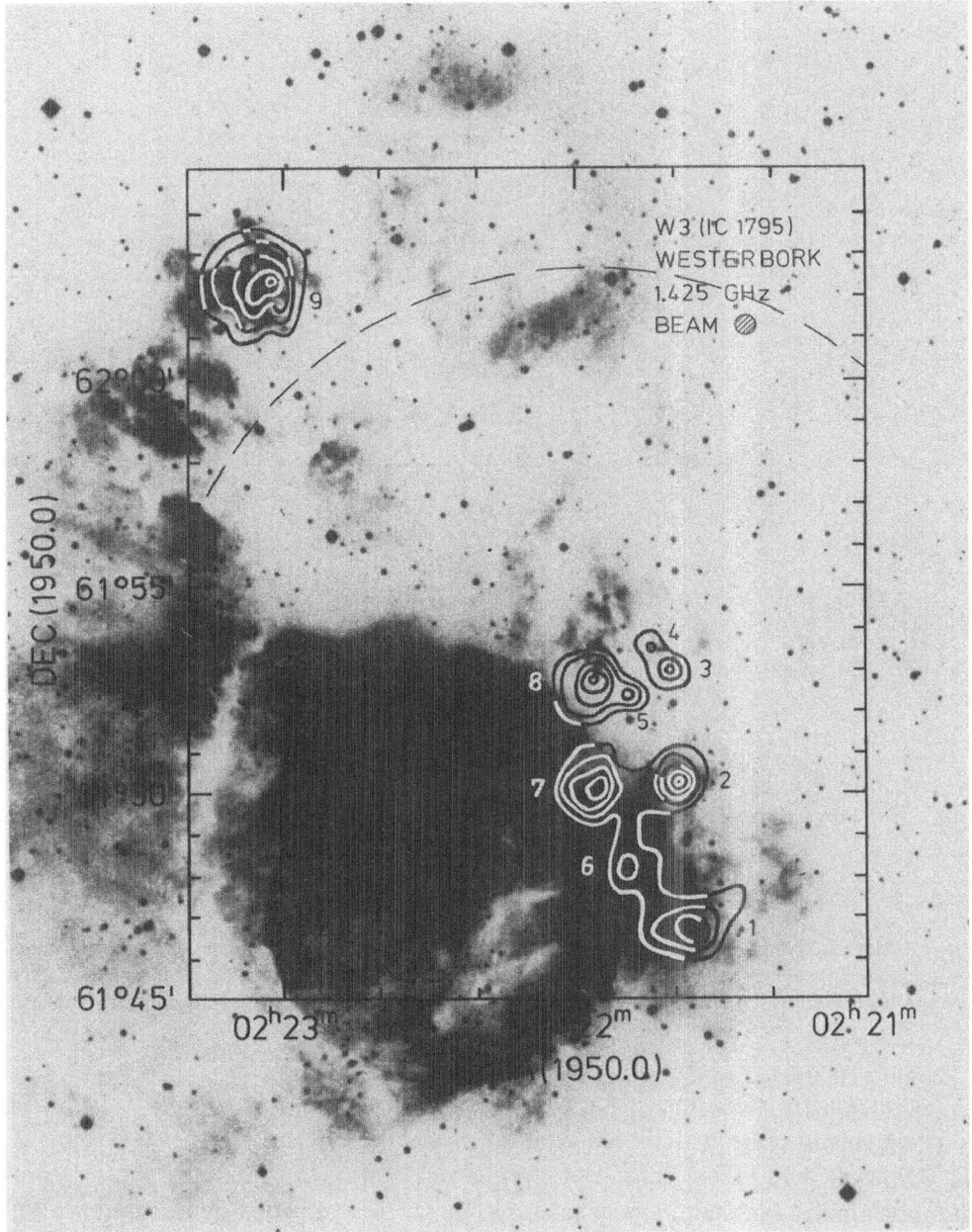

Fig. 6. Aperture-synthesis map of W3 by Sullivan and Downes (1973) at 1.4 GHz (continuum) with a resolution of 25″ × 28″. (Copyright by National Geographic Society – Palomar Observatory Sky Survey. Reproduced by permission from the Hale Observatories.)

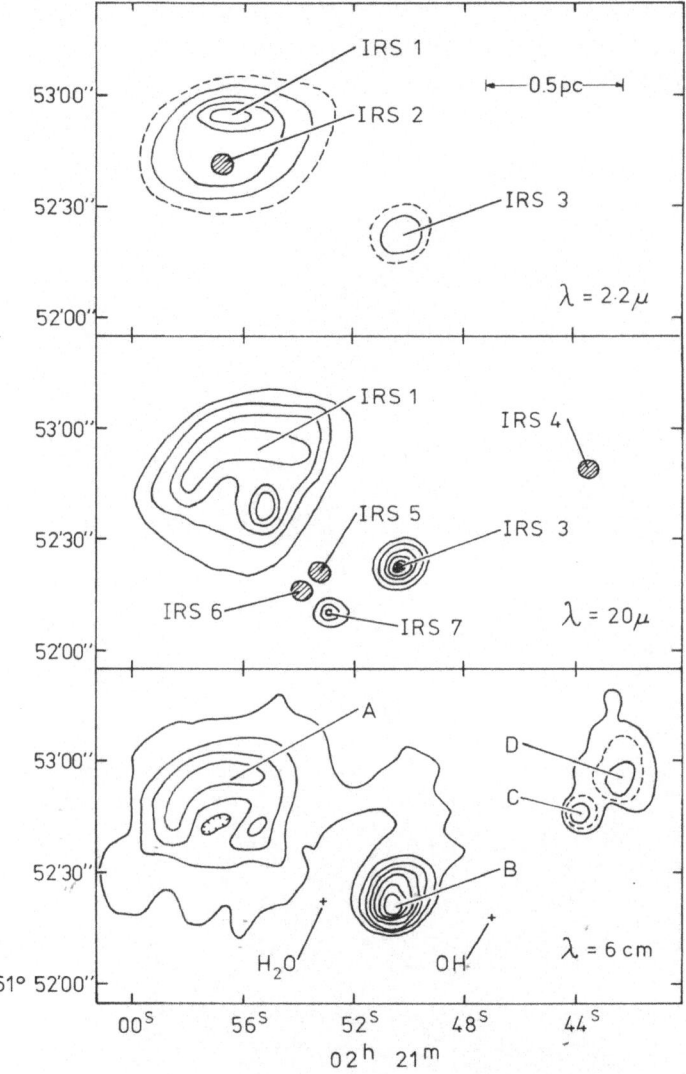

Fig. 7.   Maps by Wynn-Williams *et al.* (1972) at 2.2 μm (top) and 20 μm (middle), with a resolution of ~ 10″ drawn at the same scale as the 5.0-GHz map (bottom). The 1720-MHz OH emission source position (Wynn-Williams, 1973) has been added for comparison with the IR, radio, and H$_2$O emission sources.

The measured excitation parameters, U_{obs}, with the implied spectral type of the ionizing star in brackets (column 2), and the sizes and positions of components A, B, C, and D were taken from Wynn-Williams (1971). The L$_c$-photon fluxes N_c required to ionize each component (column 3) and that provided by the assumed O5 star (column 5) were calculated from the relation

$$\left[\frac{N_c}{s^{-1}}\right] = 3.806 \times 10^{43} \left[\frac{U}{pc\ cm^{-2}}\right],$$

where it is assumed that $T_e = 8000$ K. The effect of dust is ignored in all cases.

In no case can the O5 star account for the ionization of all components (which make up no more than 50% of the total radio emission from G133.7 + 1.2). Component A can be easily ionized with an O5 or even an O6 star at its center, but then

TABLE III

Ionization of compact components in W3 (G133.7 + 1.2)

Component	U_{obs} (pc cm^{-2})	Required N_c $\times 10^{48}$ (s^{-1})	Geometrical dilution factor	N_c provided by O5 star $\times 10^{48}$ (s^{-1})
Case 1: O5 star at the center of component A				
A	92 ($\sim$O6)	30	1	30.0
B	63 (O8–O7)	9.5	0.026	1.1
C	25 ($\sim$B0)	0.6	<0.0006	<0.03
D	38 (O9.5–O9)	2.1	0.0075	0.3
Case 2: O5 star at the center of component B				
A	92 ($\sim$O6)	30	0.15	8.6
B	63 (O8–O7)	9.5	1	9.5
C	25 ($\sim$B0)	0.6	<0.002	<0.1
D	38 (O9.5–O9)	2.1	0.002	0.1
Case 3: O5 star at the geometrical center between components A, B, C, and D				
A	92 ($\sim$O6)	30	0.11	6.3
B	63 (O8–O7)	9.5	0.09	5.2
C	25 ($\sim$B0)	0.6	<0.006	<0.3
D	38 (O9.5–O9)	2.1	0.04	2.3

the other components could not be ionized. Similarly with an O5 star at the center of component B, only component B can be ionized. In case 3 only component D can be accounted for.

I conclude from Table III that it is very difficult to ionize dense small components in H II regions from the outside, even with very hot stars. When it is possible, the star must be very close to the compact component. Each of the components A, C, and D, therefore, probably has a single ionizing star at its center; likewise each of the five subcomponents in B probably has a single ionizing star at its center. Column 2 of Table III shows that the compact components in W3 do not require very hot stars to ionize them if the star is located at the center of each component.

(d) ARE COMPACT COMPONENTS SURROUNDED BY DENSE NEUTRAL SHELLS?

Theoretical analyses by Davidson and Harwit (1967) and Larson (1969a, b) predict the formation of a cocoon or dense, thin shell of neutral gas and dust around an inner sphere of gas ionized by a newly formed star. Interferometric observations of small bright OH and H_2O emission-line sources (Burke et al., 1970; Johnston et al., 1972: Hills et al., 1972) have shown that such sources tend to be closely associated both in position and velocity with compact H II regions. Since OH and H_2O can only exist for reasonable times in neutral gas clouds (Hesser and Lutz, 1970; Stief et al., 1972) it is believed by some that compact H II regions are surrounded by neutral shells. An alternative interpretation is that the OH/H_2O maser sources are themselves protostar clouds which lie in the neighborhood of a slightly more evolved star with its own compact H II region.

The best observational evidence for a dense neutral shell is the appearance of OH emission sources around the edge of W3 (OH) (Baldwin et al., 1973; Wink et al., 1973). The VLBI OH positions are only relative, however, and the apparent clustering of OH sources around the edge of W3 (OH) has yet to be confirmed by accurate measurements. From a comparison of the IR and radio observations of W3, Wynn-Williams et al. (1972) conclude that the extinction varies from component to component; on the basis of this they argue that each condensation is associated, at least in part, with its own obscuring matter. From their synthesis of the 21-cm hydrogen line toward W3, Sullivan and Downes (1973) find no variation of the absorption over the whole of W3 (G133.7 + 1.2). They indicate that one possible way to reconcile the IR and hydrogen line data is for the compact H II components to be surrounded by a considerable mass of H_2.

W3 B has the largest visual extinction of all the components in W3; also, OH and H_2O emission sources appear in projection on opposite sides of it. However, the H_2O source apparently originates from IRS5 and not from component B. Also, according to Wink (1973), component B is composed of at least five subcomponents. Thus no simple neutral shell model for W3 B can be supported by the present data.

In my opinion the data support both the neutral shell and separate-protostar models of OH and H_2O emission. The best case for the former is W3(OH) and for the latter is W3 B.

(e) THE MASS DISTRIBUTION IN GIANT H II COMPLEXES

Let us first consider the large-scale mass distribution in W3. Mezger and Henderson (1967) derive:

$$\text{total } M_{H II} \simeq 4.2 \times 10^2 \ M_\odot .$$

In section IIIc, we found that for W3 (G133.7 + 1.2) $N_c^* \gtrsim 7.3 \times 10^{49}$ s^{-1}. The average integrated Lyman continuum flux-to-mass ratio for O star clusters is: $\langle N_c \rangle / \langle M \rangle \simeq 2 \times$ $\times 10^{46}$ s^{-1} $M_\odot^{-1}$ (Mezger et al., 1974). We, therefore, infer a stellar mass in W3 of:

$$\text{total } M_* = N_c^* / (\langle N_c \rangle / \langle M \rangle) \simeq 4 \times 10^3 \ M_\odot .$$

From the column density of 14×10^{13} cm^{-2} for H_2S toward W3 (Thaddeus et al., 1972) and an assumed ratio of $N_{H_2S}/N_{H_2} \simeq 3 \times 10^{-9}$, as in Orion A (Thaddeus et al., 1972), we infer a column density of H_2 toward W3 of $\simeq 5 \times 10^{22}$ cm^{-2}. If we further assume that the H_2 cloud has the same extent as the H_2CO cloud toward W3 (about $14' \times 5'$, see Dickel (1973), then the mass of the extended molecular cloud is

$$M_{mol} \simeq 2 \times 10^5 \ M_\odot .$$

Thus the large scale distribution of mass in the W3 region goes approximately as

$$M_{mol} \gtrsim 10 \ M_*,$$

and

$$M_* \gtrsim 10 \ M_{H II}.$$

Similar analyses for Orion A and W49A yield approximately the same results (Mezger, 1973). An obvious and not new conclusion indicated by these numbers is that very massive clouds are required for star formation to occur at least for large O star clusters. The ratio of M_*-to-$M_{H II}$ has been used to infer that the efficiency of star formation is very high; however, such a ratio probably overestimates the efficiency because the low mass stars may be embedded in neutral gas.

Let us now consider what proportion of the ionized gas is in compact components. For the diffuse southern extension mapped at 15 GHz and synthesized at 0.4 GHz, Wynn-Williams (1971) gives a size of ~ 4.5 pc and a density $n_e \simeq 100$ cm^{-3}, which implies a mass of $M_{H II} \simeq 120 \ M_\odot$. The sum of the flux densities of components A, B, C, and D is ~ 42.1 Jy (Wynn-Williams, 1971), and the integrated flux density of G133.7+1.2 is 76.5 Jy (Mezger and Henderson, 1967); therefore the flux density of the *diffuse gas* in G133.7+1.2 is ~ 34.4 Jy. The size of G133.7+1.2 is 4.3', which in conjunction with a flux density of 34.4 Jy implies a mass of $\sim 400 \ M_\odot$ for the diffuse ionized gas in G133.7+1.2. Excluding the southern extension, this reduces to $\sim 280 \ M$. The ionized gas in components A, B, C, and D together is $\sim 18 \ M_\odot$ (Wynn-Williams, 1971), therefore less than 10% of the ionized gas in G133.7+1.2 is in compact components.

Let us consider the ionized mass in some of the most compact H II regions. Table IV lists several of the densest H II regions observed to date, with their derived densities, masses and excitation parameters.

From Table IV we see that the more dense the sources are, the less ionized gas mass is observable, even though all these components require hot stars for their ionization. This partly reflects the fact that with increasing resolution, the smaller source

TABLE IV

Ionized gas mass in some of the most compact H II regions

Source	n_e (cm^{-3})	$M_{H II}/M_\odot$	U (pc cm^{-2})	References
W3 C	2.5×10^4	0.09	27	Wink (1973)
NGC 7538 C	8.8×10^4	0.002	12	Wink (1973)
W3 B_5	4.6×10^4	0.006	14	Wink (1973)
W3 (OH) A	1.8×10^5	0.084	54	Wink *et al.* (1973)
M17 A	1.5×10^5	0.0008	13	Wink (1973)
ON-1 [a]	$(0.9–8.8) \times 10^5$	0.00003–0.03	4.9–31	Winnberg *et al.* (1973)
IRS 5			> 30	Wynn-Williams *et al.* (1972)

[a] Distance ambiguity

sizes which are being resolved have lower masses and higher densities because they depend on the source size to the $+1.5$ and -1.5 power respectively. Also with the high emission measures found in these sources (typically $\sim 10^8$ pc cm^{-6}), they are optically thick at frequencies of 5 GHz and lower, hence detection becomes a limitation. A third factor which limits the amount of ionized gas in dense components is the presence of dust. At high electron densities Mezger *et al.* (1974) have shown that the dust competes with the gas for L_c-photons, and in the densest regions it may even absorb the major portion of the available L_c-flux. In this regard it should be emphasized that the excitation parameters listed in Table IV are only lower limits.

IV. Radio Recombination-Line Observations in H II Regions: Helium Abundances

In the past three years observations of radio recombination lines have been most active in the following areas: (1) the detection and interpretation of H recombination lines in the diffuse interstellar gas (see for example Gottesman and Gordon, 1970; Jackson and Kerr, 1971; Gordon and Cato, 1972; Davies *et al.*, 1972); (2) the observation of recombination lines of carbon and heavier atoms (with ionization potentials < 13.6 eV) which are believed to originate from cold and dense H I clouds associated with H II regions (see for example Palmer *et al.*, 1967; Zuckerman and Palmer, 1970; Simpson, 1970; Pedlar, 1970; Gordon and Churchwell, 1970; Ball *et al.*, 1970; Menon, 1970; Chaisson *et al.*, 1972; Chaisson, 1972; Chaisson, 1973b, c); (3) the mapping of H- and (in the brightest parts) He-line emission from a few bright H II regions (see for example Mezger and Ellis, 1968; Rubin and Mezger, 1969; Gordon, 1969; Gordon and Meeks, 1968; Chaisson, 1973a, b, c); and (4) the determination of helium abundances in galactic H II regions from observations of H and He radio recombination lines. Important progress in the identification of the spectra of several nebulae as thermal by the detection of radio recombination lines has been made by Milne *et al.* (1969), Wilson and Altenhoff (1970), Wilson and Altenhoff (1972), and Dickel and Milne (1972). Time limitations do not permit a discussion of all these areas of research. Since the first two areas are only indirectly related to H II regions and the third requires a detailed consideration of individual objects, I have chosen to restrict my my discussion to the fourth topic only.

(a) DO THE MEASURED INTENSITY RATIOS OF He- AND H-LINES REPRESENT TRUE ABUNDANCE RATIOS?

To answer this question it is necessary to consider non-LTE effects on both H and He lines, the geometry of the H II regions concerned, and the broadening of H and He lines. Goldberg (1966) first pointed out the importance of departures from LTE and the possibility of maser enhancement of radio recombination lines. A series of papers mostly by Hjellming and his coworkers (Hjellming and Churchwell, 1969; Hjellming and Davies, 1970; Hjellming and Gordon, 1971; Goldberg and Cesarsky, 1970) made the first non-LTE analyses of observed H-line intensities. Unfortunately all of these analyses assumed constant-density H II regions and ignored pressure broadening altogether. Griem (1967) found an analytical expression for pressure broadening of radio recombination lines in a plasma for given T_e and N_e values. However, when this expression was applied to H II regions, no attempt was made to include it properly in the transfer equation, and therefore it predicted much wider lines than were observed. Brocklehurst and Seaton (1972) have included in their solution of the equation of transfer non-LTE effects, a variable density model, and Griem's (1967) solution for pressure broadening. A constant electron temperature was assumed. In my opinion, this is the best theoretical treatment of the intensities of radio recombination lines to the present time. Observations of high-frequency radio recombination lines by Sorochenko and Berulis (1969), Churchwell *et al.* (1970), and Waltman and Johnston (1973) all find somewhat higher values of the line-to-continuum ratio than predicted by Brocklehurst and Seaton (1972). Whether this is due to observational difficulties or to problems with the theory remains to be seen. Batchelor (1974) has calculated the effect on the measured H- and He-line intensities when the He$^+$ and H$^+$ Strömgren spheres do not coincide. He used the Brocklehurst and Seaton (1972) model of Orion A for both LTE and non-LTE cases and varied the telescope beamsize and the relative size of the He$^+$ Strömgren sphere as independent variables. Among the important conclusions drawn by both Brocklehurst and Seaton (1972) and by Batchelor (1974) is that at high enough frequencies (i.e., $v \gtrsim 5$ GHz) the observed line intensity ratios of H and He should represent quite accurately the actual abundance ratio. In other words, at higher frequencies where the line and continuum optical depths are $\tau_L \ll \tau_c \ll 1$, the line ratios are largely insensitive to non-LTE effects, pressure broadening, and even to some extent the non-coincidence of Strömgren spheres. However, at lower frequencies where the optical depths are higher the line intensity ratios may bear almost no relation to the actual abundances.

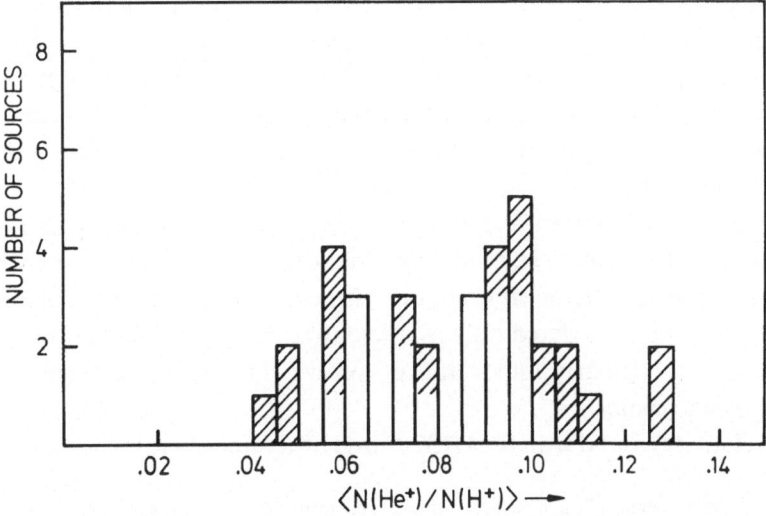

Fig. 8.   Histogram showing the distribution of observed ionized He abundances (Churchwell *et al.*, 1974).

(b) MEASURED HELIUM ABUNDANCES

Helium abundances derived from radio recombination lines have been published for only a few of the brightest H II regions (see for example Palmer *et al.*, 1969; Mezger *et al.*, 1970; Churchwell and Mezger, 1970).

Recently He- and H-line intensities have been measured at frequencies $\geqslant 5$ GHz with He-line signal-to-noise ratios $\geqslant 5$ for 39 galactic H II regions (Churchwell *et al.*, 1974). Twenty-eight of these nebulae are giant H II regions; they have intrinsic luminosities at least four times that of the Orion nebula $(N_c \geqslant 3.5 \times 10^{49} \text{ s}^{-1})$ and should therefore, have coincident He^+ and H^+ volumes. Figure 8 shows a histogram of the observed helium abundances; shaded areas indicate unconfirmed measurements (i.e., only one independent measurement is available). Three giant H II regions in the galactic center (G0.2 – 0.0, G0.5 – 0.0, and G0.7 – 0.0) are not included in this figure.

From Figure 8 we conclude that the credible range of ionized helium abundances in galactic H II regions outside the galactic center is

$$0.06 \leqslant \langle N(\text{He}^+)/N(\text{H}^+)\rangle \leqslant 0.10\,*.$$

Attempts to measure the He^+ 173α line in G0.2 – 0.0, G0.7 – 0.0, W43 and W49 *A* show that

$$\langle N(\text{He}^{++})/N(\text{H}^+)\rangle < 0.01.$$

For the giant H II regions in the galactic center we find the limits given in Table V (Churchwell and Mezger, 1973).

The low abundance of ionized helium in the galactic center has been independently confirmed by Mezger *et al.* (1970), Chaisson (1973c), and Robinson (1973).

(c) CORRELATIONS OF MEASURED He ABUNDANCES WITH GEOMETRICAL AND PHYSICAL PARAMETERS

A check for systematic trends in the measured helium abundances with several geometrical and physical parameters was undertaken. Plots of $\langle N(\text{He}^+)/N(\text{H}^+)\rangle$ as a function of distance from the Sun, distance from the galactic plane, and the fraction of the nebular volume intercepted by the telescope main beam reveal no systematic trend. An apparent correlation does exist with distance from the galactic center (see Figure 9).

The abundance of ionized helium apparently increases with galactic radius to at least 7 kpc from the center; beyond 7 kpc the scatter is too large to say whether the increase continues or whether it levels off. This conclusion is based on a minimum of data and should be considered only as a tentative result until further measurements are made. It is particularly important to observe H II regions between 1 and 5 kpc from the galactic center.

No correlation is found with the H137β-to-H109α intensity ratio, which is a measure

* Angular brackets surrounding measured abundance ratios denote average values weighted by the telescope beam.

TABLE V

Ionized helium in the galactic center

Source	$N(He^+)/N(H^+)$	$N(He^{++})/N(H^+)$
G0.2–0.0	0.025	0.008
G0.5–0.0	0.01 [a]	
G0.7–0.0	0.021	0.005

[a] Huchtmeier and Batchelor (1973)

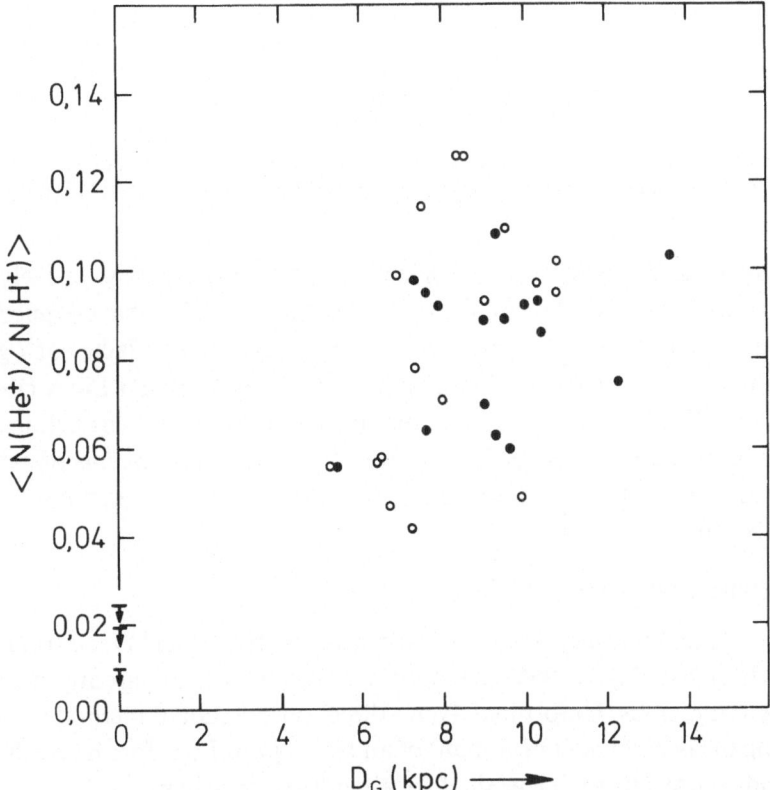

Fig. 9.   Observed ionized He abundances as a function of distance from the galactic center.

of departures from LTE. This seems to confirm Brocklehurst and Seaton's (1972) prediction that the He abundance should be independent of non-LTE effects at frequencies where $\tau_L \ll \tau_c \ll 1$.

Mezger et al. (1974) have defined the IR excess as the difference, normalized to the Ly-α luminosity, between the integrated IR luminosity of the heated dust in H II regions and that supplied by trapped Ly-α photons. The IR excess derived from the 100 μm flux measurements of Hoffmann et al. (1971) is plotted against $\langle N(He^+)/N(H^+)\rangle$ in Figure 10.

The IR excess apparently increases with decreasing ionized helium abundance in

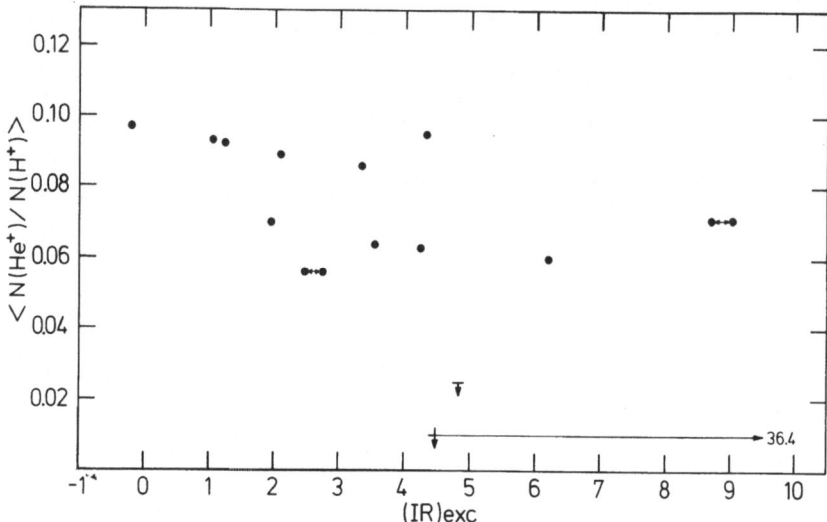

Fig. 10.   Observed ionized He abundances as a function of the IR-excess radiation, i.e., IR radiation
in excess of that supplied by the absorption of trapped Ly-α photons by dust in H II regions.

giant H II regions. We believe this is an indication that the dust selectively absorbs
stellar photons capable of ionizing helium. Mezger *et al.* (1973) have argued that the
absorption cross-section of dust must be on the average about a factor of 7 greater at
wavelengths $\lambda < 504$ Å than that found at $\lambda \sim 1500$ Å by Witt and Lilley (1973).

Leibowitz (1973) has also concluded that dust selectively absorbs helium-ionizing
photons; unfortunately NGC 2024, the source on which his conclusion was based,
shows no ionized helium because its ionizing star is too cool and not because of
selective absorption by dust.

(d) THE PROBLEM WITH LOW He$^+$ ABUNDANCES

More than 75% of our sample of H II regions have $\langle N(H^+)/N(H^+) \rangle < 0.10$, and $\sim 10\%$
have $\langle N(He^+)/N(H^+) \rangle < 0.065$. Doubly ionized helium is apparently negligible in
all galactic H II regions. Unfortunately, no direct observational method is known for
determining the neutral helium content of an H II region. Therefore it is only possible
through indirect means to derive the true abundance of helium.

Churchwell *et al.* (1974) have argued that the initial helium abundance in our
Galaxy was $N(He)/N(H) \sim 0.08$. Observed values of $\langle N(He^+)/N(H^+) \rangle > 0.08$ would
then indicate helium enrichment in the course of galactic evolution. Present-day
observations imply an average helium enrichment of ~ 0.02 by number. This value is
in good agreement with that found by Talbert and Arnett (1973) for stellar evolution
processes. In no case should the actual helium abundance be less than 0.08, and in
general it should not be less than 0.09–0.10 after enrichment. How then can this value
be reconciled with the measured He$^+$-abundances, which are largely less than 0.09?

We believe that dust may be responsible for the low He$^+$ abundances. Selective
absorption by dust of photons with wavelengths $\lambda < 504$ Å could easily cause such
an effect. The gas density may be as high as 10^4–10^5 cm^{-3} in bright compact H II

regions, thus the density of grains is probably proportionally high. As is shown in Figure 10, the He^+ abundance decreases with increasing IR excess. In this regard, we note that Sgr B2 (G0.7 − 0.0) has the highest IR excess in our sample and one of the lowest upper limits on ionized helium. We have shown also that $\langle N(He^+)/N(H^+) \rangle$ appears to decrease with decreasing distance from the galactic center for $D_G < 7$ kpc (Figure 9). We therefore argue that: (1) the actual helium abundance in the Galaxy (including the nucleus) is constant with $N(He)/N(H) \simeq 0.10$; and (2) the ionized helium abundance in the giant H II regions in the galactic center is lowered by the selective absorption of He-ionizing photons by dust. Such an effect could be caused by an increase in the dust-to-gas ratio, by a change in composition, by a change in the distribution of grain sizes, or some combination of these toward the galactic center.

The observations could also be explained if H II regions toward and in the galactic center are ionized by O9 or cooler stars. We favor the dust interpretation for three reasons: (1) very large numbers of O9 stars would be required to ionize the gas in the galactic center; (2) it is questionable whether cooler stars could account for the inferred IR excess in the galactic center; and (3) the molecules in the massive molecular clouds in the galactic center probably depend on dust for their formation and subsequent protection from the intense radiation field implied by the ionized gas. It may be that the observed increase of N and O from the edge to the center in M33 and M101 by Searle (1971) and in M31 by Rubin et al. (1972) is also related to an increase in the dust-to-gas ratio or to a change in the properties of dust toward the nuclei in these galaxies.

It is, of course, possible that the observed variation in He^+ abundances reflects a a variation in the actual He abundance; however, this is, in my opinion, fraught with many more difficulties than the explanation which we offer here.

Acknowledgements

I thank Dennis Downes, Peter Mezger and Malcolm Walmsley for critically reading this manuscript. Substantial improvements over the original version was made due to their comments. In addition I extend my thanks to Lindsey Smith for numerous discussions and to Jörn Wink for the use of his data before publication.

References

Allen, C. W.: 1963, *Astrophysical Quantities*, The Athlone Press, University of London, London, 2nd ed., p. 238.
Altenhoff, W. J., Downes, D., Goad, L., Maxwell, A., and Rinehart, R.: 1970, *Astron. Astrophys. Suppl.* **1**, 319.
Auer, L. H. and Mihalas, D.: 1972, *Astrophys. J. Suppl.* **24**, 205.
Baars, J. W. M. and Wendker, H. J.: 1974, this volume, p. 219.
Baldwin, J. E., Harris, C. S., and Ryle, M.: 1973, *Nature* **241**, 38.
Ball, J. A., Cesarsky, D., Dupree, A. K., Goldberg, L., and Lilley, A. E.: 1970, *Astrophys. J.* **162**, L25.
Batchelor, A. S. J.: 1974, *Astron. Astrophys.* **32**, 343.
Beard, M.: 1966, *Australian J. Phys.* **19**, 141.
Beard, M. and Kerr, F. J.: 1969, *Australian J. Phys.* **22**, 121.

Beard, M., Thomas, B. MacA., and Day, G. A.: 1969, *Australian J. Phys. Astrophys. Suppl.* No. 11, 27.
Benvenuti, P., D'Odorico, S., and Peimbert, M.: 1973, *Astron. Astrophys.* **28**, 447.
Brocklehurst, M. and Seaton, M. J.: 1972, *Monthly Notices Roy. Astron. Soc.* **157**, 179.
Burke, B. F., Papa, D. C., Papadopoulos, G. D., Schwartz, P. R., Knowles, S. H., Sullivan, W. T., Meeks, M. L., and Moran, J. M.: 1970, *Astrophys. J. Letters* **160**, L63.
Carranza, G., Courtès, G., Georgelin, Y., Monnet, G., and Pourcelot, A.: 1968, *Ann. Astrophys.* **31**, 63.
Chaisson, E. J.: 1973a, *Astrophys. J.* **182**, 767.
Chaisson, E. J.: 1973b, *Astrophys. J.* **186**, 545.
Chaisson, E. J.: 1973c, *Astrophys. J.* **186**, 555.
Chaisson, E. J.: 1972, *Nature Phys. Sci.* **239**, 83.
Chaisson, E. J., Black, J. M., Dupree, A. K., and Cesarsky, D. A.: 1972, *Astrophys. J. Letters* **173**, L131.
Churchwell, E. and Mezger, P. G.: 1970, *Astrophys. Letters* **5**, 227.
Churchwell, E. and Mezger, P. G.: 1973, *Nature* **242**, 319.
Churchwell, E. and Walmsley, C. M.: 1973, *Astron. Astrophys.* **23**, 117.
Churchwell, E., Mezger, P. G., and Huchtmeier, W.: 1974, *Astron. Astrophys.* **32**, 283.
Churchwell, E., Mezger, P. G., Reifenstein, E. C., III, Rubin, R., and Turner, B.: 1970, *Astrophys. Letters* **5**, 157.
Courtès, G.: 1960, *Ann. Astrophys.* **2**, 115.
Cruvellier, P.: 1967, *Ann. Astrophys.* **30**, 1059.
Dachler, M., Mack, J. E., Stoner, J. O., Jr., Clark, D., and Ring, J.: 1968, *Planetary Space Sci.* **16**, 795.
Davidson, K. and Harwit, M.: 1967, *Astrophys. J.* **148**, 443.
Davies, R. D., Matthews, H. E., and Pedlar, A.: 1972, *Nature Phys. Sci.* **238**, 101.
Day, G. A., Caswell, J. L., and Cooke, D. J.: 1972, *Australian J. Phys. Astrophys. Suppl.* No. 25, 19.
Day, G. A., Thomas, B. MacA., and Goss, W. M.: 1969, *Australian J. Phys. Astrophys. Suppl.* No. 11, 11.
Day, G. A., Warne, W. G., and Cooke, D. J.: 1970, *Australian J. Phys. Astrophys. Suppl.* **13**, 11.
Dickel, H. R.: 1973, in J. M. Greenberg and H. C. van de Hulst (eds.), 'Interstellar Dust and Related Topics', *IAU Symp.* **52**, 277.
Dickel, J. R. and Milne, D. K.: 1972, *Australian J. Phys.* **25**, 539.
Fomalont, E. B. and Weliachew, L.: 1972, *Mem. Roy. Soc. Sci. Liège III*, 453.
Felli, M. and Churchwell, E.: 1972, *Astron. Astrophys. Suppl.* **5**, 369.
Goldberg, L.: 1966, *Astrophys. J.* **144**, 1225.
Goldberg, L. and Cesarsky, D.: 1970, *Astrophys. Letters* **6**, 93.
Gordon, M. A.: 1969, *Astrophys. J.* **158**, 479.
Gordon, M. A. and Cato, T.: 1972, *Astrophys. J.* **176**, 587.
Gordon, M. A. and Churchwell, E.: 1970, *Astron. Astrophys.* **9**, 307.
Gordon, M. A. and Meeks, M. L.: 1968, *Astrophys. J.* **152**, 417.
Goss, W. M. and Day, G. A.: 1970, *Australian J. Phys. Astrophys. Suppl.* No. 13, 3.
Goss, W. M. and Shaver, P. A.: 1970, *Australian J. Phys. Astrophys. Suppl.* No. 14, 1.
Gottesman, S. T. and Gordon, M. A.: 1970, *Astrophys. J.* **162**, L93.
Griem, H.: 1967, *Astrophys. J.* **148**, 547.
Hardebeck, E. G.: 1971, *Astrophys. J.* **170**, 281.
Harper, D. A. and Low, F. J.: 1971, *Astrophys. J. Letters* **165**, L9.
Hesser, J. E. and Lutz, B. L.: 1970, *Astrophys. J.* **159**, 703.
Hills, R., Janssen, M. A., Thornton, D. D., and Welch, W. J.: 1972, *Astrophys. J. Letters* **175**, L59.
Hjellming, R. M. and Churchwell, E.: 1969, *Astrophys. Letters* **4**, 165.
Hjellming, R. M. and Davies, R. D.: 1970, *Astron. Astrophys.* **5**, 53.
Hjellming, R. M. and Gordon, M. A.: 1971, *Astrophys. J.* **164**, 47.
Hoffmann, W. F., Frederick, C. L., and Emery, R. J.: 1971, *Astrophys. J. Letters* **170**, L89.
Huchtmeier, W. K. and Batchelor, R. A.: 1973, *Nature* **243**, 155.
Jackson, P. D. and Kerr, F. J.: 1971, *Astrophys. J.* **168**, 29.
Johnson, H. M.: 1971, *Astrophys. J.* **164**, 379.
Johnston, K. J., Knowles, S. H., Sullivan, W. T., III, Burke, B. F., Lo, K. Y., Papa, D. C., Papadopoulos, G. D., Schwarz, P. R., Knight, C. A., Shapiro, I. I., and Welch, W. J.: 1971, *Astrophys. J. Letters* **166**, L21.
Larson, R. B.: 1969a, *Monthly Notices Roy. Astron. Soc.* **145**, 271.
Larson, R. B.: 1969b, *Monthly Notices Roy. Astron. Soc.* **145**, 297.
Leibowitz, E. M.: 1973, *Astrophys. J.* **181**, 369.

Lemke, D. and Low, F. J.: 1972, *Astrophys. J. Letters* **177**, L53.
Martin, A. H. and Downes, D.: 1972, *Astrophys. Letters* **11**, 219.
Menon, T. K.: 1970, *Astrophys. Letters* **7**, 55.
Mezger, P. G.: 1973, private communication.
Mezger, P. G. and Ellis, S. A.: 1968, *Astrophys. Letters* **1**, 159.
Mezger, P. G. and Henderson, A. P.: 1967, *Astrophys. J.* **147**, 471.
Mezger, P. G., Smith, L. F., and Churchwell, E.: 1974, *Astron. Astrophys.* **32**, 269.
Mezger, P. G., Wilson, T. L., Gardner, F. F., and Milne, D. K.: 1970, *Astrophys. Letters* **6**, 35.
Milne, D. K., Wilson, T. L., Gardner, F. F., and Mezger, P. G.: 1969, *Astrophys. Letters* **4**, 121.
Monnet, G.: 1971, *Astron. Astrophys.* **12**, 379.
Morton, D. C.: 1969, *Astrophys. J.* **158**, 629.
Palmer, P., Zuckerman, B., Penfield, H., Lilley, A. E., and Mezger, P. G.: 1969, *Astrophys. J.* **156**, 887.
Palmer, P., Zuckerman, B., Penfield, H., Lilley, A. E., and Mezger, P. G.: 1967, *Nature* **215**, 40.
Pedlar, A.: 1970, *Nature* **226**, 830.
Raimond, E. and Eliasson, B.: 1969, *Astrophys. J.* **155**, 817.
Robinson, B. J.: 1973, private communication.
Robinson, B. J., Goss, W. M., and Manchester, R. M.: 1970, *Australian J. Phys.* **23**, 363.
Reay, N. K. and Ring, J.: 1969, *Planetary Space Sci.* **17**, 561.
Reifenstein, E. C. III, Wilson, T. L., Burke, B. F., Mezger, P. G., and Altenhoff, W. J.: 1970, *Astron. Astrophys.* **4**, 357.
Reynolds, R. J., Roesler, F. L., and Scherb, F.: 1973, *Astrophys. J.* **179**, 651.
Rubin, R. H. and Mezger, P. G.: 1970, *Astron. Astrophys.* **5**, 407.
Rubin, V. C., Kumar, C. K., and Ford, W. K., Jr.: 1972, *Astrophys. J.* **177**, 31.
Salpeter, E. E.: 1955, *Astrophys. J.* **121**, 161.
Savage, B. D. and Jenkins, E. B.: 1972, *Astrophys. J.* **172**, 491.
Schraml, J. and Mezger, P. G.: 1969, *Astrophys. J.* **156**, 269.
Searle, L.: 1971, *Astrophys. J.* **168**, 327.
Sharpless, S.: 1959, *Astrophys. J. Suppl.* **4**, 257.
Shaver, P. A. and Goss, W. M.: 1970, *Australian J. Phys. Astrophys. Suppl.* No. 14, 76.
Simpson, J. P.: 1970, *Astrophys. Letters* **7**, 43.
Sorochenko, R. L. and Berulis, J. J.: 1969, *Astrophys. Letters* **4**, 173.
Stief, L. J., Donn, B., Glicker, S., Gentieu, E. P., and Mentall, J. E.: 1972, *Astrophys. J.* **171**, 21.
Sullivan, W. T. III and Downes, D.: 1973, *Astron. Astrophys.* **29**, 369.
Talbot, R. J., Jr. and Arnett, W. D.: 1973, *Astrophys. J.* **186**, 51.
Thaddeus, P., Kutner, M. L., Penzias, A. A., Wilson, R. W., and Jefferts, K. B.: 1972, *Astrophys. J. Letters* **176**, L73.
Thomas, B. MacA. and Day, G. A.: 1969a, *Australian J. Phys. Astrophys. Suppl.* No. 11, 3.
Thomas, B. MacA. and Day, G. A.: 1969b, *Australian J. Phys. Astrophys. Suppl.* No. 11, 19.
Waltman, E. B. and Johnston, K. J.: 1973, *Astrophys. J.* **182**, 489.
Wendker, H. J.: 1970, *Astron. Astrophys.* **4**, 378.
Wilson, T. L. and Altenhoff, W. J.: 1970, *Astrophys. Letters* **5**, 47.
Wilson, T. L. and Altenhoff, W. J.: 1972, *Astron. Astrophys.* **16**, 489.
Wink, J. E.: 1973, unpublished Ph.D. Thesis, University of Münster.
Wink, J. E., Altenhoff, W. J., and Webster, W. J., Jr.: 1973, *Astron. Astrophys.* **22**, 251.
Winnberg, A., Habing, H. J., and Goss, W. M.: 1973, *Nature Phys. Sci.* **243**, 78.
Witt, A. N. and Lillie, C. F.: 1973, *Astron. Astrophys.* **25**, 397.
Wynn-Williams, C. G.: 1971, *Monthly Notices Roy. Astron. Soc.* **151**, 397.
Wynn-Williams, C. G., Becklin, E. E., and Neugebauer, G.: 1972, *Monthly Notices Roy. Astron. Soc.* **160**, 1.
Wynn-Williams, C. G., Werner, M. W., and Wilson, W. J.: 1973, preprint.
Zuckerman, B. and Palmer, P.: 1970, *Astron. Astrophys.* **4**, 244.

E. Churchwell
Max-Planck-Institut für Radioastronomie,
Auf dem Hügel 69,
53 Bonn 1, F.R.G.

DISCUSSION

Menon: Photographs exist in the literature which show that surrounding the Orion nebula there is a highly symmetric faint H II region of extent of about 3°. This H II region can be attributed to escaped Lyman continuum photons from the density-bounded Orion nebula. The molecular clouds attributed to the CO emission are presumably within this H II region.

Habing: To some extent I agree and to some extent I disagree with your statement that star formation is not an efficient process. If one looks at the W3 map that you showed it follows that only on a large scale a lot of matter is left over after the star formation. However, on a small scale, this is not true – see, for example, the ionized hydrogen mass in W3 *B*. So it appears that in the later phases of the collapse, when the fragmentation has led to the formation of objects like W3 *B*, the stellar formation is extremely efficient.

Churchwell: This is what I meant to point out with regard to the analysis of mass in W3 *B* where ∼60% of the ionized gas is in compact components.

Gordon: Sometimes recombination lines from H II regions have asymmetrical shapes. These have been interpreted as blends of lines from the H II region and lines from foreground H I. Could the asymmetrical shape be due entirely from the small-scale component within the H II regions now seen with interferometers?

Churchwell: I think the analysis of Brocklehurst and Seaton (1972) has shown that except at relatively high frequencies the compact components in H II regions will contribute very little to the line intensities. On this basis I would not expect large line asymmetries in recombination line profiles in H II regions due to high density components except at quite high frequencies where few line measurements have been made.

Jenkins: In support of the idea that there is an observable emission measure between known H II regions you mentioned the diffuse Hα emission recorded by Courtès and others. Is there any likelihood that interstellar dust grains have a high enough albedo and are plentiful enough to scatter or reflect Hα photons from the H II regions and produce this diffuse contribution?

Churchwell: I suspect it is possible that some of the Hα emission in the spiral arm regions of the galactic disk is due to scattered Hα photons from stars, but I have no idea how much. I believe that Peimbert and his co-workers have argued that most Hα emission seen in the disks of galaxies is largely due to recombinations.

Monnet: There are in fact two different kinds of diffuse emission – one in the spiral arms, one between the spiral arms in the disk. It is completely impossible that the emission in the disk can come from scattered light from the classical H II regions, as the ratios of Hα/[N II], Hα/[S II], Hβ/[O II] in the disk are very different than those in the classical H II regions. On the other hand, the diffuse arm emission which has a filamentary structure with a typical size of 100 pc can be – at least partly – due to scattered light.

Guélin: The amount of scattered light has been estimated by others at between 0 and 30%. On that kind of estimation, you can be off by a factor of 3 or 4; that means that you cannot determine it.

Van Woerden: At Westerbork, Sullivan and Downes have not only measured the continuum but also looked for the recombination lines. He166α and C166α were below their detection limit. H166α was detected in one concentration (the strongest one), at a velocity agreeing with single-dish observations, with $T_b = 22 \pm 5$ K, that is, 0.4% of the continuum. The line flux in this concentration is only 1% of that measured with single-dish antennas. The latter must almost completely come from structures with scales > 5′. This is in agreement with the Brocklehurst and Seaton (*Monthly Notices Roy. Astron. Soc.* **157** (1972), 179) theory.

Churchwell: I point out also with regard to these measurements that with a beam of ∼25″ the carbon line was not seen in absorption. Therefore one wonders about the beam dilution spoken of by Zuckerman.

HIGH-RESOLUTION RADIO OBSERVATIONS OF A
GIANT H II REGION IN CYGNUS

JACOB W. M. BAARS

Netherlands Foundation for Radio Astronomy, Radiosterrenwacht, Dwingeloo, The Netherlands

and

HEINRICH J. WENDKER

Hamburger Sternwarte, Hamburg-Bergedorf, F.R.G.

Abstract. Westerbork observations of several large H II regions in the Cygnus-X region are described. IC 1318 appears to be a giant H II region, and two nearby sources may be genetically related to it.

I. Introduction

The highest angular resolution for observations of H II regions with a single antenna was obtained a few years ago by Schraml and Mezger (1969). They used the NRAO 43-m telescope at 2-cm wavelength, which resulted in a half-power-width of the main beam of about 2′. The results, particularly those on the nebula W49 (Mezger *et al.*, 1967), indicated that, given a sufficiently narrow antenna beam, many interesting details would be seen in H II regions of small overall angular size.

Consequently such observations have been undertaken with interferometers, e.g., the Cambridge 'One-Mile-Telescope'. (Wynn-Williams, 1971) and the NRAO three-element interferometer (Webster *et al.*, 1971). These two aperture-synthesis instruments are either extremely slow or incapable of covering the aperture plane completely. Therefore the observations have been restricted mainly to bright regions of small angular size (in most cases significantly smaller than the beam of the antenna pattern of the interferometer elements).

These observations have shown the existence of compact condensations ('bright knots'). They are thought to be excited from a star within and are considered as very young H II regions. In almost all cases the total flux density in the interferometer map is less than the total flux density measured with a single antenna. This indicates that the source contains angular structure on a scale large enough to be resolved by the shortest interferometer spacing.

The Synthesis Radio Telescope (SRT) at Westerbork, The Netherlands, delivers a completely sampled aperture plane from 36 m to 1458 m with 18-m intervals in 4 half-days of observing time. This enables the study of sources of relatively large angular size with a very good resolution.

We therefore used this instrument for observations of several large H II regions in the Cygnus-X region. One aim was to study the hierarchy of sub-structures in these presumably old sources. The choice of the general region of the sky was influenced by the availability of plates in the Hα line, enabling the study of the relation between radio and optical structures and the absorption across the nebulae.

F. J. Kerr and S. C. Simonson, III (eds.), Galactic Radio Astronomy, 219–226. All Rights Reserved.

II. The Radio Observations

Figure 1 shows part of the Cygnus-X region as observed by Wendker (1970) at 11-cm wavelength with an angular resolution of 11'. The SRT observations have been done in the region of IC 1318b and c, the companion sources north (G79.3 + 1.3) and south (G78.1 + 0.6) of these, the so-called γ-Cygni source, and the area around G78.2 − 0.4 and G78.0 + 0.0. Note that the northern and southern companions are much smaller and appear more symmetrical than the regions IC 1318b and c. Figure 2 is a copy of the Hα plate of the region; IC 1318b and c are separated by a large lane of dust, while the northern and southern sources are highly obscured. We adopt a distance of 1.5 kpc to all the sources (Dickel *et al.*, 1969).

We turn now to the Westerbork observations of IC 1318b, at 21 cm with a synthesized beam of 30" × 45". Figure 3 shows the resulting contour map. Integration of the map indicates that only 15% of the flux density measured with a single antenna is present. The map appears to be a blend of about 10 substructures of the size of a few

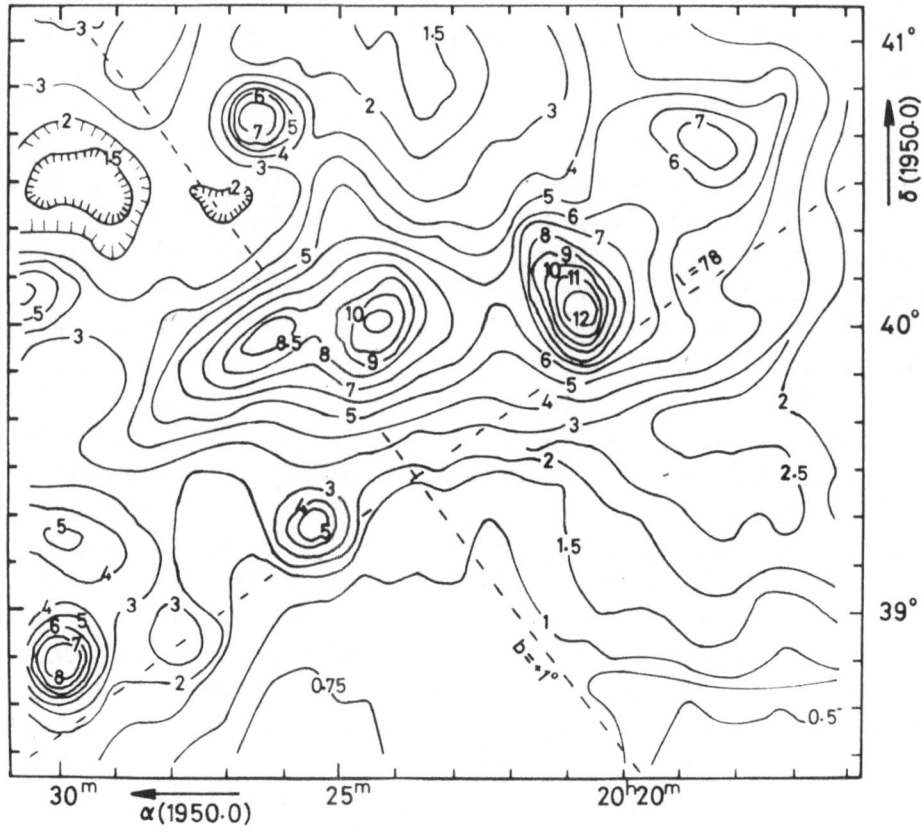

Fig. 1. Part of 11-cm map of Cygnus-X region obtained by Wendker at $\lambda = 11$ cm with a resolution of 11'. The brightest source is the γ-Cygni source; to the east of it are IC 1318b and c with their northern and southern companions.

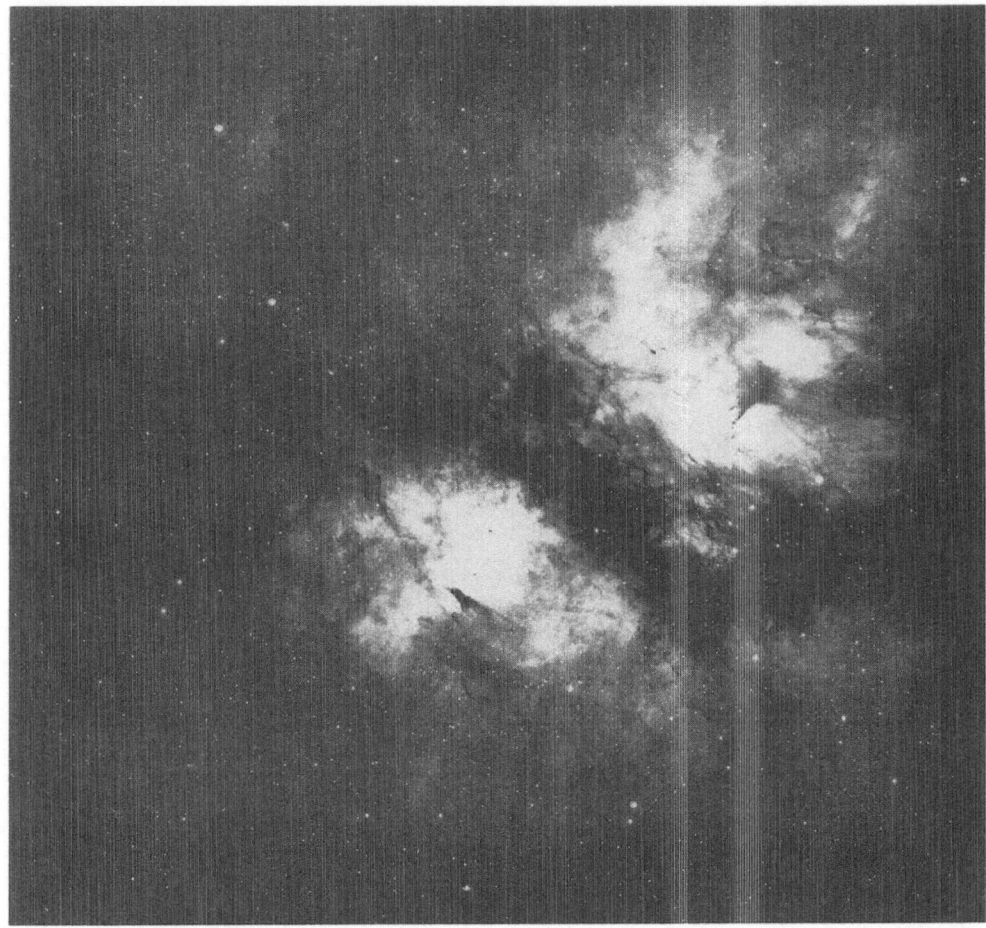

Fig. 2.   A photograph in Hα of IC 1318b and c. Note the conspicuous dust lane crossing the region. Photograph by H. R. Dickel and H. J. Wendker. Courtesy of Hale Observatories.

arcmin. The half-power angular width of the source, as deduced from the single-antenna observations is $20' \times 30'$, beyond the region where the SRT sees significant radiation. An obvious conclusion is that the greater part of the source's flux density comes from a large, smooth component which is at best only partially seen by the smallest interferometer spacing of 36 m.

To check this quantitatively we have used the source parameters from the 11-cm map as starting parameters. Since the sources are known to be optically thin at 21 and 11 cm no serious error can arise in doing this. By an iterative least-squares fit to the complex visibilities we found the parameters of those sources which dominate the 36-m and 54-m spacings of the SRT observations. In comparing the final fit with the single dish data we find that the SRT sources fit only the upper part of the single antenna scans. This suggests that a component exists with size $> 1°$ which is thus completely absent in the 36-m spacing but contributes heavily to the single-antenna

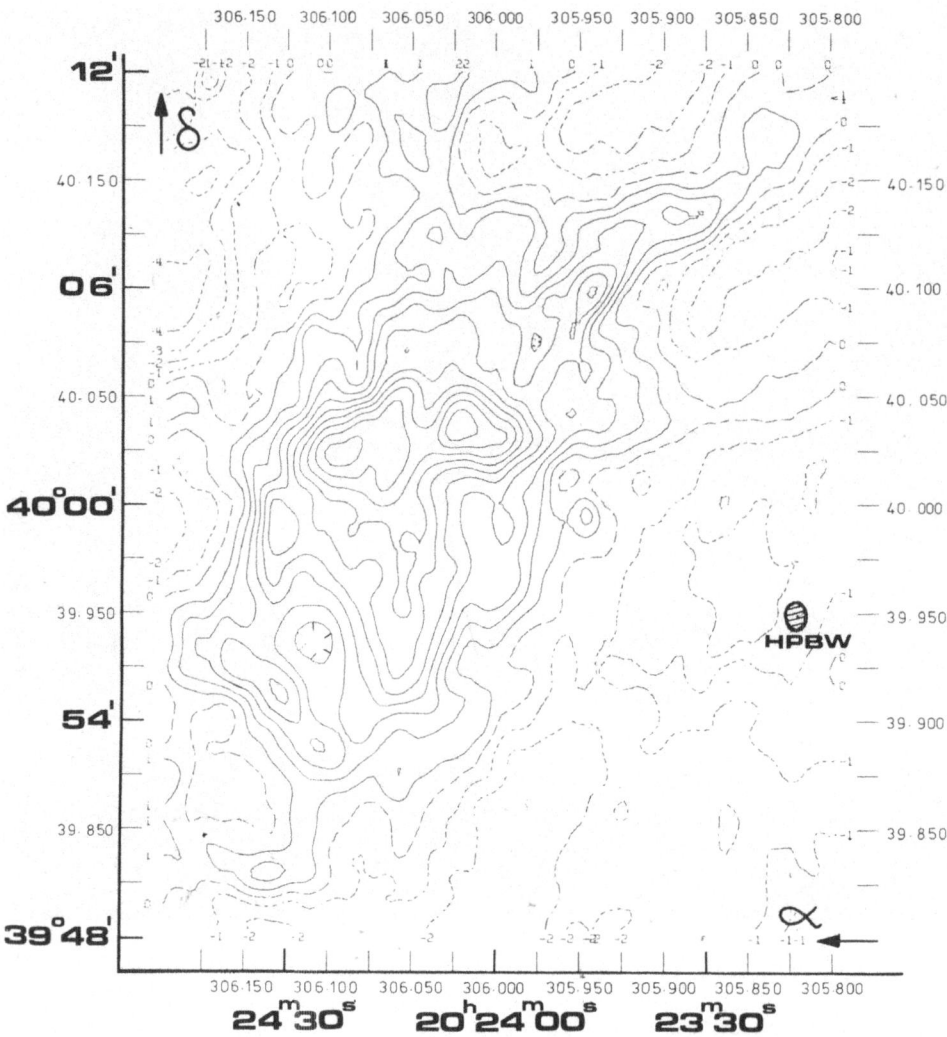

Fig. 3.   IC 1318b as observed with the SRT at $\lambda = 21$ cm with a synthesized beam of $30'' \times 45''$. The full contours are 1 (1) 10 K in brightness temperature, the dashed contours are negative responses. The observed size of the source is much too small, because of the loss of information on the large-scale structure in the interferometer output.

flux density. Introducing such a component in the single-dish data together with the components found from the two smallest spacings of the interferometer we obtained a reasonable fit to the single-dish data. We find a half-width of 75' and a flux density of about 300 Jy for this large extended component. Almost all of this component was considered to belong to the unresolved background contained in the 11-cm data by Wendker (1970).

We conclude that we see the following hierarchy of angular structures in IC 1318b:

   (i) an extended envelope of size 75';

(ii) the intermediate sources, dominating the 11-cm data, of size 15'–30';

(iii) the finer scale substructure, seen only with the SRT, of size less than 5'.

The ratios of flux densities are about: (i):(ii):(iii) $= 10:3:10^{-2}$.

Quite a different picture emerges in the case of the source G79.3 + 1.3, the northern companion of IC 1318b. SRT maps at 6 cm and 21 cm are available for this source. Figure 4 shows the 6-cm map. Here small-scale components stand out clearly. They

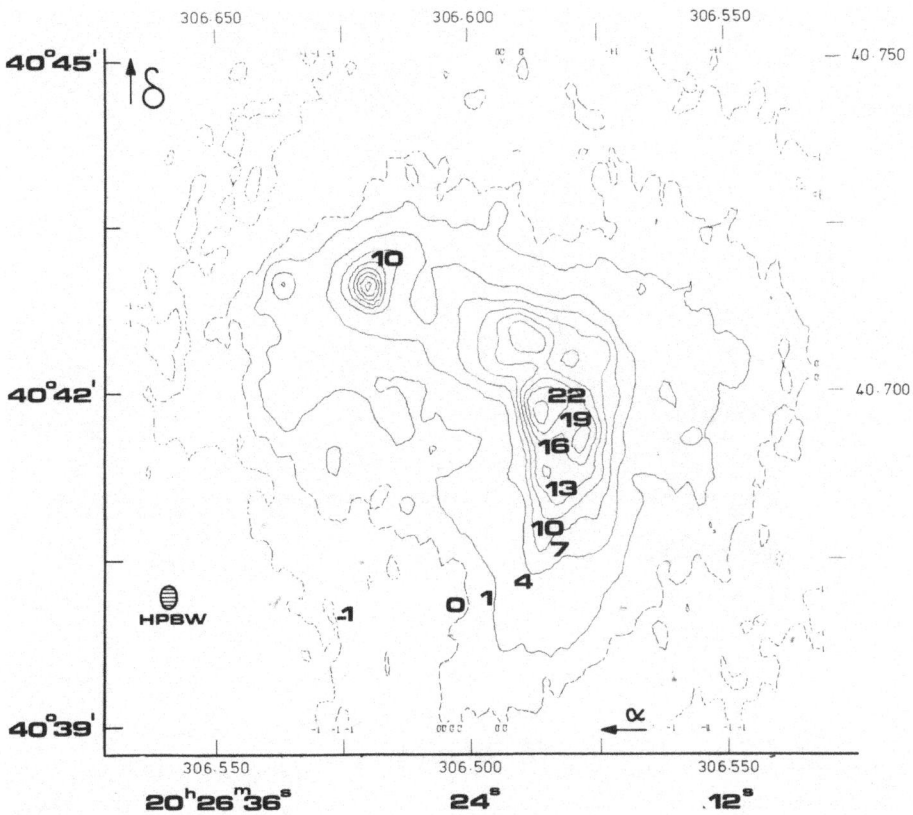

Figs. 4a–b. Contour map of G79.3 + 1.3, observed at $\lambda = 6$ cm with a resolution of $6'' \times 10''$. Contour values are approximately in kelvins in equivalent brightness temperature. There are at least seven small-scale components.

are rather symmetrical in appearance and of small angular size. A comparison with the single-antenna flux density shows that by far the largest percentage of the flux density is contained in the condensations. This is very different from IC 1318. However, an analysis similar to that described above shows that at 21 cm there is a weak extended component which is smooth and covers the total source area. This component is lost at 6 cm, but the 6-cm data do show a smooth component underlying the condensations which is of intermediate size.

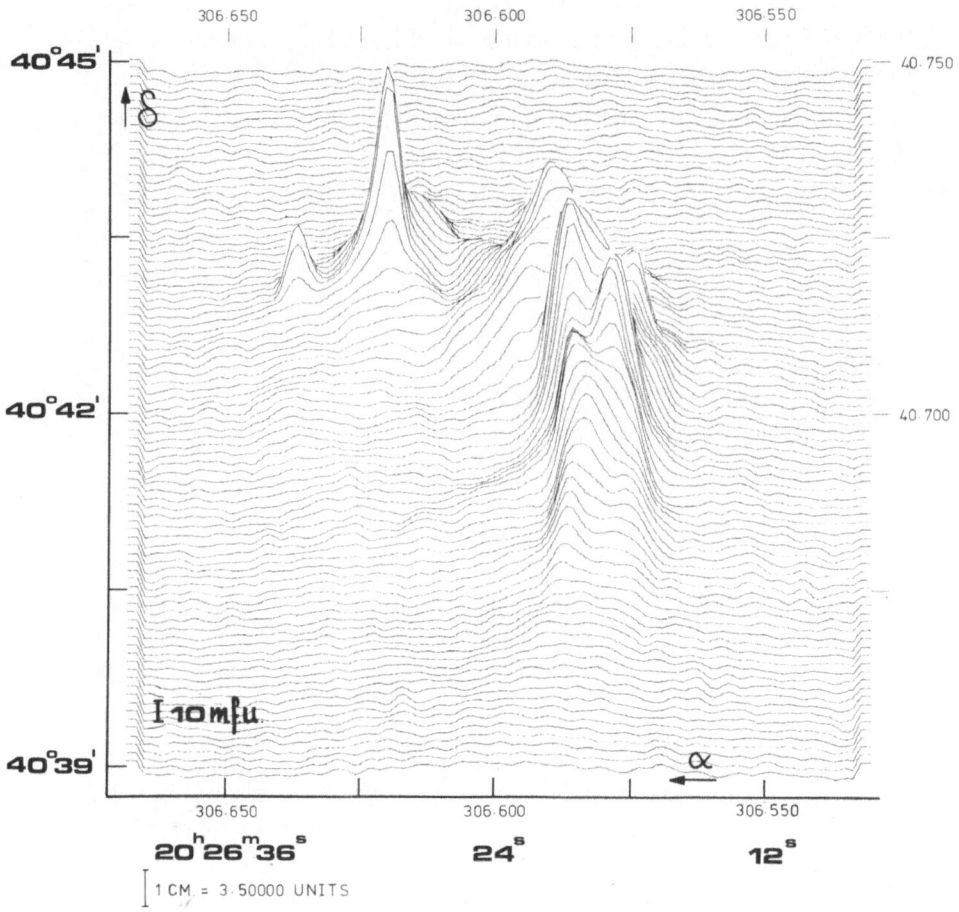

Fig. 4b.   The same observation represented in a profile map.

III. Discussion

Results similar to IC 1318b have been obtained for IC 1318c and in a preliminary form also for G78.2−0.4. In the same sense the southern companion and G78.0+0.0 resemble the northern companion of IC 1318b. Therefore the appearance of the two sources discussed above is not unique.

A main point from the above analysis is that the extended component of size 75′ covers both sources IC 1318b and c. This leads us to suggest that both nebulae IC 1318b and c are in fact one HII region. This source then belongs to the class of giant HII regions, as defined by Mezger (1970). If this suggestion is accepted it is probably of interest to point out that IC 1318 is then the giant HII region closest to the Sun. Although it is intrinsically weaker than W49, the prototype of a giant HII region, it contains all the hierarchy of substructures found in W49.

Physical parameters for the components, derived under the usual assumptions for HII regions, are listed in Table I.

TABLE I

Physical parameters of IC 1318b and c components

Parameter	Envelope	Intermediate component	Small component	Total source
$T_b(K)$ at 11 cm	4	8	1	12
Size (arcmin)	75	25	3	75
Size in pc	32	11	1.3	32
E (pc cm^{-6})	8400	17000	2100	25200
n_e (cm^{-3})	16	40	40	28
U (pc cm^{-2})	100	65	8	150
Type of exciting star	O6/7	O8 V	B0.5–1 V	O4–O5

If the total source has one excitation center, a star of type O5 would be required for a smooth distribution of the gas. Since the observations clearly show this not to be the case, a somewhat hotter star, O4, is needed. There are no O4 stars observed as yet. The intermediate components in IC 1318b and c can be excited by an O8 V star alone. The small scale components are quite different from the compact H II regions studied, e.g., by Webster *et al.* (1971) or Habing *et al.* (1972). Their bright 'knots' require an O star as excitation center, while for our small components an early B star would suffice. In the present case however, we feel that the small-scale structure should be considered as an inherent part of the intermediate component, its appearance resulting from density gradients in the gas.

Now we discuss the sources of smaller angular size. As a prototype we take G79.3 + + 1.3, the northern companion of IC 1318b. The physical parameters of the components mentioned above are summarized in Table II.

TABLE II

Physical parameters of G79.3 + 1.3 components

Parameter	Envelope	Intermediate component	Condensation (average)	Total source
$T_b(K)$ at 11 cm	5	17	50	50
Size (arcmin)	5	1.8	0.3	1.8
Size in pc	2.2	0.77	0.13	0.77
E (pc cm^{-6})	10300	35700	105000	105000
n_e (cm^{-3})	70	215	900	370
U (pc cm^{-2})	38	14	6	20
Type of exciting star	O9	B0.5 V	B1 V	O9

It is seen that the condensations dominate the source G79.3 + 1.3. Certainly an O9 star could excite the total H II region. However, it is hard to see how sharp small scale structure could develop in such a case.

We suggest on the basis of the morphology, as described in the previous section (see also Figure 4) that the condensations each have their own excitation center. This is contrary to our suggestion for IC 1318b and c. A typical condensation here can be excited by a B1 V star. Thus this H II region could represent a very young association of B stars, so young that the stars have not yet lost their surrounding ionized shells. The situation is analogous to the young O-star association thought to be responsible for the appearance of W49.

At this point it is tempting also to suggest that there is a genetic relationship between the giant H II region IC 1318b and c and their northern and southern companions. In projection both companions are contained in the very extended envelope discussed in connection with IC 1318b and c. The whole complex would then be the first stage of the generation of an OB-star association, where the H II components of different size and density represent the formation of different members at different ages. A similar relationship is suggested by Israel *et al.* (1973) for a group of H II regions around $l = 111°$. A similar group may be represented by the H II regions G81.6+0.0 and G81.7+0.5 also situated in the Cygnus-X region (see Wendker, 1970).

References

Dickel, H. R., Wendker, H. J., and Bieritz, J. H.: 1969, *Astron. Astrophys.* **1**, 270.
Habing, H. J., Israel, F. P., and de Jong, T.: 1972, *Astron. Astrophys.* **17**, 329.
Israel, F. P., Habing, H. J., and de Jong, T.: 1973, *Astron. Astrophys.* **27**, 143.
Mezger, P. G.: 1970, in W. Beckers and G. Contopoulos (eds.), 'The Spiral Structure of Our Galaxy', *IAU Symp.* **38**, 107.
Mezger, P. G., Schraml, J., and Terzian, Y.: 1967, *Astrophys. J.* **150**, 807.
Schraml, J. and Mezger, P. G.: 1969, *Astrophys. J.* **156**, 169.
Webster, W. J., Altenhoff, W. J., and Wink, J. E.: 1971, *Astron. J.* **76**, 677.
Wendker, H. J.: 1970, *Astron. Astrophys.* **4**, 378.
Wynn-Williams, C. G.: 1971, *Monthly Notices Roy. Astron. Soc.* **151**, 397.

Jacob W. M. Baars
Netherlands Foundation for Radio Astronomy,
Radiosterrenwacht
Dwingeloo, The Netherlands

Heinrich J. Wendker
Hamburger Sternwarte,
Hamburg/Bergedorf, F.R.G.

DISCUSSION

Terzian: What distance did you adopt for the source? Does it correspond well with the kinematic distance from recombination lines?

Baars: We adopted 1.5 kpc; I think it agrees.

RADIO OBSERVATIONS OF H II REGIONS IN M33*

F. P. ISRAEL and P. C. VAN DER KRUIT**

Sterrewacht, Leiden, The Netherlands

Abstract. The Westerbork Synthesis Radio Telescope was used to map the continuum radiation of M33 at 1415 MHz. Of 67 radio sources with fluxes $S > 1.8$ mJy (3σ), 60% coincide with Hα sources. These are all intrinsically stronger than $4 \times$ the Orion nebula, i.e., they are giant H II regions. The two strongest sources, NGC 604 (58 mJy) and NGC 595 (20 mJy), are similar to W51.

A map of the continuum radiation of M33 at 1415 MHz has been obtained using the Westerbork Synthesis Radio Telescope. The object was followed across the sky by 4×20 interferometers during 4×12 h. The resulting synthesized beam is $23'' \times 45''$, the field of view 0.5 deg, corresponding to 80 pc $\times$ 160 pc and 7 kpc, respectively, if one adopts a distance of 720 kpc. The rms noise σ was 0.6 mJy per synthesized beam.

The map shows a large number of small sources, virtually all point sources. It shows neither a continuous arm structure (such as in M51), nor a nucleus, indicating that perhaps little nonthermal action is going on in M33. The radio map was compared to a high quality, narrow filter Hα photograph, obtained by J. Boulesteix and G. Comte from the Marseille Observatory at the Obervatoire de Haute Provence. Out of the 67 sources in the radio map with fluxes $S > 3\sigma(=1.8$ mJy$)$ 60% appear to coincide with objects in the Hα photographs. It is assumed that these sources are all H II regions; they will be called 'the identified sources.' Even a significant fraction (27%) of the 100 sources with $2\sigma < S < 3\sigma$ show a good correlation in position with H II regions. All identified sources appear unresolved or hardly resolved with the exception of the two strongest sources: NGC 604 (58 mJy) and NGC 595 (20 mJy). These two sources will be dealt with separately. Some of the major conclusions about the other sources are as follows.

(1) The number of sources, $n(S)\,dS$, increases strongly with decreasing flux, S, approximately as $S^{-3.1}\,dS$. Extrapolating to lower flux limits it appears possible to explain most of the flux from M33 in the 21-cm continuum as being of a thermal origin.

(2) To obtain the diameter of the point sources it was assumed that the optical diameters equal the radio diameters. From the diameter and the flux one then derives n_e (rms) and the linear diameter d and all the quantities that follow therefrom (excitation parameter $u \propto n_e^2 d^3$, mass of ionized gas $M \propto n_e d^3$; emission measure $EM \propto n_e^2 d$). The masses of ionized gas run up to $2 \times 10^5\ M_\odot$, the diameters up to 120 pc, and the required ionizing flux to the equal of five O5 stars. (It should be pointed out that all identified sources are intrinsically stronger than $4 \times$ the Orion nebula, which classifies them as giant H II regions.)

(3) There is a clear correlation between the surface density (by number) of H II regions and the surface density of H I.

---

* Paper presented by H. J. Habing.
** Presently Fellow of the Carnegie Institution of Washington at Hale Observatories.

F. J. Kerr and S. C. Simonson, III (eds.), Galactic Radio Astronomy, 227–228. All Rights Reserved.
Copyright © 1974 by the IAU.

(4) The surface density (by number) of H II regions decreases outward. It is highest between 1 and 2 kpc. (However, the surface density within 1 kpc is poorly determined.) The decrease in surface density is probably even more pronounced in the large H II regions.

The two outstanding nebulae, NGC 595 and NGC 694, have ionized gas masses of 5×10^5 and $22 \times 10^5\ M_\odot$, sizes of 140 and 270 pc, and they require for their excitation 11 and 33 O5 stars, respectively. These values are rather similar to those of the (super) giant galactic H II region W51. However, they appear small compared to complexes such as 30 Doradus in the Large Magellanic Cloud and object 44 in NGC 2403.

The full text of this paper has been published in *Astronomy and Astrophysics* **32** (1974), 363.

F. P. Israel
Sterrewacht, Huygens Laboratorium, Wassenaarseweg 78,
Leiden 2405, The Netherlands

P. C. van der Kruit
Hale Observatories,
813 Santa Barbara Street,
Pasadena, California 91101, U.S.A.

(Discussion follows after paper by J. H. Spencer and B. F. Burke, p. 230.)

THE H II REGIONS OF M31

J. H. SPENCER and B. F. BURKE

Dept. of Physics and Research Laboratory of Electronics,
Massachusetts Institute of Technology, Cambridge, Mass., U.S.A.

Abstract. The luminosity of the brightest H II regions of M31 was determined with the NRAO 3-element interferometer at 3.7 cm and 11 cm wavelength. It is unlikely that M31 has any superbright H II regions such as W51 or W49; our Galaxy has between 10 and 20 H II regions that are more luminous than any in M31.

The great nebula in Andromeda, M31, is often thought of as a model for our own Galaxy, and even though there is evidence that the two systems are dissimilar, quantitative data on the differences is sparse. The occurrence of giant H II regions is related to spiral type, and as a rule the most luminous H II regions occur in the later spiral types, S_c and Magellanic Cloud-type irregulars. In order to provide a quantitative measure of the luminosity of the H II regions in M31 compared to those of our own Galaxy, a study was undertaken with the NRAO 3-element interferometer to determine the luminosity of the brightest H II regions of M31.

A complete synthesis was made of five fields using all available interferometer spacings, at both 3.7 cm and 11 cm wavelength. These five fields each include a known 5C3 source. In addition a 'quick look' synthesis of 29 fields was performed using tle single antenna configuration that gave spacings of 900, 1800, and 2700 m baselines. The catalog of Baade and Arp (1964) was used as a finding list of H II regions for comparison with the radio data.

The five 'complete synthesis' fields included the nuclear region, 5C3.97, 5C3.126, 5C3.141a, and a field partly displaced from 5C3.97. Only one example was found of a coincidence between a visible H II region and a radio source, the coincidence being close for Baade and Arp's Source 298, which agrees in position with a 4 mJy source at 11 cm.

For purposes of comparison with our own Galaxy, the surveys of Reifenstein *et al.* (1970) and of Wilson *et al.* (1970) can be used to estimate the appearance of the H II regions in our Milky Way Galaxy if they were at the distance of M31. The flux of W49 would be 25 mJy, and W51, M17, and M42 would exhibit fluxes of 13, 5, and 0.25 mJy, respectively.

In our own Galaxy there are at least seven H II regions whose flux would exceed 10 mJy, and more than 22 would exceed 2.1 mJy. The complete synthesis fields found only one source that might be an H II region, of the 167 H II regions in the fields, and that one is not as luminous as M17.

The 'quick look' synthesis of 29 fields was not as sensitive, but only seven sources stronger than the 3σ level were found, and none of them were found to coincide with catalogued H II regions. The 3σ sensitivity limit varied from field to field, but in most cases varied between 9.5 and 10.5 mJy. Seven fields had 16.5 mJy 3σ-limits.

F. J. Kerr and S. C. Simonson, III (eds.), Galactic Radio Astronomy, 229–231. All Rights Reserved.

We conclude, therefore, that it is most unlikely that M31 has *any* superbright H II regions such as W51 or W49, and while it is possible that a few of the sources are in the M17 class, it appears probable that our Galaxy has between 10 and 20 H II regions that are more luminous than any in M31.

The lack of very bright H II regions in M31 has two important implications:

(1) It is strong evidence that our Galaxy is considerably later in spiral type (and probably considerably smaller) than M31, and

(2) The use of H II regions as distance indicators for distant galaxies is a method that may be subject to order-of-magnitude errors, if it is applied without certain knowledge of the expected H II-regions luminosity class in a given galaxy.

References

Baade, W. and Arp, H. C.: 1964, *Astrophys. J.* **139**, 1027.
Reifenstein, E. C., III, Wilson, T. L., Burke, B. F., Mezger, P. G., and Altenhoff, W. J.: 1970, *Astron. Astrophys.* **4**, 357.
Wilson, T. L., Mezger, P. G., Gardner, F. F., and Milne, D. K.: 1970, *Astron. Astrophys.* **6**, 364.

J. H. Spencer
B. F. Burke
Dept. of Physics and Research Laboratory of Electronics,
Massachusetts Institute of Technology,
Cambridge, Mass. 02139, U.S.A.

DISCUSSION

Wielebinski: On an 11-cm continuum map made by Berkhuijsen and myself with the 100-m telescope we find that spectrum in the direction of clumps of optical H II regions is less than the spectrum elsewhere in the spiral arms. In fact, three radio sources in the direction of optically detected groups of H II regions have nearly thermal spectra. These sources are 5C3.98, 5C3.77 and 5C3.94. The conclusion of Burke that there are no giant H II regions in M31 of the type W49 or 30 Doradus is correct, since they would be seen as single strong point sources on our map.

Baldwin: What is the total flux density of the sources associated with H II regions in M33? From the spectrum of the total flux density it seems likely that the thermal component is quite small.

Habing: The flux is mostly nonthermal. The amount of thermal flux is very uncertain, but it is small.

Van Woerden: In M101, Allen, Ekers and Goss find several H II regions (or rather *complexes*) which, at the distance of M31 or M33, would have a flux of about 1000 mJy. These complexes have sizes of the order of 1000 pc and masses $\sim 10^7 M_\odot$.

Habing suggests that the whole radio flux from M33 could be thermal. Do none of the sources found have nonthermal spectra? One should expect to find remnants of type II supernovae.

Habing: The study of M33 by Israel and van der Kruit is based on a map at only one frequency (1415 MHz). A second map at 5 GHz is in preparation. It can be used to sort out nonthermal sources and its use could lead to identification of some supernova remnants.

Price: What is the distribution in radius of the bright H II regions with respect to the center of M33?

Habing: The largest surface density of H II regions is between 1 and 2 kpc from the center. Within 1 kpc the data are unreliable; beyond 2 kpc there is a very clear decrease in the surface density.

Greenberg: Would you be willing to conjecture that the lack of nonthermal emission in M33 could be used to imply that there is no magnetic field to produce synchrotron radiation from relativistic electrons and that – going very far indeed – the lack of continuous arms is a result of this lack of magnetic field?

Habing: You just phrased a nice speculation.

Mathewson: I would urge that Northern Hemisphere observers should, by narrow band photography, determine the ratio of $S_{II}/H\alpha$ in these H II regions to detect supernova remnants. This technique has been used successfully by Clarke and myself in the Magellanic Clouds, where we have detected 15 SNR.

Mezger: Optical observers of H II regions in external galaxies quote sizes ranging from 50 to 200 pc. Radio observations in our Galaxy never show H II complexes of such large dimensions, with the exception of the extended region in the Galactic center. Do your radio observations of H II regions in M33 yield sizes which are comparable with the optical dimensions?

Habing: In this study the dimensions the authors used were derived from the photographic plates. If the optical sizes of H II regions in other galaxies are indeed larger than in our Galaxy, as you suggest, then this is perhaps caused by blending. Complexes of H II regions are quite common in our Galaxy.

Mills: The Magellanic Clouds are other examples in which emission of giant H II regions is detected. These agree perfectly in position with the optical regions and fit smoothly on to the top-end of the galactic distributions with larger sizes and lower mean electron densities.

THE BRIGHTNESS DISTRIBUTION OF ORI A
AT 4.1 mm IN THE 1_{01}–0_{00} LINE OF PARA-FORMALDEHYDE
AND IN THE CONTINUUM

NORIO KAIFU and TETSUO IGUCH

University of Tokyo, Tokyo, Japan

and

MASAKI MORIMOTO

Tokyo Astronomical Observatory, Tokyo, Japan

Abstract. The 4.1-mm continuum emission shows a peak at the Trapezium stars and an extension to the west, where the molecular and IR sources are. The 4-mm line of para-formaldehyde has a similar distribution to that of the 3-mm line of HCN and extends more to the north than the 2-mm line of ortho-formaldehyde. This probably means a lower temperature in the northern part of the cloud.

Figure 1 shows a continuum map of Ori A at 4.1 mm. A contour unit corresponds to 1.6 K in brightness temperature. The half power beam width is 2.4'. The emission peak is found to coincide with the Trapezium stars, but there is an extension of emission in the western side where the molecular and IR sources are situated.

The 1_{01}–0_{00} line of formaldehyde (72.8381 GHz) was observed at ten positions over the molecular cloud, and the line was detected in all positions. In Figure 2 these observed points are shown as crosses surrounded by squares superimposed on the continuum map (dotted contours). Also superimposed in solid controus is the distribution of 2 mm line emission of orthoformaldehyde, observed by Thaddeus *et al.* (1971). Figure 3 shows the line profiles on the corresponding positions shown in Figure 2. The 4 mm line is found to be more extended to the north than the 2 mm line and has a very similar distribution to that of the 3 mm line of HCN (Snyder and Buhl, 1971).

Thaddeus *et al.* also observed the 2_{02}–1_{01} line of para-formaldehyde at points 2, 1 and 4 with antenna temperatures of 1.9, 1.5 and 0.7 K, respectively, and found the northern point (point 4) exhibiting much weaker emission than the southern points. For the 4 mm line, however, the emission continues to be strong to point 5 and shows very different intensity ratios across the source.

Either low density or low temperature in the northern part of the cloud can explain the difference, but we regard the latter as being more likely. Density variations alone can explain the intensity ratios among the formaldehyde and HCN lines but cannot explain the brightness temperature of the CO lines unless the temperature should be allowed to vary.

A fuller account of this work will be published in the *Astrophysical Journal*.

F. J. Kerr and S. C. Simonson, III (eds.), Galactic Radio Astronomy, 233–237. All Rights Reserved.
Copyright © 1974 by the IAU.

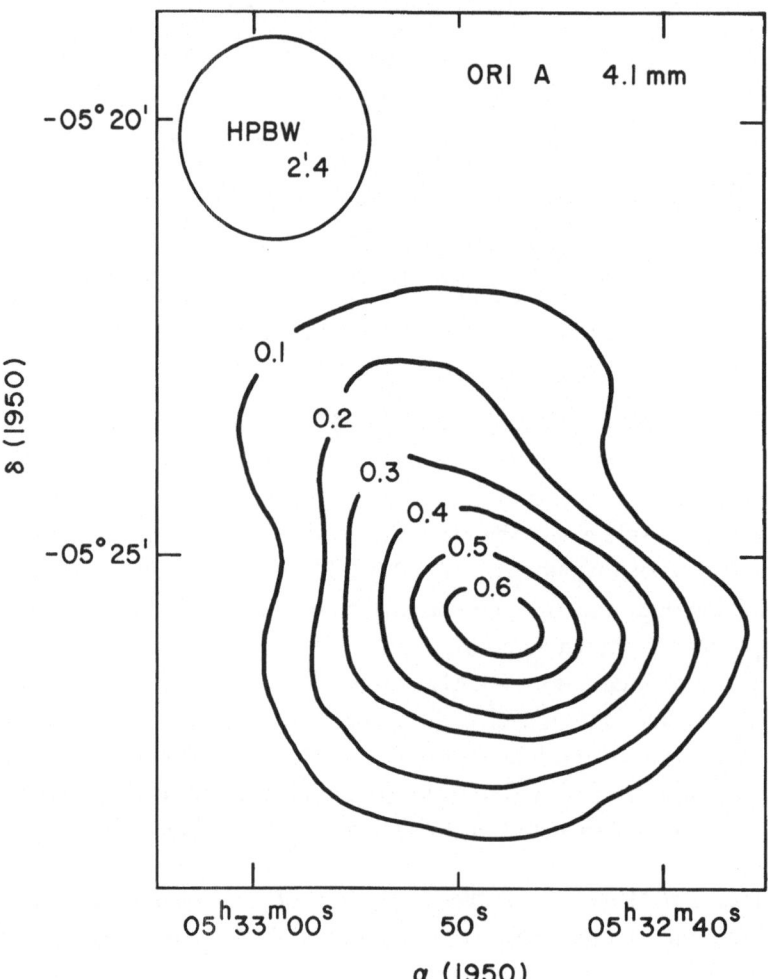

Fig. 1.   Contour map of Ori A at 4.1 mm. A contour unit corresponds to 1.6 K in brightness
temperature.

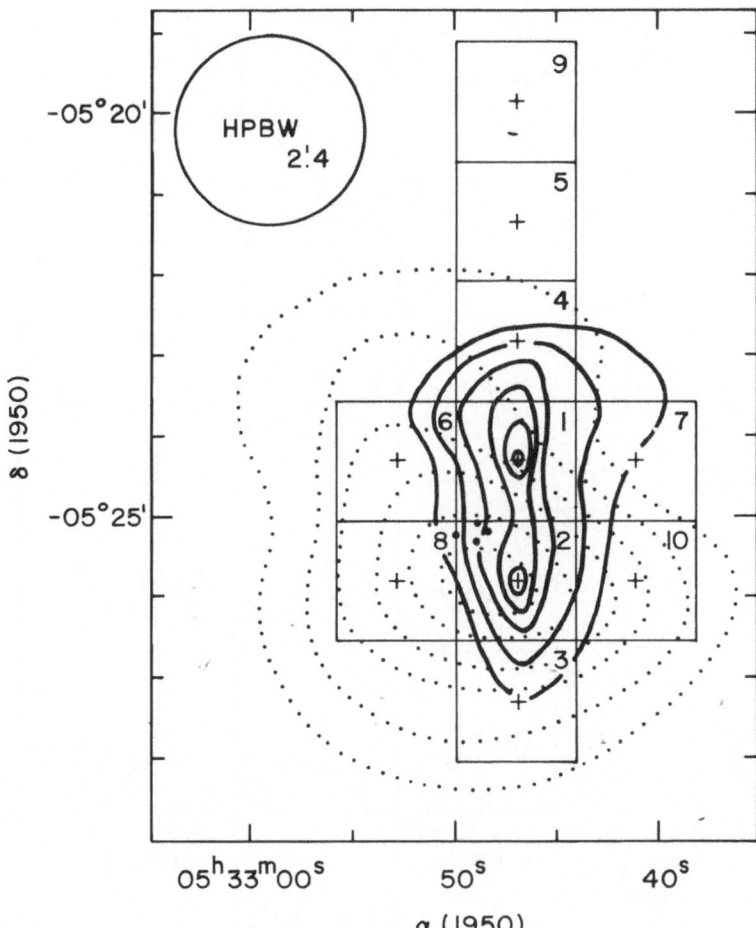

Fig. 2.   Observed points for 1_{01}–0_{00} line of para-formaldehyde. Points are shown as crosses surrounded by squares superimposed on the continuum map (in dotted contours). Solid contours represent the distribution of the 2 mm line of ortho-formaldehyde from Thaddeus *et al.*

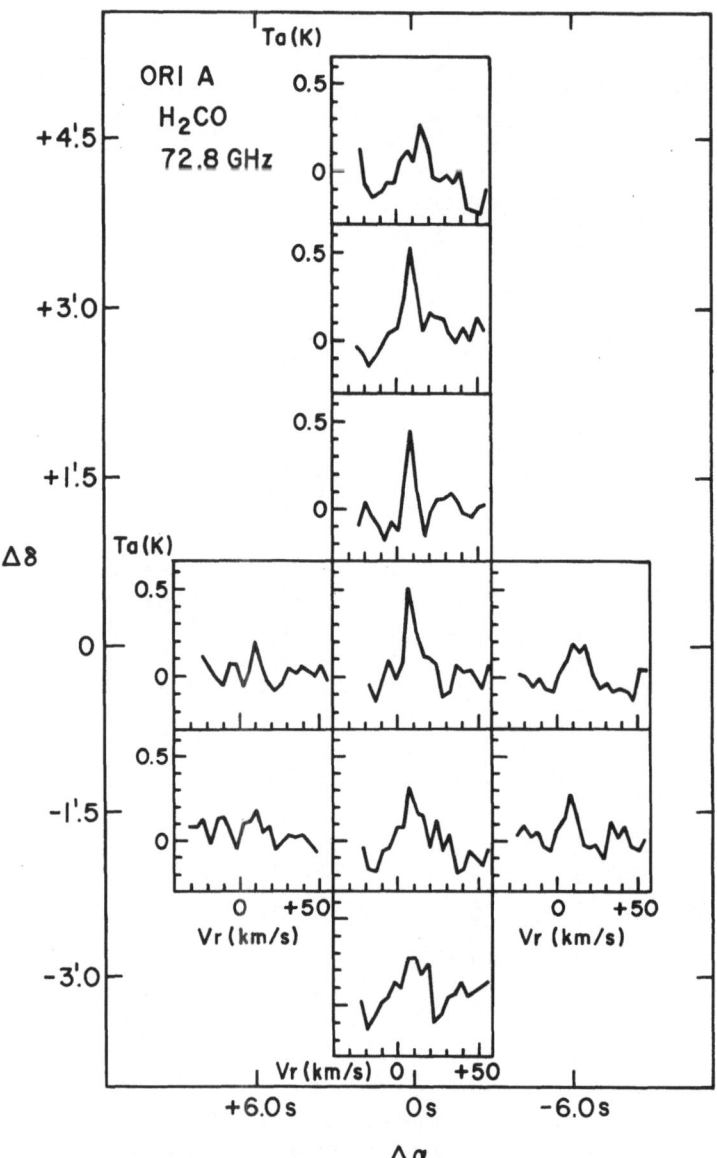

Fig. 3.   Profiles of 1_{01}–0_{00} line of para-formaldehyde at the corresponding points.

References

Snyder, L. E. and Buhl, D.: 1971, *Astrophys. J. Letters* **163**, L47.

Thaddeus, P., Wilson, R. W., Kunter, M., Penzias, A. A., and Jefferts, K. B.: 1971, *Astrophys. J. Letters* **168**, L59.

Norio Kaifu

Tetsuo Iguch

Dept. of Astronomy,

University of Tokyo,

Tokyo, Japan

Masaki Morimoto

Tokyo Astronomical Observatory,

Mitaka, Tokyo, Japan

DISCUSSION

Robinson: You state that you need very high densities and low temperatures. What are the numbers?

Morimoto: Particle densities I don't know. The temperature is exactly what you see at the CO line, that is, something like 28 K.

Zuckerman: The 4-mm continuum map of Orion that you showed peaked up at the position of the Trapezium. This is also true at 3 mm as described in a recent paper in the *Astrophysical Journal*. However, recent 1-mm measurements at Cal Tech with the 200-in. telescope show that the peak is at the Kleinmann-Low nebula, and the shape of the continuum contours agrees very well with the shape of the background molecular cloud as deduced from 2-cm H_2CO emission profiles. Thus, at wavelengths $\lesssim 1$ mm it appears that the bulk of the continuum radiation originates in the molecular cloud rather than the H II region whereas at wavelengths $\gtrsim 3$ mm the situation is reversed. It should be interesting to obtain an ~ 2 mm continuum map of Orion.

Morimoto: Our 4-mm contour shows a skirt stretched to and covering over the IR sources; it does not show much in the total flux.

LARGE-SCALE EXPANSION IN THE CARINA NEBULA

W. K. HUCHTMEIER and G. A. DAY

Division of Radiophysics, CSIRO, Sydney, Australia

Abstract. Radio recombination line observations of the Carina nebula at 6 cm and 3.4 cm were made at Parkes. Around Carina II the profiles are separated by as much as 45 km s^{-1}. This is interpreted as expansion in the H II region.

Radio recombination line observations of the Carina nebula at 6 cm and 3.4 cm were obtained with the 64-m Parkes radiotelescope. The splitting into two overlapping profiles is seen all around Carina II, whereas Carina I shows single features. The greatest separation of the overlapping profiles is 45 km s^{-1}, a value comparable to radial motions in a number of H II regions as observed optically. There are strong arguments in favour of interpreting this phenomenon as an expansion in one H II region rather than as a chance coincidence of two H II regions along the line of sight. Radial velocities from Hα observations over the nebula and the median velocities of the radio recombination lines agree well and lead to a kinematic distance of 2.7 kpc, which is in good agreement with recent distance determinations of stars in the Carina nebula and especially the two open clusters Trumpler 14 and 16. Two areas of largest linewidth and greatest separation of the two overlapping profiles are imbedded into a region of profiles with smaller separation that finally runs out into single features; the mean radial velocity of the observed lines are about the same all over the nebula. Centered to those expanding spheres are the two young open galactic clusters Trumpler 14 and 16 and the peculiar object η Carinae, suggesting that massive stellar winds are responsible for the observed expansion.

A fuller account will be published in *Astronomy and Astrophysics*.

W. K. Huchtmeier
G. A. Day
Division of Radiophysics, CSIRO,
P.O. Box 76, Epping, N.S.W. 2121, Australia

DISCUSSION

H. R. Dickel: M. Smith and T. Bohuski have optical data which show similar line splitting of ∼40 km s^{-1} between the red and blue components (i.e., comparable to radio results). They also find large splitting well to the south of the two main radio continuum peaks but still well within the nebula. The [N II]/Hα intensity ratio for the red component is different than that found for the blue component of the lines.

Huchtmeier: R. Louise, too, has observed line splitting in the Hα over large parts of the Carina nebula.

F. J. Kerr and S. C. Simonson, III (eds.), Galactic Radio Astronomy, 239. All Rights Reserved.

DECIMETER-WAVELENGTH STUDIES OF RECOMBINATION LINES

YERVANT TERZIAN and VERNON PANKONIN

National Astronomy and Ionosphere Center, Cornell University, Ithaca, N.Y., U.S.A.

Abstract. The results of new studies of decimeter wavelength Hnα and Cnα lines toward galactic nebulae are reported. The observations were made with the 305-m radio telescope of the Arecibo Observatory at 430 MHz (247α and 248α) and 606 MHz (221α). Selected positions in W51, W49, and the Rosette nebula were observed. The observed lines exhibit no pressure broadening; however, this is consistent with the interpretation of the data.

I. Introduction

Observations of radio recombination lines are well established as an important means of studying the properties of H II regions, especially when a nebula is optically obscured. However, present analyses of hydrogen lines are based almost exclusively on data obtained at frequencies greater than ~ 1 GHz. Low frequency observations are necessary in order to determine whether the source parameters derived can account for hydrogen line emission characteristics over an extended frequency range as well.

Current theories of pressure broadening in recombination lines predict a very strong dependence on the principal quantum number of the transition. Therefore, low frequency (high quantum number n) observations seem to be ideal tests for the significance of these effects.

Recombination line observations also promise to provide information which is vital to an understanding of the physical processes in predominantly neutral hydrogen regions. In this case the factors which influence the recombination line emission are not well known, and observational data are relied upon to indicate the types of source models which must be considered. In particular there are no convincing predictions of the frequency behavior of carbon recombination emission, but one interesting fact was noticed: The carbon lines become stronger relative to the corresponding hydrogen lines as the frequency of the observations decreases. The behavior of carbon line emission at frequencies below 1 GHz must be determined experimentally to help limit the possible source models.

Successful recombination line observations have been performed at frequencies below ~ 0.7 GHz. Penfield *et al.* (1967) reported the detection of H253α (~ 0.4 GHz), but only from W80. Pedlar and Davies (1972) have obtained 220α (~ 0.61 GHz) spectra toward 12 positions in galactic H II regions. In the present paper we will outline the major results of observations at the Arecibo observatory which extend the frequency range to 0.43 GHz (248α) for several selected nebulae (Parrish *et al.*, 1972; Parrish *et al.*, 1974; Pankonin and Parrish, 1974).

F. J. Kerr and S. C. Simonson, III (eds.), Galactic Radio Astronomy, 241–248. All Rights Reserved.
Copyright © 1974 by the IAU.

II. Observations at Arecibo

To date recombination line observations at Arecibo have only been performed in the lines of sight to known H II regions. 247α and/or 248α (~ 0.43 GHz) and 221α (~ 0.61 GHz) spectra were obtained toward five positions in three well-known galactic nebulae. The relevant instrumental parameters are given in Table I. The positions observed are listed in Table II together with the hydrogen line parameters derived from these observations. (The 247α and 248α spectra were averaged before the line parameters were determined.) The parameters of detected carbon lines are listed in Table III. The 247, 248α spectra of W51 A and W49A are shown in Figures 1 and 2.

The intensity of helium recombination emission from H II regions is expected to be $\sim 10\%$ that of the corresponding hydrogen line (Palmer *et al.*, 1969; and Churchwell, 1970). There was no indication of helium lines in any of the 221α and 247, 248α spectra observed at Arecibo (Pankonin, 1973). However, since the peak intensities of the hydrogen lines in this study were $\lesssim 10$ times the rms noise in the spectra this result is

TABLE I

Instrumental parameters

Frequency (MHz)	Transition	Sensitivity (K Jy^{-1})	HPBW (arcmin)	System temp. (K)	Max. freq. resol. (km s^{-1})
606	221α	4.5	9	170, 350	3.3[a]
430	247α, 248α	14.7	9	130	4.8[a]

[a] These correspond to 7 kHz.

TABLE II

Hnα Observations

Source	W51A	W51B	W51C	W49A	Rosette
Gal. Coord.	G49.5–0.4	G49.0–0.3 G49.1–0.4	G49.0–0.6	G43.2+0.0	G206.0–2.1
R.A. (1950)	$19^h21^m28^s$	$19^h20^m18^s$	$19^h21^m25^s$	$19^h07^m52^s$	$06^h28^m30^s$
Dec. (1950)	14°24′28″	14°02′05″	13°52′00″	08°59′08″	05°13′20″
			H247, 248α		
T_L/T_c	$(1.8\pm0.1)\,10^{-3}$	$(3.7\pm0.4)\,10^{-3}$	$(3.2\pm0.8)\,10^{-3}$	$(1.8\pm0.2)\,10^{-3}$	$(2.8\pm0.3)10^{-3}$
Δv_L (km s^{-1})	31.4±2.3	30.9±3.7	15.3±4.1	15.0±1.7	22.1±3.2
V_{LSR} (km s^{-1})	54.6±0.9	64.1±1.6	56.3±2.4	11.2±0.8	10.8±1.3
			H221α		
T_L/T_c	$(1.9\pm0.2)\,10^{-3}$	$(3.9\pm0.5)\,10^{-3}$	$<3.0\times10^{-3}$	$(3.4\pm0.5)\,10^{-3}$	$(4.8\pm0.6)10^{-3}$
Δv_L (km s^{-1})	31.6±3.2	32.8±4.5	–	8.4±1.4	17.2±2.3
V_{LSR} (km s^{-1})	58.9±1.4	58.4±1.8	–	9.7±0.7	15.1±1.0

Note: Errors quoted here and in Table III represent one standard deviation.

TABLE III
Cnα Observations

Source	W51A	W51B	W49A[a]
		C247, 248α	
T_L/T_c	$(1.4 \pm 0.1)\, 10^{-3}$	$(2.2 \pm 0.7)\, 10^{-3}$	$\leqslant 1.0 \times 10^{-3}$
Δv_L (km s^{-1})	15.2 ± 1.8	8.4 ± 3.6	–
V_{LSR} (km s^{-1})	61.6 ± 0.8	62.2 ± 1.6	–
		C221α	
T_L/T_c	$(1.1 \pm 0.2)\, 10^{-3}$	$\leqslant 2.4 \times 10^{-3}$	–
Δv_L (km s^{-1})	16.2 ± 3.7	–	–
V_{LSR} (km s^{-1})	62.0 ± 1.8	–	–

[a] A recombination line has been detected in the spectrum of W49A, which may be due to carbon and is probably not associated with the W49A source; see text for discussion.

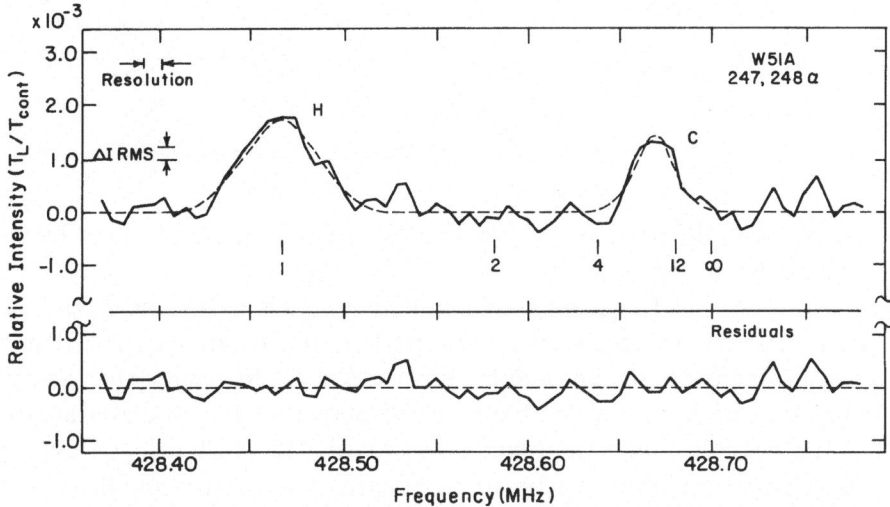

Fig. 1.   247, 248α spectrum of W51A and the residuals after subtraction of the fitted function (dashed line in the upper part). The numbers beneath the spectrum indicate the expected position of recombination lines of hydrogenic emitters of that mass if they have the same radial velocity as the H line.

not surprising. Pedlar and Davies (1972) conclude that in certain H II regions there may be a deficiency of ionized helium in the low density parts of the nebulae which are responsible for the major contribution to the low frequency hydrogen lines. This would imply that the Strömgren spheres of hydrogen and singly ionized helium do not coincide.

III. The Hnα Lines

The general problem in interpreting the low frequency hydrogen line observations is to explain the frequency dependence of the emission between 221α and 248α. The integrated line intensity at 248α relative to 221α was usually larger than could be explained by assuming uniform (in density and temperature) source models. A density and/or temperature distribution in the source or nonthermal continuum radiation in

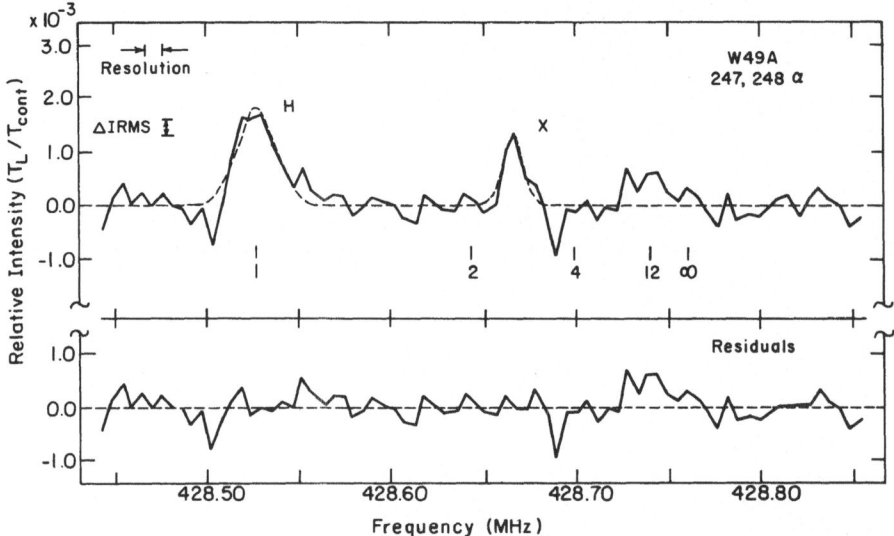

Fig. 2.   247, 248α spectrum of W49A. Figure format is described in the Figure 1 caption.

the beam provides the most straightforward explanation of the observed behavior. (The latter applies to W51C.)

In analyzing the low frequency results it is not possible to assume *a priori* either an optically thin source or LTE emission. Both optical depth and LTE departure effects must be considered, and both are functions of the frequency and the density and temperature of the source. With the limited amount of data available at low frequencies it is possible to infer approximate source models which will account for the observed emission, and thus to estimate the electron temperature and density of the emitting regions.

The electron temperatures for the H II regions considered here are fairly well known and fall within the range $5000 \lesssim T_e \lesssim 10\,000$ K. These temperatures are typical of H II regions in general. Normally the density structure is more pronounced than temperature variations in H II regions, hence to a first approximation isothermal source models can be assumed. Only in the case of W49A must there be a temperature distribution as well as a density distribution.

The Doppler temperature, calculated from the measured width of a Doppler broadened line, can be considered an upper limit to the electron temperature in the emitting region. In this way the widths of the H247, 248α and H221α lines from W49A impose an upper limit to the electron temperature of ~ 2700 K (see Figure 4) in the part of the nebula which emits those low frequency lines.

If current pressure broadening theories are correct, it should be possible to use the measured widths of low frequency recombination lines to estimate the electron density in the emitting region, or at least obtain an upper limit. Griem (1967) indicates pressure broadening effects are very sensitive to the principal quantum level of the transition and should dominate the line width for high level transitions. For the ratio of the pressure broadened to Doppler broadened line width, Griem gives

$$\frac{\Delta v_p}{\Delta v_D} \propto \frac{N_e}{\sqrt{T_e T_D}} n^7,$$

where T_e and T_D are the electron and Doppler temperatures and N_e is the electron density of the emitting region.

There was no conclusive evidence of pressure broadening in any of the lines observed at Arecibo (see also Parrish *et al.*, 1972). Figures 3 and 4 show the measured

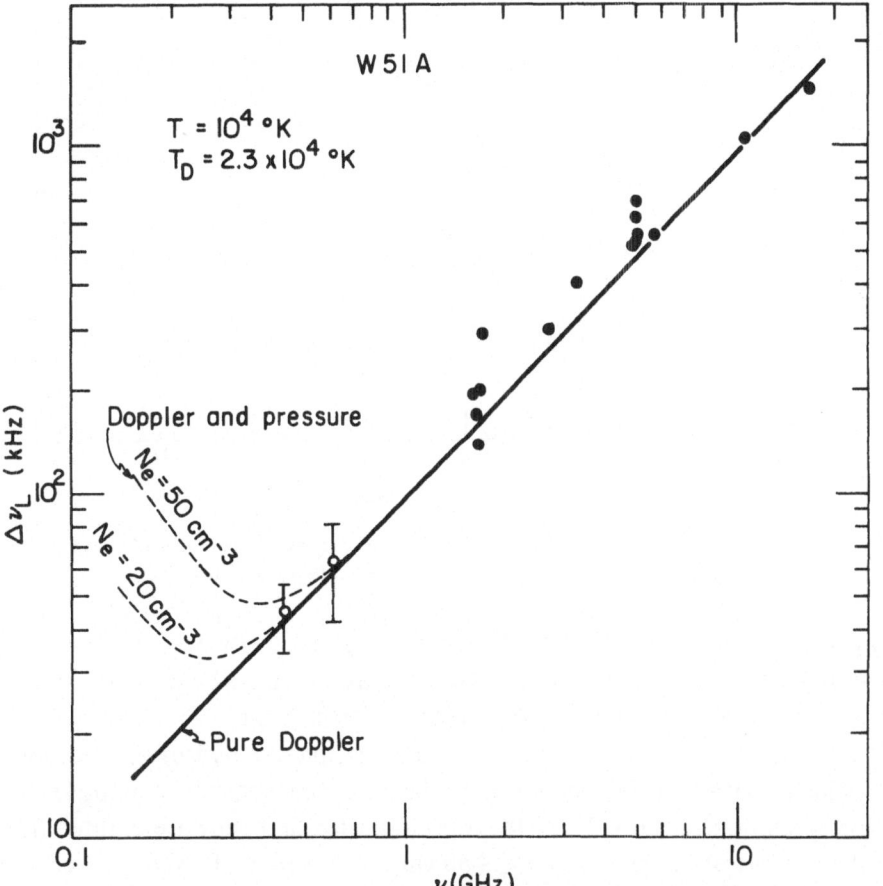

Fig. 3. Widths of Hnα lines from W51A. Open circles indicate present results. Pressure broadening effects (dashed curves) are compared to expected pure Doppler broadened line widths for the physical conditions indicated.

widths of Hnα lines plotted vs. frequency of the observations from W51A and W49A. In W49A the low frequency line widths are narrower than expected for Doppler broadening, in part indicating low temperatures as discussed above. It should be pointed out that the attenuation of the line radiation by continuum absorption increases with decreasing frequency, and pressure broadening redistributes the line

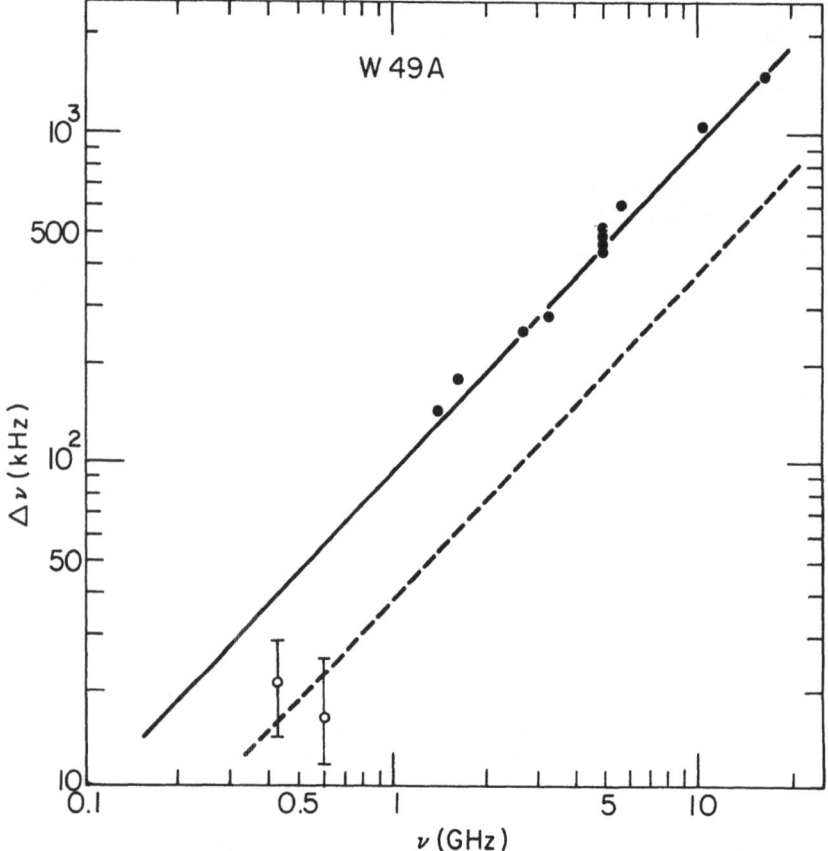

Fig. 4.   Widths of Hnα lines from W49A. Open circles indicate present results. Solid and dashed lines represent expected widths of Doppler broadened lines.

energy into the line wings. The resulting lower peak intensity becomes more difficult to detect. A combination of these two factors may explain why pressure broadening effects have not been observed in high quantum level α lines.

In summary, the non-LTE analysis of high frequency hydrogen recombination lines yields physical parameters typical of the most dense structure in the H II region. The interpretation of the H247, 248α lines indicates that they are emitted from the least dense parts of the nebulae – $N_e \lesssim 50$ cm^{-3}. These low densities are compatible with the upper limits imposed by pressure broadening considerations.

It appears that a better test of pressure broadening theories is to make observations of a series of higher order lines at nearly the same frequency. This is most practical at $v \gtrsim 1.5$ GHz, and these higher frequency lines are probably emitted from the higher density regions of the nebulae. Higher order observations have been performed by numerous investigators; generally with a negative result on pressure broadening effects. However, recently Simpson (1973) reports that the H211β profiles of Orion A, M17, and W3 are pressure broadened. She also states that the H168β and H211β intensity ratios agree with LTE predictions.

IV. The Cnα Lines

Current theories of carbon recombination lines predict that under certain conditions the lines should be observed in absorption at some low frequency (Dupree, 1974). However, this has not been observed. In fact for W51A the carbon line intensity continues to increase relative to hydrogen as the frequency decreases at least down to 0.43 GHz.

The W51A carbon lines are unusual because their observed widths are at least twice as wide as carbon lines in general. The best spectral resolution to date (3.5 km s^{-1}, Chaisson, 1973) has not resolved the Cnα lines from W51A into more than one component.

An unidentified line was detected in the 247, 248α spectrum of W49A. That spectrum is displayed in Figure 2. Pankonin et al. (1973) discuss that this is due to electronic recombination onto atomic carbon located in an intervening spiral arm between the nebula W49A and the observer.

With the possible exception of a carbon line toward W3A (Chaisson, 1972) previous observations have tended to associate the detected carbon recombination emission with H I clouds which may be physically related to H II regions. If the present identification is correct, then there are important implications that conditions exist for detectable carbon recombination emission to originate far from normal H II regions.

Acknowledgements

The active collaboration of Dr A. Parrish in many phases of the Arecibo recombination line study is gratefully acknowledged. The Arecibo Observatory, National Astronomy and Ionosphere Center, is operated by Cornell University under contract to the U.S. National Science Foundation.

References

Chaisson, E. J.: 1972, Nature, Phys. Sci. 239, 83.
Chaisson, E. J.: 1973, in press.
Churchwell, E.: 1970, Ph.D. Thesis, Indiana University.
Dupree, A. K.: 1974, Astrophys. J. 187, 25.
Griem, H. R.: 1967, Astrophys. J. 148, 547.
Palmer, P., Zuckerman, B., Penfield, H., Lilley, A. E., and Mezger, P. G.: 1969, Astrophys. J. 156, 887.
Pankonin, V.: 1973, Ph.D. Thesis, Cornell University.
Pankonin, V. and Parrish, A.: 1974, in preparation.

Pankonin, V., Parrish, A., and Terzian, Y.: 1973, *Astrophys. J. Letters* **180**, L113.
Parrish, A., Pankonin, V., Heiles, C., Rankin, J. M., and Terzian, Y.: 1972, *Astrophys. J.* **178**, 673.
Parrish, A., Pankonin, V., and Terzian, Y.: 1974, in preparation.
Pedlar, A. and Davies, R. D.: 1972, *Monthly Notices Roy. Astron. Soc.* **159**, 129.
Penfield, H., Palmer, P., and Zuckerman, B.: 1967, *Astrophys. J.* **148**, L25.
Simpson, J. P.: 1973, *Astrophys. Space Sci.* **20**, 187.

Yervant Terzian
Vernon Pankonin
National Astronomy and Ionosphere Center,
Cornell University,
Ithaca, New York 14850, U.S.A.

DISCUSSION

Zuckerman: You mentioned the possible detection of the H274α line at Arecibo. I would like to point out that this would set a record in radio astronomy as the lowest frequency spectral line yet observed. Previously the H253α line observed in 1966 by Penfield, Palmer and myself had the lowest frequency.

Terzian: We have recently made observations of the H274α line at 318 MHz, and we have good indications of a detection; however, we need a better signal to noise ratio before we say anything more on this.

Mezger: Apart from breaking an all-time record on quantum number, the thing has serious repercussions on the nature of the carbon recombination line.

OPTICAL OBSERVATIONS OF
IONIZED GAS IN EXTERNAL GALAXIES

G. MONNET

Observatoire de Marseille, Marseille, France

Abstract. This paper reviews recent optical results on the large scale distribution of ionized gas in spiral galaxies, including our own. There is a diffuse, inhomogeneous emission in the arm region in spirals, including our Galaxy, and in gas-rich galaxies a fainter diffuse emission between the arms.

I. Ionized Gas in External Galaxies

Surveys of the ionized gas in external galaxies show three major distributions on a large scale:

(i) The classical H II regions, mainly located in the spiral arms, as is well known. The emission measures (computed throughout this paper from the Hα line intensity for a temperature of 7500 K) are typically a few times 10^3 cm^{-6} pc.

(ii) A diffuse arm emission which connects the H II regions. It can be seen easily on plates taken in Hα light through an interference filter for a number of gas-rich galaxies, i.e., of S_c to S_d Hubble types, for instance in M33, M101, M83,... (Monnet, 1971). On the nearest S_{cd} galaxy, M33, this emission exhibits a clear filamentary structure (Figure 1), which a scale of ∼100 pc.

The same phenomenon – with an emission measure of ∼100 cm^{-6} pc – has been found in the nearest S_b galaxy, M31, by Deharveng and Pellet (1970) (Figure 2).

The intensity ratios between Hα 6563 Å and [N II] 6584 Å have been studied in M33 by Benvenuti *et al.* (1973) and by Comte (1974) (in preparation) with similar conclusions. Results by Comte are shown in Figure 3; the Hα/[N II] ratio is fairly normal in the diffuse arm emission, ∼5. I have made a rough study of the Hα/[N II] ratio in M31 from a comparison of the Hα plate shown in Figure 2 and a [N II] plate taken in the same field. The ratios appear also quite normal in the diffuse arm emission. These results indicate that the ionization is probably normal, similar to that of the more condensed H II regions.

(iii) A diffuse emission in the disk, between the arms. The first detection of a disk emission was made by Mayall and Aller (1939) in M33, in the [O II] 3727 Å doublet, and the first two dimensional study by Carranza *et al.* (1968) likewise in M33, but in the Hα and [N II] 6584 Å lines. So far disk emissions have been detected – with emission measures around 30 cm^{-6} pc – only in the central part (2 to 3 kpc diameter) of S_c to S_d type galaxies, marginally detected in M51 by Tully (1974) and not at all in the S_b spiral M31. In M33, the Hα to [N II] ratio is abnormal, ∼1.5, which can be explained in terms of ionization by late blue stars – either normal B stars or the dwarf blue stars hypothetized by Hills (1972) – or with more exotic ionization sources, as hard UV, soft X-rays or MeV protons. For a clearcut answer one needs a sufficient number of line intensity ratios, as the theoretical values in each case have been com-

F. J. Kerr and S. C. Simonson, III (eds.), Galactic Radio Astronomy, 249–256. All Rights Reserved.
Copyright © 1974 by the IAU.

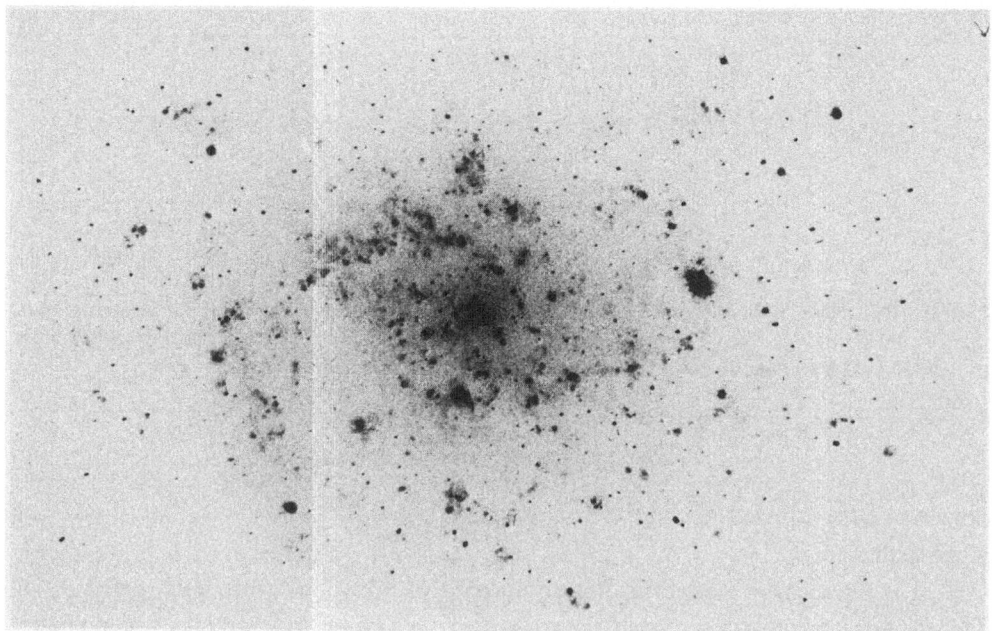

Fig. 1. Hα filter photograph of M33 over a 40' field made with the 193 cm telescope at Haute Provence Observatory. North is at right. Note the diffuse filamentary emission in the arms. The disk emission is too weak to be seen on a photograph and can be only detected with spectrographs.

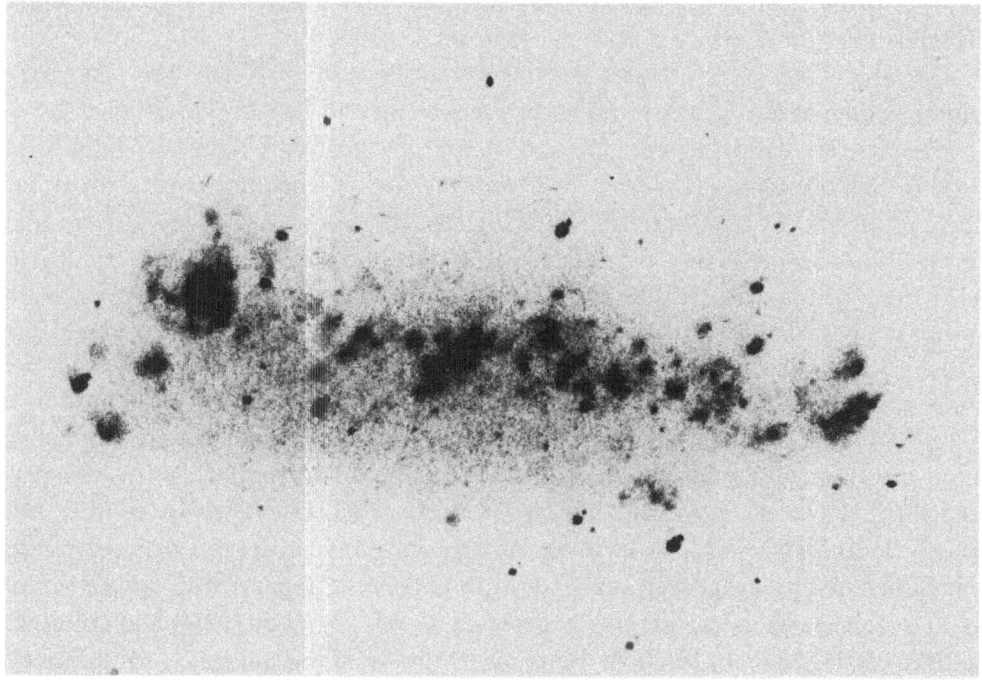

Fig. 2. Hα filter photograph of M31 showing a 15' length of the N_4 Spiral arm (Baade and Arp's notation). Note the diffuse filamentary emission in the arm between the classical H II regions. The H II regions are much fainter than in Sivan's atlas of our Galaxy made with a roughly similar scale in parsecs, showing that our Galaxy is more likely a $S_{bc} - S_c$ than a S_b.

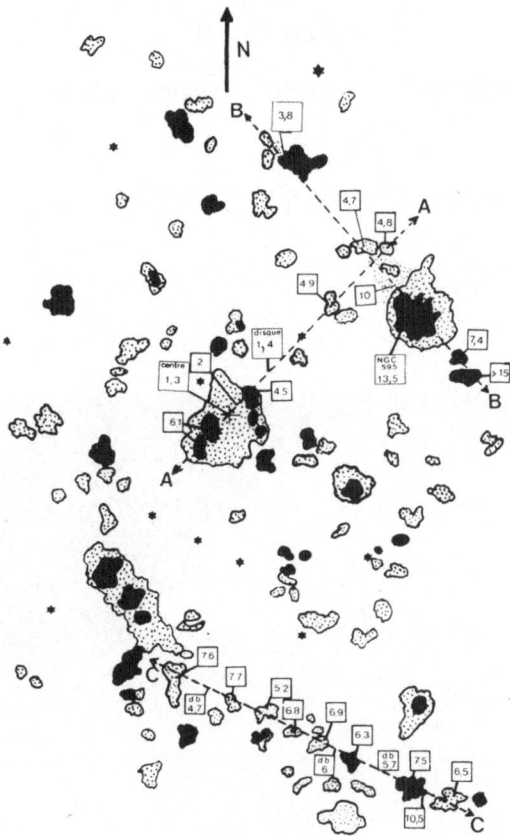

Fig. 3.   Map of the Hα/[N II] ratios on a 15′ field centered on M33, obtained by Comte with a Cassegrain spectrograph attached to the 193-cm telescope at Haute Provence Observatory.

Spectra are labeled *A*, *B* and *C*. Note the low values 1.4 in the disk emission (disk) and in the very central part of the galaxy (Spectrum A). Note the high values 4.7, 5.7, and 6 on the diffuse arm emission (db) on Spectrum C.

puted for the most common lines by Bergeron and Souffrin (1971). Benvenuti *et al.* (1973) have discovered a faint [S II] emission in the disk of M33, with an abnormal Hα/[S II] ratio ∼ 1.5 (instead of 10 for classical H II regions). A similar result has been obtained by Comte (1974) (in preparation), who has also observed Hβ and an exceedingly faint [O III] 5007 Å emission, but with a probability of positive detection of only 80%. These data are summarized in Table I. The faintness of the [O III] line definitely excludes ionization by hard UV or soft X-rays and leaves moderately blue stars or MeV protons as the only candidates.

II. Ionized Gas in Our Galaxy

Judging from the large scale distribution in external galaxies, one can expect in our Galaxy – whose Hubble type is most probably in the S_{bc}, S_c range – a diffuse arm

TABLE I

Ionized gas in M33

	Emission measure	Hα/N II	Hα/S II	Hβ/O III	Ionization
Classical H II regions	> 10^3 cm^{-6} pc		8 to 12	0.3 to 1	O stars
		4 to 8			
Diffuse arm emission	100 cm^{-6} pc		> 5	not measured	probably O stars
Disk emission	30 cm^{-6} pc	1.5	1.5	⩽ 4	B stars or MeV protons

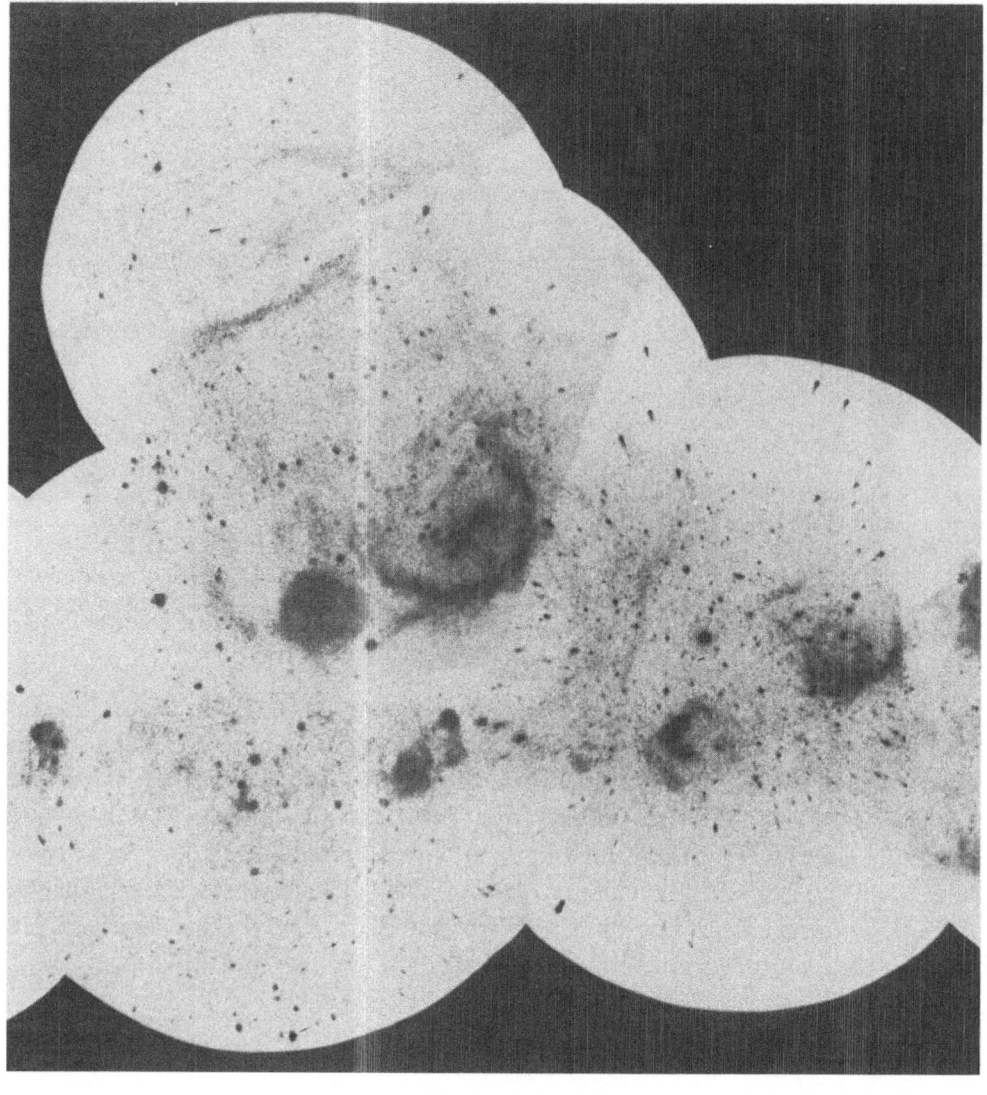

Fig. 4.   Hα photograph over an 85° field in the Orion region. One sees filaments over a 60° field. The strongest filament at top has been discovered by Meaburn (1965) and interpreted by him as an optical counterpart of the Cetus arc. The two large nebulae at the center of the plate are λ Orionis and the Orion loop.

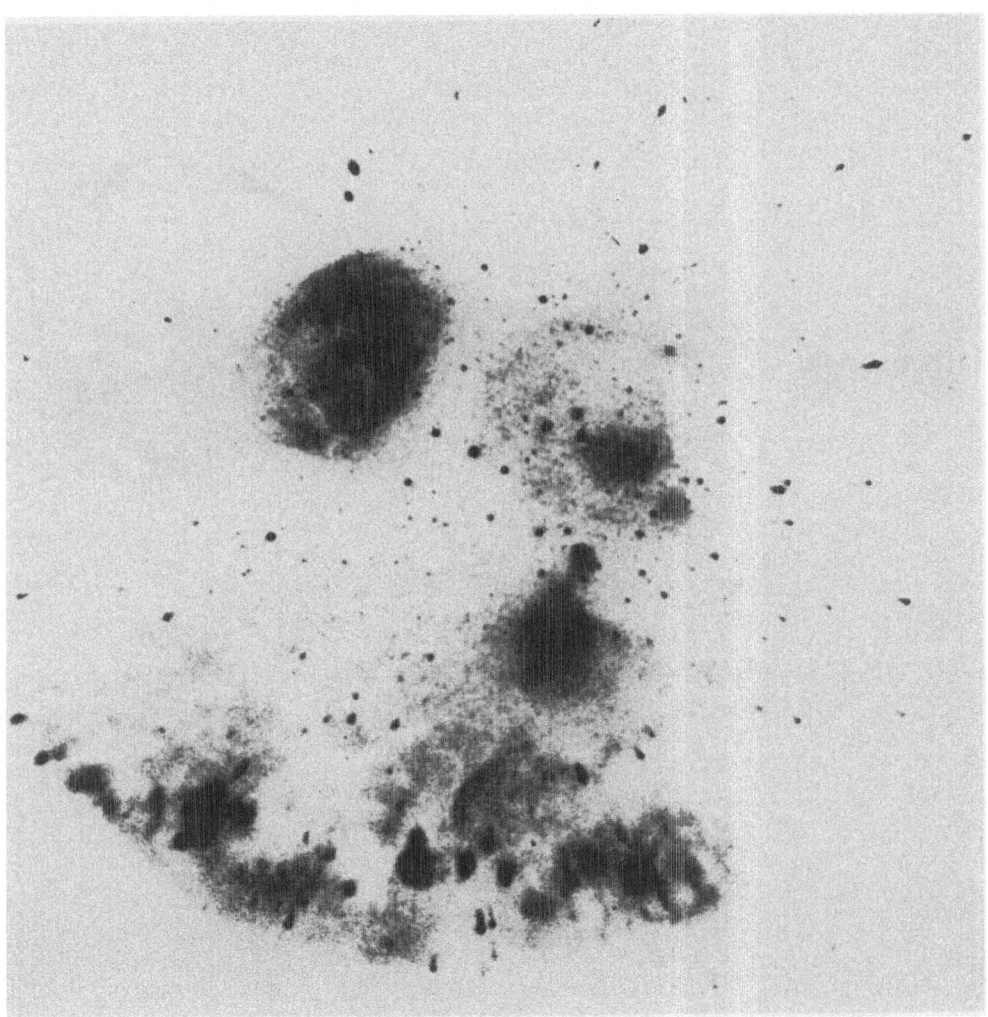

Fig. 5.   Hα photograph over a 40° field in Sagittarius. At bottom a portion of the Sagittarius arm. Top left, the ζ Ophiuchi nebula. Top right a new shell nebula in Scorpius. The position and the shape of this nebula are strikingly similar to a neutral hydrogen shell found by Sancisi and presented during this symposium.

emission ~ 100 cm^{-6} pc (perpendicular to the plane). This emission has been observed by various authors in radio recombination lines (see Guelin, 1974) and quite recently in optical lines by Reynolds and Scherb (1973).

On the other hand, if there is any disk emission it is probably very faint (a few cm^{-6} pc), quite close to the center, and of course severely blended with the stronger emission in the arm.

A near whole-sky survey in Hα has been made by Sivan with a small diameter (1 cm), large field (55°) and high aperture ratio ($F/1$) telescope. Part of this survey is shown in Figures 4, 5 and 6. There is a clear diffuse arm emission with some filament-

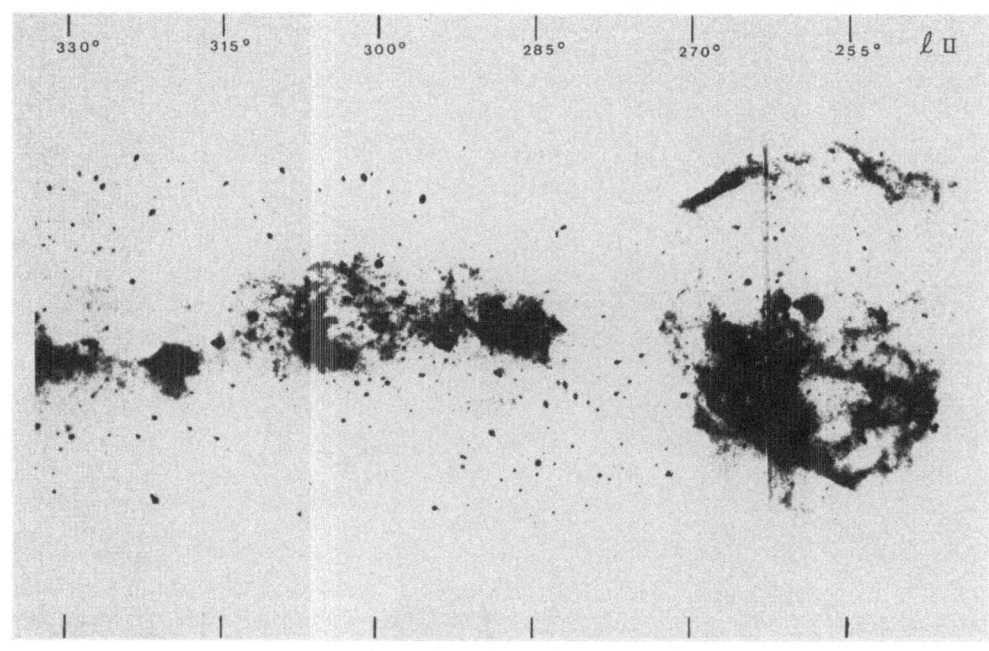

Fig. 6.   85° × 40° field in the Southern Milky Way from Puppis to Norma. The Gum nebula (40° diameter) is at right. North is at top.

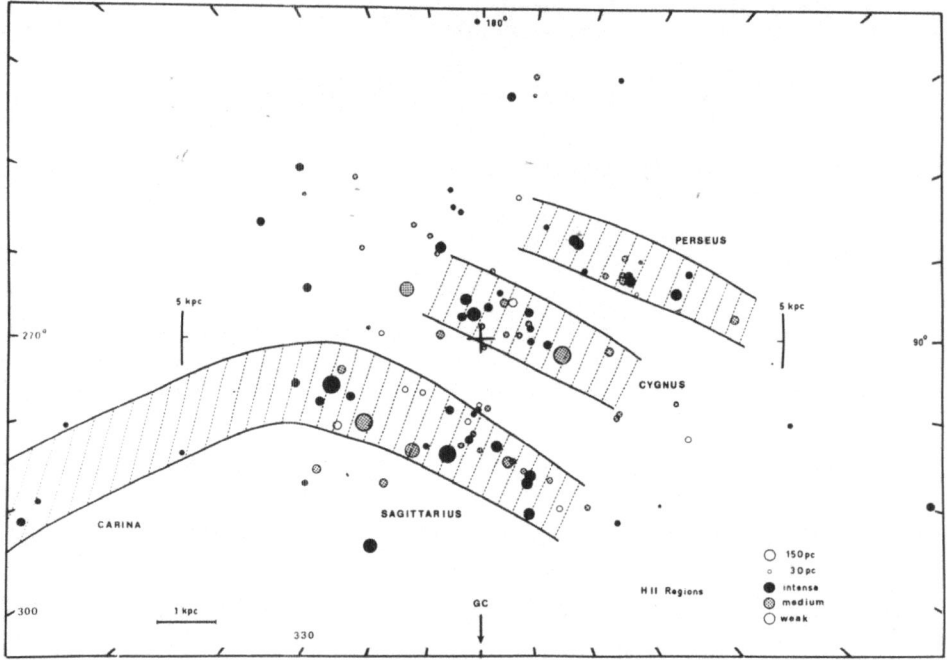

Fig. 7.   Map of the optically identified H II regions. Strong to weak H II regions are indicated by circles with three different intensities and with their real diameter. The regions of continuous arm emission are shaded.

ary structure in all the major nearby spiral arms, i.e., in Orion, in Perseus, and in the Carina-Centaurus region.

Of special interest is the Orion region where nearby filaments cover a 60° field on the sky. This structure is probably closely related to the filaments in neutral gas at high galactic latitudes shown by Heiles during this symposium, but obviously this point deserves a careful study.

These diffuse emissions have been put on a map of the optically identified H II regions in our Galaxy (Figure 7). The distances of the H II regions have been computed by Georgelin, using the photometric distance of the exciting stars *and* the radial velocities measured with a Fabry-Pérot interferometer.

Of special interest is the connection between the Carina and the Sagittarius arm between longitudes 292°–304° at a distance ~ 1.5 kpc from the Sun. One sees a continuous Hα emission over all this range and a strong H II region at longitude 295° (Figure 6), opposite to the conclusion reached by Kerr and Kerr (1970) from a study of that region in the thermal radio continuum.

On the other hand there is no connection between the Carina arm and the Vela arm between longitudes 282°–268°. This supports the Weaver (1970) model of our Galaxy, but not the Leiden-Sydney model.

References

Benvenuti, P., D'Odorico, S., and Peimbert, M.: 1973, *Astron. Astrophys.* **28**, 447.
Bergeron, J. and Souffrin, S.: 1971, *Astron. Astrophys.* **14**, 167.
Carranza, G., Courtès, G., Georgelin, Y., Monnet, G., and Pourcelot, A.: 1968, *Ann. Astrophys.* **31**, 61.
Deharveng, J. and Pellet, A.: 1970, *Astron. Astrophys.* **9**, 181.
Hills, J.: 1972, *Astron. Astrophys.* **17**, 155.
Kerr, F. J. and Kerr, M.: 1970, *Astrophys. Letters* **6**, 175.
Mayall, N. and Aller, L.: 1939, *Publ. Astron. Soc. Pacific* **51**, 112.
Meaburn, J.: 1965, *Nature* **208**, 575.
Monnet, G.: 1971, *Astron. Astrophys.* **12**, 379.
Reynolds, R., and Scherb, F.: 1973, preprint.
Tully, B.: 1974, *Astrophys. J. Suppl.* **27**, 415.
Weaver, H.: 1970, in W. Becker and G. Contopoulos (eds.), 'The Spiral Structure of Our Galaxy', *IAU Symp.* **38**, 126.

G. Monnet
Observatoire de Marseille,
2, Place Le Verrier,
F-13004 Marseille, France

DISCUSSION

Baldwin: What is the value of $\langle n_e^2 \rangle$ which you deduce from your observations of the distributed Hα emission in the arms of our Galaxy? And is this value consistent with the values of $\langle n_e \rangle$ obtained from pulsar dispersion measures?

Monnet: With an emission measure of about 100 cm^{-6} pc and a typical size of about 100 pc, the value of $\langle n_e^2 \rangle^{1/2}$ is of the order of 1 cm^{-3}. If it was homogeneously distributed $\langle n_e \rangle$ would be equal to

1, too much to fit with the values given by the dispersion measures on pulsars. But it is clear that the ionized gas distribution is highly nonhomogeneous and there is probably no large conflict.

Burton: What is the scale height of the distributed emission in our Galaxy and in M33?

Monnet: In our Galaxy, the continuous arm emission height can be measured directly from the plates. It gives about 200 pc, similar to that of the classical H II regions. In M33 both continuous arm emission and continuous disk emission heights can only be inferred from the rms radial velocity dispersion, which is about 6 km s^{-1} – even smaller than that of the classical H II regions. The scale height must be thus similar, ~ 200 pc.

Mezger: Could you define what you mean by "low excitation ionized gas in the disk?"

Monnet: We see only line emission from ions whose ionization potentials are very similar to hydrogen, i.e., N^+, S^+ and probably O^+ but practically no O^{++}.

Terzian: How do you arrive at the lower limit of the electron density from your emission measure values? Do you assume a filamentary structure for the diffuse interstellar medium?

Monnet: If there was no filling factor n_e would be ~ 1 cm^{-3} just by dividing the emission measure by the scale height. This, of course, is only a lower limit, especially because we know that there is filamentary structure.

DISTRIBUTION OF DUST IN THE ORION NEBULA

LINDSEY F. SMITH

Max-Planck-Institut für Radioastronomie, Bonn, F.R.G.

Abstract. The distribution of the dust is found by comparing 3-cm brightness temperatures, Hα surface brightness, and Hβ surface brightness. The dust is probably mixed with the gas. The Balmer radiation originates in front of the dark bay.

Comparison is made between 3-cm brightness temperatures determined by Wink (1973) with the NRAO aperture synthesis interferometer and Hα and Hβ surface brightnesses determined from narrow band interference photographs obtained in conjunction with the Marseille group with the Haute Province Observatory 193-cm telescope.

The Hα intensity is $\sim 1/5$ that expected at the positions of peak radio brightness and $\sim 1/2$ that expected in the neighbouring regions. Yet the Hα/Hβ ratio is approximately constant over the whole central region. This implies that the dust is mixed with the gas rather than being a foreground phenomenon.

The intensity of the Balmer radiation from the darkest parts of the dark bay is $\sim 1/5$ of that of adjacent regions, but the Hα/Hβ ratio is not greater than in adjacent regions. Radio observations with resolution as high as 2' (Schraml and Mezger, 1969) have not detected any structure in the region of the bay. This implies that most of the observed Balmer radiation originates in front of the bay. The gas in the bay must be neutral and the extinction by the dust must be high enough to prevent significant contribution to the observed Balmer radiation from regions behind the bay.

References

Schraml, J. and Mezger, P. G.: 1969, *Astrophys. J.* **156**, 269.
Wink, J.: 1973, private communication.

Lindsey F. Smith
Max-Planck-Institut für Radioastronomie,
Auf dem Hügel 69,
53 Bonn 1, F.R.G.

DISCUSSION

Habing: To what extent do your results agree with those of Münch and Persson?

L. F. Smith: Since Münch and Persson have not used radio observations, I have not studied the paper carefully enough to answer the question.

Van Woerden: Do I understand that you observe Hα and Hβ in front of the dark bay? If so, how can they be reddened?

F. J. Kerr and S. C. Simonson, III (eds.), Galactic Radio Astronomy, 257–258. All Rights Reserved.
Copyright © 1974 by the IAU.

L. F. Smith: Reddening would then be due only to dust mixed with the emitting gas.

Zuckerman: Could half the reddening be due to foreground material?

L. F. Smith: Only if the foreground material covered just the central part of the nebula.

Greenberg: In estimating the ratio of Hα to Hβ would an anomolous extinction curve characterized by different total to selective extinction strongly modify your physical picture?

L. F. Smith: The ratio of 1.5 that I assumed between Hα and Hβ optical depths is for the Whitford reddening curve with $R = A_V/E_{B-V} = 3$. It is possible in Orion that R is greater than 3. While it would not affect the simple conclusion made here that most of the dust is inside the nebula, the possibility must be allowed for in the final reduction.

Baars: Now that you have these beautiful high angular resolution data in radio and Hα, Hβ wavelengths, are you able to map the distribution of the dust by your method?

L. F. Smith: That is the aim of the program. For this initial report I have only calculated the numbers at key points.

Pishmish: A qualitative confirmation for the existence of dust in the Orion region is provided by the Fabry-Pérot interferograms which we have obtained on that region. These show that in between the interference rings of Hα there is emission due to the continuum. This continuum may arise from the re-emission by the dust of the radiation from the embedded Trapezium stars. The interferograms may also help in mapping the dust distribution in regions where the continuum emission is observable.

INFRARED EMISSION FROM HII REGIONS

C. G. WYNN-WILLIAMS*

California Institute of Technology, Pasadena, Calif., U.S.A.

Abstract. Most HII regions emit large amounts of energy at infrared wavelengths. Dust grains absorb a major fraction of the total energy emitted by the O stars, and this is reradiated in the infrared. The dense HII condensations seen by aperture synthesis radio observations are strong 20-μm infrared sources. The dust is probably mixed with the ionized gas in these sources. There are also infrared sources that are not radio sources; these may be protostars. Infrared emission has also been found from many OH/H$_2$O maser sources associated with HII regions.

In the past two or three years it has become clear that most HII regions emit large quantities of energy at infrared wavelengths. Dust grains within and around the HII region absorb stellar and nebular photons and are thereby heated to temperatures in the range 50 to 200 K. Figure 1 shows the overall energy distribution of W3, a source with features common to many HII regions. It may be seen that the flux density at 100 μm is about three orders of magnitude greater than is produced by free-free and free-bound transitions in the ionized plasma.

The infrared energy distribution deviates from that of a black-body at both long and short wavelengths. At short wavelengths this effect is probably due to real variations in the dust temperatures of different parts of the source, while at long wavelengths the effect is due to the decrease in the emissivity of the dust particles with increasing wavelength. The deficit of emission shortward of 3 μm as compared to the expected free-free flux density must be attributed to extinction by dust grains. Very large amounts of extinction have been found this way, including a value in excess of 50 mag in the visible for part of W3.

Because of the opacity of the Earth's atmosphere at wavelengths near 100 μm, measurements of the total energy emitted at infrared wavelengths have to be made from small telescopes carried to high altitude by aircraft or balloons (Harper and Low, 1971; Emerson *et al.*, 1973). The results of these measurements indicate that the total infrared luminosity of an HII region is closely related to its total rate of ionization and, moreover, that the dust grains absorb and re-emit a major fraction of the total energy emitted by the O stars inside the HII region. In general, however, the dust grains do not absorb a significant number of Lyman-continuum photons direct from the star (Wynn-Williams and Becklin, 1974).

Infrared maps of HII regions may be made at wavelengths of 20 μm or shorter using large optical telescopes and beamwidths of a few arcseconds. These maps indicate that most of the dense HII condensations seen by aperture synthesis radio observations are strong 20-μm infrared sources (Figure 2). This result strongly sug-

* Now at Mullard Radio Astronomy Observatory, Cambridge, United Kingdom.

F. J. Kerr and S. C. Simonson, III (eds.), Galactic Radio Astronomy, 259–266. All Rights Reserved.

gests that the dust giving rise to the emission at this wavelength is coextensive with the ionized gas; there are, however, indirect arguments that indicate that much of the emission at around 100 μm comes from cooler dust grains associated with atomic or molecular hydrogen exterior to the ionized region.

Figure 2 shows that in W3 there also exist infrared sources which are not radio

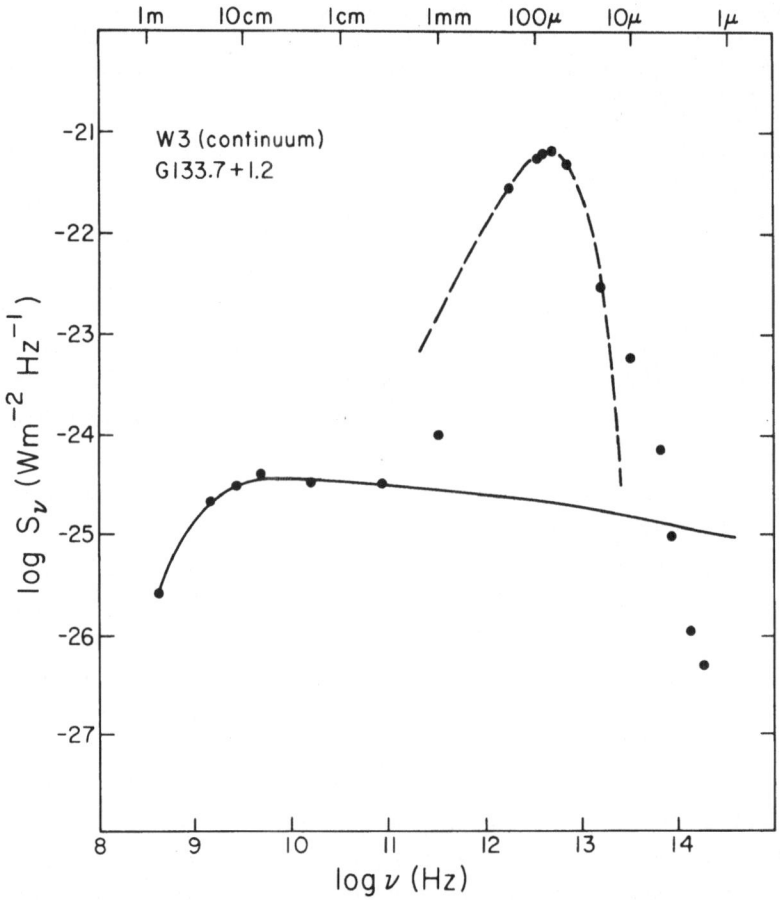

Fig. 1. The radio and infrared energy distribution of W3. Radio data are from Webster and Altenhoff (1970), Wynn-Williams (1971), Schraml and Mezger (1969) and Hobbs, *et al.* (1971). The 1-mm point is by Harvey (1973) using the Hale 200-in. telescope, the 30- to 300-μm data is by Harper (1973), and the remainder of the infrared data is from Wynn-Williams, *et al.* (1972). The solid line shows the emission expected from transitions in the H II plasma, calculated according to the relationships given by Willner *et al.* (1972), and the dashed line is a black-body curve at 70 K.

sources. IRS2, for example, seen at 2.2 μm in W3, is probably the early-type exciting star of W3(A)/IRS1. IRS5, on the other hand, is associated with the H_2O maser source and has the energy distribution shown in Figure 3. It resembles that of a black-body at 350 K except that it has a strong 'silicate' absorption at 10 μm (Aitken and Jones, 1973). This object has a luminosity of at least $3 \times 10^4 L_\odot$ and a black-body diameter

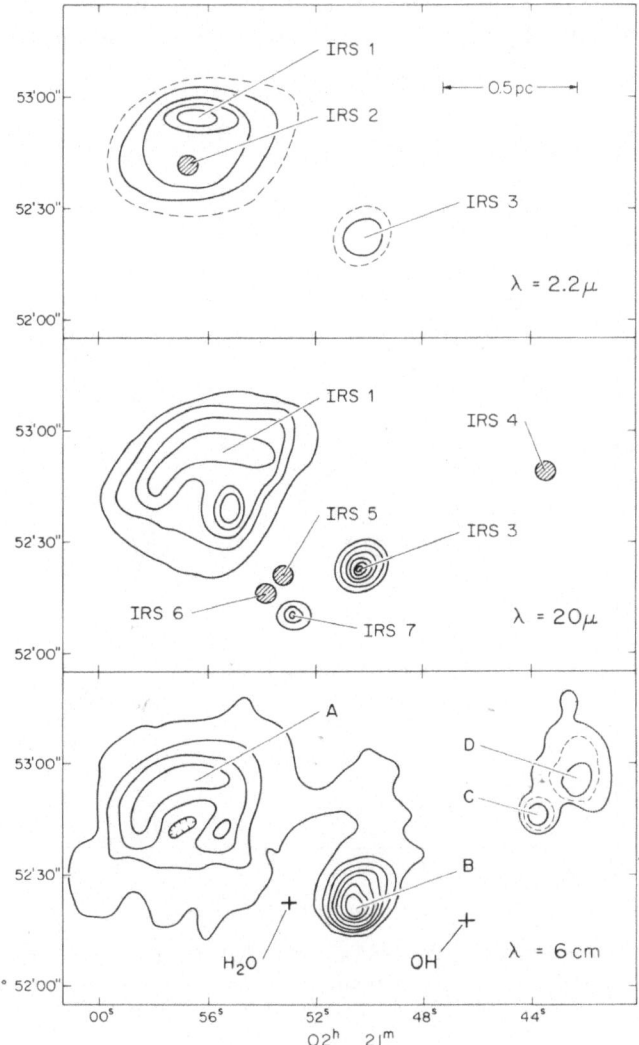

Fig. 2.   W3 (G133.7 + 1.2) mapped at 2.2 μm, 20 μm and 6 cm wavelength (Wynn-Williams, *et al.*, 1972; Wynn-Williams, 1971) with resolutions in the range 5 to 10″. The H_2O maser position is from Hills *et al.* (1972) and the OH position is from Wynn-Williams *et al.* (1974).

of 600 AU. It is tempting to speculate that W3-IRS5 is a massive protostar since an O star with this luminosity should produce a detectable H II region. Several other rather similar objects have been found in other H II regions, including one in NGC 7538, although the latter appears to be associated with faint continuum radio emission (Wynn-Williams *et al.*, 1973; Martin, 1973).

Infrared emission has been found from many, but not all, of the OH/H$_2$O maser sources associated with H II regions. In contrast to the situation found for the 1612-

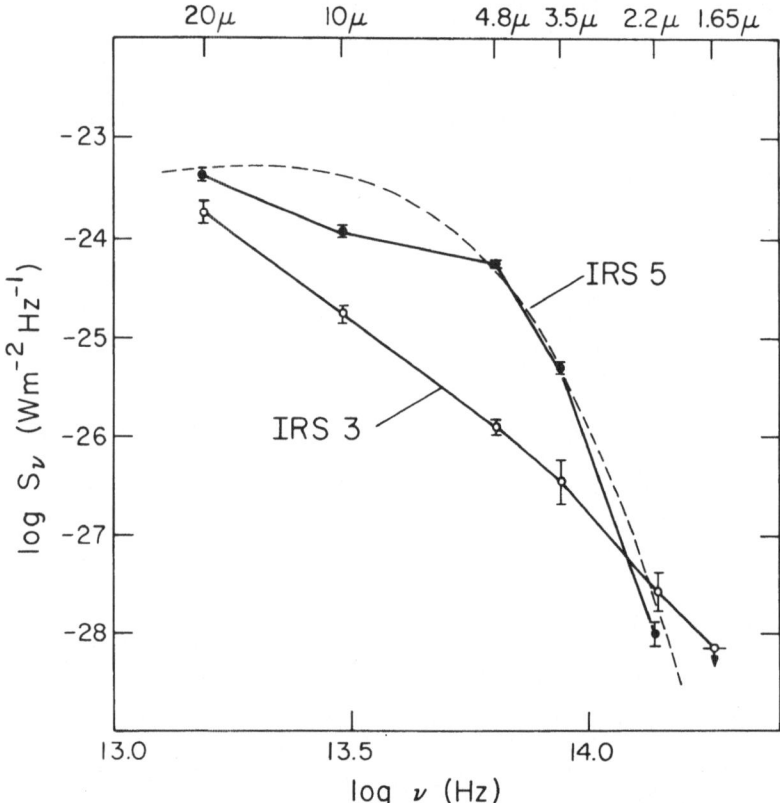

Fig. 3.   Infrared energy distributions of W3-IRS5 and W3-IRS3. (from Wynn-Williams *et al.*, 1972) The dashed line is a black-body curve at a temperature of 350 K.

MHz OH/IR stars, no quantitative relationships have yet been found between the infrared and microwave properties of these sources, except that there is perhaps a tendency for infrared sources which are associated with OH/H$_2$O masers to be hotter and more compact than are most H II components (Wynn-Williams and Becklin, 1974).

Figure 4 shows the Orion nebula at 20 μm. Most noticeable is the cluster of small sources about 1′ from the Trapezium O star cluster. These objects, which include the Becklin-Neugebauer and Kleinmann-Low infrared sources, are closely associated with the OH and H$_2$O masers, show deep silicate absorption features in their spectra, are highly polarized at infrared wavelengths and lie at the center of a more extensive cool dust cloud which is seen by 1-mm continuum and by 2-cm formaldehyde observations (Figure 5). The infrared observations indicate that the temperature of the cloud increases inwards towards the most compact part of the infrared cluster; it

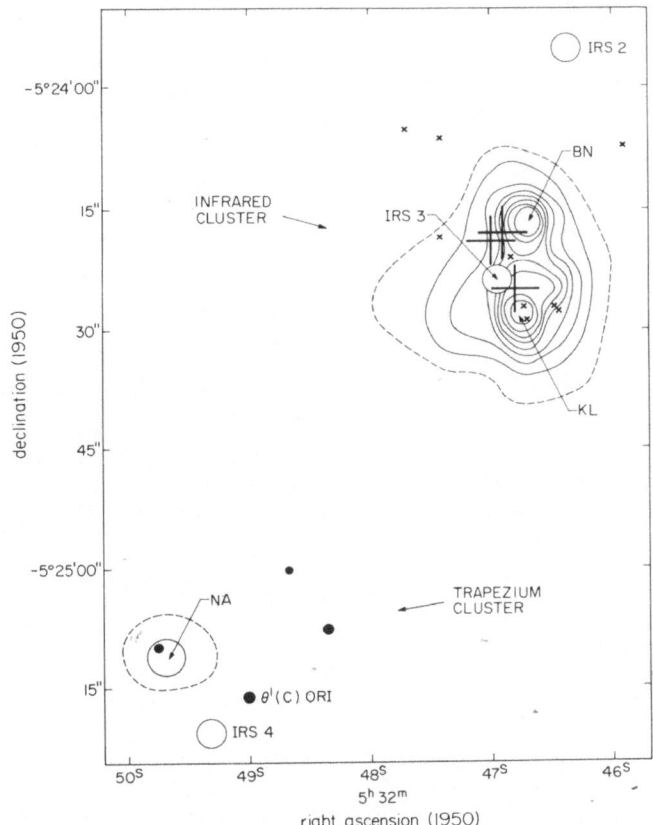

Fig. 4.   The Orion nebula mapped at 20 μm with a 5″ diaphgram on the Hale 200-in telescope by Becklin, Neugebauer and Wynn-Williams (inpublished). The contour interval is 3×10^{-15} W m^{-2} Hz^{-1} sr^{-1}, with the dashed line at half this interval. The three large crosses are the OH positions of Raimond and Eliasson (1969), while the nine small crosses are the H$_2$O maser positions of Moran et al. (1973); the whole group of nine is subject to a possible 5″ absolute position error, which is not shown in the figure.

therefore again seems very plausible that these objects are other protostars or stars formed so recently that they are still accreting matter from the surrounding clouds of gas and dust.

A much more extensive review of the infrared properties of H II regions is currently in press (Wynn-Williams and Becklin, 1974).

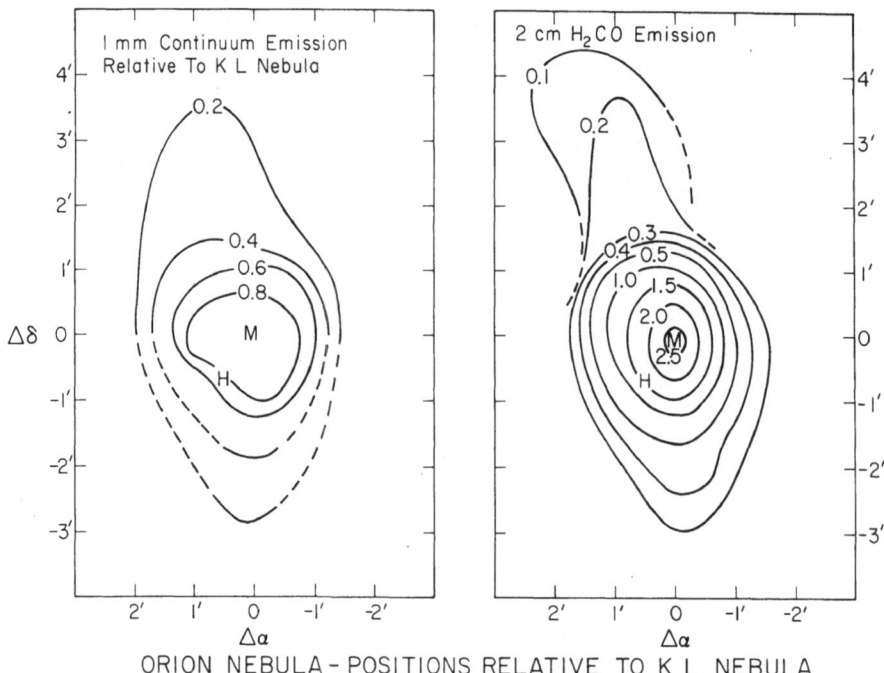

ORION NEBULA – POSITIONS RELATIVE TO K L NEBULA

Fig. 5.   The Orion nebula at 1 mm (750 to 1500 μm) as mapped with 100″ resolution on the 200-in. telescope by Harvey *et al.* (1973), and in the formaldehyde line by Evans *et al.* (1973). The peak of the H II emission, which coincides with the Trapezium cluster, is denoted H, while the center of the molecular emission, which coincides with the infrared cluster, is denoted by M.

References

Aitken, D. K. and Jones, B.: 1973, *Astrophys. J.* **184**, 127.

Emerson, J. P., Jennings, R. E., and Moorwood, A. F. M.: 1973, *Astrophys. J.* **184**, 401.

Evans, N., Morris, J. Sato, T., and Zuckerman, B.: 1973, private communication.

Harper, D. A.: 1973, private communication.

Harper, D. A. and Low, F. J.: 1971, *Astrophys. J. Letters* **165**, L9.

Harvey, P. M.: 1973, private communication.

Harvey, P. M., Werner, M. W., Gatley, I., and Elias, J.: 1973, *Astron. Soc. Pacific* meeting, Los Angeles.

Hobbs, R. W., Modali, S. B., and Maran, S. P.: 1971, *Astrophys. J. Letters* **165**, L87.

Martin, A. H. M.: 1973, *Monthly Notices Roy. Astron. Soc.* **163**, 141.

Moran, J. M., Papadopoulos, G. D., Burke, B. F., Lo, K. Y., Schwartz, P. R., Thacker, D. L., Johnston,
    K. J., Knowles, S. H., Reisz, A. C., and Shapiro, I. I.: 1973, preprint.

Raimond, E. and Eliasson, B.: 1969, *Astrophys. J.* **155**, 817.

Schraml, J. and Mezger, P. G.: 1969, *Astrophys. J.* **156**, 269.

Webster, W. J. and Altenhoff, W. J.: 1970, *Astron. J.* **75**, 896.

Willner, S. P., Becklin, E. E., and Visvanathan, N.: 1972, *Astrophys. J.* **175**, 699.

Wynn-Williams, C. G.: 1971, *Monthly Notices Roy. Astron. Soc.* **151**, 397.

Wynn-Williams, C. G., Becklin, E. E., and Neugebauer, G.: 1972, *Monthly Notices Roy. Astron. Soc.* **160**, 1.

Wynn-Williams, C. G., Becklin, E. E., and Neugebauer, G.: 1974, *Astrophys. J.* **187**, 473.

Wynn-Williams, C. G. and Becklin, E. E.: 1974, *Publ. Astron. Soc. Pacific* **86**, 5.

Wynn-Williams, C. G., Werner, M. W., and Wilson, W. J.: 1974, *Astrophys. J.* **187**, 41.

C. G. Wynn-Williams

Mullard Radio Astronomy Observatory, Cavendish Laboratory,
Free School Lane, Cambridge, CB2 3RQ, United Kingdom

DISCUSSION

Churchwell: Is there any unambiguous way to distinguish between IR radiation from protostars and from evolved M supergiants?

Wynn-Williams: Not really, but coincidences of late type M supergiants with the most interesting parts of H II regions must be rather unlikely.

Townes: In answer to the question whether there is a definite way of distinguishing between a new star and a red giant surrounded by dust, I would like to note the possibility of discrimination on the basis of the $^{12}C/^{13}C$ ratio in circumstellar material. Geballe, Wollman, and Rank have recently applied this technique in examining IRC + 10216. Old stars characteristically show a low ratio (4–10), whereas new stars presumably would show a high ratio (50–100) since interstellar gases do. In particular, it appears practical to determine this ratio from the CO absorption spectrum for such sources as the brighter Orion sources.

Van Woerden: The 50 magnitudes of extinction refer, I suppose to the visual and are estimated from the infrared on the basis of a particular model?

Wynn-Williams: The ratio of the measured 2.2 μm flux density to the expected free-free or free-bound transition radiation as calculated on the basis of the radio flux density leads to a lower limit to the extinction at 2.2 μm. An assumed interstellar extinction law, such as Whitford's, is then used to estimate the visual extinction.

F. D. Kahn: An IR source without a detectable H II region might also be produced in the following way. An early-type star, already on the main sequence, may still be accreting material at a higher rate. This leads to a high gas density around the star. Any H II region forming in this gas will have a very small surface area and will be strongly self absorbing in the radio range. The H II emission would therefore not be detectable at radio wavelengths.

Mezger: It is unfortunately the case that the density where the dust absorbs most of the Lyman continuum is also the density where the H II region becomes optically thick in the free-free continuum. So you cannot discriminate between the two possibilities that the H II region is so compact that the turn-over frequency is in the millimeter wavelength range or that the dust soaks up all the Lyman continuum.

H. J. Habing: I do not like the term 'protostar' for an object that emits sufficient UV photons to ionize a detectable H II region. This object is probably already rather close to the main sequence; I would prefer to call 'protostars' those objects still in their contraction phase, without nuclear burning.

Mezger: The best explanation for the use of 'protostar' in the context of H II regions is that everything that you can't explain in ordinary terms is called a 'protostar'. This has certainly turned out to be the case.

Price: You note that at a few hundred microns the radiation falls off more rapidly than you would expect for black bodies. Can you infer something about particle sizes from this?

Wynn-Williams: All it says is that particle sizes are less than 100 μm.

Baldwin: The present upper limit to the 5-GHz flux density of IRS5 in W3 of 10 mJy from the 5-km telescope does not yet put a very tight limit on the absence of a compact H II region. An optically thick region of 0.2″ diameter would have the flux density close to this limit, and 0.2″ is only a little smaller than compact regions we already know.

Habing: We may have detected the H_2O source at the level of 5–7 mJy at 6 cm, but this is uncertain. We are fairly certain that the OH measure in the same W3 area is below 10 mJy. In the Orion case one should be extremely careful in stating that there is no radio phenomenon associated with the Kleinman-Low nebula and all the rest because really good data are lacking.

Zuckerman: What fraction of the IR emission in Orion comes from the Kleinman-Low nebula, and what is its temperature?

Wynn-Williams: It's not the Kleinman-Low nebula itself but a region centered on the Kleinman-Low nebula that is doing the emitting. Its diameter depends somewhat on the wavelength. The peak temperature is about 75–100 K, and it cools off outwards. Rieke reckons that two-thirds of the emission cannot be explained as coming from this small area but must come from a larger region.

Terzian: It seems to me that the similar brightness distribution of a source in the radio and infrared is a necessary but not sufficient condition for saying that the gas and dust are well mixed together. Is there other evidence on this problem?

Wynn-Williams: It is unnecessary to assume any special characteristics or concentration for the dust within the ionized region; an interstellar concentration of interstellar-type particles appears to be more than sufficient to cause the infrared emission. The assumption that the dust and ionized gas are mixed is therefore the simplest assumption you can make, and the observations seem to support it.

Donn: At infrared wavelengths up to 100 μm, when grains are smaller as you indicated, probably less than 1 μm, emissivity depends strongly on wavelength. How is this taken into account in deriving temperatures from the observed energy distribution?

Wynn-Williams: Since the H II regions contain dust grains at a variety of temperatures, the derived color temperatures in the 10 μm to 20 μm range are uncertain anyway. Changing the emissivity law to $\varepsilon \propto \lambda^{-1}$ or $\varepsilon \propto \lambda^{-2}$ still gives temperatures broadly in the 100 to 200 K range for these wavelengths.

Radhakrishnan: Do you think it is possible that the shape of the long wavelength end of the spectrum has something to do with the type of a particle?

Wynn-Williams: Yes, I think it is just the emissivity decreasing with increasing wavelength, and you can easily get up to a slope of up to 4 on the $d\varepsilon_\nu$ versus ν curve.

MASER SOURCES IN HII REGIONS

B. F. BURKE

Massachusetts Institute of Technology, Cambridge, Mass., U.S.A.

Abstract. When OH and H_2O sources are found in H II regions, a compact source of radio continuum and infrared emission is usually, perhaps always, found nearby, $\lesssim 10^{17}$ cm away (approximately the conjectured size of a prestellar condensation). Whether the masers are self-excited, are amplifying the background radiation, or are themselves associated with an exciting star is clearly model-dependent, and observations have not yet given the answer. H_2O masers occur less frequently than OH masers. When they occur together they appear as a cluster of compact sources of line radiation, exhibiting the same range of radial velocities, The OH sources nearly always are strongly polarized, while the H_2O maser is not. Both classes of maser are probably saturated, and some care is needed in inferring the maser geometry from the observations. The period of stability of neither OH nor H_2O can be much more than 10^3 yr, and this fact can be used as an estimator of stellar formation rate.

I. Introduction

The OH and H_2O emission-line sources associated with H II regions exhibit such high surface brightness that it is virtually certain they arise from regions where natural maser processes can occur. This means that the upper levels of the molecular transition have a larger population than the lower states, i.e., the state temperature, T_s, defined by the Boltzmann equilibrium $n_2/n_1 = g_2/g_1 \exp(-h\nu/kT_s)$, must be negative. The stimulated emission rate predominates over the stimulated absorption rate, and amplification rather than absorption occurs. In other words, the emission lines are characterized by a negative optical depth.

The present status of our observational knowledge is in some ways sufficiently detailed to allow one to draw a few conclusions about the nature of the maser sources. Very-long-baseline interferometry (VLBI) has provided quantitative information on the frequency of occurrence, angular sizes and distributions, and association with phenomena observed at other wavelengths. There still are many more questions than answers about their nature, however, and this review will consist mostly of an attempt to formulate these questions. The OH and H_2O masers will be treated in parallel, although it is not clear to what extent the two phenomena are related.

II. What Is the Geometry of the Sources, and Where Do They Occur?

Surveys of dense, highly excited H II regions, where only those H II regions with emission measures greater than 10^6 cm^{-6} pc are considered, have demonstrated that OH maser sources are more common than H_2O sources. Studies of OH masers have shown that about one third of the dense H II regions have associated OH line sources, although the actual probability of occurrence may be somewhat higher if weak lines are searched for. The H_2O masers are less common, and a survey of the literature reveals that the probability of maser occurrence is about 10%.

F. J. Kerr and S. C. Simonson, III (eds.), Galactic Radio Astronomy, 267–273. All Rights Reserved.
Copyright © 1974 by the IAU.

The maser sources do not occur independently of other objects in H II regions. Wynn-Williams *et al.* (1972) have shown that there is a high probability of finding an infrared source within a few arcseconds of a line maser, and this probability may be 100%. It is not yet certain whether the coincidence must be exact, and in fact the infrared sources in the Orion nebula seem to lie to one side, although they are probably not separated by more than 10^4 AU, more or less. Furthermore, the maser sources are also associated with compact knots of continuum radio emission, as first shown by Baldwin *et al.* (1973), for the case of W3, and then in great detail for a number of sources by Matthews *et al.* (1973). It has been suggested that the maser sources lie at the outer edges of the continuum condensation, but the generality of this rule is not clear. In the example cited by Baldwin *et al.* (1973), not all of the maser sources of the W3 complex were included, and a substantial number do in fact lie in the interior. A more careful statistical study must be made before the suggestion is acceptable.

Both the OH and H_2O maser complexes in H II regions share the common characteristic of occurring in compact groups. The emission line shape typically exhibits a number of components, each of which comes from a different position. The diameter of the entire group is 20″ for the Orion nebula and 1″ for W49, which corresponds in linear dimensions to 1.3×10^{17} cm and 2×10^{17} cm, respectively, as determined from VLBI work by Moran *et al.* (1973).

The individual apparent sizes are much smaller. In general, the OH masers have larger apparent sizes, and there is a strong suspicion that scattering by the interstellar medium may not be negligible for OH. The H_2O masers have apparent sizes in the range 0″.001 to 0″.003 for M42, while the sizes are of the order of 0″.0002 to 0″.0005 for W49. The apparent diameters, then, correspond to linear dimensions of the order of 1/2 to a few astronomical units.

The apparent source shapes may not be round. Typically, in M42, the fringe visibilities vary with hour angle in a way that suggests lack of circular symmetry, as reported by Moran *et al.* (1973). The effect may be due to a blending of separate features, however, and so the conclusion is not yet certain. It should be emphasized that the *apparent* shape may not be the true shape. Since the maser is basically an amplifier, it is capable of amplifying the plane wave from a distant emitter, and thus one might observe merely the amplified image of a distant radio source. The problem of the excitation of the maser is such an important matter that it must be treated as a separate question later.

The positions of the OH and H_2O masers do not change greatly with time, although the intensity of the H_2O masers varies dramatically with a characteristic time-scale of a few months. The turn-on time may, in fact, be much shorter. Some of the OH sources have been reported to vary slowly in intensity, but the well-studied W3 OH source has been essentially constant over the last four years (Moran, private communication).

The question of velocity variation with time is an interesting one. Although there have been velocity changes reported, the line velocity never has changed by more than a line width, so varying intensity within a blend of several lines cannot be excluded.

The observations provide a useful upper limit, however, that has direct physical significance since change of velocity with time, dv/dt, is an acceleration. The observations give an upper limit of 0.03 cm s^{-2} for all cases, and for most of the lines the upper limit is 3×10^{-3} cm s^{-2}. From this one may conclude that if each masering object is in the vicinity of a star of one solar mass, it is typically at least 45 AU distant. Thus the masering objects are not likely to be proto-planets in solar systems, although one cannot exclude the possibility that there are pressure forces that balance the effects of gravitation.

III. What Is the Pump Mechanism?

Despite the large amount of theoretical work expended on the problem, it is not easy to draw definite conclusions. The OH and H_2O molecules have very different energy level structures, OH having a set of hyperfine levels in a Λ-doublet spectrum, while the H_2O line arises from a pure rotational transition high above the ground state. It does seem probable, however, that ultraviolet pumping is not very likely, both because of the difficult energy requirement (spending one ultraviolet photon for each radio photon) and because of the dissociation of the molecules in the presence of the ultraviolet radiation unless a strong cutoff between the pumping wavelength and the photo-dissociation wavelength can be devised.

The OH problem is very complex, because of the large number of angular momenta that determine each state. In principle, it is a very rewarding problem because of the large number of observables. All ground-state hyperfine components are in principle observable, and the transitions between the Λ-doublets of the excited states of both the $^2\Pi_{3/2}$ and $^2\Pi_{1/2}$ ladders are observable. Infrared pumping appears to be very promising, especially in view of the observational fact that the masers occur in conjunction with infrared sources.

The H_2O masers present a severe problem, since the energy requirements are heavy. The energy output of 10^{31}–10^{33} erg s^{-1} actually observed requires a large supply of infrared photons. Since we know, observationally, the infrared flux in the vicinity of the sources, it has been shown that there are not enough infrared photons to supply the maser energy. The most promising mechanism appears to be collisional pumping. De Jong's (1973) collisional pumping model requires a molecular hydrogen density of 10^8–10^{10} H cm^3 and a kinetic temperature of 100–1000 K, although he had to use estimated cross-sections that may be very far from the actual value.

It is interesting to note that, for H_2O, in which the transition lies 465 cm^{-1} above the ground state, frequent collisions are absolutely necessary to populate the states, and one has a potentially valuable estimate of the local density and temperature if only one can understand the pump mechanism thoroughly.

IV. Are the Masers Saturated or Unsaturated?

The answer to this question has important implications for the next two questions. By definition, an unsaturated maser is one in which the induced transition rate is small compared to the other rates by which the molecules enter and leave the upper state. This means that the excitation temperature is independent of the radiation field, and the specific intensity will grow exponentially. In a saturated maser, on the other hand, the transition rate from the upper state is determined primarily by the induced radiative transition, and so the rate at which the specific intensity grows is controlled by the pump rate. Thus, the intensity grows linearly.

The observational evidence is not conclusive, but the strongest maser sources, such as W49, W3, and M42, probably are saturated. Each source exhibits 10 or so lines with intensity 5% or greater than that of the strongest line. If the masers are unsaturated, the (negative) optical depths are between 20 and 25, which would require the brightest 10 lines to have optical paths that differ from one another by only 10%. (In a domain of exponential growth, even small path differences become greatly exaggerated.) This does not seem to be a very likely situation, while, on the other hand, if the growth is linear the optical path lengthens greatly, even for the strongest lines. Unless the bright masers form an exceptionally uniform class, therefore, the evidence appears to favor saturated masers, at least for the strong ones.

V. How Are the Masers Excited?

The central point that must be addressed is the relation between the observations and the real physical objects. Following the suggestion of T. Gold that the physical size of the masers might be very different from the apparent size, both Litvak (1973) and Goldreich and Keely (1972) worked out simple models that demonstrated very clearly that this is the case. In this discussion, I shall follow the latter, who treat the development of the radiation field in a homogeneous, spherical, uniformly pumped two-state system. While the model is certainly an idealized one, the important difference between the apparent size and physical size is amply demonstrated.

The observed physical properties, derived from the considerations of the preceding questions, can be summarized in Table I. In all cases, entries refer to *observed* properties, although the luminosity is inferred under the assumption that the radiation is

TABLE I

Observed properties of OH and H_2O masers

Quantity	OH	H_2O
Occurrence in H II regions	1/3	1/10
Time scale	10^8 s	10^6 s
Velocity	$V(\text{H II}) \pm 10$ km s^{-1}	
Line widths	0.5–2 km s^{-1}	
Polarization	Circular	None (or linear in Orion)
Energy	10^{-5}–10^{-3} $L_\odot$	10^{-4}–1 $L_\odot$
Brightness	10^{13} K	10^{15} K
Group size	10^3–10^4 AU	
Acceleration	$< 10^{-2}$ cm s^{-2}	
Apparent size	5–10 AU	0.5–2 AU

effectively isotropic. (For individual lines this may not be so, but since a number of different velocity components are usually observed simultaneously in a given maser, the average power output must be effectively isotropic.)

The maser properties themselves are not directly observed, and are of course model dependent. The Goldreich-Keely model assumes the following are given: a differential pumping rate for the two levels, $\Delta R/R$, a relaxation rate Γ which includes all non-radiation channels, and a molecular density n_x for the molecule under consideration. These quantities will be functions of the total (molecular hydrogen) density, the temperature, and the radiation field, and in general these relations are not well known, although estimates can be made. Goldreich and Keely then solve simultaneously the equations of radiative transfer and of detailed balance, in a dimensionless form where the natural unit is the scale length L required to give a factor e amplification in the unsaturated regime.

Interestingly, the ratio of apparent size to actual size of such a homogeneous maser is given by $(1/\alpha)^{1/4}$, α being the dimensionless quantity $(A/\Gamma)(R/\Delta R)$, where A is the Einstein spontaneous radiation probability. A typical set of physical parameters, consistent with the observations, is given in Table II. These masers consist of an un-

TABLE II

Typical parameters for maser regions

Quantity	OH	H_2O
L (unsaturated length)	2×10^{13} cm	5×10^{12} cm
T_{ex}	45 K	2500 K
$\alpha = RA/\Delta R\Gamma$	2.5×10^{13}	6×10^6
Size	50 AU	200 AU

saturated core, in which all rays are amplified, outside of which there is a transition region and then, for the bulk of the cloud, only the outgoing rays from the unsaturated core are amplified. Thus, most of the maser is saturated, and the induced transitions dominate all other radiation processes. Note that the apparent maser sizes, of the order of 1–10 AU imply physical sizes of about 100 AU, which corresponds very closely to the observed separations between different velocity components. The question of saturation of the masers is evidently a crucial one. Note, however, that the maser size to apparent size ratio is not sensitive to the assumptions about physical conditions because the fourth root of the parameter α is taken.

Goldreich and Keely also considered models with cylindrical geometry. In this case, for a long cylinder, the apparent size is the same as the diameter of the cylinder. For H_2O masers, we see the spots turning on and off, turning on rather suddenly and off more slowly. It does not seem likely that this is caused by the narrow beams of filaments sweeping past us, and it therefore seems likely that the physical shapes of the masers are more nearly spheres than filaments.

VI. What Is the Magnetic Field?

From the observed polarization it should be possible to derive the magnetic field. Unfortunately, this cannot be done uniquely. The Goldreich theory for the observed polarization predicts that one should see the full Zeeman pattern. But since components are observed at a large number of velocities and since one is free to specify both the radial velocity and the Zeeman effect for each line, there are twice as many free parameters as observables. Solutions can be found, but not unique solutions.

One wants to have a sharper test than that, and the H_2O masers do give a sharper test. Goldreich and Keely found that there is a regime where linear polarization can arise. This is where the rate of radiation, r, is less than the gyrofrequency, ω_z, which in turn is less than the natural line width, $\delta\omega$, i.e., $r < \omega_z < \delta\omega$. For Orion this predicts a magnetic field $B \sim 10^{-2}$ G, a very high field indeed. For $B > 10^{-3}$ G the field would already separate out the Zeeman components for OH, but complete OH Zeeman patterns are not observed in a convincing way. One should note that Moran *et al.* (1973) have demonstrated that the H_2O masers in Orion are quite separate from the OH masers, and so the magnetic fields may not, in fact, be the same for both.

VII. What Use Are They?

The natural time scale seems to be something on the order of 300 years for an entire complex of these objects. This follows from noting that, for the observed velocities, a group cannot stay together in the sky for more than about 300 to 1000 yr. However, the lifetime of the H II regions is generally believed to be 10^5 to 10^6 yr. Therefore, we are dealing with an object that lives for only about 10^{-3} the lifetime of a given H II region, an observation recently made by de Jong (private communication).

We are seeing a transient phenomenon, but we also see something that is more common than the appearance of an O6 star. The apparent rate of about 1000 of these masers in the lifetime of one of these density-limited H II regions is more appropriate to the entire formation rate of all O and B stars. Therefore, in addition to the obvious interest in understanding the physics of these peculiar masering regions, it seems very likely that by surveying the Galaxy for all its OH and H_2O masers we may hope to locate all the new B stars in the Galaxy.

Acknowledgements

This research was supported in part by a grant from the National Science Foundation.

References

Baldwin, J. E., Harris, C. Stella, and Ryle, M.: 1973, *Nature* **241**, 38.
Goldreich, P. and Keely, D. A.: 1972, *Astrophys. J.* **174**, 517.
de Jong, T.: 1972, Thesis, University of Leiden.
Litvak, M. M.: 1973, *Astrophys. J.* **182**, 711.

Matthews, H. E., Goss, W. M., Winnberg, A., and Habing, H. J.: 1973, *Astron. Astrophys.* **29**, 309.
Moran, J. J., Papadopoulos, G. D., Burke, B. F., Lo, K. Y., Schwarz, P. R., Thacker, D. L., Johnston, K. J., Knowles, S. H., Reisz, A. C., and Shapiro, I. I.: 1973, *Astrophys. J.* **185**, 535.
Wynn-Williams, C. G., Becklin, E. E., and Neugebauer, G.: 1972, *Monthly Notices Roy. Astron. Soc.* **160**, 1.

B. F. Burke
Dept. of Physics,
Massachusetts Institute of Technology,
Cambridge, Mass. 02139, U.S.A.

DISCUSSION

Kaufmann: The physical parameters you assume for H_2O masers, such as molecular densities and pump rates, are they convenient because they provide adequate solutions to Goldreich equations, or because there are some other evidence or thoughts on physical conditions at the sources?

Burke: There is a wide range of acceptable values for the various parameters.

Van Woerden: What is the evidence that masers are related to the formation of O- and B-type stars only, and not to that of less massive stars?

Habing: I think one good reason to assume that OH masers are associated with early-type stars is that a sizeable fraction of the Class I OH masers (possibly up to 100%, but most likely more than 30%) is associated with H II regions, as Winnberg, Goss, Mathews and I discovered recently at Westerbork.

Townes: There is a variety of arguments which lead to the conclusion that many of the interstellar masers are probably saturated. The lack of weaker masers in a given field is not to me a very convincing one, since not many careful searches and statistical studies of weaker lines have been carried out.

An important feature which must be considered is dust; in fact, its presence seems to be required for some masers in order to allow any reasonable pumping scheme to function. Furthermore, the dust may be quite important in producing feedback and scattering which affects both saturation and polarization.

Burke: There are several strong maser sources with a large number of strong lines whose intensities are comparable.

Robinson: In type I OH masers, satellite line emission or absorption is commonly observed only an order of magnitude in intensity below the main line emission. Given the lower transition probabilities for the satellite lines, I take this as supporting evidence for saturated maser emission. For unsaturated maser emission we would, on the average, expect intensity ratios of 1665 MHz to satellite line emission of exp (20)/exp(4) for comparable degrees of inversion.

MAGNETIC FIELDS IN OH MASER CLOUDS

R. D. DAVIES

University of Manchester, Nuffield Radio Astronomy Laboratories,
Jodrell Bank, Cheshire, United Kingdom

Abstract. Observations of Class I OH maser sources show a range of features which are predicted on the basis of Zeeman splitting in a source magnetic field. Magnetic field strengths of 2 to 7 mG are derived for eight OH maser sources. The fields in all the clouds are directed in the sense of galactic rotation. A model of W3 OH is proposed which incorporates the magnetic field data. It is shown that no large amount of magnetic flux or angular momentum has been lost since the condensation from the interstellar medium began.

I. Introduction

Compact OH maser sources show strong polarization – especially those associated with H II regions (Class I OH sources). A characteristic of these sources is their strong (nearly 100%) circular polarization and weaker linear polarization. When circular polarization was first detected, it was attributed to Zeeman splitting by magnetic fields of a few milligauss in the OH cloud (Davies *et al.*, 1966). Subsequently it became clear that these sources were masering and this simple explanation was not sufficient, and other explanations were sought.

Shklovskii (1969) suggested that twin source components of opposite circular polarization might be produced by a rotating OH cloud in which the lines are split by a Zeeman effect. His method of producing circular polarization is similar to that suggested earlier by Cook (1966).

Linear polarization can result from directional pumping by unpolarised radiation which produces irregularities in the Zeeman sublevel populations, even in the absence of significant magnetic fields (e.g., Perkins *et al.*, 1966). The more frequently observed circular polarization cannot be formed in this way.

Proposals for generating circular polarization have been advanced (Heer and Graft, 1965; Culshaw and Kanneland, 1966; Heer and Settles, 1967) which are based upon saturated masers where one of the senses of polarization is suppressed. These theories predict patterns of emission and polarization for the other lines of $^2\Pi_{3/2}$, $J = 3/2$ quartet which are not confirmed by observations (Litvak, 1969; Turner, 1970). The suppression theory has been criticized by Bender (1967), who showed that the likelihood of producing a resultant circular polarization was much less than had been calculated. Further, the Heer and Settles theory required a magnetic field of $\lesssim 10^{-10}$ G, a value considerably less than any likely interstellar value. In a similar theory Hall and ter Haar (1973) have recently discussed the circular and linear polarization resulting from saturated maser mode competition occurring in a single-line model of Class I OH sources.

The possibility of a Zeeman interpretation of the circular polarization from some OH maser sources has again found favour with some recent authors (Coles *et al.*,

F. J. Kerr and S. C. Simonson, III (eds.), Galactic Radio Astronomy, 275–292. All Rights Reserved.
Copyright © 1974 by the IAU.

1969; Rydbeck *et al.*, 1970; Gardner *et al.*, 1970; Litvak, 1971; and Zuckerman *et al.*, 1972). Goldreich *et al.* (1973) have given a theoretical discussion of the role of Zeeman splitting in producing polarization in astrophysical masers. Further it is recognized that in an interstellar maser there is unlikely to be a strict one-to-one correspondence between LH and RH components (Davies, 1967).

The present paper examines the Zeeman splitting interpretation in some detail and shows that for a number of sources the predictions are verified by observations of single telescope spectra for a wide range of transitions and also by interferometer data. Magnetic field strengths are derived for a number of OH maser sources, and these are compared with the values expected for the compression of the interstellar field. The dynamical stability of the OH sources is also discussed.

II. The Expected Zeeman Splitting of the OH Molecule

(a) THE ZEEMAN PATTERN

In this section the normal Zeeman pattern of OH will be described for the various transitions which have been observed in OH maser sources. These include the $^2\Pi_{1/2}$, $J=1/2$ state triplet, $J=3/2$ and $5/2$ quartets and the $^2\Pi_{3/2}$, $J=3/2$ and $5/2$ quartets.

Each of the Π ground electron states mentioned above is split into two states by Λ doubling, and these in turn are doubled by hyperfine splitting, resulting in four states for each value of J. In a magnetic field each of these states is further split into $2F+1$ magnetic energy sublevels separated from the original energy level by

$$\Delta W = h\Delta v = -\mu_n g_I \cdot \frac{M_F B}{2F(F+1)} \cdot [I(I+1)+F(F+1)-J(J+1)],$$

$$-\mu_0 g_J \cdot \frac{M_F B}{2F(F+1)} \cdot [J(J+1)+F(F+1)-I(I+1)],$$

where μ_n and μ_0 are the nuclear and Bohr magneton ($\mu_n = \mu_0/1836$)
    $g_1 \sim g_J \sim 1$ (see below for actual values of interest);
    $\mu_0/h = 1.39967$ MHz G^{-1};
        J = quantum number for the total angular momentum including rotation but excluding nuclear spin;
        F = total angular momentum;
        I = nuclear spin;
    M_F = the magnetic quantum number giving the projection of F on the direction of the magnetic field; it takes the values $F, F-1 \ldots -F+1, F$;
        B = the magnetic field strength.
For the $^2\Pi_{3/2}$ states of OH the first (nuclear and magnetic moment) component of the splitting is negligible compared with the second (electronic) component. However, in the $^2\Pi_{1/2}$ states the spin and orbital components of the electron magnetic moment are oppositely directed and largely cancel, and the nuclear spin component is a significant contribution; the resultant splitting is $\sim 10^{-3}$ of the $^2\Pi_{3/2}$ states.

The Zeeman splitting pattern expected for a particular OH line results from transitions between the magnetic sublevels of the two states. Allowed transitions are $\Delta M_F = 0$ (π components) and $\Delta M_F = \pm 1$ (σ components). The resulting Zeeman patterns for the $^2\Pi_{3/2}$, $J = 3/2$ and $5/2$ states are shown in Figure 1. When the field is directed parallel to the line of sight the LH and RH σ components are seen; when

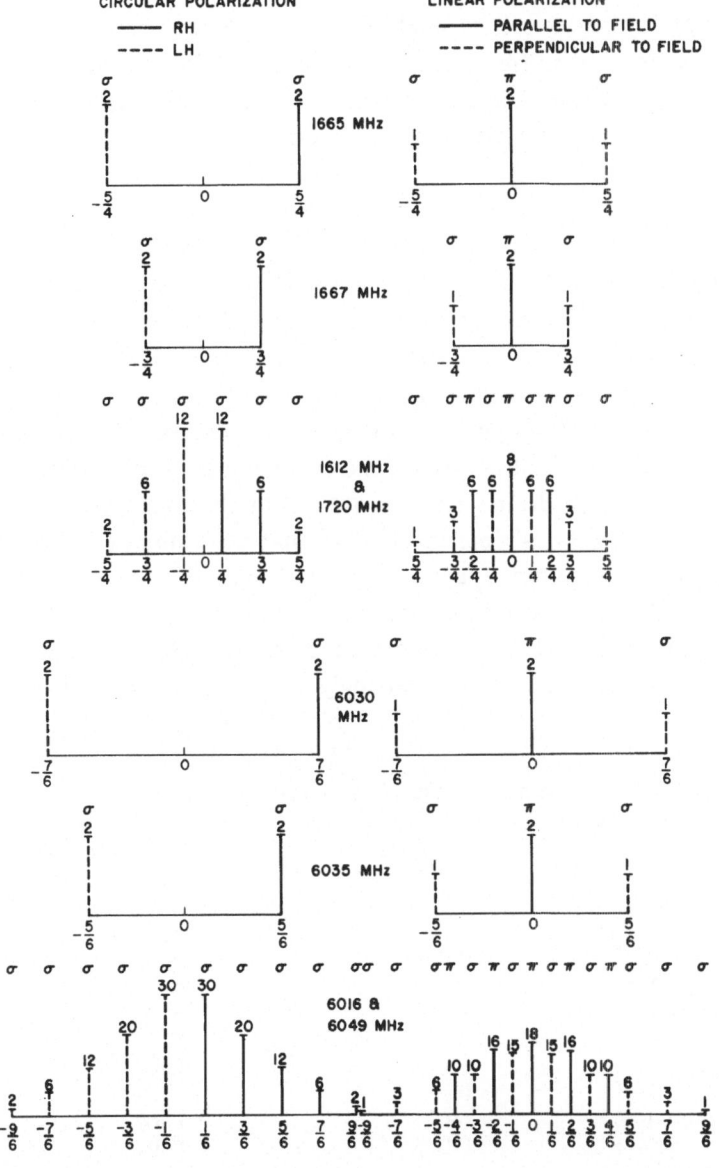

Fig. 1.   Zeeman splitting patterns for the $^2\Pi_{3/2}$, $J = 3/2$ and $J = 5/2$ quartets of OH. The frequency splitting is plotted in units of $g_J \cdot \mu_0/h$; this is 1.31 MHz G^{-1} for the $J = 3/2$ lines and 0.679 MHz G^{-1} for the $J = 5/2$ lines. Relative intensities of the Zeeman components are shown for each transition. The patterns in the left-hand side of the figure result from observations parallel to the field and those on the right-hand side are from observations perpendicular to the field.

perpendicular to the line of sight the π components are linearly polarized parallel to the field while the σ components are perpendicular. The intensities of the various σ and π components have been estimated using Table 10.1 of Townes and Schawlow (1956); relative intensity values are given in Figure 1.

The amount of Zeeman splitting in the commonly observed OH transitions is given in Table I. These refer to the σ components and have been estimated from the values of g_J ($=0.935$, 0.485 and 0.325 for $J=3/2$, $5/2$ and $7/2$) for the $^2\Pi_{3/2}$ state given by Radford (1961). In the case of the satellite lines which have many σ components, the tabulated Zeeman split is the average split weighted by the component intensity. Table I also emphasizes the smaller Zeeman splitting measured in terms of velocity for the higher frequency excited hydrogen lines.

(b) APPLICATION TO AN OH MASERING SOURCE

In any astronomical situation the polarized components shown in Figure 1 will be broadened by thermal and mass motions of the OH molecules. When the broadening is comparable to the Zeeman splitting the observed polarization will be reduced. A typical cloud is illustrated in Figure 2, where the emission spectra on a given line of sight show the effects of differing mass motions and magnetic fields; the field *direction* is assumed to have a component away from the observer at each point. If the OH were not masering and it had a constant excitation temperature its emission spectrum (proportional to $\int n_{OH}(v)\,dl$) in each polarization might appear as in (a). However, if the molecular energy states were inverted so that the emission column was an unsaturated maser the resultant spectra might look like (b), where each spectrum will be proportional to $\exp[\text{const} \times \int n_{OH}(v)\,dl]$. In (a) there are only small fractional differences in the LH and RH spectra, whereas in (b) these differences have been amplified by the exponential term to give strongly polarized signals. For example, in the

TABLE I

Zeeman splitting of OH lines (σ components)

Transition		Frequency (MHz)	Zeeman split between LH and RH	
			(MHz G^{-1})	(km s^{-1} G^{-1})
$^2\Pi_{1/2}J=1/2$	$F=0\rightarrow1$	4660.2	$\sim10^{-3}$	$\sim10^{-2}$
	$1\rightarrow1$	4750.6	$\sim10^{-3}$	$\sim10^{-2}$
	$1\rightarrow0$	4765.5	$\sim10^{-3}$	$\sim10^{-2}$
$^2\Pi_{3/2}J=3/2$	$F=1\rightarrow2$	1612.2	1.308	236
	$1\rightarrow1$	1665.4	3.270	590
	$2\rightarrow2$	1667.3	1.964	354
	$2\rightarrow1$	1720.5	1.308	236
$J=5/2$	$F=2\rightarrow3$	6016.7	0.678	33.8
	$2\rightarrow2$	6030.7	1.582	79.0
	$3\rightarrow3$	6035.0	1.132	56.4
	$3\rightarrow2$	6049.0	0.678	33.8

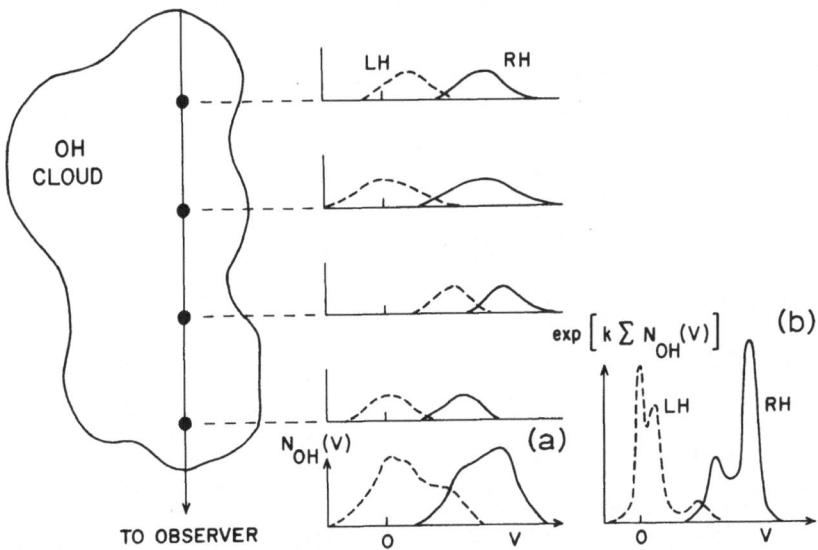

Fig. 2.   The masered spectrum produced by contributions along a line of sight through an OH cloud. The distribution of molecules as a function of velocity, $N_{OH}(v)$, at each point is shown. The sum of all contributions to N_{OH} is $\Sigma N_{OH}(v)$. The resultant masered spectrum $T_B(v)$ is of the form $\exp[\text{const} \times \Sigma N_{OH}(v)]$. The two circularly polarized components are shown by full and broken lines.

stronger OH sources the exponent is 20–30, so a fractional difference of 0.2 in the LH and RH line integral spectra would give a difference of e^4–e^6 (i.e., 50–400) in the masered LH and RH spectra. Moreover, the LH and RH integrals are not identical, nor are they frequency-shifted versions of one another; accordingly the LH and RH masered spectra are in general very different. In addition the pumping may produce different amounts of inversion along the line of sight, and this will enhance the differences between the LH and RH spectra.

If the Zeeman splitting is comparable with (say 20% or more) the total Doppler broadening of a source, a high percentage of circular polarization of the masered spectrum is expected. The amount of linear polarization in this situation will be less than the circular polarization both in terms of the number of masered lines and their percentage polarization. This is evident from Figure 1, where the orthogonal linear components have only half the frequency separation of the circular components, and, further, the outer linear components have the same polarization. Further, Faraday depolarization along the emission column and propagation path in the source can reduce or remove linear polarization but will not affect circular polarization.

The above statements, based on a superposition approach, are essentially justified by Goldreich *et al.* (1973), who have shown in a detailed discussion of the transfer equations for the Stokes parameters that there is no polarization unless the Zeeman splitting exceeds the bandwidth of the amplified radiation when the emitted radiation is just the amplified Zeeman pattern. If the Faraday rotation across the region is large they show that the linear polarization is destroyed while the σ components of the Zeeman pattern are 100% circularly polarized.

TABLE II

Zeeman splitting measurements in OH sources

Line frequency (MHz)	Observed splitting (km s^{-1})	Magnetic field[a] (mG)	
W3 (OH) ($l=134.0, b=1.0$)			
Mean displacement			
1612	1.14	+4.8	
1665	2.43	+4.3	
1667	1.80	+5.1	+4.95
1720	0.84	+3.5	
6030	0.55	+6.9	
6035	0.29	+5.1	
Major component displacement			
1612	1.21	+5.1	
1665	1.31	+2.2	
1667	2.20	+6.2	+4.70
1720	0.83	+3.5	
6030	0.54	+6.8	
6035	0.25	+4.4	
Source map			
1665	2.76	+4.7	
Adopted value		+4.8	
W75B ($l=81.9, b=0.8$)			
Mean displacement			
1665	4.67	+7.9	
6035	0.33	+5.9	
Major component displacement			
1665	2.75	+4.7	
	(8.0)	(+13.6)	
6035	0.33	5.9	
Adopted value		+6.9	
NGC 6334A ($l=351.4, b=0.7$)			
Mean displacement			
1665	2.40	−4.1	
1667	1.30	−3.7	−3.9
1720	0.86	−3.6	
6035	0.24	−4.2	
Major component displacement			
1665	3.50	−5.9	
1667	1.98	−5.6	−4.8
1720	0.86	−3.6	
6035	0.24	−4.2	
Adopted value		−4.3	
G285.2−0.0 ($l=285.2, b=0.0$)			
1665	1.50	−2.5	

Table II (Continued)

Line frequency (MHz)	Observed splitting (km s^{-1})	Magnetic field[a] (mG)	
G291.6−0.4 (l=291.6, b=−0.4)			
1665	3.6	−6.1	
NGC 6334B (l=351.1, b=0.7)			
1665	1.1	−1.9	−2.2
1667	0.8	−2.3	
Sgr B2 (l=0.7, b=0.0)			
1665	1.95	+3.3	
1720	0.90	+3.8	+4.1
6035	0.29	+5.1	
W49 (l=43.2, b=0.0)			
Interferometer			
1665	1.68	+2.8	

[a] A positive magnetic field is directed away from the observer.

(c) PREDICTIONS OF THE ZEEMAN SPLITTING EXPLANATION

The polarization properties of the maser emission from OH sources will now be summarized. It is supposed that the Zeeman splitting is comparable with the Doppler broadening for the lowest frequency lines ($^2\Pi_{3/2}$, $J=3/2$) lines at ~ 1660 MHz.

(i) A strict Zeeman pattern like Figure 1 is not expected for each emitting component for the reasons given above. If there is a dominant velocity component in the source the main polarized features (generally with unequal intensity) will be displaced by an amount given by the Zeeman split relation in Table I.

(ii) The masered spectra observed in different transitions will appear to have quite different component velocities because of the different ratio of Zeeman to Doppler shift in the various transitions.

(iii) Circular polarization will dominate the spectra. The number of linear components and their percentage polarization will be small because they overlap.

(iv) All the $^2\Pi_{3/2}$ lines will show the same sense of overall displacement between RH and LH spectra provided there is a consistent field direction in the cloud.

(v) The *mean* velocity displacement of the LH and RH spectra will approximate the values listed in Table I and will be greatest for the 1665 and 1667 MHz transitions.

(vi) The Zeeman component of the velocity displacement of the LH and RH spectra in $^2\Pi_{3/2}$, $J=5/2$ transitions will be a factor of ~ 7 less than in the $J=3/2$ transitions.

(vii) Transitions in the $^2\Pi_{1/2}$ state will be unpolarized.

(viii) Interferometer observations will show the LH and RH emission components from the same general area of the source (i.e., having the same systematic velocity) with the appropriate Zeeman split for the field at that point.

III. The Zeeman Splitting Interpretation of Observed OH Spectra

A number of OH maser sources of Class I have been studied in sufficient transitions to test the validity of the Zeeman interpretation. The magnetic fields in these sources will have to be sufficiently large that the Zeeman splitting in at least the 1665 and 1667 MHz lines is comparable with the Doppler broadening. For a few brighter sources long baseline interferometry gives the relative spatial positions of the stronger LH and RH components; these can be compared with the predictions. Eight sources (W3 OH, W75 B, NGC 6334 A and B, G258.2−0.0, G291.6−0.4, Sgr B2 and W49) show some of the Zeeman effects described above.

(a) *W3 (OH)* (R.A. $= 02^h23^m17^s$, Dec. $= 61°38'54''$ [1950])

W3 is the brightest and the most extensively observed Class I OH maser source. Figure 3 plots the integrated LH and RH spectra of the stronger transitions taken from published material. They show all the characteristics (a) to (g) of Zeeman splitting discussed in Section IIc. Since there are approximately 20 components in the 1665 MHz spectrum a statistical determination of the magnetic field responsible can be obtained from the mean displacement of the LH and RH spectra in the transitions. The velocity displacements for the six transitions and their corresponding magnetic fields are listed in Table III. These give a mean field of 4.95 mG directed away from the observer. Another estimate of the magnetic field is provided by the displacement between the brightest LH and RH features in each spectrum. This gives a mean field of 4.7 mG. Although this method gives a similar value to the previous method, in general it is more liable to error because the brightest masered component is not necessarily representative of the mean spectrum.

Interferometry gives the third method of determining magnetic fields in an OH masering source by indicating directly the Zeeman displacement between the LH and RH components in each region of the source. 1665 MHz observations of W3 OH by Harvey *et al.* (1973) give a mean field of 4.7 mG in separated regions of the object. The three methods are consistent in indicating a field in W3 OH of 4.8 mG directed away from the observer.

(b) *W75B* (R.A. $= 20^h36^m54^s$, Dec. $= 42°25'40''$ [1950])

Observations at 1665 and 6035 MHz of adequate sensitivity and frequency resolution are available for W75B ($=$W75N). The 6035 MHz spectra show the same LH to RH displacement in the three major components. At 1665 MHz it is not clear which major LH components should be combined with the two strongest RH components without reference to the 6035 MHz splitting. Two splittings are suggested in Table II. The adopted field is 6.9 mG directed away from the observer.

(c) *NGC 6334A* (R.A. $= 17^h17^m32^s$, Dec. $= -35°44'15''$ [1950])

Figure 5 summarizes the published data on NGC 6334*H* ($=$NGC 6334*N*), which strongly supports the Zeeman interpretation. The splittings of four strong lines give a consistent field of 4.3 mG directed towards the observer.

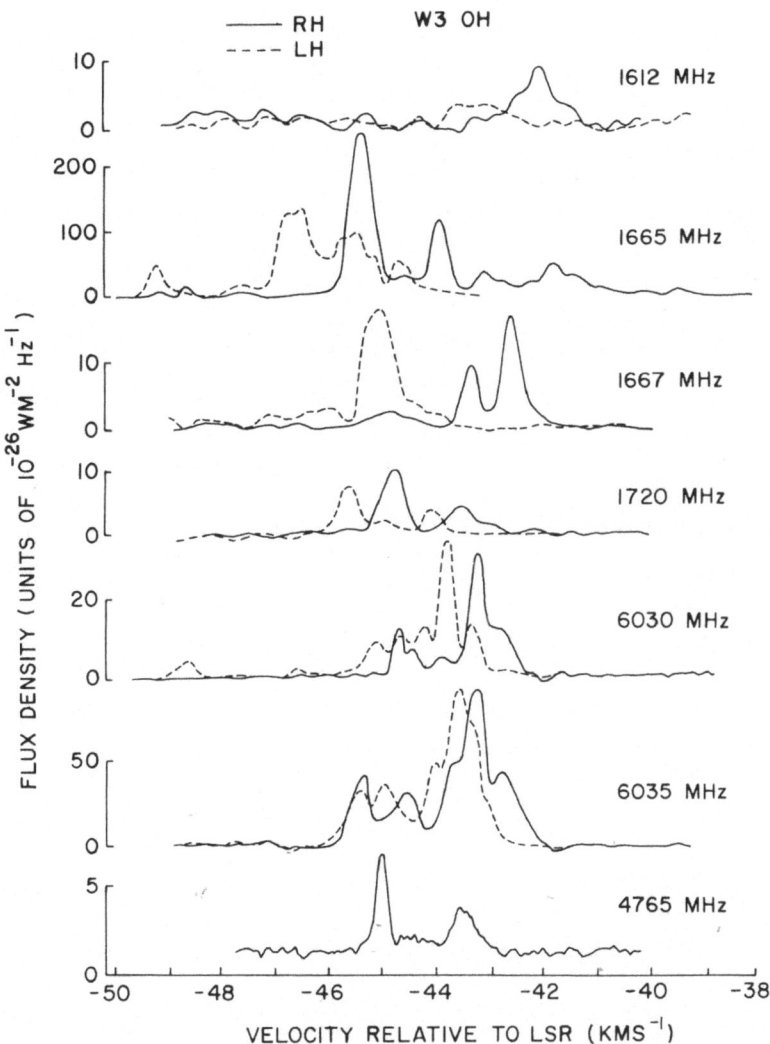

Fig. 3.   Observed spectra of W3 OH. RH polarization is shown by full lines and LH by broken lines. The
4765 MHz spectrum is the sum of orthogonal polarizations; it is believed to be unpolarized.

(d) OTHER OH SOURCES

A number of other OH sources have been observed with adequate frequency resolution and show spectral line displacements that indicate Zeeman splitting. The clearest
examples are G285.2−0.0, G291.6−0.4, NGC 6334B and Sgr B2. The LH and RH
displacements and the derived magnetic field strengths and directions are given in
Table II. Of these sources Sgr B2 is the most complex; consistent splittings in three
transitions are suggested. Another source, W49, shows strong frequency splitting of
the LH and RH components in many of its sub-sources when mapped at 1665 MHz
with an interferometer (Harvey *et al.*, 1973). The corresponding fields in the five sub-

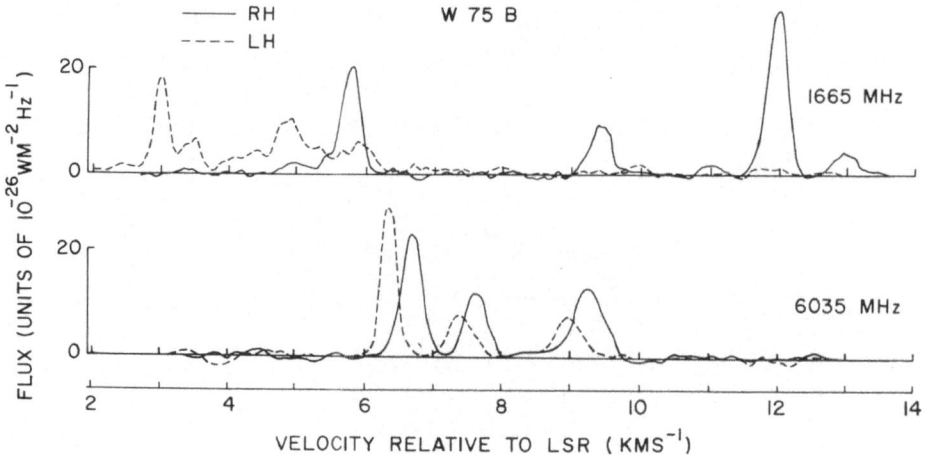

Fig. 4.   Observed spectra of W75B(= W75N). RH polarization is shown by full lines and LH
by broken lines.

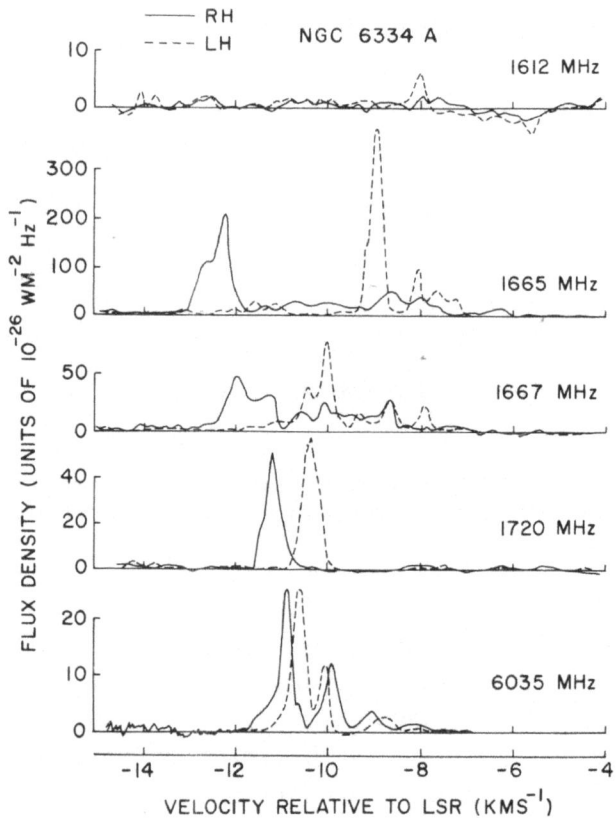

Fig. 5.   Observed spectra of NGC 6334A. RH polarization is shown by full lines and LH by broken lines.

sources of W49 range from 0.7 to 5.2 mG directed away from the observer in each case.

IV. Magnetic Field Amplification in Interstellar Clouds

Class I maser sources are closely associated with H II regions and areas of contemporary star formation. It is therefore to be expected that the OH masers are themselves gas clouds collapsing towards the protostar phase. It is quite possible that in many of these objects the central star is in the nuclear burning phase and has already ionized the gas in its immediate vicinity. This situation probably exists in W3 OH where an H II region and IR source are closely associated with the OH source. A number of similar associations are suspected. In this section we consider the effect of the collapse of the interstellar cloud on the magnetic field strength and alignment. Larson's (1969) models of collapsing protostars will be used.

If a uniform magnetic field B_0 permeates the average interstellar cloud with a hydrogen density $n_0(H)$ and the cloud is of such a configuration that it will collapse gravitationally then the field will also be compressed. Conservation of magnetic flux in an isotropic collapse will lead to a field B at a density $n(H)$ given by

$$B = B_0 \left[\frac{n(H)}{n_0(H)} \right]^{2/3} .$$

Measurements of the Zeeman effect in normal interstellar H I clouds (Verschuur, 1971; Davies *et al.*, 1974) show that this relationship is obeyed. The available H I data show that $B_0 = 4\sqrt{2}\mu G$ for $n_0(H) = 10 \text{ cm}^{-3}$. For the OH maser sources estimates of $n(H)$ can be made from the measured OH maser gain which suggest that $n(H) \sim 3 \times 10^6$ cm^{-3}. The uncertainty in this is no more than a factor of 10 either way (see Section V for a discussion of $n(H)$ for W3 OH). The corresponding value of B would be 25 mG if magnetic flux is conserved in an isotropic collapse. Reference to Table II shows that the highest observed values are only a factor of 4 less than this predicted magnetic field value.

An alternative estimate of the magnetic field amplification can be made by including the effect of the unimpeded infall along the magnetic axis of the cloud. If the radius in this direction is a fraction $\eta < 1$ of the radius perpendicular to the field then the compressed field is $\eta^{2/3}$ of that calculated above. For $\eta = 0.5$, a value consistent with the observed dimensions of W3 OH, the value of B would be reduced from 25 to 16 mG. In view of the uncertainty in estimating $n(H)$ it is clear that the observed values are consistent with magnetic flux conservation of the general interstellar field during gravitational cloud collapse. Only a small flux-loss is permitted by the data. It would therefore be surprising if the effects of Zeeman splitting were *not* detectable in those collapsing clouds which have reached a density of $\sim 10^6$ cm^{-3}.

The relationship between B and $n(H)$ given above will be preserved throughout the depth of the cloud even if the density varies with depth, providing the assumption of conservation of magnetic flux still holds. Thus in W3 OH, where the

OH lies within an outer shell surrounding ionized gas, we would expect the field to be still given by the expression since this gas is a part of the original gas cloud.

In the foregoing discussion it has been assumed that the magnetic field is frozen into the gas. This assumption can be justified on the following arguments. For a fully ionized gas the time for the Ohmic losses to dissipate an amount of energy comparable with the magnetic energy is given by Spitzer (1962) as

$$\tau = 2 \times 10^{-13} \ T^{2/3} \ L^2 \ \text{seconds},$$

where T is the gas temperature and L is a length characteristic of the system. Putting $T = 100$ K and $L = 10^{15}$ cm, the smallest length scale found in the OH masers, $\tau = 6 \times 10^{12}$ yr. This dissipation time far exceeds the collapse time of the cloud, which is $\sim 10^5$ yr (Larson, 1969). In a partially ionized gas, which is more appropriate for the OH cloud, Cowling (1957) gives a dissipation time of

$$\tau_B = 1.2 \times 10^{-3} \ n_e n R^2 B^{-2} \ \text{yr}$$

where $n_e (\sim 10^{-4} \ n)$ and n are in cm^{-3}, R the cloud radius in pc, and B is in G. With a cloud radius of 0.012 pc, $n = 10^6$ cm^{-3}, we find $\tau_B = 2 \times 10^6$ yr. This is still greater than the lifetime of the cloud. It is therefore evident that the magnetic field is frozen into the gas of the OH cloud.

A surprising pattern is found in the magnetic field directions within the eight OH masers listed in Table II. All the fields are aligned in the direction of galactic rotation, which is also the direction of the general galactic magnetic field determined from Faraday rotation measurements of pulsars and extragalactic sources (Gardner et al., 1967; Manchester, 1972). The neutral hydrogen clouds for which a significant Zeeman splitting has been measured also fit this pattern (Davies et al., 1974). The locations of the OH and H I clouds in the Galaxy are shown in Figure 6. This seems to indicate that these dense clouds which are widely spread throughout the Galaxy have preserved the direction of the general magnetic field as they have collapsed from densities of a few atoms per cubic centimeter to $\sim 10^6$ cm^{-3}. The conditions under which such a cloud can collapse in diameter by a factor of ~ 70 and still preserve its alignment relative to the weak external field are of interest but will not be discussed further here.

It is pertinent to consider in more detail the implication that the Class I OH maser sources are associated with collapsing protostars. The number of OH maser sources would therefore be expected to equal the rate of formation of O and B stars multiplied by 10^5 yr, the gravitational collapse time of the protostar; the OH maser action occurs only when the density is $\sim 3 \times 10^6$ cm^{-3}. Estimates of the rate of formation of O stars in the Galaxy suggest $\sim 10^{-3}$ yr^{-1} (Spitzer, 1968). This leads to an expected number of 10^2 OH sources in the Galaxy, a value which is consistent with the observations.

V. A Model of the Magnetic Field in W3 OH

In this section we will discuss a model of the magnetic field in the W3 OH maser source. This source has been observed extensively at a wide range of frequencies, and

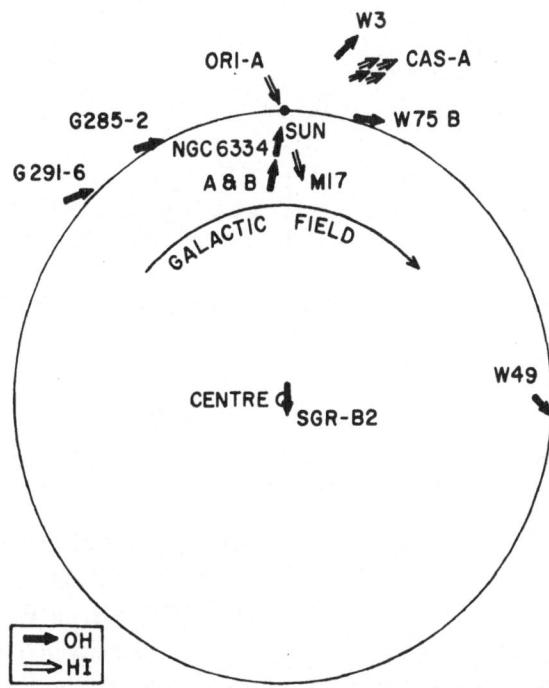

Fig. 6.   The (line of sight) magnetic field direction in eight OH maser sources (this paper) and seven H I clouds (Davies *et al.*, 1974). The general interstellar field of the Galaxy is directed in the clockwise sense. All the cloud fields are seen to be in the same sense.

it has been mapped in most detail. Its measured properties are as follows: (a) Its dimensions are 2.5″ × 1.4″ (0.036 × 0.020 pc) with a major axis at position angle ∼0°. (b) Its magnetic field is 4.8 mG. (c) When the effect of Zeeman splitting is allowed for in each region of the source a velocity gradient is found which indicates a rotational velocity of ∼2 km s^{-1} about an axis perpendicular to the major axis; the more negative velocities are at the northern end.

A sketch of the proposed model for W3 OH is shown in Figure 7. The magnetic field is taken to be parallel to the rotation axis for the following reason. The fields in the OH maser sources are found to be parallel to the general galactic field and consequently will be parallel to the galactic plane; for W3 OH this field will be directed to p.a. ∼270°. This is also the rotation axis inferred from the observations.

The W3 OH source is associated with an H II region with dimensions 2.3″ × 1.9″ which is at the same position within the errors of measurements (Baldwin *et al.*, 1973). An IR source is also associated; its dimensions are unknown. The rms density of the H II region is 1.25×10^5 cm^{-3}. An O7 star at the centre of W3 OH is required to produce the ionization and to heat circumstellar dust to produce the IR emission (Wynn-Williams *et al.*, 1972). The mass of the O7 star is ∼20 $M_\odot$, while the H II region is 10^{-1} $M_\odot$ and the OH cloud is ∼1.5 $M_\odot$ (see below). The OH cloud, which also contains a magnetic field of ∼5×10^{-3} G, is a cool shell surrounding the H II region. This outer gas which has a longer free-fall time is probably still collapsing inwards

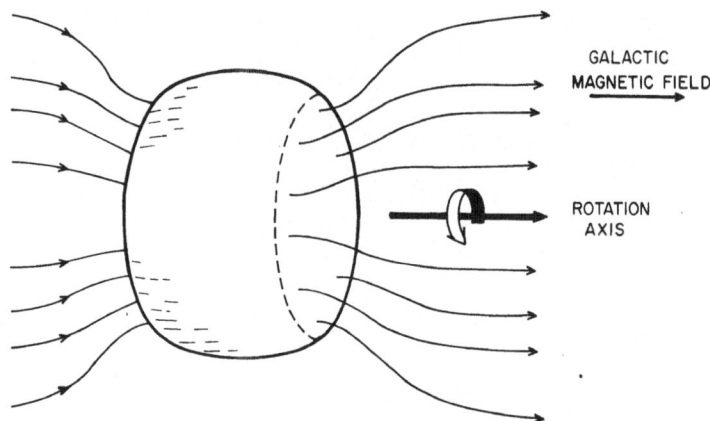

Fig. 7.   A model of the W3 OH source showing the sense of rotation and the magnetic field direction. The OH source is a shell of gas surrounding an H $_{II}$ region and an O7 star.

while the inner ionized gas is heated by the star formed from the greater mass of gas which collapsed first (Larson, 1969). The hot inner gas and the star are probably the source of the pumping mechanism of the Class I OH maser in W3 OH.

An important parameter in this discussion is the gas density in the molecular masering cloud; this can be estimated from the observed maser gain. Various authors (e.g., Litvak, 1969; Litvak *et al.*, 1969) have shown that the molecular line integral is given for a source similar to W3 OH by

$$\int n(\text{OH}) \, dl \cdot \frac{v}{\varDelta v} \approx 10^{22} \text{ cm}^{-2},$$

where $v/\varDelta v \sim 3 \times 10^5$ is the ratio of the line frequency to the line-width before amplification. This gives $\int n(\text{OH}) \, dl = 3 \times 10^{16} \text{ cm}^{-2}$. With a masering depth of $\sim 10^{16}$ cm – a typical line of sight through the shell – we obtain $n(\text{OH}) = 3 \text{ cm}^{-3}$. Taking $n(\text{H})/n(\text{OH}) = 10^6$ (Litvak, 1969) we find $n_{\text{H}} = 3 \times 10^6 \text{ cm}^{-3}$ with an uncertainty of a factor of 10 either way arising from uncertainties in the various steps of the calculation. This gas density would be in pressure equilibrium with the hot H $_{II}$ region gas if the temperature of the OH cloud were $\sim 10^3$ K. The mass of a circumstellar gas cloud with this density and a thickness of 20% of the radius derived for the OH maser source would be 1.5 $M_{\odot}$.

The dynamical stability of the OH maser source can be checked with the data available. If the collapse of the object were entirely stabilized by rotation a point on the equator would satisfy the relationship

$$\frac{v^2}{R} = \frac{GM}{R^2}, \text{ i.e., } v^2 = \frac{GM}{R}.$$

Using a total system mass $M \sim 20 \, M_{\odot}$, $R = 4.5 \times 10^{16}$ cm, the RH side of the equation leads to $v \sim 2.5 \text{ km s}^{-1}$, whereas the observed rotational velocity is 2 km s^{-1}. Thus the rotation is capable of largely stabilizing the source. In fact the flattening of the object may, in part, be due to the significant centrifugal force at the equator.

The other likely contributor to the flattening of the source will be magnetic rigidity of the cloud in the plane perpendicular to the magnetic axis of the cloud. Estimates using the values of $n(H)$, B and velocity spread (1 km s^{-1}) given above show that the magnetic and gravitational pressures have a similar magnitude of ~ 3 to 10×10^{-7} dyn cm^{-2} while the thermal gas pressure is about one tenth of this value. This magnetic pressure, acting orthogonally to the field direction will result in a significant flattening of the gas cloud parallel to the magnetic (and rotational) axis.

VI. Conclusion

A study of the expected properties of Zeeman splitting of OH maser emisssion and a comparison with observed properties indicate that a number of the sources show all the properties expected for either saturated or unsaturated masers (Goldreich *et al.*, 1973). These sources evidently have a Zeeman splitting comparable to or larger than the velocity dispersion of the OH gas. In other sources the spectra are so complex that even though the Zeeman effect is still producing the strong circular polarization it is impossible to decide which LH and RH components are related. Long baseline interferometry is shown to be an effective method of demonstrating the Zeeman splitting of such sources (e.g., W49). The main effects observed here are predicted in the models discussed by Goldreich *et al.* (1973), where the Zeeman splitting is greater than the (non-masered) line-width for unsaturated masers and for small Faraday rotation. Smaller fields than this could still produce strong circular polarization, but the Zeeman splitting patterns would be confused by random motions in the sources. It is interesting that the Class I OH maser sources have collapsed to such a density that the frozen-in magnetic fields are amplified so that they produce a Zeeman splitting comparable to the random motions and therefore give strong circular polarization.

In the sources where the field is sufficiently strong to separate the LH and RH circularly polarized components, estimates can be made of the magnetic field strength. These are in the range 2–7 mG. It is conjectured that the OH sources represent a much later phase of the collapse of the most dense clouds studied in neutral hydrogen Zeeman observations. The magnetic fields found in the OH clouds are of the magnitude expected for the conservation of magnetic flux. Only a small proportion, if any, of the magnetic flux appears to have been lost in collapsing from a density of atoms of ~ 10 cm^{-3} to $\sim 3 \times 10^6$ cm^{-3}. In a further isotropic collapse to stellar densities $(3 \times 10^{24}$ cm$^{-3})$ these fields would be amplified to 2×10^9 G; photospheric fields would be $\sim 10^5$ G. Since such high fields are not seen in normal stars, a mechanism is required for removing magnetic flux from the contracting cloud (see for example Spitzer, 1968). Most of this flux loss must therefore occur between OH maser densities and stellar densities, namely when the density is greater than 10^6 cm^{-3} and the radius is less than 5×10^{16} cm. The magnetic field will be largely dissipated by hydromagnetic instabilities with a timescale equal to the dynamical collapse time of the cloud of $\sim 10^5$ yr (Mestel, 1972).

The effect of magnetic braking on the cloud angular momentum due to any coupling between the cloud magnetic field and the interstellar field can also be investigated for the W3 OH source. The present angular momentum per unit mass for the OH cloud as determined from its rotation velocity is $\sim 30\%$ of the angular momentum of the cloud from which it collapsed if we take the interstellar turbulent velocity to be 0.1 km s^{-1} (Spitzer, 1968). Within the uncertainties of the parameters used there is no strong evidence for any large-scale loss of angular momentum. Like magnetic flux, the angular momentum loss occurs principally when the cloud collapses to higher densities than are represented by the OH sources. As a consequence the timescale for both loss processes must be $\lesssim 10^5$ yr, the collapse time of the OH cloud.

The recent measurements of magnetic fields in H I and OH clouds are now providing a corpus of observational data which can be related to theoretical studies of the kind carried out by Mestel (1965) and Parker (1973) for example.

Acknowledgements

I wish to thank Professors F. D. Kahn and L. Mestel and Dr R. S. Booth for discussions on many of the topics of this paper.

References

Baldwin, J. E., Harris, C. S., and Ryle, M.: 1973, *Nature* **241**, 38.
Bender, P. L.: 1967, *Phys. Rev. Letters* **18**, 562.
Coles, W. A., Rumsey, V. H., and Welch, W. J.: 1968, *Astrophys. J. Letters* **154**, L61.
Cook, A. H.: 1966, *Nature* **211**, 503.
Cowling, T. G.: 1957, *Magnetohydrodynamics*, Interscience Publishers, New York, p. 110.
Culshaw, W. and Kanneland, J.: 1966, *Phys. Rev.* **145**, 257.
Davies, R. D.: 1967, in H. van Woerden (ed.), 'Radio Astronomy and the Galactic System', *IAU Symp.* **31**, 82.
Davies, R. D., de Jager, G., and Verschuur, G.: 1966, *Nature* **209**, 974.
Davies, R. D., Booth, R. S., and Wilson, A. J.: 1974, in preparation.
Gardner, F. F., Ribes, J. C., and Goss, W. M.: 1970, *Astrophys. Letters* **7**, 51.
Gardner, F. F., Whiteoak, J. B., and Morris, D.: 1967, *Nature* **214**, 371
Goldreich, P., Keeley, D. A., and Kwan, J. Y.: 1973, *Astrophys. J.* **179**, 111.
Hall, L. M. and ter Haar, D.: 1973, *Monthly Notices Roy. Astron. Soc.* **162**, 97.
Harvey, P. J., Booth, R. S., Davies, R. D., Whittet, D., and McLaughlin, W.: 1974, *Monthly Notices Roy. Astron. Soc.*, in press.
Heer, C. V. and Graft, R. G.: 1965, *Phys. Rev.* **140**, A1088.
Heer, C. V. and Settles, R. A.: 1967, *J. Mol. Spectrosc.* **23**, 448.
Larson, R. B.: 1969, *Monthly Notices Roy. Astron. Soc.* **145**, 271.
Litvak, M. M.: 1969, *Astrophys. J.* **156**, 471.
Litvak, M. M.: 1971, *Astrophys. J.* **170**, 71.
Litvak, M. M., Zuckerman, B., and Dickinson, D. F.: 1969, *Astrophys. J.* **156**, 875.
Manchester, R. N.: 1972, *Astrophys. J.* **172**, 43.
Mestel, L.: 1965, *Quart. J. Roy. Astron. Soc.* **6**, 161 and 265.
Mestel, L.: 1972, in P. Schindler (ed.), *Cosmic Plasma Physics*, Plenum Publishing Corporation, New York, p. 203.
Parker, D. A.: 1973, *Monthly Notices Roy. Astron. Soc.* **163**, 41.
Perkins, F., Gold, T., and Salpeter, E. E.: 1966, *Astrophys. J.* **145**, 361.

Radford, H. E.: 1961, *Phys. Rev.* **122**, 144.
Rydbeck, O. E. H., Kollberg, E., and Elldér, J.: 1970, *Astrophys. J. Letters* **161**, L25.
Shklovskii, I. S.: 1969, *Soviet Astron.* **13**, 1.
Spitzer, L.: 1962, *Physics of Fully Ionized Gases*, Interscience Publishers, New York.
Spitzer, L.: 1968, *Stars and Stellar Systems* **7**, 1.
Townes, C. H. and Schawlow, A. L.: 1956, *Microwave Spectroscopy*, McGraw-Hill, London.
Turner, B. E.: 1970, *J. Roy. Astron. Soc. Can.* **64**, 221.
Verschuur, G. L.: 1971, *Astrophys. J.* **165**, 651.
Wynn-Williams, C. G., Becklin, E. E., and Neugebauer, G.: 1972, *Monthly Notices Roy. Astron. Soc.* **160**, 1.
Zuckerman, B., Yen, J. L., Gottlieb, C. A., and Palmer, P.: 1972, *Astrophys. J.* **177**, 59.

R. D. Davies
University of Manchester,
Nuffield Radio Astronomy Laboratories,
Jodrell Bank, Cheshire SK11 9DL, United Kingdom

DISCUSSION

Turner: Use of the Litvak (1969) IR pumping theory to deduce total particle densities in W3(OH) is inappropriate for two reasons:

(i) That theory applies to type II(a) and type II(b) OH masers, whereas the W3(OH) source is a type I source.

(ii) Even if the Litvak (1969) theory did apply, the IR line strengths for the 2.8 μm OH transition have been recently revised downward by a factor of 10 to 100. This lowers the particle density estimates by a similar factor, for the same pumping action.

Davies: There is no generally accepted pumping theory for the Type I sources. In my estimates of gas density I have used the conclusion of a number of authors who have examined both collisional and radiative pumping mechanisms. To obtain the observed maser gain an OH column density of $\sim 3 \times 10^{16}$ cm^{-2} is required. This can then be used to derive a gas density and a mass for the cloud.

Robinson: If the LH and RH components of the OH emission are produced by Zeeman splitting, I am puzzled that they are observed to come from spatially separated points in the sky. In the interferometer maps each spot has a characteristic velocity and sense of circular polarization.

Burke: I cannot agree that one should expect to see different Zeeman components from different parts of the sky. On the basis of any theory I know of, the full Zeeman pattern from a given cloud should be seen if the Zeeman splitting is bigger than the line width.

Davies: The model which discuss in my paper is of an emitting region with small fluctuations in velocity, magnetic field and density. In such a situation you certainly would not expect the LH and RH emission features to be coincident in apparent position.

Burke: I agree that the general tendency of one hand of circular polarization to be shifted in frequency with respect to the opposite hand is good evidence for the several milligauss field. The main point is that the existing theories fail to explain the absence of the full pattern from each cloud. Some nonlinear theory is needed, such as the one proposed by Townes.

Townes: Presently described theories do not allow for the very large polarizations apparently observed in a single cloud. A possible way of obtaining such large polarization is to allow for scattering and feedback by dust particles. Because of the feedback, very large advantage can be given to one polarization over another, and thus very large polarizations obtained.

Davies: Large percentage circular polarizations are expected for masering in the presence of Zeeman splitting where the splitting is comparable with the line width. The actual maser gain on any line of sight will depend upon the particular fluctuations of magnetic field and velocity along that line. Left and right hand circular polarizations will then have different maser gains, as described in my paper.

Zuckerman: The left and right should be coming from the same point if you want to call it a Zeeman doublet. You can't wipe out the left in one doublet, wipe out the right in another doublet coming from another point and then say those two things left over are Zeeman effect.

Townes: I don't think that's quite the point. We need only, it seems, have the same value of magnetic field. In other words, it is close enough to have the same magnetic field; it need not come from exactly the same batch of neutral gas.

Burke: I think the principal point is that the evidence the left and right are generally separated is in fact good evidence for a magnetic field there. The fact that you don't see the whole Zeeman pattern is good evidence that existing theory has something seriously wrong, and one must probably have to go to a non-linear theory such as Townes just described.

ON THE FIRST SOUTHERN-SKY H₂O MASERS OBSERVED
AT ITAPETINGA, BRAZIL

PIERRE KAUFMANN, W. G. FOGARTY, E. SCALISE, JR., and R. E. SCHAAL

Centro de Rádio-Astronomia e Astrofísica, Universidade Mackenzie, São Paulo, Brazil

Abstract. The first spectra of southern H_2O masers obtained with the 14-m Itapetinga radio telescope are described. A new H_2O source was found in H2-3 (G345.4 − 0.9).

The first southern-sky H_2O-maser spectra were obtained at Itapetinga Radio Observatory, Atibaia, São Paulo, Brazil, in the middle of 1973, with a 45-ft radome-enclosed radio telescope. A preliminary description of the instrument was given by Kaufmann and D'Amato (1973). Based on measurements of Jupiter at 22.2 GHz, the antenna has an aperture efficiency of 60% for the dish alone (without radome), and thus for point sources 1 K of corrected antenna temperature corresponds to 31 Jy. A load at 330 K served as a calibration source for the observations. The relative pointing and tracking accuracies of the radio telescope were 4″ rms. The absolute accuracies were not yet completely determined, but an upper limit of less than 20″ is known for the repeatability – negligible in comparison to the antenna HPBW at K-band.

A simplified spectrometer was used during these first experiments. The IF output from a dual beam front end was fed into a spectrum analyzer which scans in frequency. After synchronous detection the measurements were stored and averaged by an on-line digital integrator. The integration time corresponding to a multichannel system was $T = N\Delta t (\Delta f/\Delta F)$ seconds, where N is the number of sample scans, Δt the duration of each scan, Δf the filter bandwidth and ΔF the frequency interval analyzed. Most of the measurements were taken with $\Delta f = 43$ kHz, $\Delta F = 2$ MHz and $\Delta t = 1$ s. The actual integration times, $N\Delta t$, of thousands of seconds corresponded to relatively small T. These first spectra were somewhat noisy but repeatable with respect to intensities and structures found.

Antenna temperatures were carefully corrected for atmospheric and radome attenuation, following the relation

$$T_a = T_a^* \exp(\tau \sec z)/\eta_r,$$

where T_a is the corrected antenna temperature, T_a^* the measured antenna temperature, τ the optical depth of the troposphere, z the zenith angle, and η_r the radome transmission at K-band. During the period of observation, τ ranged from 0.1 to 0.2 and was frequently measured.

The main results can be summarized as follows:

(a) Well known sources measured by northern hemisphere observatories were observed. The spectrum of the Orion A H_2O maser is shown in Figure 1 for comparison purposes, since it is well known in position and typical strength of the lines (Hills *et al.*, 1972).

F. J. Kerr and S. C. Simonson, III (eds.), Galactic Radio Astronomy, 293–297. All Rights Reserved.

(b) Southern hemisphere H_2O sources discovered by Johnston *et al.* (1971) using the 9-m Carnavon antenna in Australia and by Johnston *et al.* (1972) using the Parkes telescope (with its inner reflector resurfaced) were then observed and confirmed at Itapetinga. Among these sources we should comment that G285.3−0.1,

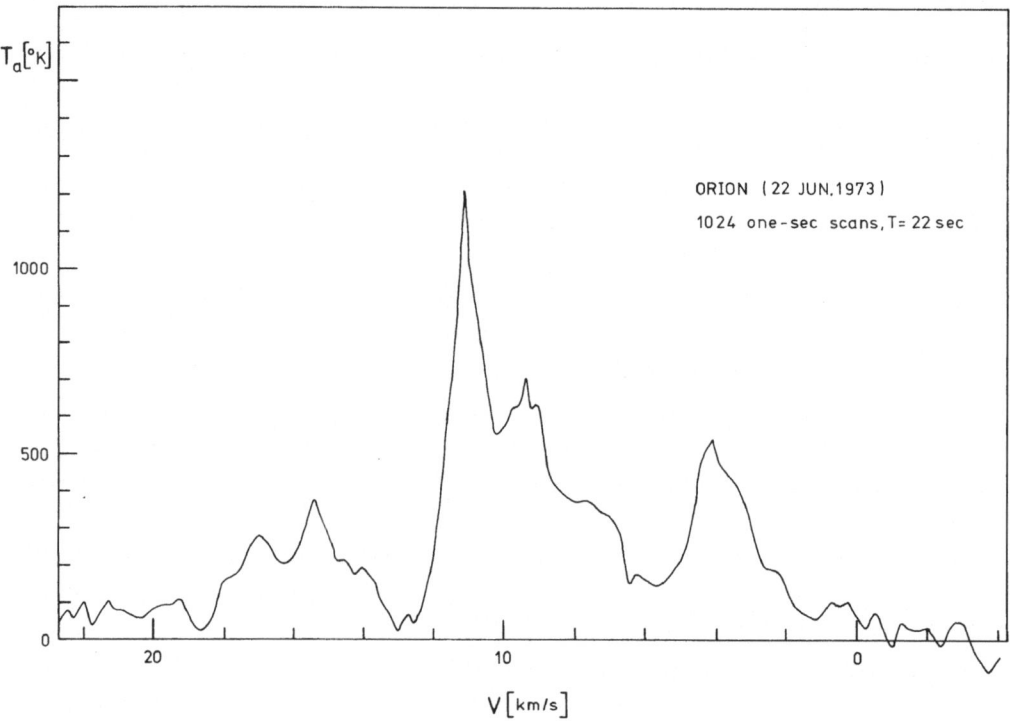

Fig. 1.   The principal H_2O lines observed in Orion, centered on $+10$ km s^{-1}.

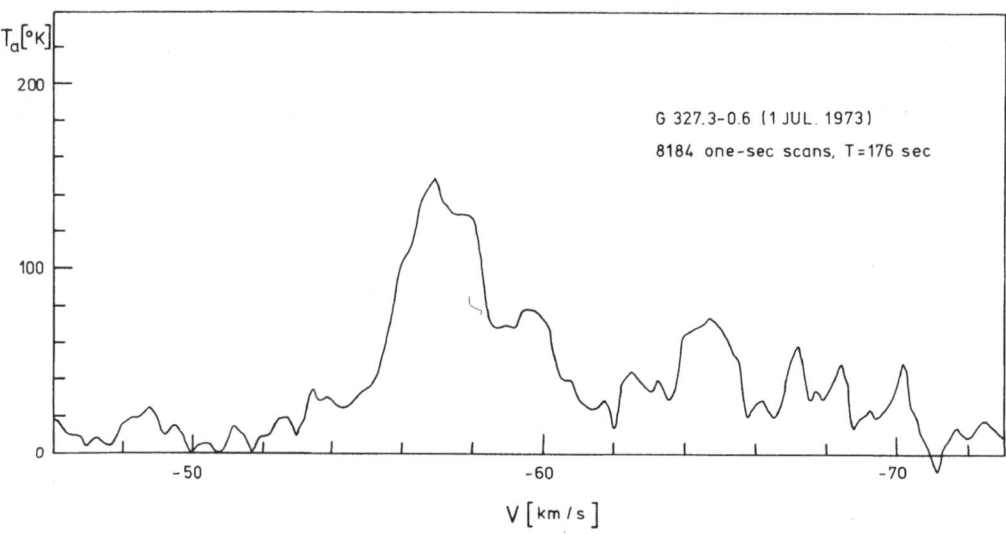

Fig. 2.   A more complex spectrum of water vapour lines from the galactic source G327.3−0.6.

G327.3 − 0.6, G331.5 − 0.1, G333.6 − 0.2 and G345.7 − 0.1 have shown stronger lines, with some changes in their original structures, and gave antenna temperatures ranging from 150 K to 950 K. The source G327.3 − 0.6 is shown in Figure 2 with an additional line appearing at −65 km s^{-1}. The source G331.5 − 0.1 is shown in Figure 3, and it is the strongest so far known in this series of new H_2O masers. It shows a −97 km s^{-1} line that is much stronger and well resolved than the broader emission between −86 km s^{-1} and −94 km s^{-1}.

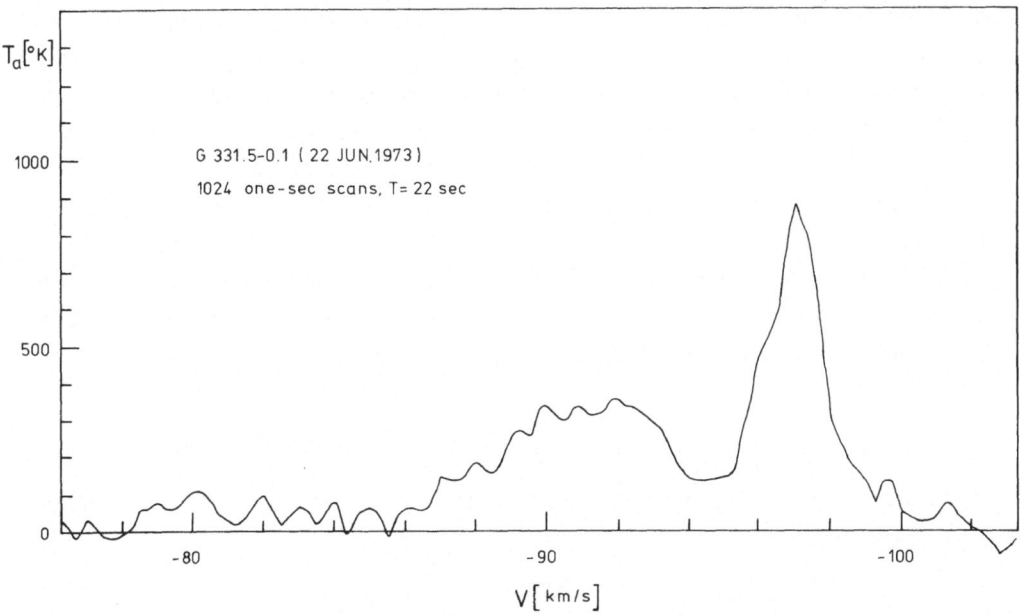

Fig. 3.   This source, G331.5 − 0.1, is the strongest from the new series of H_2O sources discovered by Johnston *et al.* (1972) and has shown line intensity variations.

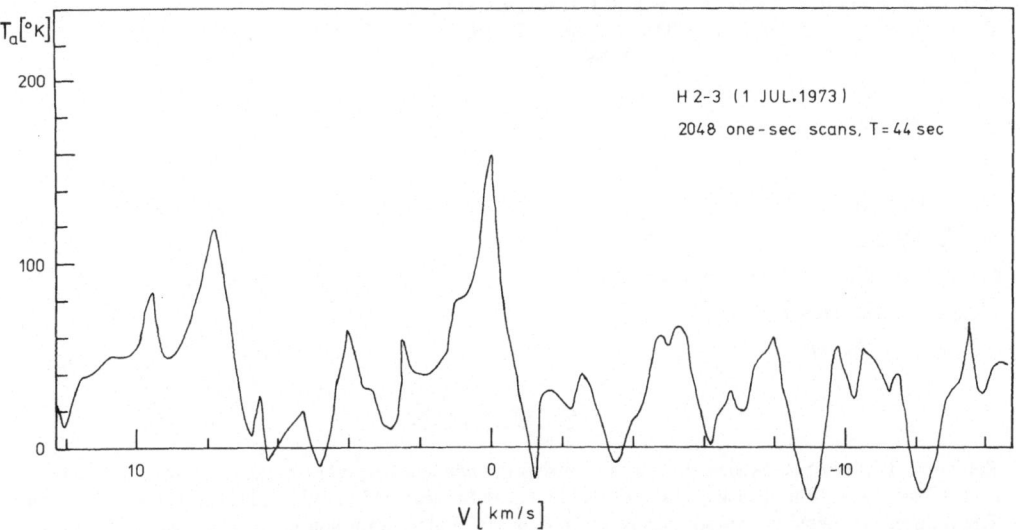

Fig. 4.   The presence of H_2O found in the compact H II object H2-3.

For all these sources, including Orion A, the Itapetinga results in antenna temperature are bigger than Parkes' by a factor of about ten. Although it is known that H_2O sources show important time variations in intensity, this does not seem to be the case for all the sources observed in this study. It is then suggested that the sources discovered by Johnston *et al.* (1972) typically present lines with flux densities of about ten times larger than previously reported.

(c) Quite surprisingly VY CMa was probably 'off' during the epoch of the observations (end of June 1973) with an upper limit in antenna temperature of 35 K rms, suggesting that drastic variability might be happening in that well known and usually strong source.

(d) A new H_2O source was found in the compact H II object H2-3 (also G345.4 − 0.9), as shown in Figure 4, with one line close to 0 km s^{-1} and possibly another weaker one at $+8$ km s^{-1}. OH in absorption was known for H2-3 at -14 km s^{-1} and $+20$ km s^{-1}, as well as a H109α recombination line at -23 km s^{-1} (Turner, 1970; Rubin and Turner, 1971). The presence of an H_2O maser in H2-3 favours the discussion of Rubin and Turner (1971) on the non-planetary nebula nature of this object and supports the possibility that it is more likely a young compact H II object like the well studied nebulosity K3-50.

Acknowledgements

This work has been supported by the Brazilian research agencies BNDE-FUNTEC, CNP_q and FAPESP.

References

Hills, R., Janssen, M. A., Thornton, D. D., and Welch, W. J.: 1972, *Astrophys. J. Letters* **175**, L59.
Johnston, K. J., Knowles, S. H., and Sullivan, W. T.: 1971, *Astrophys. J. Letters* **167**, L93.
Johnston, K. J., Robinson, B. J., Caswell, J. L., and Batchelor, R. A.: 1972, *Astrophys. Letters* **10**, 93.
Kaufmann, P. and D'Amato, R.: 1973, *Sky Telesc.* **45**, 144.
Rubin, R. H. and Turner, B. E.: 1971, *Astrophys. J.* **165**, 471.
Turner, B. E.: 1970, *Astrophys. Letters* **6**, 99.

Pierre Kaufmann
W. G. Fogarty
E. Scalise, Jr.
R. E. Schaal
Centro de Rádio-Astronomia e Astrofísica,
Universidade Mackenzie,
São Paulo, Brasil

DISCUSSION

Robinson: It is most surprising that for all sources the *antenna temperatures* measured with the Universidade Mackenzie 14-m dish are a factor of 10 higher than measured in 1971 at Parkes using a 34-m dish. The ratio of flux densities would then be more than 20. Water vapor sources are very variable, but it is most unlikely that the flux from *all* sources would have increased by 20 times.

The published Parkes measurements were made with the NRL 25-m dish (using Jupiter as a flux calibrator). The Parkes observations of W49, Orion A and YV CMa agreed to within a factor of 2 with earlier NRL measurements; the factor of 2 uncertainty arises from the variability of the H_2O sources.

Kaufmann: We are interested in cooperating in order to understand why Parkes values for H_2O sources were so low, since we do not see any factor that could change Itapetinga values by ± 10 to 20%. Our values are just compatible with the antenna performance.

Johnston: After seeing your antenna temperatures, I would estimate the discrepancy to be about a factor of 3.

Note added in proof *(Kaufmann):* The H_2O sources shown in this paper were again measured in November 1973 and April 1974. A negative feedback in the radiometer has been found and eliminated, and the Moon has also been used for calibration, and this reduced the originally quoted antenna temperatures by a factor of 2. The discrepancy discussed in the paper reduces to a factor of 5. On the other hand, another set of measurements of H_2O southern masers by Caswell, Batchelor, Haynes, and Huchtmeier (*Australian J. Phys.*, in press, 1974) with the Parkes telescope, taken in the same period as the results shown in this paper but with higher resolution (15.6 kHz per channel), shows smaller discrepancies on the stronger sources.

TIME SCALES AND MASSES OF OH-EMISSION REGIONS

T. K. MENON

Tata Institute of Fundamental Research, Bombay, India

Abstract. The velocity structure of OH profiles of OH maser regions is used to estimate the time scales of the source distributions. Time scales are found to be of the order of a few thousand years in typical cases. The velocity dispersions and intensity distribution of components are interpreted in terms of a collapsing cloud model of the emission regions. The masses of the clouds are found to be in the range from a few solar masses to a few hundred solar masses for the well-known emission regions.

(The following summary was prepared from the tapes by the Editors.)

A number of OH maser sources were investigated, in which the scale size is known from VLBI or high-resolution observations. The velocity structure of the OH profiles was used to estimate the time scales of the source distributions. For all groups of OH sources that are as wide in angular size as the Orion or W3 groups, the time scales are found to be on the order of a few thousand years in typical cases.

While investigating this time scale a peculiarity was noticed in the intensity distribution of the spectral components. Orion shows a typical spectrum of the multi-component OH sources. W3 and W49 are other examples. The Orion OH spectrum consists of 20 or more well-separated emission components, covering the velocity range -11 to $+23$ km s^{-1}. (The velocity of the local standard of rest at Orion is $\sim +8$ km s^{-1}.) Starting at the more negative velocities, the intensity of the components increases, building up to very large intensities toward positive velocities. The component at the highest velocity on the positive side has the largest intensity. At still higher velocities nothing more is seen. All the low intensity components are to the negative velocity side of the high intensity components.

Since this happens in a large number of multi-component sources it may imply that there is a cutoff velocity at the positive velocity side. This may be a velocity of escape, or it may be a terminal velocity of infalling material. In either case one can estimate the mass. I have made a collapsing cloud model of the emission regions. The masses of the clouds are found to be in the range from a few solar masses to a few hundred solar masses for the well known emission sources.

T. K. Menon
Tata Institute of Fundamental Research,
Homi Bhabha Road,
Bombay 400 005, India

DISCUSSION

Mezger: Do you apply Zeeman analysis to the W49 region? Can you estimate a mass?

Menon: Not Zeeman analysis, but just the fact that the line profile has its highest intensity at the

F. J. Kerr and S. C. Simonson, III (eds.), Galactic Radio Astronomy, 299–300. *All Rights Reserved.*
Copyright © 1974 by the IAU.

positive velocities, with a falling off of intensity toward lower velocities. I have not computed a mass for W49.

Morimoto: Davies stated the W3 OH cluster can be supported by a star of 20 $M_\odot$. Your mass of 600 $M_\odot$ seems to contradict this.

Menon: My estimate of 600 $M_\odot$ refers to the Orion OH cluster. The mass for W3 OH turns out to be about 15 $M_\odot$ since the velocity spread is only about 2 km s^{-1}.

Davies: You based your estimates of cloud masses on infall velocities of approximately 20 km s^{-1}. In the case of W3 OH, where the structure of the OH object has been investigated in detail by interferometry, the spread in velocity is ± 2 km s^{-1} when account is taken of the Zeeman splitting. This velocity leads to a much lower estimate of the object.

Menon: My estimate of mass refers to the Orion OH cluster. In the case of W3 OH since the velocity spread is small the corresponding mass will also be small since the mass is proportional to the square of the velocity.

Townes: In some of these profiles I think I can see several weak components at higher velocities than that of the strongest peak.

Menon: There may be a few very weak components beyond, but there is a piling up of material at this terminal velocity.

OH OBSERVATIONS NEAR THE REFLECTION NEBULAE
NGC 2068 AND NGC 2071

L. E. B. JOHANSSON and B. HÖGLUND

Onsala Space Observatory, Onsala, Sweden

A. WINNBERG

Max-Planck-Institut für Radioastronomie, Bonn, F.R.G.

NGUYEN-Q-RIEU

Observatoire de Meudon, Meudon, France

and

W. M. GOSS

Kapteyn Astronomical Laboratory, Groningen, The Netherlands

Abstract. Narrow OH emission lines at 1667 MHz, apparently from a Class I source, have been observed near the reflection nebula NGC 2071. The region contains many T Tauri stars. OH emission corresponding to the dust cloud north and east of NGC 2024 is also seen. At 1720 MHz the dust cloud component appears in absorption; presumably the isotropic 2.7 K cosmic background is being absorbed.

(Summarized by the Editors.)

I. Dust Cloud Emission

We surveyed OH emission in the direction of the dust cloud near NGC 2024 (W12 or Orion B) between limits of $-2\overset{\circ}{.}0 < \delta < +1\overset{\circ}{.}5$, $5^h40^m < \alpha < 5^h50^m$, using the 25-m radio telescope of the Onsala Space Observatory.

We found dust cloud emission at ~ 9 km s^{-1} over the whole survey area north and east of NGC 2024. The antenna temperatures were 0.10 K at 1665 MHz and 0.21 K at 1667 MHz. Because of uncertainties in gain and noise fluctuations we cannot rule out the LTE radio, but the satellite lines show clear evidence for nonthermal emission. The 1612 MHz line is stronger than the 1665 MHz line, and the 1720 MHz line appears in absorption. Presumably the 2.7 K isotropic background is being absorbed; the situation is similar to H_2CO absorption in dark clouds. A concentration of T Tauri stars in the vicinity may provide near-IR flux to pump the OH molecules to the nearest vibrationally excited state, enhancing both the 1612 MHz emission and the 1720 MHz absorption.

II. The Class I Source

We found unusual OH emission, apparently from a Class I source, near the reflection nebula NGC 2071. A narrow emission line appears at 1667 MHz only. It has two components, at 13.45 and 13.75 km s^{-1}. The half-power widths are 0.3 km s^{-1}, *implying a kinetic temperature of ~ 33 K if due to thermal broadening. We saw no*

F. J. Kerr and S. C. Simonson, III (eds.), Galactic Radio Astronomy, 301–302. All Rights Reserved.
Copyright © 1974 by the IAU.

time variation between August and December 1972. We measured its position at Nançay with a beamwidth of $3.5 \times 18'$: $\alpha = 5^h44^m30^s \pm 1^s$, $\delta = +0°20'.5 \pm 1'$ (1950). Its galactic designation is OH 205.1−14.1.

The Class I source is quite close to a group of T Tauri stars but does not seem to coincide with any one of them. We used the 100-m telescope at Effelsberg to search for a compact H II region near the source and found none to limits of 0.08 Jy at 2.695 GHz and 0.02 Jy at 10.7 GHz. We suggest the source may represent an early phase in the development of a T Tauri star.

We give a fuller account in *Astrophysical Journal* **189** (1974), 455.

L. E. B. Johansson
B. Höglund
Onsala Space Observatory,
S-430 34 Onsala, Sweden

A. Winnberg
Max-Planck-Institut für Radioastronomie,
Auf dem Hügel 69,
53 Bonn 1, F.R.G.

Nguyen-Q-Rieu
Observatoire de Meudon,
92 Meudon, France

W. M. Goss
Kapteyn Astronomical Laboratory,
Postbus 800, Groningen 8002, The Netherlands

DISCUSSION

Turner: The relative line brightnesses shown on the slide are consistent with near-IR pumping (2.8 μm) but not far-IR pumping. These lines were said to arise over an extended area, probably a dust cloud. Are there enough T Tauri stars over this area to maintain near-IR pumping over the entire area? Are the relative line brightnesses constant over this entire area? It is important to decide these questions for dark dust clouds in general, to explain the many similar OH anomalies found in these clouds by Turner and Heiles (1971) and by Turner (1973).

Höglund: Herbig and Kuhi listed 45 T Tauri stars in the area, so I think there will be sufficient 2.8 μm flux available. We observed all four ground state lines only at the position of the Class I source.

HII REGIONS: SOME THEORETICAL DEVELOPMENTS

F. D. KAHN

University of Manchester, Manchester, United Kingdom

Abstract. Observations have revealed the presence of inhomogeneities in both density and velocity in HII regions. Dense non-ionized condensations can persist for a long time, developing bright rims of dense ionized gas around them. Attempts have been made to understand the large ~ 50 km s^{-1} motions seen in some HII regions by invoking stellar winds or the flow of cosmic ray particles through the plasma.

I. Introduction

It has long been known that the structure of HII regions is not homogeneous. In particular, the electron density n_e is often uneven, and there are frequently considerable differences in velocity between adjacent parts of an HII region.

The unevenness in electron density shows up when attempts are made to determine $\bar{n}_e$ by means of different techniques. To understand how this happens, suppose that a fraction α of the available space contains electrons, at a density of N_e per unit volume, and that the rest of the space is empty. Then a measurement of the electron density by the technique of line ratios will find the average value

$$\bar{n}_{e,1} \equiv \frac{\int n_e \, dn_e}{\int dn_e} = N_e, \tag{1}$$

since the mean is, in this case, weighted by the local ion, and therefore electron, density. A determination based on the emission measure of the HII region will find the value

$$\bar{n}_{e,2} \equiv \left\{ \frac{\int\int n_e^2 \, dl}{\int dl} \right\}^{1/2} = \left\{ \frac{\int\int n_e^2 \, dv}{\int dv} \right\}^{1/2} = \alpha^{1/2} N_e. \tag{2}$$

Finally, a determination based on pulsar dispersion measure would, if it could be carried out unambiguously, find the value

$$\bar{n}_{e,3} \equiv \frac{\int n_e \, dl}{\int dl} = \frac{\int n_e \, dv}{\int dv} = \alpha N_e. \tag{3}$$

Discrepancies between known results show that the filling factor is around 0.03 to 0.05 (Osterbrock and Flather, 1959; Danks and Meaburn, 1971).

The velocity structure of HII regions is also noticeably uneven. Thus the observations of the Orion nebula by Wilson *et al.* (1959) revealed velocity differences between adjacent regions of the order of 10–20 km s^{-1}. More recently Meaburn and his associates have found rather larger velocity differences in various nebulae, ranging up to 50 km s^{-1} (see, for example, Meaburn, 1971). In this case the linear scale on which the variation occurs may be rather larger.

F. J. Kerr and S. C. Simonson, III (eds.), Galactic Radio Astronomy, 303–308. All Rights Reserved.

II. The Effect of Non-Ionized Condensations

The temperature does not vary much from point to point in an H II region, so that any fluctuations in electron density must be strongly correlated with pressure fluctuations. It is therefore natural that irregular gas flows should occur in H II regions, typically with velocities of the order of the speed of sound. This is just what Wilson and Münch observed.

But the inhomogeneities in density and pressure occur on a small linear scale, perhaps 10^{16} or 10^{17} cm. If the resulting motions have speed of about 10^6 cm s^{-1} then the density of the ionized gas will even itself out in a short time, of the order of 1000 yr. This is clearly much less than the age of the typical H II region. The suggestion has therefore been made that the spaces where there is high electron density are associated with condensations of non-ionized gas, embedded in the H II region. Around the boundary of each condensation there is an ionization front, itself surrounded by a shell containing dense ionized gas (a bright rim). The gas in the bright rim streams away from the condensation into the lower density H II region.

Quite a long time ago Dyson (1968) worked out these ideas in some detail and found reasonable agreement with the observed phenomena. His model for a non-ionized globule with its ionized jacket was later reconsidered (Kahn, 1969); the main change in this revised version was that it gave a different treatment of the mechanics of the ionization front. Qualitatively the results were not changed much. Both treatments were restricted to condensations which had spherical symmetry throughout. Naturally the incident radiation field must then be isotropic. Such an idealisation makes the theory easier to handle, but obviously it cannot give a realistic description of the radiation field in an actual H II region.

It is therefore interesting that Dyson (1973b, 1974) has made a first attempt to extend his description to cases where the incident radiation field is anisotropic. His treatment is linearized, in the sense that it allows for small distortions from spherical symmetry. Once again the problem is to calculate the structure of the non-ionized part of the globule, and to fit the solution to the appropriate boundary conditions at the ionization front. Dyson concludes that the distortions in the globule are usually rather smaller than the anisotropy of the radiation field.

However, the discussion is rather a delicate one. The boundary conditions applicable to the anisotropic case have to be found by making an analogy with the isotropic situation. It is arguable whether Dyson has used the proper analogy, but there will probably be no qualitative change in his overall conclusion even if he has to revise his boundary conditions. So we conclude, provisionally, that the existence of a globule does not depend on its being spherically symmetrical; globules can therefore exist in much the same way in real radiation fields.

If a globule does not collapse under its own self-gravitation, it will provide a source of high density ionized gas until it has been eaten up by the incident Lyman continuum radiation. The complete ionization of a globule would typically require a few hundred thousand years.

III. Other Causes of Inhomogeneity

The presence of non-ionized condensations within H II regions probably explains the existence of density fluctuations in the ionized gas, and produces the irregular velocity pattern described by Wilson and Münch. But it is rather less likely that the 50 km s^{-1} motions observed by Meaburn can be explained in the same way.

To understand why there may be a difficulty, let us consider the steady flow of ionized gas from a globule. It is most realistic to assume that the gas remains isothermal, at a temperature of about 10^4 K. With the usual cosmic mixture of elements, the corresponding speed of sound is $c_0 = (kT/m)^{1/2} = 1.2 \times 10^6$ cm s^{-1}. For a steady flow, Bernoulli's theorem takes the form

$$\tfrac{1}{2}u^2 + c_0^2 \log \varrho = \text{constant}, \tag{4}$$

where u is the speed and ϱ the density of the gas. If the density is ϱ_I and the speed $u_I \equiv \mu_I c_0$ at the ionization front, then it follows from (4) that elsewhere in the flow

$$\frac{\varrho}{\varrho_I} = \exp\left\{-\frac{u^2 - \mu_I^2 c_0^2}{2c_0^2}\right\}. \tag{5}$$

With $\mu_I = 2$, $u = 5 \times 10^6$ cm s^{-1} and $c_0 = 1.2 \times 10^6$ cm s^{-1},

$$\frac{\varrho}{\varrho_I} \doteq e^{-7} \doteq 10^{-3},$$

so that the gas flow reaches a high velocity only after the density has fallen by a large factor. At very low density the gas contributes only a little to the emission measure. It would therefore be quite hard to explain Meaburn's observations in this way. On the other hand, compared with the luminous output of an O5 star, it does not require an extravagant amount of energy to set up the 50 km s^{-1} motions. Much of the energy radiated by the early type stars in H II regions is, in fact, given to the gas, but the gas re-radiates it in turn, and only a little remains available to drive any mass motions. A good mechanism for the acceleration of the gas requires that energy be stored, for a long enough time, in a form where the particles or photons carrying it have a momentum density which can couple with the gas in the H II region. The problem is not that the energy requirement is excessive, but rather that it is not easy to retain the energy in a suitable form.

Dyson and de Vries (1972) (see also Dyson, 1973a) have investigated whether a powerful stellar wind can drive high speed motions through an H II region. A speed of 1500 km s^{-1} is adopted for the wind; it turns out that the wind will form an expanding hole in the interstellar gas. Beyond the boundary of the hole there is a highly compressed layer of gas, which is itself enclosed on the outside by a shock, S_1. Inside the hole there is a thick shell of hot gas, bounded on the interior side by an inward facing shock, S_2. As the wind from the star passes through this shock, its kinetic energy is converted into thermal energy. For a wind speed of 1500 km s^{-1}, the shocked gas acquires a temperature of the order of 10^8 K. Little cooling is expected to occur in so hot a gas. Instead the shock-heated gas pushes against the surrounding interstellar gas. In short, the stellar wind can be regarded as a source of thermal energy which pumps pressure into the hole in the interstellar gas and causes it to expand.

The result of the calculation is that about 25% of the energy released in the stellar wind becomes available in the form of mass motion in the interstellar gas. The rest is lost as thermal energy behind shock S_1, and is later radiated. As an example, if the stellar wind relases energy at a rate $Q = 10^{38}$ erg s^{-1}, and blows into interstellar gas having a density $\varrho_0 = 10^{-22}$ g cm^{-3}, then the interstellar gas will move at a speed $V = 5 \times 10^6$ cm s^{-1} after time $t = 10^5$ yr. For different values of Q, ϱ_0 and t, V scales like $(Q/\varrho_0 t^2)^{1/5}$. Dyson considers a rather smaller value for Q, and is somewhat pessimistic about the likely significance of stellar winds in this context; nevertheless, it seems that this process may actually be quite important. In particular it is worth noting that Wendker, Smith, Habing, Israel, Dickel and Price consider that a stellar wind, originating in the WR star HD 192163, may be responsible for forming the ellipsoidal nebulosity NGC 6888 (Lindsey Smith, private communication). The nebulosity would here be identified with the compressed interstellar gas lying between the contact discontinuity and shock S_1. The ellipsoidal shape of the nebulosity is interesting, and doubtless arises from an anisotropy in the stellar wind itself. It will be interesting to modify Dyson's calculations to allow for departures from spherical symmetry, and to see what restrictions this change places on the model.

There are other possible ways of storing the energy, for example in the form of cosmic ray particles. As is well known, cosmic rays readily lose energy to free electrons via electrostatic interactions. If cosmic ray protons are trapped in an ionized gas with density 10^{-22} g cm^{-3} then the time scale for this energy loss exceeds 10^5 yr, provided the protons have a kinetic energy greater than 160 MeV. For other densities the minimum energy scales like $\varrho^{2/3}$. Clearly suprathermal particles, whose energies are supposedly around 5 MeV, will not last long enough to provide any important acceleration, unless they can be prevented from mixing with the interstellar gas. But cosmic rays in the more usual energy range will last long enough. The mechanical coupling between the cosmic ray plasma and the thermal plasma is due to the well-known Alfvén wave instability. This effect prevents any streaming between the two plasmas with relative speeds substantially greater than the Alfvén speed (Kulsrud and Pearce, 1969; Skilling, 1970; Kahn, 1971). It is now thought that this effect is not quite as universally important as had once been believed, largely because damping of the Alfvén waves by ion-neutral collisions restricts the range of conditions where the wave instability occurs (Wentzel, 1974). But this certainly does not apply in H II regions.

Finally, to complete the list of energy sources, there may be mechanical effects caused by very low frequency radiation trapped in the H II region. The vlf radiation is thought to originate in pulsars, and on emission the period $2\pi/\omega$ of a wave is the same as the pulsar period. The wave cannot propagate through a plasma of electron density n_e unless the amplitude A of its vector potential exceeds $A_* = 4\pi n_e ec^2/\omega^2$. Even when a wave does propagate, it is still unstable on a short time-scale and is therefore expected to break up into secondary waves having a different frequency (Max, 1973). Non-linear interactions occur once the plasma contains waves with a variety of frequencies. The interactions impede propagation, so that once again energy is trapped in the H II region, and becomes available for accelerating the interstellar gas. But here we are on very speculative ground, for these ideas have not really been worked out yet.

All these processes should have about the same efficiency for the conversion of energy into kinetic energy of the ionized gas. In other words, to explain the 50 km s^{-1} motions we must look for a source which will provide energy at the rate of about 10^{38} erg s^{-1}, in a form which does not rapidly get lost as radiation, and will maintain the supply for some 10^5 yr.

References

Danks, A. C. and Meaburn, J.: 1971, *Astrophys. Space Sci.* **11**, 398.
Dyson, J. E.: 1968, *Astrophys. Space Sci.* **1**, 388; **2**, 461.
Dyson, J. E.: 1973a, *Astron. Astrophys.* **23**, 381.
Dyson, J. E.: 1973b, *Astron. Astrophys.* **27**, 459.
Dyson, J. E.: 1974, *Astron. Astrophys.* **32**, 349.
Dyson, J. E. and de Vries, J.: 1972, *Astron. Astrophys.* **20**, 223.
Kahn, F. D.: 1969, *Physica* **41**, 172.
Kahn, F. D.: 1971, *Quart. J. Roy. Astron. Soc.* **12**, 384.
Kulsrud, R. M. and Pearce, W. P.: 1969, *Astrophys. J.* **156**, 445.
Max, C. E.: 1973, preprint.
Meaburn, J.: 1971, *Astrophys. Space Sci.* **13**, 110.
Osterbrock, D. E. and Flather, E.: 1959, *Astrophys. J.* **129**, 26.
Skilling, J.: 1970, *Monthly Notices Roy. Astron. Soc.* **147**, 1.
Wentzel, D. G.: 1974, *Ann. Rev. Astron. Astrophys.* **12**, 71.
Wilson, O. C., Münch, G., Flather, E., and Coffeen, M.: 1959, *Astrophys. J. Suppl.* **4**, 199.

F. D. Kahn
Astronomy Department,
University of Manchester,
Manchester 13, United Kingdom

DISCUSSION

J. R. Dickel: The high velocity wings all appear on one side. Is it positive or negative radial velocity? And won't the spherical expanding shocks or stellar winds produce a symmetric profile with wings on both sides?

Kahn: The early observations by Meaburn showed high velocity streaming on the negative side only. I believe that it is not yet quite certain whether this asymmetry is real. If it is, then a spherically symmetrical stellar wind will not offer any explanation of this phenomenon.

Yuan: Those profiles which you showed all have a narrow central peak and broad tails. But your stellar wind model seems only capable of producing a simple Gaussian. How do you explain?

Kahn: The H II region at large produces a narrow spectral line. The broad wings would be due to the emission from the region where the gas is expanding at ~ 50 km s^{-1}.

Habing: What is the evolution of the kind of globules you showed? Provided that they do not collapse, will they gradually dissolve, forming thin shell sources of ionized gas, or will they form dense ionized blobs?

Kahn: The massive blobs collapse under gravitation. In the less massive ones the gas becomes ionized and streams away into the surrounding H II region.

Parijskij: We may simplify the picture of the OH (and H_2O) emission by a suggestion that many observable features are due to propagation effects, through a moving ionized (and neutral) medium.

Example: a strong stellar wind is between the primary single source at a single frequency and the observer. We should expect in this case a multi image structure and a separation of R and L-handed polarized spots. The physical conditions in that case do not seem extreme.

Swarup: Would you be able to identify these compact sources in the microwave region?

Kahn: No, I don't think so. No star has get condensed inside these globules. When they are illuminated from inside they look quite different, they become infrared sources, perhaps eventually compact H II regions.

Dickel: For a 10 $M_\odot$ globule what is the size?

Kahn: The size is 10^{16} cm.

Pishmish: In the region of NGC 2467 over a very small region of 0.5 pc there is a radial velocity difference of 50 km s^{-1} (that makes it 50 again).

Terzian: Do we expect to see these sources in radio recombination lines?

Kahn: I don't see why not.

Menon: Some years ago photographs were taken of the Orion nebula by Fisher at the University of Hawaii with very narrow band filters which could be tuned around Hα. Photographs were taken every 2–3 km s^{-1} over a wide range of velocities, and again they showed a component at about −50 km s^{-1}.

PART 3

SUPERNOVA REMNANTS

"Supernovae are called on to explain everything, and indeed they may, but I await further studies."

D. W. Goldsmith, in the discussion following his paper

OBSERVATIONAL ASPECTS OF SUPERNOVA REMNANTS

B. Y. MILLS

University of Sydney, Sydney, Australia

Abstract. A supernova event may lead to four observable features: a pulsar, an expanding nebulosity, a radio source and an X-ray source. The great majority of supernovae do not produce observable pulsars, and discussion is restricted largely to the other features. An increasing number of X-ray sources is now being detected and the structure and spectrum of the stronger sources investigated; these observations yield information about the physical state of the remnant. Recently, 11 new optical and radio remnants have been found in the Magellanic Clouds. These have led to a good determination of the $\Sigma - D$ relation, thus providing a more reliable distance scale for galactic SNR, but have also shown that a one-to-one correspondence between a radio source and a supernova event is questionable. When these remnants are combined with corrected earlier catalogues and a new southern catalogue containing a high proportion of distant old remnants the number of known SNR is about 150. The evolution of galactic SNR may be investigated empirically, and although the derived rate of occurrence is very uncertain a rate of about 2 supernovae per century is consistent with most determinations. The galactic SNR are distributed rather like the radio disc emission, but more closely confined to the galactic plane, and selected SNR show evidence of a spiral pattern.

I. Introduction

The observable remnant of a supernova comprises all or some of a number of features. The radio source is most obvious and is conventionally assumed to be common to all events but, in addition, the remnant may exhibit an optical nebulosity, an X-ray source and a pulsar. Other less certain features which may be associated comprise 'fossil' Strömgren spheres and expanding shells of neutral hydrogen, but there will be no time to discuss these. Pulsars will be considered only in their relation to the expanding remnants. From about 150 radio sources which have been classified as supernova remnants (SNR) with some degree of certainty, only two exhibit all four of the main features, the Crab nebula and Vela X. Of these, only the Crab nebula is also recorded as an actual supernova event. Nevertheless, there is little doubt that we are dealing with a common morphological class.

My task is made easier by the recent appearance of a comprehensive review by Woltjer (1972). A brief summary is therefore adequate for many of the observational features, and it is possible to discuss the most recent developments at some length. Woltjer's review also serves as a source of references for much of the earlier work.

II. Optical Features

Optical emission from SNR is observed only under favourable circumstances when the interstellar absorption is not too great. The remnants of 25 galactic SNR have been observed as filamentary emission nebulae (e.g., Woltjer, 1972; van den Bergh *et al.*, 1973) and, just recently, the numbers recognized in the Magellanic Clouds have increased dramatically from three to 14 (Mathewson and Clarke, 1972, 1973a, b, c). However, because of their great distance little detailed information is available on the latter. The filamentary structures of galactic supernovae display a wide variety of appearances, which are discussed and classified by van den Bergh *et al.* (1973).

Expansion of the filaments is observed for nine objects: Tycho's SN, the Crab nebula, IC 443, the Monoceros nebula, Pup A, Kepler's SN, the Cygnus Loop, HB21 and Cas A. No measurements of radial velocities or proper motions are available for the remaining nebulosities. Expansion velocities reported range from about 5500 km s^{-1} for Cas A (van den Bergh, 1971) to about 22 km s^{-1} for HB21 (Losinskaya, 1972). Cas A has a second, low velocity, filamentary system apparently formed from the stationary interstellar medium. The filamentary system associated with Kepler's SN also has radial velocities much lower than expected from the other characteristics of the supernova (Minkowski, 1959), and there are some similar difficulties with the Cygnus Loop (e.g., Tucker, 1971).

Woltjer (1972) has summarized the measurements of relative emission line intensities for nine SNR and compared them with the corresponding intensities in a typical H II region, the Orion nebula. A notable feature is the great strength of the [S II] doublet (λ6716-31), which is comparable or greater than Hα for the SNR but at least an order of magnitude weaker for the Orion nebula and the other H II regions. Other forbidden lines, particularly [N II] (λ6548-84), are also relatively much stronger in most SNR. However, Woltjer concludes that, after allowing for the great variations in T and N_e, most of the data are compatible with 'normal' compositions. Notable exceptions are the Crab nebula with an apparent over-abundance of He, and Cas A in which the H, N and O abundances appear anomalous. The relatively high strength of the [S II] lines is an excellent but not infallible indication of a remnant and has been used to identify SNR in the Magellanic Clouds, first by Westerlund and Mathewson (1966) and later by Mathewson and Clarke (1972, 1973a, b, c).

Observations of the emission spectrum have been used to estimate masses of the expanding gaseous envelopes, but these masses are lower limits because not all the gas is necessarily in a radiating condition. The overall picture is that the physics of the optical emission is reasonably well understood. Further observations are possible, particularly of the largely neglected southern objects, and future improved instrumentation should allow more detailed analysis of the SNR in the Magellanic Clouds.

III. X-Ray Emission

Soft X-ray emission from the hot gas behind the shock front of the expanding shell is expected and has been detected in all the well known SNR. Unfortunately, it is not yet possible to measure the spectrum with sufficient accuracy to decide whether the origin is thermal exept in one case, the Crab nebula. This, however, has a power law spectrum characteristic of synchrotron radiation and, moreover, it is linarly polarized (see Woltjer, 1972). The other remnants have spectra which can be equally well represented by power law or exponential spectra, corresponding to synchrotron and thermal emission respectively, and in most cases the agreement is not particularly good with either. Uncertainties arise because of interstellar absorption and the expected presence of emission lines in a thermal emitter, distorting the simple bremsstrahlung spectrum.

A detailed calculation of the effects of line emission and comparison with the measured soft X-ray spectrum of Pup A yields a very good fit at a temperature of 4×10^6 K (Burginyon et al., 1973b). A similar temperature is obtained by a simple exponential fit, but the detailed agreement is not so good. The temperatures of this and four other remnants listed by Woltjer all lie within the range 2.5×10^6 K to 2.5×10^7 K whereas a much greater spread of temperatures would be expected from shock wave theory if the observed filament radial velocities are equated to the shock velocity. For slowly moving filaments, implying lower temperatures than those observed, there is reason to suppose that the shock is actually faster, as discussed by Woltjer (1972), but in Cas A the fitted temperature (1.4×10^7 K) would imply a shock velocity much slower than the velocities of the filaments. Possibly the synchrotron mechanism is dominant in young SNR, the relativistic electrons being supplied by a neutron star as for the Crab nebula.

Woltjer listed six SNR from which X-ray emission had been detected: the Crab nebula, Cas A, Tycho's SN, Pup A, Vela X and the Cygnus Loop. As a result of some recent surveys (Schwartz et al., 1972a, b; Palmieri et al., 1972; Giacconi et al., 1972; Burginyon et al., 1973a) a further seven identifications have been suggested with catalogued galactic SNR, although some of these are uncertain. In addition X-ray emission has been found from the region of the North Polar Spur (Bunner et al., 1972), a possible SNR, as discussed later. An identification has also been suggested between one of the five X-ray sources in the LMC and an extended radio source (Byrne and Butler, 1973), but this does not have SNR characteristics. These new identifications are all comparatively weak X-ray sources and the spectra ill-defined.

A significant advance has been the detailed mapping of the X-ray emission from several SNR. The Cygnus Loop (Stevens and Garmire, 1973) has an associated X-ray source which closely corresponds in position with the radio and optical object. It displays marked limb brightening, suggesting confinement to a relatively thin shell, but the detailed correlation with other features is not good; the X-ray distribution suggests a more complete shell. Cas A (Fabian et al., 1973) presents a similar picture within the limits of resolution. The most probable result is a shell source similar in size to the radio source, but a disc distribution cannot be excluded; a point source is most unlikely. Pup A (Zarnecki et al., 1973) presents a different picture with the X-ray emission contained well within the radio source and strongly peaked in the NE quadrant; significant emission from a compact object cannot be excluded.

IV. Radio Remnants

The radio source is usually the basis for the recognition of a past supernova event. Radio sources come in two types, the extended, often shell-like object associated with the expanding remnant and characteristic of 'recent' supernova events, which we will subsequently refer to as the SNR, and the pulsar, believed to be the neutron star remnant, and usually associated with 'old' events.

I do not propose to discuss pulsars as such, although it seems very likely that the neutron star, which may not be an observable pulsar, could have an important relationship with the expanding remnant by providing a continuing source of energy in the early stages. Observational association of a pulsar and SNR is certain in only one case, the Crab nebula. However, the Vela X association is very probable, while a third recently suggested by Large and Vaughan (1972), the Crux SNR with PSR 1154−62, appears reasonable if one is prepared to accept a transverse velocity of about 600 km s^{-1} for the pulsar. Davies *et al.* (1972) have associated IC443 with PSR 0611+22, which is within 0°6 of the centre of the SNR although well outside its outer boundary. The estimated distances of the objects are compatible but the ages appear incompatible; the association is discussed in Section VIII.

An attempt to associate all SNR with pulsars on a statistical basis (Tsarevsky, 1973) is clearly without foundation (Combe and Large, 1973). It seems necessary to accept the view that few supernova events produce a detectable pulsar although the proportion producing neutron stars may, of course, be much larger (even 100%).

One other possible association has been noted, the pulsar CP1919, which is located just within the boundary of a weak extended non-thermal source (Caswell and Goss, 1970). However, both the estimated distances and ages of the assumed SNR and pulsar are in gross discordance. It has been suggested by Blandford *et al.* (1973) that the extended source is a 'ghost remnant' of an old pulsar formed from the outward diffusion of the relativistic electrons. If so, other examples might be expected, but Schönhardt (1973) failed to find any such remnants in the fields of 19 'old' pulsars.

Let us now turn to the SNR proper. Characteristic morphological features are that the source forms a ring or arc-like structure, that it has a non-thermal spectrum (although this may be very flat and mimic a thermal spectrum), that the radio recombination lines are absent or at least very weak and that the radio emission exhibits linear polarization. These properties are well known and discussed by Woltjer (1972); some very brief comments are therefore adequate.

(a) STRUCTURE

All the catalogued SNR have been mapped with varying degrees of resolution. The most common characteristic is that of an irregular arc, but almost complete rings are not uncommon, and some quite amorphous structures have been classified as SNR. Generally, it has been found that every increase in resolution has led to an increase in the amount of detailed and irregular structure recorded, and it is clear that a model of a uniform expanding shell source is quite inappropriate. When shell sources can be recognized the thickness of the shell is typically of the order of 15% of the radius, but old SNR may have relatively thinner shells.

It is well known that the Crab nebula displays a central condensation rather than a ring structure, but it is difficult to understand how much older and larger remnants could have similar structures. It appears possible that such objects may be the ejected sources from nearby supernova events which have by chance met with very low densities in the interstellar medium or else the scattered remnants of super-supernovae.

There is evidence for both such processes in the Magellanic Clouds (Section VII).

Some evidence has been presented by Shaver (1969) that the emission in a sample of SNR tended to be greater in the directions perpendicular to the galactic plane. This would be expected if a significant proportion of the emission resulted from the compression of a predominantly longitudinal interstellar magnetic field. However, some recent analysis of other remnants has not provided such a clear picture (Willis, 1972).

Ring- and arc-like structures are also very common among H II regions and have resulted in many misidentifications. The application of other criteria are necessary to confirm that an object is an SNR.

(b) POLARIZATION

All well-resolved SNR exhibit linear polarization, which is usually distributed rather irregularly over the object (for recent work see Downes and Thompson, 1972; Milne, 1972a; Willis, 1972; Kundu and Velusamy, 1972; Kundu and Becker, 1972). The polarization is usually strongest at the highest frequencies, and there is evidence that the rotation of the plane of polarization is non-linear in λ^2, suggesting that internal Faraday depolarization may be effective (Velusamy and Kundu, 1973). The polarization is usually quite small, of the order of 5%, although values up to 25% have been quoted (e.g., Kundu and Becker, 1972).

The general low value of polarization, even at high frequencies, indicates a disordered magnetic field, but quite large-scale regularities in direction are found. For example, about half the SNR studied appear to possess largely radial fields, while a substantial minority give evidence for circumferential fields. However, observations at several frequencies are necessary to correct for the effects of Faraday rotation and obtain the true orientation, which may be in doubt in some of the published maps. There appears to have been no real correlation established between SNR and field direction or amount of polarization (e.g., Willis, 1972). Clearly more high frequency, high resolution observations are needed to clarify the situation.

(c) SPECTRA

All catalogued SNR display a nonthermal spectrum with a mean spectral index (defined by $S \propto v^{\alpha}$) of $\alpha \simeq -0.5$ with a dispersion of about 0.2. There is no indication that the spectral index is correlated with any other features of the SNR.

Generally, the evidence points to a simple power law spectrum at high frequencies, although there are some indications of a high frequency steepening for the Cygnus Loop (Kundu and Becker, 1972) and HB9 (Willis, 1972). Also many SNR spectra have a marked low-frequency turnover around 100 MHz or lower. Analysis appears to support the view that this is the result of free-free absorption (e.g., Dulk and Slee, 1972) but the origin of the absorption is not at present clear. According to Dulk and Slee it is likely to occur either in dense partially ionized H I clouds or in the intercloud medium.

V. Catalogues of SNR

During the last few years several comprehensive catalogues of SNR have been prepared by Milne (1970, 1971), Downes (1971) and Ilovaisky and Lequeux (1972a). These have become progressively more complete and reliable as accumulated observations lead to the recognition of more candidates and the more accurate classification of previously known objects. However, some uncertainties remain. For example, the most recent catalogue (Ilovaisky and Lequeux, 1972a) contains 116 galactic SNR but four are rejected in a footnote because radio recombination lines had subsequently been observed and, later, eight others have been similarly confirmed as H II regions (Caswell, 1972; Dickel and Milne, 1972; Dickel *et al.*, 1973). There is also the likelihood of contamination by extra-galactic objects because all the radio features common to SNR have been observed in well-identified radio galaxies. Several nonthermal sources close to the galactic plane are, in fact, believed to be extragalactic and have not been included in their catalogue by Ilovaisky and Lequeux, but it would be surprising if all had been recognized, particularly among the weaker sources of small angular size.

Two partial catalogues have recently increased the number of well classified SNR significantly. Clark, Caswell and Green (1973) list 27 mainly southern galactic SNR of high reliability; these are, on average, older and more distant objects than those in earlier catalogues. Mathewson and Clarke (1972, 1973a, b, c) have increased the number of well identified SNR in the Magellanic Clouds from three to 14; the latter are of great importance because of their accurately known distances and will be discussed separately.

Finally, at a much lower level of reliability we have the large galactic loops, such as the North Polar Spur, for which evidence is increasing of their similarity to SNR. The properties of five such loops ranging in size from about 5° to 120° are catalogued by Berkhuijsen (1973).

VI. Distances of Galactic SNR

The distances of relatively few galactic SNR have been determined, and none of these may be considered as precise measurements. The majority are kinematic distances derived from H I and molecular absorption lines; Ilovaisky and Lequeux (1972a) list 13 such distances and six in which lower limits can be obtained by this method. Other estimates are obtained by comparison of the radial velocities and proper motions of optical filaments, by association with stars or nebulosities where distances have been measured independently and by estimates of the brightness of associated historical supernovae.

Kinematic distances appear to be the most reliable but are available for only a small selection of SNR. The usual uncertainties arise because of the uncertain distance to the galactic centre, ambiguities in interpretation and the possibility of non-circular motions but, in addition, the irregular distribution of the absorbing matter is a serious complicating factor. Uncertainties by a factor of two or more are not unusual.

Estimates of the distances of Cas A, the Crab nebula and the Cygnus Loop have been made by combining the proper motions of filaments near the edge of the nebulosities and the radial velocities of more central filaments (cf. Woltjer, 1972). However, the possibility must be considered that because of selection effects low radial velocities are obtained leading to underestimates of the distances. There is also some evidence, as mentioned earlier, that the radial velocity of the filaments in the Cygnus Loop may be considerably less than that of the shock front.

The linking of SNR with other features such as OB associations, H II regions and dark nebulae may lead to accurate distances or to completely irrelevant results. Finally, estimates made from the brightness of supernovae deduced from historical records are very uncertain and do little better than separate the SNR into close and distant objects.

Although individual measurements are often unreliable, statistical improvement can be sought by making use of the empirical surface brightness-diameter relation (the $\Sigma - D$ relation). There is a correlation between these parameters implying similar evolutionary trends for all supernovae, as discussed in Section VIII. Various authors have derived empirical power law relations of the form $\Sigma = AD^\beta$ (e.g., Woltjer, 1972; Ilovaisky and Lequeux, 1972a), which may be used not only as a check on the calibrating SNR but to provide a scale for other SNR for which no other distance estimation is possible. However, the whole process is subject to the choice of calibrators and the weights assigned to them; marked differences exist between estimates of the constants A and β. Before discussing this relation we must look at the SNR detected in the Magellanic Clouds, which are unique in having accurate distances assignable.

VII. SNR in the Magellanic Clouds

Recently the results of two major investigations of the Magellanic Clouds have been published, one using the Parkes 64-m radiotelescope (McGee et al., 1972a, b; McGee and Newton, 1972; Broten, 1972; Milne, 1972b), the other using a combination of radio measurements with the Molonglo Cross and optical measurements using the 1-m reflector at Siding Spring Observatory (Mathewson and Clarke, 1972, 1973a, b, c).

Three SNR in the LMC have been known for some time (Mathewson and Healey, 1964; Clarke, 1971). A list of eight new SNR candidates has been prepared by Milne (1972b) based on the Parkes work, but for the reasons discussed by Mathewson and Clarke (1973b) the measurements on these objects are misleading. SNR tend to occur in regions of recent star formation rich with H II regions, and the resolutions attained in the Parkes surveys appear to be inadequate to identify with certainty the different objects. Mathewson and Clarke confirm only two of Milne's SNR identifications and, moreover, find that the sizes given for these, based on the broadening of a 4' beam, grossly exceed the size of the optical remnants. To make progress with SNR identifications in the Clouds based solely on radio data it appears that the beamwidth on one frequency at least should be $\lesssim 1'$, while to determine SNR structure a beam width $\lesssim 20''$ is desirable. Radio galaxy contamination is also a significant problem because the Cloud SNR are much fainter than their galactic counterparts, so that supporting optical data would always seem very desirable.

The technique of Mathewson and Clarke has proved very fruitful and the radio and optical emission of nine new remnants have been detected in the LMC and two in the SMC. The relative strengths of the [S II] doublet and Hα have been used as a criterion to identify an optical remnant. Two further identifications with emission nebulosities in the LMC were also suggested, but in these the characteristic [S II] remnant was not visible. Comparison of the Molonglo 408 MHz data (Le Marne, 1968) with a map prepared using the Fleurs Compound Grating Interferometer at a frequency of 1415 MHz and a resolution of 40″ (Christiansen, private communication) indicates that one of these is an H II region (N157B).

The well known SNR, N49, is interesting, as a second, weaker, nonthermal source having an associated patch of strong [S II] emission is found to be connected to N49 by filamentary structures, suggesting that it was ejected in the original supernova explosion (Mathewson and Clarke, 1973a). Another less obvious example is NIIL where there is a similar and much closer patch of [S II] nebulosity also with an associated weak radio source. The projected linear separation from N49 is 110 pc, consistent with the ejected material encountering an exceptionally low density in the interstellar medium. If similar features occur in the Galaxy they would not be recognized as physically related objects.

Two very extended nonthermal sources showing no associated [S II] emission are also discussed by Mathewson and Clarke (1973b). If these are located in the LMC they have linear sizes ∼200 pc, rather like the galactic loops, but their mean surface brightness is greater by an order of magnitude and no ring structure is apparent. Mathewson and Clarke consider that, together with the nonthermal source underlying the 30 Doradus nebula, they may represent fragmentary remains of very old super-supernovae. Westerlund and Mathewson (1966) suggested that there is evidence for such rare events in the distribution of Population I objects in the Clouds.

Mathewson and Clarke (1973b) have constructed a $\Sigma - D$ diagram from their reliably identified SNR in the LMC and have added a selection of the most reliable galactic SNR. This diagram is reproduced in Figure 1 with the addition of the SNR in the SMC (Mathewson and Clarke, 1972, 1973c), the other nonthermal sources discussed above and the two galactic loops for which direct distance determinations are available (Berkhuijsen, 1973). The solid line representing the best fit to the Cloud data is defined by $\Sigma_{408} = 10^{-15} D^{-3}$ W m^{-2} Hz^{-1} sr^{-1}. Among the Cloud objects significant departures from this relation occur only for the small nebulosity possibly ejected from NIIL and the two large extended sources. Among the well defined galactic objects the Crab nebula and the Cygnus Loop stand out as exceptions; it may be significant that the distances to both these objects have been obtained by their filament expansion rate. However, even among the sources with kinematic distances there is a slight discrepancy in the sense that the diameters of the galactic objects are about 15% less than those of the same surface brightness in the Clouds.

Many reasons can be advanced for the apparent difference between the galactic

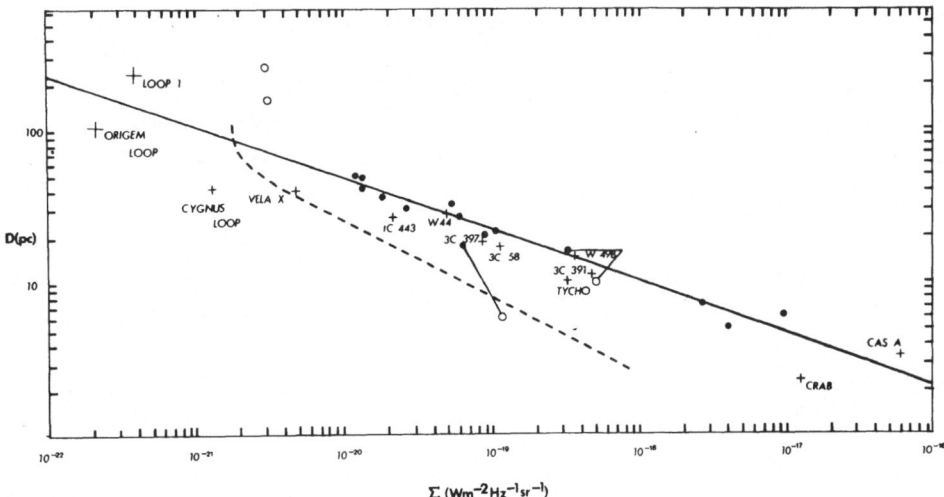

Fig. 1. The $\Sigma - D$ relation. Filled circles represent SNR in the Magellanic Clouds; open circles represent separate ejecta from the associated SNR plus the two low-brightness extended sources discussed in the text. Galactic SNR are indicated by crosses and named. The dashed line represents the approximate sensitivity limits of the radiotelescope.

and Cloud SNR. The important question, however, is whether the difference is real or the result of observational selection due to inadequate radio sensitivity in the Cloud searches. This seems unlikely. Although there is evidence of a strong selection in the detection of SNR in the Cloud this selection is based primarily on the strength of the [S II] doublet. A large number of nebulosities was examined but in no case was an optical remnant observed without a corresponding radio source. It therefore seems probable that the slope of the $\Sigma - D$ relation is quite well defined but that in applying it to determine the distance of galactic objects the constant multiplying factor should be reduced; this would obviously be necessary in comparing the distribution of SNR in the Galaxy with other features defined by kinematic distances.

In the remaining discussion I will adopt the relation suggested by Clark *et al.* (1973), $\Sigma_{408} = 7 \times 10^{-16} D^{-3}$ W m^{-2} Hz^{-1} sr^{-1}. Usually the relation is expressed at a frequency of 1000 MHz where, assuming a mean spectral index of -0.5, we have $\Sigma_{1000} = 4.5 \times 10^{-16} D^{-3}$ W m^{-2} Hz^{-1} sr^{-1}. This may be compared with the result of Ilovaisky and Lequeux (1972a), $\Sigma_{1000} = 4.6 \times 10^{-15} D^{-4.0}$ W m^{-2} Hz^{-1} sr^{-1}. The constant given by Ilovaisky and Lequeux has been corrected by the factor $4/\pi$ (e.g., Berkhuijsen, 1973). Other estimates based on the galactic SNR give similarly high exponents caused by the high weight given to Cas A and the Cygnus Loop in defining the slope of the $\log \Sigma - \log D$ relation (e.g., Woltjer, 1972). The relation may also be converted to a distance scale thus, $d = 49 (S \langle \phi \rangle)^{-1/3}$ kpc where S is the flux density of the SNR in Jy at 1000 MHz and $\langle \phi \rangle$ its geometric mean diameter in minutes. This scale derived from the Cloud SNR gives closely similar distances to the various galactic scales for young SNR but increasingly greater distances for old SNR.

VIII. SNR Evolution

The evolution of SNR is largely the preserve of theory. Nevertheless, it is useful to discuss briefly the conclusions which can be reached by statistical analysis of the available observational data. As a starting point we assume that the data may be represented by the following set of equations:

$$\Sigma = AD^\beta, \tag{1}$$

$$\Sigma = Bt^\gamma, \tag{2}$$

$$D = Ct^\delta, \tag{3}$$

$$N = P\Sigma^\xi, \tag{4}$$

$$N = QD^\zeta. \tag{5}$$

In these, Σ and D have already been defined, t is the time in years elapsed since the outburst, N is the number of SNR in the Galaxy with surface brightness greater than Σ or diameter smaller than D, and the remaining symbols represent constants to be determined by observation. While it is known that these equations cannot represent the evolution of SNR over their whole lifetimes, we have seen already that Equation (1) appears to hold over a wide range of Σ and D. The equations are not independent, and if we assume a constant rate of supernova events there are three independent multiplying constants and two independent exponents.

Equations (1) and (4) are most directly accessible to observation. The former has already been discussed and the constants obtained at an observing frequency of 1000 MHz, $A = 4.5 \times 10^{-16}$, $\beta = -3$.

Equation (4) would seem to be easily obtained by a simple counting process, once we have a reliable and complete catalogue, but there are some difficulties. Firstly, the determination of Σ can only be performed satisfactorily when the remnant forms a ring or well-defined arc to which an angular diameter can be unambiguously assigned. Partially resolved or amorphous structures may represent the complete remnant, or only portions of a larger object. Secondly, the effects of selection are severe and tend to flatten the slope. Ilovaisky and Lequeux (1972a) have given careful consideration to the selection effect and derive a relation for a limited range of SNR diameters in the form $N \propto D^{3.15 \pm 0.70}$. Their $\Sigma - D$ relation ($\Sigma \propto D^{-4.0}$) may be removed from this result using the relation $\xi = \zeta/\beta$ to give the direct observational result, $\xi = -0.79 \pm 0.17$. The number of SNR with large diameters ($\gtrsim 30$ pc) falls drastically below this relation, which they ascribe to the difficulty of detecting such objects in the Galaxy. I have estimated the numerical constant P in Equation (4) from their Equation (5b) by assuming the total number of SNR in the Galaxy is 3 times the number in their sample area of radius 7.6 kpc centred on the Sun and, by using their $\Sigma - D$ relation and the number actually listed in their catalogue within this area (40). Thus $P \simeq 2.5 \times 10^{-14}$, with considerable uncertainty. The above estimates of ξ and P could be improved using the additional catalogue information now available, but a rough check indicates that the changes would be relatively small compared with the statistical and other uncertainties.

We have now enough information to derive the exponents of Equations (2) and (3), which describe the time evolution of SNR, thus: $\gamma = 1/\xi = -1.27 \pm 0.32$, and $\delta = 1/\beta\xi = 0.42$ with an uncertainty which is greater than ± 0.1 but which cannot be accurately specified because of the possibility of systematic effects in β. It is interesting that the latter exponent is close to the value of 0.4, the well-known Sedov adiabatic solution for an explosion in a medium of constant heat capacity, which should represent the expansion rate of SNR over a wide evolutionary range (cf. Woltjer, 1972).

Determination of the related constants B and C is not satisfactory as wide variations are expected in individual SNR and calibrating data are difficult to obtain. The expansion rate of young historically recorded SNR for which t is known is not likely to be described by our derived parameters. For example, it is known by direct observation of the filaments in the Crab nebula that $\delta \simeq 1$, as expected for an explosion unretarded by the interstellar medium. Similarly, the observed rate of decrease of the flux density of Cas A is not predicted by our equations, from which it is easily shown that $(1/S)(\mathrm{d}s/\mathrm{d}t) = (2+\beta)/t'$ where t' is the 'expansion age' obtained by extrapolating the observed proper motions of the expanding filaments back to the origin. Inserting the value of β obtained above and assuming an expansion age of 270 yr, the predicted rate of decrease is 0.37% per year, about three times less than the observed rate (cf. Woltjer, 1972; Findlay, 1972). Other historical supernovae cannot be checked so directly, but the combination of low age and locations far from the galactic plane, where the interstellar gas has a low density, makes them unreliable calibrators.

Older SNR must be used and their ages derived indirectly. There appear to be only two possibilities, Vela X and the Cygnus Loop, but both these are suspect because they lie below the adopted $\Sigma - D$ relation. If this displacement results from errors in the distances it will not affect our determination of B; if the estimated distances are correct the derived B will be too low. The age of the pulsar associated with Vela X has been estimated from the rate of increase of its period as 1.1×10^4 yr (Reichley et al., 1970). When this age is inserted in Equation (2) we find $B \simeq 9 \times 10^{-16}$. The age of the Cygnus Loop may be obtained from the relation $t = \delta t'$, where t' is the expansion age, which is equal to 150 000 yr (Minkowski, 1958). Taking $\delta = 0.42$ we find $t = 6.3 \times 10^4$ yr, leading to $B \simeq 9 \times 10^{-16}$ as before! The agreement is clearly nothing but a coincidence as both determinations are very uncertain. In particular there is some possibility that the Cygnus Loop is entering a late stage of evolution characterized by isothermal expansion for which $\delta = 0.25$ (e.g., Moffat, 1971); use of our equations would then lead to an overestimate of B. Taking $B \simeq 9 \times 10^{-16}$ as above we find $C = (A/B)^{1/3} = 0.79$, so that the expansion of a typical SNR is defined by $D \simeq 0.79\, t^{0.42}$. It is interesting to compare this with the Sedov solution, which gives $D = 0.65(E/n)^{1/5}\, t^{0.4}$ where E is the energy of the supernova explosion in units of 7.5×10^{50} erg (Shklovskii, 1962). Shklovskii's estimate that $E/n \sim 1$ for a typical supernova explosion is not very different from our result.

Finally, we may estimate the rate of occurrence of supernovae in the Galaxy using the relation $R = N/t = PB^\xi$. Inserting the values derived above we find $R \simeq 0.02$ yr^{-1}, or the mean time between supernova events $\tau \simeq 50$ yr. If, however, we assume that direct distance determinations of Vela X and the Cygnus Loop are correct and that the brightnesses are anomalous, we may calculate the constant C from Equation (3). Thus for Vela X, $C = 0.58$ and for the Cygnus Loop, $C = 0.48$. Adopting the mean value, $C = 0.53$, we find $B = AC^{-3} \simeq 3 \times 10^{-15}$, leading to a mean time between supernova events $\tau \simeq 130$ yr.

It is interesting to compare these results with other determinations. Using Sedov's relation directly, with assumed values of E and n, Downes (1971) gives $\tau \simeq 45$ yr and Milne (1970) gives $\tau \simeq 75$ yr. Ilovaisky and Lequeux (1972b) base their calibration mainly on the historical remnants, with a theoretical correction for the density of the interstellar medium; they find $\tau \simeq 50$ yr. Recent direct determinations based on the observed rate of occurrence of supernovae in the Galaxy have been made by Katgert and Oort (1967) who find $\tau \simeq 26$ yr. Based on the rates in external galaxies Tammann (1970) finds $\tau \simeq 26$ yr for an Sbc and $\tau \simeq 65$ yr for an Sb. The range of estimated values is therefore about 5:1, with a most likely rate of about 2 per century. The estimates based on SNR should now be amenable to considerable improvement using new and improved catalogues and a combination of empirical and theoretical approaches.

The association of a pulsar with IC 443 by Davies et al. (1972) does not fit this general evolutionary picture. For the pulsar, $P/\dot{P} \simeq 125000$ yr, suggesting an age of about 60000 yr (e.g., Reichley et al., 1970) but the SNR appears too small and bright for such an age. If its evolution had been typical, as defined above, the age should be about 6000 yr, but the corresponding transverse velocity for the pulsar would then be about 4000 km s^{-1}. If the pulsar is unassociated and formed in a supernova event 60000 yr ago its own remnant should be observable with properties similar to the Cygnus Loop. It may prove possible to reconcile these apparent contradictions without violating any observational data by assuming extreme values for all the parameters. An alternative hypothesis is that pulsars may be formed by direct contraction of a star without a typical supernova explosion (e.g., Cameron, 1970).

IX. Distribution of SNR

Several authors have discussed the distribution of SNR throughout the Galaxy (e.g., Milne, 1970; Ilovaisky and Lequeux, 1972a, b; Clark et al., 1973). The most thorough statistical analysis is that of Ilovaisky and Lequeux, and although there is now much additional data which warrants further analysis it seems probable that their main conclusions will not be affected. Adoption of the present distance scale leads to an addition of between 20% and 25% to their mean distances. This hardly affects the overall shape of the derived radial distribution, but it does lead to a correction to the z distributions.

Ilovaisky and Lequeux found a scale height of about 90 pc for SNR within 6 kpc of the Sun. Using the present distance scale this should be increased to 110 pc, which, however, makes little difference to their discussion. The z distribution of SNR is very similar to that of the Cepheids, OB associations and OB stars and suggests a largely Population I origin. The detection of a few SNR, including remnants of some of the historical supernovae, at very high z distances is consistent with the generally accepted picture of a type I supernova associated with Population II and type II supernovae associated with Population I. It is clear that the great majority of galactic SNR must be of type II, and this would therefore be expected also in the Magellanic Clouds.

When the z distribution is derived as a function of distance from the galactic centre an increase of scale height is found as the distance increases. The same effect is observed in the z distribution of the nonthermal radio emission and H I (e.g., Mills, 1971) although the scale height of the SNR is less than either, in fact almost an order of magnitude less than that of the radio emission.

The surface density of SN remnants projected onto the galactic plane was also derived as a function of the distance from the galactic centre, assuming radial symmetry. When compared with the derived radial distributions of H I, H II and nonthermal radio emission it was found that only the latter displayed similarities. Both are restricted to little more than the solar distance and both have a fairly flat distribution out to this distance, although the radio emission does show more of a central concentration.

Clark *et al.* (1973) have prepared a map of the galactic distribution of SNR in their catalogue, together with those listed by Woltjer (1972) as good SNR identifications (Figure 2). They have used the distance scale presented here and limited the selection to SNR with $z < 250$ pc. Inspection of their map confirms that the SNR are distributed more or less uniformly out to 10 kpc from the galactic centre with a sharp decrease in numbers beyond. Within 10 kpc of the Sun the SNR are mainly confined to the regions of high H I density according to Kerr's (1970) map; in particular the Norma-Scutum arm is well delineated. Although the Sagittarius arm also shows up clearly there are some anomalies and the results do not really help to resolve the uncertainties in this region. However, the general appearance is that of a Population I distribution; in view of the uncertain distances of individual SNR little more can be expected.

X. A New SNR

Finally, the SNR associated with the supernova 1970g in M101 must be mentioned. This was a type II supernova discovered at the end of July 1970 in the southern part of M101 by Detre (1970). Weak radio emission (3.8 mJy) was reported at λ 11 cm by Gottesman *et al.* (1972) several months later, but by February 1972 the same group could detect nothing exceeding 3 mJy. Meanwhile, measurements at λ 21 cm using the Westerbork radiotelescope had detected a radio source having a flux density of 5.4 mJy in December 1970 and 11.3 mJy in December 1971 and at λ 6 cm a flux density of 4.0 mJy in January 1973 (Goss *et al.*, 1973).

Clearly, radio emission from the SNR has been observed, but the picture is complicated by the close proximity of an H II region (NGC 5455) in M101. Measurements of the Hβ flux for the nebula by Searle (1971) predict a radio flux density of only

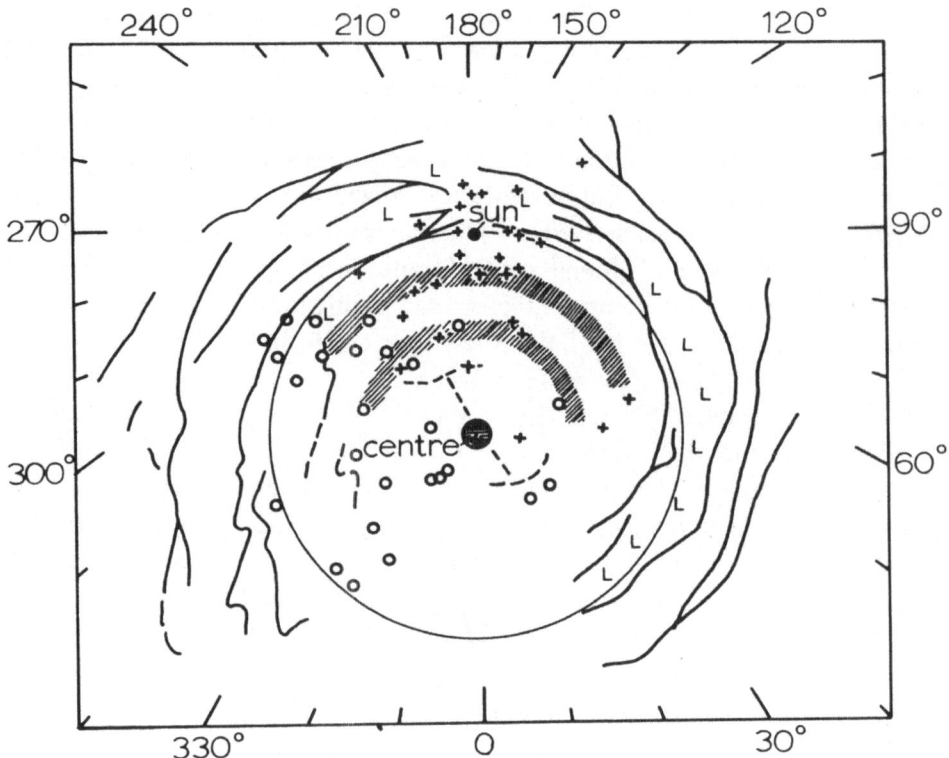

Fig. 2. The distribution of SNR in the galactic disk according to Clark *et al.* (1973) superimposed on the H I map of Kerr (1970). Open circles represent SNR from their catalogue; crosses represent well established SNR listed by Woltjer (1972). Selection has been restricted to those with $|z| < 250$ pc.

about 1.4 mJy, but there is evidence that Hβ measurements substantially underestimate the radio emission of H II regions in external galaxies (see also Mills and Aller, 1971). Most of the observed emission may be originating in the H II region, but even if one attributes to the SNR only the 6 mJy difference between the Westerbork 1970 and 1971 observations, the object would have several times the radio luminosity of Cas A. No convincing explanation for the observed emission has been advanced, and other recent extra-galactic supernovae studied at Westerbork have not been detected. One awaits its subsequent evolution with interest.

References

Bergh, S. van den: 1971, *Astrophys. J.* **165**, 457.
Bergh, S. van den, Marscher, A. P., and Terzian, Y.: 1973, *Astrophys. J. Suppl.* **26**, 19.
Berkhuijsen, E. M.: 1973, *Astron. Astrophys.* **24**, 143.
Blandford, R. D., Ostriker, P. J., Paccini, F., and Rees, M. J.: 1973, *Astron. Astrophys.* **23**, 145.
Broten, N. W.: 1972, *Australian J. Phys.* **25**, 599.
Bunner, A. N., Colemand, P. L., Kraushaar, W. L., and McCammon, D.: 1972, *Astrophys. J. Letters* **172**, L67.

Burginyon, G., Hill, R., Seward, F., Tarter, B., and Toor, A.: 1973b, *Astrophys. J. Letters* **180**, L75.

Burginyon, G., Hill, R., Palmieri, T., Scudder, J., Seward, F., Stoering, J., and Toor, A.: 1973a, *Astrophys. J.* **179**, 615.

Byrne, P. B. and Butler, C. J.: 1973, *Nature Phys. Sci.* **244**, 6.

Cameron, A. G. W.: 1970, *Ann. Rev. Astron. Astrophys.* **8**, 179.

Caswell, J. L.: 1972, *Australian J. Phys.* **25**, 443.

Caswell, J. L. and Goss, W. M.: 1970, *Astrophys. Letters* **7**, 142.

Clark, D. H., Caswell, J. L., and Green, A. J.: 1973, *Nature* **246**, 28.

Clarke, J. N.: 1971, *Proc. Astron. Soc. Australia* **2**, 44.

Combe, V. and Large, M. I.: 1973, *Astrophys. Letters* **14**, 59.

Davies, J. G., Lyne, A. G., and Seirdakis, J. H.: 1972, *Nature* **20**, 229.

Detre, L.: 1970, *IAU Circ.* No. 2269.

Dickel, J. R. and Milne, D. K.: 1972, *Australian J. Phys.* **25**, 539.

Dickel, J. R., Milne, D. K., Kerr, A. R., and Ables, J. G.: 1973, *Australian J. Phys.* **26**, 379.

Downes, D.: 1971, *Astron. J.* **76**, 305.

Downs, G. S. and Thompson, A. R.: 1972, *Astron. J.* **77**, 120.

Dulk, G. A. and Slee, O. B.: 1972, *Australian J. Phys.* **25**, 429.

Fabian, A. C., Zarnecki, J. C., and Culhane, J. L.: 1973, *The Soft X-ray Structure of Cassiopeia A*, preprint.

Findlay, J. W.: 1972, *Astrophys. J.* **174**, 527.

Giacconi, R., Murray, S., Gursky, H., and Kellogg, E.: 1972, *Astrophys. J.* **178**, 281.

Goss, W. M., Allen, R. J., Ekers, R. D., and Bruyn, A. G. de: 1973, *Nature Phys. Sci.* **243**, 42.

Gottesman, S. T., Broderick, J. J., Brown, R. L., Balick, B., and Palmer, P.: 1972, *Astrophys. J.* **174**, 383.

Ilovaisky, S. A. and Lequeux, J.: 1972a, *Astron. Astrophys.* **18**, 169.

Ilovaisky, S. A. and Lequeux, J.: 1972b, *Astron. Astrophys.* **20**, 347.

Katgert, P. and Oort, J. H.: 1967, *Bull. Astron. Inst. Neth.* **19**, 239.

Kerr, F. J.: 1970, in W. Becker and G. Contopoulos (eds.), 'The Spiral Structure of Our Galaxy', *IAU Symp.* **38**, 95.

Kundu, M. R. and Becker, R. H.: 1972, *Astron. J.* **77**, 459.

Kundu, M. R. and Velusamy, T.: 1972, *Astron. Astrophys.* **20**, 237.

Large, M. I. and Vaughan, A. E.: 1972, *Nature Phys. Sci.* **236**, 117.

Le Marne, A. E.: 1968, *Monthly Notices Roy. Astron. Soc.* **139**, 461.

Lozinskaya, T. A.: 1972, *Soviet Astron.* **20**, 219.

McGee, R. X. and Newton, L. M.: 1972, *Australian J. Phys.* **25**, 619.

McGee, R. X., Brooks, J. W., and Batchelor, R. A.: 1972a, *Australian J. Phys.* **25**, 581.

McGee, R. X., Brooks, J. W., and Batchelor, R. A.: 1972b, *Australian J. Phys.* **25**, 613.

Mathewson, D. S. and Clarke, J. N.: 1972, *Astrophys. J. Letters* **178**, L105.

Mathewson, D. S. and Clarke, J. N.: 1973a, *Astrophys. J.* **179**, 89.

Mathewson, D. S. and Clarke, J. N.: 1973b, *Astrophys. J.* **180**, 725.

Mathewson, D. S. and Clarke, J. N.: 1973c, *Astrophys. J.* **182**, 697.

Mathewson, D. S. and Healey, J. R.: 1964, in F. J. Kerr and A. W. Rodgers (eds.), 'The Galaxy and the Magellanic Clouds', *IAU Symp.* **20**, 283.

Mills, B. Y.: 1971, 12th International Conference on Cosmic Rays, Invited and Rapporteur Papers, p. 72.

Mills, B. Y. and Aller, L. H.: 1971, *Australian J. Phys.* **24**, 609.

Milne, D. K.: 1970, *Australian J. Phys.* **23**, 425.

Milne, D. K.: 1970, in R. D. Davies and F. G. Smith (eds.), 'The Crab Nebula', *IAU Symp.* **46**, 248.

Milne, D. K.: 1972a, *Australian J. Phys.* **25**, 307.

Milne, D. K.: 1972b, *Astrophys. Letters* **11**, 167.

Minkowski, R.: 1958, *Rev. Mod. Phys.* **30**, 1048.

Minkowski, R.: 1959, in R. N. Bracewell (ed.), 'Paris Symposium on Radio Astronomy', *IAU Symp.* **9**, 315.

Moffat, P. H.: 1971, *Monthly Notices Roy. Astron. Soc.* **153**, 401.

Palmieri, T. M., Burginyon, G. A., Hill, R. W., Scudder, J. K., and Seward, F. D.: 1972, *Astrophys. J.* **177**, 387.

Reichley, P. E., Downs, G. S., and Morris, G. A.: 1970, *Astrophys. J. Letters* **159**, L35.

Schwartz, D. A., Bleach, R. D., Boldt, E. A., Holt, S. S., and Serlemitsos, P. J.: 1972, *Astrophys. J. Letters* **173**, L51.

Schwartz, D. A., Bleach, R. D., Boldt, E. A., Holt, S. S., and Serlemitsos, P. J.: 1972, *Rep. NASA-TM-X-65825*, NASA, Greenbelt, Maryland.

Schönhardt, R. E.: 1973, *Nature Phys. Sci.* **243**, 62.

Searle, L.: 1971, *Astrophys. J.* **168**, 327.

Shaver, P. A.: 1969, *Observatory* **89**, 227.

Shklovskii, I. S.: 1962, *Soviet Astron.* **6**, 162.

Stevens, J. C. and Garmire, G. P.: 1973, *Astrophys. J. Letters* **180**, L19.

Tammann, G. A.: 1970, in W. Becker and G. Contopoulos (eds.), 'The Spiral Structure of Our Galaxy', *IAU Symp.* **38**, 236.

Tsarevsky, G. S.: 1973, *Astrophys. Letters* **10**, 71.

Tucker, W. H.: 1971, *Science* **172**, 372.

Velusamy, T. and Kundu, M. R.: 1973, *Polarization of Supernova Remnants: Internal Faraday Effects and Derived Magnetic Fields*, preprint.

Westerlund, B. E. and Mathewson, D. S.: 1966, *Monthly Notices Roy. Astron. Soc.* **131**, 371.

Willis, A. G.: 1972, Ph.D. Thesis, University of Illinois.

Woltjer, L.: 1972, *Ann. Rev. Astron. Astrophys.* **10**, 129.

Zarnecki, J. C., Culhane, J. L., Fabian, A. C., Rapley, C. G., Silk, R., Parkinson, J. H., and Pounds, K. A.: 1973, *A Low Energy X-Ray Map of the Puppis-A Supernova Remnants*, preprint.

B. Y. Mills
School of Physics,
University of Sydney,
Sydney, N.S.W., Australia

DISCUSSION

Lequeux: It is interesting to see that the galactic distribution of SNR is not much affected by the choice of the $\Sigma - D$ relation used for determining distances. I doubt that distances to SNR are accurate enough to be able to establish a genuine association with galactic spiral arms.

Mills: The differences between the Ilovaisky and Lequeux $\Sigma - D$ relation and the one presented here do not affect the distances of the younger SNR very much, so I believe the radial distribution derived by Ilovaisky and Lequeux would not be greatly altered by the new scale. Since the true position of the galactic spiral arms is so uncertain perhaps the SNR will help determine them. There is clear evidence of spirality in the distribution.

J. R. Dickel: The shell thickness of old SNR is not necessarily thin. The plot of the thickness-to-radius ratio versus surface brightness shows tremendous scatter (A. G. Willis: 1973, *Astron. Astrophys.* **26**, 237). For example, the old remnant HB9 has a $\Delta R/R$ of at least 0.4.

Smith: It would be useful to measure the velocity of the pulsar close to IC 443, using the scintillation drift method. This would give an age as well as a test of the association, which at the moment rests on the fact that the pulsar is unusually young.

Mills: It is very important to pursue this possible association because we have so little information on the time evolution of SNR.

Oort: In a previous session, Goldsmith has stressed that we do not know where interstellar clouds come from. I think it is worth considering the possibility that they are generally generated by supernovae. If the radius of the disk is taken to be 12 kpc, its thickness 0.2 kpc, and if it is assumed a third of this volume is occupied by the spiral arms, the total volume is 3×10^{10} pc^3. If there is one supernova per 30 yr, we find that in a sphere of 100 pc radius there would on the average be one supernova explosion per 2×10^5 yr. This is of the same order as the probable life time of an SNR. It appears therefore possible that the 'clouds' in the interstellar medium as well as their velocities are created by supernovae.

Kerr: If the $\Sigma - D$ relationship is different for the Large Cloud and the Galaxy, could this be interpreted as a recalibration of the distance to the Large Cloud?

Mills: Either that or a recalibration of the distance to the galactic center. Either this distance would have to be increased by 10% to 15% or the distance to the Clouds reduced by this amount, provided the SNR in both places are the same. Alternatively it might be suggested that the Cloud SNR are brighter at the same diameter.

Lequeux: The apparent difference between galactic SNR and the SNR in the Magellanic Clouds can first be due to a selection effect: it is natural that one sees only the brightest SNR in the Clouds, whilst the sample of galactic SNR used to calibrate the galactic $\Sigma - D$ relation seems more random (these are just the SNR for which distances are known).

Mills: As discussed in my paper, the SNR in the Magellanic Clouds are strongly selected, but on the basis of the [Sɪɪ] line brightness, *not* the radio emission. I would be surprised if only the stronger radio sources had been found, but of course this is not impossible.

Menon: You mentioned that the mean spectral index of supernova remnants is 0.5. Is it then possible to account for the nonthermal radio spectrum of the galactic background as being due to electrons from the remnants?

Mills: I do not know. Although there is clear evidence that the spectra of SNR do not evolve with time this applies only to the earlier phases. It is conceivable that changes may occur in the final stages of their evolution when merging into the galactic background.

Terzian: I like to report that very recently van den Bergh, Marscher, and myself published an Optical Atlas of Galactic Supernova Remnants (*Astrophys. J. Suppl.* **26**, (1973), 19), which should assist in making comparisons of the radio, X-ray and other maps with the optical ones.

Mathewson: All the supernova remnants in the Large Cloud I saw came in Hɪɪ regions. Smith and Weedman determined the radial velocities of 70 Hɪɪ regions in the Large Cloud and found that they agree to within a few kilometers per second with the H ɪ. For two small supernova remnants which could be put right over the slit, the systemic velocities differed by about 70 km s^{-1} of the H ɪ. Perhaps the supernova occurs in stars moving at high velocities, or it could be the sling-shot effect where the pulsar goes off one way and the remnant goes the other way.

SUPERNOVA REMNANTS: CAN A FOSSIL HII REGION BE FORMED AS THE RESULT OF A SUPERNOVA EXPLOSION?

F. D. KAHN

University of Manchester, Manchester, United Kingdom

Abstract. An estimate of the amount of ionizing radiation available in a supernova explosion falls far short of the amount needed in the standard model of the Gum nebula. The only possibility left is radiation from the pulsar.

I. Physics of a Fossil HII Region

The Gum nebula is widely thought to be the best example of a fossil HII region. Its properties were discussed at a conference at Greenbelt in 1971 (Maran *et al.*, 1971). The basic data for the model of the nebula were given by Brandt (1971). They are

total number of free electrons	$\approx 2 \times 10^{62}$,
average electron density	$\approx 0.16 \text{ cm}^{-3}$,
energy required to produce the ionization	$\approx 5 \times 10^{51}$ erg.

The Gum nebula is believed to have been formed by the supernova that also produced the Vela pulsar. Its age is therefore estimated to be 10000 yr $\approx 3 \times 10^{11}$ s. Now the recombination rate in the nebula is βn_e, where $\beta = 2 \times 10^{-13}$ cm^3 s^{-1} is the usual recombination coefficient for ionized hydrogen in interstellar space. With $n_e \doteq 0.16$ cm^{-3} the rate becomes 3.2×10^{-14} s^{-1}, and during the lifetime of the nebula only about one per cent of the protons could possibly have recombined. This justifies the estimate for the energy required to ionize the nebula: each H atom needs to be ionized only once, and only negligible energy is required to balance recombinations.

There is another important difference between the physics of fossil HII regions and that of standard HII regions. In the standard case the ionizing radiation comes from early type stars, with surface temperatures around 50000 K. Only a fraction of the photons are beyond the Lyman limit, and only a small proportion of these have energies more than a few eV above the ionization potential of hydrogen. To estimate how many ionizations are produced by the star one then simply works out the rate at which it produces photons in the Lyman continuum.

But the calculation is different for the case of a supernova remnant. Here the radiation comes, for a time at least, from a surface which is much hotter than any stellar photosphere. The ionizing photons are therefore much more energetic, on average. If a photon with, say, 100 eV energy ionizes a hydrogen atom, then the electron so produced has a kinetic energy of about 86 eV. This electron can, and will, produce further ionization by collision. By this argument Brandt derives his estimate for the energy required to ionize the nebula. Almost all the radiant energy absorbed by the gas goes to ionize hydrogen atoms.

But there are some limitations on this argument. The number of protons along the

F. J. Kerr and S. C. Simonson, III (eds.), Galactic Radio Astronomy, 329–334. All Rights Reserved.

line of sight through the nebula is of the order of 10^{20} per cm^2. This is also the number of hydrogen atoms per unit area along the line of sight before ionization. If the ionizing radiation at a given frequency is to be efficiently absorbed then the mixture of elements in the interstellar gas must present a cross-section of at least 10^{-20} cm^2, per H atom, for absorption at that frequency. From the data given by Brown and Gould (1970) it follows that $v = 5 \times 10^{16}$ Hz is the largest frequency for effective absorption. This corresponds to a minimum wavelength $\lambda = 60$ Å; the maximum wavelength is, of course, $\lambda = 912$ Å, at the edge of the Lyman continuum. Photons with $\lambda = 60$ Å are typical of radiation with a colour temperature of 500 000 K. We are therefore not interested here in the very hard radiation that might be produced when the initial shock from the supernova explosion reaches the surface of the star in which it occurs. The radiation which would possibly produce the ionization must come from the hot remnant, in the early phase of its expansion.

At a later stage of evolution the typical supernova remnant shares its kinetic energy with the ambient interstellar medium via a shock interaction (Woltjer, 1972). Typically, at time t, the speed of sound in the gas behind the shock is a, the radius of the shocked region is of order at and its energy content is of order $\varrho_{is} a^5 t^3$. Here ϱ_{is} is the interstellar gas density before the shock wave overtakes it; in the present case we estimate $\varrho_{is} \sim 2 \times 10^{-25}$ g cm^{-3}. The radiant losses from the shocked region become important only when the temperature has fallen below about 5×10^6 K, or $a \lesssim 2 \times 10^7$ cm s^{-1}. Let E_0 be the energy of the supernova explosion. We must then clearly wait until

$$a \sim \left(\frac{E_0}{\varrho_{is} t^3} \right)^{1/5} \lesssim 2 \times 10^7$$

or

$$t \gtrsim 6 \times 10^{-13} \left(\frac{E_0}{\varrho_{is}} \right)^{1/3} \tag{1}$$

before the kinetic energy given to the supernova remnant is converted back into radiation. In the present case E_0 must exceed 5×10^{51} erg. Therefore the lower limit on the time exceeds 1.8×10^{12} s or 6×10^5 yr. This is very much too late for the Gum nebula. The most likely source of the radiation is therefore the fireball which expands away from the star after the supernova explosion.

II. The Fireball

We now discuss the properties of the expanding fireball. The supernova explosion releases an amount of energy E_0, say 10^{52} or 10^{53} erg, into a mass M_*, say 10^{34} g. Some of this energy is carried to the surface by shock waves, but most of it is trapped in the gas. It cannot be radiated immediately, because the opacity of the gas is too high. Therefore a considerable pressure builds up in the gas, and the fireball accelerates outwards. The acceleration is largest immediately after the explosion.

It is readily shown that radiation pressure will dominate in the fireball. We have, for a fireball of initial radius R_0, that the temperature T is given by

$$E_0 = \frac{3}{2} \frac{M_*}{m} kT + \frac{4\pi}{3} R_0^3 aT^4, \tag{2}$$

where m is the mean molecular weight, about 10^{-24} g, and a is Stefan's constant. The two terms on the right hand side are, respectively, the gas kinetic and the radiant energy content. With $R_0 = 10^{12}$ cm, a typical value, they are in the ratio $1:10^3$ (when $E_0 = 10^{53}$ erg) or $1:250$ (when $E_0 = 10^{52}$ erg). In either case the radiant energy content and, therefore, radiation pressure dominate. The ratio of radiation pressure to gas pressure remains unchanged as long as the subsequent expansion is adiabatic, that is as, long as radiative losses from the fireball have a negligible effect on its energy content.

We therefore model the radiative loss process of the fireball as follows. A mass M_* is given energy E_0 which is largely converted into kinetic energy. The mass therefore expands with speed $V \equiv (2E_0/M_*)^{1/2}$. At time t we suppose that it is spread over the surface of a sphere with radius $R = (2E_0/M_*)^{1/2} t$. The mass per unit area of the sphere is then

$$\sigma = \frac{M_*^2}{8\pi E_0} t^{-2}. \tag{3}$$

If electron scattering is the main cause of opacity in the material, then the optical depth of the shell is

$$\tau \equiv \kappa_e \sigma = \frac{\kappa_e M_*^2}{8\pi E_0} t^{-2}. \tag{4}$$

Here the electron opacity is denoted by κ_e ($=0.34$ cm^2 g^{-1} in the usual cosmical mixture of elements). Let E be the total radiant energy enclosed by the shell at time t. The rate of change of E is then given by

$$\frac{dE}{dt} = -\frac{E}{R} \frac{dR}{dt} - \frac{4\pi R^2 c}{3\tau} \frac{E}{(4\pi/3) R^3}. \tag{5}$$

The first term on the right-hand side gives the adiabatic energy loss rate; the second gives the rate of flow of energy through the surface. On substituting for R and τ in terms of t we get that

$$\frac{dE}{E} = -\frac{dt}{t} - \frac{4\pi\sqrt{2}}{\kappa_e} \left(\frac{E_0}{M_*^3}\right)^{1/2} ct \, dt. \tag{6}$$

On integration

$$E = \frac{\text{const.}}{t} \exp\left\{-\frac{2\pi\sqrt{2}}{\kappa_e} \left(\frac{E_0}{M_*^3}\right)^{1/2} ct^2\right\}. \tag{7}$$

For $E_0 = 10^{52}$ or 10^{53} erg and $M_* = 10^{34}$ g we get a value of 10^{-13} or 3×10^{-13} for the coefficient of t^2 in the exponent. This means that the exponential term does not become small until after time $t = 3 \times 10^6$ or 1.7×10^6 s, respectively. Until that time the energy content of the shell varies inversely as the time or the radius of the shell – in other words, its variation is purely adiabatic.

In order to fix the constant in equation (7) we set $E = E_0$ at time $t = t_0$, when the radius of the shell was $R = R_0$. This gives the relation

$$V t_0 = \left(\frac{2E_0}{M_*} \right)^{1/2} t_0 = R_0,$$

so that t_0 is of the order of a few hundred seconds, much less than the time required for radiation to leak through the shell. At this time the exponential term is very nearly equal to unity, and we find that Equation (7) becomes

$$E = \frac{R_0}{t} \left(\frac{M_* E_0}{2} \right)^{1/2} \exp \left\{ - \frac{2\pi\sqrt{2}}{\kappa_e} \left(\frac{E_0}{M_*^3} \right)^{1/2} c t^2 \right\}. \tag{8}$$

From (6) and (8) we get that the total radiant energy lost through the shell during its expansion is

$$Q = \int_0^\infty \frac{4\pi\sqrt{2}}{\kappa_e} \left(\frac{E_0}{M_*^3} \right)^{1/2} c t \, E \, dt$$

$$= \frac{4\pi}{\kappa_e} \left(\frac{E_0}{M_*} \right) R_0 c \int_0^\infty \exp \left\{ - \frac{2\pi\sqrt{2}}{\kappa_e} \left(\frac{E_0}{M_*^3} \right)^{1/2} c t^2 \right\} dt$$

$$= 2^{1/4} \pi \frac{R_0 c^{1/2} E_0^{3/4}}{\kappa_e^{1/2} M_*^{1/4}}. \tag{9}$$

We insert numerical values and find that $Q = 3.5 \times 10^{48}$ or 2×10^{49} erg, for $E_0 = 10^{52}$ or 10^{53} erg, and with $M_* = 10^{34}$ g and $R_0 = 10^{12}$ cm. We see that the radiant energy output of the fireball is totally inadequate and cannot possibly result in the formation of a fossil H II region with the parameters that are customarily quoted for the Gum nebula.

In fact the discrepancy is even wider than these arguments suggest. Before radiant losses through the shell become serious, the energy density of the radiation within is, at time t,

$$\frac{E}{\frac{4\pi}{3} R^3} = \frac{3}{16\pi} \frac{M_*^2 R_0}{E_0 t^4}, \tag{10}$$

with the help of our formulae. The equivalent temperature of the radiation field within the shell is

$$T = \left(\frac{3}{16\pi} \right)^{1/4} \frac{M_*^{1/2} R_0^{1/4}}{E_0^{1/4} a^{1/4}} t^{-1}. \tag{11}$$

This should also be the colour temperature of the radiation which leaks through the shell.

But in making the estimate for the energy required to form the Gum nebula it is assumed that most of the photons emitted by the fireball are harder than the Lyman limit. This means, effectively, that the colour temperature of the radiation must exceed 60 000 K. With the usual values substituted in relation (11) we find that t must therefore be less than 3×10^5 or 1.7×10^5 s, for E_0 equal to 10^{52} or 10^{53} erg. We found earlier that the shell goes on radiating energy, at an almost constant rate, until time 3×10^6 or 1.7×10^6 s in these two cases. But we see now that the photons emitted are hard enough to cause appreciable ionization only during the earliest one-tenth of this time. Therefore, only 3.5×10^{47} or 2×10^{48} erg are available, in these two cases, in the form of ionizing radiation.

III. Discussion

Obviously there is a great discrepancy between our estimate of the amount of ionizing radiation available and the amount that is needed for the standard model of the Gum nebula. The shortfall is so large that it is hardly worthwhile to make such improvements as would replace our approximate calculation by a more exact one; the result would still be qualitatively the same. It seems that only one possibility remains. It is that the ionizing radiation was produced by the Vela pulsar itself. Let us see what this implies.

The likely mass of a pulsar is $M = 10^{33}$ g and its radius $R = 10^6$ cm. The moment of inertia of a neutron star is of the order of 2×10^{44} g cm^2 (Ruderman, 1972), and its limiting angular velocity (to avoid break-up under centrifugal force) is given by

$$\omega_c^2 = \frac{GM}{R^3} = 7 \times 10^7 \text{ s}^{-2}.$$

Hence the maximum kinetic energy of rotation is 7×10^{51} erg. This is also the maximum possible amount of energy available for radiation. Note that ω_c comes out to be about 100 times the present angular velocity of the pulsar (cf. Smith, 1972). In the simplest version of the oblique rotator theory the slow-down time for a pulsar varies like ω^{-2}, so that the bulk of the original energy of the neutron star would have been emitted as radiation over a period of a year. Our estimate for the opacity of the fireball shows that radiation can begin to escape freely once the supernova remnant is more than about 3×10^6 s (or 40 days) old. Therefore the bulk of the pulsar radiation would be available for the production of the fossil H II region.

This seems to be the only possibility. If it fails then there is no way in which one can associate the Gum nebula with the Vela supernova.

References

Brandt, J. C.: 1971, in S. P. Maran, J. C. Brandt, and T. P. Stecher (eds.), *The Gum Nebula and Related Problems*, Goddard Space Flight Center, Greenbelt, Maryland, p. 15.
Brown, R. L. and Gould, R. J.: 1970, *Phys. Rev.* **1**, 2252.

Maran, S. P., Brandt, J. C., and Stecher, T. P.: 1971, *The Gum Nebula and Related Problems*, Goddard Space Flight Center, Greenbelt, Maryland.
Ruderman, M.: 1972, *Ann. Rev. Astron. Astrophys.* **10**, 427.
Smith, F. G.: 1972, *Rep. Prog. Phys.* **36**, 399.
Woltjer, L.: 1972, *Ann. Rev. Astron. Astrophys.* **10**, 129.

F. D. Kahn
Astronomy Department,
University of Manchester,
Manchester 13, United Kingdom

DISCUSSION

Caswell: I would like to point out that the Gum nebula, the experimental prototype of a fossil Strömgren sphere, can be quite satisfactorily interpreted as a normal H II region. A detailed analysis reaching this conclusion was published recently (Beuermann, K. P.: 1973, *Astrophys. Space Sci.* **20**, 27.)

Mathewson: Several weeks ago M. Clearly and I made a detailed H I survey of the Gum nebula using the 18-m reflector at Parkes. High velocity H I was found around the nebula, and the correlation between the start of the H I emission and the optical limits of the nebula was so good that we concluded that the expanding shell of the Gum nebula was producing the high velocity H I. Radial velocities of 70 km s^{-1} were generally recorded although there were areas with velocities as high as 200 km s^{-1}. As there is no known mechanism that can produce such high gas velocities other than SNR we are forced to the conclusion that the Gum nebula is a SNR. The velocity of expansion must be at least several hundred km s^{-1}, which implies a diameter of about 40 pc, which places the Gum nebula at a distance of 60 pc from the Sun, much closer than the Vela 10 SNR which lies in the same direction. The ionization of the Gum nebula would then be due to collisional excitation. (However, optical spectra taken recently by us at five points in the Gum nebula are not characteristic of SNR nor do they show the same radial velocities as the H I, which casts doubt on the conclusion that this nebula is an SNR. It may be that the high-velocity H I belongs to a more distant spiral arm which by chance has an edge coincident with the closer edge of the Gum nebula.)

Baldwin: Can the theoretical models be calculated for expansion into an inhomogeneous medium, and can they account for the observed features, say in Cas A and the Cygnus Loop?

Kahn: The calculation would probably be feasible for small perturbations. But gross inhomogeneities might be very hard to deal with.

Burke: Stringent upper limits have now been placed on the flux from SN 1885 (S And). I wish to report a result obtained by J. H. Spencer with the NRAO interferometer. At 11 cm and 3 cm, a limit of 0.6 mJy was obtained. All the phenomenological theories predict several orders of magnitude more flux. The failure to observe radiation is consistent with the recent theory of Gull, and it appears that S And has not yet reached the 'turn on' stage.

POLARIZATION OBSERVATIONS OF
SUPERNOVA REMNANTS

D. K. MILNE

Division of Radiophysics, CSIRO, Sydney, Australia

and

JOHN R. DICKEL

University of Illinois Observatory, Urbana, Ill., U.S.A.

Abstract. Polarization and total power at 5000 MHz and 2700 MHz have been obtained for 30 supernova remnants (SNR) using the 64-m radio telescope at Parkes. This large sample includes a range of SNR from the young bright objects (such as Kepler's SNR) through the old faint sources such as MSH 14-63. Among the old remnants, the individual SNR have very varied properties, but several general conclusions emerge from our study. (1) In many cases, the polarization of the galactic background is as strong as that of the SNR and can vary significantly over the angular extent of the SNR. (2) The Faraday rotation is generally small and varies quite uniformly across the source. (3) The magnetic field patterns, although associated with the SNR, do not show any particularly characteristic pattern or relations to the total-power structure of the source.

I. Introduction

We have obtained 6- and 11-cm polarization and total power maps of 30 supernova remnants (SNR) using the 64-m radio telescope at Parkes. This large sample includes a range of SNR from the young bright objects (such as Kepler's SNR) through the old faint sources such as MSH 14-63. Among the old remnants, the individual SNR have very varied properties but several general conclusions emerge from our study: (i) In many cases, the polarization of the galactic background is as strong as that of the SNR and can vary significantly over the angular extent of the SNR; (ii) the Faraday rotation is generally small and varies quite uniformly across the source; and (iii) the magnetic field patterns, although associated with the SNR, do not show any particularly characteristic pattern or relations to the total power structure of the source. These points will be described below.

II. Equipment and Observations

At 2700 MHz we used a two-channel parametric correlation receiver, giving a system noise temperature of 100 K over a 400-MHz band, and at 5000 MHz a switched parametric receiver with an 80-K noise temperature and 500-MHz bandwidth was used. Both of these receivers compared the signals from probes of opposite polarization in circular waveguide feed horns. The beamwidths were 8.4′ and 4.4′ at 2700 and 5000 MHz, respectively.

The observing procedures have been described in detail by Milne (1972) and consisted, briefly, of scans in declination spaced at approximate half-beamwidth intervals in right ascension. At each right ascension a total of four scans were made, at position angles separated by 45°. The polarization **E** vectors were then obtained

by combining these scans in the usual way. At 2700 MHz two instrumental effects
were recognized and removed; (i) a 'gain' effect yielding a spurious polarization vector
always at the same feed angle and (ii) a spurious polarization proportional to the
total power intensity gradient and directed along this gradient. No significant
instrumental effects were detected at 5000 MHz.

III. Results

(a) BACKGROUND POLARIZATION

A good example of the extremely large background contribution which can occur
in the direction of some sources is seen in Figure 1. This shows the polarization **E**

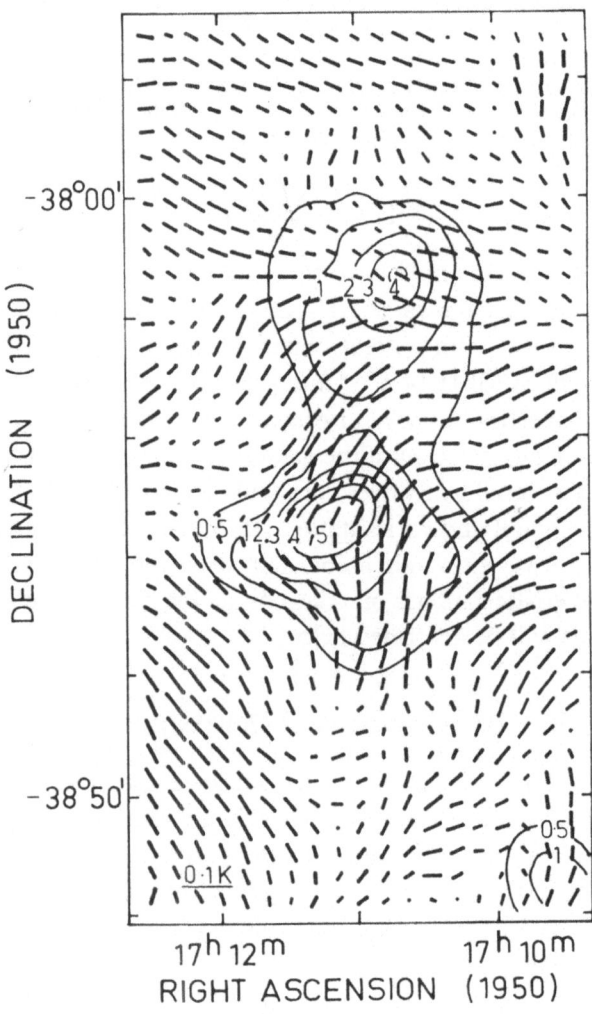

Fig. 1.   The high level of background polarization is seen here in the 5000-MHz **E** vectors around CTB
37*A* and *B*. The total power contour unit is 1 K in brightness temperature in all of these figures.

vectors at 6 cm in the field of CTB 37*A* and *B*, two separate SNRs (G348.5 + 0.1 and G348.7 + 0.3). It is virtually impossible to distinguish where the SNRs should be from this figure; the polarization and the SNRs bear no relation. Presumably the strong galactic magnetic fields and perhaps intervening Faraday rotation have masked the source polarization. The source MSH 11 − 61*A* (G290.1 − 0.8) is an example of an SNR whose polarization clearly stands out above the background, although the latter is still significantly polarized (Figure 2). It can also be seen that the background polarization changes significantly across the 15′ subtended by the SNR. In this case the H II region MSH 11 − 61*B* just to the west of the SNR shows negligible polarization, as would be expected for such an object. A final example of the background polarizations is shown in Figure 3a. Although the SNR MSH 14 − 63 (G315.4 − 2.3) is faint its polarization can be seen to just stand out above the noise as well as the background. Even without the total power contours the general position of the source can be picked out from the map of the magnetic fields alone in Figure 3b. In this case we plot the directions of the projected magnetic field, i.e., orthogonal to the observed **E** vectors after removal of the Faraday rotation.

(b) FARADAY ROTATION

The Faraday rotation was evaluated from the two-frequency data after convolution to the same effective resolution for both; the lowest value of |*RM*| generally was

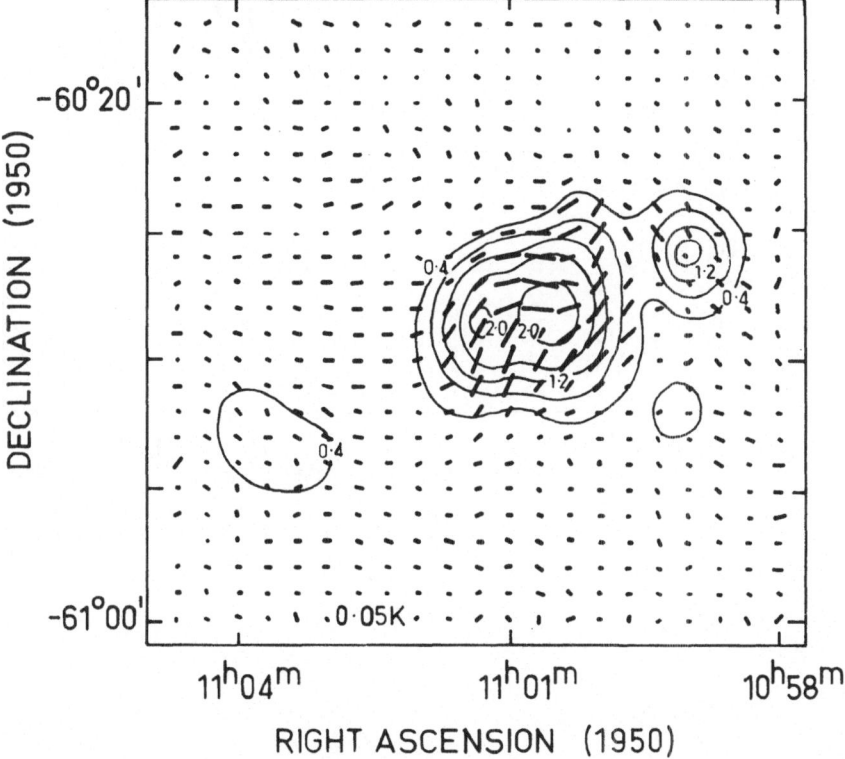

Fig. 2.   5000-MHz polarization **E** vectors and total power contours for MSH 11 − 61*A* and *B*.

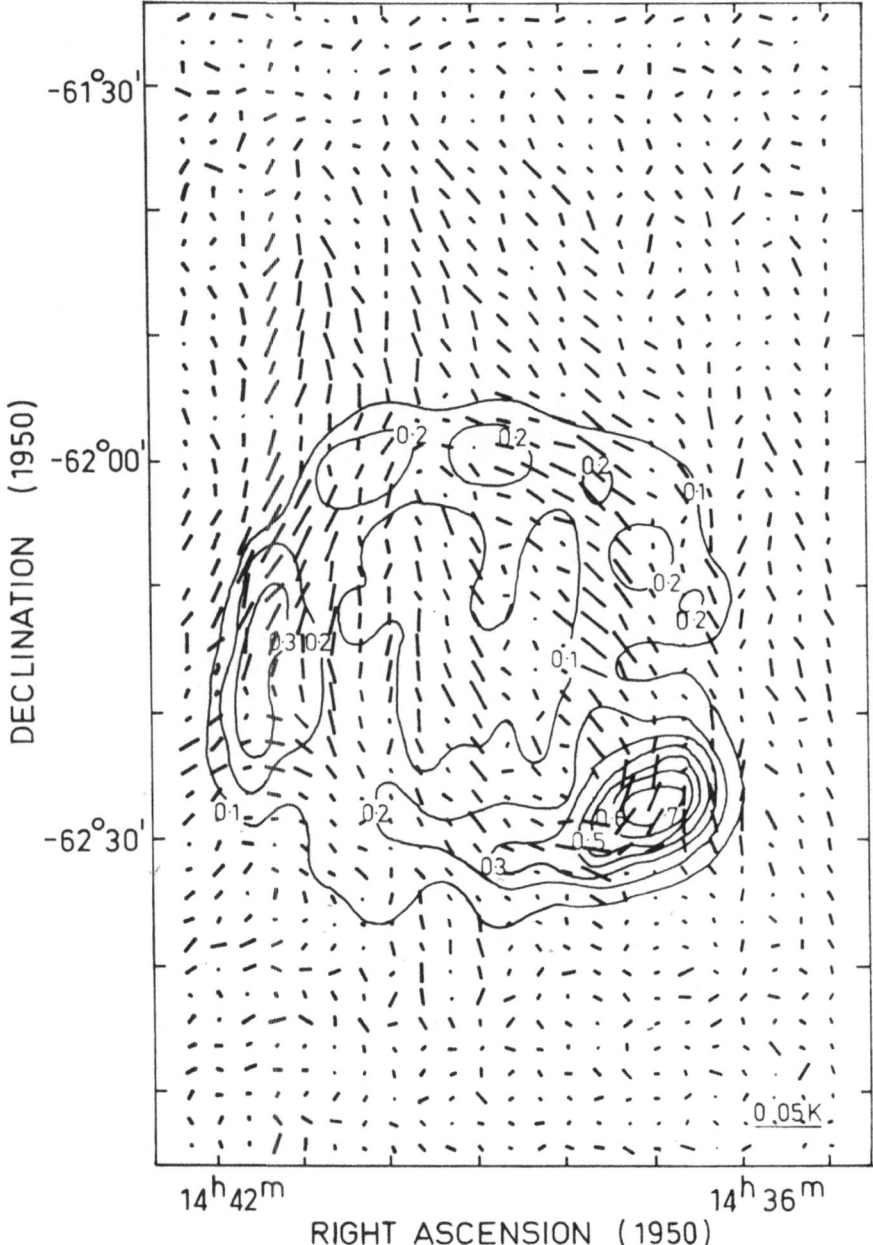

Fig. 3a.   5000-MHz polarization **E** vectors for MSH 14−63.

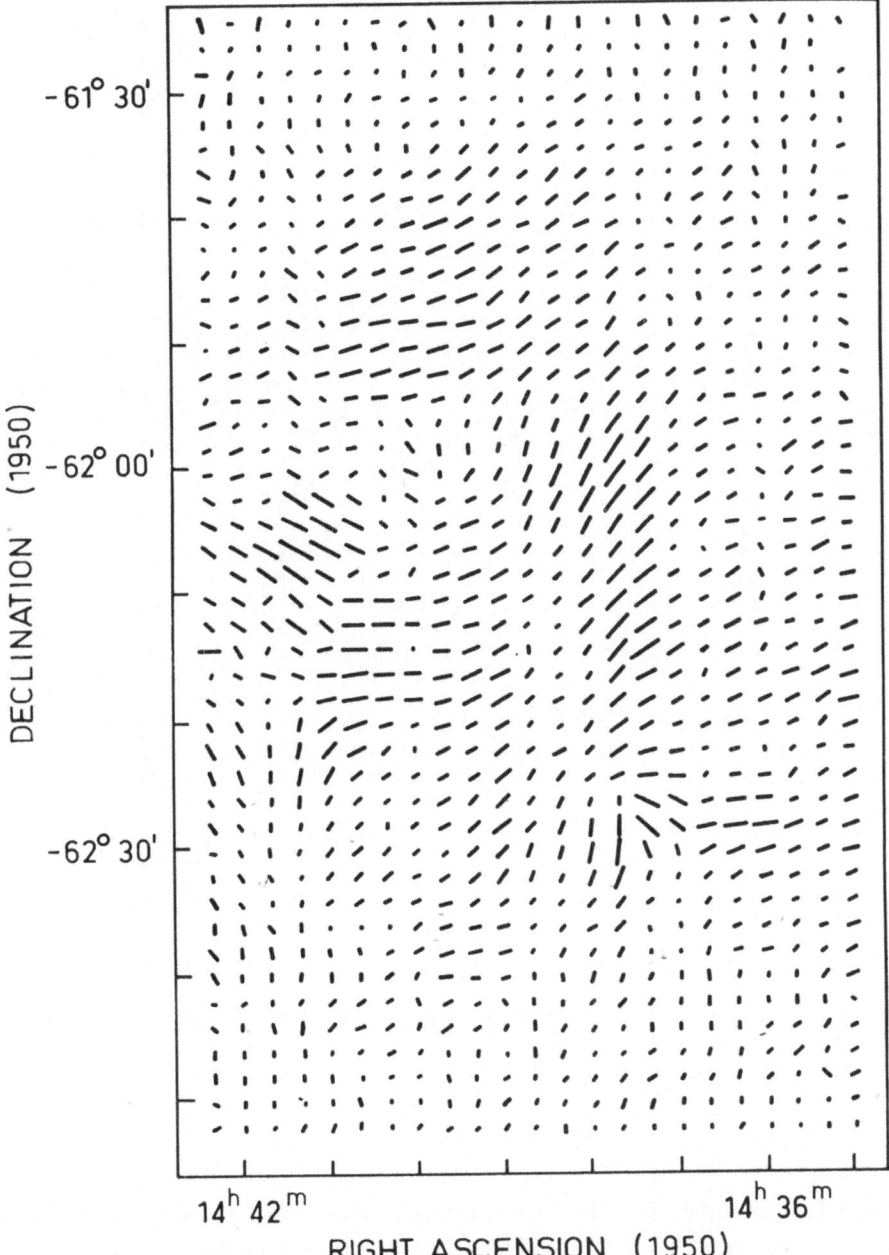

Fig. 3b. Directions of magnetic field for MSH 14−63.

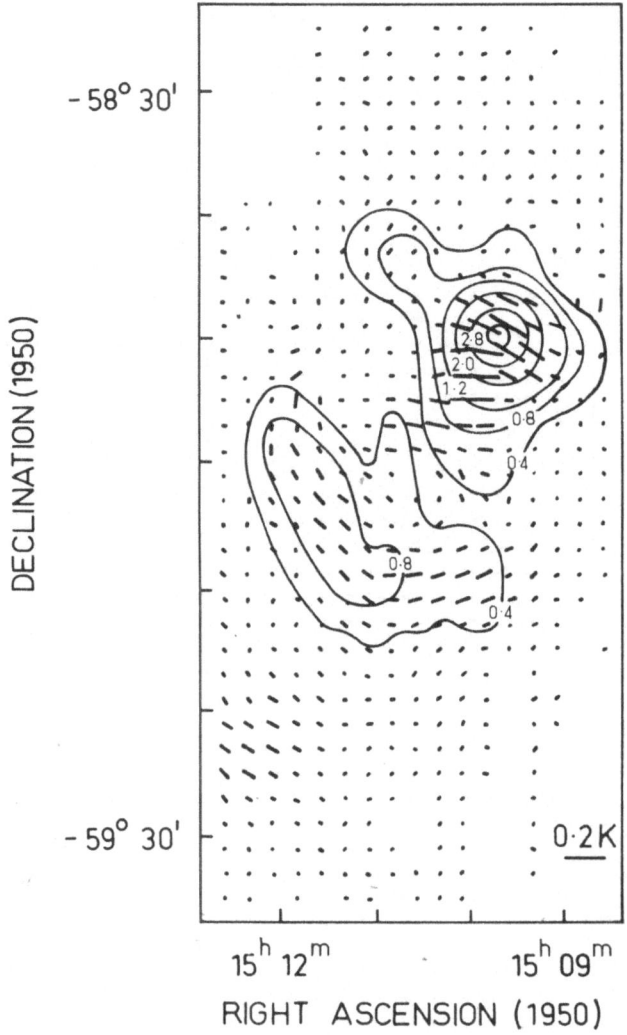

Fig. 4a.   5000-MHz **E** vectors for MSH 15−52.

accepted. It is of course possible that additional rotations by multiples of π radians could occur between 2700 MHz and 5000 MHz, but each of these would change the rotation measure by 350 rad m^{-2} and in no case where a third frequency is available has this been observed within the errors imposed on the measured position angles. In the case of MSH 14−63 discussed above, the Faraday rotation was quite small and varied only very gradually across the source. This was confirmed by a relatively small depolarization ratio between the two frequencies.

Comparison of the **E** vectors at 6 and 11 cm for the extended Vela X SNR (G263.4−3.0) (Milne, 1973) shows again only a very small Faraday rotation. The average value was $+56$ rad m^{-2} with an extreme range from -100 rad m^{-2} to

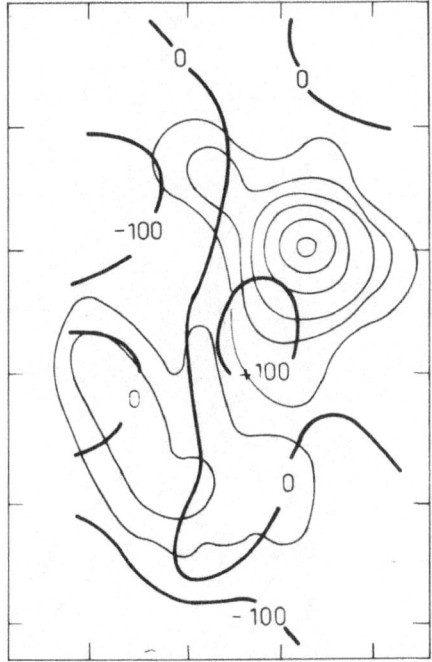

Fig. 4b. Distribution of rotation measure (rad m^{-2}).

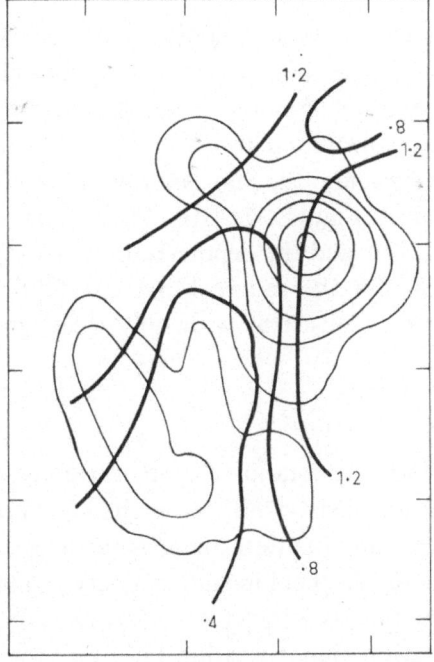

Fig. 4c. Depolarization ratio ($P_{2700\,\mathrm{MHz}}/P_{5000\,\mathrm{MHz}}$).

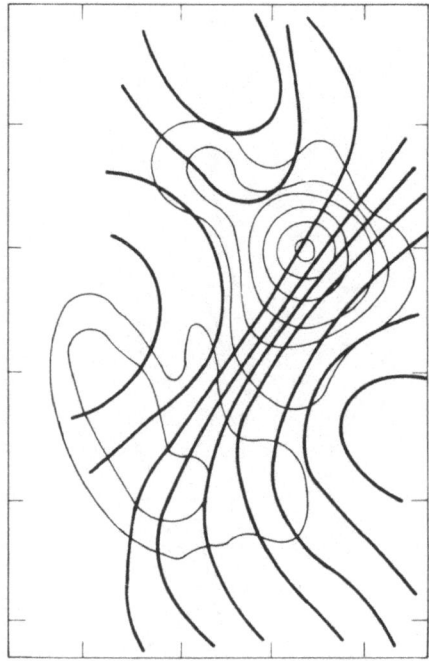

Fig. 4d.   The directions of magnetic field.

$+150$ rad m^{-2}. This amount of rotation could reasonably be internal to the source since for a source with a diameter of 20 pc and a longitudinal magnetic field strength of 10^{-5} G a rotation measure of 56 rad m^{-2} implies a density of only 0.33 cm^{-3} for the thermal electrons in the medium. However it is equally acceptable that all of this rotation is between the object and the Sun. Another illustration of the small general Faraday rotation is given in Figures 4a, b, c, which show respectively the 6-cm **E** vectors, the Faraday rotation, and the depolarization ratio (per cent polarization at 11 cm / per cent at 6 cm) for the source MSH 15−52 (G320.4−1.2). The small depolarization is, of course, consistent with only weak gradients in the Faraday rotation across the beam.

(c) MAGNETIC FIELD CONFIGURATION

After correcting for the Faraday rotation shown above, the orientations of the projected magnetic field lines for MSH 15−52 are as shown in Figure 4d. Although some areas show a more or less radial pattern the general structure is quite confused. A better example of a more regular field including a very significant radial component is shown in our results for Puppis A (G260.4−3.4) (Figures 5a, b, c and d, the 6 cm **E** vectors, the Faraday rotation, depolarization ratio and the magnetic field orientations, respectively). We note in passing that, perhaps fortuitously, the position of most significant Faraday rotation northwest near the position of the extended X-ray

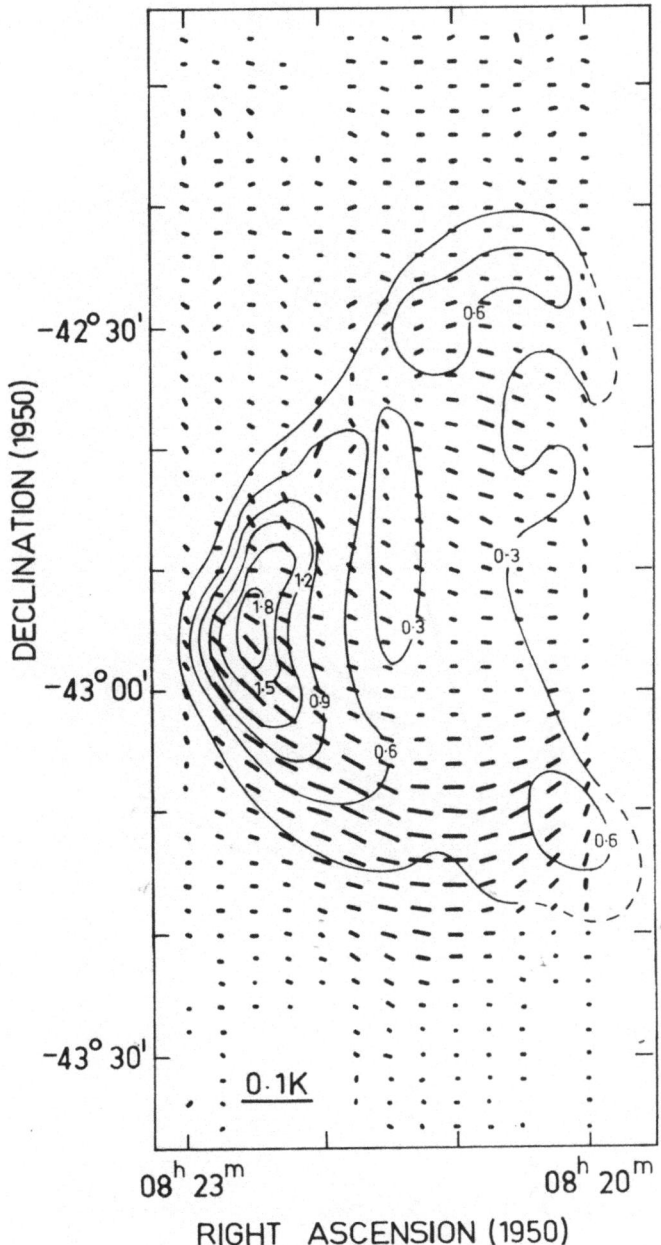

Figs. 5a–d.   The same as Figure 4 for Puppis A.

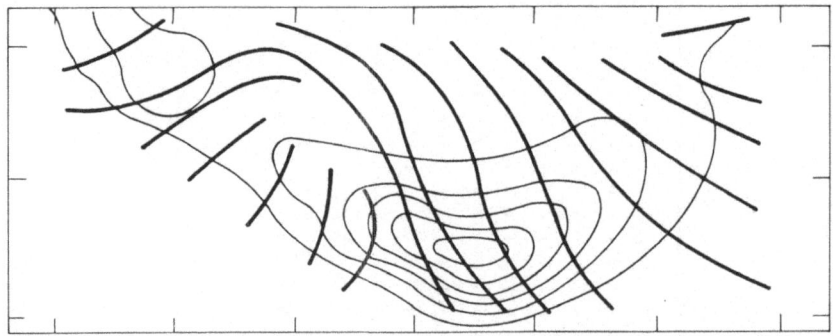

Fig. 5d.

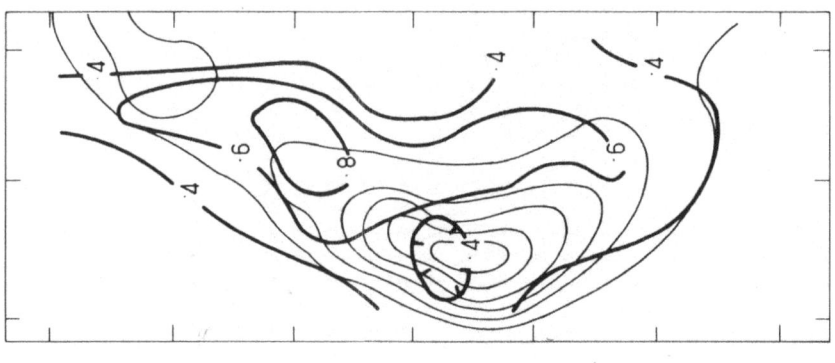

Fig. 5c.

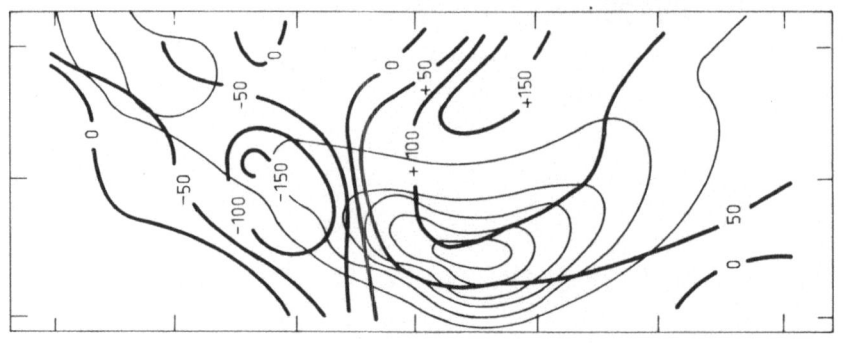

Fig. 5b.

source associated with the SNR (Zarnecki *et al.*, 1973). As a final figure (Figure 6) we show the magnetic field vectors of Vela X as determined from the data shown above superimposed upon an optical photograph of the region. The vector lengths are the mean of the polarized intensities at 6 and 11 cm, and the circle marks the position of the pulsar. Clearly there is polarization associated with the SNR, but its pattern including such features as the whorl in the south centre will be very difficult to evaluate even with three-dimensional models.

In addition to the results shown above we recall the tangential field found for 1209 − 51/52 (G296.3 + 10.0) (Whiteoak and Gardner, 1968, confirmed by us at 6 cm) and the nearly uniform field across W44 (G34.6 − 0.5) (Kundu and Velusamy, 1972; Milne and Dickel, 1973). Thus from this preliminary analysis of a few sources, no

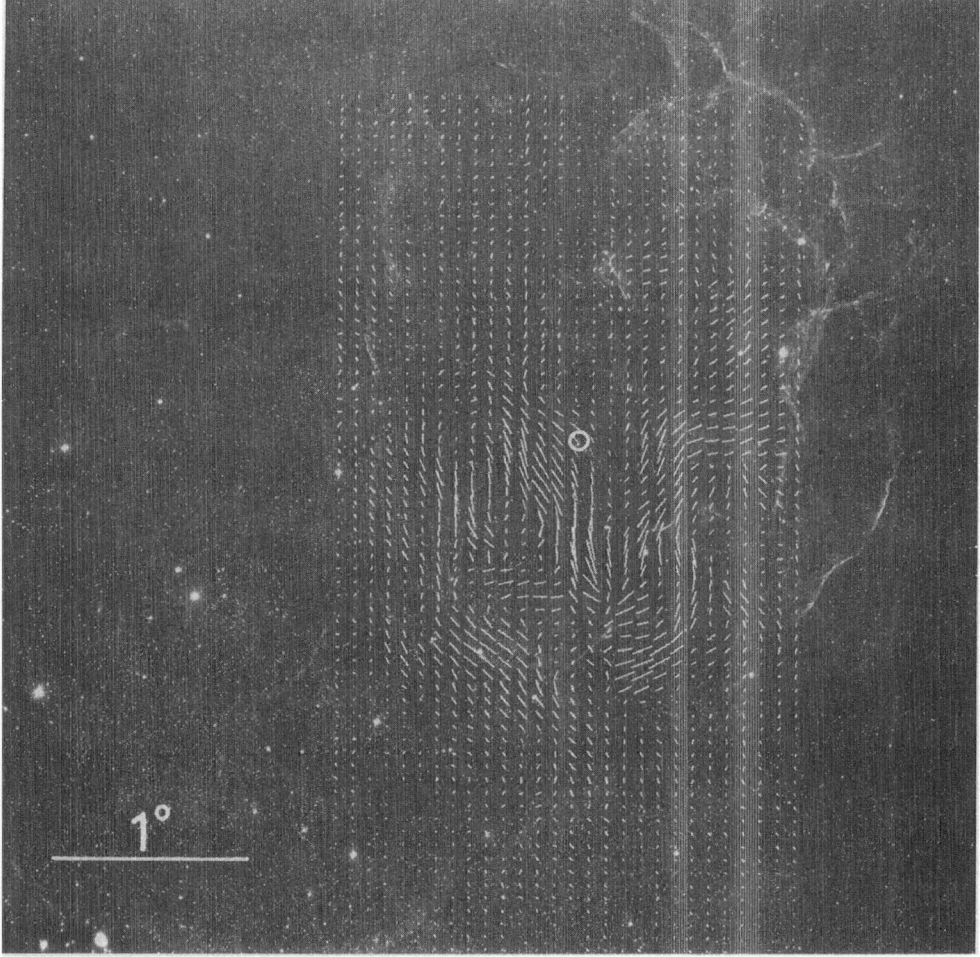

Fig. 6.   Vela X showing the optical features (courtesy of B. J. Bok) overlaid with vectors representing the direction of the projected magnetic field. The length of these vectors is the average of the 5000-MHz and 2700-MHz polarization intensity.

clear picture of the magnetic field configurations has emerged, and it would appear that the individual SNR had very different starting conditions and/or very varied interaction with the surrounding interstellar medium. If any general trend is seen it would be that the mean percentage polarization increases with decreasing surface brightness, i.e., the older remnants perhaps have more orderly fields.

IV. Conclusions

In most directions along the galactic plane there is strong polarization of the background emission with a scale comparable with the angular size of many of the SNR. Studies of these objects should therefore include sufficient measurements of the background to assess its effect.

Working within the error in position angle we find that rotation measures generally lie within the range ± 200 rad m^{-2} and vary uniformly across the source. Except in very few objects the magnetic fields delineate patches with uniform direction but bearing no relation to the source structure.

References

Kundu, M. R. and Velusamy, T.: 1972, *Astron. Astrophys.* **20**, 237.
Milne, D. K.: 1972, *Australian J. Phys.* **25**, 307.
Milne, D. K.: 1973, *The Vela Supernova Remnant*, in preparation.
Milne, D. K. and Dickel, J. R.: 1973, submitted to *Australian J. Phys.*
Whiteoak, J. B. and Gardner, F. F.: 1968, *Astrophys. J.* **154**, 807.
Zarnecki, J., Culhane, J. L., Fabian, A. C., Rapley, C. G., Silk, R., Parkinson, J. H., and Pounds, K. A.: 1973, *Nature Phys. Sci.* **243**, 4.

D. K. Milne
Division of Radiophysics, CSIRO,
P.O. Box 76, Epping, N.S.W. 2121, Australia

John R. Dickel
University of Illinois Observatory,
Urbana, Ill. 61801, U.S.A.

(Discussion follows after paper by R. M. Duin et al., p. 361.)

HIGH-RESOLUTION OBSERVATIONS OF
GALACTIC NONTHERMAL SOURCES AT 160 MHz

O. B. SLEE and G. A. DULK*

Division of Radiophysics, CSIRO, Sydney, Australia

Abstract. Preliminary maps of the brightness distributions of 11 supernova remnants are presented. The sources, with diameters $\lesssim 10'$, were observed at 160 MHz with resolution better than $2'$. Ten of the supernova remnants are significantly resolved, and at least six show evidence of a shell-like structure. For those sources where high-frequency maps are available with comparable resolution the brightness distributions are generally very similar.

I. Introduction

This paper presents some preliminary maps of the brightness distributions of 11 suspected galactic supernova remnants (SNRs). The observations were made at 160 MHz with the radio-heliograph operated by the Division of Radiophysics, CSIRO, at Culgoora, N.S.W. At this frequency the radioheliograph's beamwidth, $1.85' \times 1.85'$ secant $(30°18' + \delta)$, is small enough to partially resolve many of the SNRs near the galactic equator. Each observation consisted of a series of drift scans of the source. A detailed description of the observing method and reductions is given by Dulk and Slee (1972). The flux density scale was established from observations of several strong sources with well-known spectra from Kellermann *et al.* (1969).

Each of the sources was observed on several different occasions; the maps selected for presentation here show the features commonly found on the different maps of a particular source. No two maps of a source are exactly alike because of the effects of noise, ionospheric refraction changes, and, in some cases, perturbations caused by other sources in the grating lobes of the instrument. Later we plan to publish more accurate brightness contours obtained by averaging many separate maps taken at a variety of hour angles; the resulting partial synthesis will reduce the effects of grating responses. In addition, we will publish the flux densities of these and other galactic sources and discuss them from the point of view of interstellar absorption.

II. Comments on 160-MHz Maps

G348.5 + 0.1. This source (CTB 37*A*) has long been recognized as a probable SNR. The present 160-MHz map shows a definite shell-like structure which is particularly enhanced to the north-east. Dickel *et al.* (1973) have observed this source with 2.8′ resolution at 8.8 GHz, where it shows a remarkably similar brightness distribution. At 80 MHz, if we take into account the lower resolution, the brightness distribution is again similar (Dulk and Slee, 1972). Two important conclusions flow from this similarity:

* Exchange visitor from the Department of Astro-Geophysics, University of Colorado, Boulder, Colo., U.S.A.

F. J. Kerr and S. C. Simonson, III (eds.), Galactic Radio Astronomy, 347–351. All Rights Reserved.

(i) The low-frequency absorption reported for this source by Dulk and Slee (1972) is fairly uniformly distributed over the $9 \times 6'$ area occupied by the remnant.

(ii) There is little change in the radio spectrum over the area of the source, implying that the electron energy distribution is similar in various parts of the remnant.

G349.7+0.2. This source is barely resolved by the 1.85′ beam so that no structural detail is seen.

G357.7−0.1. The source is well resolved by the 160-MHz observations. The crowding of the contours, especially in the south-west quadrant, is suggestive of a shell-like structure. Dickel et al. (1973) published a map of this source at 8.8 GHz with a 2.8′ beam showing a similar general elongation. Once again it appears that the electron energy distribution remains constant over the remnant.

G04.5+6.8. Kepler's SNR is barely resolved by the 1.85′ beam. It shows a weakly elliptical distribution with major axis at $PA \sim 60°$. Dickel et al. (1973) published a map at 8.8 GHz with 2.8′ resolution which shows a similar distribution, but with the major axis at $PA \sim 0°$.

G11.2−0.4. The source is significantly resolved at 160 MHz with the 1.85′ beam. The flat top suggests that the source has a shell-like structure which is approximately symmetrical. The source was observed by Shaver and Goss (1970) at 408 MHz with a 2.8′ beam, but the shell-like structure of the source was not apparent at their resolution.

G25.3−0.1. The source is barely resolved in declination but is significantly resolved in hour angle. Because this source is quite weak the reliability of the contour map is questionable. Shaver and Goss's (1970) map at 408 MHz with a 2.8′ beam showed no structural detail; on that basis, they suggested that the source is an extragalactic object.

G29.7−0.2. The source is significantly elongated in hour angle, but there is little evidence of shell structure or other brightness details. Shaver and Goss (1970) published a 408-MHz map in which the source was not significantly resolved.

G31.9+0.1 (3C391). The $1.85' \times 2.12'$ beam at 160 MHz resolves the source and shows an apparent shell-like structure, particularly in the north and west. The 8.8-GHz map of this source given by Dickel et al. with a 2.8′ beam does not match the 160-MHz map very closely; in particular the tongue-like projection to the south in the 160-MHz map is not a feature of the 8.8-GHz map. If the differences in the two maps are real, they imply that there are spectral changes over the SNR, either because of the presence of a separate component with a different spectrum, or because the electron energy distribution for this particular remnant is not constant throughout. Patchy, low-frequency absorption can be ruled out in this case because the optical depth is small at 160 MHz. (Caswell et al. [1971] showed that $\tau \approx 1$ at 80 MHz.)

G39.2 − 0.3 (3C396). The source is significantly resolved by the $1.85' \times 2.27'$ beam at 160 MHz. The distribution is approximately spherical with some indication of a shell-like structure, especially to the north-east and south-west of the peak. The 408-MHz map of Shaver and Goss (1970) with 2.8′ resolution is generally similar and indicates a shell structure extending from the north through west.

G41.1 − 0.3 (3C397, CTB 67). The source is significantly extended at $PA - 57°$ but is barely resolved in the perpendicular direction. There is weak evidence for a shell structure.

G43.3 − 0.2 (3C398, W49B). The $1.85' \times 2.32'$ beam of the radioheliograph has partial-

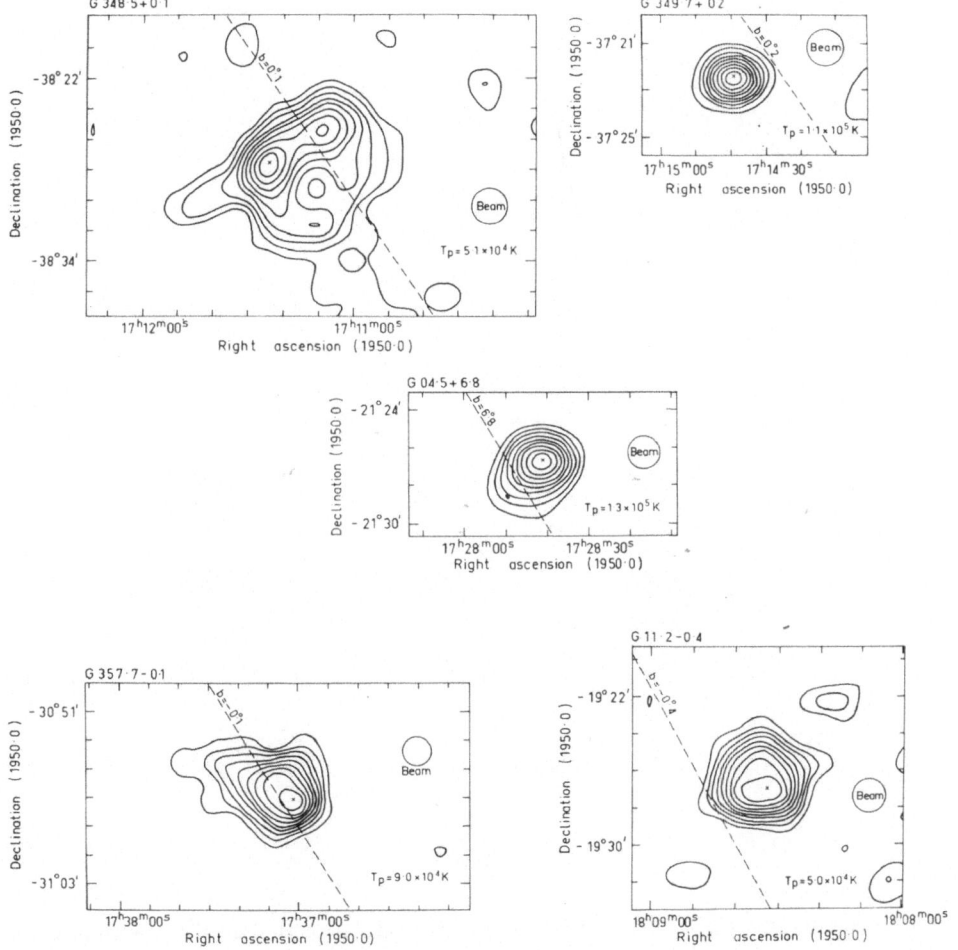

Figs. 1a–b.   Representative contour maps showing the brightness distributions of 11 galactic nonthermal sources at 160 MHz. The contour levels are 10, 20, ..., 90% of T_p, the peak beam brightness temperature measured by the radioheliograph. The value of T_p given is the average value found from all maps of the source; it is only approximately correct for the representative map presented here.

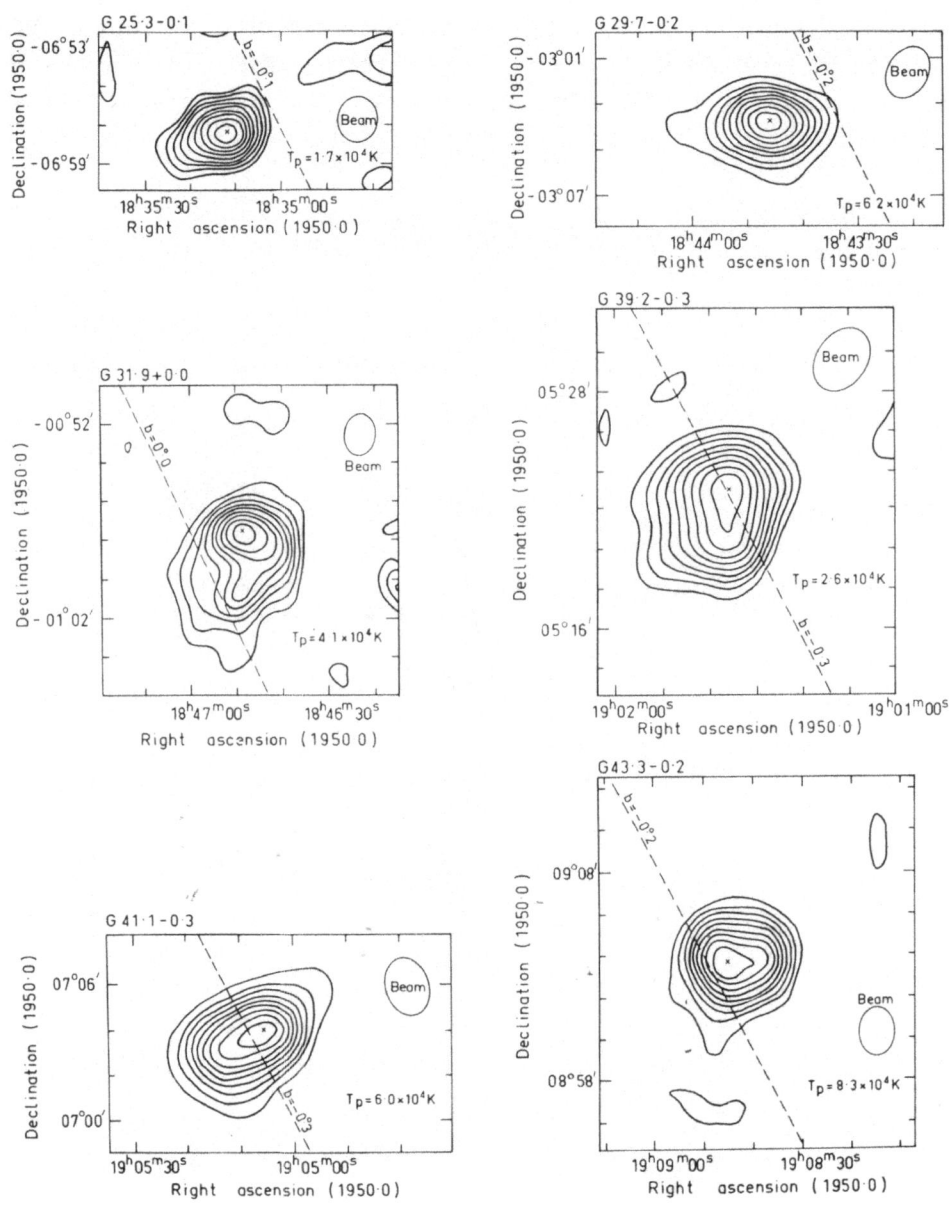

Fig. 1b.

ly resolved this source. The symmetrical radio brightness distribution with a flat top is indicative of a shell structure.

III. Discussion

The present preliminary maps of 11 nonthermal galactic sources have demonstrated the feasibility of high-resolution observations at metre wavelengths, provided care is taken to make enough observations to ensure that the effects of noise, ionospheric refraction and grating responses can be recognized. Shell-like structure appears to be a common characteristic of these small-diameter SNRs. For four of the five sources for which high-frequency maps of similar high resolving power are available, the radio brightness distributions at the high and low radio frequencies are generally very similar. This agreement indicates, first, that the energy distribution of the relativistic electrons responsible for the synchrotron emission from the SNR is similar throughout the remnant. Secondly, for those galactic sources showing pronounced absorption at low frequencies, the agreement between the radio brightness distributions at high and low frequencies shows that the optical depth varies on a scale which is larger than the angular extent of the source. The details of the brightness distributions (for example, the orientation of the major axis of the deconvolved elliptical brightness distribution) does not show any consistent relationship with the direction of the galactic equator. Such a relationship might be expected if the structure of the source were determined by the interaction of the expanding gas cloud with the general field of the Galaxy.

References

Caswell, J. L., Dulk, G. A., Goss, W. M., Radhakrishnan, V., and Green, A. J.: 1971, *Astron. Astrophys.* **12**, 271.

Dickel, J. R., Milne, D. K., Kerr, A. R., and Ables, J. G.: 1973, *Australian J. Phys.* **26**, 379.

Dulk, G. A. and Slee, O. B.: 1972, *Australian J. Phys.* **25**, 429.

Kellermann, K. I., Pauliny-Toth, I. I., and Williams, P. J.: 1969, *Astrophys. J.* **157**, 1.

Shaver, P. A. and Goss, W. M.: 1970, *Australian J. Phys. Astrophys. Suppl.* **14**, 77.

O. B. Slee
G. A. Dulk
Division of Radiophysics, CSIRO,
P.O. Box 76,
Epping, N.S.W. 2121, Australia

(Discussion follows paper by R. M. Duin et al., p. 361.)

A 160-MHz MAP OF CAS A

H. HIRABAYASHI

Tokyo Astronomical Observatory, Tokyo, Japan

Abstract. A rotational synthesis map of Cas A was obtained at 160 MHz with a resolution of $2 \times 3'$.

I should like to report my preliminary results of a rotational synthesis map of Cas A at 160 MHz. The results were obtained using the compound grating interferometer of the Nobeyama Solar Radio Station, Tokyo Astronomical Observatory. The resolution was $2 \times 3'$. Thus the frequency and the resolution are much the same as that used by Slee and Dulk on the Culgoora Solar Radioheliograph. The interferometer consists of two independent one-dimensional arrays aligned in the E-W and N-S directions.

The data were obtained as a series of one-dimensional images, and to make the weightings uniform on the two-dimensional u, v plane, the observed images were convolved with the function whose Fourier transform has an amplitude proportional to the absolute value of the spatial frequency vector. They were then integrated on the two-dimensional image plane, taking into consideration the variation of base-line vector (i.e., the orientation and elongation of the images) with respect to the hour angle.

One day was used for the observation, and the observed hour angle ranged over 9 h. At this time only the E-W arm was used. The figure shows the synthesized map of Cas A and the beam on the same scale. The overall appearance of the map is similar to the higher frequency results. Although high resolution maps have been made at higher frequencies, those at such low frequency are important to determine the structure of the supernova remnants in a wider frequency range.

This is the first preliminary result. By adding the N-S arm data one can get much better coverage on the u, v plane. Also, the 'CLEAN' procedure would restore the images of the supernova remnants to reveal more plausible structure in many cases. I hope to extend the observation to some of the other supernova remnants accessible from the Northern Hemisphere.

H. Hirabayashi
Tokyo Astronomical Observatory,
Mitaka, Tokyo, Japan

(Discussion follows paper by R. M. Duin et al., p. 361.)

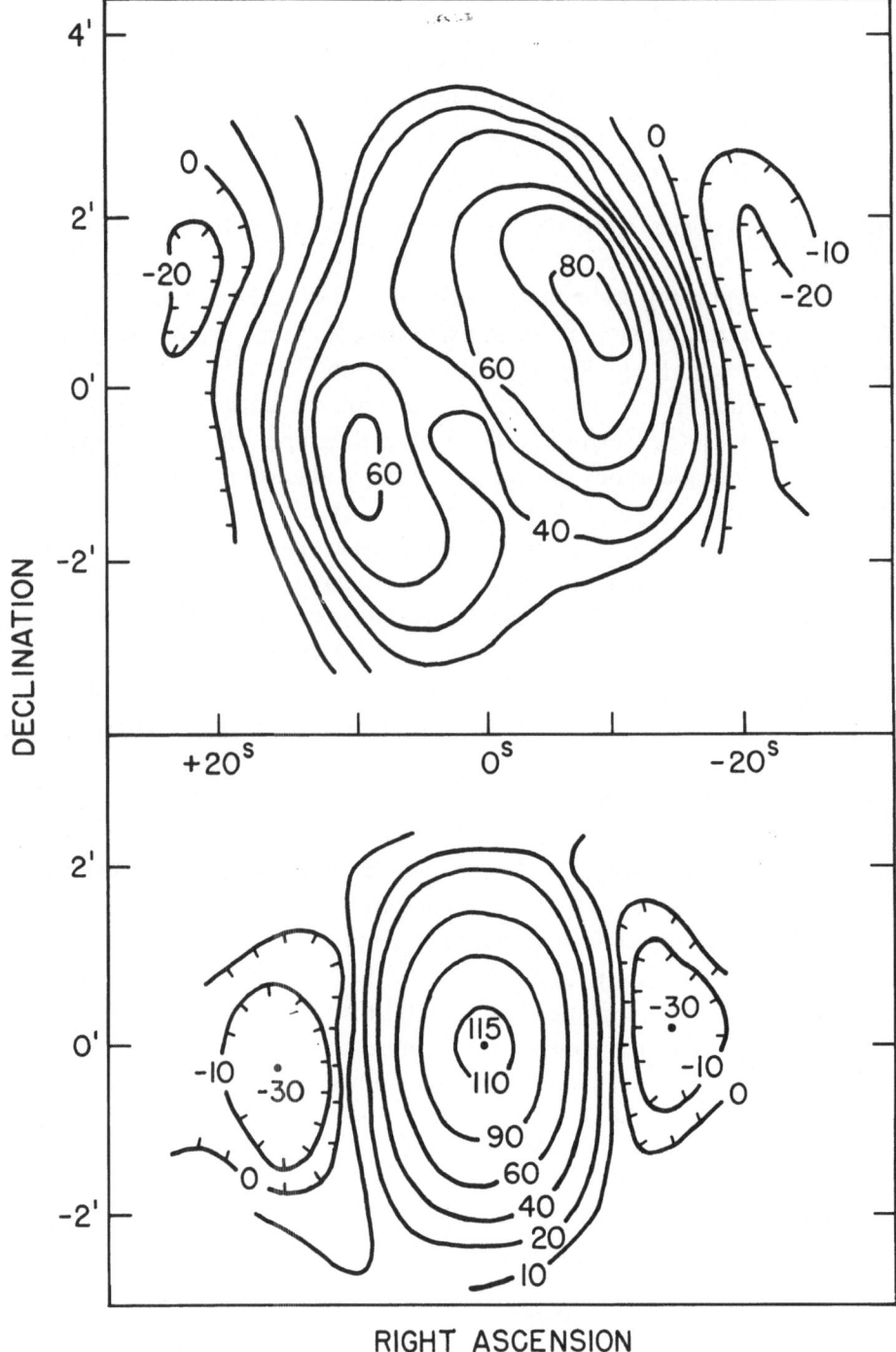

Fig. 1.   Brightness contour map of Cas A (above) and the corresponding two-dimensional beam (below). Brightness scales are in arbitrary units. The map was obtained by the E-W arm and is not yet 'CLEANED.'

WESTERBORK OBSERVATIONS OF THE
GALACTIC SUPERNOVA REMNANTS IC 443 AND TYCHO*

R. M. DUIN, R. G. STROM, and H. VAN DER LAAN

Sterrewacht, Leiden, The Netherlands

Abstract. Continuum observations of IC 443 were made at 21 cm with a resolution of $23'' \times 59''$. The appearance in the radio continuum is very similar to the optical appearance. Since the radio emission is nonthermal, this requires a mechanism that can enhance the synchrotron volume emissivity within thermal filaments, which emit strong optical radiation. We suggest compression of gas and magnetic fields in a shock.

Tycho's remnant has been observed both at 21 cm (resolution $24'' \times 27''$) and at 6 cm (resolution $7'' \times 8''$). Most of the fine scale structure is resolved. The similarity at the two wavelengths indicates a remarkable constancy in the spectral index of features down to a scale of $20''$.

I. Introduction

High resolution studies of extended radio sources require an aperture synthesis instrument of great sensitivity. The Westerbork Synthesis Radio Telescope (WSRT), with its twelve 25-m paraboloids, admirably fulfills this requirement. Here we report on observations of two supernova remnants, IC 443 and 3C 10 (Tycho's remnant), made with sufficient resolution that about 5000 synthesized beams are required to fully scan each object.

II. Observations of IC 443

IC 443 is the remnant of a probable type II supernova of several tens of thousands of years ago. In the bright NE part of the remnant, filaments aligned in the tangential direction are readily apparent on the Palomar Sky Survey print (Figure 1). Temperatures of about 10^4 K and densities on the order of 400 cm^{-3} are indicated for these filaments (Parker, 1964). To the west and southwest the filaments are weaker and define a more irregular morphology. It is likely that in these regions the interstellar medium is more tenuous than to the northeast, where H I observations suggest $n_H = 10$ cm^{-3} for the interstellar density (Locke *et al.*, 1964).

A short period ($P = 0.34$ s) pulsar, PSR 0611+22, has been discovered just to the west of the remnant (Davies *et al.*, 1972). The period and slowing down rate define a time, $P/\dot{P} = 190000$ yr, which by analogy with the Crab nebula pulsar suggests an age of 70000 yr. If a Sedov solution is applicable to the present shock front in IC 443, the average expansion rate should be 2.5 times the present speed of 65 km s^{-1} (Lozinskaya, 1969). Combined with the pulsar age, this would place the remnant at 2.5 kpc rather than the much quoted distance of 1.5 kpc.

Continuum observations of IC 443 were made with the WSRT at a wavelength of 21 cm. The high resolution ($23'' \times 59''$) map in Figure 2 reveals an overwhelming amount of fine scale structure, especially in the northeastern sector. To the southwest,

* Presented by H. J. Habing.

F. J. Kerr and S. C. Simonson, III (eds.), Galactic Radio Astronomy, 355–361. All Rights Reserved.

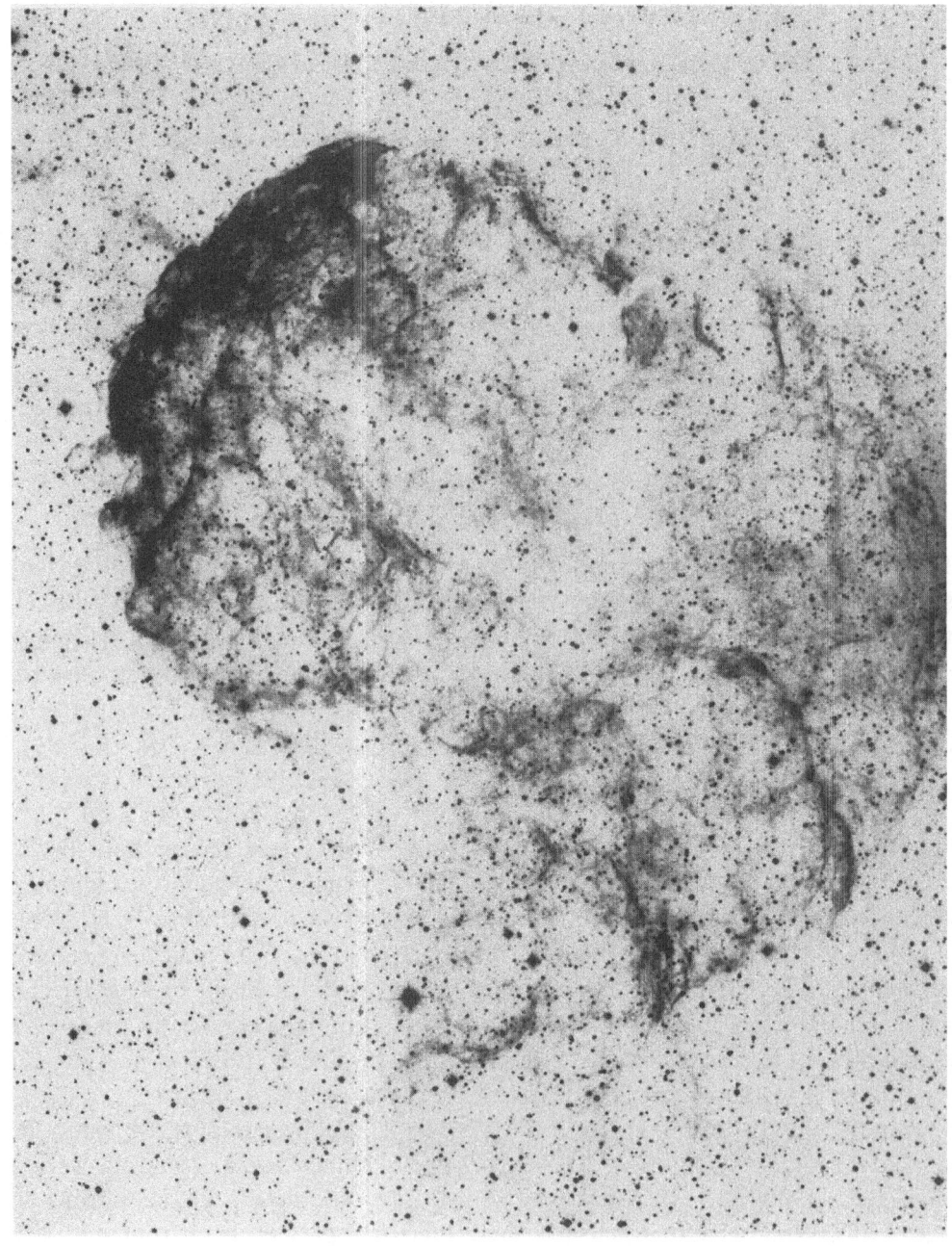

Fig. 1.   A copy of the red Palomar Sky Survey plate of IC 443. The main contribution to the optical picture comes from line radiation (Hα). (Copyright National Geographic Society – Palomar Observatory.)

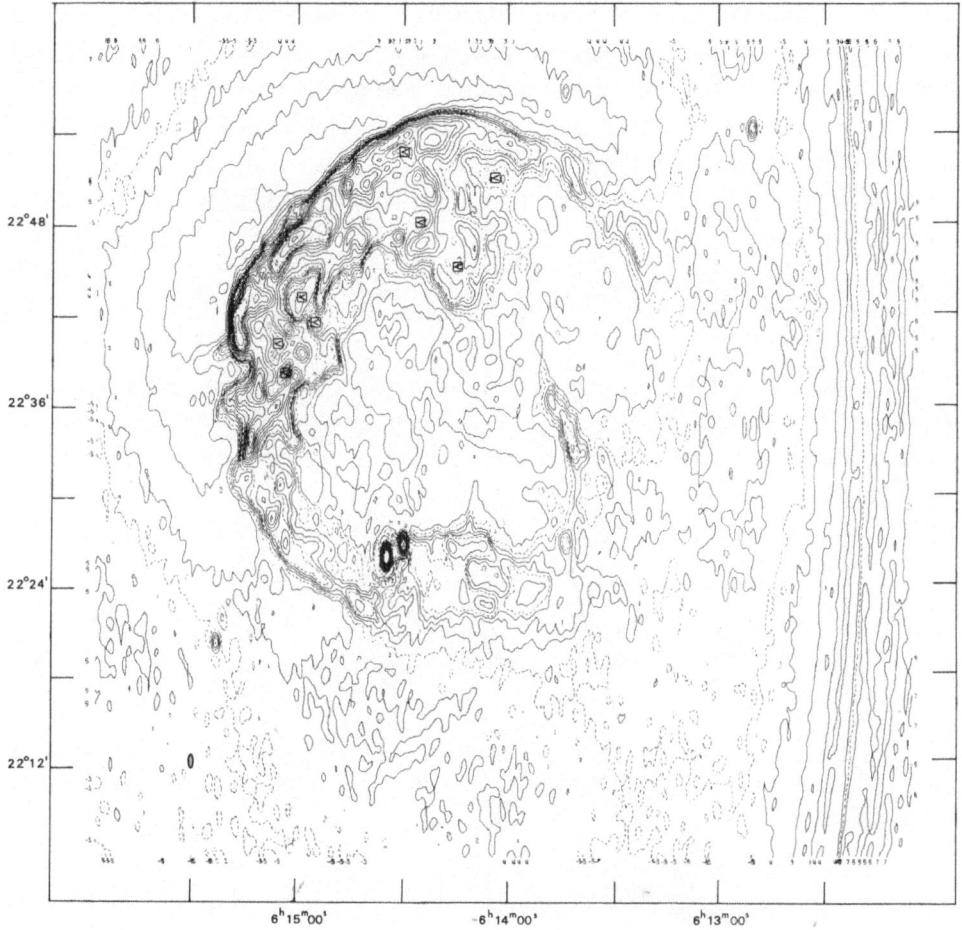

Fig. 2.   Contour map of the total intensity of IC 443 at 21-cm continuum. The outermost dashed contour is the apparent zero level. From that level the contour intervals between pairs of dashed contours are 1.1, 2.2, and 4.4 K, respectively. The symbol '<' enclosed in a square indicates a local minimum. The shaded ellipse in the SE corner represents the half-power beam.

the signal to noise ratio is considerably degraded by primary beam attenuation (the field center is at $RA = 6^h14^m36^s$, $Dec = 22°42'6''$). It should be noted that because of the shortest baseline (36 m) a considerable amount of large scale structure is absent from the map.

Comparing the optical and radio morphology, a great similarity is apparent. This spatial coincidence has been noted by Hill (1971) but our smaller beam confirms it in still greater detail. Such a close correlation would be expected in an H II region, for example, where the (optical) line emission and (radio) free-free emission originate in the same material. It is surprising, however, in a supernova remnant whose radio emission almost certainly originates in a low density gas of relativistic particles radiating in a magnetic field. It is unlikely that the radio emission is thermal in origin:

no H109α recombination line has been detected (Dickel and Milne, 1972) from IC 443, and the spectrum of the northeastern part is even steeper than the decidedly non-thermal spectrum of the remnant as a whole. The observations therefore require a mechanism which can enhance the synchrotron volume emissivity within thermal filaments which emit strong optical radiation.

The passage of a 65 km s^{-1} shock front through the interstellar medium will increase its temperature to 10^5 K and compress it by a factor of four. If this hot gas then undergoes unstable cooling to the observed temperature of 10^4 K, under pressure equilibrium, a density similar to that in the filaments results: 4×10 cm$^{-3} \times 10^5$ K = $= 400$ cm$^{-3} \times 10^4$ K.

If the magnetic field also is compressed during the cooling, then the strength will be increased to a value of the order of 10^{-4} G. Then the synchrotron volume emissivity is sufficiently enhanced to explain the correspondence between the optical and 21 cm continuum features.

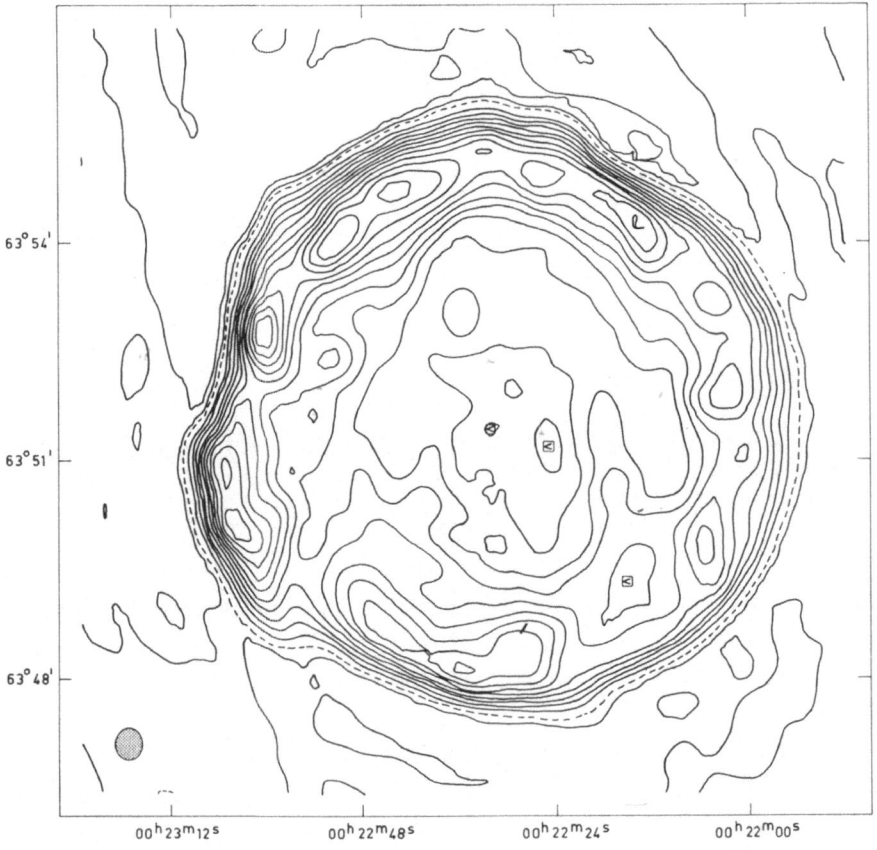

Fig. 3. Contours of total intensity for Tycho's remnant at 21-cm continuum. Above the map zero level, which is dashed, the contour interval is 27.5 K. The symbol '<' enclosed in a square indicates a local minimum. The shaded ellipse represents the synthesized half-power beam.

III. Observations of 3C 10 (Tycho's Remnant)

3C 10 has now been observed at two wavelengths with the WSRT. A 21-cm continuum map (with a $24'' \times 27''$ synthesized beam) of the total intensity and polarized emission has been discussed by Strom and Duin (1973). The total intensity map is reproduced in Figure 3. Among its features are the remarkable circularity of the ring of emission, the steep outer gradients and the fine scale structure. Although the peaks of emission inside the main ring (many of which are not apparent because of the contour interval) are generally weaker than those superimposed on top of it, the number in each group is about the same. Special attention is drawn to the interior ridge at $RA = 0^h22^m19^s$ which extends northwards from the main ring to at least $Dec = 63°52'$.

The 6-cm map (resolution $7'' \times 8''$) is shown in Figure 4, and it is apparent that most of the fine scale structure is resolved, although the outer edge still is unresolved in some places. The aforementioned interior ridge now appears as a string of emission peaks, and many of the other components observed at 21 cm split up into numerous sub-components. In some cases it is clear from the 6-cm map that the more inward position of the outermost peaks in the 21-cm map is the result of the larger beam at the latter wavelength.

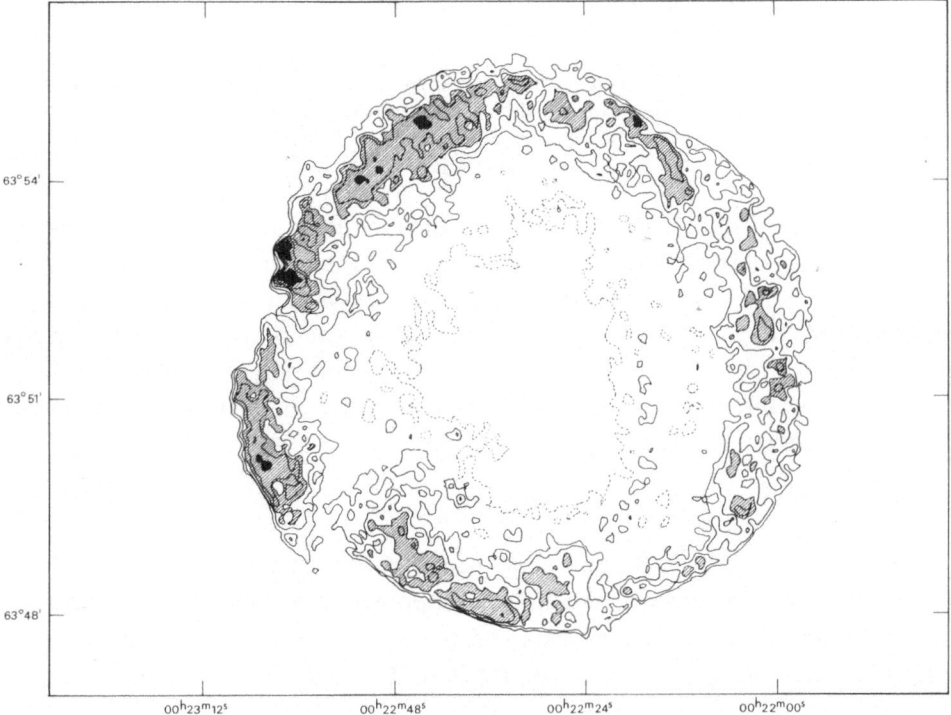

Fig. 4.   Contour map of Tycho's remnant at 6 cm. The resolution is $7'' \times 8''$. For the sake of clarity only some positive contours are shown. Regions above 8.25 K are shown black, while everything above 4.13 K is shaded. The outermost contour is 1.38 K above the apparent zero level (see text). The map is uncorrected for the primary beam, which has a HPBW of 10.6'.

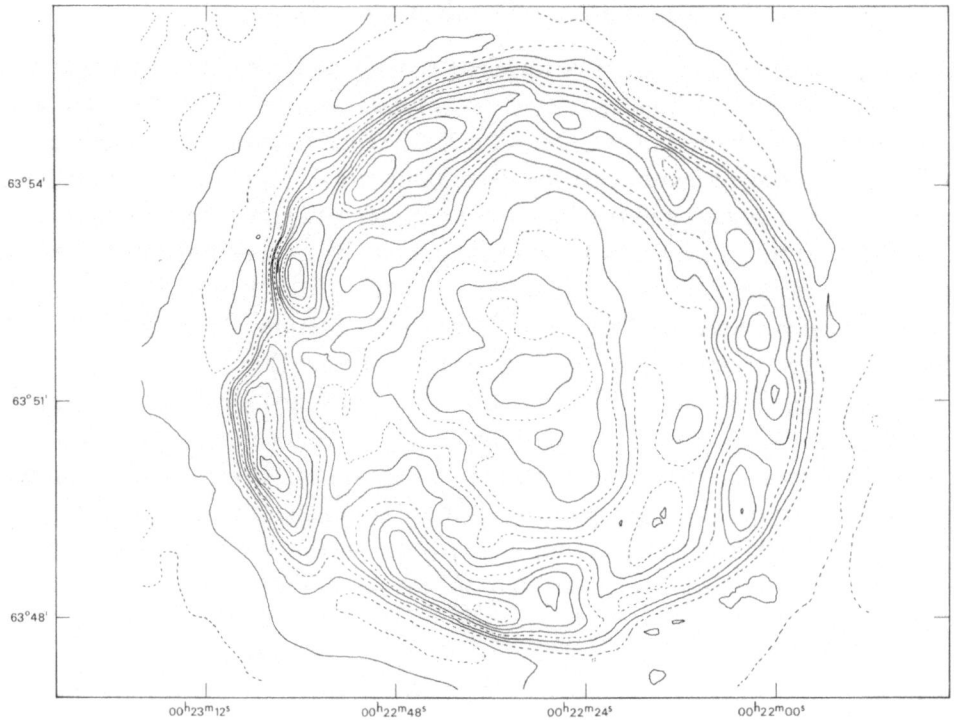

Fig. 5.   Contour map of Tycho's remnant at 6 cm, convolved to the 24″ × 27″ resolution of the 21-cm map. Above the dashed apparent zero level, the first dashed contour that encloses the whole source, the contour interval is 0.67 K. Below the zero level the contour interval is 0.9 K. The map is uncorrected for the primary beam, which has a HPBW of 10.6′.

The 6-cm map is distorted by missing zeroth and first spacings. This causes a negative shift of the contours which is greatest near the remnant center. Apart from this effect, the 6-cm map, when smoothed to the same resolution as the 21-cm map (Figure 5), is virtually indistinguishable from the 21-cm map. This indicates that the spectral index of features down to a scale of 20″ in all parts of the remnant is remarkably constant. We conclude that relativistic particles throughout the remnant must have identical acceleration histories.

Acknowledgements

The Westerbork Synthesis Radio Telescope is operated by the Netherlands Foundation for Radio Astronomy with the financial support of the Netherlands Organisation for the Advancement of Pure Research (Z.W.O.).

R.M.D. acknowledges receipt of a grant from Z.W.O.

References

Davies, J. G., Lyne, A. G., and Seiradakis, J. D.: 1972, *Nature* **240**, 229.
Dickel, J. R. and Milne, D. K.: 1972, *Australian J. Phys.* **25**, 539.
Hill, I. E.: 1972, *Monthly Notices Roy. Astron. Soc.* **157**, 419.

Locke, J. L., Galt, J. A., and Costain, C. H.: 1964, *Astrophys. J.* **139**, 1071.
Lozinskaya, T. A.: 1969, *Soviet Astron.* **13**, 192.
Parker, R. A. R.: 1964, *Astrophys. J.* **139**, 493.
Strom, R. G. and Duin, R. M.: 1973, *Astron. Astrophys.* **25**, 351.

R. M. Duin, R. G. Strom, and H. van der Laan
Sterrewacht, Huygens Laboratorium, Wassenaarseweg 78,
Leiden 2405, The Netherlands

DISCUSSION

Menon: I would like to know whether that steep gradient is resolved or still unresolved at 6-cm resolution?

Habing: It is unresolved.

J. R. Dickel: Hirabayashi's map agrees very closely with the 2-cm map (with 2′ beam) by Mayer and Hollinger (*Astrophys. J.* **151**, (1968), 53). Our 11-cm map of Tycho (Hermann and Dickel: 1973, *Astron. J.* No. 9) shows the same cell structure in the polarization as Strom and Duin find at 21 cm, supporting their conclusion of a thin-screen Faraday rotation in front of the source. 3C 391 is so poorly resolved by us at 8.8 GHz that I would be leery of attaching much significance to any spectral features found by Slee and Dulk based on that map. In CTB 37A the amount of significant fall-off in the spectrum at low frequencies depends critically upon the point at 408 MHz by Kesteven, which has recently been revised downward by D. Clarke. If real, the very uniform absorption mentioned by Slee must have a very abrupt boundary because CTB 37B only 20′ to the north (but presumably more distant based upon the $\Sigma - D$ relation) does not show significant absorption. A uniform cloud which falls abruptly at the edge of a source must lie very close to the source itself.

Shakeshaft: Although we have seen the remarkable similarity of maps of several supernova remnants over very wide ranges of frequency, there is good evidence in IC 443 of changes with frequency, the bright rim at the north-east having a spectrum steeper than that of the rest of the source. (Hill, I. E.: 1973, *Monthly Notices Roy. Astron. Soc.* **164**, 398). Further, there is a good opportunity to obtain spectral information at low frequencies from observations of the lunar occultations occurring during the next year.

J. R. Dickel: But for integrated spectrum be careful of the point sources on the south (as Duin presented to Lecce symposium on supernovae, May 1973) which might be unrelated and of different spectral index.

Kerr: Kundu and Velusamy, at University of Maryland, have done rather detailed polarization observations on several dozen northern sources from Green Bank at 3 and 11 cm. Their work gives results which support Dickel's main conclusions.

Mills: Some of Slee's supernova remnants are very, very small. Could these be extragalactic objects? At very small angular size, it is difficult to decide whether it is a supernova remnant or an extragalactic object. Has there been any independent evidence, say on whether the spectrum is very flat or something like that?

Slee: In most of these cases the spectral index is 0.5 or 0.6 or less. I don't think any polarization measurements have been made on many of them.

Oort: I understood from Dickel's communication that the 'cell' structure he found in the polarization measurements in the Vela SNR could not be due to Faraday rotation. This is interesting in connection with the much more pronounced cellular structure found by Strom and Duin at Leiden in Tycho's SNR at 21 cm and 6 cm. This structure was suggested by the authors and van der Laan to be due to structure in a screen around the radio SNR. Dickel's observations in the Vela object may indicate that the thin shield might not be the correct explanation, but that the structure would be inherent to the shell itself.

Dickel: The Vela structure looks more like a continuous whorl rather than the small discrete cells we see in Tycho. See also my comment on Habing's report of Strom and Duin's work.

Parijskij: I would like to report that observations by Soboleva at Pulkovo at 2 cm with 40″ resolution of the Crab nebula give evidence of the existence of a cavity around the Crab pulsar but of much larger size than we may expect from low frequency radiation pressure.

Jenkins: In connection with our guest investigator program on Copernicus, we have observed through the Vela supernova several stars. They have interstellar line components at about —80 km s^{-1} and — 150 km s^{-1}; I don't think we'll see much difference in the composition of the gas. It's probably mostly interstellar material swept up. We can get a measure of the ionization balance for the gas in this region. We have not yet had a chance to analyze the results, but a quick look seems to show that it is a higher than usual level of ionization.

21-CM LINE ABSORPTION STUDY
OF EXTENDED SUPERNOVA REMNANTS

FUMIO SATO

Chiba Prefecture Education Center, Chiba, Japan

Abstract. By using the Maryland-Green Bank Galactic 21-cm Line Survey, a series of equal-velocity contour diagrams has been constructed for the regions of the extended sources W51 and W44. The nonthermal component of W51 has no absorption features above 53 km s^{-1}, but its thermal components have clear absorption features up to much higher velocities. From this fact, the association of the thermal and nonthermal components of W51 is rejected. Absorption features toward W44 cannot be seen above 50 km s^{-1}. Kinematic distances to the nonthermal component of W51 and to W44 are 4.4 kpc and 3.5 kpc, respectively. From peculiar brightness distributions in the equal-velocity contour diagrams I conclude that a dense H$_I$ cloud, as well as clouds of OH and H$_2$CO, is associated with W44.

Many studies of the neutral hydrogen 21-cm line in absorption have been using profiles in the special directions of continuum sources (e.g., Kerr and Knapp, 1970; Radhakrishnan *et al.*, 1972). They are based on the emission profiles expected in the absence of the sources. But high-resolution observations reveal many small irregularities in the brightness distribution, and it is hardly possible to get plausible expected profiles, with a few exceptions such as pulsars and some strong nonthermal sources. In the cases of extended sources and those with low surface brightness, the expected profile method is particularly ineffective. In order to inspect the absorption features on such sources, the use of an overall view of the sources and their surroundings in the two-dimensional equal-velocity contour diagrams is more effective (Sato, 1968, 1969, 1970, 1973). The high-resolution survey made with the NRAO 300-ft telescope at Green Bank (Westerhout, 1969) is available for the construction of equal-velocity contour diagrams in the regions of the extended sources W51 and W44.

The source W51 consists of several thermal components and an extended nonthermal one. The 21-cm line in absorption due to the thermal components has been studied by Kerr and Knapp (1970) and Radhakrishnan *et al.* (1972). Kinematic distances to the thermal components were estimated from a hydrogen recombination line (Wilson *et al.*, 1970). But no 21-cm line work on the nonthermal component has been done. Some of the constructed equal-velocity contour diagrams for radial velocities of 53 km s^{-1} and lower are very similar to each other and to the map of the continuum intensity. All the components of W51 show clear absorption features at those radial velocities. At higher velocities, however, no absorption feature due to the nonthermal component is seen, while those due to the thermal components are clearly seen. Thus, association of the thermal and nonthermal components is definitely rejected. The distance to the nonthermal component is 4.4 kpc from the Sun, assuming the Schmidt (1965) model for the velocity-distance relation.

The source W44 is a well-known supernova remnant. Some authors (Goss, 1968; Goss *et al.*, 1971; Whiteoak and Gardner, 1972) observed the OH 1720-MHz line in emission and OH and H$_2$CO lines in absorption near the edge of the shell at radial velocities of about 42 km s^{-1}. Goss *et al.* (1971) and Whiteoak and Gardner (1972) suggested that the dense clouds of OH and H$_2$CO are physically associated with the source.

In equal-velocity contour diagrams of the 21-cm line brightness temperature, clear absorption features are seen at $V=10$ to 20 km s^{-1} and $V=28$ to 50 km s^{-1}, but no trace of absorption features is seen above $V=50$ km s^{-1}. From this fact, W44 is located at a distance of 3.5 kpc from the Sun, assuming the Schmidt (1965) model for the velocity-distance relation.

To examine whether a dense H I cloud is associated with any part of the source, expected profiles obtained by interpolation of the absorption dip on the constant-declination scans or by averaging the intensities at several points outside the source are not sufficiently reliable. Therefore, I have constructed expected equal-velocity contour diagrams by using the 1414-MHz continuum survey (Altenhoff *et al.*, 1970) made with the NRAO 300-ft telescope. I add a constant multiple of the continuum brightness to the observed 21-cm brightness, which will give the brightness distribution with the source subtracted, assuming constant optical depth. Figure 1 is such an expected equal-velocity contour diagram for $V=20$ km s^{-1}, assuming a constant optical depth of 0.29 over the source W44. No feature associated with the source is

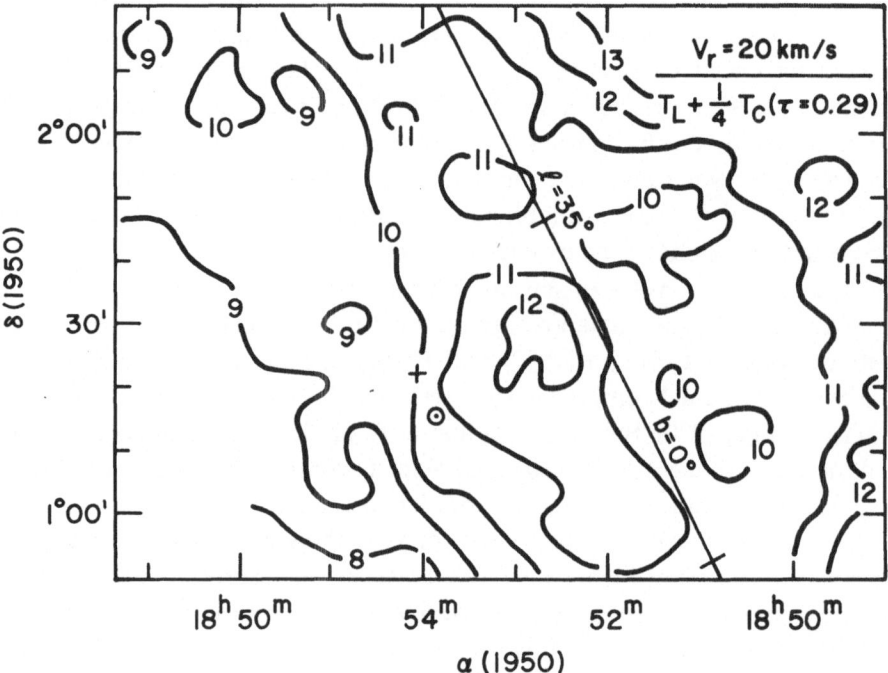

Fig. 1. Equal-velocity contour diagram at $V=20$ km s^{-1} assuming a constant optical depth of 0.29 over the source W44, derived from the 'Maryland-Green Bank Galactic 21-cm Line Survey', 2nd edn. (Wester-hout, 1969) and the 1414-MHz map observed with the NRAO 300-ft telescope (Altenhoff *et al.*, 1970). Contour unit is 5 K in T_a. A circled dot shows the position of the continuum maximum, and a cross shows that of the maximum opacity of the OH 1667-MHz line (Goss *et al.*, 1971).

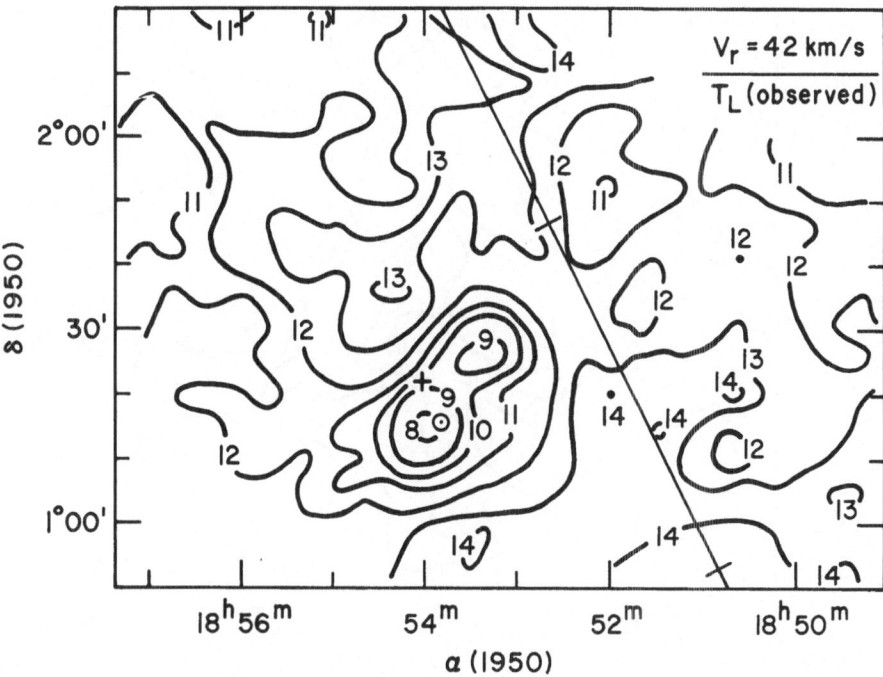

Fig. 2. Equal-velocity contour diagram for a region of W44 at $V = 42$ km s^{-1} derived from the 'Maryland-Green Bank Galactic 21-cm Line Survey', 2nd edn. (Westerhout, 1969). See Figure 1.

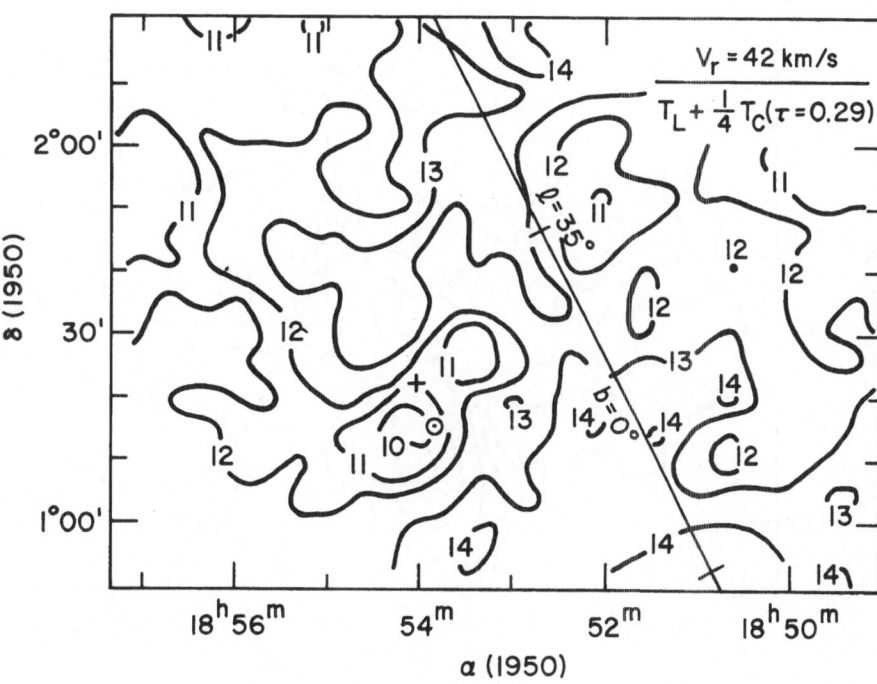

Fig. 3. Equal-velocity contour diagram at $V = 42$ km s^{-1} assuming a constant optical depth of 0.29 over W44. See Figure 1.

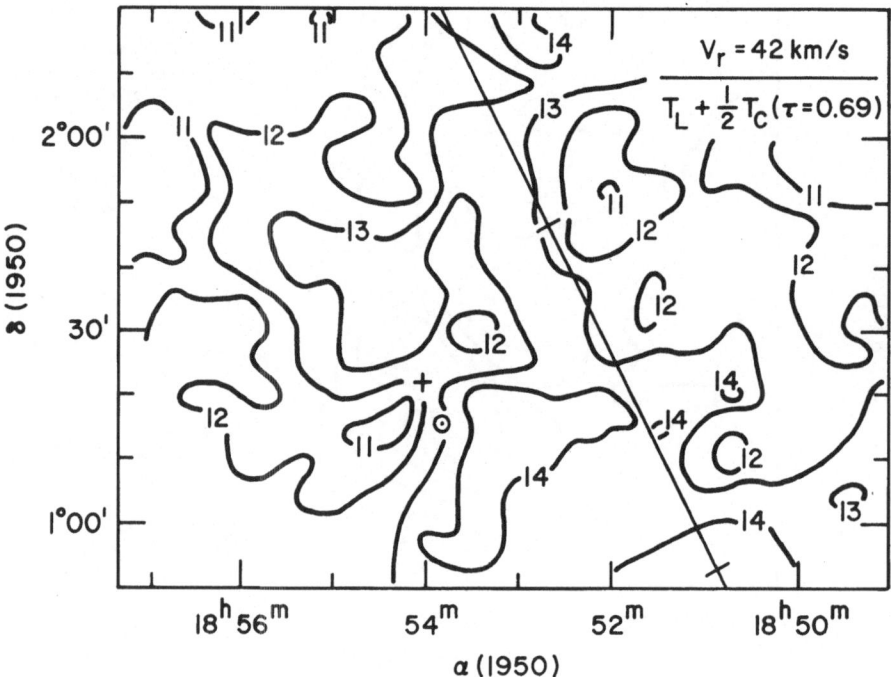

Fig. 4.   Equal-velocity contour diagram at $V = 42$ km s^{-1} assuming a constant optical depth of 0.69 over W44. See Figure 1.

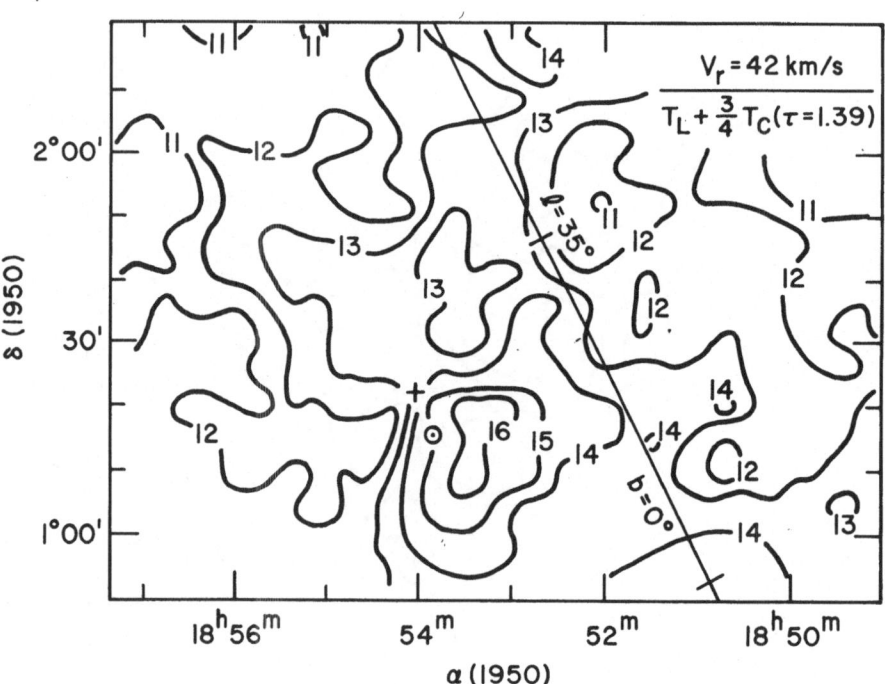

Fig. 5.   Equal-velocity contour diagram at $V = 42$ km s^{-1} assuming a constant optical depth of 1.39 over W44. See Figure 1.

seen, and the smooth distribution of the brightness shows that the assumption of constant optical depth is reasonable.

Figure 3 is an expected equal-velocity contour diagram for $V = 42$ km s^{-1}, assuming a constant optical depth of 0.29. It resembles well the observed intensity contour diagram in Figure 2 and still shows a rather clear absorption feature, but its bottom is displaced to the northeast of the source from the vicinity of the continuum peak. This phenomenon is more clearly seen in Figure 4, which is for the same velocity assuming a constant optical depth of 0.69 over the source. The southwestern part of the source begins to emerge, but in its northeastern part there can be seen a trough along the shell through the region of large optical depth of the absorbing OH and H_2CO. In Figure 5 for $V = 42$ km s^{-1}, assuming a constant optical depth of 1.39, the continuum emission of the source added to the observed 21-cm line intensity is quite strong, and the western half of the source appears. In the eastern half, however, the trough is barely filled, and there is no enhancement of emission due to the added continuum radiation of the source. These facts imply that at $V = 42$ km s^{-1}, H I gas in the southwestern part of W44 has a small optical depth of 0.29 or so, but in the northeastern part its optical depth is as large as 1.39 or so. Such features are seen also in the contour diagram for the velocity range from 40 to 46 km s^{-1}. The dense H I gas agrees well in position and velocity with OH and H_2CO concentrations which have been suggested to be associated with the source (Goss *et al.*, 1971; Whiteoak and Gardner, 1972). Thus, I have come to the conclusion that a dense H I gas cloud with a radial velocity of 40 to 46 km s^{-1} is associated with the shell of the source W44.

References

Altenhoff, W. J., Downes, D., Goad, L., Maxwell, A., and Rinehart, R.: 1970, *Astron. Astrophys. Suppl.* **1**, 319.

Goss, W. M.: 1968, *Astrophys. J. Suppl.* **15**, 131.

Goss, W. M., Caswell, J. L., and Robinson, B. J.: 1971, *Astron. Astrophys.* **14**, 481.

Kerr, F. J. and Knapp, G. R.: 1970, *Australian J. Phys. Astrophys. Suppl.*, No. 18, 9.

Radhakrishnan, V., Goss, W. M., Murray, J. D., and Brooks, J. W.: 1972, *Astrophys. J. Suppl.* **24**, 49.

Sato, F.: 1968, *Publ. Astron. Soc. Japan* **20**, 303.

Sato, F.: 1969, *Ann. Tokyo Astron. Obs. Second Ser.* **11**, 67.

Sato, F.: 1970, *Ann. Tokyo Astron. Obs. Second Ser.* **12**, 1.

Sato, F.: 1973, *Publ. Astron. Soc. Japan* **25**, 135.

Schmidt, M.: 1965, in A. Blaauw and M. Schmidt (eds.), *Galactic Structure*, University of Chicago Press, Chicago, p. 513.

Westerhout, G.: 1969, 'Maryland-Green Bank Galactic 21-cm Line Survey', 2nd edn., University of Maryland, College Park, Maryland.

Whiteoak, J. B. and Gardner, F. F.: 1972, *Astron. Astrophys.* **21**, 159.

Wilson, T. L., Mezger, P. G., Gardner, F. F., and Milne, D. K.: 1970, *Astrophys. Letters* **5**, 99.

Fumio Sato

Chiba Prefecture Education Center,
2-10-1 Katsuragi, Chiba, Japan

DISCUSSION

J. R. Dickel: Maps of the 21-cm line at various radial velocities in the vicinity of IC 443 show a hole right at the position of the source at all velocities, which suggests that the SNR is at or behind all the significant hydrogen in that direction. Depending on how one continues the Perseus arm around to that area through the anticenter of the Galaxy, we can obtain various estimates of the distance, but the closest values are about 2.2 kpc (using Roberts' shock model). This would put the distance well beyond that of the H II region to the northeast of IC 443, which several people have suggested is responsible for the retardation of the shell's expansion. (The distance estimate of 550 pc is by Racine from photometry of a B8 star which illuminates the reflection nebula.) At positive velocities the neutral hydrogen appears to be part of the general galactic emission, but at $V = -6$ to -10 km s^{-1} there is a definite cloud to the northeast of IC 443, and it is probably this which retards the expansion and may make it old enough to match the pulsar age.

Guélin: Minkowski has derived a distance of $\gtrsim 2$kpc for IC 443 from the motion of the filaments. (*IAU Symp.* **9** (1959), 315.)

Mills: The distance of IC443 on the $\Sigma - D$ relation I presented also is about 2–5 kpc. It would be hard to accept 550 pc.

Caswell: How do you determine a kinematic distance for the neutral hydrogen in the direction of IC 443, near the anticenter?

Dickel: From the shock model by Roberts (*Astrophys. J.* **173** (1972), 259), who continued the Perseus arm around to this direction.

THE BIRTHPLACES OF PULSARS

A. G. LYNE

University of Manchester, Nuffield Radio Astronomy Laboratories, Jodrell Bank, Cheshire, United Kingdom

Abstract. With the refinement of the estimates of electron density in the Galaxy it is clear that pulsars have z distances away from the galactic plane of a few hundred parsec. There is now a very strong indication that young pulsars lie at small z distances, suggesting that pulsars originate in a small z population and are given velocities of about 100 km s^{-1} at birth.

Earlier in this symposium it was demonstrated that the z distribution of pulsars is well-defined and substantially independent of distance from the Sun out to at least 5 kpc. The electron distribution is smooth enough for the dispersion measure to be quite a good measure of distance, and the mean electron density in the plane cannot be far different from 0.03 cm^{-3}. Figure 1 shows the distribution of pulsars off the plane in units of dispersion measure. The mean distance off the plane is about seven of these units, or 230 pc using a value of 0.03 for the mean electron density.

How did they get there? Do pulsars originate from a stellar population having a similar distribution to the one we see now, or from a much narrower one? The ideas

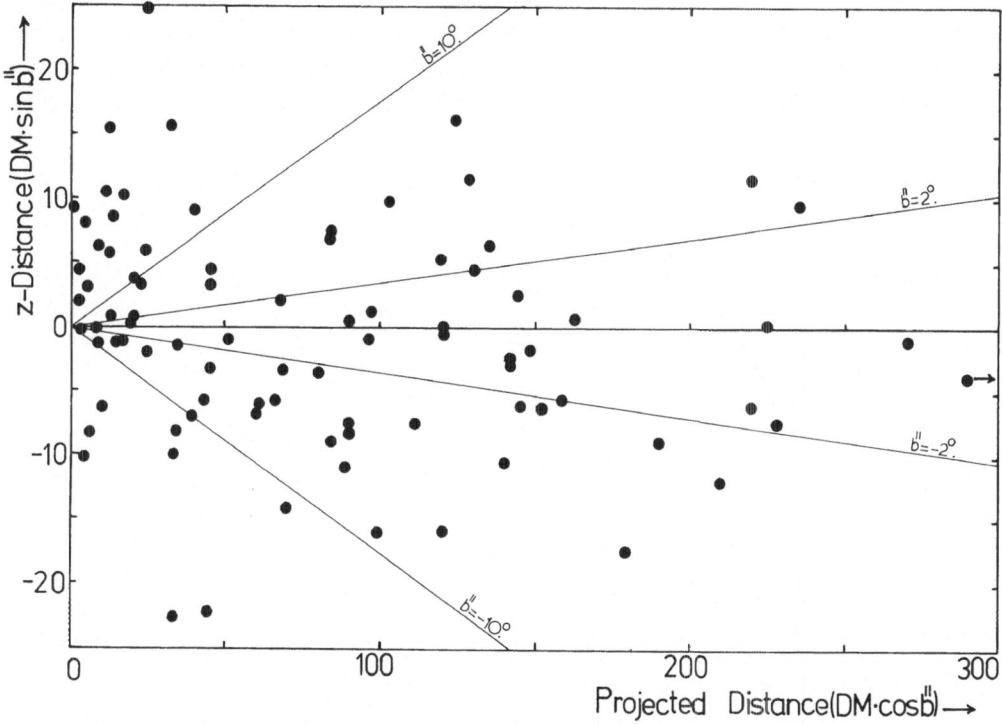

Fig. 1.   The derived z distances of pulsars plotted against the distances along the galactic plane from the Sun. The units of both quantities are in units of dispersion measure, pc cm^{-3}.

F. J. Kerr and S. C. Simonson, III (eds.), Galactic Radio Astronomy, 369–372. All Rights Reserved.
Copyright © 1974 by the IAU.

I am going to present are not entirely new (see, for instance, Gold and Newman, 1970 and Prentice, 1970) but I think we can now investigate them more critically with the present data.

The notion that pulsars originated from very close to the plane can be investigated from the ages of pulsars, young pulsars appearing closer to the plane than old ones. The age as measured from the derivative of the period is available at the moment only for a minority of pulsars. However, we can use period as a crude measure of age. If we plot a pulsar in a graph of period, P, against time, t, it will follow one of a family of curves. These curves will be parabolae in the case of magnetic braking and will take the general form $P \propto t^{1/2}$ as indicated in Figure 2a. Which of the curves a pulsar follows is determined by the strength of the braking.

If the pulsar at birth is given a kick, for instance due either to the disruption of a binary system or an asymmetric supernova explosion, the velocity normal to the plane will remain essentially constant during the lifetime of the pulsar, since this lifetime of about 10^7 yr is small compared with the period of oscillation across the plane (Oort, 1965). Thus z would be a crude measure of time, the crudeness depending upon the dispersion in velocity and direction of the velocity.

Because of the dispersion in braking, velocity and direction we should not expect to find a very strong relationship. Whether the effect exists can be judged from Figure 2b, which shows period plotted against z distance for the 100 pulsars which have reliable dispersion measurements. The mean z of the dozen or so shortest period objects is about 3 pc cm^{-3} while that of the rest is something over 7 pc cm^{-3}. I think that this lends some support to the model I have described, but the evidence is by no means concrete. By early next year, period derivatives for the majority of pulsars

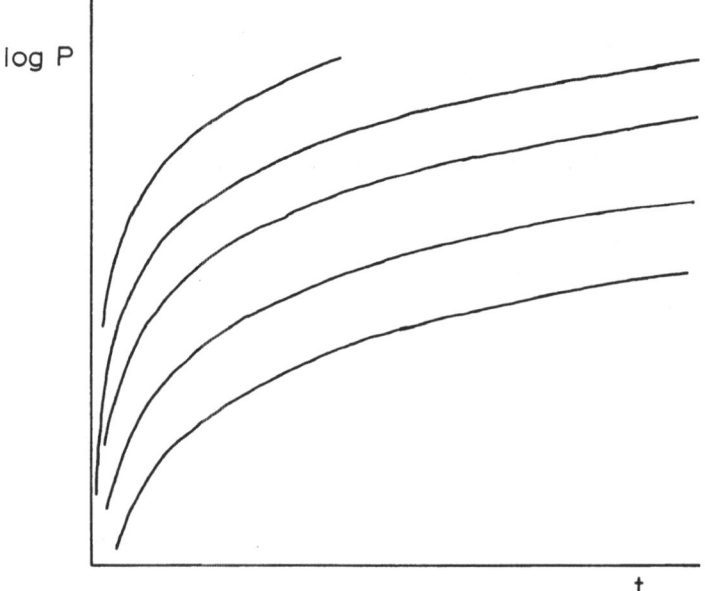

Fig. 2a.   A schematic evolution of the periods of pulsars with time.

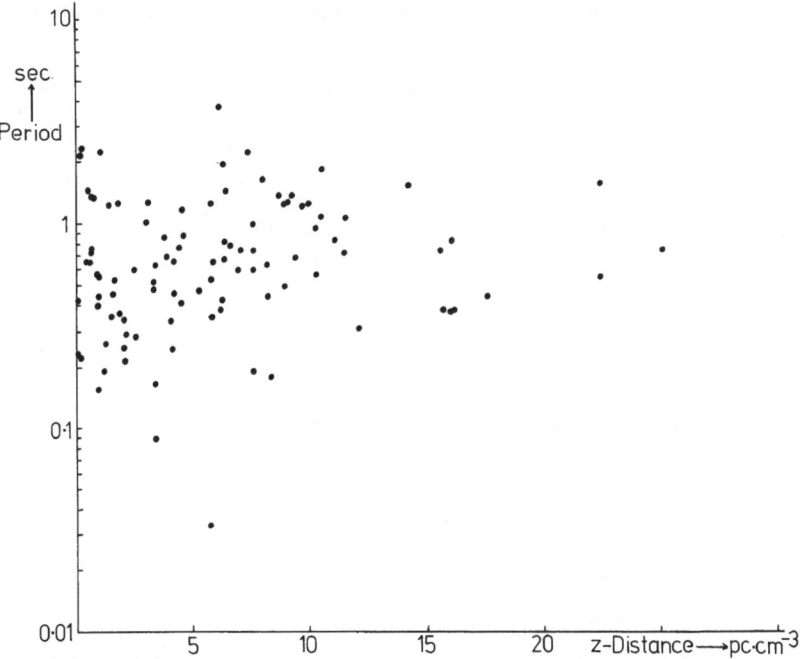

Fig. 2b.   The observed distribution of pulsars in period and z distance.

should be available and this should remove one of the main decorrelating influences in this diagram.

This suggests that the pulsars are born in a population having a scale height of very much less than the estimated 230 pc of the observed distribution. Possibly they originate in population I objects. Supernova remnants (Poveda and Woltjer, 1968) as well as many other populations have scale heights of less than 100 pc. The distribution in longitude lends some support to this.

Let me pursue the argument further and now assume that pulsars are born on the plane and attempt to compute the velocity they must have in order to have reached their present heights off the plane during their lifetime as pulsars. The mean age of the two dozen or so objects with measured derivatives is about 3×10^6 yr. This leads to an estimate of the mean velocity normal to the plane of 70 km s^{-1}. Provided that the velocity distribution is isotropic in space, then from the Earth we will observe pulsars to have velocities normal to the line of sight of 95 km s^{-1} and total mean velocities of 125 km s^{-1}.

That pulsars may have high velocities has been suggested by Gunn and Ostriker (1970) and by Prentice (1970) and others. The velocities of the order of 100 km s^{-1} suggested here are an order of magnitude lower than suggested by these authors. However, they are consistent with estimates of velocity from observations of the interstellar scintillation patterns (Rickett, 1970; Downs and Reichley, 1971; Ewing *et al.*, 1970; Lang and Rickett, 1970; Galt and Lyne, 1972).

The present indications are therefore that the pulsars originate in a small *z* distance galactic population, possibly type II supernovae, and are given velocities of the order of 100 km s^{-1} at birth.

References

Downs, G. S. and Reichley, P. E.: 1971, *Astrophys. J. Letters* **163**, L11.
Ewing, M. S., Batchelor, R. A., Friefield, R. D., Price, R. M., and Staelin, D. H.: 1970, *Astrophys. J. Letters* **162**, L169.
Galt, J. A. and Lyne, A. G.: 1972, *Monthly Notices Roy. Astron. Soc.* **158**, 281.
Gold, T. and Newman, H. M.: 1970, *Nature* **227**, 151.
Gunn, J. E. and Ostriker, J. P.: 1970, *Astrophys. J.* **160**, 979.
Lang, K. R. and Rickett, B. J.: 1970, *Nature* **225**, 528.
Oort, J. H.: 1965, *Stars and Stellar Systems* **5**, 455.
Poveda, A. and Woltjer, L.: 1968, *Astron. J.* **73**, 65.
Prentice, A. J. R.: 1970, *Nature* **225**, 438.
Rickett, B. J.: 1970, *Monthly Notices Roy. Astron. Soc.* **150**, 67.

A. G. Lyne
University of Manchester,
Nuffield Radio Astronomy Laboratory,
Jodrell Bank, Cheshire SK11 9DL, United Kingdom

DISCUSSION

Sutton: For the 24 or so pulsars with measured dP/dt, have you made any preliminary diagrams of age vs z distance?

Lyne: No; however, a few months ago with 18 pulsars there was only a very weak effect in such a diagram.

Milne: When you extrapolate these values back toward the plane, do you find a supernova remnant there? For example, IC 443 – is the pulsar moving in the right direction?

Lyne: From this data all we know is the component of velocity normal to the plane, and normal to the line of sight. We don't know what the other two components are, so it is difficult to do that. I'm not saying that the velocities are always moving away from the plane. The simple assumption that I've made here is that they were born at zero height off the plane. In fact, they are presumably born a few tens of parsecs away from the plane.

Smith: You might make it clear though, that it is possible to measure the velocities, including the velocity vector. Exactly what you suggest could be done for IC 443. I made that remark earlier on.

Van Woerden: Goss and Schwarz (*Nature Phys. Sci.* **234** (1971), 52) have shown that the weak extended radio source near pulsar CP 1919 has a shell structure. They now find, from more sensitive observations, a second shell source overlapping the first and again having the pulsar close to its edge. The discrepancies of age and of distance remain as severe as they were. Maybe several events have happened in this region?

NEW RESULTS ON SUPERNOVA REMNANTS*

I. S. SHKLOVSKII

Sternberg Astronomical Institute, Moscow, U.S.S.R.

Abstract. Estimates of the masses and energies of the supernova explosions that resulted in the Cygnus Loop and similar remnants are revised downward on the basis of recent soft X-ray observations.

(The following summary of Parijskij's report was prepared from the tapes by the Editors.)

I would like to just give some information. Shklovskii (1974) has taken a new point of view about supernova remnants.

Very roughly, the main result of his new research was connected with recent soft X-ray observations. He tried to recalculate once more the mass and energies of, first of all, the Cygnus Loop remnant. His main conclusion is that the old values for mass and energy were overestimated. In the real case the energy is closer to the case of a type I supernova, about 10^{50}–10^{51} erg, and the mass is comparable with 1 $M_\odot$. Moreover, he tried to consider other supernovae of that type, IC 443, Tycho, and Vela 10, for example. His feeling is that they are also closer to type I supernovae. The masses of the envelopes of the Cygnus Loop and also these supernovae are ~ 1 $M_\odot$. Practically all these sources have some kind of pulsar inside, a radio pulsar or an X-ray point source, which Shklovskii suggests is just the situation when the axis of the magnetic field of the pulsar or neutron star is oriented in the proper way. So by a combination of these two facts he came to the conclusion that the total mass of the exploding original star is not very high, a few solar masses, and again it supports his feeling that for the majority of well-known supernova remnants, the supernova belongs to type I, or closer to type I than type II.

There is a by-product of this consideration. He stresses greatly that there should be a strong selection effect. According to his approach the lifetime of supernova shells or remnants which exploded in dense regions of the Galaxy will be larger than those sources which exploded at higher latitude. So the statement that many supernova shells belong to the extreme flat population of original stars should be considered once more with that kind of selection effect taken into account.

Reference

Shklovskii, I. S.: 1974, *Astron. Zh.* **51**, 3 (= *Soviet Astron.* **18**, 1).

I. S. Shklovskii
Sternberg Astronomical Institute,
Prospect 13,
Moscow, U.S.S.R.

* Reported by Yu. Parijskij.

F. J. Kerr and S. C. Simonson, III (eds.), Galactic Radio Astronomy, 373–374. All Rights Reserved.
Copyright © 1974 by the IAU.

DISCUSSION

Mills: In preparing my review I looked for some indication of differences between the radio properties of type II and known type I SN, but could not find anything significant. However, if all those suggested by Shklovskii are actually type I, I note that these are all well below the average $\Sigma - D$ relation.

PART 4

STELLAR AND CIRCUMSTELLAR SOURCES

"I asked for the physics, not the empirical model."
G. Westerhout, in the discussion following the paper
by Yervant Terzian

RADIO CONTINUUM OBSERVATIONS OF
STELLAR SOURCES

L. L. E. BRAES

Sterrewacht, Leiden, The Netherlands

Abstract. Thirty optically visible stellar sources have been observed in the radio continuum: four red dwarf flare stars, three novae, two red supergiants, eight binary systems, and 13 related peculiar objects. The observations of red dwarfs, novae, and supergiants are briefly reviewed. Their emission seems reasonably well understood. Binary systems such as Algol show erratic flaring; consistent explanations have not yet been given. The related peculiar objects, such as V1016 Cyg and MWC 349, are even less well understood. Some of them may have just arrived on the main sequence; others may be planetary nebulae in an early stage of formation.

(The following summary was prepared from the tapes by the Editors.)

I. Introduction

In this review on stellar radio sources, I shall confine myself to radio continuum observations of optically visible stellar sources. Apart from the Sun, 30 such objects have been detected (see Table I): four red dwarf flare stars, three novae, two red supergiants, eight binary systems, and 13 that I call related peculiar objects.

For only 22 objects, however, can the detection be regarded as definite. For these objects there is either a close positional coincidence, say to within a few arcseconds with the radio source, or there is a correlation between the radio and optical variability. For the eight remaining sources there is strong evidence that they are radio emitters, but definite detection must await accurate measurements of the radio position. These eight objects have so far only been observed with single dish instruments, and their positional accuracy is not high enough to conclude that there is definite radio detection.

The radio emission from all these objects is almost always very faint and highly variable. Their detection has not only been a matter of using high sensitivity instruments with high resolving power, but it has also been a matter of luck and patience. Many more objects have been looked at, of course, at radio wavelengths, but with negative results. In particular, people have looked at early-type stars, Of stars, Wolf-Rayet stars, magnetic variables, but all with negative results. I know of 50 published negative results, but I estimate the total number of negative results to be at least 150. That indicates radio stars might be quite rare, at least down to current sensitivity levels, say 5 mJy. You will see that all the radio stars detected up to now are real pathological cases. You will find most of them discussed in Section IV of the *Eighth Liège Astrophysical Colloquium on Emission-Line Stars*, entitled, 'Étoiles particulières et anormales'.

F. J. Kerr and S. C. Simonson, III (eds.), Galactic Radio Astronomy, 377–381. All Rights Reserved.

II. Flare Stars, Novae, and Supergiants

The red dwarf flare stars detected so far are located in the solar neighborhood, say
within 10 pc from the Sun. Observations of the radio emission of these stars have
been made from Jodrell Bank and in Australia since 1958. Several cases of simultaneous optical and radio flares have been observed, but spectral information is
still limited. In order to find out what the mechanism responsible for the radio emission is, many more measurements of flares at different frequencies are required. We
may be dealing here with nonthermal phenomena like radio bursts on the Sun. For
a review paper on radio observations of these objects I refer to Wilson (1971).

The novae have been studied extensively by the NRAO group. The spectra they
obtained are entirely consistent with thermal bremsstrahlung from the expanding
ionized nova envelopes. These novae are easily detectable because the envelopes
quickly attain angular sizes of the order of a few tenths of an arcsecond while
remaining opaque to radio waves (see, e.g., Herrero *et al.*, 1971).

The red supergiant Betelgeuse, α Orionis, was first detected by Kellermann and
Pauliny-Toth (1966) at 15.8 GHz. They found this result only on one night; on the
11 following nights it was not detected. Low also discovered intense and variable
infrared emission from that star. Later, Seaquist (1967), using the Algonquin Radio
Observatory telescope concluded that he may have detected radio emission from α Ori

TABLE I

Stellar sources detected at radio frequencies

Red dwarf flare stars		
UV Cet (dM5.5e)	YZ CMi (dM4.5e)	
V371 Ori (dM3e)	EV Lac (dM4.5e)	
Novae		
FH Ser (N Ser 70)	HR Del (N Del 67)	
V368 Sct (N Sct 70)		
Red supergiants		
α Ori (M2 Iab)		
π Aur (M3 II)		
Peculiar binary systems		
β Per (B8 V + G8 III)	β Lyr (Bpe + B7 V)	
b Per (A2 III + ?)	Cyg X-1 (B0 Ib + ?)	
α Sco (M1 Ib + Be)	AR Lac (K2 III + F8)	
RY Sct (B0ep + late)	R Aqr (M7e + Pec)	
Related peculiar objects		
HD 37806 (A0e, Dn)	V1016 Cyg (e [ ], D)	MWC 137 (Be [ ], Dn)
M1–11 (e [ ], D)	P Cyg (B1e, F)	
Sco X-1 (Pec)	MWC 349 (e [ ], D)	
M2–9 (e [ ], D)	HBV 475 (Pec)	
HD 167362 (WC [ ], Dn)	Cyg X-2 (sdGpec)	
Vy2-2 (e [ ], D)	LkHα 101 (e [ ], Dn)	

and π Aur. These attempts at detection were made following a suggestion by Weymann and Chapman that the free-free emission from the corona formed by ejected matter from these stars might be detectable at radio frequencies. The stars show circumstellar lines in their spectra, indicating expanding gas over several hundred stellar radii.

Recently, Altenhoff and Wendker (1973) using the Bonn 100-m dish detected α Ori at a flux density of 7 ± 1 mJy. Efforts to detect the star with the Westerbork Synthesis Radio Telescope at 1.4 GHz and with the Green Bank three-element interferometer at 2.7 and 8.1 GHz did not yield a positive result. So strictly speaking, it remains uncertain whether α Ori has been detected at radio wavelengths. Certainly it remains uncertain whether it is a variable radio star as suggested by the data. Maybe it occasionally undergoes brief radio flares.

III. Binary Stars

I've also listed all the radio stars that are known to be binary. For all these binaries there are indications of mass loss or mass exchange in the system. It is tempting to believe there is a relationship between such gas-streaming phenomena and the production of radio emission. The radio emission from most of these objects is very faint.

Only Algol, β Persei, is bright enough to be studied in some detail. This has been done at Green Bank. It appears from the Green Bank observations that the radio behavior of Algol is erratic, showing both long quiescent periods and occasional periods of flaring. During quiescent periods the spectrum remains flat, but during most of the flares it increases at the higher frequency. This led Hjellming (1972) to propose that these flares are due to bremsstrahlung from a partially self-absorbed hot plasma with the following parameters: $T_e \sim 10^8$ K, $n_e \sim 10^{10}$ cm^{-3}, $D \sim 0.3$ AU. The value of the diameter he gets, 0.3 AU, is of the order of the separation of the close binary pair of the system (Algol is a triple system consisting of A B, and C components). This indicates that the radio emission originates from that close pair and not from the C component. Pooley and Ryle (1973), using the Cambridge 5-km telescope, came to the same conclusion. They found a variation of the right ascension of the radio source with time, suggesting orbital motion with respect to the C component.

There's one difficulty with the Hjellming model, and that's the high temperature of 10^8 K. From such high temperatures one would expect Algol to be an X-ray source during the radio flares, and that has not been observed yet. So this failure to detect Algol in the X-ray region, together with the occurrence of occasional nonthermal events, led Canizares et al. (1973) to reject Hjellming's model. They suggest that most of the radio emission must be ascribed to complex nonthermal phenomena.

Since its detection in October 1971, Algol has shown progressively stronger and more frequent radio flaring. Hjellming et al. (1972) suggest there may be a connection between this recent radio activity and a change in period of the binary system. The period of Algol is variable. It is known that period changes occur in the system on the average once every 25 yr. Hjellming believes that these period changes indicate sudden mass losses in the system, what he calls 'starquakes'. He believes that the radio flares we are detecting are the direct result of such starquakes. He suggests the rare occurrence of such activity might explain why radio emission has been detected in only a few objects out of roughly 50 binaries observed at Green Bank. There are more objects of variable period in Table I, and from all these objects Hjellming found radio emission similar to that from Algol.

Antares, α Scorpii, is a quite different system. It's clearly variable at 8.1 GHz, but at lower frequencies there is no significant, if any, change with time. So maybe here we are dealing with a two-component radio source, one component due to the envelope surrounding the supergiant in the system and a variable component associated with the early-type star. We know from the Green Bank observations that this 8.1 GHz emission originates from this B component. RY Scuti and R Aquarii may be similar systems, though Hjellming et al. (1973) prefer to call them radio nebulosities because of the lack of short time scale flux variations. They may, however, represent the same case as α Sco, with constant radio emission from a circumstellar envelope and variable radio emission at higher frequencies from the binary system.

IV. Related Peculiar Objects

The related peculiar objects include the well-known X-ray sources Sco X-1 and Cyg X-2, nova-like variables like P Cygni, and so-called stellar planetaries, mainly from Minkowski's lists, Minkowski 1–11, 2–9, etc. These objects show the spectral characteristics of early-type emission-line stars and compact planetary nebulae. The list also contains very peculiar systems like V1016 Cygni, which in the literature has been described as a symbiotic star, a slow nova, or a planetary nebula in an early stage of formation. Most of these stars show forbidden lines in their spectra. Most of them occur in Allen's (1973) infrared catalog of early-type emission stars, from which the types in Table I have been taken. 'F' means that the observed infrared color indices fall close to the range of values expected from an early-type star combined with optically thin free-free radiation from the star's ionized hydrogen envelope. 'D' means that the color indices are too large to be accounted for by hot free-free radiation. The IR excess is ascribed to re-radiation from circumstellar dust clouds. The 'n' means that there is a visible nebulosity associated with the star. Probably we are dealing here with objects that have just arrived on the main sequence and are still surrounded by part of the parent interstellar material.

The best-studied object is MWC 349 (Greenstein, 1973). According to Olnon (preprint) its radio spectrum ($\alpha \sim 0.8$) can be explained by adopting an inhomogeneous density distribution. With a distance of 2.1 kpc and $T_e = 10^4$ K, he finds $n_e = 0.73 \, r^{-2.15}$ cm^{-3}, where r is the distance from the center of the object in pc. He estimates the observed radio flux originates in layers at a distance of more than 0.002 pc from the center, radiation from the deeper levels being effectively absorbed. So we do not get any information about the density inside the sphere with that radius. But using the volume emission measure computed from the Hβ flux he derives a mean electron density of 7×10^6 cm^{-3}.

References

Allen, D. A.: 1973, *Monthly Notices Roy. Astron. Soc.* **161**, 145.

Altenhoff, W. J. and Wendker, H. J.: 1973, *Nature* **241**, 37.

Canizares, C. R., Neighbours, J. E., Clark, G. W., Lewin, W. H. G., Schnopper, H. W., and Sprott, G. F.: 1973, *Astrophys. J. Letters* **183**, L91.

Greenstein, J. L.: 1973, *Astrophys. J. Letters* **184**, L23.

Herrero, V., Hjellming, R. M., and Wade, C. M.: 1971, *Astrophys. J. Letters* **166**, L19.

Hjellming, R. M.: 1972, *Nature Phys. Sci.* **238**, 52.

Hjellming, R. M., Webster, E., and Balick, B.: 1972, *Astrophys. J. Letters* **178**, L139.

Hjellming, R. M., Blankenship, L. C., and Balick, B.: 1973, *Nature Phys. Sci.* **242**, 84.

Kellermann, K. I. and Pauliny-Toth, I. I. K.: 1966, *Astrophys. J.* **145**, 953.

Pooley, G. G. and Ryle, M.: 1973, *Nature* **244**, 270.

Seaquist, E. R.: 1967, *Astrophys. J. Letters* **148**, L23.

Wilson, A. J.: 1971, *IAU Colloq.* **15**, 114.

L. L. E. Braes
Sterrewacht, Huygens Laboratorium, Wassenaarseweg 78,
Leiden 2405, The Netherlands

DISCUSSION

Hughes: I can confirm that our observations to date on RY Sct at 2.8 cm show that the radio emission is constant to within about 20%. There was a numerical error in one of the values for flux density as listed in a recent publication in *Nature*. Some caution needs to be exercised when selecting radio sources according to their optical properties. We have observed a number of stars for radio emission at 2.8 cm. Under one classification, we chose Bep stars showing P Cyg profiles and an infrared excess, as contained in the list by Geisel. This included β Lyr, MWC 349 and RY Sct. β Lyr shows large radio variability and small angular size, while the latter two are constant to within about 20%, the radiation originating in a radio nebulosity which surrounds the star. Clearly we have two types of radio source.

Macrae: A comment about Algol. If one looks into the history books one finds a very surprising thing – that this star was not known to be variable in light until about 400 yr ago. This was only about 150 yr before Goodricke's recognition of Algol as an eclipsing pair. β Per was a well-known star in all of the principal constellations and there had been many assiduous observers over the centuries. Perhaps we have evidence that pronounced changes took place in Algol as recently as four centuries ago. If so, the time scale of the phenomena we are discussing may be shorter than generally assumed.

Van Woerden: If MacRae's suggestion is correct, then certainly the same case should be made for Mira Ceti, which at maximum is roughly as bright as Algol but has a much larger amplitude, and still was only discovered as variable around 1600. I doubt whether the arguments for profound changes around 1600–1700 would be strong. In medieval (and earlier?) times one tended to believe that the stellar heavens were unchangeable. The discovery of variable stars after 1600 might rather be due to the breakthrough of an experimental approach to physical science.

STELLAR X-RAY SOURCES

HARVEY D. TANANBAUM*

American Science & Engineering, Cambridge, Mass., U.S.A.

Abstract. Data are presented for Cygnus X-1, Cygnus X-3, and Scorpius X-1 from radio to X-ray wave-lengths. The evidence for Cygnus X-1's being a black hole is now quite convincing. New data for Cygnus X-3 show the presence of X-ray activity at the time of the giant radio outburst. The data for Scorpius X-1 show a close correlation between the X-ray and optical behavior, but coverage was not sufficiently complete to fully assess the relationship (if any) between the X-ray and optical emission and the radio emission.

I. Introduction

At the IAU General Assembly the subject of binary X-ray sources was reviewed. The prevalence of intensity variations often on time scales as short as 0.1 s was established for the galactic X-ray sources, and the evidence for the binary nature of the galactic sources was presented.

For this meeting I have chosen for discussion three galactic X-ray sources: Cygnus X-1, Cygnus X-3 and Scorpius X-1. The reasons for choosing these sources were that all three have radio counterparts, all three X-ray and radio sources have been fairly extensively studied, and the three objects have different appearances. Without attempting to link the behavior of these three sources, what I would like to do here is to look at them with the objective of assessing whether the radio emission and the X-ray emission are in any way related.

As I began to work on the paper, I approached it with the belief that the X-ray and radio emission were essentially unrelated. About the only relationship that had been observed was the appearance of a detectable radio flux that coincided within a week's time with a factor of 4 decrease in the 2–6 keV emission from Cygnus X-1. As we shall see there is now some additional evidence suggesting that the radio emission is not totally decoupled from the X-ray, although there are still very many unanswered questions.

II. Cygnus X-1

Perhaps the most significant of the Uhuru results for the galactic X-ray sources has been the discovery of pulsations from Cygnus X-1. These data led to further study of this object and to the present belief that we are dealing with a black hole. I would like to present the data we have on this object and consider the status of the black hole identification. Figure 1 contains data showing substantial variations in X-ray intensity on time scales from 100 ms to 10 s. Some 80 s of data are shown here summed on four time scales from 100 ms up to 14 s. I should point out that similar X-ray variability also reported by Rappaport *et al.* (1971a), Holt *et al.* (1971) and Shulman *et al.* (1971)

* Now at Center for Astrophysics, Harvard College Observatory/Smithsonian Astrophysical Observatory, Cambridge Massachusetts.

F. J. Kerr and S. C. Simonson, III (eds.), Galactic Radio Astronomy, 383–399. All Rights Reserved.
Copyright © 1974 by the IAU.

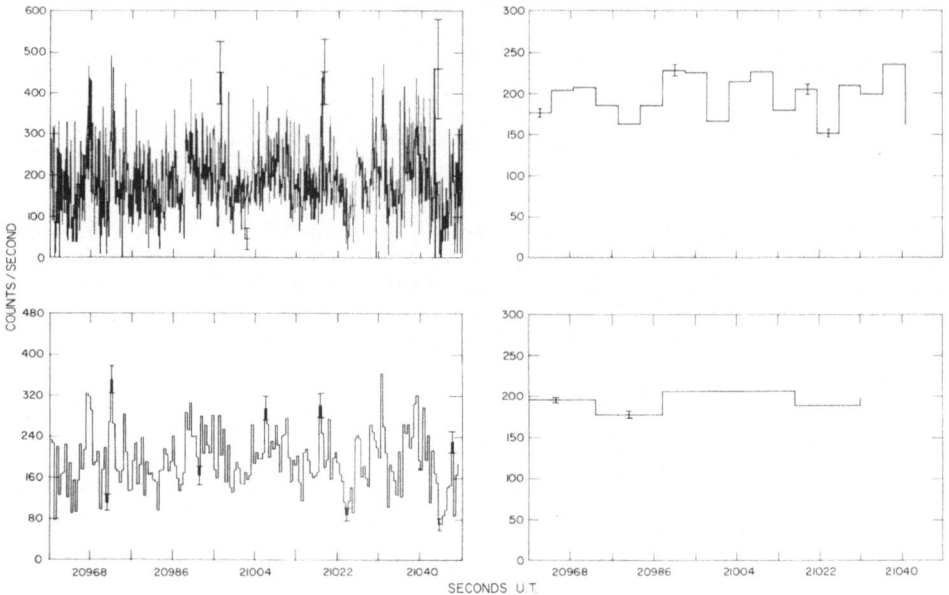

Fig. 1. Observations of Cygnus X-1 on 1971, June 10. Data have been corrected for the triangular collimator response. Data are summed over 0.096 s, 0.48-s, 4.8-s, and 14.4-s intervals. Typical 1-σ error bars are shown.

compels us to consider a source region of 10^9 cm or less. Figure 2 shows the X-ray location obtained from an MIT rocket flight (Rappaport *et al.*, 1971b) and from Uhuru which led to the discovery of a radio source by Braes and Miley (1971) and by Hjellming and Wade (1971). It is this precise radio location that led to the optical identification by Webster and Murdin (1972) and by Bolton (1972a) of Cygnus X-1 with the 5.6-day spectroscopic binary system HDE 226868. The central object of this system is a 9th magnitude star, most likely a B0 supergiant, and first mass estimates for the primary led to a mass in excess of 3 $M_\odot$ for the unseen companion. This came from the mass function determined from the absorption line velocities

$$\frac{M_s^3 \sin^3 i}{(M_s + M_{B0})^2} \simeq 0.2.$$

Even at $i = 90°$, which gives a minimum value of the mass of the secondary,

$$M_{B0} = 20 \rightarrow M_s = 5 \ M_\odot$$

If the secondary is the compact X-ray source, then it could well be a black hole, a point to which we will return shortly.

We attempted to confirm the identification of the X-ray source with HDE 226868 by looking for a 5.6-day effect on the X-ray light curve. In December 1971 and January 1972 we observed Cygnus X-1 continuously for 35 days. We folded the X-ray data with many different periods including 5.6 days, and the results are shown in Figure 3. Data at 2–6 keV are shown folded modulo 3.0, 5.6, and 6.2 days; the average is indi-

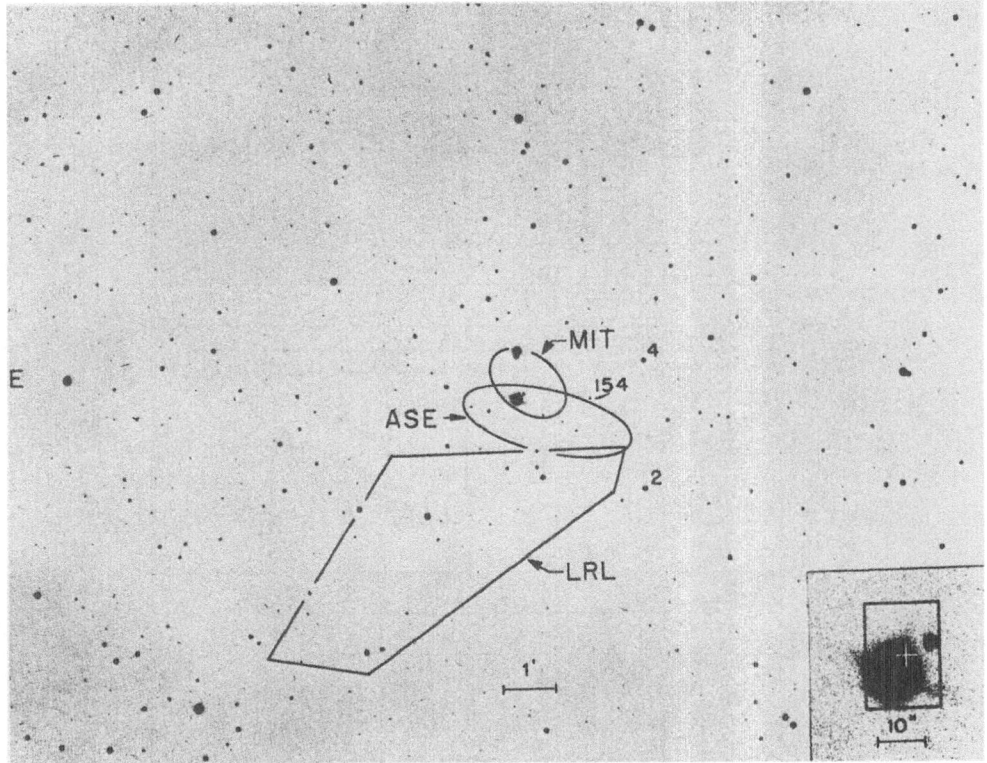

Fig. 2.   X-ray location of Cygnus X-1. HDE 226868 is the bright star in the overlap between the MIT and ASE error boxes. The insert shows the radio location, which was reduced to an uncertainty of less than 1″ after this figure was drawn and is coincident with HDE 226868.

cated by the dotted lines, and 2-σ error bars are indicated by the solid lines. We conclude that there is no evidence for a 5.6-day eclipse here and believe that previous reports from balloon observations of such an effect at higher energies were caused by the large-scale time variability and not by a 5.6 day effect. The report by Professor Boyd at Sydney of a possible 5.6-day effect at energies below 3 keV may be due to a real effect or may be due to the variability of the source and the presence of only two data points. In any case, the absence of a 5.6-day effect does not rule out the identification and can be understood in terms of an appropriate inclination angle for the orbital plane of the binary system.

With the use of Uhuru as an observatory, we have analyzed 16 months of data on Cygnus X-1, which are shown in Figure 4. We have plotted the 2–6 keV intensity versus day of 1970. The vertical lines for a given day are not error bars, but rather show the range of variability observed on that day. For some days we have only the average intensity shown by a dash available in our analyzed results. We see that a remarkable transition occurred in March and April 1971, with the source changing its average 2-6 keV intensity level by a factor of 4. We have also indicated in the figure the 6–10 keV and 10–20 keV X-ray intensities and see that the average level of the

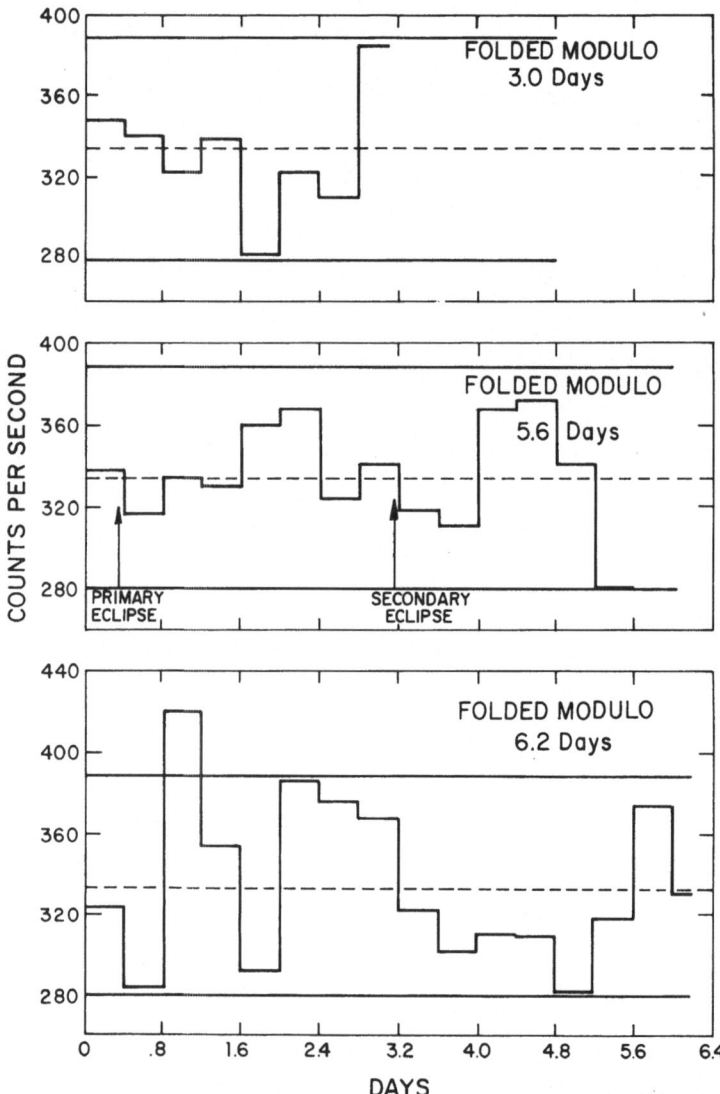

Fig. 3.   Cygnus X-1 folded intensity, 17 December 1971 to 21 January 1972; 35 days of 2–6 keV data are shown folded modulo 3.0, 5.6, and 6.2 days. The dotted lines give the average intensity, and the solid lines are 2-σ error bars.

10–20 keV flux increased by a factor of 2. The figure also shows that at the same time the X-ray intensity changed, a weak radio source appeared at the Cyg X-1 location and was detected by the Westerbork and NRAO groups. Hjellming (1973) has recently reported analysis of additional radio data which shows the radio source first appeared some time between March 22 and March 31, essentially the time during which the 2–6 keV X-ray intensity first headed downward. This correlated X-ray-radio behavior is the major evidence in addition to the positional data that Cyg X-1 is in fact identified with the optical and radio object.

   With respect to the arguments concerning the mass of the secondary and thereby

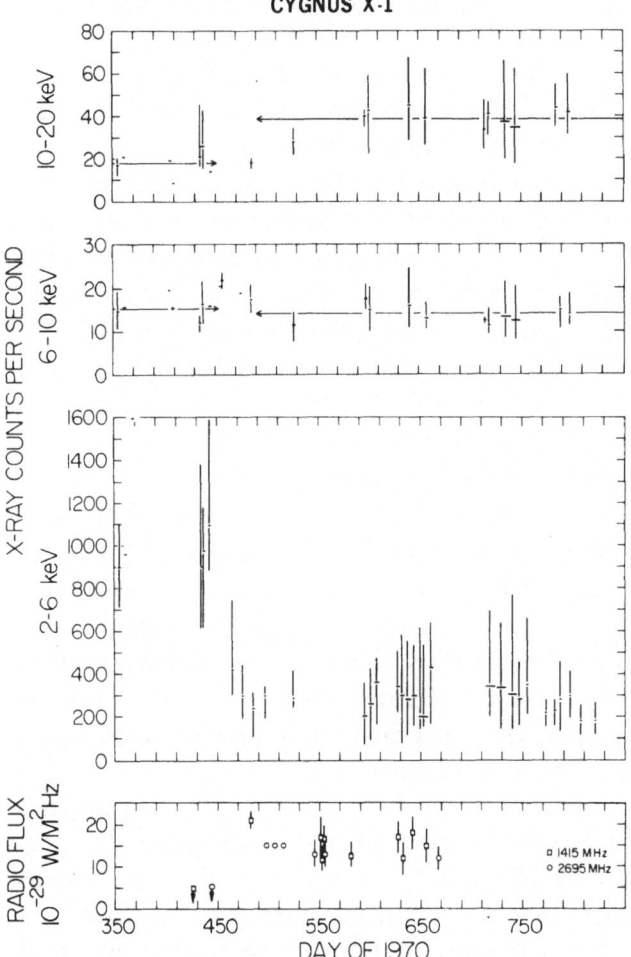

Fig. 4.   16 months of observations of Cygnus X-1. X-ray data are shown for three energy bands, 2–6 keV, 6–10 keV, and 10–20 keV, plotted vs. day of 1970. The transition discussed in the text occurred near day 450. The radio data are shown at the bottom of the figure, to which should be added the additional positive sighting on day 455 reported by Hjellming (1973).

the possibility of Cygnus X-1's being a black hole, there are three recent papers that approach the question. One paper by Bolton (1972b) uses the absorption line velocities, the He II 4686 Å emission line velocities, the 2.2 kpc distance derived from the observed reddening and the equivalent width and velocity of the interstellar lines to determine values of 20 $M_\odot$ for the primary, 13 $M_\odot$ for the secondary, and an inclination angle of 26°. Cherepashchuk et al. (1973) use the absorption line velocities, the He II emission line velocities with some allowance that the emission region may not belong to the X-ray star, but may lie between the two stars (the possible flaw in Bolton's work), and the photoelectric observations showing 0.07 mag changes due to a tidally distorted system. Then taking into account limb darkening and gravity darkening, and making the not necessarily correct assumption that the primary fills its Roche lobe, they determine a primary mass between 10.7 and 22 $M_\odot$, a secondary mass between 7.8 and 17 $M_\odot$, and a distance of order 5 kpc – although they appear to have neglected interstellar absorption effects and could therefore have overestimated the distance by a factor of 2. H. Mauder (1973) uses the mass function determined from the absorption line velocities, the possible distances allowed by the observed reddening and the absence of a bright infrared source which would be produced by an absorbing circumstellar shell around the source, the absence of any substantial reflection effects as demonstrated by the photoelectric observations and the X-ray to visible light energy ratio, and the photoelectric observations plus the assumption that the star cannot be any larger than its Roche lobe. He determines a self-consistent set of parameters that gives for a distance of 2 kpc, a primary mass of 25 $M_\odot$ and a secondary mass between 6.0 and 7.3 $M_\odot$.

What is the point of this rather lengthy, yet sketchy description? Since we are talking about the first possible observation of a black hole, we want to consider all the possibilities. The first optical papers suggested a mass of $\sim 20\ M_\odot$ for the primary star based on a B0 Ib spectral typing. Among others, Trimble *et al.* (1973), and Faulkner (1973) pointed out that a helium burning star of much less mass could give a temperature and effective surface gravity identical to that observed. Since such a source would be at a distance of 1 kpc at most, a knowledge of the distance is crucial for deciding the masses of the stars. Now the recent distance measurements that have been made by Bregman *et al.* (1973) and Margon *et al.* (1973) rule out the more exotic helium stars, determine that the primary is a massive supergiant, and thereby yield masses from 6 to 17 $M_\odot$ for the secondary.

There are three links in the chain of arguments leading to our conclusion about Cygnus X-1 – (1) the identification, where location agreement plus the X-ray-radio correlation suggest the HDE 226868 secondary is the X-ray source; (2) the compactness, where the short time variations, accretion as an energy source, and the absence of substantial visible light emission from the secondary indicate that this object is compact; and (3) the mass estimates of $\geqslant 6\ M_\odot$, which therefore indicate that Cygnus X-1 is a black hole.

In the way of future work on this source, we have the promise of a 1 arcsec X-ray location from the HEAO-B X-ray telescope in the late 1970's to make the identification absolute, the large-area NRL experiment planned for the HEAO-A mission to study the intensity variations on very short time scales to pursue the compactness, and the continuing optical efforts to confirm the distance measurements.

III. Cygnus X-3

If we now turn to Cygnus X-3, we find a different looking X-ray picture. Figure 5 shows the 2–6 keV Uhuru observation of Cygnus X-3 for 9 days in May 1972. The error bars contain both the statistical and systematic uncertainties. The data show intensity variations of about a factor of two with a period of 4.8 h first reported by Parsignault *et al.* (1972). Figure 6 shows the 9 days' data in three different energy bands from 1.8 to 10.0 keV plotted folded with the 4.8-h period. We see that the

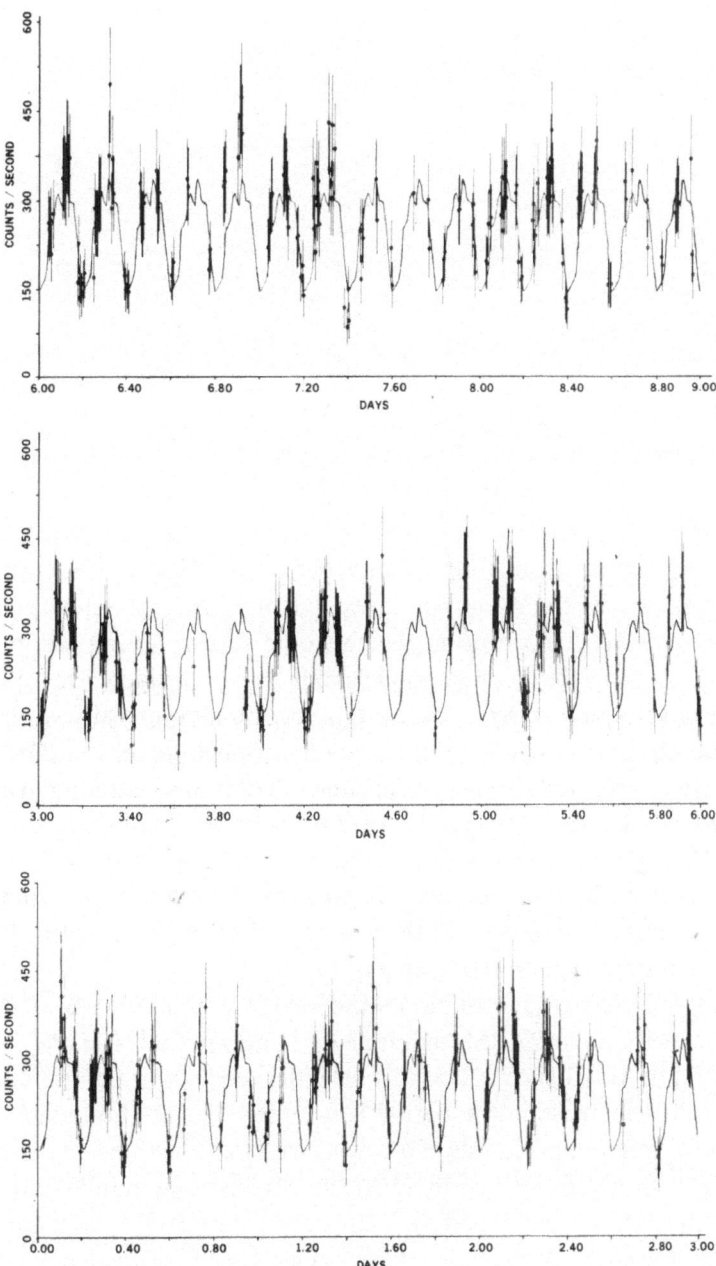

Fig. 5.   Observations of Cygnus X-3 from 8 May to 17 May are plotted as a function of time. The 2–6 keV intensities are given in counts s^{-1} and have been corrected for aspect; error bars include both statistical error and systematic error due to aspect correction. Also shown is the average light curve obtained by folding all the data modulo 4.80 h and finding the average intensity every 0.24 h.

CYGNUS X-3

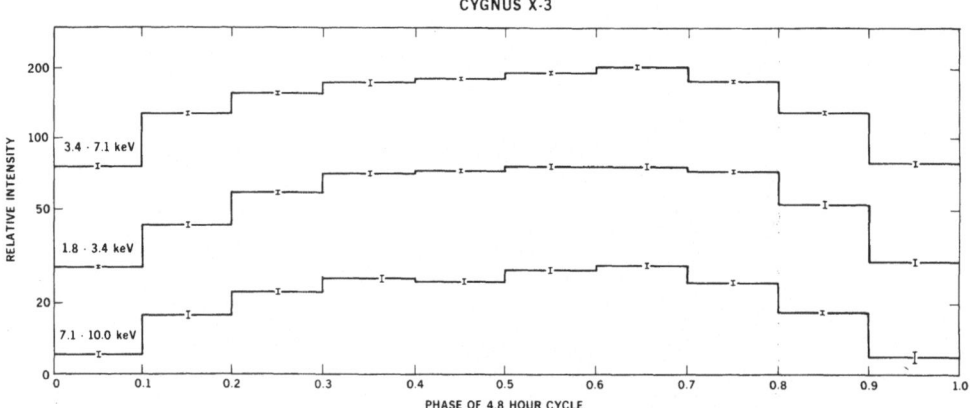

Fig. 6.   Counting rate data for Cygnus X-3 for the same 9 days in May 1972 folded modulo 4.8 h for three energy bands. The intensity scale is logarithmic, and the data show that the 4.8-h intensity variations are essentially energy independent.

minimum is observed in all three energy bands and that the intensity variations across the 4.8-h cycle are essentially independent of energy. Thus the minimum is not due to photoelectric absorption, which would show a strong energy dependence $(e^- (Ea/E)^{8/3})$. Among the possible models still being considered for this 4.8-h light curve are a binary system with a 4.8-h orbital period and intensity variations caused either by geometrical obstruction or by different optical depths in different viewing directions due to only weakly energy dependent Compton scattering in a hot ionized cloud, an idea suggested by Gursky (1973). Compton scattering calculations by Tucker (1973) suggest an emitting region that would be of order 10^8 cm, a size consistent with short time scale fluctuations now observed on at least some occasions, indicating a compact component of the source. Other possibilities for the 4.8-h cycle would be pulsation or rotation of a single star.

For the 4.8-h cycle we have a new, tentative period of 0.1996515 ± 0.0000032 days constant from December 1970 through June 1971 and a period of 0.1996787 ± 0.0000012 from June 1971 to May 1972, an 8σ statistical difference. A word of caution concerning this 1 part in 10^4 increase in the period is in order. The method used to determine the phase of the minimum consists of fitting a sine function to the intensity variations. The period is then determined by dividing the time between minimum phases by the appropriate integer number of periods. Canizares et al. (1973) have reported evidence for changes in the shape of the light curve and we must investigate the effects of such a change on our analysis technique before considering the period change as definite. Changes in this period could prove very important in choosing among various models for the 4.8-h cycle.

Figure 7 shows several points over a 1.5-yr period where we have determined the average intensity of Cygnus X-3. The points with the smallest error bars are data for which the average intensity was determined by the sine fitting technique; other points are obtained as daily averages of randomly selected points or as averages of selected

CYG X-3: AVERAGE INTENSITY

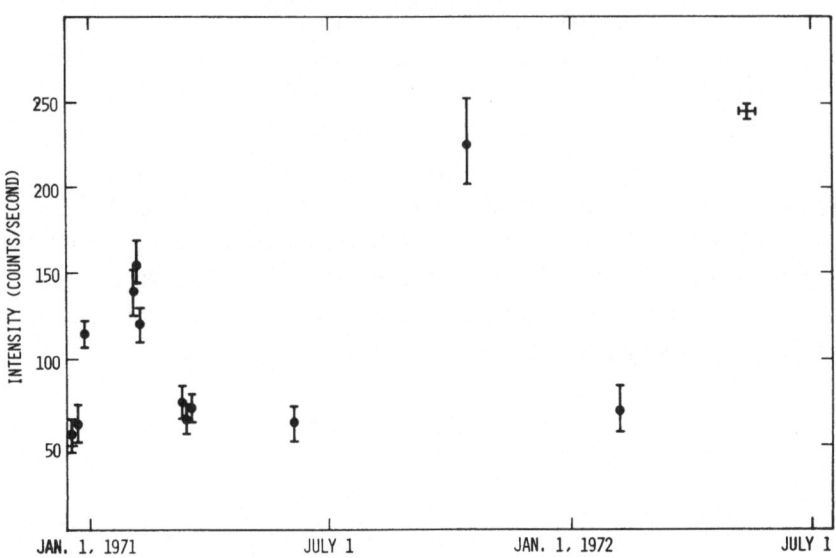

Fig. 7. Average 2–6 keV intensity for Cygnus X-3 on various days from January 1971 until July 1972. Data are corrected for aspect and the concept of 'average' is discussed in the text.

quick look data points at various phases of the 4.8-h cycle. The data suggest that Cyg X-3 may have average intensity levels which persist for times of months with transitions between levels sometimes observed. Average intensity levels of 60, 125 and 230 cts s^{-1} are seen at various times. This picture is somewhat similar to that observed for Cygnus X-1 and suggests that the radio data be checked against the X-ray observations, if possible, for possible long term correlated changes in average intensity.

Figure 8 shows new results for the X-ray intensity for 6 days at the time of the first September 1972 radio flare. We now include production data whereas our earlier report of no significant X-ray changes was based on quick look data only. We see that there are several points of high intensity on September 1, the day before the radio flare was first reported, although the radio data allow for the possibility of an earlier start up since the flare was first observed already in progress. The presence of X-ray intensities of at least 600 cts s^{-1}, an intensity greater than ever previously observed for Cygnus X-3, and at least a factor of 2 higher than any intensities measured on August 30 and 31 indicates a connection between the X-ray and radio behavior. This is also supported by preliminary observations of higher temperature energy spectra for the high intensity data points. This observation is important since it confirms the X-ray radio identification previously only suggested by positional coincidence of a few arcminutes. This identification is also confirmed as reported by Professor Boyd at the IAU General Assembly in Sydney by the very recent observation by Becklin *et al.* (1973) of a 4.8-h period in the infrared object identified with the Cyg X-3 radio source, with the infrared minimum in phase with the X-ray minimum.

Wallace Tucker and I have performed some very rough, back of the envelope calculations to see what the parameters of a system might be in which 10^{38} erg s^{-1} is produced in a synchrotron process X-ray flare and 10^{33} erg s^{-1} is produced in a radio flare. We basically have a model in which the electrons are injected and first produce X-rays in a small region with an intense magnetic field and then expand into a larger region with smaller B and produce the radio emission. We assume an X-ray emitting region of order 10^9 cm and a radio emitting region of order 10^{15} cm.

Assuming a radio lifetime of 10 days we find as already determined by Gregory *et al.* (1972) a magnetic field of $\lesssim 5$ G in the radio emitting region; we now can determine a magnetic field of 5×10^9 G in the X-ray region, electron energies of ~ 10 MeV, and number densities of $\sim 5 \times 10^{-2}$ cm^{-3} for the radio emitting region and $\sim 5 \times 10^3$ cm^{-3} for the X-ray region.

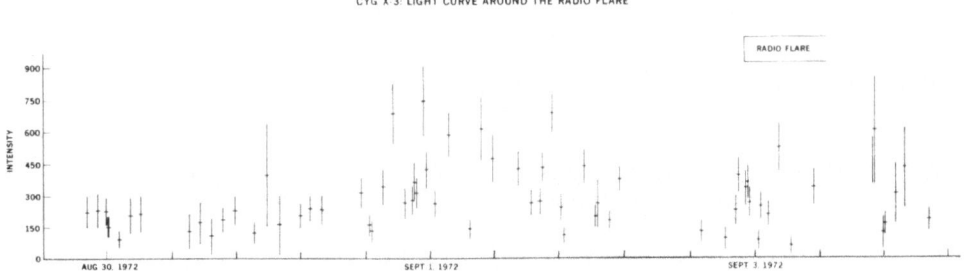

Fig. 8. 2–6 keV X-ray intensity for Cygnus X-3 in late August and early September 1972. The time of the first observation of the giant radio flare is indicated at 2200 on September 2. Note the difference in the X-ray behavior around September 1 compared to the preceding days.

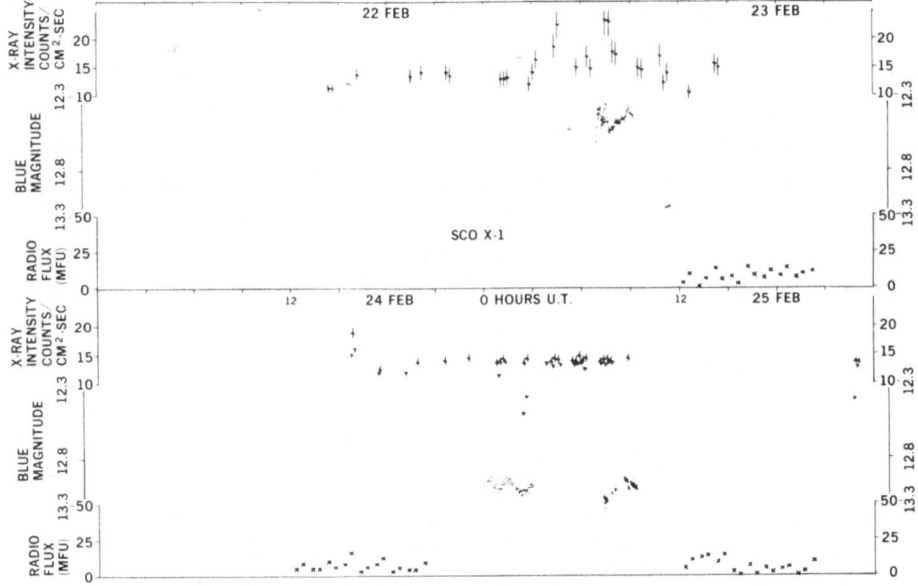

Fig. 9. Simultaneous X-ray, optical, and radio observations for Sco X-1 on 22–25 February 1971. The data are discussed in the text.

IV. Scorpius X-1

Turning to Sco X-1 we have considerably more detailed observations, particularly those obtained by collaborative efforts of a number of radio and optical observers together with Uhuru in February and March 1971.

Figure 9 shows the X-ray intensity in counts cm^{-2} s^{-1}, the blue magnitude, and the radio intensity observed at NRAO and Westerbork in mJy.

The radio source is relatively quiet and weak on February 23, 24 and 25; the optical intensity is relatively bright and variable on February 23 and the X-ray source is also highly variable. On February 24 and 25 the blue-magnitude is fainter than 13 mag and the X-ray emission is relatively low and quiet.

Figure 10 shows a continuation of this optical and X-ray behavior on February 26, with an enormous radio flare beginning around 18 h on February 26. Unfortunately, the X-ray and optical observations are notably absent on the 2nd half of February 26 and on February 27, except for three hours of relatively faint blue observations. When the radio data resume at 12 h on February 27, the flare has totally subsided, and then the X-ray and optical data for February 28 show low intensity and not much variability. Data such as these have been used to infer that the radio behavior is not related to the optical and X-ray emission, but the scarcity of X-ray and optical observations during much of the time on February 26 and 27 makes this conclusion unwarranted.

In Figure 11 we pick up Sco X-1 on March 23, 1971. Much of the data for March 23 and 24 show the blue magnitude brighter than 12.6 mag and the X-ray source intense and varying (although there are some quiet X-ray times). When the data are

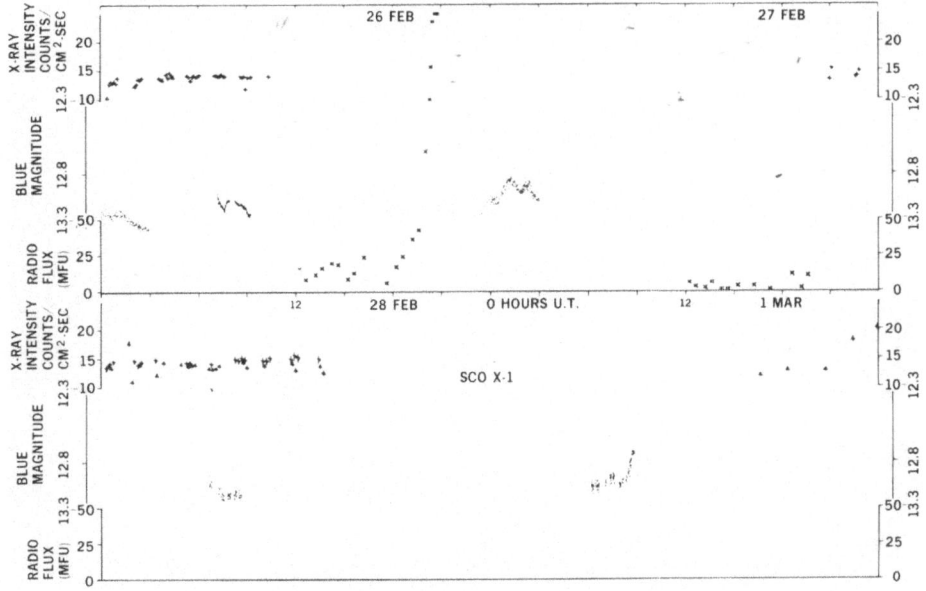

Fig. 10.   Same as Figure 9 except for 26 February–1 March 1971.

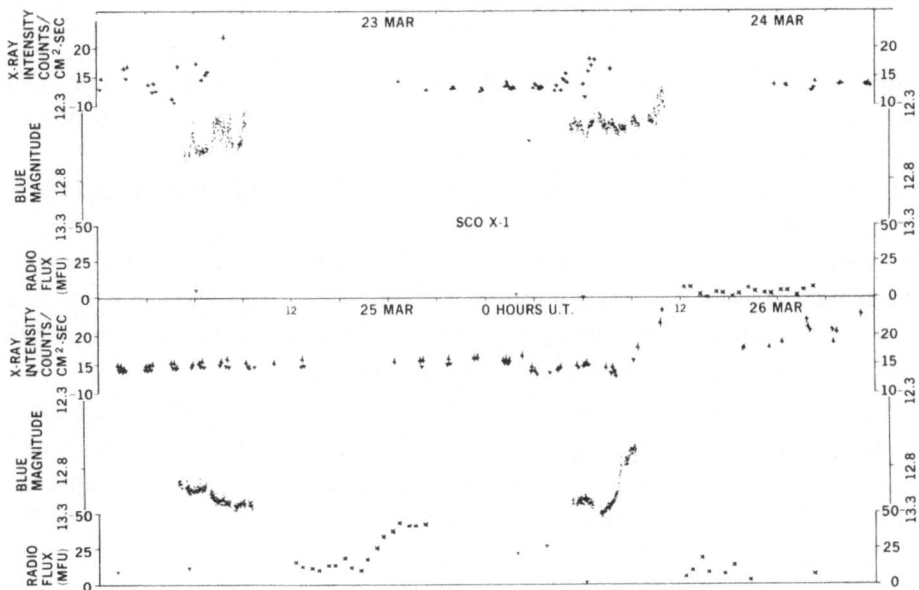

Fig. 11.   Same as Figure 9 except for 23–26 March 1971.

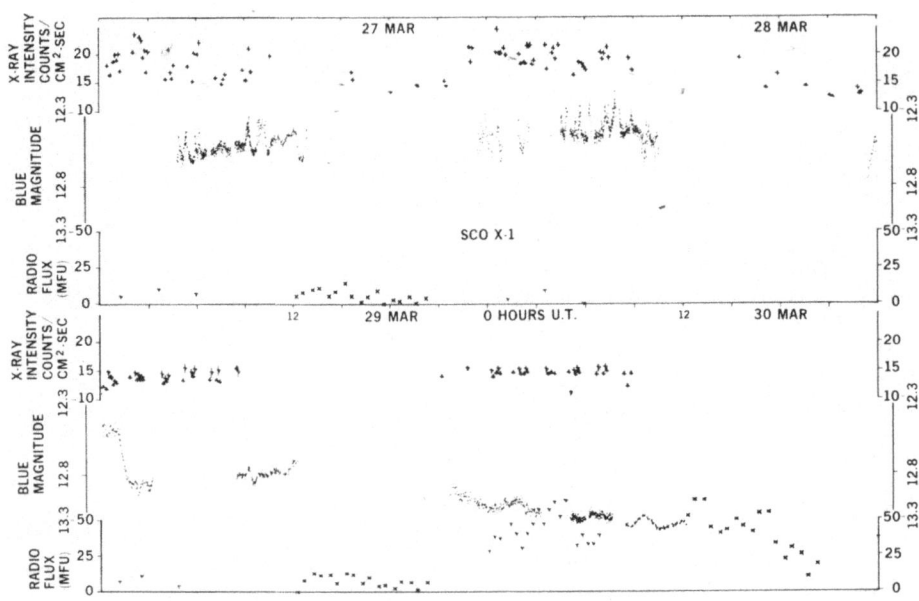

Fig. 12.   Same as Figure 9 except for 27–30 March 1971.

picked up on March 25, the X-ray intensity is very quiet and the optical is fainter than 13 mag. What would be helpful in attempting to understand how the source quieted down would have been more complete coverage on March 24 and 25.

Late on March 25, the radio data show a flare to about 500 mJy; the X-ray source remains low and quiet during this time, although there is a 6 hour gap in the data during the actual radio increase. On March 26 some 15 h after the radio flare, the optical brightness increases steeply, by several tenths of a magnitude in about an hour. About 1 h after the optical increase began, the X-ray intensity increases substantially and a bright, variable X-ray state follows. A few scattered radio points suggest that the radio flare may have disappeared some 2 to 4 h before the optical increase.

In Figure 12 we pick up the optical data showing the source still bright and highly variable on March 27 and 28 with a large decrease early on March 29. The X-ray points are also bright and variable on March 27 and most of March 28 and are definitely faint and relatively quiet on March 29. The March 28 data suggest that the X-ray decrease may have preceded the optical decrease by as many as 6 to 8 h. The X-ray and optical data remain low and quiet on March 29 and 30, with the radio data showing another flare, beginning between 21 and 24 h on March 29. The previous data from March 25 and 26 suggest that we either look for an optical and X-ray brightening some 15 hours after the start of the radio activity or some 2 to 4 h after the radio activity ends. Unfortunately, there are no X-ray data after 9 h on March

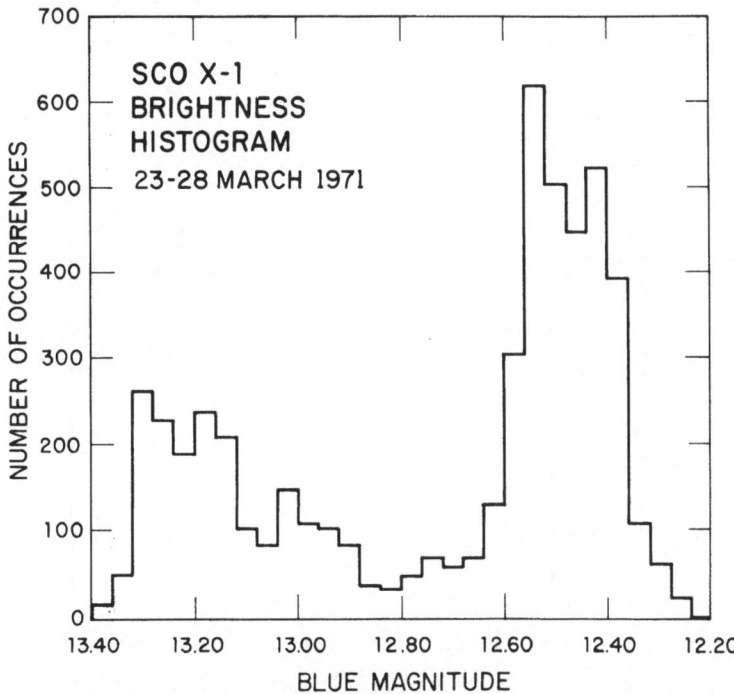

Fig. 13. Sco X-1 brightness histogram 23–28 March 1971. The number of occurrences of each blue magnitude is plotted vs. blue magnitude. The data indicate the clustering around 13.2 mag and 12.5 mag as discussed in the text.

30 nor any optical data at all on March 30, although optical data later on March 31 show the source faint. Thus, no conclusions can be drawn concerning repeated relationships between the radio and the optical and X-ray data.

Figure 13 shows a blue-magnitude occurrence histogram for March 23–28 suggesting two states for Sco X-1, with most of its time spent either bright or faint with little time in the middle. In Figure 14 we see the X-ray intensity versus blue magnitude for these days. This figure shows the correlation between the X-ray and optical. When the optical source is faint the 2–6 keV X-ray intensity is low and varying only slightly. When the blue magnitude passes 12.6 mag, the X-ray intensity is greater by up to a factor of 2 and much more variable. This is also related to observations by Hiltner and Mook (1970) that Sco X-1 only flares in the optical when it is brighter than this same 12.6 mag. Figure 15 is 2 days of these same data where we show the X-ray temperature vs the 2–6 keV intensity and find the source cooler when it is weaker and hotter and more variable when it is more intense. A better approach would have been

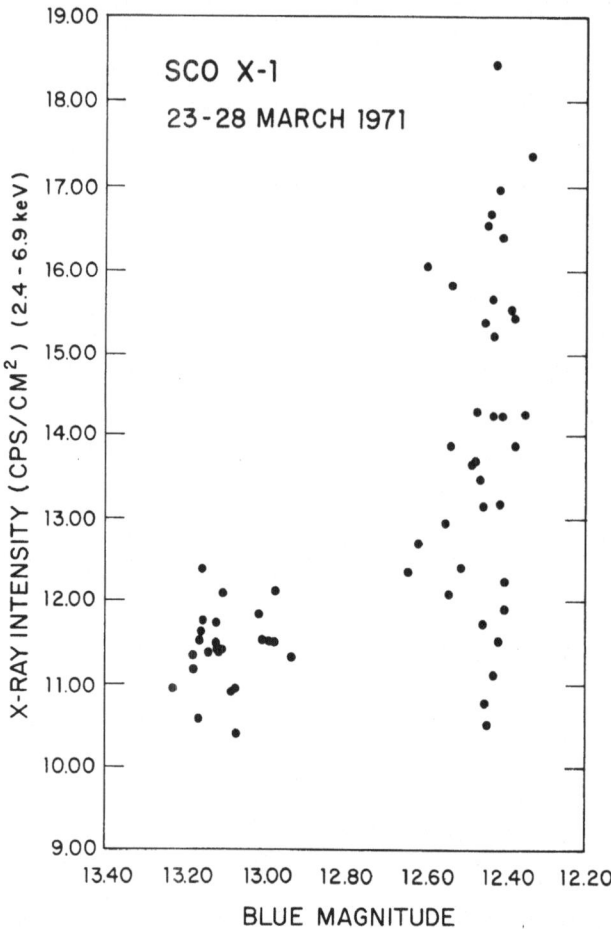

Fig. 14.   Simultaneous X-ray-optical variability of Sco X-1. Each point represents a single, simultanous measurement of the X-ray and optical intensity of Sco X-1. Note the relatively quiet, faint state and the highly variable, bright state.

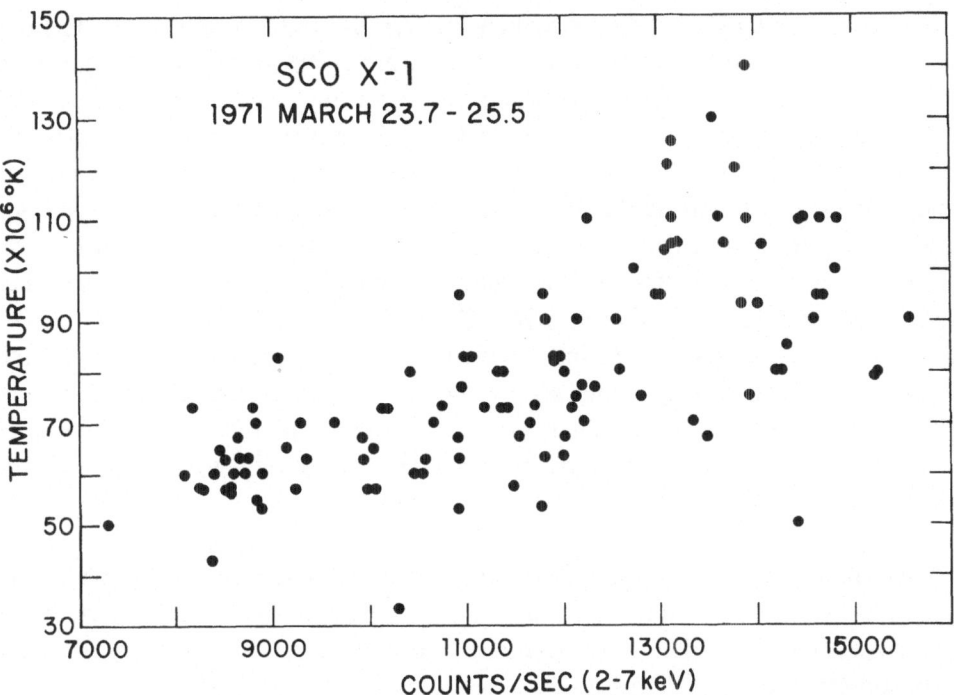

Fig. 15.   Temperature versus X-ray intensity for Sco X-1. Each point represents a single temperature and X-ray intensity observation. We see that as the source intensity increases the temperature increases and is more variable.

to plot temperature versus emission measure here, since the 2–6 keV intensity is of course greatly affected by the exponential temperature factor.

An effort of this type was undertaken by Kitamura *et al.* (1971) with rocket data and optical observations which suggested a size of 10^8 to 10^9 cm and a density of 10^{15-16} cm^{-3} for the X-ray emitting region. These conclusions were based on the observation that the emission measure decreased as the source became hotter. The Uhuru and MIT OSO 7 results suggest that this is not the case, but rather the emission measure increases as Sco X-1 becomes hotter. It may be that different laws govern this relationship depending on whether the source is in the bright flaring state or in the quiet, faint state. Present models suggest the radio emission occurs relatively far out from the X-ray emitting region, with the optical emission coming from a region also outside of but much closer to the X-ray emitting region, resulting in the correlation between the optical and X-ray emission.

V. Conclusion

It should now be perfectly clear that not too much is perfectly clear. What is new is that at least some of the time the radio emission is more closely related to the X-ray emission than previously thought for these objects. Yet the time scales of a day or two that may be involved require future round-the-clock coordinated observations, such as suggested by Dr Hughes in Sydney for Cyg X-3 radio observations. There is also, no doubt, a wealth of information on the nature of the sources and the emission process still to be extracted in interpreting observations already made. One final interesting point recently made by Tucker (1973) is that the sources with intrinsic, variable low energy cut-offs – the eclipsing binaries – have not yet been observed as radio sources; this might be due to the high density gas near the source along our line of sight to the source with the high plasma frequency preventing the transmission of low frequency radio emission which may be present at the source. On the other hand, the X-ray data analyzed for Cyg X-1, Cyg X-3, and Sco X-1, as well as GX17 + 2 and GX9 + 1, do not require intrinsic low energy absorption, which may mean the absence of the high density, radio absorbing gas near the source along the line of sight, and therefore the ability for us to observe radio emission. This suggests that the most viable candidates for future radio searches are those sources without large intrinsic cutoffs in their X-ray emission.

Note added. Further analysis has shown that the possible change in 4.8-h period for Cygnus X-3 cannot be isolated from effects due to changes in the shape of the X-ray light curve. Hence, we determine a period of 0.19967 ± 0.00001 days for all of the Uhuru observations of Cygnus X-3.

Acknowledgements

I gratefully acknowledge the contributions to this paper of several members of the Uhuru group at the HCO/SAO Center for Astrophysics and at AS&E including Riccardo Giacconi, Ethan Schreier, Christine Jones, Bill Forman, Edwin Kellogg, Herb Gursky, Steven Murray, Dan Parsignault, Robert Leach and Wallace Tucker. I also appreciate the efforts of Hale Bradt and his co-workers in obtaining and analyzing the data from the coordinated observations of Sco X-1, and for allowing me to use the soon-to-be-published results.

References

Becklin, E. E., Hawkins, F. J., Mason, K. O., Matthews, K., Neugebauer, G., Sanford, P. W., and Wynn-Williams, C. G.: 1973, preprint.
Bolton, C. T.: 1972a, *Nature* **235**, 271.
Bolton, C. T.: 1972b, *Nature Phys. Sci.* **240**, 124.
Braes, L. and Miley, G. K.: 1971, *Nature* **232**, 246.
Bregman, J., Butler, D., Kemper, E., Koski, A., Kraft, R. P., and Stone, R. P. S.: 1973, preprint.
Canizares, C. R., McClintock, J. E., Clark, G. W., Lewin, W. H. G., Schnopper, H. W., and Sprott, G. F.: 1973, *Nature Phys. Sci.* **241**, 28.
Cherepaschuk, A. M., Lyutiy, V. M., and Sunyaev, R. A.: 1973, *Astron. Zh.* **50**, 3.
Faulkner, J.: 1973, Tucson Workshop on Compact X-Ray Sources.
Gregory, P. C., Kronberg, P. P., Seaquist, E. R., Hughes, V. A., Woodsworth, A., Viner, M. R., and Retallack, D.: 1972, *Nature Phys. Sci.* **239**, 440.
Gursky, H.: 1973, private communication.
Hiltner, W. A. and Mook, D. W.: 1970, *Ann. Rev. Astron. Astrophys.* **8**, 139.
Hjellming, R. M.: 1973, *Astrophys. J. Letters* **182**, L29.

Hjellming, R. M. and Wade, C. M.: 1971, *Astrophys. J. Letters* **168**, L21.

Holt, S., Boldt, E., Schwartz, D., Serlemitsos, P., and Bleach, R.: 1971, *Astrophys. J. Letters* **166**, L65.

Kitamura, T., Matsuoka, M., Miyamoto, S., Nakagawa, M., Oda, M., Ogawara, Y., Takagishi, K., Rao, U. R., Chitnis, E. V., Jayanthi, U. B., Prakasa-Rao, A. S., and Bhandari, S. M.: 1971, *Astrophys. Space Sci.* **12**, 378.

Margon, B., Bowyer, S., and Stone, R.: 1973, *Astrophys. J. Letters*, to be published.

Mauder, H.: 1973, *Astron. Astrophys.*, to be published.

Parsignault, D. R., Gursky, H., Kellogg, E. M., Matilsky, T., Murray, S., Schreier, E., Tananbaum, H., Giacconi, R., and Brinkman, A. C.: 1972, *Nature* **239**, 123.

Rappaport, S., Doxsey, R., and Zaumen, W.: 1971a, *Astrophys. J. Letters* **168**, L43.

Rappaport, S., Zaumen, W., and Doxsey, R.: 1971b, *Astrophys. J. Letters* **168**, L17.

Shulman, S., Fritz, G., Meekins, J., and Friedman, H.: 1971, *Astrophys. J. Letters* **168**, 449.

Trimble, V., Rose, W. K., and Weber, J.: 1973, *Monthly Notices Roy. Astron. Soc.* **162**, 1.

Tucker, W.: 1973, private communication.

Webster, B. L. and Murdin, P.: 1972, *Nature* **235**, 37.

Harvey Tananbaum

Center for Astrophysics,
Harvard College Observatory/Smithsonian Institution,
60 Garden Street,
Cambridge, Mass. 02138, U.S.A.

DISCUSSION

Kaufmann: During the Cyg X-3 bursts in the first days of September 1972 we observed at Itapetinga (CRAAM), Brazil, very clear VLF propagation anomalies in the terrestrial lower ionosphere. Changes in diurnal phase variation were larger than 10%. Since at nighttime in the ionospheric E-region the maximum electron production rate is somewhere in the 6–12 keV range, the observed VLF anomalies would require X-ray emission in excess from Cyg X-3 or, even better, a steeper energy spectrum, with excesses at energies greater than 12 keV, as has been reported in this communication.

Lequeux: You showed variations in Cygnus X-3 in three different energy bands, and they look rather similar. Doesn't that contradict the statement that the low-energy cut-off in the source moves with the material?

Tananbaum: No, my statement was that the low energy cut-off does not vary across the 4.8-h cycle, precisely the point that rules out photo-electric absorption as a possible explanation of this 4.8-h variation.

RADIO OUTBURSTS FROM CYGNUS X-3

V. A. HUGHES, M. R. VINER, and A. WOODSWORTH

Queen's University, Kingston, Ontario, Canada

Abstract. The variation in flux density obtained at 10522 MHz for Bursts 2, 3, and 4 is compared with that obtained by others at frequencies down to 365 MHz. The bursts appear to have a quasi-periodic modulation with a period of 3–4 h, which is different from the 4.8-h periodicity observed at X-ray and in-frared wavelengths. The modulation is attributed to a fluctuation in the size of the expanding cloud of particles produced by either an instability in the atmosphere of Cygnus X-3 or by a built-in instability in the cloud itself.

The first observed giant radio outburst from Cygnus X-3 (Burst 1) occurred on September 2, 1972 (Gregory *et al.*, 1972a). In the period September 20–30, three further outbursts (Bursts 2, 3, 4) were observed by a large number of observatories and provide the most comprehensive frequency coverage yet obtained for any comparable event. The variation in flux density obtained by us at the Algonquin Radio Observatory at 10522 MHz is shown in Figure 1 and is compared with the flux densities as published by others in Figure 2. Since the maximum period for continuous monitoring at any one observatory is typically about 12 h, an attempt has been made to interpolate the values for flux density between observing periods by means of broken lines. Though no great reliance can be placed on detailed values for flux density for these latter periods, they nevertheless show the general features of the bursts. Typical absolute values as shown by the solid lines are better than about 5%, or ±0.1 Jy, though in most cases relative variations in intensity are far less uncertain.

Burst 1 was originally attributed to the result of an expanding cloud of relativistic particles (Gregory *et al.*, 1972b). On this assumption, when the source was first ob-

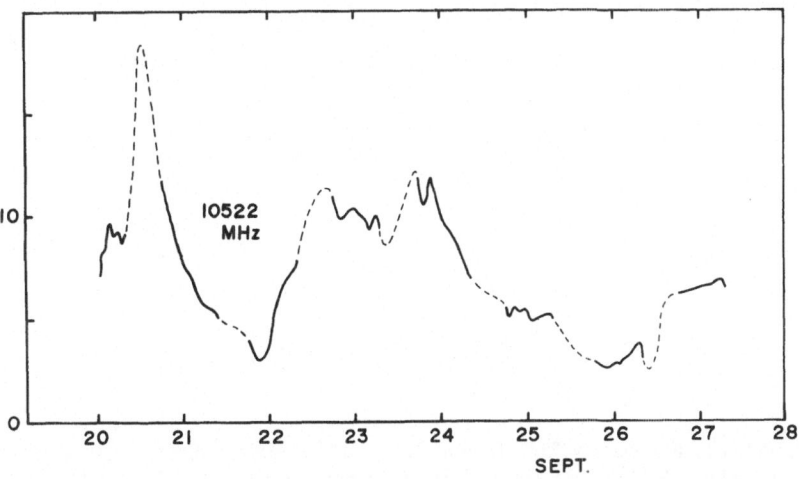

Fig. 1.   Flux density from Cygnus X-3 at 10522 MHz for Bursts 2, 3, and 4.

F. J. Kerr and S. C. Simonson, III (eds.), Galactic Radio Astronomy, 401–405. All Rights Reserved.

served at 10522 MHz, its diameter was about 50 AU on the assumption that the magnetic field was about 14 G, though more recent estimates of the magnetic field could reduce this to about 10 AU.

There are a number of differences between the second series of bursts and Burst 1.

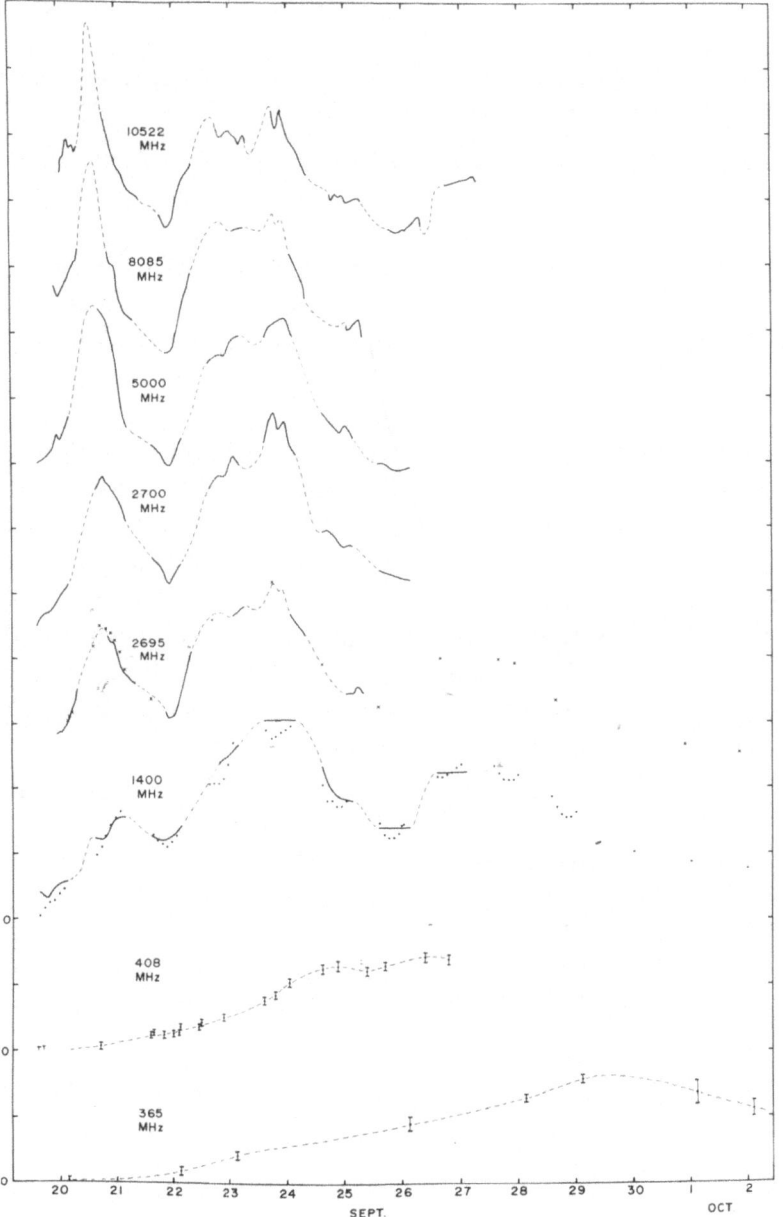

Fig. 2. Variation in flux density from Cygnus X-3 at 10522 MHz compared with that obtained by Hjellming and Balick (1972) at 8085 and 2695 MHz, Bransen *et al.* (1972) at 5000, 2700 and 1400 MHz, Braes *et al.* (1972) at 2695 and 1400 MHz, Anderson *et al.* (1972) at 408 MHz, and Bash and Ghigo (1973) at 365 MHz.

Burst 2 is similar in shape, but its duration is a factor of 2–3 times shorter. Bursts 3 and 4 are more complicated, showing a superimposed modulation. In addition, the rate of drift from high to low frequencies for Bursts 2, 3, and 4 is a factor of about 2 smaller than for Burst 1.

It was suggested by Peterson (1972) that an alternative explanation for Burst 1 was a process of injection of relativistic particles of specific duration such that at no time was the region optically thick to synchrotron radiation. However, attempts to fit this model to Burst 2 using a least squares fit were not satisfactory. It has not been found possible to reproduce the high peak, in particular at 5000 MHz, while still maintaining the observed decay time. The original model of the expanding cloud still seems attractive since it explains both the decay phase of the burst and the change in spectral index for both Bursts 1 and 2, and there is insufficient evidence to show that the later bursts could not be explained in this way. In addition, further confirmation comes from long-baseline observations of the burst on September 24 which show that the source had a size of at least 100 AU, assuming a distance of 10 kpc (Hinteregger *et al.*, 1972).

A further characteristic which was evident in the initial outburst, but which is far more pronounced in the later bursts, is a fluctuation or modulation having a quasi-period of 3–4 h, and which is apparent in Figure 1. The amplitude of the modulation amounts to about three times the uncertainty in the relative value for flux density and, as can be seen in Figure 2, has a counterpart at the lower frequencies. On the assumption that the radiation is produced by a cloud of relativistic particles, such a modulation could be the result of a change in the total number of particles, as would be produced by a series of injections or by an acceleration mechanism acting over the cloud, or of an oscillation in the size of the cloud with corresponding change in magnetic field.

If the fluctuation is the result of a series of injections, then it seems a little strange that, for instance, in Burst 3 the individual injections would just be sufficient to maintain the overall amplitude to within $\pm 10\%$ over a period of about 2 days, when large individual injections such as produce Bursts 1 and 2 are known to exist. It is also difficult to see how an injection mechanism can act almost instantaneously over the whole of the cloud, which must be the case if the estimates of the size of the cloud when seen initially are correct. One possibility is that the expanding cloud is situated in a region of a fluctuating magnetic field, as could be produced by an instability, and that the cloud itself then undergoes oscillation in size. In this case, it is easy to show that under the normal assumptions that the relativistic gas behaves adiabatically, that magnetic flux is conserved and that the cloud has radial oscillations, the flux density, S, will vary with the radius, r, of the cloud as

$$S \alpha r^3, \qquad \tau > 1,$$
$$S \alpha r^{-2\gamma}, \qquad \tau < 1,$$

where it is assumed that the electrons have a power law distribution of energies with an exponent of $-\gamma$. It is also assumed that any acceleration of particles due to the betatron process is negligible. Hence it is expected that at some frequency, namely where $\tau = 1$, the modulation will reverse in phase. There is some evidence that the fluctuations at 1400 MHz are in antiphase with those at higher frequencies, suggesting that at these points τ becomes equal to unity at a frequency of about 2000 MHz. If this is correct, then the radio outburst can be explained in terms of an event which causes to be emitted a comparatively dense cloud of relativistic particles, carrying with it some of the background magnetic field. The cloud moves outwards through a region of instability which takes the form of fluctuations in pressure or magnetic field or both, and which produces a small quasi-periodic modulation in the radius of the cloud, or the cloud itself has its own built-in instability. At the same time, the cloud is expanding such that after a certain time radiation starts to appear, first at the higher frequencies, as in the case of the van der Laan model (van der Laan, 1966), but with superimposed modulation. After a certain distance, the cloud will have moved through the region of instability, or the built-in instability becomes damped out. In either case, the modulation disappears and the cloud continues to expand. It is of interest that the presence of a region of instability suggests that Cygnus X-3 may have a comparatively extensive atmosphere with parameters such that it can sustain oscillations with a period of 3–4 h; more likely the cloud itself has the instability.

Both the X-ray and infrared observations show the presence of a 4.8 h periodicity (Parsignault *et al.*, 1972; Canizares *et al.*, 1973). It is suggested in the two preceding papers that they are both associated with one component of a binary system which consists of a hot plasma of radius about $1R_\odot$ and that the magnetic field associated with the X-ray source is about 5×10^9 G. Attempts have been made to extract a 4.8 h periodicity from the radio modulation, but without success. It appears that the radio emission is not related directly to the X-ray and infrared emissions, though minor variations in the intensity of the latter may indicate that a radio outburst may occur later. The proposed model is consistent with these facts when we also consider the dimensions associated with the different events. The X-rays and infrared appear to originate from a small region with dimensions much less than about 10^{11} cm. The radio emission comes from a cloud of particles which has a size when first seen of at least 10^{14} cm and hence must be at least this distance from the X-ray and infrared source.

Acknowledgements

We thank H. M. Bradford and D. S. Retallack for help in taking the 10 522 MHz observations and W.-Y. Chau and R. N. Henriksen for helpful discussions.

This work has been carried out under an Operating Grant from the National Research Council of Canada. The latter also operate the Algonquin Radio Observatory.

References

Anderson, B., Conway, R. G., Davis, R. J., Packham, R. J., Richards, A. C. S., Spencer, R. E., and Wilkinson, P. N.: 1972, *Nature Phys. Sci.* **239**, 117.

Bash, F. N. and Ghigo, F. D.: 1973, *Nature Phys. Sci.* **241**, 93.

Braes, L. L. E., Miley, G. K., Shane, W. W., Baars, J. W. M., and Goss, W. M.: 1973, *Nature Phys. Sci.* **242**, 66.

Bransen, N. J. B. A., Martin, A. H. M., Pooley, G. C., Readhead, A. C. S., Shakeshaft, J. R., Slingo, A., and Warner, P. J.: 1972, *Nature Phys. Sci.* **239**, 133.

Canizares, C. R., McClintock, J. E., Clarke, G. W., Lewin, W. H. G., Schnopper, H. W., and Sprott, G. F.: 1973, *Nature Phys. Sci.* **241**, 28.

Gregory, P. C., Kronberg, P. P., Seaquist, E. R., Hughes, V. A., Woodsworth, A., Viner, M. R., and Retallack, D. S.: 1972a, *Nature* **239**, 440.

Gregory, P. C., Kronberg, P. P., Seaquist, E. R., Hughes, V. A., Woodsworth, A., Viner, M. R., Retallack, D. S., Hjellming, R. M., and Balick, B.: 1972b, *Nature Phys. Sci.* **239**, 114.

Hinteregger, H. F., Catuna, G. W., Counselman, C. C., Ergas, R. A., King, R. W., Knight, C. A., Robertson, D. S., Rogers, A. E. E., Shapiro, I. I., Whitney, A. R., Clark, T. A., Hutton, L. K., Marandino, G. E. Perley, R. A., Resch, G., and Vandenberg, N. R.: 1972, *Nature Phys. Sci.* **240**, 159.

Hjellming, R. M. and Balick, B.: 1972, *Nature Phys. Sci.* **239**, 135.

van der Laan, H.: 1966, *Nature* **211**, 1131.

Parsignault, D. R., Gursky, H., Kellogg, E. M., Malitsky, T., Murray, S., Schreier, E., Tananbaum, H., Giacconi, R., and Brinkman, A. C.: 1972, *Nature Phys. Sci.* **239**, 123.

Peterson, F. W.: 1973, *Nature* **242**. 173.

V. A. Hughes
M. R. Viner
A. Woodsworth
Astronomy Group, Department of Physics and Astronomy,
Queen's University,
Kingston, Ontario, Canada

(Discussion follows the paper by Wynn-Williams, p. 409.)

INFRARED AND X-RAY VARIABILITY OF CYG X-3

C. G. WYNN-WILLIAMS*

California Institute of Technology, Pasadena, Calif., U.S.A.

Abstract. The 2.2 μm flux from the infrared counterpart of Cygnus X-3 has been found to vary in phase with the X-ray flux with a 4.8 h period. The infrared observations imply a radius $\sim 1\ R_\odot$ for the X-ray object and do not favor models involving a compact object such as a neutron star or a black hole.

A 4.8 h periodicity has been detected in the 2.2 μm flux density from the infrared counterpart of Cygnus X-3. Coordinated measurements (Becklin *et al.*, 1973) by the Caltech infrared group on the 200-in. telescope and by the Mullard Space Sciences Laboratory's X-ray package on the Copernicus satellite on 1973 July 9 showed that the infrared and X-ray fluxes vary in phase over the 4.8 h period, although their light curves are not identical (Figure 1). The infrared periodicity has been observed on several occasions; the ratio of maximum to minimum flux density appears to stay constant, even though the mean infrared flux density was seen to vary between observing periods. A similar behaviour has been noted in the X-ray emission.

The simplest and most attractive explanation for the 4.8 h periodicity is that Cyg X-3 is an eclipsing binary system. From the period it may be deduced that the separation of the stellar centres is $1.4(M/M_\odot)^{1/3} R_\odot$, where M is the total mass of the system. Since the radius of the infrared source cannot be greater than this and since a lower limit on the distance to the source is known from radio observations (Braes *et al.*, 1973), a lower limit to the 2.2 μm brightness temperature may be obtained:

$$T_{\mathrm{B}} > 6 \times 10^6 \left(\frac{M}{M_\odot}\right)^{-2/3} \mathrm{K}.$$

It therefore may be deduced that for any reasonable total mass of the Cyg X-3 binary system the object seen at 2.2 μm has a surface temperature higher than that of the photosphere of any normal star. Furthermore, given the very high estimated value for the infrared surface brightness, and given the similarity of the X-ray and infrared eclipse curves, it seems extremely probable that both the X-ray and infrared fluxes originate from the same hot object. For any reasonable value for the total mass of the system, this object has a radius of the order of $1\ R_\odot$. The observations described in this paper therefore definitely do not favour models for Cyg X-3 involving a compact object such as a white dwarf, a neutron star or a black hole.

Infrared data taken on 1972 July 11 are shown on two time scales in Figure 2. The data on the left show a sharp cut off at about 0715 attributable to the 4.8 h periodicity. Half an hour later the outburst shown on an extended time scale on the right of Figure 2 occurred. It may be seen that this outburst lasted only about five

---

* Now at Mullard Radio Astronomy Observatory, Cambridge, United Kingdom.

F. J. Kerr and S. C. Simonson, III (eds.), Galactic Radio Astronomy, 407–409. All Rights Reserved.
Copyright © 1974 by the IAU.

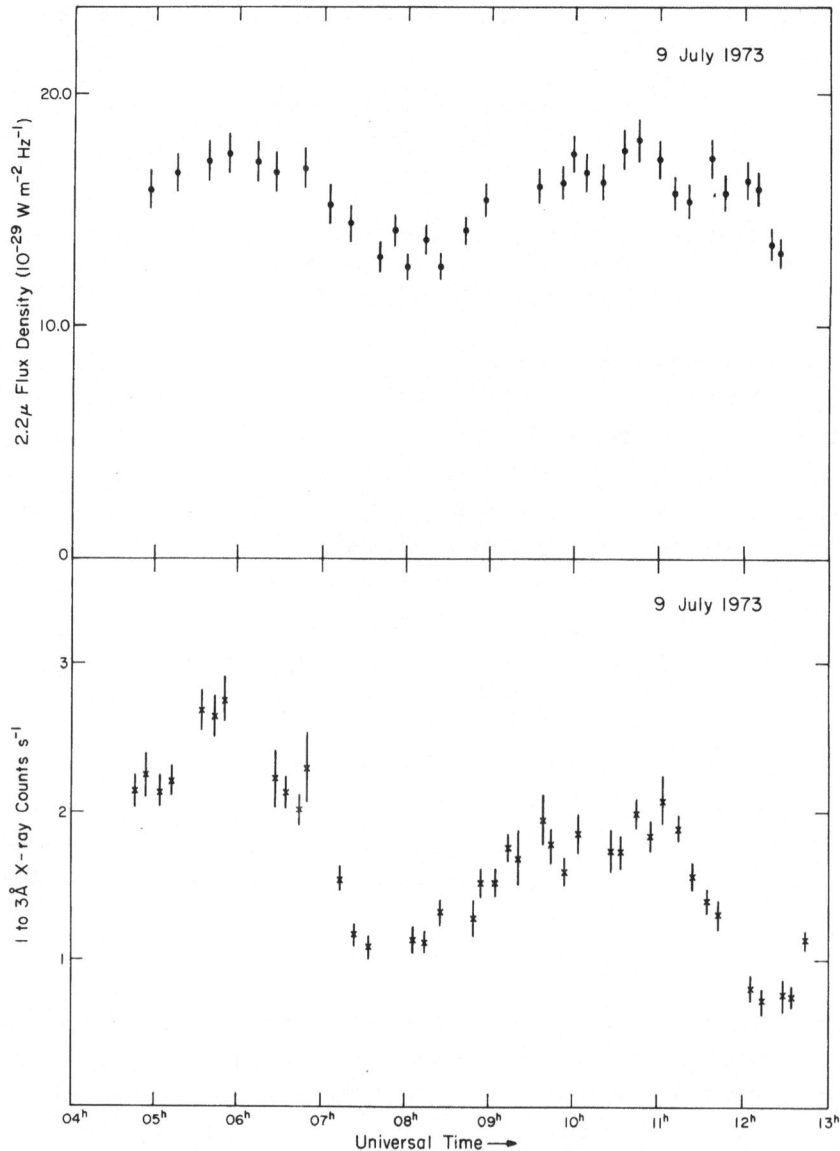

Fig. 1. The 4.8 h periodicity of Cyg X-3. The X-ray data have been averaged over the 6 min integration times of the infrared data. During the time period shown the infrared flux from a comparison star located within 5″ of Cyg X-3 appeared constant within ±2%.

minutes. Outbursts qualitatively similar to this one have been observed on other occasions; they do not appear to be correlated with the 4.8 h periodicity.

Although this erratic behaviour is highly reminiscent of the radio outbursts for which Cyg X-3 is notorious, it is not yet established how, if at all, the radio and infrared flares are related. Further observations to correlate X-ray, infrared and radio behaviour are in progress.

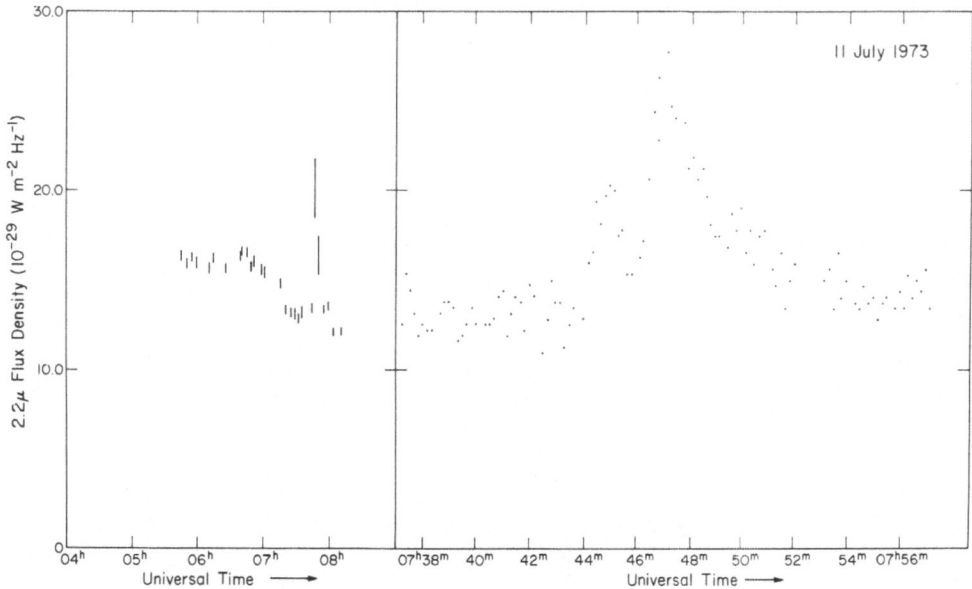

Fig. 2.   The infrared flare of 1973 July 11. On the left hand side, the points represent 3 min integrations. A portion of the data at the left is shown broken up into 10 s samples. The statistical accuracy of each point is about 10%.

References

Becklin, E. E., Neugebauer, G., Hawkins, F. J., Mason, K. O., Sanford, P. W., Matthews, K., and Wynn-Williams, C. G.: 1973, *Nature* **245**, 302.
Braes, L. L. E., Miley, G. K., Shane, W. W., Baars, J. W. M., and Goss, W. M.: 1973, *Nature Phys. Sci.* **242**, 66.

C. G. Wynn-Williams
Mullard Radio Astronomy Observatory,
Cavendish Laboratory, Free School Lane,
Cambridge, CB2 3RQ, United Kingdom

DISCUSSION

Van Woerden: Would you say that from the fact that the flare does not come to a specific phase in the 4.8 h period, one would think of hot spots on the surface.

Wynn-Williams: No, one would not think of hot spots on the surface.

Fourikis: I would like to report that we observed radio emission from Henize 1044, yet another early-type emission line object. Using the Parkes telescope at 3 cm the flux density is 242 ± 34 Jy. This peculiar object is in many ways similar to HD 16362. This work was done in collaboration with C. R. Purton, P. A. Feldman, and A. E. Wright.

Westerhout: Hughes' picture of time variations as a function of frequency certainly reminds one of dynamic spectra of solar bursts, except for the time scale.

Hughes: This is true, but the three Cygnus X-3 bursts shown appear to be of different character, though all have the same rate of change of frequency with time. The different types of solar burst appear to have different values for $d\nu/dt$.

MEASUREMENTS ON PLANETARY NEBULAE

D. K. MILNE

Division of Radiophysics, CSIRO, Sydney, Australia

and

L. H. ALLER

University of California, Los Angeles, Calif., U.S.A.

Abstract. 165 planetary nebulae south of $+27°$ declination have been observed with the Parkes 64 m telescope. These nebulae were chosen to include most of those for which Hβ flux densities are available. Radio positions and flux densities (generally with probable errors of only a few millijanskies) obtained from the survey must represent the most reliable and homogeneous radio data currently available for the class of object. The data have been used to obtain distances and optical extinctions for these nebulae.

I. Introduction

One hundred sixty-five planetary nebulae south of $+27°$ declination have been observed with the Parkes 64 m telescope. These nebulae were chosen to include most of those for which Hβ flux densities are available. Radio positions and flux densities (generally with probable errors of only a few mJy) obtained from the survey must represent the most reliable and homogeneous radio data currently available for the class of object. The data have been used to obtain distances and optical extinctions for these nebulae.

II. Observations

At 5000 MHz the beamwidth of the Parkes telescope is 4.5', and with the cryogenic parametric receiver a system noise temperature of ~ 80 K was obtained over a 500 MHz band.

The telescope was scanned in right ascension and declination over the optical position of each nebula many times and the intensity averaged over each arcminute and then integrated over all of the scans. The number of scans varied from six in each coordinate for the stronger sources to 90 for the very weakest. After 20 scans the peak to peak noise level was typically 0.015 Jy. In Figure 1 we show 10 integrated scans in each coordinate through the nebula M1-46, for which we obtain a flux density of 0.095 ± 0.015 Jy. The confusion which limited us at 2700 MHz (Aller and Milne, 1972) is apparent in the declination scan but is not a serious limitation here.

III. Results

The flux densities obtained here are of much greater accuracy than those from any previous survey. More important, these results are for a much greater sample than previously obtained at this level of accuracy and should tell us something of the interstellar absorption. So firstly, we derived a radio distance scale. This follows from

F. J. Kerr and S. C. Simonson, III (eds.), Galactic Radio Astronomy, 411–416. All Rights Reserved.
Copyright © 1974 by the IAU.

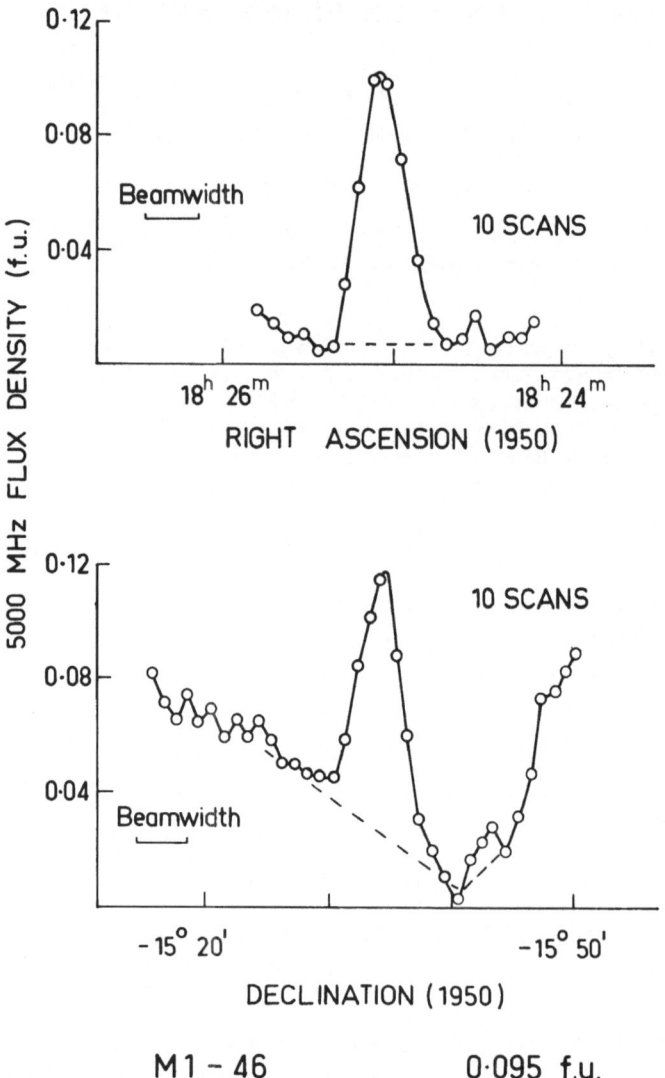

Fig. 1.   Averages of 10 scans through the optical position of the moderately weak planetary nebula M1-46 (0.095 Jy) The base levels chosen for each set of scans are shown. The zero point on the flux density scale is arbitrary.

the method of Minkowski and Aller (1954) and Shklovskii (1956) – the constant mass method – in which it is assumed that the same fixed mass of gas is ejected to form the shell in all of the nebulae. One then calculates the free-free radio emission from this gas, arriving at the formula for the distance in parsecs,

$$d = 6180 \, \theta^{-3/5} S^{-1/5},$$ (1)

where θ is the angular radius of the nebula in arcseconds and S is the optically thin radio flux density in janskies. In this derivation a nebular mass of 0.16 $M_\odot$ and an electron temperature of 10^4 K were assumed.

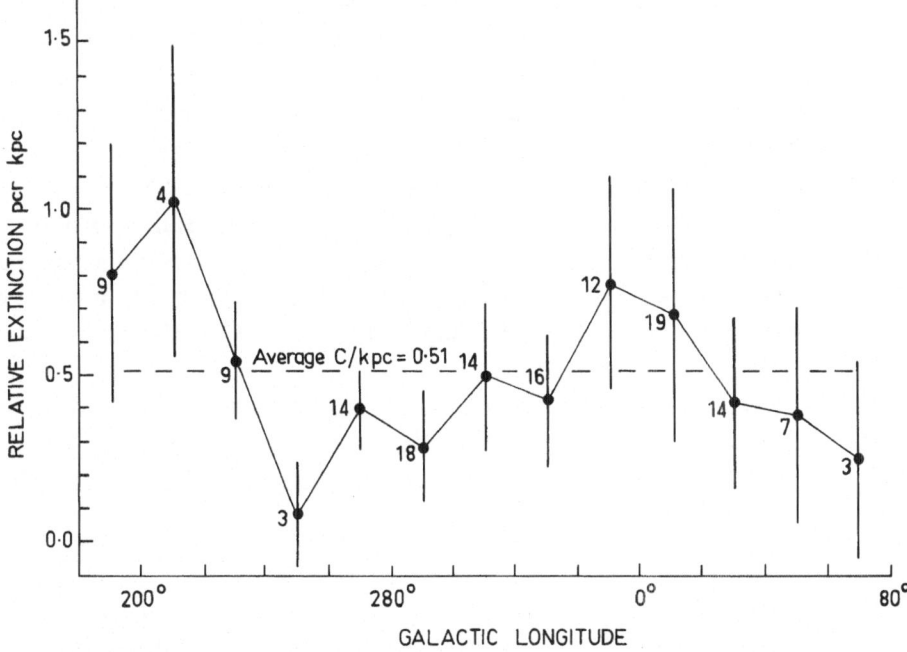

Fig. 2. The extinction per kiloparsec averaged over each 20 deg of galactic longitude. This diagram is based on a galactic disk model 300 pc thick with no absorption outside this disk. The vertical bars represent the error in the average absorption at each 20° point, the numbers of nebulae averaged in each 20° sample are shown. The average extinction per kpc is indicated.

NGC 7293

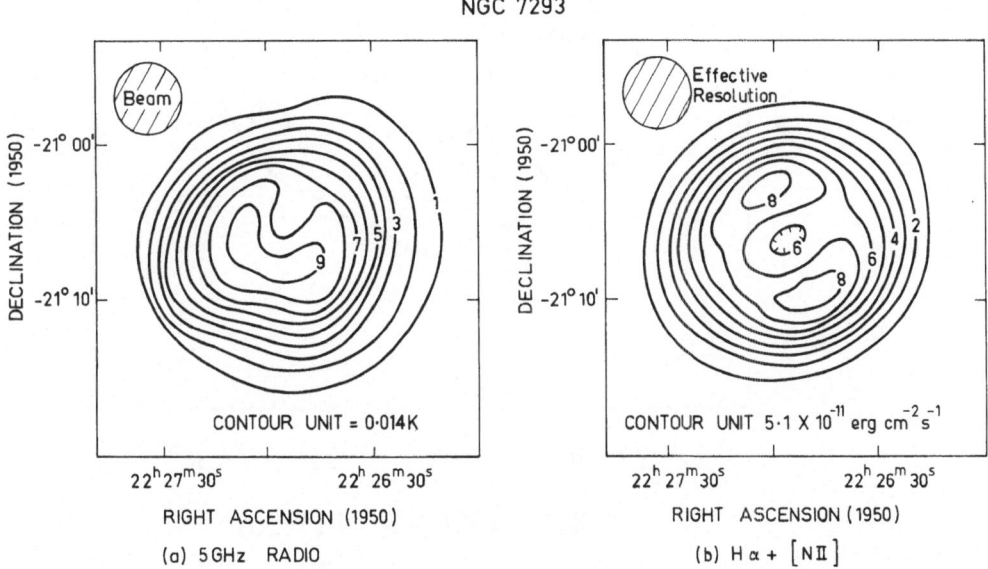

Fig. 3. (a) 5000 MHz isotherms of the radio emission from NGC 7293. The contour unit is 0.014 K full beam brightness temperature. (b) Hα + [N II] isophotes of NGC 7293 (Gurzadyan 1963) smoothed to the same resolution as the 5000 MHz contours in Figure 3a. The contour unit is 5.1×10^{-11} erg cm^{-2} s^{-1}.

Extinction coefficients were calculated from comparison with the Hβ flux densities. The extinction coefficients obtained here show none of the irregularities seen in previous surveys; at least they are, with two exceptions, all positive within the errors.

Finally, since the distances obtained from Equation (1) are independent of absorption we have calculated values for the mean absorption per kiloparsec. For this we adopted, after some consideration, a uniformly absorbing galactic disk 300 pc thick with no absorption outside this disk. In Figure 2 we show the mean absorption per kpc averaged each 20° along the galactic plane. Significant increases in the mean absorption are seen in the direction of Orion and the galactic centre. The average extinction coefficient per kpc is 0.15±0.3 in log F (Hβ), i.e., 1.28±0.75 mag of absorption per kpc.

There are suggestions (Mottmann, 1972) that the nebular masses increase with time and that a mass of 0.16 $M_\odot$ is typical of an older nebula, in which case our distances would be upper limits, yielding lower limits on the mean extinction per kpc. How-

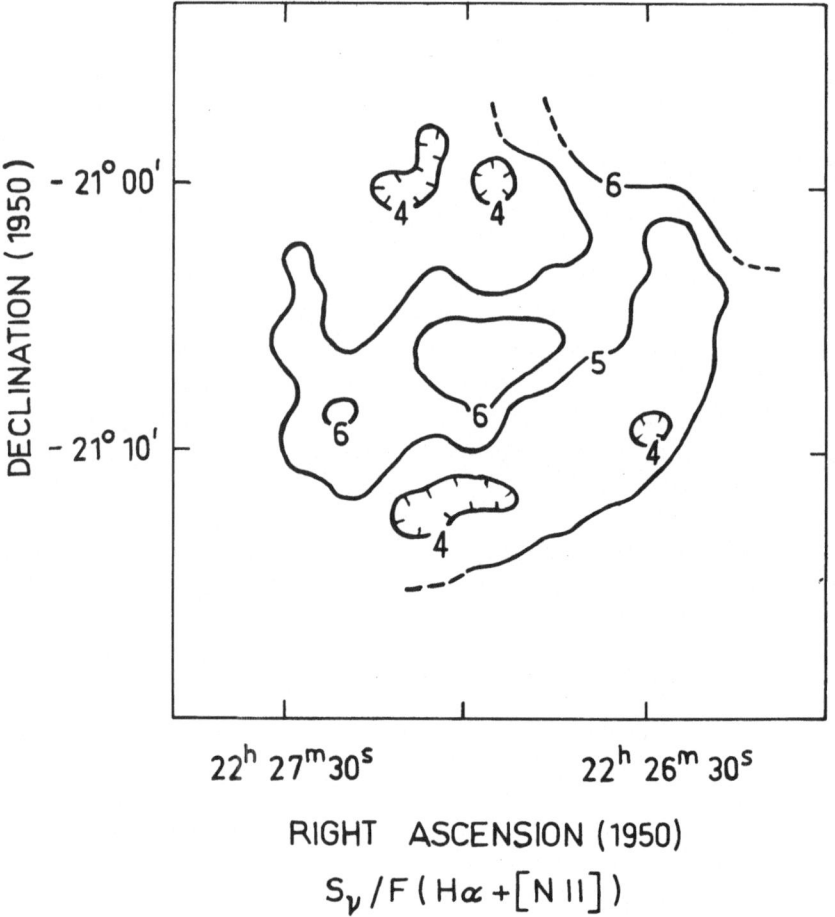

Fig. 4.   Contours of the ratio of the radio brightness to the Hα+[N II] intensity over the nebula NGC 7293. The contour unit is 10^{+8} Jy/erg cm^{-2} s^{-1}, and the effective resolution is 4.4′.

ever one should bear in mind that many of these objects are more than 150 pc from the galactic plane, so that only part of the distance obtained here is in the absorbing layer.

Only one of the nebulae in this study was resolved in our 4.5' beam. This nebulae, NGC 7293 (the Helix nebula), was mapped with integrated scans at several declinations. The radio contours in Figure 3a show the dipole structure of this object. In Figure 3b we have convolved Gurzadyan's (1963) Hα ($+$[NII]) isophotes to the same resolution as the radio contours. The contour unit in this diagram is 5.1×10^{-11} erg cm^{-2} s^{-1} and was obtained from Metik and Gershberg's (1964) value for the integrated Hα+[NII] flux density for NGC 7293.

The main difference between Figures 3a and 3b is in the greater angular depth of the indentation in the NW side of the optical features: an absence of optical emission where there is a maximum in the radio contours. This is shown quantitatively in Figure 4, where we have drawn contours of the ratio radio flux density/(Hα+[NII]) flux density. Changes in this ratio could be explained either by increased absorption or an increase in the electron temperature over the region of high S_v/F(Hα). Assuming that about 20% of the optical emission near Hα is in the [NII] lines and adopting an electron temperature of 10^4 K, a value of $4.4 \times 10^{+8}$ in Figure 4 would imply an extinction coefficient of zero. Further the range in S_v/F(Hα) in Figure 4 would total a range of ~ 0.25 in extinction coefficient and an average value of 0.02, close to the value of 0.01 obtained from the comparison of radio and Hβ flux densities. If these variations were to be satisfied by changes in the electron temperature, then temperature differences of 15000 K would be necessary.

Structure in the interstellar extinction with this angular scale ($\sim 10'$) is not unexpected. However, Terzian, in this symposium, has shown from his fine synthesis maps of much smaller planetary nebulae that absorption on the angular scale of the smaller objects exists, implying that this absorption is associated with the nebula. If this is so then the mean interstellar absorption would be less than the values that we derive.

We are continuing these observations to include fainter objects and to include those for which optical fluxes are not yet available.

References

Aller, L. H. and Milne, D. K.: 1972, *Australian J. Phys.* **25**, 91.
Gurzadyan, G. A.: 1963, *Soobshch. Byurakansk. Obs. Akad. Nauk Arm. S.S.R.* **34**, 59.
Metik, L. P. and Gershberg, R. E.: 1964, *Izv. Krymsk. Astrofiz. Obs.* **31**, 112.
Minkowski, R. and Aller, L. H.: 1954, *Astrophys. J.* **120**, 261.
Mottmann, J.: 1972, Thesis, University of California, Los Angeles.
Shklovskii, I. S.: 1956, *Astron. Zh.* **33**, 222.

D. K. Milne
CSIRO Division of Radiophysics,
P.O. Box 76,
Epping, N.S.W. 2121, Australia

L. H. Aller
Astronomy Department,
University of California,
Los Angeles, Calif. 90024, U.S.A.

DISCUSSION

H. R. Dickel: What is the excitation class or type of the NGC 7293 planetary nebula? Would a variation of [N II] also be able to explain optical-radio variation, rather than extinction?

Milne: I don't remember. I think that the rough sum suggested that it would take approximately a 15 000 K variation across the nebula to explain it as an [N II] effect. I believe that it is all differential absorption.

Terzian: I think we should be careful and not overinterpret the interstellar extinction data in deriving 'accurate' distances. Since recent work shows that many planetary nebulae are strong infrared sources, part of the extinction (absorption) may be directly associated with the nebulae.

Milne: I think that the presence of differential absorption on the scale of NGC 7293 and also on the scale of the much smaller objects that you have observed also suggests that this absorption is close to the nebula.

Greenberg: Terzian's comment on the existence of dust in the planetary nebula affecting the extinction raises the important point that one should carefully distinguish between extinction and absorption. Even if there were dust in the nebula it would show up quite differently than if it were somewhere along the line of sight. Roughly speaking, it would tend to exhibit its extinction only as absorption (without scattering), whereas in interstellar space the extinction measures the *sum* of absorption and scattering. I have suggested this effect in *Mem. Soc. Roy. Sci. Liège* (1972), p. 197.

FINE STRUCTURE IN PLANETARY NEBULAE

YERVANT TERZIAN

National Astronomy and Ionosphere Center, Cornell University, Ithaca, N.Y., U.S.A.

Abstract. A short report is presented on the aperture synthesis radio observations of planetary nebulae performed by Terzian *et al.* (1974). Examples of radio maps at 8085 MHz are shown with angular resolutions of the order of 2″ to 3″.

Fine Structure in Planetary Nebulae

Recently aperture synthesis techniques at centimeter wavelengths have made possible the study of planetary nebulae with angular resolutions as good as 2″ (Terzian *et al.*, 1974; Balick *et al.*, 1973; and Scott, 1973). Such observations have in many cases angular resolutions comparable to optical photographs.

In this short report I should like to present the most recent results of aperture synthesis observations which I have completed with the collaboration of B. Balick and C. Bignell (Terzian *et al.*, 1974). Observations were made using the National Radio Astronomy Observatory (NRAO) three-element interferometer at observing frequencies of 2695 MHz (λ 11 cm) and 8085 MHz (λ 3.7 cm). This instrument consists of three 85-ft (25.4 m) telescopes, two of which can be moved along a linear track to form up to 16 different telescope separations from 100 to 2700 m.

The synthesis maps of the observed planetary nebulae were generated from the data by the standard Fourier inversion programs available at NRAO. The maps generated in this fashion contain sidelobes because of the unfilled nature of the synthesized aperture. In order to remove the sidelobes, these maps were 'cleaned' using a pattern recognition technique described by Fomalont (1973) and Högbom (1974).

More than 40 planetary nebulae were observed in this program. Complete observations on the available telescope spacings have been performed for 14 nebulae, for which radio maps have been produced. Figures 1 to 4 show examples of radio maps at 8085 MHz for NGC 6543, NGC 7027, NGC 7354 and NGC 7662.

The results of this study show that several of the observed planetary nebulae have a general double source structure – this is more evident in the lower-resolution 2695 MHz maps. The higher-resolution 8085 MHz maps show substantial fine structure. In addition, spherical or elliptical shells are present for most of the observed nebulae, and central intensity depressions appear in the nebulae for which the synthesized beam is smaller than the source dimensions.

NGC 6543, shown in Figure 1, is a helical type nebula studied by Münch (1968) and has a complex optical structure. There exists some correlation between the bright optical features of the nebula and the radio intensity peaks. Woolf (1969) has reported a large infrared excess radiation from NGC 6543.

Figure 2 shows the radio structure of NGC 7027. The optical appearance of this nebula is irregular with bright condensations. Significant local obscuration must exist

F. J. Kerr and S. C. Simonson, III (eds.), Galactic Radio Astronomy, 417–421. *All Rights Reserved.*
Copyright © 1974 by the IAU.

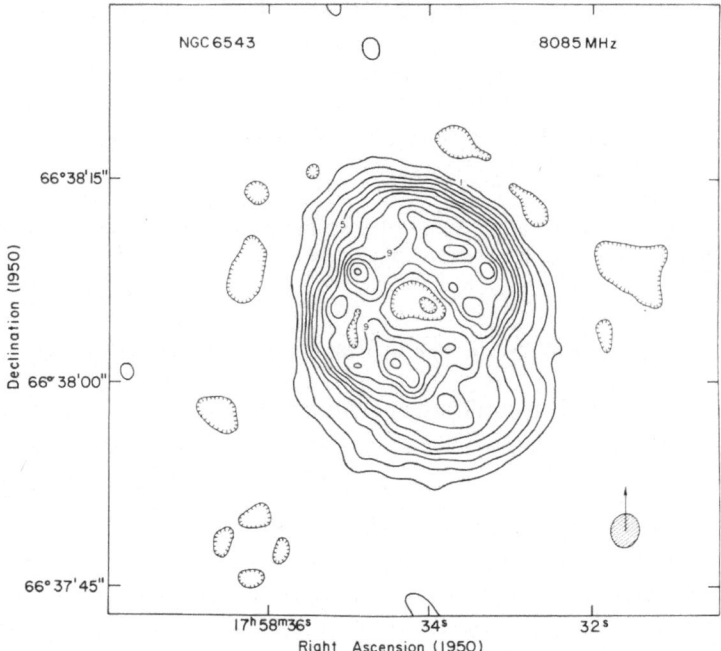

Fig. 1.   The radio brightness distribution for NGC 6543 at 8.1 GHz. The dimensions of the synthesized beam are 2.5″ × 2.2″. The contour spacings correspond to 6.7 K of brightness temperature.

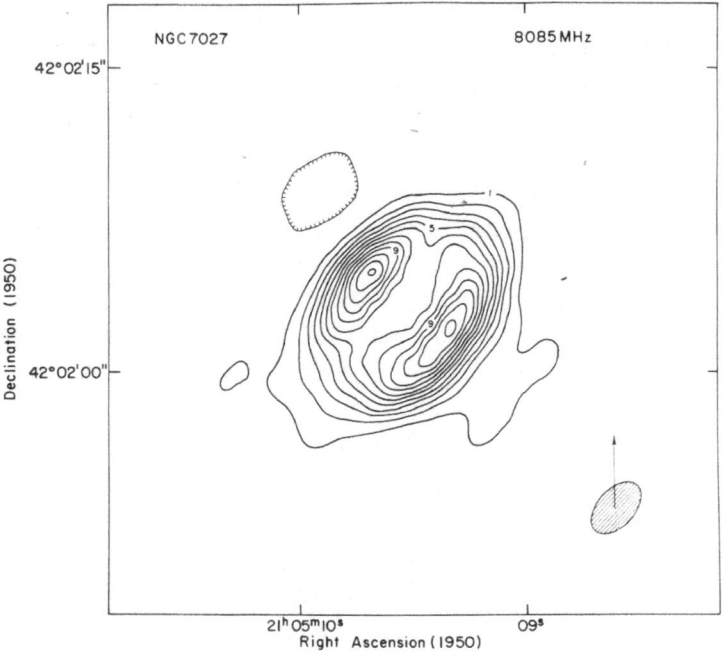

Fig. 2.   The radio brightness distribution for NGC 7027 at 8.1 GHz. The dimensions of the synthesized beam are 3.0″ × 1.9″. The contour spacings correspond to 200 K of brightness temperature.

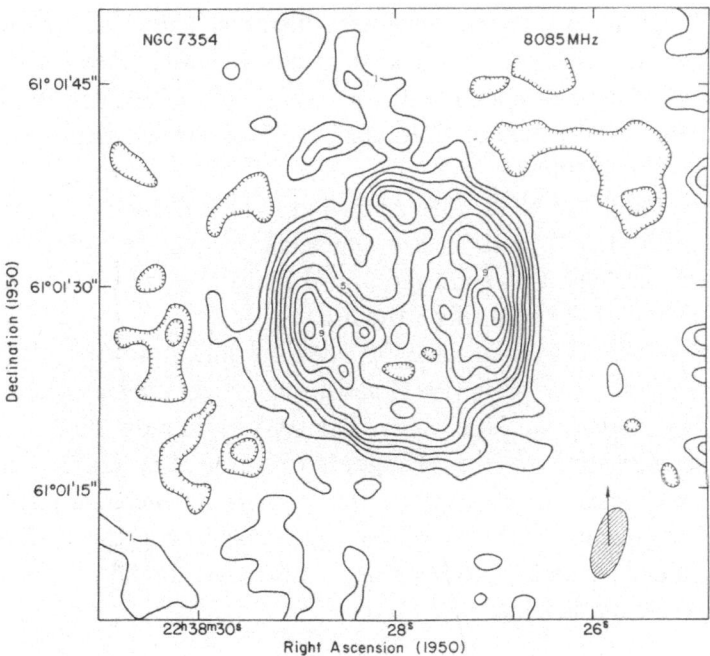

Fig. 3.   The radio brightness distribution for NGC 7354 at 8.1 GHz. The dimensions of the synthesized beam are 2.7″ × 2.1″ (note drawing error in beam major axis). The contour spacings correspond to 2.9 K of brightness temperature.

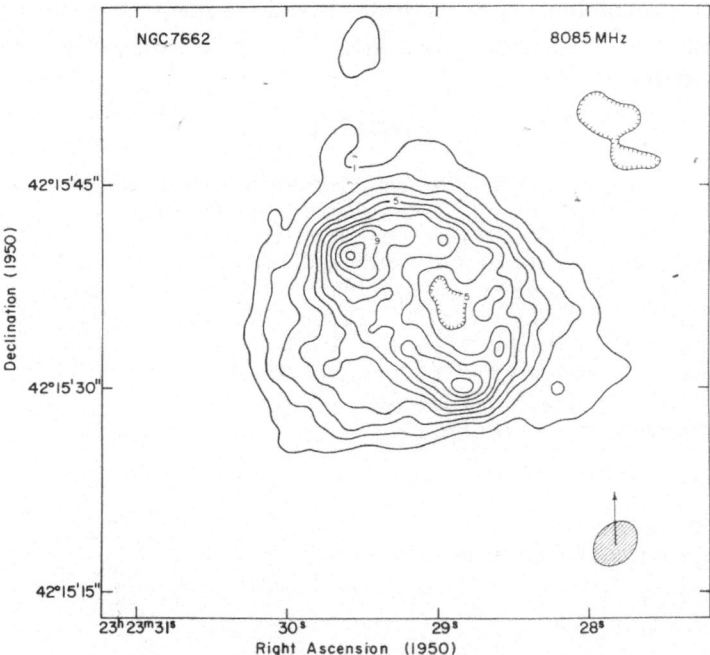

Fig. 4.   The radio brightness distribution for NGC 7662 at 8.1 GHz. The dimensions of the synthesized beam are 3.5″ × 2.0″. The contour spacings correspond to 5.8 K of brightness temperature.

in NGC 7027, particularly in the north-east direction. The north-east major radio component is hardly visible at optical wavelengths. It is well known that NGC 7027 has a very large infrared excess (Gillett *et al.*, 1972). Recently Wynn-Williams (1973) has produced an intensity map of NGC 7027 at $\lambda 11$ μm which greatly resembles the radio map shown in Figure 2.

Optical photographs of NGC 7354 (shown in Figure 3) show that it is composed of an irregular oval ring $22'' \times 18''$ in size, which is surrounded by an outer fainter envelope $\sim 32''$ in size. The two main radio peaks have a good correspondence with the optical maxima of the oval ring structure.

Figure 4 shows the radio map for NGC 7662. At optical wavelengths this nebula shows two very thin rings. Aller (1971) has given optical intensity contours superimposed on a photograph of NGC 7662. The 8085 MHz radio map is very similar to these results, indicating no large changes in obscuration across the nebula.

The aperture synthesis radio observations of planetary nebulae give the spacial distribution of matter in these objects, since interstellar extinction and/or local absorption is negligible at radio wavelengths. Such studies should help us understand the origin of planetary nebulae.

Acknowledgements

The aperture synthesis work was supported by the National Astronomy and Ionosphere Center, which is operated by Cornell University under contract with the National Science Foundation, and by the National Radio Astronomy Observatory, which is operated by Associated Universities, Inc., under contract with the National Science Foundation.

References

Aller, L. H.: 1971, *The Planetary Nebulae*, Sky and Telescope Monographic Series I.
Balick, B., Bignell, C., and Terzian, Y.: 1973, *Astrophys. J. Letters* **182**, L117.
Fomalont, E.: 1973, *Proc. Inst. Elec. Electron. Engrs.* **61**, 2111.
Gillett, F. C., Merrill, K. M., and Stein, W. A.: 1972, *Astrophys. J.* **172**, 367.
Högbom, J. A.: 1974, *Astron. Astrophys. Suppl.* **15**, 417.
Münch, G.: 1968, in D. E. Osterbrock and C. R. O'Dell (eds.), 'Planetary Nebulae', *IAU Symp.* **34**, 259.
Scott, P. F.: 1973, *Monthly Notices Roy. Astron. Soc.* **161**, 35.
Terzian, Y., Balick, B., and Bignell, C.: 1974, *Astrophys. J.* **188**, 257.
Wynn-Williams, C. G.: 1973, private communication.
Woolf, N. J.: 1969, *Astrophys. J. Letters* **157**, L37.

Yervant Terzian
National Astronomy and Ionosphere Center,
Cornell University,
Ithaca, N.Y. 14850, U.S.A.

DISCUSSION

Shakeshaft: I suggest that it is misleading to describe these radio sources as having mainly double structures, since the features you refer to as double are generally just peaks upon the broader components. The Cambridge 5 km map of NGC 7027 at 6 cm has such a double feature, and Paul Scott was able to fit this map very well by a model with a thick cylindrical shell, the axis of which is at about 30° to the line of sight.

Terzian: Many optical photographs of planetary nebulae show distinct brightness maxima at opposite directions; it is not surprising for the radio observations to suggest a similar pattern. In our work so far we have not made any attempt to fit cylindrical models, mainly because we have seen that at the higher resolutions the sources have a great deal of fine structure and cylindrical models will be unrealistic.

Dickel: If these planetaries really are shells with two more or less symmetrically opposite condensations, is there any correlation of the orientation of these blobs with galactic coordinates, magnetic field, or anything else?

Terzian: I think the number of sources studied in detail are still very small to be able to do such statistical studies.

Wynn-Williams: Observations of NGC 7027 by Becklin, Neugebauer and myself at 10 μm with 2″ resolution show that the dust giving rise to the infrared emission has a symmetrical structure similar to that at radio wavelengths. The hot dust is, therefore, almost certainly unrelated to the cold dust causing the extinction seen in part of the source, since the extinction is seen to be so uneven.

Westerhout: Can Shakeshaft explain the *physics* of a 'cylindrical' shell?

Shakeshaft: I do not know what might be the physical origin of a cylindrical shell, but it is a model that has been used successfully to account for the distribution of optical emission in various planetary nebulae.

Habing: D. George and M. Kaftan-Kassim have observed three planetary nebulae with the Westerbork Synthesis Radio Telescope at 6 cm with 6″ resolution. In the two well resolved cases (NGC 6720 and NGC 40) they find structures that resemble cylindrical models.

Menon: It is rather curious that in the majority of cases you have studied, the planetary nebulae seem to select the orientation of the maximum of the interferometer resolution to show their double structure.

Terzian: This is not so. In the cases where the synthesized beam is nearly circular the orientation of the double structure is random.

Ekers: This effect could be caused by the elliptical cross section of the beam. For example, a ring of emission observed with an elliptical beam will have two peaks in the position angle of the minor axis of the beam. I suggest that you should convolve these maps to a circular beam in order to obtain a better estimate of the brightness distribution.

Terzian: I certainly agree. This will be helpful in the cases where the beam is very elongated.

Milne: If these objects were supernova remnants, we would have no hesitation interpreting these observations in terms of spherical shells.

Zuckerman: Have your results been able to shed any light on the question of recombination lines?

Terzian: I believe only three planetary nebulae show the recombination line so far – NGC 7027, IC 418, and NGC 6543. The recombination lines in planetary nebulae are exceedingly weak and one has to integrate over 30 h with the best telescopes in order to get any detection. However, even the few existing results are difficult to explain, and we haven't tried to work out any relationship yet.

RADIO OBSERVATIONS OF IR/OH/H₂O STARS

J. L. CASWELL

Division of Radiophysics, CSIRO, Sydney, Australia

Abstract. A small percentage of infrared stars, selected on the basis of their IR brightness at 2.2 μm and their redness ($I-K$ colour index), show 18 cm OH emission, and some exhibit 1.35 cm H₂O emission; further investigations identify some of these objects as M-type supergiants and the remainder as M-type giant long-period variable stars.

For the strongest 1612 MHz OH emitters, long-baseline interferometry shows that the emission arises from a large number of small-diameter features distributed over a much larger, approximately elliptical region corresponding to a circumstellar cloud. Models for the geometry of the emitting regions are discussed.

The OH characteristics with respect to the ratios of line intensities for different transitions and the velocity structure are distinctive, not being found in other types of OH source; consequently a large number of new OH emitters can be recognized as probably IR/OH objects despite the absence of IR or optical data; the precise classification of these sources is considered together with their relevance in understanding the identified sources.

I. Introduction

Shortly after the discovery of cosmic OH masers, infrared radiation was invoked as a possible source of the maser pumping; it was natural therefore to search for OH emission from the large number of objects found in the Caltech infrared survey at 2 μm (Neugebauer and Leighton, 1969). As a result, Wilson and Barrett (1968) detected OH emission with intensity greatest in the 1612 MHz transition from four IR objects, one of which, NML Cyg, proved to be the strongest OH emitter discovered to date. Subsequently a further class of OH emitter associated with IR objects was found with emission chiefly in the main lines at 1665 and 1667 MHz (Robinson *et al.*, 1971b). We will refer to both types of objects as IR/OH sources; they correspond to oxygen-rich M-type supergiants, and M-type giants, the latter group comprising long-period (Mira) variables (as shown by Hyland *et al.*, 1972).

H₂O emission at 1.35 cm has also been detected in the IR/OH sources, VY CMa showing strong H₂O emission and a number of other sources showing weaker emission (Schwartz and Barrett, 1970).

In the following sections we discuss the major characteristics of the OH and H₂O emissions, together with some of the closely related optical and infrared properties. The unidentified OH emitters of this type also give significant clues to our understanding of the objects.

II. Characteristics of the OH and H₂O Emission

It is convenient to follow Wilson and Barrett (1972) in treating the IR/OH objects as forming three classes according to their readily observed properties; in addition, we have treated as a separate class those OH sources with no IR identification but whose OH properties suggest they are IR/OH objects.

F. J. Kerr and S. C. Simonson, III (eds.), Galactic Radio Astronomy, 423–438. All Rights Reserved.

(a) CLASS A. M-TYPE SUPERGIANT 1612 MHz OH EMITTERS

Only four objects in this class are known; they are listed in Table IA. The common properties, several of which can be seen in Figure 1, are:

(i) OH emission is strongest in the 1612 MHz line and occurs in two velocity ranges separated by about 50 km s^{-1}; in each velocity range the emission generally shows a gradual decrease of intensity in the direction of the other peak and a sharper decrease at the outer side. Little or no polarization is detectable.

(ii) Weaker 1667 and 1665 MHz (main-line) OH emission is present and may be highly polarized.

(iii) H$_2$O emission is observed in all four stars.

(iv) The 1720 MHz OH transition has not been detected.

(b) CLASS B. WEAKER 1612-MHz OH EMITTERS ASSOCIATED WITH M-TYPE GIANT LONG-PERIOD VARIABLE STARS

The 1612 MHz emission is similar to that of the supergiants but is 'scaled down' in

TABLE I

Positions and microwave properties of IR/OH stars

Name	Equatorial coordinates (1950.0)				Galactic coordinates		Peak intensity[a] OH emission		H$_2$O	References[b]
	R.A.			Dec.	l	b	1612 MHz	1665/7 MHz	22235 MHz	
	h	m	s	° ′	°	°		(Jy)		
A. M-type supergiants										
VY CMa	07	20	55	−25 40.2	239	−5.1	230	140	10000	1, 2
VX Sgr	18	05	03	−22 14.0	8	−1.0	25	14	√	3, 4, 5
NML Cyg	20	44	34	+39 55.9	80	−1.9	520	13	80	6, 7
PZ Cas[c]	23	41	41	+61 31.0	115	−0.0	2.3		√	8, 9
B. M-type giant 1612 MHz OH emitters										
IRC+10011	01	03	48	+12 19.8	129	−50.1	37	3	–	10, 4
NML Tau	03	50	46	+11 15.7	178	−31.4	3	2.5	80	10, 5
IRC+50137	05	07	20	+52 48.9	156	+7.8	12	0.7	–	10, 4
IRC+40156	06	29	45	+40 44.9	174	+14.1	4.6	1.3	–	10
IRC−20197	09	42	56	−21 48.1	256	+23.3	7	0.7	–	10
R Crt	10	58	09	−18 03.6	269	+37.2	2	2	290	5
WX Ser	15	25	32	+19 44.1	30	+53.5	3	3	–	10
IRC+30292	16	25	59	+34 54.6	56	+43.5	3	–	–	10
IRC−10381	17	48	28	−08 00.7	19	+9.5	2.5	–	–	10
IRC−10434	18	30	30	−07 29.0	24	+0.6	2.5	0.8	–	10
IRC+10365	18	34	59	+10 23.0	41	+7.8	1.5	–	–	10
IRC−10450	18	37	35	−05 45.8	27	−0.1	3.4	–	–	10
R Aql	19	03	58	+08 09.2	42	+0.4	84	7.5	110	10, 4, 7
IRC−20540	19	05	56	−22 19.2	15	−13.6	12	√	–	10
RR Aql	19	54	58	−02 01.2	39	−15.6	5	2	19	10, 5
IRC−10529	20	07	46	−06 24.7	36	−20.4	35	8	–	10
UX Cyg	20	53	00	+30 13.4	74	−9.4	3	–	–	10
V Mic	21	20	37	−40 55.2	1	−45.5	8.4	1.9		11
IRC+40483	21	25	23	+36 29.0	84	−10.2	3.5	–	–	10

Table I (Continued)

Name	Equatorial coordinates (1950.0)			Galactic coordinates		Peak intensity[a] OH emission		H₂O	References[b]
	R.A.		Dec.	l	b	1612 MHz	1665/7 MHz	22235 MHz	
	h	m	s	° ′	°	°	(Jy)		

C. Unidentified 1612 MHz OH emitters

Name	h	m	s	Dec.	l	b	1612	1665/7	22235	Refs
	16	37	30	−46 13.9	338	+0.1	5.1	–		11
	16	49	52	−41 43.8	343	+1.3	12	–		10, 4
	17	35	58	−32 10.3	356	−0.6	29	4	–	10, 4
d	17	50	02	−26 45	2	−0.3	4	–		12
	18	19	48	−12 53	18	+0.4	8			13
	18	21	17	−12 28.0	19	+0.4	26	2		10, 4, 14
	18	25	36	−10 13	21	+0.4	11			13
	18	26	48	−12 44	19	−1.0	16			13
W 41	18	31	27	−09 00.9	23	−0.3	12	–	–	4
	18	37	42	−05 00.6	27	+0.2	30	–		10, 4
	18	44	48	−02 39	30	−0.3	13			13
W 43A	18	45	05	−01 48.3	31	+0.0	55	2.5		4, 15
	18	45	54	−02 54	30	−0.6	55			13
	18	46	00	−01 51	31	−0.2	10	1		13
	18	48	48	−01 05	32	−0.4	7			13
	18	49	48	−00 18	32	−0.3	30	2		13
	18	54	56	+02 08	36	−0.3	18	1	–	4, 16
	19	11	54	+11 05	45	+0.1	11	–		13
ON 4	20	26	40	+38 57.0	78	+0.2	9	–	–	4, 17

D. M-type giant main line OH emitters

Name	h	m	s	Dec.	l	b	1612	1665/7	22235	Refs
R Hor	02	52	12	−50 05.6	265	−57.4	–	3.3	–	18
IRC+20082	04	26	07	+24 37.6	173	−16.3	–	0.7	–	19
U Ori	05	52	51	+20 10.1	189	−2.5	–	6	80	20, 21
R LMi	09	42	35	+34 44.6	191	+49.8	–	0.6	–	19, 22
W Hya	13	46	12	−28 07.1	318	+32.8	–	13	230	20, 7, 21
RU Hya	14	08	42	−28 38.4	323	+30.8	–	0.4	150	22, 5
S CrB	15	19	21	+31 32.8	49	+57.2	1.4	3	57	20, 21
U Her	16	23	35	+19 00.3	35	+40.3	–	3.5	330	7, 20, 21
IRC−20424	18	00	58	−20 19.5	10	+0.8	1.7	4	160	20, 5, 21
GY Aql	19	47	20	−07 44.3	33	−16.5	–	0.2	–	22
R Peg	23	04	08	+10 16.4	85	−44.6	–	0.4	70	22, 5
R Cas	23	55	53	+51 06.6	115	−10.6	–	3.5	56	20, 23

Notes to Table I

[a] Representative values from published profiles. Variability can be considerable – e.g., reported intensities for H₂O emission from VY CMa range from 1000 to 20000 Jy. No entry indicates absence of data; '–' indicates null result; '√' indicates positive detection but no details available.

[b] Basic radio data (positions, flux densities, line profiles) are referenced; more detailed studies are referred to in the text.

[c] Data from Kukarkin *et al.* (1969) indicate similarity with VX Sgr; wide separation of OH peaks (∼55 km s⁻¹) is also characteristic of supergiants rather than giants.

[d] This OH source was found while pointing at the 100 μm IR source Hoffman 39 (no shorter wavelength IR counterpart has been reported); a chance coincidence is likely at this low galactic latitude, and the OH source is probably 'unidentified'.

References to Table I
 1. Eliasson and Bartlett (1969).
 2. Knowles *et al.* (1969).
 3. Caswell and Robinson (1970).
 4. Hardebeck (1972).
 5. Dickinson *et al.* (1973).
 6. Wilson *et al.* (1970).
 7. Schwartz and Barrett (1970).
 8. Dickinson and Chaisson (1973).
 9. Dickinson (1973).
10. Wilson and Barrett (1972).
11. Caswell *et al.* (1971).
12. Chaisson and Dickinson (1972).
13. Winnberg *et al.* (1973).
14. Turner (1970a).
15. Robinson *et al.* (1970b).
16. Downes (1970).
17. Elldér *et al.* (1969).
18. Robinson *et al.* (1971b).
19. Wilson and Riegel (1973).
20. Wilson *et al.* (1972).
21. Hardebeck and Wilson (1971).
22. Fillit *et al.* (1972).
23. Turner and Rubin (1971).

both intensity and in velocity width (both in the separation of the peaks and widths of each feature). The main-line emission is also weaker and is sometimes not seen above the sensitivity limits achieved to date; H_2O emission is rather less frequently detectable and apparently confined to those sources showing main-line emission. Table IB lists sources in this category.

(c) CLASS C. UNIDENTIFIED 1612 MHz OH SOURCES WITH CHARACTERISTICS OF
      IR/OH OBJECTS

This group of sources we define as having the OH properties of Classes A and/or B, but they have been detected independently of known stars or of any prior IR search. It might be expected that a subsequent IR search would unambiguously group these with A or B, but in the few cases where such a search has been made IR emission is either absent or surprisingly weak; accordingly it seems desirable to define a separate category for these sources pending further study. Table IC lists these sources. With the addition of many new unpublished detections (Section VI), these now comprise the largest single group.

(d) CLASS D. IR/OH SOURCES WITH OH STRONGEST IN THE MAIN LINES
      (1665 AND 1667 MHz)

This class of source is associated with an IR object/M-type long-period variable star and shows OH emission strongest at 1665 or 1667 MHz. In some instances 1612 MHz emission is detectable, indicating no clear-cut division between Classes B and D (see Section IVc). A complex, possibly double-peaked, velocity structure is often recognizable, the velocity separation being less than that at 1612 MHz. H_2O emission is

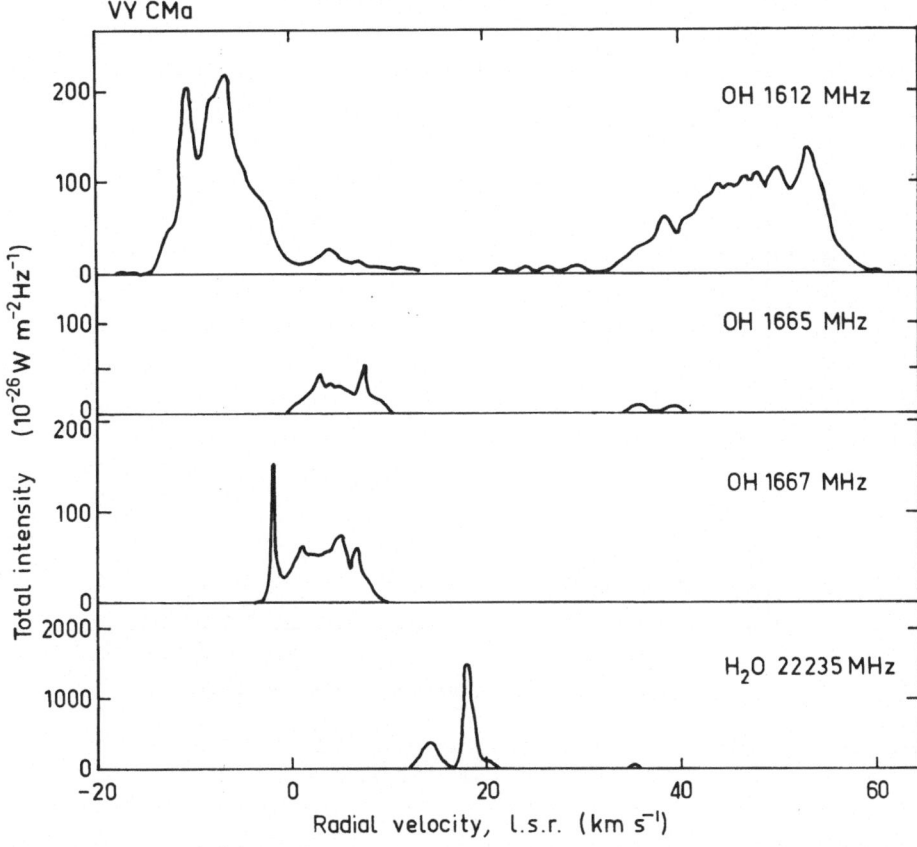

Fig. 1.   OH and H$_2$O emission associated with the M-type supergiant VY CMa.

frequently detectable; again, the 1720 MHz transition has not been detected. Table ID lists the sources in this category.

III. More Detailed Study of Radio Properties

(a) THE CHARACTERISTIC 1612 MHz VELOCITY STRUCTURE

This is well shown in VY CMa (Figure 1). In this particular example the low-velocity feature has a greater intensity (both peak and integrated) than the higher (more positive velocity) feature. This is not a general characteristic, and of the 52 sources in Classes A, B and C for which data are available, approximately one-half (27) have the high-velocity feature more intense than the low-velocity feature, and within each individual group the fraction does not differ significantly from one-half. The histogram of Figure 2 shows the frequency distribution of the intensity ratio of the two features:

$$R = \frac{\text{Intensity of stronger feature}}{\text{Intensity of weaker feature}}.$$

We conclude that most commonly the features are approximately equal in intensity

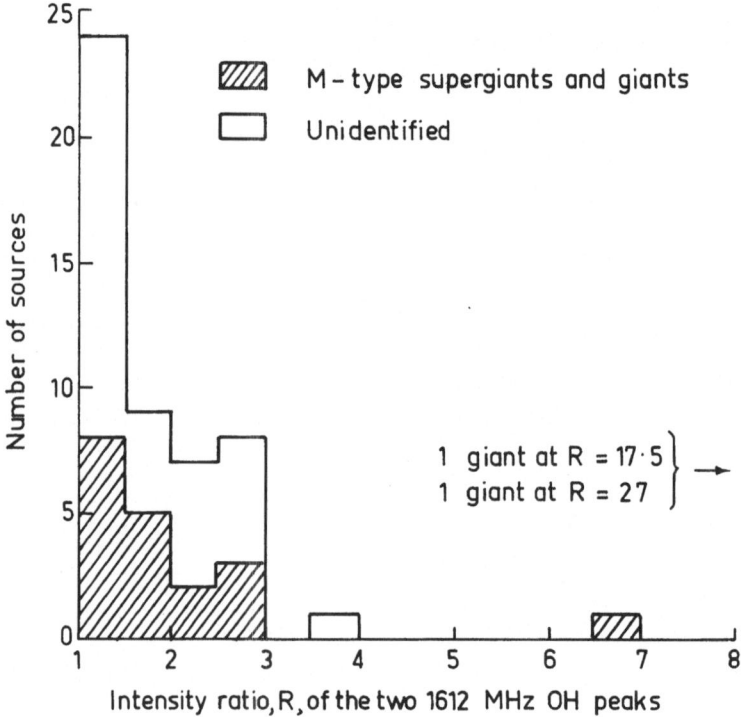

Fig. 2.   The distribution of the intensity ratio, R, of the two 1612 MHz OH peaks for 52 sources.
The median value of R is 1.7.

(both peak and integrated) with a steadily smaller fraction showing large ratios. Apart from UX Cyg, the distribution is complete for the identified sources and indicates that large ratios are rare. For UX Cyg only a single feature has been detected, probably because the sensitivity and/or velocity range of the observations was inadequate to detect the other feature. A number of single-feature unidentified 1612 MHz emitters have also been detected but probably only a few of these are double with high intensity ratios and at least some such objects do not belong to Class C; accordingly all single-feature unidentified objects are omitted from further study here (for example, OH 284.2–0.8 probably belongs to Class C, but OH 331.5–0.1 differs in several characteristics – see Caswell *et al.* (1971)).

(b) 1612 MHz ANGULAR STRUCTURE

Davies *et al.* (1972) have used long-baseline interferometry to derive the map of NML Cyg reproduced in Figure 3. In more recent work, Masheder *et al.* (1973) discuss further the NML Cyg results and present similar data for VY CMa; for both sources it is found that many individual emission features with diameters of only about 0.05″ are spread over a larger elliptical region of about 2″ and the positions of the low-velocity peaks are contained within a slightly smaller area than that covered by the high-velocity peaks. Over the total extent of the source, a gradient of velocity is observed which can be interpreted as indicative of rotation; however, it is important

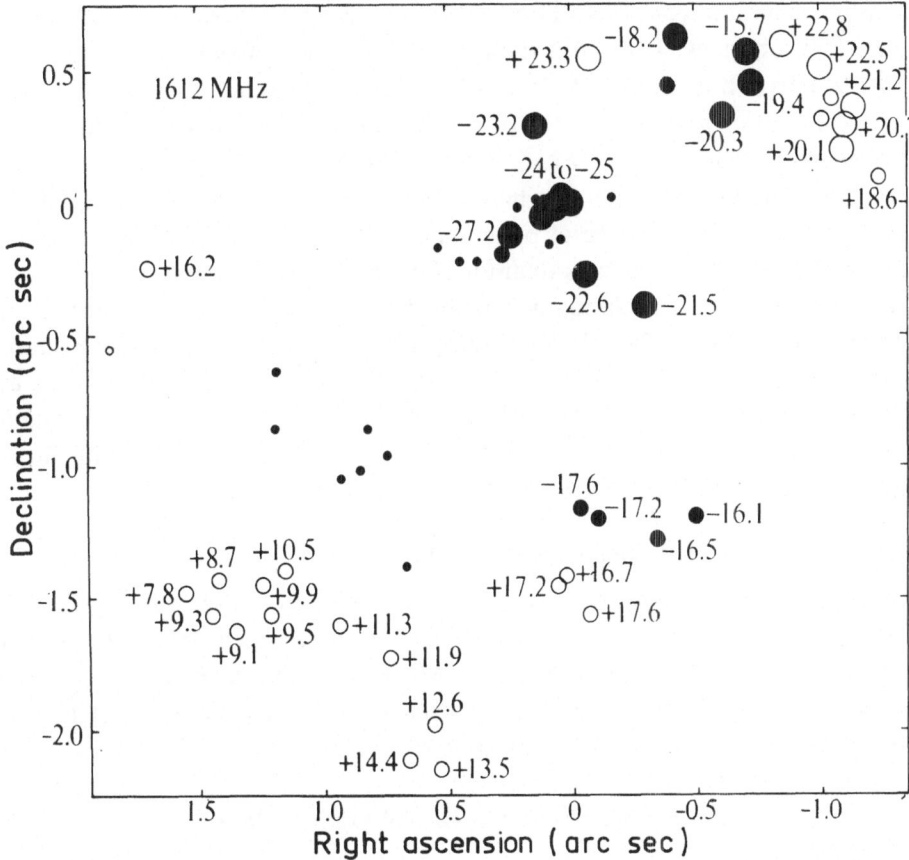

Fig. 3. The positions of the principal 1612 MHz OH features in NML Cygnus. Open circles indicate more positive velocity features; filled circles indicate more negative velocity features. Numbers indicate radial velocities (km s^{-1}) relative to the LSR (from Davies et al., 1972).

to note that the map rules out rotation as the major cause of the double-peaked velocity structure. These data will be discussed further in connection with proposed models of the source (Section V).

(c) MAIN-LINE EMISSION AND WATER VAPOUR EMISSION

For Class A, B and C sources, two main-line emission peaks are sometimes present, with velocities usually just within the range of velocities of the 1612 MHz emission. Where only one main-line peak is present, it shows no preference for occurring at the high or low velocity, nor does it systematically accompany the stronger or the weaker 1612 MHz peak. The overall angular extent of the region emitting at 1665 or 1667 MHz is (for VY CMa, the only object studied to date) much smaller than at 1612 MHz (Harvey et al., in preparation, quoted by Masheder et al., 1973).

The 1665 and 1667 MHz emission from the supergiants VY CMa and VX Sgr contrasts with the 1612 MHz emission by showing both circular and linear polarization and also erratic variability in both intensity and polarization characteristics (Booth, 1969; Robinson et al., 1970a; Robinson et al., 1971a).

Main-line emission from Class D objects often displays a double-peaked velocity structure (Wilson *et al.*, 1972), but the central null is less distinct than that of the 1612 MHz emission in Classes A, B and C. Whether the main-line emission (in all four classes) is strictly similar to that of H II region emitters is not clear; for IR/OH objects the intensity of 1667 MHz emission is commonly comparable with or greater than that of 1665 MHz emission, whereas the converse is true of H II region OH emitters; in addition, the degree of polarization of 1665 and 1667 MHz emission is perhaps lower for IR/OH sources than for H II region OH emitters.

H_2O emission usually accompanies main-line emission in Classes A and D and sometimes also in Class B, but H_2O has not yet been detected in any Class B sources without detectable main-line emission. A number of long-period variable stars and IR sources have been found to show H_2O emission prior to any search being made for OH emission (Dickinson *et al.*, 1973; Dickinson, 1973); these are probably further examples of Classes A, B and D rather than a new class of object.

(d) TRANSITIONS OF OH IN THE EXCITED STATE

Several IR/OH objects have been searched for transitions in both the $^2\pi_{3/2}$, $J=5/2$ state and the $^2\pi_{1/2}$, $J=1/2$ state; a positive detection has been obtained only from the $^2\pi_{3/2}$, $J=5/2$ state for a single source, NML Cyg (Zuckerman *et al.*, 1972). With increased sensitivity, further study of the excited state transitions might assist in defining physical conditions within the sources.

IV. Comparisons with IR and Optical Data

(a) POSITIONS

The identifications of the OH/H_2O emitters with optical and IR stars are based on good positional agreements. Hyland *et al.* (1972) summarize new IR data and radio positions (the latter chiefly from Hardebeck, 1972) which indicate that for Class A and many Class B objects the agreement is very good; Hardebeck and Wilson (1971) show that positional agreement is good for several Class D objects.

Masheder *et al.* (1973) point out that, unfortunately, the absolute positional accuracy of their detailed OH maps is not sufficient to allow precise positioning of the stellar or IR object relative to the individual OH components.

(b) INTENSITY VARIABILITY

In the infrared, some sources such as NML Cyg show little or no variability, while others such as CIT 3 show variation of several magnitudes. The infrared variability allows some objects with no detectable optical counterpart (because of obscuration) to be unambiguously classified as Mira variables. The 1612 MHz OH emission exhibits slow periodic variations apparently paralleling the optical and infrared changes (Bechis *et al.*, 1971; Hardebeck and Wilson, 1971) and corroborating the identifications. The percentage modulation of the OH intensity is always less than that of the infrared or optical emission.

The main-line OH emission and the H_2O emission have also been reported to show systematic long-term variability (in addition to the more erratic variability noted earlier) for all classes of source.

Dickinson (1973) concludes, on the basis of a new search for water-vapour emission, that the Mira variables most commonly showing H_2O emission are those which show the greatest change in visual magnitude.

(c) INTENSITY AND COLOUR OF ASSOCIATED IR OBJECTS

We will merely summarize an interpretation and list the supporting evidence.

For OH emission at 1612 MHz and on the main lines, and for water-vapour emission, an appropriate cool M-type star is required – an inference made from the correlation of the presence of all three emissions with the colour index $I-K$ (Hyland et al., 1972).

The main-line OH emission often found in Class B objects will be considered together with that of Class D objects; its detectability is correlated with K magnitude (Wilson et al., 1972), and since K magnitude is a satisfactory distance indicator for this type of star the correlation indicates that only nearby objects are detected and that the luminosity of main-line OH emission is generally low. Furthermore there must presumably be a quite low scatter in maximum main-line luminosity in order for the correlation to be recognizable. The fact that objects with detectable main-line OH emission are local is reflected in the large scatter in galactic latitude which they display (the mean $|b|$ for main-line emitters in Tables IB and ID is $\sim 28°$).

The detectability of H_2O emission is correlated with that of main-line emission, showing that it requires similar physical conditions. However, the intensity of H_2O emission is more sensitive to small changes in these conditions (as is also indicated by its generally more extreme time variability); it thus has a large range of luminosity which weakens any correlation with K magnitude (Dickinson et al., 1973).

1612 MHz emitters are best regarded as a sub-class of the main-line emitters; the very wide range of 1612 MHz luminosity which they exhibit results from a strong dependence on the presence of a circumstellar dust cloud (inferred from an IR excess in the wavelength range 10–20 μm). The wide range of luminosity destroys any recognizable correlation with distance, i.e., with K magnitude (Wilson and Barrett, 1972). Because of its wide range of luminosity 1612 MHz emission is not detectable in many of even the nearby main-line emitters (Class D objects), while at the other extreme it may be an order of magnitude stronger than main-line emission so that for high-luminosity 1612 MHz emitters at large distances *main-line* emission is not detectable; the mean value of $|b|$ for Class B emitters with no detectable main-line emission is only 13°, consistent with their being objects more distant than those showing main-line emission.

(d) OH VELOCITY COMPARISONS WITH OPTICAL AND IR DATA

The optical spectra of M-type supergiants and M-type giant long-period variables usually show an emission line spectrum. For about half the stars studied an absorption spectrum has also been measured, displaced from the emission to more positive velocities. In the few IR/OH sources for which optical data exist, the emission and absorption velocities correspond approximately with the more negative and more positive velocity 1612 MHz peaks, respectively (see Table 3 of Wilson and Barrett, 1972). A similar correlation is found for main-line emission (Wilson *et al.*, 1972). However, the optical velocities in some cases change considerably during the light cycle (cf., e.g., Feast, 1963), whereas to a high degree of precision (better than 1 km s^{-1}) no velocity changes have been detected in the OH velocities, so that the correlation must be treated with caution.

Optical observers of long-period variable stars have noted a correlation of $A - E$ (the separation of the extreme velocities) with period (cf., e.g., Feast, 1963); since $A - E$ appears to correspond with the separation of the peaks of 1612 MHz OH emission, it is to be expected that a plot of the latter against period will show a similar correlation; furthermore, for this investigation it is possible to use infrared periods even where no optical data are available. Dickinson and Chaisson (1973) verify that such a relation appears to hold.

The separation in velocity of the OH peaks is also correlated with redness (or dust-shell thickness) for both 1612 MHz and main-line peak separation (Wilson *et al.*, 1972), i.e., the shell is expanding with a velocity approximately proportional to the dust-shell thickness. It might be expected that since the dust shell appears to be a prime requirement for 1612 MHz emission, the luminosity of 1612 MHz emission would in turn be proportional to the peak separation, but this is not readily seen above the scatter, although there is some suggestion of its validity in that intense 1612 MHz emission from supergiants has a wider velocity separation than that from the weaker giants.

V. Models of the Sources

The models proposed so far are very preliminary. They have been designed to fit the supergiants for which more data are available but would be expected with little change to apply to Class B sources also.

For VY CMa, a source for which a large amount of optical and IR data exist, Hyland *et al.* (1969) explain the infrared excess in terms of a circumstellar dust cloud and give appropriate model parameters. The OH emission can be fitted into this picture in a variety of ways. However the more negative velocity ('blue-shifted') OH peak is generally assumed to arise from the near side of an expanding shell.

In both the models to be mentioned here, the more positive velocity OH peak is assumed to be at the stellar velocity, by comparison with optical absorption data (Merrill, 1940). Davies *et al.* (1972) suggest that it arises in the same shell as the more negative velocity emission, but at the periphery as projected along the line of sight, where the expansion is merely a tangential motion. The intensity distribution as a function of radial velocity (showing a central minimum) is not an immediate consequence of the model but the angular distribution of the emission is accounted for satisfactorily. Dickinson *et al.* (1973) suggest the data could be better accounted for if a shock-front is invoked with the smaller area of 'blue-shifted' emission being due to its containment within the shock-front and the 'stationary' (more positive velocity) gas being external to it.

Essentially Davies et al. (1972) emphasize that the area covered by the more negative velocity peak is smaller than that covered by the more positive velocity peak, while Dickinson et al. (1973) emphasize that it is not very much smaller. The absence of high-velocity emission in the central region (in angular extent) is necessary to the Davies interpretation but is irrelevant to the Dickinson interpretation. For neither model is it clear why the far side of the shell, with a still greater red shift, should not be seen; Davies et al. suggest free-free absorption by ionized gas, optically thick at 18 cm, within the OH emitting shell; this hypothesis might be contradicted if no continuum emission from the absorbing cloud could be detected at high frequencies. In the direction of ON-4, an upper limit to the continuum flux density of 1.5×10^{-3} Jy at 1400 MHz has been reported by Baars and Wendker (1973). Turner (1970b) cites continuum interferometer measurements at 2700 MHz yielding upper limits of 10×10^{-3} Jy for NML Cyg and VY CMa. At 5000 MHz an upper limit of 20×10^{-3} Jy for VY CMa has been obtained (Caswell, unpublished); for this source Wilson (1971) has reported a tentative detection $(200 \pm 70 \times 10^{-3}$ Jy) at 3.5 mm which still lacks confirmation. Goss et al. (1974) have recently reported the detection of a continuum source with $S(10.7 \text{ GHz}) = 12 \pm 2 \times 10^{-3}$ Jy coinciding with NML Cyg to within 30″. Information on its angular extent and spectrum is necessary to determine whether it possesses the properties required to fit the model of Davies et al. (1972)

General OH and H$_2$O maser pumping mechanisms have been discussed by Turner (1970b) whereas Litvak and Dickinson (1972) present a pumping mechanism designed to account specifically for IR/OH objects; any refinements will probably require a better understanding of the physical geometry of the sources.

VI. The Unidentified Sources

The unidentified sources of Table IC constitute 'accidental' discoveries (while studying other objects) together with objects found in a *survey* at Onsala (Winnberg et al., 1973).

None of the objects has been found to coincide with a known late M-type star or with a previously catalogued IR source. Although an IR source has subsequently been discovered in the direction of ON-4, apparently it is weak and has properties unusual for OH/IR objects (Wilson and Barrett, 1972). The IR source which Glass and Feast (1973) propose as an identification with OH 338.5 + 0.1 is probably not associated, since a remeasurement of the OH position (Caswell and Haynes, unpublished data) confirms a separation of more than 2′. It is clearly very important to achieve the maximum possible positional accuracy for the OH sources in order to allow sensitive searches for related optical and IR objects.

The mean value of $|b|$ for unidentified sources in Table IC is 0°.4. These sources have been discovered largely in searches confined to low galactic latitudes, but it should be noted that no unidentified sources have been found in several hundred search positions at higher latitudes during either the Wilson and Barrett (1972) or the Caswell *et al.* (1971) search of known IR objects. We infer that the small value of |b| for unidentified sources would not be increased much if systematic surveys were extended to higher latitudes and we note that such a small value of |b| is characteristic of distant supergiants. In contrast, objects of Class B have a median value of |b| ~ 20°, characteristic of Mira variables, as noted by Wilson and Barrett (1972).

The velocity separations of the 1612 MHz emission peaks in the case of unidentified sources are characteristic more of the long-period variable giants than of the supergiants (Winnberg *et al.*, 1973).

Overall, we conclude that although the unidentified objects have OH properties similar to both the supergiants (Class A) and the giants (Class B), they are not merely unidentified members of either class but probably represent a new class in their own right.

Figure 4 shows a velocity-longitude diagram for Class A, B and C objects in which the velocity plotted is that of the more red-shifted peak. The Class B objects (M giants)

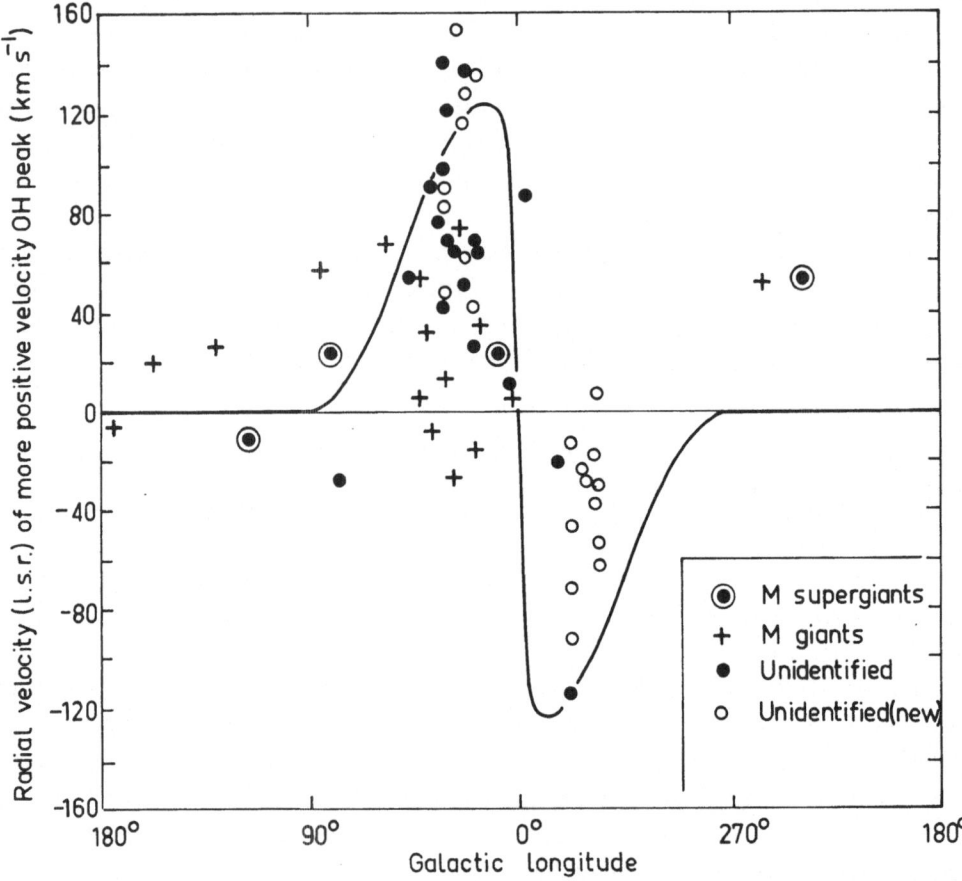

Fig. 4.   Velocity-longitude plot for 1612 MHz IR/OH emitters; the radial velocity plotted is that of the more positive velocity OH peak. The curve indicates the maximum velocity 'allowed' on the Schmidt galactic rotation model. Unidentified (new) sources are unpublished results from the Onsala survey and a new survey by Caswell and Haynes (see text).

are relatively nearby and have no significant velocity component arising from galactic rotation but show a quite high scatter in their random velocities – both positive and negative. For the unidentified objects published to date in the longitude range 0° to 50° no objects show negative velocities while three show velocities greater than the maximum 'allowable' on the Schmidt (1965) galactic rotation model. Wallerstein (1973) infers from the distribution in velocity of the published Onsala unidentified sources in this region that (a) they are more like supergiants at large distance and (b) the true velocity representing the stellar motion may be midway between the OH peaks.

We have added a number of new unpublished sources to study this situation further. In the longitude range 18° to 33° there are nine additional sources from the Onsala search (Winnberg et al., private communication) and 12 sources in the range 327° to 340° have been added from a new low-latitude survey of 1612 MHz OH emission by Caswell and Haynes (unpublished). In the region 18° to 33° the absence of negative velocities results partly from less coverage of this velocity range, but the survey by Caswell and Haynes extended to velocities of $+130 \text{ km s}^{-1}$ so that the striking absence of positive velocity sources in the region 327° to 340° is real and significant. The velocities range from zero to the maximum 'allowed'; since at least some of the objects would presumably be nearby, the absence of positive velocities at longitudes 327° to 340° (and the absence of negative velocities at longitudes 18° to 50°) implies a small velocity dispersion. Where large velocities are observed, they are presumably due to galactic rotation and indicate large distances. We conclude that the objects are more like supergiants in their kinematic properties and it is reasonable to use the characteristic velocities to infer kinematic distances. This important conclusion appears inescapable.

In the longitude range 18° to 33° six sources exceed the maximum 'allowed' velocity whereas none does so in the longitude range 327° to 340°. No such striking asymmetry is evident in the HI distribution (Kerr and Hindman, 1970). In Figure 5 we have replotted the velocity-longitude diagram using the mean velocity of the two OH peaks, instead of the more positive of the two velocities. It is seen that this produces a much greater degree of symmetry (or more precisely, antisymmetry) about $l=0°$, particularly for the unidentified OH sources. This plot suggests interpreting the mean velocity as representing the motion of the star; such an interpretation would not be consistent with the models considered previously but would allow a much simpler explanation of the two OH emission peaks in which they arise from near and far sides of an expanding shell. It was shown in Section III that, statistically, in their widths and intensities and their accompaniment by main-line emission, the more positive and less positive velocity 1612 MHz OH peaks are not distinguishable; this symmetry also favours the simple expanding-shell model. (We might also note that a contracting rather than an expanding shell would be equally acceptable for the OH interpretation.) Problems arise chiefly with the optical interpretation because, for a few of the identified objects, the more positive velocity OH peak corresponds fairly well with the optical absorption velocity near maximum light which in turn is customarily assumed to represent the characteristic stellar motion (see Section IVd).

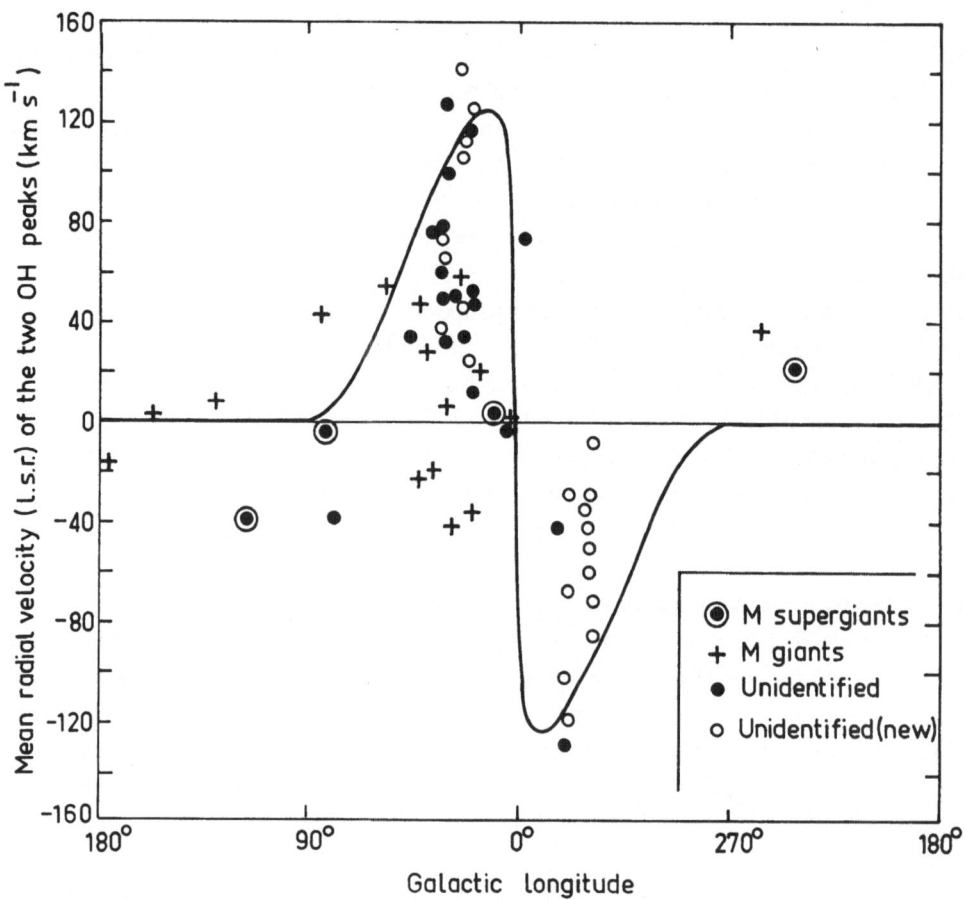

Fig. 5. Velocity-longitude plot for 1612 MHz IR/OH emitters; the radial velocity plotted is the mean of the two OH peaks.

VII. Conclusion

IR/OH stars commonly show weak main-line OH emission: the 1612 MHz OH emitters appear to represent a sub-class satisfying more stringent conditions – in particular a long-wavelength IR excess which probably provides the maser pump and which arises from a prominent circumstellar dust shell. The 1612 MHz emitters have well-defined general characteristics, and some sources have been studied in detail, but the correct geometrical model for the 1612 MHz emitting region remains a problem. A solution to this problem is suggested by the unidentified sources which support a simple expanding shell model in which the two major velocity peaks of 1612 MHz emission arise in near and far sides of the shell. The apparent conflict with optical interpretations which caused an earlier rejection of this model now merits further investigation.

The unidentified 1612 MHz OH sources constitute the largest class of IR/OH objects and represent a rapidly growing area of study; identification of these objects with IR stars is of major importance, since preliminary unsuccessful searches for identifications suggest that the amount of IR radiation needed to produce the 1612 MHz emission is less than for the identified objects.

With a better understanding of these aspects of 1612 MHz emission, viz. the geometrical model and the role of the IR radiation, there would be good prospects for solving the long-standing problem of a detailed pumping mechanism for OH maser emission.

Acknowledgements

I am grateful to colleagues who graciously sent recent reprints and contributed preprints and data in advance of publication; in particular the Onsala group kindly supplied information on their new sources. The unpublished data in the longitude range 327° to 340° I obtained in collaboration with Dr R. F. Haynes.

References

Feast, M. W.: 1963, *Monthly Notices Roy. Astron. Soc.* **125**, 27.
Fillit, R., Gheudin, M., Nguyen-Quang-Rieu, Paschenko, M., and Slysh, V.: 1972, *Astron. Astrophys.* **21**, 317.
Glass, I. S. and Feast, M. W.: 1973, *Astrophys. Letters* **13**, 81.
Goss, W. M., Winnberg, A., and Habing, H. J.: 1974, *Astron. Astrophys.* **30**, 349.
Hardebeck, E. G.: 1972, *Astrophys. J.* **172**, 583.
Hardebeck, E. G. and Wilson, W. J.: 1971, *Astrophys. J. Letters* **169**, L123.
Hyland, A. R., Becklin, E. E., Frogel, J. A., and Neugebauer, G.: 1972, *Astron. Astrophys.* **16**, 204.
Hyland, A. R., Becklin, E. E., Neugebauer, G., and Wallerstein, G.: 1969, *Astrophys. J.* **158**, 619.
Kerr, F. J. and Hindman, J. V.: 1970, *Australian J. Phys. Astrophys. Suppl.* No. 18, 1.
Knowles, S. H., Mayer, C. H., Cheung, A. C., Rank, D. M., and Townes, C. H.: 1969, *Science* **163**, 1055.
Kukarkin, B. V., Kholopov, P. N., Efremov, Yu. N., Kukarkina, N. P., Kurochkin, N. E., Medvedeva, G. I., Perova, B. N., Federovich, V. P., and Frolov, M. S.: 1969, *General Catalog of Variable Stars*, 3rd ed., Astron. Council Acad, Sci. U.S.S.R., Moscow.
Litvak, M. M. and Dickinson, D. F.: 1972, *Astrophys. Letters* **12**, 113.
Masheder, M. R. W., Booth, R. S., and Davies, R. D.: 1974, *Monthly Notices Roy. Astron. Soc.* **166**, 561.
Merrill, P. W.: 1940, *Spectra of Long-Period Variable Stars*, University of Chicago Press, Chicago.
Neugebauer, G. and Leighton, R. B.: 1969, *Two Micron Sky Survey, A Preliminary Catalog*, NASA SP-3047, Washington, D.C.
Robinson, B. J., Caswell, J. L., and Goss, W. M.: 1970a, *Astrophys. Letters* **7**, 79.
Robinson, B. J., Goss, W. M., and Manchester, R. N.: 1970b, *Australian J. Phys.* **23**, 363.
Robinson, B. J., Caswell, J. L., and Dickel, H. R.: 1971a, *Astrophys. Letters* **8**, 171.
Robinson, B. J., Caswell, J. L., and Goss, W. M.: 1971b, *Astrophys. Letters* **7**, 163.
Schmidt, M.: 1965, *Stars and Stellar Systems* **5**, 513.
Schwartz, P. R. and Barrett, A. H.: 1970, *Astrophys. J. Letters* **159**, L123.
Turner, B. E.: 1970a, *Astrophys. Letters* **6**, 99.
Turner, B. E.: 1970b, *J. Roy. Astron. Soc. Can.* **64**, 221.
Turner, B. E. and Rubin, R. H.: 1971, *Astrophys. J. Letters* **170**, L113.
Wallerstein, G.: 1973, *Astrophys. Letters* **15**, 83.
Winnberg, A., Goss, W. M., Höglund, B., and Johansson, L. E. B.: 1973, *Astrophys. Letters* **13**, 125.
Wilson, W. J.: 1971, *Astrophys. J. Letters* **166**, L13.
Wilson, W. J. and Barrett, A. H.: 1968, *Science* **161**, 778.
Wilson, W. J. and Barrett, A. H.: 1972, *Astron. Astrophys.* **17**, 385.
Wilson, W. J. and Riegel, K. W.: 1973, *Astron. Astrophys.* **22**, 473.
Wilson, W. J., Barrett, A. H., and Moran, J. M.: 1970, *Astrophys. J.* **160**, 545.

Wilson, W. J., Schwartz, P. R., Neugebauer, G., Harvey, P. M., and Becklin, E. E.: 1972, *Astrophys. J.*
  177, 523.
Zuckerman, B., Yen, J. L., Gottlieb, C. A., and Palmer, P.: 1972, *Astrophys. J.* **177**, 59.

J. L. Caswell
CSIRO Division of Radiophysics,
P.O. Box 76,
Epping, N.S.W. 2121, Australia

DISCUSSION

Palmer: First, how do you sort out which of the many main line emitters fit into your classification D; and second, does Orion fit into your classification scheme?

Caswell: I define Class D OH emitters as being associated with known IR objects, generally identified as M-type giant long period variables. Since the OH characteristics of Class D sources do not differ markedly from those of main line emitters associated with H II regions, no attempt has been made to see if any of the latter fit better into Class D. Orion does not fit into any of the classes discussed here.

Batchelor: How do the OH velocities compare with the stellar velocity?

Caswell: The more negative velocity peak corresponds approximately with the optical emission velocity and the more positive velocity with the optical absorption velocity; the latter is usually interpreted as representing the characteristic velocity of the star.

Barrett: A unique OH source was recently discovered by Lo and Bechis. The source appears coincident with V1057 Cyg and radiates at 1720 MHz in two features of flux density 48 and 34 Jy. The width of the features is ~ 0.3 km s^{-1}, and the separation is ~ 0.35 km s^{-1}. Both features show elliptical polarization of $\sim 25\%$. Emission or absorption at 1612, 1665 and 1667 MHz was not detected at a level of 0.7 Jy.

Zuckerman: Because the OH emission was observed with a single antenna it is dangerous to associate it with V1057 Cyg. A similar identification of an unusual OH source with the planetary nebula NGC 2438 turned out to be incorrect after an interferometric position was obtained for the OH source.

THE INFRARED PROPERTIES OF CIRCUMSTELLAR OH/IR
AND H₂O/IR SOURCES

A. R. HYLAND

Mount Stromlo and Siding Spring Observatory, Canberra, Australia

Abstract. Anomalous OH and H_2O microwave emission is a common occurrence in circumstellar shells surrounding cool oxygen-rich M supergiants and Mira variables. Infrared spectra have shown that the atmospheric conditions in these stars are such that the concentration of OH molecules is a maximum. For those with H_2O microwave emission, strong H_2O 1.9 μm absorption is present.

There exists a growing number of typical OH/IR sources discovered by radio surveys, for which no associated IR object has yet been found. Preliminary results of TV photography to discover IR counterparts are shown.

Several interesting correlations have been defined between the IR colour, period and amplitude of infrared variability, and the velocity separation of the OH emission peaks. The relationship of these to pulsationally driven mass loss and the observed characteristics of circumstellar dust shells is investigated.

The maser emission from OH/IR and H_2O/IR sources is believed to be pumped by IR radiation. Combined studies of the variability of OH and H_2O emission and the IR continuum of the Mira sources confirm that IR pumping is probably the dominant mechanism, although the data do not appear to favour any specific pump scheme. The OH masers are at least partially saturated; the degree of saturation of the H_2O masers is unclear.

I. Introduction

Since the association of certain microwave OH emission sources with stellar infrared sources was discovered several years ago (Wilson and Barrett, 1968; Eliasson and Bartlett, 1969), microwave, infrared and optical studies have established the general pattern of relationships which exist between the properties observed in different parts of the spectrum. The basic microwave characteristics of the OH/IR sources have been derived from the investigations of Wilson *et al.* (1970), Wilson and Barrett (1972) and Robinson *et al.* (1970, 1971). These may be summarized as follows:

(i) The OH emission may be strongest in (a) the 1612 MHz satellite line (the Turner [1970] type IIb sources, often called simply the 1612 MHz sources) or (b) the 1665/ 1667 MHz mainlines (the type I or mainline emitters).

(ii) The emission occurs in two main ranges of velocity separated by between 5 and 60 km s⁻¹. Where optical data are available, the more positive velocity peak corresponds reasonably well with the stellar absorption-line velocity, while the more negative peak agrees well with velocities derived from stellar emission lines. The OH profile peaks are usually narrow, though in a small number of cases (specifically, those identified with M supergiant stars) the 1612 MHz emission occurs in two broad features.

(iii) The OH emission is weakly polarized.

(iv) No emission has been detected at 1720 MHz, nor (with the exception of NML Cyg) has any continuum source been detected at the location of OH/IR sources.

As a result of a very recent investigation by Harvey *et al.* (1974), a further important characteristic should be added to the above.

(v) The OH emission often varies in a regular manner by factors of 2–4 in intensity, in phase and with the same period as the infrared variations.

F. J. Kerr and S. C. Simonson, III (eds.), Galactic Radio Astronomy, 439–458. All Rights Reserved.
Copyright © 1974 by the IAU.

It has also been shown (Schwartz and Barrett, 1970) that a number of microwave H_2O sources are also associated with infrared sources and late M stars, and that in several cases these are also OH sources. For these the H_2O emission velocity peaks generally lie between those of the OH emission, and the H_2O emission is often highly variable.

Combining the microwave characteristics with the optical and infrared studies of Hyland *et al.* (1969, 1972) and Wilson *et al.* (1972), a simple picture has been developed (relevant to both OH/IR and H_2O/IR sources) which encompasses the basic observations and is a convenient starting place for a review of the known infrared properties of such systems. Each system is considered as being composed of three main parts (shown schematically in Figure 1).

(i) A central stellar source which is either an M supergiant star or long-period variable M star with an effective temperature lying in the range $1800\,\text{K} < T_* < 2600\,\text{K}$.

(ii) A circumstellar shell of gas and dust which emits predominantly in the infrared and which has a characteristic temperature $T_s \sim 600$–700 K. (For simplicity all circumstellar shells are considered here to be spherical.)

(iii) A further circumstellar shell of gas which includes the OH and H_2O emitting regions and which may be partly or wholly co-extensive with the dust shell (ii), but which at least in two cases (NML Cyg and VY CMa) is approximately 10 times larger (see, e.g., Davies *et al.*, 1972). Dynamically it is currently thought that the circumstellar shells are expanding at a steady rate indicative of the rate of mass loss from the central

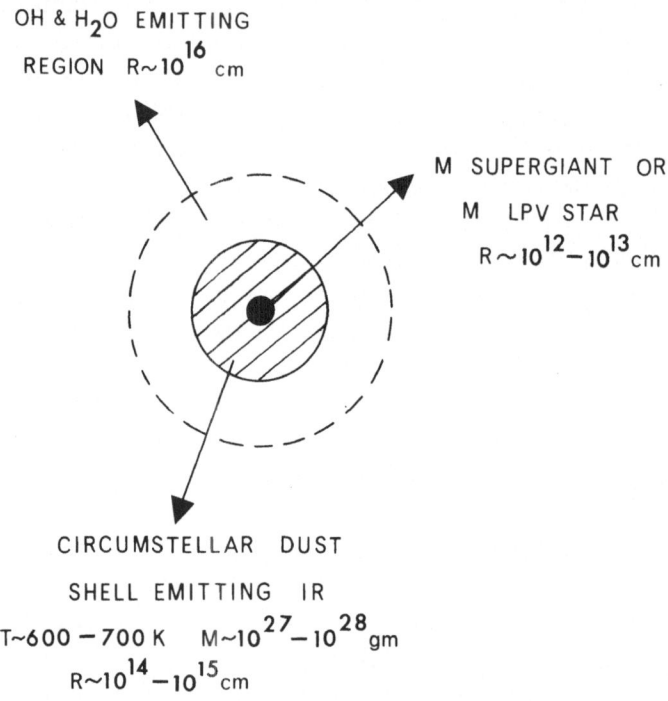

Fig. 1.   A schematic representation of the main components of an OH/IR source with typical dimensions.

source, while from considerations of the possible maser processes it is widely believed that the OH and H_2O molecules are pumped by infrared photons.

In this review the available infrared observations are discussed insofar as they relate to (a) OH and H_2O systems as a whole, (b) any one of the individual components described above or to (c) questions of radiative pumping and any correlations of properties which may help to bring the phenomena together in a coherent manner.

II. The Identification of OH/IR Systems

For historical reasons most surveys for OH/IR sources have been conducted by searching at the positions of known infrared sources, and these have proved eminently successful (see, e.g., Wilson *et al.*, 1970; Wilson and Barrett, 1972). More recently OH and H_2O emission has been sought successfully by searching at the positions of long period variable M stars (Schwartz and Barrett, 1971; Caswell *et al.*, 1971; Nguyen-Quang-Rieu *et al.*, 1971).

The identity of the microwave and infrared sources was established for several cases by the interferometric observations of Hardebeck (1972) and Hardebeck and Wilson (1971), positional agreement being obtained to better than 15″. A few accidentally discovered OH sources with the right characteristics have remained unidentified with any known infrared source.

Discoveries of OH/IR and H_2O/IR sources certainly differ from other radio surveys because of the unusual search technique which has been used, and consequently there are large selection effects inherent in the discoveries. The situation is presently being rectified somewhat by comprehensive sky surveys along the galactic plane. Winnberg *et al.* (1973) and Caswell and Haynes (1973) have already discovered a large number of sources with characteristics similar to the supergiant OH/IR sources. The most interesting feature of these recent results is that in no case is there any obvious IR source which can be identified with the microwave source. The possibility exists that these may be a new class of OH emission source, or at the very least the relationship of the intensity of microwave to infrared emission differs considerably from that of the well known sources.

To settle the question as to the nature of these new sources, it is important for infrared surveys to be undertaken in conjunction with the OH surveys to search for possible infrared counterparts. The common method of scanning the microwave error boxes at infrared wavelengths is time consuming unless the OH positional accuracy is high. This method has been used to locate infrared identifications for ON-4 (Neugebauer, quoted by Winnberg *et al.*, 1973) and G338.5+0.1 (Glass and Feast, 1973). However, no details are known regarding the characteristics of the former, and the latter identification has been challenged because of the lack of positional and velocity agreement.

A faster technique has been used by the author for searching for IR sources which holds promise for future identifications of OH/IR sources. The region to be searched is photographed through blue and infrared filters with a TV camera (SEC vidicon,

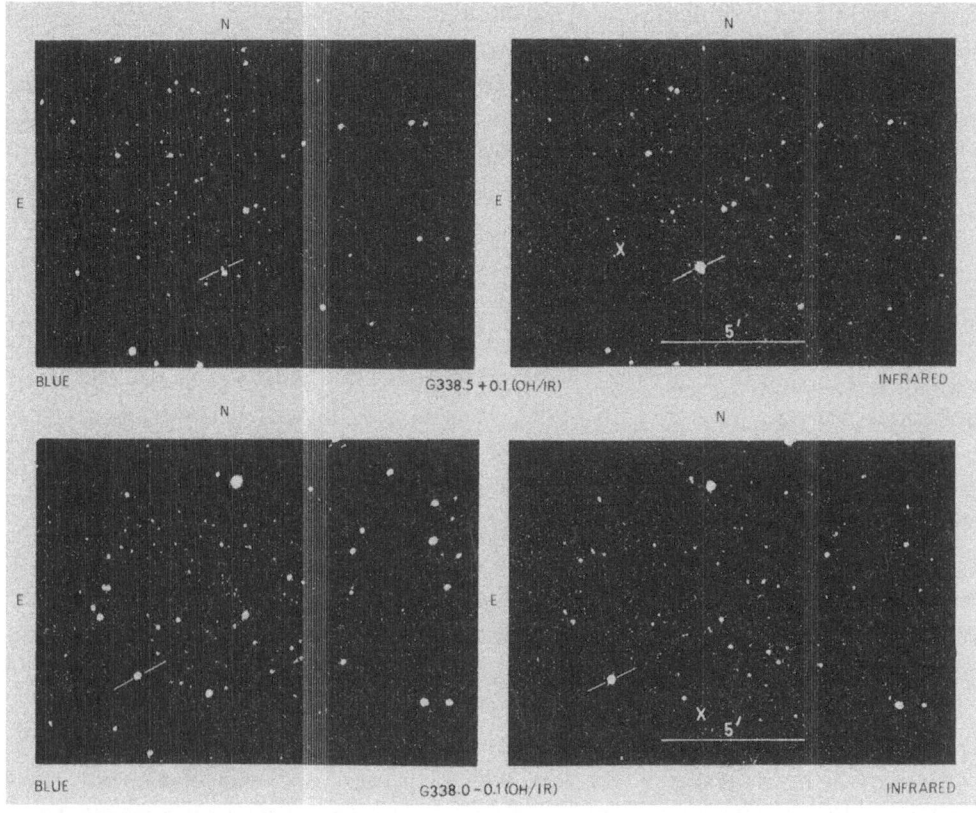

Fig. 2. Blue and infrared television photographs of the regions near the two sources OH338.5 + 0.1 and OH338.0 − 0.1 as described in the text. The OH positions are marked by a cross on the infrared frames, and the nearest infrared sources are marked.

S-25 photocathode) attached to the cassegrain focus of a 16 in telescope, and which gives a visible field of $\sim 15 \times 10'$. The filters and photo-cathode response provide effective wavelengths corresponding approximately to the $B(\lambda 4300 \text{ Å})$ and I_{Kron} $(\lambda 8000 \text{ Å})$ bands. Two examples of this technique are shown in Figure 2 where (a) the infrared source found by Glass and Feast (1973) is shown in comparison with the OH position for G338.5 + 0.1 and (b) the region near OH 338.0 − 0.1 is shown. There is an infrared source $\sim 2'$ east of the OH position in (b), but the positional agreement is too poor for this to be claimed as an identification.

This technique is clearly applicable to searches for infrared counterparts of OH sources. Typically for an integration time of 2.5 s, a limiting magnitude $I_{\text{Kron}} \sim 14$ can be reached, and this corresponds to a 2.2 μm magnitude $K \sim 6$ for the very reddest known objects. Such a search thus effectively extends the limit of the reddest known infrared sources some 3 mag fainter than that of the Caltech Two Micron Sky Survey (Neugebauer and Leighton, 1969).

III. Spectroscopic Observations and the Nature of the Central Stellar Sources

It has been firmly established from both optical and infrared spectra that the central sources of OH/IR systems are M supergiants and M-type long-period variable stars. The stars associated with microwave H_2O emission sources belong to the same categories, although irregular variables of late M spectral type have also been found to be H_2O emitters but not OH emitters (e.g., RX Boo). Because a large number of the OH/IR sources (in particular the 1612 MHz sources) are bright in the infrared, but very faint visually, the most comprehensive spectroscopic data has been obtained from infrared spectra in the 2 μm region (Hyland *et al.*, 1972; Frogel, 1970). From these spectra, which include most of the identified OH/IR and H_2O/IR sources, the following general features can be noted.

(i) Three of the strongest 1612 MHz OH sources (which are also microwave H_2O sources), NML Cyg, VY CMa and VX Sgr, have been classified as late M supergiants. The major characteristics of their spectra are the strong absorption due to the first overtone CO vibration rotation band and the presence of weak but visible H_2O absorption between 1.7 and 2.0 μm. These characteristics are illustrated by the spectra of VY CMa and VX Sgr in Figure 3. It is notable that the spectra of earlier supergiants (none of which has yet been found to have microwave OH or H_2O emission lines) show no absorption due to H_2O at the resolution available.

(ii) The majority of the 1612 MHz OH/IR sources have infrared spectra characteristic of the long-period variable (Mira) M stars; i.e., they have both very strong H_2O absorption in the broad 1.9 μm and 2.7 μm bands and moderate to strong CO absorption. The spectra of two typical examples of this class are also shown in Figure 3. In extreme cases (such as IRC+10011 shown) the continuum may be affected by absorption and re-emission of radiation by the circumstellar dust particles; however, the effect is not large enough to affect the classification of the central source to any marked degree.

(iii) Infrared and optical spectra of the mainline OH/IR sources, and the H_2O/IR sources show that these are also generally long-period variable M stars. No distinguishing spectroscopic features have been found by which one can separate these from the 1612 MHz sources. Three examples of such spectra are shown in Figure 4. S CrB is both a mainline OH source and an H_2O emitter, RX Boo is an H_2O source, and IRC−20424 is an unusual mainline OH emitter which will be discussed further in the next section. Hyland *et al.* (1972) have emphasized that the unique contribution of the infrared spectra is that estimates of the effective temperatures and abundance characteristics (especially the oxygen to carbon ratio) of the central sources can be obtained *independently* of interstellar or circumstellar reddening. The conclusions of their study of 1612 MHz sources are also applicable to the mainline OH emitters and the H_2O sources, and are:

(α) All the central stellar sources have photospheric temperatures which lie in the limited range 1800 to 2600 K. For conditions typical of the photospheres of luminous M stars this is the very temperature range at which the partial pressure of OH reaches

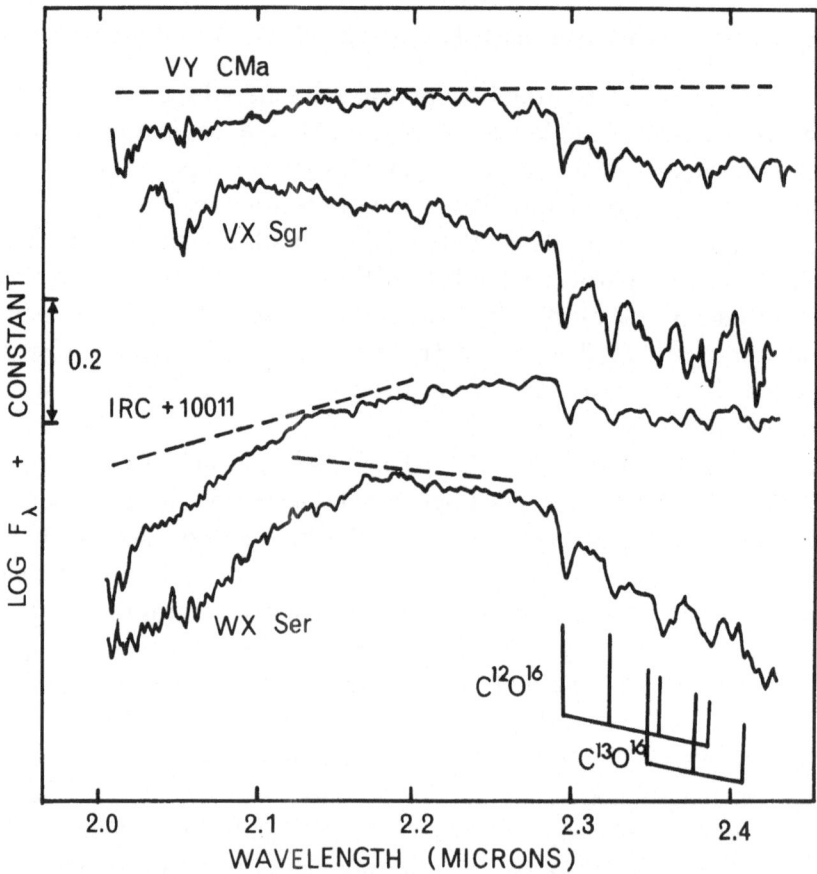

Fig. 3. Infrared spectra of the supergiant OH sources VY CMa and VX Sgr and the 1612 MHz long period variable sources IRC + 10011 and WX Ser are shown. F_λ is the normalized flux per unit wavelength interval. The dashed lines indicate the slope of the continuum as determined from broad band photometry. The spectra are notable for the strong absorption by CO beyond 2.3 μm and for absorption by H_2O molecules between 2.0 and 2.15 μm.

a maximum (see, e.g., Tsuji, 1964). At a given photospheric pressure this peak in the partial pressure of OH is very sharp; a change of 200 K either way produces a decrease in the partial pressure by more than an order of magnitude. Additionally, the partial pressure of H_2O lies within a factor of 3 of that at total association.

(β) All stars associated with OH or H_2O emission have oxygen-rich photospheres. No carbon stars, where O/C < 1, or S stars, where O/C ~ 1, have been identified with OH or H_2O sources.

Hence, the photospheric layers of all the OH and H_2O microwave sources are rich in OH molecules and H_2O molecules, and Hyland et al. have suggested that the molecules responsible for the microwave emission originate in these photospheres and remain associated as they are driven outwards into a circumstellar shell where the conditions for maser emission become favourable. Further direct evidence linking stellar mass loss and microwave emission will be discussed in the next section.

All the available infrared spectroscopic data has been obtained at medium resolu-

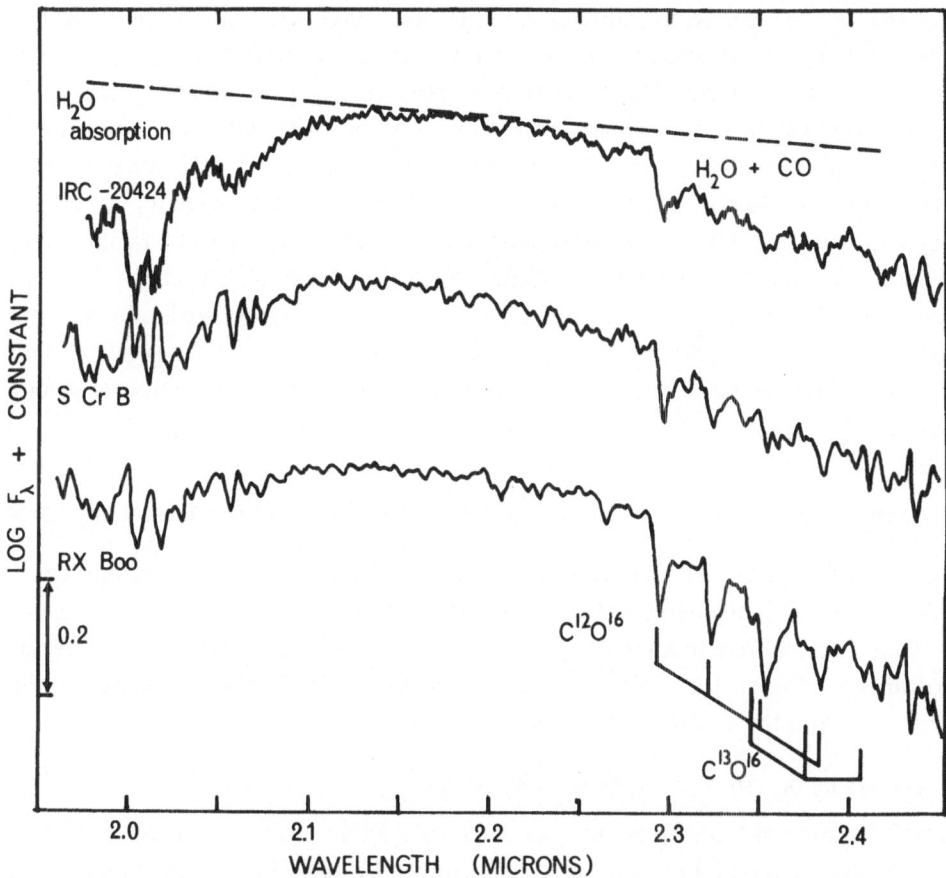

Fig. 4.   Same as Figure 3. IRC-20424 and S CrB are long-period variable 1665/1667 MHz emitters. S CrB
and the irregular M8 giant RX Boo are microwave H$_2$O sources.

tion, and no direct measurements have been made on the individual molecular lines, in particular those of CO, H$_2$O and OH, which are available for study at high resolution. Some very recent work at high resolution in the 1 and 2 μm region has been reported (Maillard, 1973) and holds great promise for detailed investigations into the photospheric properties of the M giant and supergiant OH/IR stars. Maillard obtained evidence from one star (R Leo) of a splitting of the individual CO lines into two components at one phase of its variation. The velocity difference agreed well with that found for the separation of the OH emission peaks. Curiously, he found no evidence for the splitting of the measurable infrared OH lines! The continuation of such high-resolution studies can be expected to provide exciting results necessary for the understanding of the relationships between pulsation and mass loss, and for deriving more realistic models for OH/IR sources.

Few optical spectroscopic studies of OH/IR sources have been reported in the literature (Hyland *et al.*, 1969; Wallerstein, 1971). Their major contribution has been in the determination of the optical velocity systems for comparison with the OH emission velocities. The general results obtained from these and other published velocities of long-period variable M stars (Feast, 1963) are that the more positive velocity OH peak corresponds approximately with the stellar absorption line velocity, while the more negative velocity peak is in reasonable agreement with the optical emission line velocity. A summary of the available velocity data has been given by Wilson and Barrett (1971). The situation regarding the equality of OH emission velocities and optical velocities is not, however, clear cut, since it is known that photospheric velocities vary by 5 km s^{-1} or more during the cycles of long-period variable stars. It is necessary therefore to have accurate observations around a cycle to be able to determine the stellar systemic velocity. Programmes of this nature are highly desirable. Furthermore, optical velocity studies are of prime importance for the positive identification of the stellar counterparts of the unidentified high velocity sources recently discovered along the galactic plane (Winnberg *et al.*, 1973).

IV. Infrared Photometry and the Properties of the Circumstellar Dust Component

In the simple picture of OH/IR sources described in the introduction, the central stellar source was considered to be embedded in a large circumstellar shell of dust and gas, which determined to a large extent the infrared properties of the source. Infrared photometry between 1 and 20 μm provides information both about the circumstellar component and the central stellar source.

(a) STATISTICS OF THE NEAR INFRARED COLOURS

These have been obtained from the extensive 1612 MHz survey of Wilson and Barrett (1972) which included 403 stars selected from the Caltech Two Micron Sky Survey, and have been discussed by those authors and Hyland *et al.* (1972). The main conclusion is that *all* 1612 MHz OH/IR sources have $0.8 - 2.2$ μm colours $[I - K] > 4.6$. Since 140 infrared sources included in the survey had values of $[I - K] < 4.6$ this is unlikely to be due to selection effects and essentially corroborates the spectroscopic analysis, which showed that all OH/IR sources have photospheric temperatures lower than 2600 K.

(b) INFRARED EXCESSES AND COLOUR DIFFERENCES BETWEEN THE 1612 MHz SOURCES AND MAINLINE EMITTERS

From observations covering the wavelength range 1.25–20 μm, Hyland *et al.* (1972) showed that almost all the 1612 MHz sources possess large infrared excesses and that they were among the very reddest stars found in the Two Micron Sky Survey. The majority of those stars with large infrared excesses but which did not exhibit OH emission were found to be carbon stars. The infrared excesses of the 1612 MHz OH/IR sources have been generally interpreted in terms of circumstellar dust shells with characteristic temperatures between 600 and 700 K which absorb the visual and near infrared radiation and re-radiate the energy at longer wavelengths. This widely accepted model can be used to derive useful information regarding the structure of such shells and will be discussed below.

While the 1612 MHz (supergiant and type IIb) OH/IR sources generally possess large infrared excesses, Wilson *et al.* (1972) and Hyland, Hirst *et al.* (1972) have shown that this is not the case for the mainline (type I) OH/IR sources. The distinction between the two groups of sources is illustrated in Figure 5, where histograms of the number of sources in a given 2.2–3.5 μm $(K-L)$ colour interval and 2.2–10.2 μm $(K-N)$ colour interval are shown. Both colours are representative of the dust shell structure, the former representing the stellar continuum as modified by the absorption properties of the dust particles, the latter representing the low temperature dust shell emission.

The histograms show that mainline emission predominates in those sources with the bluest colours, i.e., those for which there is little evidence for the presence of circumstellar dust particles. On the other hand 1612 MHz emission increases as the optical depth of the circumstellar dust increases and dominates the other lines when the shell is optically thick at near infrared wavelengths. Wilson *et al.* (1972) have also shown that the 10.2–20 μm colours, which depend on the extent of dust at temperatures ~ 300 K, are also correlated with the OH type in the same sense.

Mention should be made of the two unusual sources which complicate the simple picture, R Aql and IRC-20424. The former is a strong 1612 MHz source which possesses no dust shell characteristics, while the latter is a mainline emitter which has a large infrared excess, typical of the 1612 MHz sources.

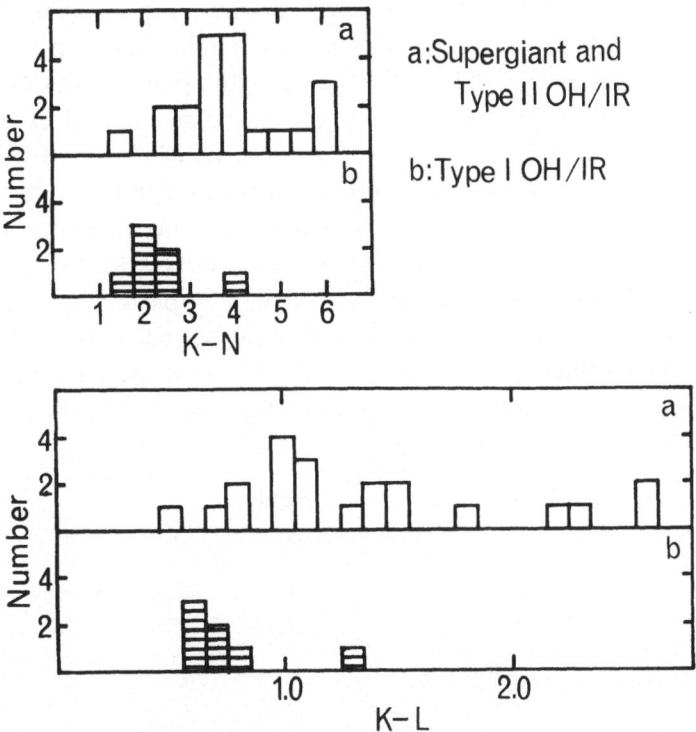

Fig. 5. Histograms of the number of OH/IR sources in a given colour interval. The distribution with colour of the type I (mainline) OH/IR sources is seen to be different from that of the type II (1612 MHz) sources in the sense that the former are significantly bluer in both the $K-N$ and $K-L$ colours.

The existence of these anomalous cases suggests that although the presence and extent of circumstellar dust particles are important in determining the characteristics of the OH emission, a further important parameter as yet unidentified must also be taken into account.

(c) SHELL STRUCTURE

It is widely accepted that the large infrared excesses such as those of the 1612 MHz sources are the result of absorption and re-emission processes by circumstellar dust particles. For a few sources (e.g., VY CMa, VX Sgr) spectroscopic data in the 8–14 μm region confirm the presence of an emission feature which can be attributed to emission by silicate-like particles (e.g., Gillett *et al.*, 1970). The observations show, however, that the excess emission clearly extends to wavelengths as short as 4.8 μm.

Only the simplest models have been applied in any systematic way in the literature. Models have been constructed by Hyland *et al.* (1972) for the case of a central star with $T = 2200$ K surrounded by an isothermal dust shell in which the absorption coefficient of the dust at short infrared wavelengths was assumed to have a λ^{-1} dependence. Figure 6 shows that the colours derived from these models adequately describe the observations of the OH/IR sources. Again a fairly clear separation between the 1612 MHz sources and the mainline emitters is seen. From these simple models the majority of sources appear to have dust shells with temperatures in the range 600–700 K. The percentage of the observed flux which is re-radiated emission from the shell is also shown in the figure. Although the implications have not been fully explored, the large number of 1612 MHz sources with optically thick dust shells and the lack of any visually unreddened sources with large infrared excesses suggest that the assumption of spherical symmetry is not grossly in error. However, further investigation of models where the assumption of spherical symmetry has been relaxed, such as that advanced by Herbig (1970), would appear to be needed at this stage.

Typical properties of the dust shells have been derived (Hyland *et al.*, 1972), and a few examples from their paper are shown in Table I. Typically the shell radii range between 5×10^{14} and 10^{15} cm, and the dust shell masses between 2×10^{27} and 1.5×10^{28} g, where these values are based on source luminosities at maximum of $10^4\ L_\odot$. Correspondingly, the optical depths at 1.65 μm range from $\tau = 0$ for R Aq1 and several mainline emitters to $\tau = 2$ for IRC + 10011.

TABLE I

Typical shell characteristics of OH/IR sources

Object		$\dfrac{L_{shell}}{L_{total}}$	T_{shell} (K)	R_{shell} (10^{15} cm)	M_{shell} (10^{27} g)
M supergiants	VY CMa	0.75	600	1.5	15
	NML Cyg	0.75	600	1.5	15
Long-period	IRC + 10011	0.85	700	0.5	3
variable M	WX Ser	0.30	500	1.0	2
stars	R Aql	0.05	–	–	–

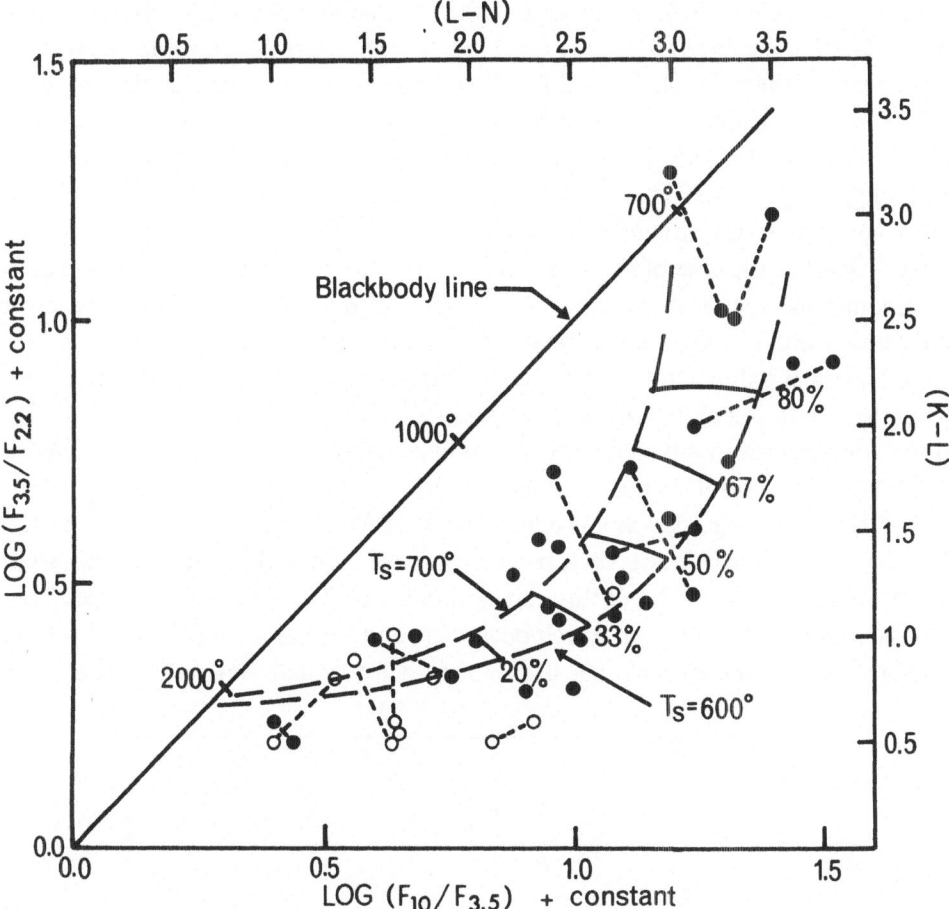

Fig. 6. Flux ratios at 10.2 and 3.5 μm and at 3.5 and 2.2 μm are compared for both 1612 MHz (filled circles) and 1665/1667 MHz (open circles) OH/IR sources. The diagonal line corresponds to the flux ratio for black bodies at the temperatures marked. The dashed lines correspond to dust shells of temperatures of 600 and 700 K surrounding a central star with photospheric temperature of 2200 K; the solid lines joining the dashed curves show the fraction of the total energy emitted by the shell. R Aql is the 1612 MHz source at the extreme lower left edge of the distribution and IRC-20424 is the lone 1665/1667 MHz source in the middle of the region dominated by 1612 MHz emitters. The small dashed lines join extreme values of the flux ratios observed during the cyclical variations of the long-period variable sources.

(d) SHELL STRUCTURE AS A CONSEQUENCE OF MASS LOSS

The association of 1612 MHz sources with thick circumstellar dust shells and the one to one correlation of the (photospheric) molecular abundances lead to the suggestion of a common mass ejection process for the dust shells and the OH and H₂O molecules responsible for the microwave emission. Although attractive, this idea has not been universally accepted, e.g. Davies *et al.* (1972) suggest that in the case of NML Cyg the circumstellar material is the remnant of a protostellar dust cloud. Evidence in favour of the common mass ejection hypothesis may be summarised as follows:

  (i) Direct evidence for mass loss from the photospheres of long-period variable M stars is derived from the differential velocities of the absorption and emission lines

(Merrill, 1960; Feast, 1963). In the case of the M supergiants VY CMa (Hyland et al., 1969) and VX Sgr (Wallerstein, 1971), blue shifted sharp circumstellar absorption components of the zerovolt K I and Ca I lines provide the evidence for mass ejection.

(ii) The rates of mass loss required to support the circumstellar dust shells are entirely reasonable, being of the order of 10^{-5}–10^{-6} $\mathcal{M}_\odot$ yr^{-1} although Davies et al. (1972) suggest that mass loss rates of 10^{-3} $\mathcal{M}_\odot$ yr^{-1} are required to maintain the circumstellar material surrounding NML Cyg.

(iii) At least in the case of the long-period variable stars, which are highly evolved, the circumstellar material can in no way be identified with the remnants of a proto-stellar dust cloud but must be of the star's own fabrication.

(iv) The most telling evidence in favour of the hypothesis comes from the correla-tion of the separation of the OH velocity peaks ΔV (OH) with the infrared 3.5–10.2 μm colour, which has been independently noted by Hyland, Hirst et al. (1972) and Wilson et al. (1972). This correlation is shown in Figure 7.

ΔV (OH) can be regarded as a crude estimate of the system's rate of mass ejection, while the 3.5–10.2 μm colour gives a good estimate of the optical depth of circumstellar dust. A further characteristic of this correlation which has not previously been noted is the break in the distribution which occurs around the values ΔV (OH) $\sim$ 10–14 km s^{-1}. Since this is precisely the range of escape velocity for material in the photospheres

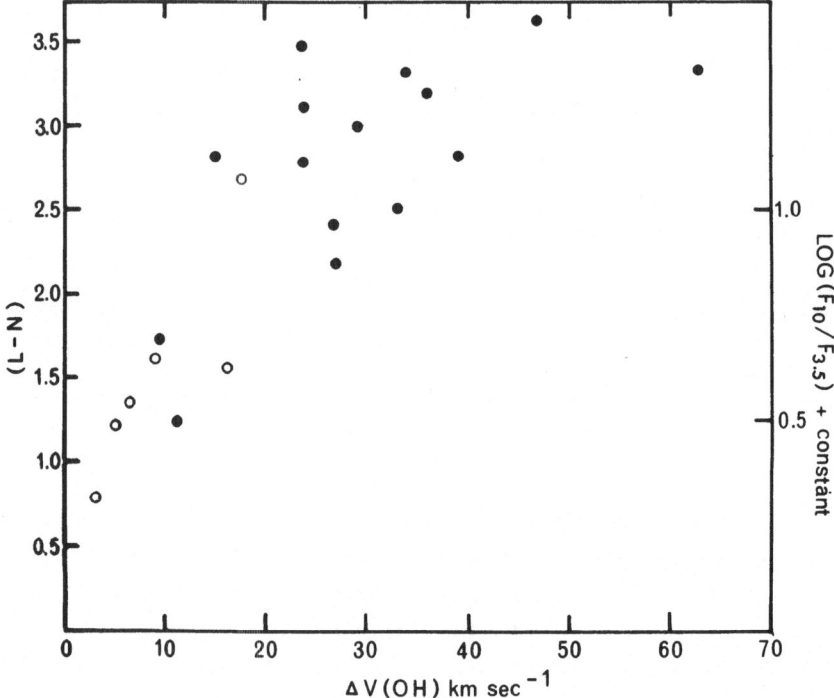

Fig. 7. The correlation of the 10.2 to 3.5 μm flux ratio (or $[L-N]$ colour) with the velocity separation of the OH emission peaks, ΔV(OH), for the 1612 MHz sources (filled circles) and the 1665/1667 MHz emitters (open circles). This correlation is discussed in the text in terms of mass loss and the formation of circumstellar dust shells.

of M giants and supergiants, the figure may be interpreted as showing empirically that if the velocity of ejection is below the escape velocity, thick circumstellar dust shells will not form, but that if the escape velocity is exceeded, then thick circumstellar dust shells almost invariably do form. The fact that the observed position of NML Cyg in this diagram is entirely consistent with the above picture contradicts the proto-stellar remnant hypothesis.

V. Infrared and Microwave Variability of the OH/IR and H₂O/IR Sources

(a) GENERAL CONSIDERATIONS

One of the most important tasks which has been undertaken regarding the OH/IR and H₂O/IR sources has been to establish the relationship between the infrared and microwave flux densities. Although Hyland et al. (1972) showed that such a relation-ship appeared to exist (see their Figure 9), it was clear that a co-ordinated study of the infrared and microwave variations was necessary to clarify the situation. The recent studies of Schwartz et al. (1974) for H₂O sources and Harvey et al. (1974) for OH sources fill this need and because of their importance will be discussed in some detail here.

(b) INFRARED VARIABILITY

The studies of variability cover eight H₂O/IR sources (both long-period variable stars and irregular variables) and 14 OH/IR sources (ten 1612 MHz sources and four main-line emitters). They thus give comprehensive results on all the common types of these sources.

The infrared variations are interesting in themselves and have the following charac-teristics:

(i) All sources investigated with the exception of the supergiants NML Cyg and VY CMa showed variations at 2.2 μm between 0.5 mag and 2.5 mag.

(ii) The infrared phases lag the visual by approximately 0.1–0.2 of a period, and the period is the same as the visual period within the errors of observation, covering the range 300–700 days.

(iii) Colour variations are seen to occur in the sense that the infrared continuum is bluer at maximum light and redder at minimum. Although data on the H₂O sources is available only at a single wavelength (2.2 μm), the colour variations should be similar.

Among the most interesting results are the correlations which are evident between the quantities: the period, the 2.2 μm amplitude, colour and velocity separation of the OH emission peaks, in particular for the long period variable M stars. These are shown in Figure 8. The colour-velocity separation correlation has been discussed in Section IV, the period-velocity separation correlation has been noted by Dickinson and Chaisson (1973), and period-amplitude relations from observations at shorter wave-lengths have been discussed in the literature (Merrill, 1960; Lockwood and Wing, 1971).

The causal relationships among the four quantities are unclear and may indeed

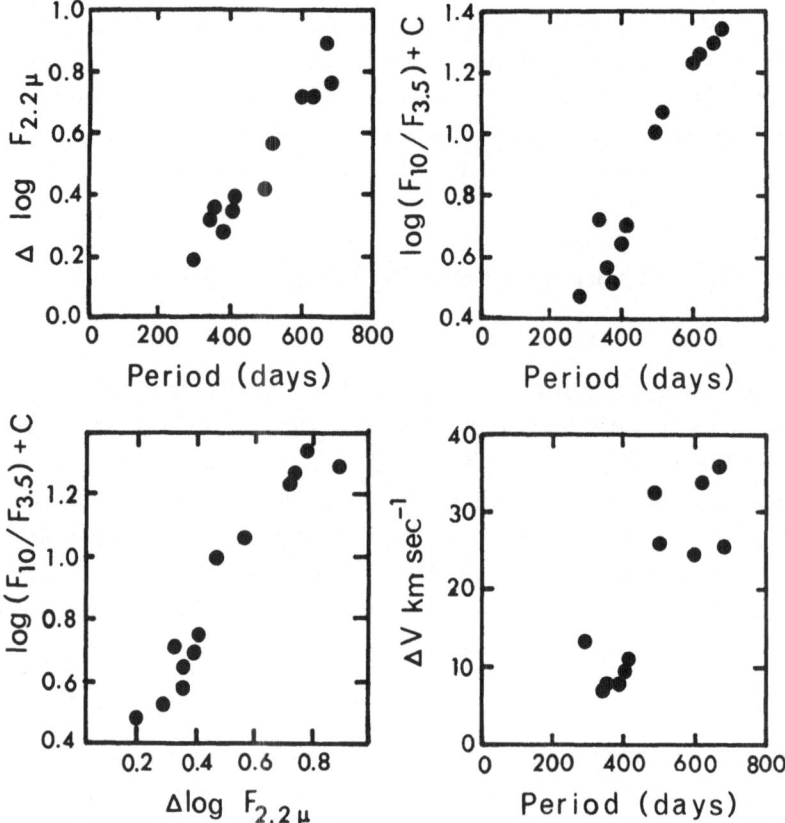

Fig. 8.   The observed correlations between the log of the amplitude of variation at 2.2 μm, the 10.2 to 3.5 μm flux ratio, the velocity of separation of the OH emission peaks (ΔV), and the period, for long-period variable OH/IR stars (Harvey *et al.*, 1974).

depend on a further parameter such as luminosity, as one expects a period-luminosity relation to exist for the long period variable stars. Certainly the association of longer periods and larger amplitudes with larger velocity separation and infrared excesses suggests that the mass ejection processes are intimately connected with the mechanical pulsation mechanisms, and that mass outflow due to radiation pressure on dust grains may be a secondary process which dominates only at large distances from the star.

(c) CORRELATIONS OF THE INFRARED AND MICROWAVE VARIABILITY

The empirical relationships which have been defined by the investigations of Schwartz *et al.* (1974) and Harvey *et al.* (1974) are:

(i) The infrared and microwave variations of the long period variable 1612 MHz OH/IR and 1.35 cm H_2O/IR stars are periodic, and have the *same period and phase* within the observational limits. Examples of such variations taken from the above papers are shown in Figure 9.

(ii) The 1665/1667 MHz variations are more random than the 1612 MHz variations and no definite correlation with the infrared variations has been established.

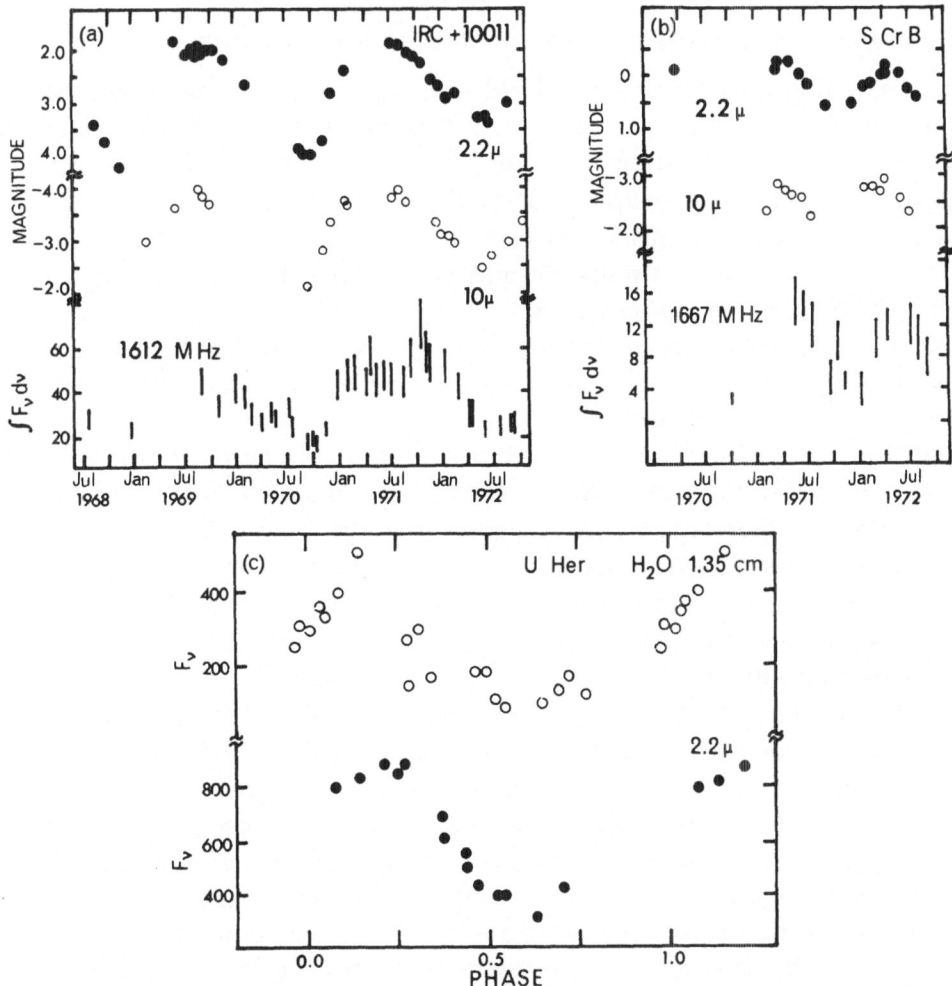

Fig. 9. (a) Observed infrared and 1612 MHz OH variations of the highly variable 1612 MHz source IRC + 10011. (b) Observed infrared and 1667 MHz OH variations for the mainline emitter S CrB. (c) Observed infrared and 1.35 cm H$_2$O variations for the H$_2$O source U Her.
The observations shown in (a) and (b) are taken from Harvey *et al.* (1974) and those in (c) from Schwartz *et al.* (1974).

(iii) The amplitude of variation in the OH microwave flux is similar or significantly smaller than the amplitude at infrared wavelengths. On the other hand, the amplitude of the 1.35 cm H$_2$O variations may be larger than the infrared amplitudes.

(iv) The 1612 MHz OH lines and the H$_2$O lines show no velocity variations within the observational limits of 0.3–0.5 km s^{-1} (which is of course different from observations of spectral lines in the optical).

The interpretation of these observations in terms of the coupling of the microwave emission regions to the other components of simple OH/IR systems as envisaged in Section I places stringent conditions on future models which may be proposed.

(d) RADIATIVE COUPLING OF THE INFRARED AND MICROWAVE EMISSION

The observational results on infrared and microwave variability show that there is strong coupling between the pulsations of the central source and the microwave emission processes. The nature of this coupling has been investigated (Harvey et al., 1974) and *radiative coupling* appears to be the only viable alternative. Mechanical coupling of the regions by mass motions has been rejected on several counts:

(i) VLBI observations show that the microwave emission may come from regions up to 10^{16} cm from the central source;

(ii) the circumstellar densities need to be as low as 10^7 cm^{-3} or quenching of the maser processes by collisions becomes important, and these low densities require the regions also to be several stellar radii from the central source;

(iii) the stability of the velocities of the OH and H_2O lines differs from the variability of line velocities in the stellar photosphere and also indicates that the microwave emission regions are mechanically decoupled from the central source; and (iv) the observed velocities of mass motion (up to ~ 50 km s^{-1}) are such that large phase lags (up to several years) would be required for the mechanical coupling of the stellar pulsations with regions situated between 10^{14} and 10^{16} cm from the central source, in disagreement with observation.

The proposed radiative coupling may be obtained by three separate mechanisms.

(i) The radiation from the central source may affect the equilibrium population density of the OH and H_2O molecules responsible for the microwave emission. This can be discounted not only because the UV radiation field required to dissociate both OH and H_2O molecules is weak in regions surrounding late M stars, but also because such a process would act in the *wrong sense*; i.e., at maximum light the dissociation would be strongest, whereas the microwave emission is seen to increase.

(ii) The variable radiation from the central source may contribute to the collision rate, which determines the extent of population inversion in the masering region. This suggestion is also unlikely since neither the microwave line widths or velocities are observed to vary, as would be expected in the case of large changes in the collision rates. Further reasons are given by Harvey et al. (1974).

(iii) The most likely explanation is that of *radiative pumping* of the OH and H_2O molecules by infrared radiation as has been suggested by Shklovskii (1967), Litvak (1969a), and most recently Litvak and Dickinson (1972).

Such a mechanism is consistent with all observations to date. In particular it has been shown to be energetically possible if the infrared pump radiation lies between 2 and 40 μm (Harvey et al., 1974), and indeed pumping by 2.8 μm or 35 μm photons has been suggested for the OH sources. It has also been suggested that H_2O sources are pumped by 2.7 μm radiation (Litvak, 1969b).

It is not the purpose of the review to detail the infrared pumping schemes which have been proposed to explain OH/IR and H_2O/IR sources. Accurate predictions from such models require a more detailed knowledge of the conditions prevailing in the masering regions than currently available, although the schemes proposed by

Litvak and Dickinson (1972) appear to qualitatively reproduce the observed OH characteristics. The degree of saturation of the masers is still a point of debate, although for the OH sources most authors favour the suggestion that the masers are at least partly saturated. There are a number of mechanisms which can account for the different observed infrared and microwave amplitudes. For the H$_2$O sources the situation is less clear, and Schwartz *et al.* (1974) have shown that either an unsaturated or saturated maser model can be used to explain the variations in R Aql.

VI. Some Other Properties

Although most of the major properties of OH/IR and H$_2$O/IR sources have been covered in this paper, it has been necessary to omit detailed discussion of a number of interesting characteristics. Two of these should however be briefly mentioned here.

(a) SPATIAL DISTRIBUTION OF OH/IR SOURCES

The spatial distribution of the OH/IR sources has been found to be typical of the long-period variable M stars and M supergiants. However, two interesting differences have been noted between the 1612 MHz and the mainline sources (Wilson *et al.*, 1972). Working on the assumption that the luminosity of all the long-period variable stars is 10^4 $L_\odot$ at maximum light, it is found that the average distance of the mainline sources is roughly one-third of that derived for the 1612 MHz sources. This correlation is also evident in the K magnitudes of the sources, which are on the average much brighter for the mainline sources. If, as suspected, a period-luminosity law exists for long-period variable M stars, one may infer from the period-colour relationship discussed earlier that the mainline emitters have intrinsically lower luminosities than the 1612 MHz sources. If this possibility is included, the differences between the spatial distribution of mainline and 1612 MHz OH/IR sources are enhanced. The probable explanation for this phenomenon is that the intrinsic luminosity of the mainline OH emission is also lower than for the 1612 MHz sources.

(b) THE RATIO (R) OF THE INTEGRATED MICROWAVE OH LUMINOSITY TO THE TOTAL OPTICAL AND INFRARED LUMINOSITY

This quantity bears directly on the question raised above. It is derivable from the observations and is independent of distance. The values obtained for both 1612 MHz and mainline OH/IR sources are plotted in Figure 10 against the 3.5–10.2 μm colour $(L-N)$, in the manner of Hyland, Hirst *et al.* (1972). There is an apparent general increase in R as a function of colour (i.e., increasing shell characteristics), though this dependence is largely enhanced by the large values attributed to the supergiant sources. While the M supergiants generally have larger values of R than the long-period variables, for the latter there appears to be an upper limit independent of colour. No explanation has been advanced for this apparent upper limit, but it may be related to the maximum efficiency of infrared pumping and to the fact that in the long-period variables the 2.8 μm OH line lies in a region of heavy photospheric absorption by

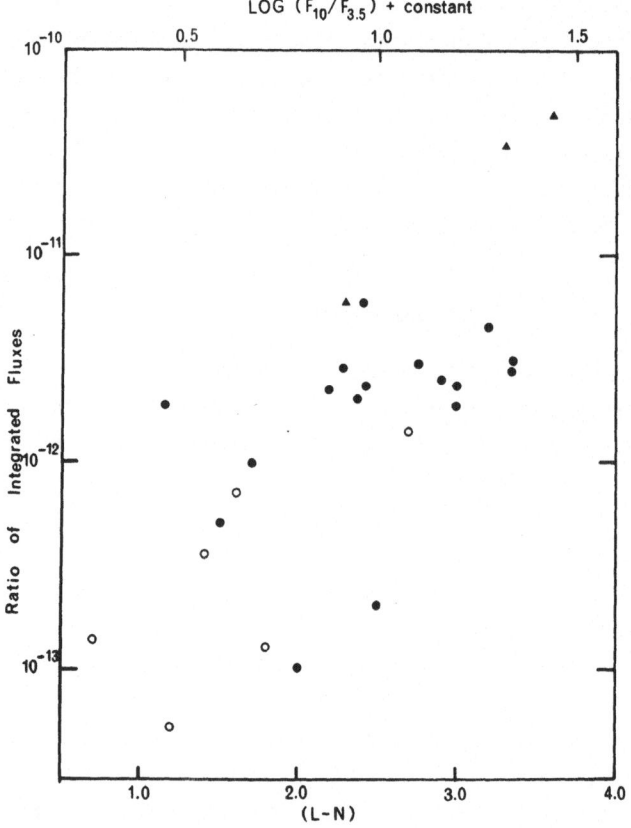

Fig. 10.   The ratio (R) of the integrated OH luminosity to the total optical and infrared luminosity as a function of the 10.2 to 3.5 μm flux ratio (or $L - N$ colour). The 1612 MHz supergiant sources are shown as filled triangles, the 1612 MHz long period variable sources as filled circles, and the 1665/1667 MHz long-period variable sources as open circles.

H_2O, which reduces the radiation field by a significant amount. This effect is worthy of further investigation.

The average value of R for the mainline emitters is almost an order of magnitude lower than for the 1612 MHz long-period variable sources (although the maximum values are similar), and this would also help to account for the different observed spatial distributions of the two classes of source.

Finally, Figures 7 and 10 can be used to predict the colour and integrated optical and infrared luminosity from the OH characteristics of the unidentified sources discussed in Section II. Such considerations indicate that bright infrared sources should be found associated with these OH sources, the present lack of which suggests that they may have properties unlike the 'classical' OH/IR stars discussed in this review.

VII. Conclusions

In this paper current knowledge on the infrared properties of the circumstellar OH/IR and H$_2$O/IR sources has been reviewed in terms of the simple schematic model proposed in the Introduction. This model still appears to be valid as a crude approximation to such systems, and the major characteristics bear repeating here:

(i) The OH/IR and H$_2$O/IR sources discussed here have a central stellar component consisting of an M supergiant or long-period variable M star. These stars are all oxygen rich and have a limited range of photospheric temperatures such that the partial pressures of OH and H$_2$O molecules are close to the maximum achievable.

(ii) All the central stars appear to be losing mass, and those with higher rates of mass loss have thicker circumstellar dust shells, a tendency towards the strongest OH emission, and a predominance of 1612 MHz emission.

(iii) The masering regions appear to be at distances up to 10^{16} cm from the central source and are radiatively coupled to the luminosity variations of that source. Radiative pumping in the 2–40 μm region appears to be consistent with all the observations.

Although the basic characteristics of the OH/IR and H$_2$O/IR sources appear to be well established, many unsolved problems await the investigator. The nature of the new unidentified OH/IR sources, the geometry of the masering and infrared emitting regions, the causal relationships between stellar pulsation and mass loss, and more realistic investigations of the infrared pumping mechanisms still have to be explored. The investigations of OH/IR and H$_2$O/IR sources discussed in this review have been a prime example of the fruitful cooperation which can be achieved between the optical, infrared, radio and theoretical astronomer, and it is to be hoped that such cooperation will continue in the field and spread to other domains of astronomical research.

Acknowledgements

I wish to thank Dr P. M. Harvey and his co-authors for providing me with preprints of their papers, and to Drs J. L. Caswell and B. J. Robinson for useful discussions and for sending me positions of unidentified OH/IR sources prior to publication.

References

Caswell, J. L. and Haynes, R.: 1973, in preparation.
Caswell, J. L., Robinson, B. J., and Dickel, H. R.: 1971, *Astrophys. Letters* **9**, 61.
Davies, R. D., Masheder, M. R. W., and Booth, R. S.: 1972, *Nature Phys. Sci.* **237**, 21.
Dickinson, D. F. and Chaisson, E. J.: 1973, *Astrophys. J. Letters* **181**, L435.
Eliasson, B. and Bartlett, J. F.: 1969, *Astrophys. J. Letters* **155**, L79.
Feast, M. W.: 1963, *Monthly Notices Roy. Astron. Soc.* **125**, 367.
Frogel, J. A.: 1970, *Astrophys. J. Letters* **162**, L5.
Gillett, F. C., Stein, W. A., and Solomon, P. M.: 1970, *Astrophys. J. Letters* **160**, L173.
Glass, I. S. and Feast, M. W.: 1973, *Astrophys. Letters* **13**, 81.
Hardebeck, E. G.: 1972, *Astrophys. J.* **172**, 583.
Hardebeck, E. G. and Wilson, W. J.: 1971, *Astrophys. J. Letters* **169**, L123.
Harvey, P. M., Bechis, K. P., Wilson, W. J., and Ball, J. A.: 1974, *Astrophys. J. Suppl.* **27**, No. 248.
Herbig, G. H.: 1970, *Astrophys. J.* **162**, 557.
Hyland, A. R., Becklin, E. E., Neugebauer, G., and Wallerstein, G.: 1969, *Astrophys. J.* **158**, 619.
Hyland, A. R., Becklin, E. E., Frogel, J. A., and Neugebauer, G.: 1972, *Astron. Astrophys.* **16**, 204.
Hyland, A. R., Hirst, R. A., Robinson, G., and Thomas, J. A.: 1972, *Astrophys. Letters* **11**, 7.

Litvak, M. M.: 1969a, *Astrophys. J.* **156**, 471.

Litvak, M. M.: 1969b, *Science* **165**, 855.

Litvak, M. M. and Dickinson, D. F.: 1972, *Astrophys. Letters* **12**, 113.

Lockwood, G. W. and Wing, R. F.: 1971, *Astrophys. J.* **169**, 63.

Merrill, P. W.: 1960, *Stars and Stellar Systems* **6**, 525.

Neugebauer, G. and Leighton, R. B.: 1969, *Two Micron Sky Survey, A Preliminary Catalog*, NASA, Washington, D.C.

Nguyen-Quang-Rieu, Fillit, R. and Gheudin, M.: 1971, *Astron. Astrophys.* **14**, 154.

Robinson, B. J., Caswell, J. L., and Goss, W. M.: 1970, *Astrophys. Letters* **7**, 79.

Robinson, B. J., Caswell, J. L., and Goss, W. M.: 1971, *Astrophys. Letters* **7**, 163.

Schwartz, P. R. and Barrett, A. H.: 1970, *Astrophys. J. Letters* **159**, L123.

Schwartz, P. R. and Barrett, A. H.: 1971, in G. W. Lockwood and H. M. Dyck (eds.), *Proc. Conference on Late Type Stars*, Kitt Peak National Obs., Tucson, Arizona, p. 57.

Schwartz, P. R., Barrett, A. H., and Harvey, P. M.: 1974, *Astrophys. J.* **197**, 491.

Shklovskii, I. S.: 1967, *Astr. Circ. U.S.S.R.* **424**, 1.

Tsuji, T.: 1964, *Ann. Tokyo Astron. Obs.* **9**, 1.

Turner, B. E.: 1970, *J. Roy. Astron. Soc. Can.* **64**, 221.

Wallerstein, G.: 1971, *Astrophys. Letters* **7**, 199.

Wilson, W. J. and Barrett, A. H.: 1971, in G. W. Lockwood and H. M. Dyck (eds.), *Proc. Conference on Late Type Stars*, Kitt Peak National Obs., Tucson, Arizona, p. 77.

Wilson, W. J. and Barrett, A. H.: 1972, *Astron. Astrophys.* **17**, 385.

Wilson, W. J., Barrett, A. H., and Moran, J. M.: 1970, *Astrophys. J.* **160**, 545.

Wilson, W. J., Schwartz, P. R., Neugebauer, G., Harvey, P. M., and Becklin, E. E.: 1972, *Astrophys. J.* **77**, 523.

Winnberg, A., Goss, W. M., Höglund, B., and Johansson, L. E. B.: 1973, *Astrophys. Letters* **13**, 125.

A. R. Hyland

Mt. Stromlo and Siding Spring Observatory,
Private Bag, Woden, A.C.T. 2606, Australia

DISCUSSION

Davies: You suggested that the OH and H_2O molecules seen in the radio masers were injected from the central star. I think there is good evidence, at least for the M supergiant OH/IR objects, that the OH and H_2O are associated with the pre-existing circumstellar cloud for the following reasons:

(i) The angular momentum derived from the radio interferometry is far too large for gas ejected from a star.

(ii) The amount of gas ($\sim 10^{-1}$ $M_\odot$) and rate of mass outflow (10^{-3} $M_\odot$ yr^{-1}) are greater than for any known group of stars.

(iii) There is already evidence in VY CMa of a large gas and dust complex surrounding the IR/OH object.

Hyland: Firstly, the observations of the long period variable OH/IR stars provide circumstantial evidence for the expulsion of the emitting OH and H_2O molecules from their own atmospheres which is hard to ignore. Further, infrared objects whose central stellar sources have high temperatures (such that OH and H_2O molecules cannot form in their atmospheres) do not appear to exhibit OH or H_2O microwave emission, regardless of whether the circumstellar material appears to be due to mass loss (such as for RV Tauri stars) or is the possible remnant of a protostellar cloud (as may be the case for Z CMa). It is true that the angular momentum of the circumstellar material surrounding NML Cyg and VY CMa appears to be embarrasingly high. However, it is marginally possible from the numbers currently available to explain such high angular momentum if those stars were rotating near break-up velocity while on the main sequence. The high rate of mass outflow does not appear to be too much of a problem given the erratic and energetic nature of the outbursts in VY CMa.

Donn: What is the wavelength band of your infrared search?

Hyland: The effective wavelength of the television system and infrared filter is 8000 Å and corresponds closely with the Kron I system.

SOME HIGH-VELOCITY SOURCES OF 1612 MHz OH EMISSION

F. J. KERR and P. F. BOWERS

University of Maryland, College Park, Md., U.S.A.

Abstract. A pilot survey was made to study the feasibility of using radio observations of OH/IR stars to investigate the galactic distribution and kinematics of these presumably intermediate-population objects. Seven new sources were found, four near the center, two near $l = 30°$, and one at $l = 128°$.

Most identified OH/IR stars are believed to be long-period oxygen-rich Mira variables. Winnberg *et al.* (1973) have recently surveyed the galactic plane from $l \simeq 20°$ to $50°$ and have found about 30 new 1612 MHz OH sources. The lack of optical or infrared counterparts to the sources indicates that some of the sources may be quite distant since many 1612 MHz OH sources are also bright infrared objects and most of these sources have the microwave characteristics of OH/IR stars.

The velocities range up to $+140$ km s^{-1}, and in several cases the velocities are higher than would be expected for HI in the same direction, implying perhaps that these sources represent a population with a higher velocity dispersion than would be characteristic of Population I objects. It seems possible then to use radio observations of the OH/IR stars to study the galactic distribution and kinematics of a disk popula-

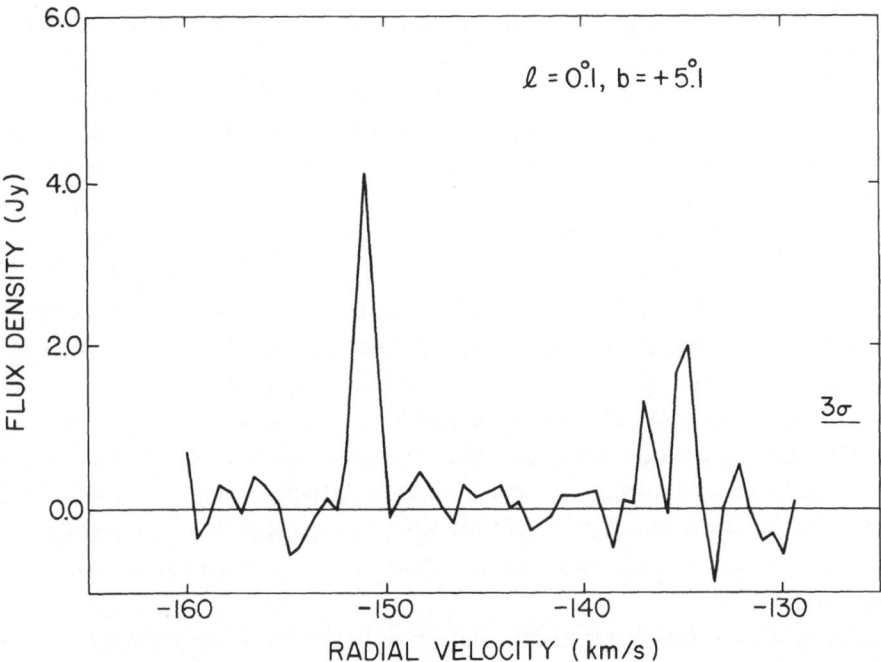

Figs. 1a–b. A 1612 MHz OH source in the general direction of the galactic center and (b) a 1612 MHz source toward the outer part of the Galaxy. The galactic coordinates of each source are indicated. The velocity resolution is 0.67 km s^{-1}, and the polarization angle was 0°. The radial velocities are with respect to the LSR.

F. J. Kerr and S. C. Simonson, III (eds.), Galactic Radio Astronomy, 459–461. All Rights Reserved.
Copyright © 1974 by the IAU.

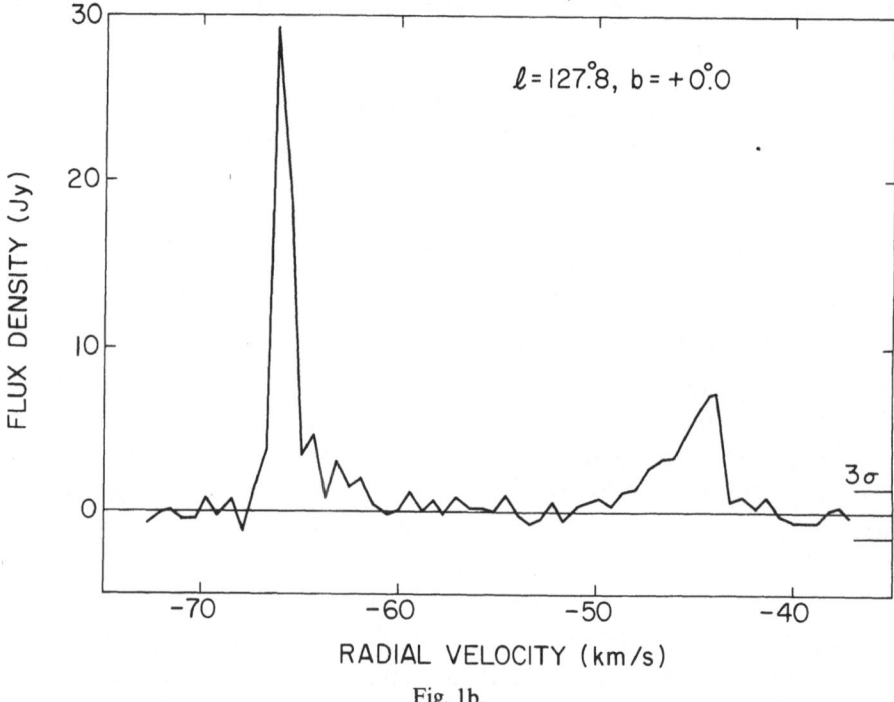

Fig. 1b.

tion. A pilot program has been completed to investigate the feasibility of such a project, using the 140-ft (43-m) radio telescope at the National Radio Astronomy Observatory* in Green Bank, West Virginia.

Observations were obtained at about 150 positions. Over half of these were in directions in the vicinity of the galactic center, where an older population might be expected to be concentrated. Positions near $l = 30°$ and $l = 130°$ were also observed. Seven new sources were detected at 1612 MHz. Most of the sources have double peaks, as with the known OH/IR stars, and may be at large distances since there are no obvious optical or infrared counterparts.

Figure 1a shows the profile for a source which is especially interesting because this high velocity observed near zero longitude must be entirely due to a noncircular component of motion. Figure 1b shows the profile of a strong source in a direction toward the outer part of the Galaxy, where little 1612 MHz work has been done so far.

We are planning a larger survey to look for the stronger sources of this class. With respect to galactic structure, radio astronomy has hitherto been concerned primarily with Population I objects. The bright microwave emission from these sources may provide a new method for studying the distribution and kinematics of an older population.

In closing, we want to report briefly that a number of RV Tauri stars, some of which have large infrared color excesses, have been observed primarily at 1612 MHz to a sensitivity of ∼1 Jy (Bowers and Cornett, 1974). No OH emission was detected.

* NRAO is operated by Associated Universities, Inc., under contract with the National Science Foundation.

References

Bowers, P. F. and Cornett, R. H.: 1974, *Astrophys. Letters*, in press.
Winnberg, A., Goss, W. M., Höglund, B., and Johansson, L. E. B.: 1973, *Astrophys. Letters* **13**, 125.

F. J. Kerr
P. F. Bowers
Astronomy Program, University of Maryland,
College Park, Md. 20742, U.S.A.

DISCUSSION

Habing: You said that this is a pilot program, so you are going to extend this to the whole sky?

Kerr: Yes, we're extending it over the whole sky at low sensitivity, just looking for the strongest objects.

Caswell: I think that for the unidentified sources the major concern is whether the velocities are due to galactic rotation and give good kinematic distances. I believe the characteristic velocity of the objects is better represented by the mean velocity of the two OH peaks than by the red-shifted one, as is conventionally believed at present. We have six objects below the galactic center quite clearly well above the maximum velocity allowed as shown by the H I velocity distribution in this area which is very similar, symmetric about the galactic center. There is just nothing above the galactic center corresponding to these. However, there is a considerable increase in symmetry if you take the mean of the two OH peaks as the characteristic velocity of the objects.

Kerr: It's true that galactic rotation is certainly the most important component, and it can be argued which feature should be taken. If we take the extreme peak, that goes a long way outside the Schmidt curve. But I would like to point out that on the southern side the hydrogen distribution is somewhat different from the northern side in velocity. That's what the asymmetry is.

Palmer: How does the distribution of the newly discovered sources near the galactic center compare with that of the high velocity gas in the direction of the galactic center?

Kerr: There are not enough known yet for good statistics. However, some of the sources are high enough in latitude to be outside the normal gas layer, *if* they are at galactic center distances.

Robinson: Is it true to say that none of the Class C unidentified sources have been found in latitudes more than 3° and they are all very close to the plane?

Caswell: That's true as far as I recollect, but there is a significant selection effect. The large number of search positions that has been done at high latitudes gives some indication that the unidentified sources are quite closely confined to the plane.

Kerr: Practically all searching has been done pretty close to the plane.

Caswell: The searching at the positions of long-period variable stars gives a mean galactic latitude of something like 6°.

PART 5

THE GALACTIC CENTER

"I have a feeling that one does not understand at all
how things happen at the center."
J. H. Oort, in the discussion following his paper

INFRARED RADIATION FROM THE GALACTIC CENTER

J. BORGMAN

Kapteyn Observatory, Roden, The Netherlands

Abstract. IR observations of the region within 0°5 of Sgr A are discussed. The total IR luminosity is $150 \times 10^6 \, L_\odot$. Most of it is stellar radiation, absorbed by low temperature dust and reradiated in the IR. There is no unique interpretation for the highly structured infrared core. It may comprise compact infrared sources like those in some H II regions, or the structure may simply result from the dust distribution. Nonthermal radiation is also feasible, but a tradeoff between source luminosity and extinction is possible and prevents a final decision on this question.

I. Introduction

Since the first papers on infrared radiation from the galactic center by Becklin and Neugebauer (1968, 1969), considerable observational work has been done. Observations are now available on the far infrared surface brightness of a $4° \times 2°$ region (Hoffmann *et al.*, 1971). Fine structure has been mapped within 20″ of the near infrared maximum surface brightness by Rieke and Low (1973). The broadband spectrum of a number of features has been determined (Becklin and Neugebauer, 1969; Rieke and Low, 1973; Harper and Low, 1973), and a broad absorption band has been detected near 9.7 μm (Woolf, 1973).

The present paper aims to briefly describe and discuss published as well as some new observations. We will largely restrict ourselves to the region within 0°5 of Sgr A. Following the usual practice the observations will be ascribed to 'sources'. This terminology is used rather loosely and is not much more than a convenient way to characterize the observations in terms of the spectral and spatial distribution of the observed radiation.

II. Observations

Table I presents a condensed summary of observations which are more fully discussed below.

(a) THE NEAR INFRARED HALO

Becklin and Neugebauer (1968) mapped the 2.2 μm radiation in a region of $30' \times 40'$ using an aperture of 1.8′; in addition, they obtained maps at 2.2 μm of smaller regions with higher resolution. For the central 110″ a study was made of the relation between the field of view and the measured radiation at 1.65, 2.2, 3.5 and 4.8 μm.

Figure 1 shows the 2.2 μm map of Becklin and Neugebauer, obtained with a 1.8′ aperture. The map is dominated by the near infrared nucleus (see Section IIb), the center of which coincides with the radio position of Sgr A. In addition several secondary sources can be seen; none of these secondary sources has been observed at longer wavelengths. The position of the source at $17^h42^m44^s$, $-29°00'$, is close to the position

TABLE I

Condensed summary of typical data

Name	Area	F_λ at λ		$P_{\lambda\lambda}$ and	L at $\lambda\lambda$		Ref.[a]
		Jy	μm	10^{-10} W m^{-2}	$10^6 L_\odot$	μm	
Near infrared halo	$\sim 40' \times 30'$	2000	2.2	11	3	1.5–2.4	1
Near infrared halo	$\sim 1° \times 1°$	1000	5.5	1	0.3	5–6	2
Near infrared nucleus	$5' \times 3'$	80	2.2	1	0.3	1.5–4.0	1
Near infrared nucleus	$\begin{cases} (10' \times 6') \\ \text{minus} \\ (5' \times 3') \end{cases}$	100	2.2	0.6	0.2	1.5–2.4	1
Near infrared nucleus	22"	5	2.2	0.1	<0.1	1.06–4.0	1, 7
Infrared core	22"	900	12.2	3	1	3.1–25	1, 3, 4, 5
Far infrared halo	$38' \times 15'$	1.5×10^6	100	240	74	75–125	6
Far infrared halo	7'	–	–	130	40	10–300	8

[a] References quoted:

1. Becklin and Neugebauer (1968, 1969)
2. Houck *et al.* (1971)
3. Woolf (1973)
4. Borgman (1972, 1973)

5. Low *et al.* (1969)
6. Hoffman *et al.* (1971)
7. Spinrad *et al.* (1971)
8. Low and Aumann (1970)

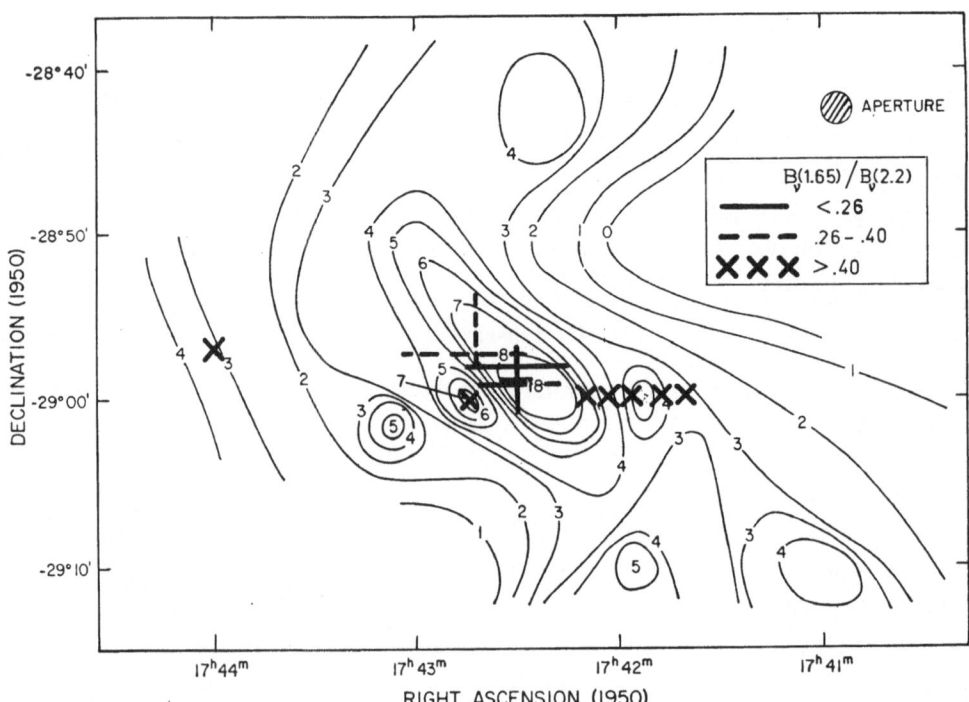

Fig. 1.   Becklin and Neugebauer's (1968) contour map at 2.2 μm observed with an aperture of 1.8' diameter. Contour lines are separated by 8.5×10^{-20} W m^{-2} Hz^{-1} s^{-1} and are uncertain by the same amount. Color information, added to the contour diagram refers to 1.65 and 2.2 μm scans (Becklin and Neugebauer, 1968) and is explained in the insert.

of the 40 km s^{-1} H$_2$CO cloud discussed, e.g., by Fomalont and Weliachew (1973). Becklin and Neugebauer made a number of scans and measurements within the area of Figure 1 which indicate that the color gets redder towards the near infrared nucleus. We will adopt $B\lambda(1.65)/B\lambda(2.2)=0.51$ for the near infrared halo, with a flux density of 2000 Jy at 2.2 μm. The flux density has been derived from a mean brightness of 25×10^{-20} W m^{-2} Hz^{-1} s^{-1} outside the near infrared nucleus over the 1000 arc-min^2 mapped in Figure 1. It is clear that the near infrared halo extends beyond the limits of this map; therefore, the actual 2.2 μm flux density will be higher.

(b) THE NEAR INFRARED NUCLEUS

This is the dominant source of Figure 1, which has been observed at wavelengths 1.65, 2.2, 3.5, 4.8 and 10.1 μm. The observations at 4.8 and 10.1 μm will be dealt with under the infrared core (see Section IIc), together with the 20 μm observations.

The near infrared nucleus has been observed in considerable detail at 2.2 μm by Becklin and Neugebauer (1968). They found that the 2.2 μm radiation distribution had a dominant source at $17^h42^m30^s \pm 1^s$, $-28°59\!'4 \pm 0\!'1$ with a full width at half maximum of $3' \times 5'$ when observed with a 1.8$'$ aperture. This source has a 2.2 μm angular distribution which is characterized by a power law such that the flux density within a circular aperture of diameter D is proportional to $D^{1.2 \pm 0.1}$, corresponding to a mean surface brightness which is proportional to $D^{-0.8 \pm 0.1}$. This power law holds in the aperture range $5''$–$10'$. The same dependence on aperture has been observed at 1.65 μm over the aperture range $16''$–$110''$. According to Becklin and Neugebauer (1968), the near infrared nucleus has a color $B\lambda(1.65)/B\lambda(2.2)=0.20$. In our discussion we will adopt this figure with the restriction that it is valid only for the central $5' \times 3'$ part. Somewhat arbitrarily we adopt for our discussion $B\lambda(1.65)/B\lambda(2.2)=0.35$ for the annulus $(10' \times 6')$–$(5' \times 3')$, a value intermediate between the color of the center and the color of the near infrared halo (see Section IIa).

Spinrad et al. (1971) detected the near infrared nucleus at 1.06 and 1.18 μm, with measurements at the position of the maximum 2.2 μm brightness with a 22$''$ diaphragm.

(c) THE INFRARED CORE

Becklin and Neugebauer (1969) and Low et al. (1969) made the first measurements of a small source $(d \approx 16'')$ which emits most of the 10 and 20 μm radiation of the galactic center. Ground-based observing techniques do not exclude the existence of a more extended 10 and 20 μm source of low surface brightness; however, rocket observations by Houck et al. (1971) make it unlikely that a $1'$ aperture would contain less than half of the total radiation at these wavelengths (cf. Rieke and Low's data for a $1'$ aperture).

The position of the 16$''$ source coincides with the center of the near infrared nucleus at 2.2 μm. In addition, Becklin and Neugebauer (1968) found a point-like source at less than 10$''$ from the center of the 16$''$ source.

Recently, Rieke and Low (1973) obtained detailed maps of $\sim 30'' \times 30''$ area at

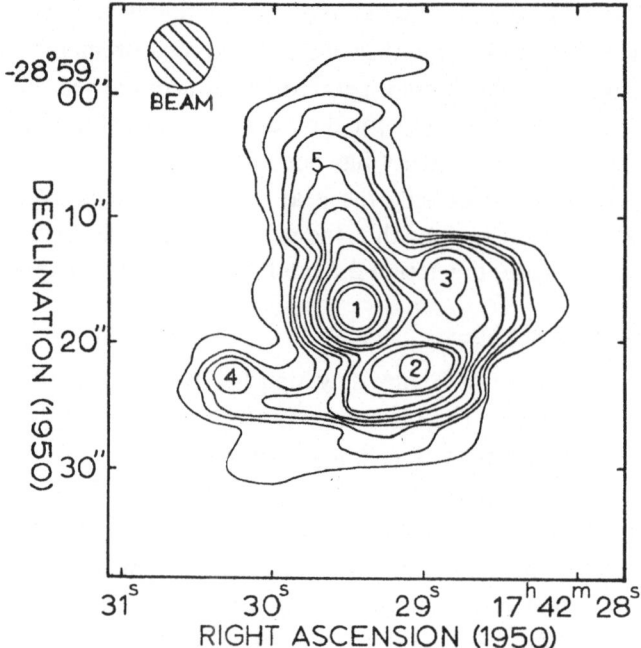

Fig. 2.   Rieke and Low's (1973) 10.5 μm map of the infrared core. The contours are at 1, 2, 3, 4, 6, 8, 10, 12, 14, 16, 18, 20, 22, 24 and 26 times 7.6×10^{-17} W m^{-2} Hz^{-1} s^{-1}. The position of source 5 has been copied from Rieke and Low's (1973) 5 μm map.

3.5, 5.0, 10.5 and 21 μm. Their maps show five sources, four of which can be seen in the 10.5 μm map reproduced in Figure 2.

Woolf (1973) published a scan of a strong and broad absorption feature at 9.7 μm. The profile was measured with a 22″ diaphragm which possibly encircles the sources 1 and 2 of Figure 2; these two sources largely determine the position of the 10 μm centroid. Borgman (1973) measured the depth of the 9.7 μm absorption feature at the position of the four sources in Figure 2 separately. The following discussion summarizes the results of these studies.

(i) *The pointlike source.* Following Rieke and Low (1973), we will identify the pointlike source (Becklin and Neugebauer, 1968) with source 3, discussed under Section IIc (iv).

(ii) *The 16″ source.* We will assume that the 16″ source (5″ south and 2″ east of the pointlike source according to Becklin and Neugebauer, 1969) is a blend of sources 1 and 2 in Figure 2. This expectation is supported by a comparison of the observed flux density (Becklin and Neugebauer, 1969) and an integration of the brightness over a circular 16″ diaphragm in Figure 2. The positional agreement between Becklin and Neugebauer (1969) and Rieke and Low (1973) is, however, not entirely convincing.

(iii) *Sources 1 and 2.* These two sources are the major features of the 10.5 μm map in Figure 2. At 21 μm Rieke and Low (1973) observed a single concentration, coinciding with the position of source 1. The contribution of source 2 to the 21 μm map is difficult to estimate. Rieke and Low correct their data for 29 mag of visual extinction, resulting in intrinsic color temperatures of approximately 350 K.

Borgman (1973), observing sources 1 and 2 with 0.7 μm bandwidth filters and a 7″ diaphragm finds [8.1–12.2] color temperatures of 275 ± 20 K and 250 ± 20 K, resp. The depth of the 9.7 μm absorption feature, when measured with a 1.0 μm bandwidth filter, is 1.7 and 2.0 mag, resp. These differences are considered real. The difference with Rieke and Low's color temperatures may be partly due to Rieke and Low's neglecting the 9.7 μm depression of approximately 0.4 mag when observing with a wide-band 10 μm filter. The influence of a correction of 29 mag of visual extinction on a [8.1–12.2] color temperature of 250 K is only 10 K.

Sources 1 and 2 are the dominant features of the infrared core at 10 μm. This suggests that the profile of the 9.7 μm depression as published by Woolf (1973) is typical for Sources 1 and 2.

(iv) *Source 3*. The broadband spectrum between 1.65 μm and 5 μm, corrected for 29 mag of visual extinction, fits a temperature of 2000 K (Rieke and Low). The 10 and 20 μm observations show a considerable excess over the 2000 K curve, suggesting that at these wavelengths Source 3 is lost in the background. This suggestion is confirmed by our 7″ data, which indicate a [8.1–12.2] color temperature of 275 ± 30 K. With the same 7″ diaphragm we measured the depth of the 9.7 μm feature, which turns out to be 2.3 mag. For a fuller discussion of the pointlike object we refer to Becklin and Neugebauer (1968).

(v) *Source 4*. Our 7″ aperture data indicate a color temperature of 200 ± 20 K at the position of source 4. The 9.7 μm depression is 2.1 mag (mean value of two discordant observations) when measured with a 1.0 μm bandwidth filter. Source 4 shows only up on the 10.5 μm map of Rieke and Low; at other wavelengths there are no closed contours at the position of this source. We suggest that Source 4 can be regarded as part of the diffuse radiation in the infrared core.

(vi) *Source 5*. This source shows only up on the 5 μm map of Rieke and Low, where it is about as strong as Source 3. At 3.5, 10.5 and 21 μm there is no evidence for the existence of Source 5.

(d) THE FAR INFRARED HALO

Hoffman *et al.* (1971) mapped the galactic center region, using a 12′ aperture and a 75–125 μm bandpass. They detected at least three discrete sources and an extended background of $4° \times 2°$. The observed luminosity in this area (which includes Sgr A, Sgr B2 and G359.4–0.1) is $3.4 \times 10^8 L_\odot$.

Hoffmann *et al.* have adjusted the position of their largest and most luminous source to the radio coordinates of Sgr A. The identification is supported by the relative positions of the two other sources, which then coincide with Sgr B2 and G359.4–0.1.

Our discussion will be limited to the source identified with Sgr A. This source measures 38′ × 15′ full width at half maximum, and its observed 75–125 μm luminosity is $7.4 \times 10^7\,L_\odot$. Low and Aumann (1970) studied the dependence of the 50–350 μm flux density on the field of view; they found that the brightness decreased according to a $D^{-1.1 \pm 0.2}$ power law in a 3 to 14′-aperture range. With a 5′ aperture Harper and Low (1973) observed at 56, 68, 91 and 105 μm; the flux density seems to peak at ∼ 80 μm.*

III. Discussion

The main conclusions of the discussion of the observations are given in Table II and Figure 3.

Following Becklin and Neugebauer (1968) we will interpret the near infrared radiation of the galactic nucleus as stellar, with a population as suggested by the broadband spectrum of the nucleus of M31 (Sandage *et al.*, 1969): blackbody radiation characterized by a temperature of 4000 K. Assuming the interstellar extinction law to be the same as in the solar neighbourhood (Johnson, 1965) they interpreted the [1.65–2.2] color index as being reddened by 1.9 mag, leading to a visual extinction of 27 mag. Essentially the same result was found by Spinrad *et al.* (1971) and Borgman (1972) when observations at 1.06 and 1.18 μm were included.

Actually the figure of $A_V = 27$ mag was derived for the near infrared nucleus. However, for the near infrared halo it has been argued in Section IIa that the [1.65–2.2] color is 1.0 mag bluer than in the near infrared nucleus. This leads to $A_V = 13$ mag and a luminosity of $120 \times 10^6\,L_\odot$. It should be noted that the 5.5 μm flux density of Table I is somewhat but not significantly below the expected value. As the 5.5 μm

TABLE II

Interpretation

Name	Intrinsic source	A_V mag	λ μm	A_λ mag	F_λ° Jy	L_{tot} $10^6\,L_\odot$
Near infrared halo	stellar, 4000 K BB	13	2.2	1.2	6600	120
Near infrared nucleus 5′ × 3′	stellar, 4000 K BB	27	2.2	2.7	1000	18
Near infrared nucleus (10′ × 6′) − (5′ × 3′)	stellar, 4000 K BB	20	2.2	2.0	600	12
[a] Near infrared nucleus, 22″	stellar, 4000 K BB	27	2.2	2.7	60	1
[a] Near infrared nucleus 22″	stellar, 4000 K BB	75	4.8	1.9	200	10
[a] Infrared core, 22″	diluted 270 K BB	27	12.2	0.2	1100	1
Far infrared halo 38′ × 15′	dust	13	100	0	1.5×10^6	150

[a] These are three alternatives to explain the radiation from the central 22″ central region.

* The agreement between the published photometric quantities of Hoffman *et al.* (1971), Harper and Low (1973), Aumann and Low (1970) and Low and Aumann (1970) is not very satisfactory; without full information on actually observed data and reduction procedures it is difficult to make a judicious assessment. A preliminary comparison indicates that the flux data of Hoffmann *et al.* (1971) may be too high, a conclusion which is supported by some recent results of the group at the University of Groningen (Olthof, private communication).

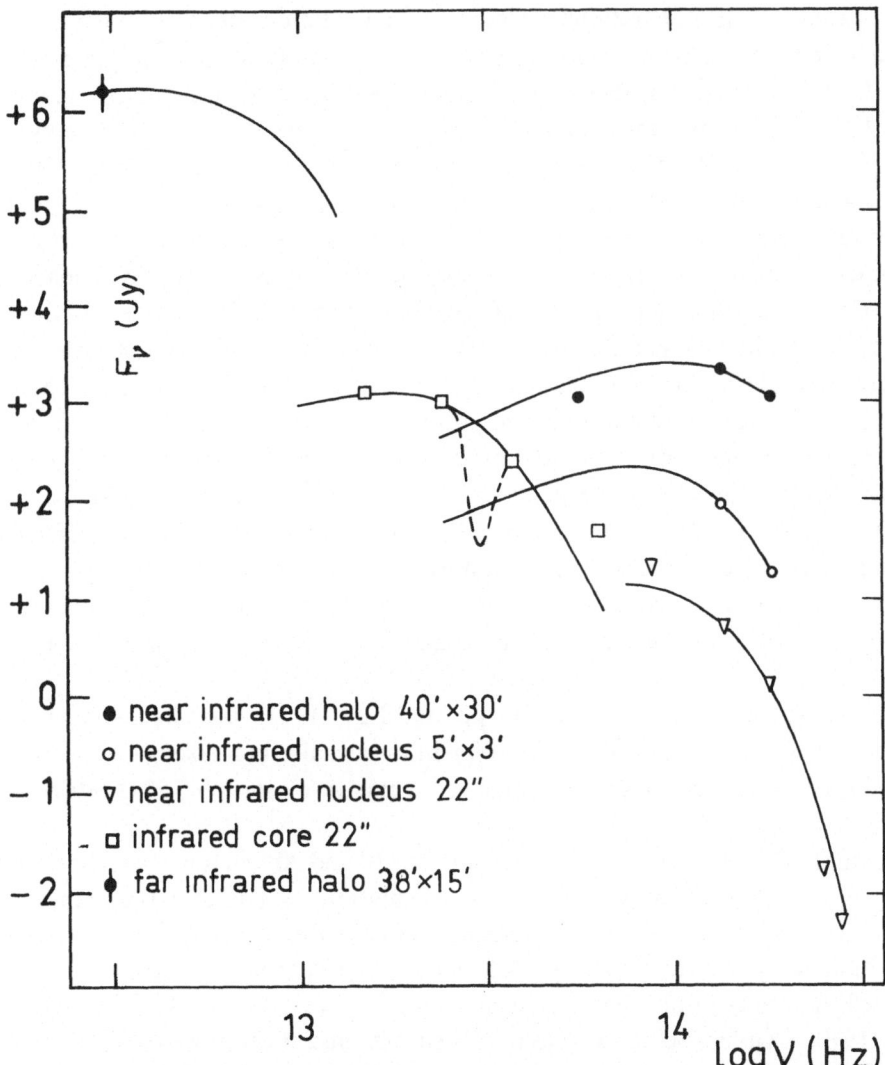

Fig. 3. Spectral energy distribution of various regions in the galactic center. Plotted points are observations taken from the literature quoted in Table I. The dashed curve is the 9.7 μm absorption feature, adapted from Woolf (1973). Full drawn curves refer to the interpretation in Table II. The curve fitted to the infrared core data is for a 270 K black-body and $A_V = 27$ mag. For reference, a 65 K black-body curve has been drawn through the 100 μm point of the far infrared halo, but this temperature is certainly too high for the outlying portions of the halo (see text).

flux has been measured for a larger area than is interpreted here we tentatively conclude that the near infrared halo does not extend much beyond the 1000 arcmin² mapped by Becklin and Neugebauer (1968); this conclusion should not be given too much weight as some unpublished uncertainties and observing constraints could not be taken into account.

The adopted $A_V = 27$ mag for the 5' × 3' near infrared nucleus leads to a luminosity of $18 \times 10^6 L_\odot$. Adopting an intermediate extinction $A_V = 20$ mag for the annulus

(10×6) arcmin^2 – (5×3) arcmin^2 we find a projected luminosity of 12×10^6 $L_\odot$. With a simple spherical mass distribution model and adopting $M/L = 3$ it can be shown that the total stellar luminosity within the 1000 arcmin^2 area is approximately 140×10^6 $L_\odot$ at a total mass of 4×10^8 $M_\odot$.* An inspection of Table I shows that only 5% of this radiation is observed as direct starlight in the near infrared; 130×10^6 $L_\odot$ are available for redistribution over other parts of the spectrum.

The far infrared halo and the near infrared halo have approximately the same brightness distribution, suggesting that they occupy the same volume in space. The total infrared luminosity of the far infrared halo may be a factor 2 larger than the 75–125 μm luminosity as follows from the data in Table I and the flux-aperture relation determined by Low and Aumann (1970). The resulting luminosity of 150×10^6 $L_\odot$ is in good agreement with the assumption that 130×10^6 $L_\odot$ of stellar energy from the near infrared halo are reradiated by dust in the far infrared. Also the agreement between the data in the smaller fields of $10' \times 6'$ (30×10^6 $L_\odot$ stellar luminosity) and $7'$ diameter (40×10^6 $L_\odot$ total infrared radiation ascribed to dust) is quite satisfactory and supports the local energy conversion hypothesis. The agreements are somewhat fortuitous (but how much so is difficult to estimate) in view of considerable uncertainties in relation to the data and the somewhat different areas which were compared.

The flux density peak at 80 μm (Harper and Low, 1973) suggests a color temperature of 65 K for the central $5'$ of the far infrared halo, but taking into account a diluted spectrum at low frequencies this value may be closer to 45 K (Andriesse and Olthof, 1973).

The infrared core, radiating predominantly at 10 and 20 μm, is not well understood and as a consequence has aroused much speculation. Okuda and Wickramasinghe (1970) have proposed a strong concentration of the dust towards the center where they situated a powerful 5×10^4 K blackbody. The origin of this source remains unspecified. A variety of nonthermal mechanisms has been discussed by Burbidge and Stein (1970). Lynden-Bell and Rees (1971) suggest accretion of gas by a black hole, resulting in an efficient conversion of the rest-mass energy into UV and optical radiation which is then available for the ionization of the thermal radio source and heating of the dust in the extended far infrared halo.

The 9.7 μm absorption feature of the infrared core has been studied in detail by Aitken and Jones (1973), who predict 75 mag of visual extinction. The extinction value depends fully on the adopted extinction in front of the Becklin/Neugebauer object in the Orion nebula and has to be revised if the criticism of Becklin et al. (1973) proves to be right.

Borgman (1972) proposed a dense 1 pc diameter star cluster with a luminosity of 5×10^8 $L_\odot$, sufficient to excite all of the observed far infrared emission in an area of $4° \times 2°$; 200 mag of visual extinction between the source and the Sun are needed to

---

* As a consequence of adopting a bluer color for the near infrared halo this mass is lower than the values discussed by Sanders and Lowinger (1972), which include those derived earlier by Oort (1971) and Becklin and Neugebauer (1968).

explain the 5–20 μm spectrum as an absorption spectrum, but this requires the intervening dust cloud to be at a distance of at least 5 pc in order to keep the dust radiation at 20 μm at a tolerable level (Visser, private communication). However, the 2.2 μm radiation shows a 1 pc wide peak at the position of the infrared core (Becklin and Neugebauer (1968)); if this is interpreted as radiation from the near infrared nucleus stellar population, which can be seen within 0.5 pc from the center, it is suggested that the postulated dense dust cloud is not larger than 1 pc and coincides with the infrared core. Taking these considerations into account we find that a reduced cluster luminosity of $10^7 L_\odot$ (rather, the central peak in the continuous distribution of stars in the near infrared halo and nucleus) embedded in 75 mag of visual extinction reproduces the 3.5 and 4.8 μm data; the excess radiation at 8–20 μm in the central 1 pc region (21″) can be ascribed to the high temperature dust in that region, adopting reasonable absorption efficiencies. It should be stressed that the choice of source luminosity and extinction is rather arbitrary and that trade-offs between these two parameters are quite feasible if dust radiation is allowed for. It is not possible, on the basis of now available observations, to decide convincingly between radiation contributions of hot dust and partially obscured sources. The only hard fact is the observed luminosity of $10^6 L_\odot$, which is a lower limit as it does not include any far infrared contribution where there are no high angular resolution data.

There is some evidence that the infrared core sources 1 and 2 (and possibly 5) as discussed in Section IIc can be identified in the 8085 MHz map of Balick and Sanders (Gordon, 1974), which would mean that these sources are real and are not the result of extinction windows. Both sources show the 9.7 μm feature in absorption, which invites a comparison with the Becklin and Neugebauer source in the Orion nebula (Gillett and Forrest, 1973) and IRS5 in W3 (Wynn-Williams et al., 1973; Aitken and Jones, 1973). However, there is at least one important difference: the luminosity of the galactic core is two orders of magnitude greater. Some 40 sources like W3–IRS5 would be required to explain the 10–20 μm data. It should be noted that the near infrared data cannot supply us with a lower estimate of extinction on the basis of an extrapolation of a flat radiospectrum, as the near infrared halo stars (and perhaps hot dust in the dense center) form a comparatively bright foreground.

It is of some interest to consider the contribution to the radiation field by planetary nebulae, which are known to be good infrared emitters. The spatial distribution of these objects has been discussed by Minkowski (1965), who finds a strong concentration towards the galactic center. Addressing ourselves to the volume of space occupied by the near infrared halo we will make an estimate of the numbers of planetaries required to explain the 10–20 μm data. Adopting $10^4 L_\odot$ and 350 K as a typical luminosity and color temperature for a planetary and assuming that the radiation is evenly divided between the UV and the infrared we would need 1000 planetaries to produce the observed 10–20 μm flux density (5000 Jy according to Houck et al. [1971]; notice that the 10–20 μm fluxes of the cold dust and the halo stars have been considered negligible). The UV luminosity would be $5 \times 10^6 L_\odot$, which is compatible with the $10^7 L_\odot$ required to keep the extended thermal region ionized (Lynden-

Bell and Rees, 1971). The numbers given here suggest a ratio of giant stars to planetaries of 1000:1, which is not unreasonable considering the possible genetic relationship between these objects and their relative life span expectations. Again, these numbers are not too meaningful, as a hot dust component and a number of O stars will explain the data just as well. It should be noted that the total luminosity of the region occupied by the dust in the far infrared halo of the galactic center can be understood quite well with the luminosity of the late type stellar population; there is no reason to believe that a large fraction of the dust radiation is caused by heating through direct absorption of UV photons, though this is probably the dominant mechanism in the nearby H II regions.

IV. Conclusion

Infrared observations of the galactic center region indicate a total luminosity of $150 \times 10^6 \, L_\odot$ within 50 pc from the center. Nearly all of this luminosity is of stellar origin, but most of it is absorbed and reradiated by low temperature dust in the far infrared. There is no unique interpretation for the highly structured infrared core. Though the deep 9.7 μm absorption feature hints at high local extinction (as seems to occur also in some nearby dense molecular clouds) it is probably premature to accept this parameter as a calibrated measure of extinction. The sources in the infrared core may well be giant examples or clusters of compact infrared sources as found in some H II regions. The observations do not rule out that the infrared core is simply the structured central peak of the stellar and dust distribution.

Assuming that the mass contribution of the 1 pc infrared core is small and adopting $M/L = 3$ we find that the mass within 50 pc from the center is $5 \times 10^8 \, M_\odot$.

Acknowledgements

It is a pleasure to thank Dr C. D. Andriesse for a critical reading of an early version of the manuscript. The final paper has benefitted from discussions with Drs Gordon, Greenberg, Mezger, Oort and Wynn-Williams.

References

Aitken, D. K. and Jones, B.: 1973, *Astrophys. J.* **184**, 127.
Andriesse, C. D. and Olthof, H.: 1973, *Astron. Astrophys.* **27**, 319.
Aumann, H. H. and Low, F. J.: 1970, *Astrophys. J.* **159**, L159.
Becklin, E. E. and Neugebauer, G.: 1968, *Astrophys. J.* **151**, 145.
Becklin, E. E. and Neugebauer, G.: 1969, *Astrophys. J.* **157**, L31.
Becklin, E. E., Neugebauer, G., and Wynn-Williams, C. G.: 1973, *Astrophys. J. Letters* **182**, L7.
Borgman, J.: 1973, in L. Mavridis (ed.), *Stars and the Milky Way System*, p. 188.
Borgman, J.: 1974, in preparation.
Burbidge, G. R. and Stein, W. A.: 1970, *Astrophys. J.* **160**, 573.
Fomalont, E. B. and Weliachew, L.: 1973, *Astrophys. J.* **181**, 781.
Gillett, F. C. and Forrest, W. J.: 1973, *Astrophys. J.* **179**, 483.
Gordon, M. A.: 1974, this volume, p. 477.

Harper, D. A. and Low, F. J.: 1973, *Astrophys. J.* **181**, 781.

Hoffman, W. F., Frederick, C. L., and Emery, R. J.: 1971, *Astrophys. J. Letters* **164**, L23.

Houck, J. R., Soifer, B. T., Pipher, J. L., and Harwit, M.: 1971, *Astrophys. J. Letters* **169**, L31.

Johnson, H. L.: 1965, *Astrophys. J.* **141**, 923.

Low, F. J. and Aumann, H. H.: 1970, *Astrophys. J. Letters* **162**, L79.

Low, F. J., Kleinmann, D. E., Forbes, F. F., and Aumann, H. H.: 1969, *Astrophys. J. Letters* **157**, L97.

Lynden-Bell, D. and Rees, M. J.: 1971, *Monthly Notices Roy. Astron. Soc.* **152**, 461.

Minkowski, R.: 1965, *Stars and Stellar Systems* **5**, 321.

Okuda, H. and Wickramasinghe, N. C.: 1970, *Nature* **226**, 134. ,

Oort, J. H.: 1971, in D. J. K. O'Connell (ed.), *Nuclei of Galaxies, Pontificiae Academiae Scientiarum Scripta Varia* **35**, North-Holland Publishing Co., Amsterdam, p. 321.

Rieke, G. H. and Low, F. J.: 1973, *Astrophys. J.* **184**, 415.

Sandage, A. R., Becklin, E. E., and Neugebauer, G.: 1969, *Astrophys. J.* **157**, 55.

Sanders, R. H. and Lowinger, T.: 1972, *Astron. J.* **77**, 292.

Spinrad, H., Liebert, J., Smith, H. E., Schweizer, F., and Kuhi, L. V.: 1971, *Astrophys. J.* **165**, 17.

Woolf, N. J.: 1973, in J. M. Greenberg and H. C. van de Hulst (eds.), 'Interstellar Dust and Related Topics', *IAU Symp.* **52**, 485.

Wynn-Williams, C. G., Becklin, E. E., and Neugebauer, G.: 1972, *Monthly Notices Roy. Astron. Soc.* **160**, 1.

J. Borgman
Kapteyn Observatory,
Mensingheweg 20,
Roden, The Netherlands

DISCUSSION

Mezger: The far IR halo has an apparent size of $38' \times 18'$. It thus comprises at least the H II region G0.2−0.0, one of the most powerful H II regions in the galactic center. You attribute all of the dust heating to radiation of Population II stars. I feel that at least part of the heating must be due to heating by O stars which ionize the H II region.

Borgman: This is only a small part, probably; it is, moreover, uncertain whether the two regions coincide spatially.

Radhakrishnan: What happens to the energy absorbed by the dust at the frequency of the absorption dip? For example, is it reradiated at longer wavelengths because the temperature of the dust is low?

Borgman: It is reradiated by the dust, both at longer and shorter wavelengths as the color temperature requires a dust temperature of the order of 200–300 K.

Maxwell: What extinction is now seen in the galactic center, say at 2.2 μm?

Borgman: The answer is whatever you want, almost. In my tables I used the 'official' 27 mag extinction, which means that the 2.2 μm extinction using the normalized van de Hulst curve is about one-tenth of that: 2.7 mag. However, you can make almost any tradeoff between source luminosity and extinction in front of it by varying the source temperature.

RADIO CONTINUUM AND RECOMBINATION LINE STUDIES
OF THE GALACTIC CENTER

M. A. GORDON

National Radio Astronomy Observatory,* Green Bank, W. Va., U.S.A.

Abstract. This review discusses unsuccessful searches for electromagnetic bursts accompanying gravitational events observed to come from the galactic center, radio observations of continuum emission and recombination lines from the region of the galactic center.

I. Introduction

In our attempts to understand the basic nature of our own Galaxy, perhaps the most important observations for us – and the most difficult – are the ones directed toward the galactic nucleus. The last few years have brought many new discoveries concerning the composition of the interstellar medium and the galactic center region. This paper presents a somewhat personal review of three areas of radio research which have been heavily pursued over the last few years.

II. Pulsed E-M Radiation from the Galactic Center

Weber (1969) has reported the detection of pulsed gravitational waves from the direction of the galactic center. His detectors have an angular resolution of 70°, so that uncertainty exists as to the precise direction from which the pulses originate. Subsequent papers establish that his detection rate is approximately two events per day and that the characteristic flux density of each event is approximately 3×10^{28} Jy at his observing frequency of 1660 Hz.

The rationale for concomitant searches for pulsed electromagnetic (E-M) radiation is as follows. Presumably, the gravitational radiation arises from mass quadrupole moments associated with catastrophic events in the direction of the galactic center. Even if the coupling efficiency were extremely low, it seems reasonable either to expect these catastrophic events would also intrinsically generate bursts of detectable E-M radiation, or the energy released in the gravitational events would stimulate E-M bursts by some secondary process from material along the line of sight. The great sensitivities of apparatus designed to respond to E-M radiation could therefore be used to explore the nature of the gravitational events in an indirect manner. For example, even long-wave receivers can detect events of 100 Jy, a factor of 3×10^{26} weaker than the gravitational bursts. Thus, the logic connecting pulsed gravitational and the expected E-M bursts is intuitive; it does not presume the detailed physical processes connecting the two events.

A large number of workers have searched for these associated E-M bursts at fre-

---

* Operated by Associated Universities, Inc., under contract with the National Science Foundation.

F. J. Kerr and S. C. Simonson, III (eds.), Galactic Radio Astronomy, 477–489. All Rights Reserved.

quencies from 151.5 MHz to (indirectly) beyond 5.6×10^{12} MHz. When available, most workers have used receivers designed to reject terrestrial interference by requiring *coincidence* detection of the E-M pulses on either more than one frequency or at more than one location. To date, no E-M pulses have been found which correspond to the gravitational events recorded by Weber.

Figure 1 shows the observational limits of some of these searches, plotted as flux density in units of Jy against frequency in units of Hz. Below 160 MHz (Dulk, 1970; Brezgunov *et al.*, 1971), the galactic center is undetectable presumably because of free-free absorption along the line of sight to the source; cross-hatching marks this region of the spectrum. These observational limits have a spectral index of −0.4, but such an index probably only describes how instrumental sensitivities vary with frequency and does not describe a characteristic of the E-M bursts.

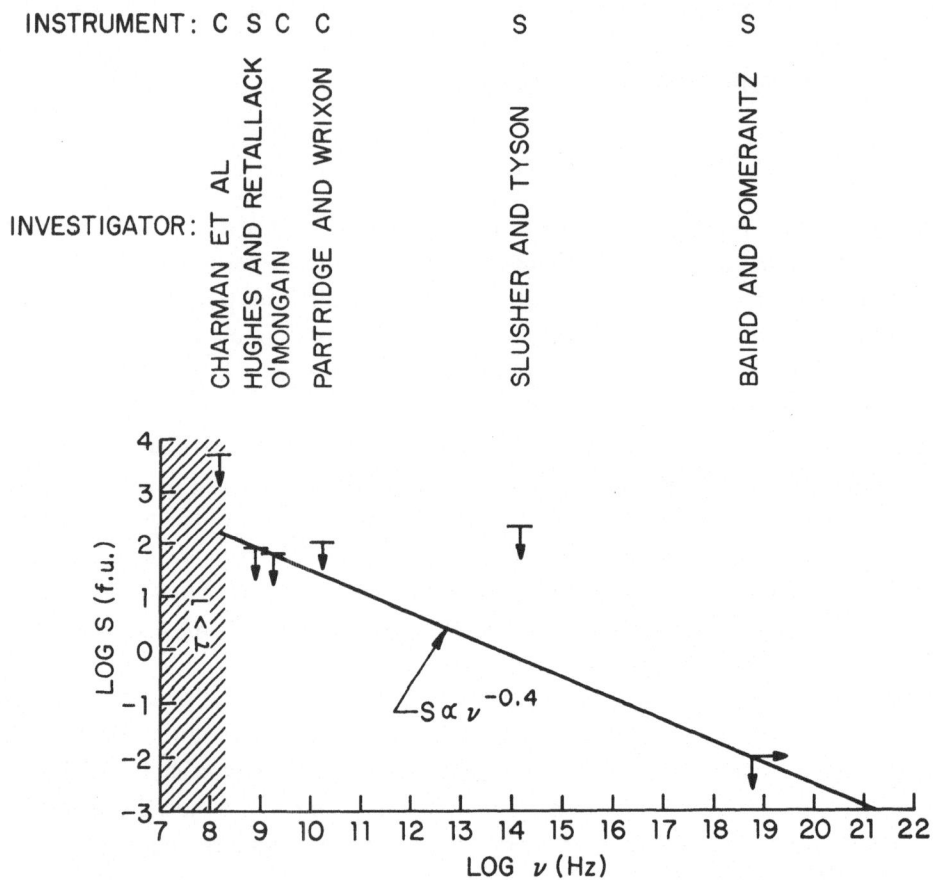

Fig. 1.   Limits to electromagnetic pulses from the galactic center associated with observed gravitational events. The hatched region marks frequencies at which Sgr A cannot be detected. The fitted spectum probably describes instrumental sensitivities. Top: the investigator(s) for each observation is shown; *C* or *S* refers to a coincidence or single receiving system, respectively.

Although no E-M pulses have been detected which correlate with the gravitational bursts, Hughes and Retallack (1973) do find discrete radio pulses of extra-terrestrial origin. Using a 60-ft telescope operating at 858 MHz, they find E-M pulses which cannot be identified with any radio, optical, infrared, X-ray, or γ-ray source within the beam during the drift scans across Sgr A, nor with solar or ionospheric events. Comparison observations made in other directions did not show the same kind of pulses. They conclude that the mysterious E-M pulses come from a region 4 min of time east of the galactic center but cannot identify their source.

The failure to find E-M pulses associated with Weber's gravitational events does not prove conclusively that they don't exist. Propagation of E-M pulses through ionized gas causes delays proportional to $v^{-2} N_e \, dl$; delays could be as large as 10^m for pulses propagating at 150 MHz through a column density of 10^{22} electrons cm^{-2}, a reasonable estimate for the ionization between earth and Sgr A (see Partridge and Wrixon, 1972). Furthermore, the radio beams used in these searches usually are much smaller than the 70° beam of Weber's apparatus; the source of the gravitational events may not be Sgr A. What *is* apparent, though, is that the flux density of E-M radiation associated with the gravitational pulses is an extremely small fraction $(< 10^{-24})$ of that radiated by the gravitational events.

III. Maps of Continuum Emission

Because of our location within the plane, absorption and scattering by interstellar gas and dust makes optical observations completely ineffective as a tool to study the nucleus of the Galaxy. This same interstellar matter also absorbs low-frequency radio waves, thereby limiting radio investigations to the centimeter and millimeter range. Furthermore, the high angular resolution required for these investigations can only be achieved (with circularly symmetric beams) by filled-aperture telescopes operated at short wavelengths; in this range, thermal sources necessarily predominate over nonthermal sources because of the intrinsic differences between the frequency dependences of the radio emission. Most interferometers are located in the northern hemisphere where synthesis of a symmetric beam to map the brightness distribution of Sgr A is extremely difficult owing to the low declination of the source; they usually produce strip scans either by name or effect. Lunar occultations also produce strip scans, but fortunately they frequently lie at considerable angles to one another according to the circumstances of each particular event. In short, the most important astronomical object within our Galaxy seems to be one of the most difficult to observe.

Figure 2 shows a recent map of the entire galactic center region made at 2 cm with a circularly symmetric beam of half-power width 2.25′ (Kapitsky and Dent, 1973). At a distance of 10 kpc, the map encloses a region 195 × 270 pc. The region is complex, containing both thermal and nonthermal sources. The best known sources in the region are the H II region Sgr B2 and the central source Sgr A, whose positions are identified in the figure. There are also many smaller discrete sources seen in the region, some blended into the larger sources Sgr A and Sgr B. Maps of the same region, made

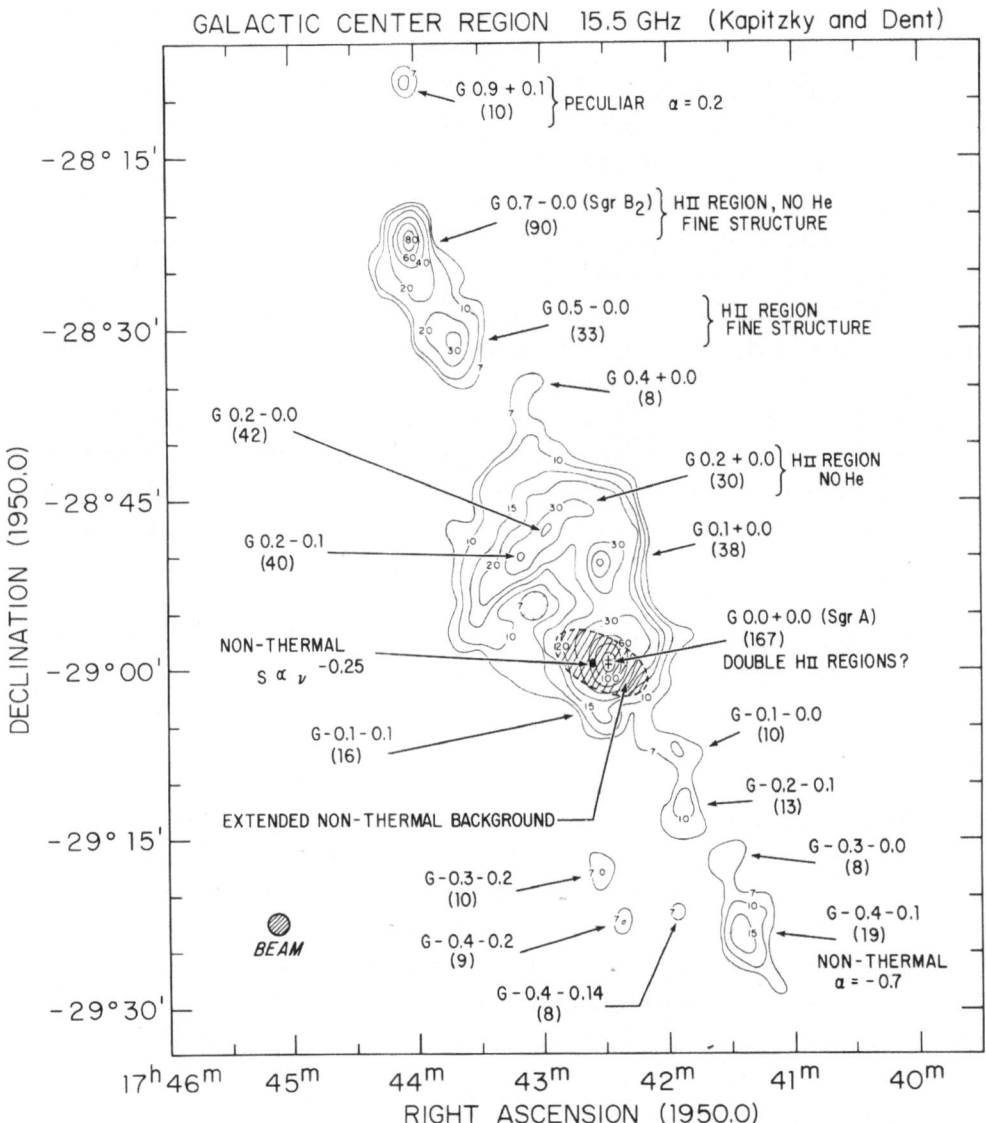

Fig. 2.   The general region of the galactic center mapped at 2 cm with a beam of half-power width 2.25′
by Kapitsky and Dent (1973). Flux densities are shown by the numbers in parentheses.

at lower frequencies (Whiteoak and Gardner, 1973), show the discrete sources to be superimposed upon an extended background several degrees in extent. An important result of the map shown in Figure 2 is that this background is no longer visible at 2 cm; that is, this extended background must be nonthermal and any radio recombination lines seen against this background must arise in the H I gas along the line of sight.

From observations at 3.5 mm, Hobbs *et al.* (1971) believe Sgr B2 to contain a dense, compact H II region. Martin and Downes (1972) interpret synthesis observations to show Sgr B2 to consist of seven discrete sources in an extended background. Recently synthesis observations by Balick and Sanders (1973) show Sgr B2 to have two discrete sources, less than 15 arcsec wide, containing 20 per cent of the flux at 13 cm. Based upon the results of Hobbs *et al.*, Balick and Sanders suggest these components have an electron density of 10^6 cm^{-3} and an electron temperature of 15000 K.

Sgr A, the radio source associated with the galactic center, is complex: it contains more than one component, and these components have different spectral indices. Evidently this complexity is the reason that, as has been known for some time, the overall source has been observed to shift in position and in angular size as a function of angular resolution and frequency. In the analysis of observations made at 327 MHz by lunar occultation, and others at frequencies up to 1665 MHz, Gopal-Krishna *et al.* (1972) incorporated many previous observations – virtually strip scans – to produce the map of Sgr A shown in Figure 3.

Figure 3 shows the radio source Sgr A to contain two discrete peaks, *A* and *S*, superimposed upon a broad background *B* (not to be confused with Sgr B) and another extended background *E*.

Component A has a width of approximately 10″ (Downes and Martin, 1971) and has the same position as the peak of the extended 2.2 μm source seen by Becklin and Neugebauer (1969). On the basis of comparison of the ratio of the 2.2 μm flux to star densities measured in the nucleus of M31, Sandage *et al.* (1969) and later Oort (1971) and Sanders and Lowinger (1972) all conclude that the infrared observations indicate that our Galaxy has a dense nucleus of approximately 10^6 $M_\odot$ in the form of population II stars within 1 pc radius, the mass density of stars tapering off toward a radius of ∼750 pc roughly as $R^{-1.8}$. Oort estimates the central star density (within a radius of 1.0 pc, or ∼2.5″) to be approximately 10^8 times that near the Sun. The radio data suggest these stars to be embedded in a nonthermal radiation field (component *A*) having a spectral index α of roughly 0.1 ± 0.01 between 327 and 5000 MHz.

Component S is less well known because of blending with the extended background in the strip scans. Its position may be associated with that of the 40 km s^{-1} features of the OH and H_2CO molecule (Sandqvist, 1971; Gardner and Whiteoak, 1972). Gopal-Krishna *et al.* suggest its spectral index is unlikely to be greater than 0.2. ('S' indicates the name of its discoverer, Sandqvist.)

Components B and E, because of their weaker intensities, are more difficult to classify as either thermal or non-thermal. They may be associated with the broad 100 μm emission sources (Hoffman *et al.*, 1971) tentatively associated with thermal emission from dust grains heated by the stars located near the galactic center (see Sanders and Lowinger [1972] for a general discussion). Gopal-Krishna *et al.* feel that the larger extended component (*E*) has been missed in earlier work because of the great difficulty of observing it and its presumably steep spectral index. The spectral index of component *B* appears to be no greater than 0.3. Both may be nonthermal, although by a similar analysis Lipovka (1971) reaches the opposite conclusion.

Figure 4 shows synthesis observations of Sgr A made by Balick and Sanders (1973) at λ 13 cm with a beam 8″ × 20″ and at λ 4 cm with a beam 3″ × 7″. The map shows two

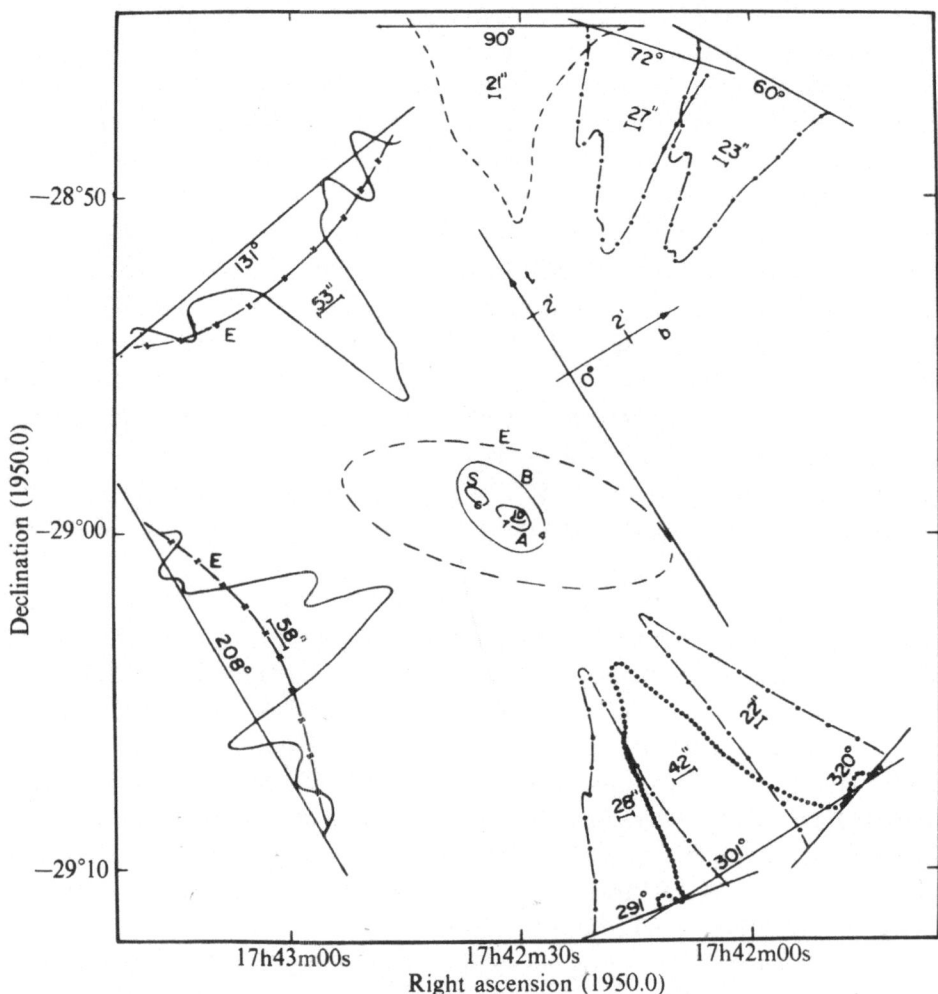

Fig. 3. A composite map of Sgr A made by Gopal-Krishna *et al.* (1972). In addition to their occultation observations made at 327 MHz (solid lines), the map includes observations at 405 MHz (dotted lines) by Thompson *et al.* (1969), at 1420 MHz (dashed lines) by Sandqvist (1971).

distinct, unresolved sources containing approximately 20% of the total flux of Sgr A. The low-level background is nonthermal. These constitute the fine structure of the *A* component of Figure 3. The *S* component seen at low frequencies does not appear in these maps and hence is either nonthermal or too broad to appear in this synthesis observation. Note that the radio map shows the same double structure seen in the 10 *μ*m IR data.

The lowest frequency at which the galactic center region has been observed is 160 MHz. Figure 5 shows the map made at this frequency by Dulk and Slee (1973) with the Culgoora spectroheliograph. The peak emission comes from a position

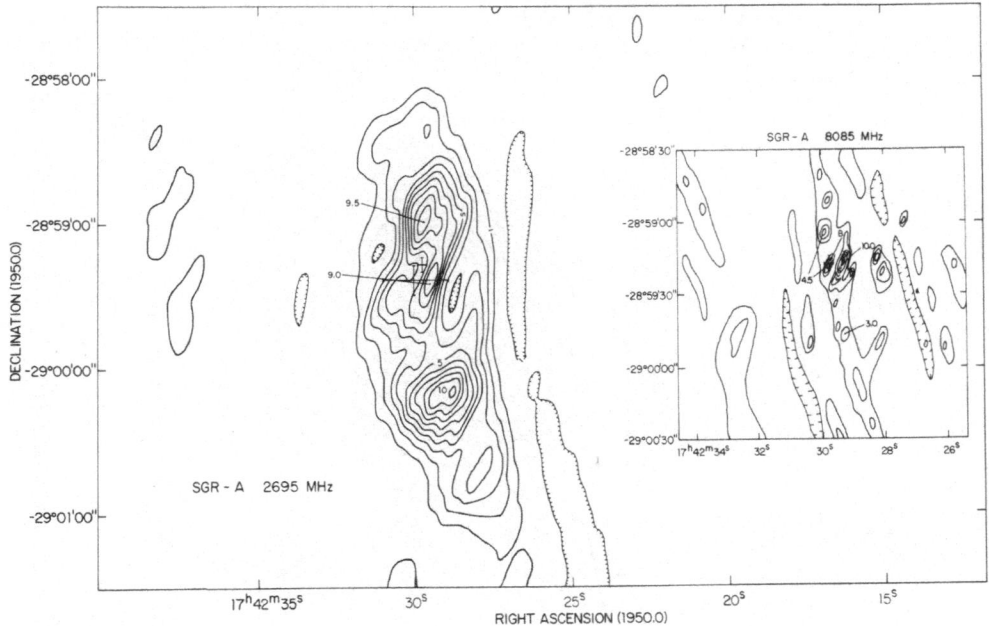

Fig. 4.   The synthesis maps of fine structure in Sgr A made at λ13 and λ4 cm by Balick and Sanders (1973). These maps have been 'cleaned'. Note the two discrete, unresolved sources superimposed upon a non-thermal background. The large cross marks the center of the extended source seen at 2.2 μm, the small cross the point source at 2.2 μm.

midway between components *A* and *S*, and the broad component may well correspond to the source *E* found at 327 MHz by Gopal-Krishna *et al.* (1972).

At frequencies less than 120 MHz, the source Sgr A cannot be observed (Dulk, 1970; Brezgunov *et al.*, 1971). The limits are consistent with free-free absorption along the line of sight such that the optical depth τ is 1 at approximately 200 MHz. As will be discussed below, this large-scale absorption is likely to be caused by partially-ionized cold gas in the interstellar medium.

IV. Radio Recombination Lines

Radio recombination lines provide still another tool with which to investigate the galactic center. First, they are emitted only by thermal sources and can be used to classify sources of flat spectra as being either thermal or nonthermal. Second, the power emitted in the lines is a measure of $T_e^{-1.5} \int n_i n_e \, dl$ subject to the (often substantial) uncertainties imposed by departures from local thermodynamic equilibrium (LTE) and by large fluctuations in the physical conditions over the path length of the emission. Third, comparison of the frequency of the line center with the known rest frequency gives a radial velocity to a thermal source. Fourth, if the actual electron temperature can be established, the line widths are a measure of the nonthermal velocity fields (microturbulence) of the gas within the beam. Fifth, comparison of

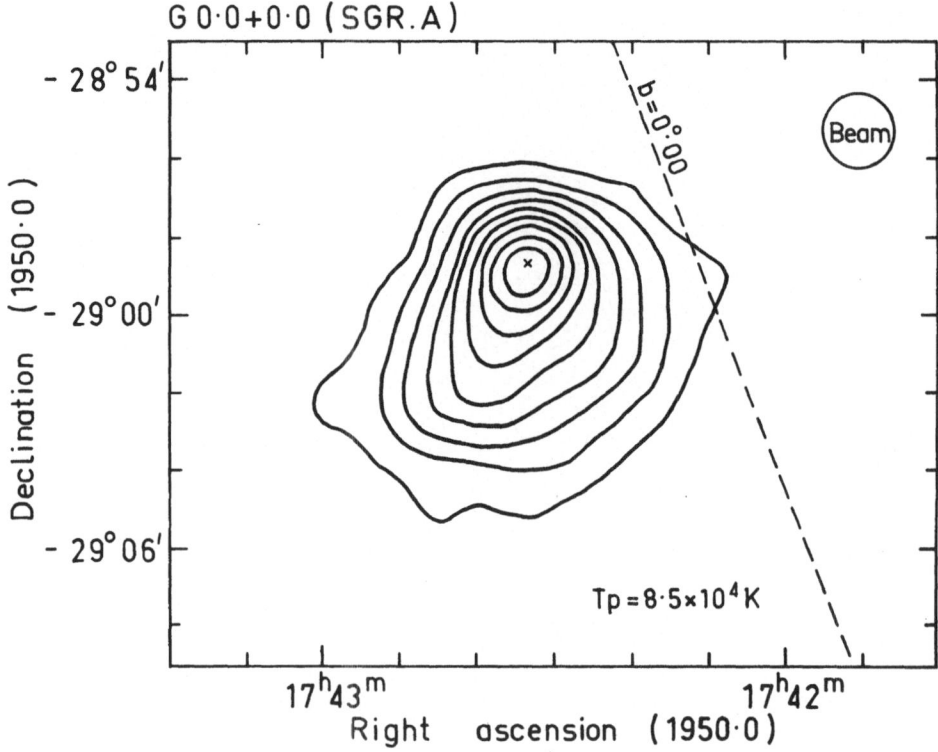

Fig. 5. Galactic center at 160 MHz mapped by Dulk and Slee (1973).

lines emitted by different elements gives relative elemental abundances, also subject to the uncertainties due to departures of the level populations from LTE.

Surveys of radio recombination lines show that H II regions emit hydrogen recombination lines out of LTE (Brocklehurst and Seaton, 1972), at 5 GHz have β lines with intensities approximately 20% of the α lines (Hjellming and Gordon, 1971), and have helium lines with intensities 5% to 12% of the corresponding hydrogen lines (Churchwell and Mezger, 1973). Line profiles tend to be gaussian with full widths at half intensity of 20 to 30 km s^{-1}, depending upon the relative angular sizes of the source and the beam (Sorochenko and Berulis, 1969; Smith and Weedman, 1970).

(a) THE Sgr B2 COMPLEX

Some sources located in the galactic center region (Figure 2) emit radio recombination lines. The giant Sgr B2 complex emits hydrogen lines having characteristics normal for H II regions. For example, the ratio of line to continuum intensities and the single measurement of the ratio of the α to β line intensity are similar to those observed in the giant H II regions lying in the plane between galactic radii 3 and 10 kpc (Mezger *et al.*, 1972; Chaisson, 1973). On this basis, the G0.5 − 0.0 and Sgr B2 components have been identified as giant H II regions themselves. Mapping these components by means of recombination lines, Chaisson (1973) finds that the velocity gradients across these sources run in opposite directions, showing the G0.5 − 0.0 and Sgr B2 components to be physically separate.

Unlike most H II regions, the Sgr B complex does not emit detectable helium recombination lines. Huchtmeier and Batchelor (1973) report no detection of the He 109α line (5 GHz) to a sensitivity of 1% of the corresponding hydrogen line. Rubin (1968) notes that a narrow range of late O and early B stars emit a radiation field capable of ionizing large amounts of hydrogen but very little helium; this situation is likely in the H II region Orion B (NGC 2024). However, the great amount of ionized hydrogen in the Sgr B2 complex requires on the order of 10^2 early-type stars, and it is highly unlikely that all of these would lie in this narrow range of spectral types. If the Sgr B2 complex is an H II region, Churchwell and Mezger (1973) concluded that either the gas does not contain neutral or ionized helium, or that the ionization is not caused by early-type stars.

The detection of a large excess of 100-μm flux may provide an alternative explanation for the deficiency of helium recombination lines in Sgr B2. Bless and Savage (1972) noted that, in general, the extinction by interstellar grains increases toward short wavelengths in the UV. Leibowitz (1973) used this fact to propose that, in H II regions showing no helium recombination lines, the grains deplete the radiation field shortward of 514 Å, reradiate the energy into the IR, and thereby prevent substantial ionization of helium within the H II region. Mezger et al. (1973) further explore the details of this mechanism. It seems possible that this mechanism can account for the lack of He II in the source Sgr B2.

Sgr B2 has some other unusual characteristics. Chaisson reports the widths of the recombination lines to be approximately 40 km s^{-1}, much larger than those seen in other H II regions having similar ratios of angular size to beamwidth. Furthermore, Sgr B2 is an unusually rich source of molecules (see Gordon and Snyder, 1973). One component of the radial velocities of the molecules generally agrees with those of the recombination lines, suggesting the molecular cloud to be closely associated with the H II region. Chaisson (1973) suggests that the molecules lie within a dense shell of cold gas surrounding the H II region.

(b) OTHER LINE SOURCES

Regions closer to the galactic center also emit radio recombination lines. The extended region immediately to the north of Sgr A shows weak, frequently asymmetric hydrogen recombination lines, which Mezger et al. (1972) attribute to the ionized gas constituting this source. However, recombination lines are also seen in the direction of the nonthermal source Sgr A itself, and from the nonthermal extended region lying to the south of Sgr A (Lockman and Gordon, 1973). These spectra are similar to those of Sgr B2 in that no helium lines have been detected and that line widths are often large. Unlike those from Sgr B2 the ration of the α-to-β line intensities is abnormally large for H II regions; that is, ⩾9 at λ18 cm (Lockman and Gordon, 1973) and ⩾17 at λ21 cm (Brown and Balick, 1973). Clearly, departures from LTE are extremely large for these recombination lines, much larger than that found for these lines in H II regions.

The line profile of the recombination lines seen in the direction of Sgr A is highly non-gaussian. It contains a main component near 0 km s^{-1} which is seen in all other lines from the region and other components which vary from one beam position to another. The intensity of the main component correlates well with that of the background continuum. Furthermore, as Figure 6 shows, the line profile seen toward Sgr A bears a remarkable similarity to the corresponding absorption profile of H I at λ21 cm, taken by an interferometer by Radhakrishnan et al. (1972).

Because of the great departures from LTE and the similarity of the line profile to that of H I Lockman and Gordon (1973) considered that the 0 km s^{-1} feature of the recombination line, at least, might arise in the cold gas lying between Sgr A and the Sun, stimulated by continuum radiation from the background sources according to the method described by Dupree and Goldberg (1969). Using the observational constraints and the non-LTE calculations of level populations (Dupree, 1972), they calculated that the lines could arise in dense, cold (20 K) clouds of a few parsecs in length, lying along the line of sight to Sgr A. The alternative hypothesis, that the lines arise in high temperature regions, requires the existence of long, small diameter 'fingers' of gas running from Sgr A directly to the Sun – a highly unlikely possibility.

The origin of these recombination lines in cold clouds also accounts for other observed phenomena. The clouds require electron densities >3 cm^{-3}, which would correspond to net ionization rates of 10^{-13} s^{-1} on the basis of ionization equilibrium. Similar rates are necessary to explain the weak recombination lines found in the plane in certain directions free of discrete radio sources (Gordon and Cato, 1972). Owing to pecularities associated with the unknown ionization mechanism and to the prob-

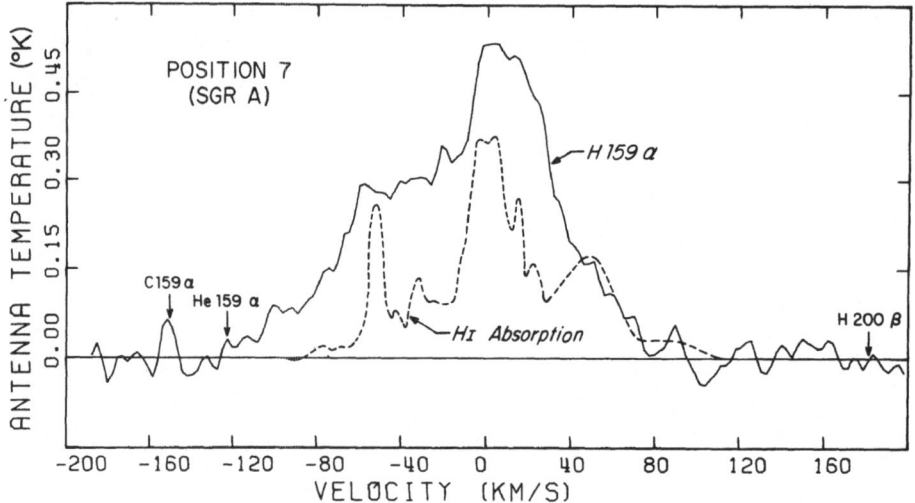

Fig. 6. Juxtaposition of the H159α-line profile (solid line) taken in the direction of Sgr A by Lockman and Gordon (1973) with an inverted absorption profile of neutral hydrogen taken in the same direction by Radhakrishnan et al. (1972).

able effects of charge-transfer reactions, helium recombination lines might not arise (Brown and Balick, 1973). Furthermore, these clouds account for the impossibility of observing Sgr A at radio frequencies less than 200 MHz because of free-free absorption (Dulk, 1970; Brezgunov et al., 1971). By independent observations and analysis, Brown and Balick (1973) reach the same conclusion, that the radio recombination lines seen at $\lambda 18$ cm and $\lambda 21$ cm in the general direction of Sgr A are emitted by dense, cold clouds along the line of sight to background sources.

The populations of levels involved in the $\lambda 3$ and $\lambda 6$ cm lines observed by Mezger et al. (1972) may be substantially different than those of the apparently maser-prone levels involved in the $\lambda 18$ and $\lambda 21$ cm lines. Therefore, the line intensities may differ greatly. The α lines seen at the shorter wavelengths involve levels near principal quantum numbers 85 and 109, where the atoms have effectively smaller collision cross-sections than at the more highly excited states seen at long wavelengths. For these levels, radiative processes may dominate so as to cause a uniform under-population and correspondingly weak recombination lines. For higher levels, the local collision and radiative processes populate in a way to make the levels highly sensitive to stimulation by background radiation. Therefore, it's possible that lines seen by Mezger et al. also come from the cold gas but are simply much weaker than those seen at lower frequencies. Clearly, additional observations of recombination lines, particularly of higher order lines, are necessary to understand the origin of the higher frequency lines from Sgr A.

Because helium recombination lines are not seen in Sgr B2 sources, it is tempting to consider the hydrogen lines seen there as arising from cold gas rather than to adopt the more radical explanations of elemental under-abundance of helium or strange new ionization mechanisms. Furthermore, the unusually wide lines observed by Chaisson (1973) could be a simple consequence of velocity dispersion (within the beam) over a long line of sight. The only impediment to such a model is the single observation of an apparently 'normal' β line reported by Churchwell (1970). Because of its importance, this line should be reobserved, and other higher order lines be observed, so as to assess fully the details of the level populations of the hydrogen atoms before the Sgr B2 sources can be reliably classified as giant H II regions deficient in helium recombination lines. Alternatively, the details of the UV extinction of grains should be quantitatively explored.

V. Summary

The radio searches for electromagnetic radiation *associated* with bursts of graviational waves have been unsuccessful, in spite of truly enormous sensitivities. By themselves, these searches cast doubt on the validity of the reported graviational bursts; together with other unsuccessful direct searches for the gravitational bursts, the existence of the gravitational bursts seems unlikely at the fluxes quoted by Weber (1969).

Observations by both filled and synthetic apertures show the sources in the galactic center to be extremely complex, always having significant structure down to the resolution limit of the telescope. Such complexity makes it difficult to determine accurate spectral indices for these components, in some cases even to answer the simple question of whether the sources are thermal or nonthermal. Unfortunately, the detection (or not) of α-type radio recombination lines does not discriminate between the two possibilities, because in principle these lines can arise in cold as well as hot gas. Lines of sight to the galactic center certainly pass through cold clouds of hydrogen, under circumstances where the amplification mechanism of Dupree and Goldberg (1969) is likely to be effective. Observations of higher order transitions can establish the nature of the level populations and hence the physical characteristics of the emitting gas.

Acknowledgements

It is a pleasure to thank Bruce Balick, Robert L. Brown, Ed Churchwell, Felix J. Lockman and Robert H. Sanders for many stimulating discussions on this subject.

References

Baird, G. A. and Pomerantz, M. A.: 1972, *Phys. Rev. Letters* **28**, 1337.
Balick, B. and Sanders, R. H.: 1973, private communication.
Becklin, E. E. and Neugebauer, G.: 1969, *Astrophys. J. Letters* **157**, L31.
Bless, R. C. and Savage, B. D.: 1972, *Astrophys. J.* **171**, 293.
Brezgunov, V. N., Dagkesamansky, R. D., and Udaltsov, V. A.: 1971, *Astrophys. Letters* **9**, 117.
Brocklehurst, M. and Seaton, M. J.: 1972, *Monthly Notices Roy. Astron. Soc.* **157**, 179.
Brown, R. L. and Balick, B.: 1973, 'Observations of the Recombination Line Region Toward Sagittarius A', *Astrophys. J.*, preprint.
Chaisson, E. J.: 1973, 'Microwave Spectroscopic Mapping of Gaseous Nebulae. IV: Excited Hydrogen in Sagittarius B_2', *Astrophys. J.*, preprint.
Charman, W. N., Fruin, J. H., Jelley, J. V., Haynes, R. F., Hodgson, E. R., Scott, P. F., Shakeshaft, J. R., Baird, G. A., Delaney, T. J., Lawless, B. G., Drever, R. W. P., and Meikle, W. P. S.: 1971, *Nature* **232**, 177.
Churchwell, E.: 1970, 'Observations of Radio Recombination Lines of Hydrogen, Helium and Carbon', Indiana University, Ph. D. thesis.
Churchwell, E. and Mezger, P. G.: 1973, *Nature* **242**, 319.
Downes, D. and Martin, A. H. M.: 1971, *Nature* **233**, 112.
Dulk, G. A.: 1970, *Astrophys. Letters* **7**, 137.
Dulk, G. A. and Slee, O. B.: 1973, private communication.
Dupree, A. K.: 1972, *Astrophys. J.* **173**, 293.
Dupree, A. K. and Goldberg, L.: 1969, *Astrophys. J. Letters* **158**, L49.
Gardner, F. F. and Whiteoak, J. B.: 1972, *Astrophys. Letters* **10**, 171.
Gopal-Krishna, Swarup, G., Sarma, N. V. G., and Joshi, M. N.: 1972, *Nature* **239**, 91.
Gordon, M. A. and Cato, T.: 1972, *Astrophys. J.* **176**, 587.
Gordon, M. A. and Snyder, L. E.: 1973, *Molecules in the Galactic Environment*, John Wiley and Sons, New York.
Hjellming, R. M. and Gordon, M. A.: 1971, *Astrophys. J.* **164**, 47.
Hobbs, R. W., Modali, S. B., and Maran, S. P.: 1971, *Astrophys. J. Letters* **165**, L87.
Hoffman, W. F., Frederick, C. L., and Emery, R. J.: 1971, *Astrophys. J. Letters* **164**, L23.
Huchtmeier, W. K. and Batchelor, R. A.: 1973, *Nature* **243**, 155.
Hughes, V. A. and Retallack, D. S.: 1973, *Nature* **242**, 105.
Kapitsky, J. E. and Dent, W. A.: 1973: 'A High Resolution Map of the Galactic Center Region', *Astrophys. J.*, preprint.
Leibowitz, E. M.: 1973, *Astrophys. J.* **181**, 369.
Lipovka, N. M.: 1971, *Astron. Zh.* **48**, 260 (English transl.: 1971, *Soviet Astron.* **15**, 203.).
Lockman, F. J. and Gordon, M. A.: 1973, *Astrophys. J.* **182**, 25.

Martin, A. H. M. and Downes, D.: 1972, *Astrophys. Letters* **11**, 219.

Mezger, P. G., Churchwell, E., and Pauls, T. A.: 1972, 'Ionized Gas in the Galactic Center' (to be published in the proceedings of the Regional Meeting of the IAU, Athens).

Mezger, P. G., Churchwell, E. B., and Smith, L. F.: 1973, 'IR-Excess Radiation and the Absorption Characteristics of Dust in Galactic HII Regions', preprint.

O'Mongain, E.: 1973, *Nature Phys. Sci.* **242**, 136.

Oort, J. H.: 1971, in D. J. K. O'Connell (ed.), *Nuclei of Galaxies*, Pontificiae Academiae Scientiarum Scripta Varia No. 35, North-Holland Publishing Co., Amsterdam, p. 11.

Partridge, R. B. and Wrixon, G. T.: 1972, *Astrophys. J. Letters* **173**, L75.

Radhakrishnan, V., Goss, W. M., Murray, J. D., and Brooks, J. W.: 1972, *Astrophys. J. Suppl.* **4**, 357.

Rubin, R.: 1968, *Astrophys. J.* **154**, 391.

Sandage, A. R., Becklin, E. E., and Neugebauer, G.: 1969, *Astrophys. J.* **157**, 55.

Sanders, R. H. and Lowinger, T.: 1972, *Astron. J.* **77**, 292.

Sandqvist, A.: 1971, Ph.D. Thesis, University of Maryland.

Slusher, R. E. and Tyson, J. A.: 1973, *Nature* **243**, 25.

Smith, M. G. and Weedman, D. W.: 1970, *Astrophys. J.* **160**, 65.

Sorochenko, R. L. and Berulis, J. J.: 1969, *Astrophys. Letters* **4**, 173.

Thompson, A. R., Riddle, A. C., and Lang, K. R.: 1969, *Astrophys. Letters* **3**, 49.

Weber, J.: 1969, *Phys. Rev. Letters* **22**, 1320.

Whiteoak, J. B. and Gardner, F. F.: 1973, *Astrophys. Letters* **13**, 205.

M. A. Gordon

National Radio Astronomy Observatory,
2010 N. Forbes Ave., Tucson, Ariz. 85705, U.S.A.

DISCUSSION

Maxwell: Have the NRAO maps been laundered? That should be indicated on the maps.

Gordon: Yes, and I agree.

Robinson: Gordon has called for β recombination line observations at Sgr B2. At Parkes we have measured the H137β line in Sgr B2. The line consists of two resolved components:

(i) A component centered on the H137β frequency (5005.033 MHz) with similar width to the H109α line and 20% of the H109α intensity.

(ii) A component at higher frequency ($\nu_0 = 5005.320$ MHz) which is produced by the 3-3 A-branch transition in methanol!

THE GALACTIC CENTRE AT 408 MHz

A. G. LITTLE

University of Sydney, Sydney, Australia

Abstract. The Molonglo radio telescope has been used to map the galactic centre region with a resolution of 2.9'. Although similar to microwave maps of comparable resolution, there are differences. In particular, the source G1.05–0.1 is not visible at microwaves, and the source G0.1–0.0 is not visible at 408 MHz. The flux values of all the other sources around Sgr A confirm that they are thermal and that the sources G0.7–0.0, G0.5–0.0, and G0.2–0.0 are optically thick.

A more extensive galactic background survey with the cross shows the existence of a peak in the background emission with a brightness temperature of about 500 K at the galactic centre. This is probably the nonthermal source observed at 85 MHz. The higher-resolution observations of this map show the narrower, hotter region which appears on the microwave map and which surrounds Sgr A. This has a brightness temperature of 2100 K, which cannot be accounted for by a simple thermal spectrum extrapolated from microwaves.

Although there are now many occultation measurements of Sgr A at low frequencies, very little data is available for the remainder of the region below about 1.4 GHz. Moreover, what high-resolution information is available is confined to frequencies above a few GHz. The Molonglo radio telescope (Mills *et al.*, 1963) operating at 408 MHz should therefore be able to add useful knowledge of temperatures and flux densities of this very complex region. If we confine the comparisons with the microwave results to the high-resolution maps of Broten *et al.* (1965), Downes *et al.* (1966), and Hollinger (1965), which have been made with beamwidths similar to that of the Molonglo telescope (2.9'), direct comparisons can be made without the uncertainties introduced by different resolving powers.

The line scans corresponding to these observations are shown in Figure 1. The side lobes associated with the strong source Sgr A can be seen, and some correction for these has been attempted by making use of the line scans for interpolation.

The resulting contour map is shown in Figure 2. The unit contour interval is 680 K, which is in terms of the main beam brightness temperature. This has been derived from the flux calibration using the Rayleigh-Jeans relation between flux and temperature. The lowest contour corresponds to a temperature of 1075 K, which was determined from a survey of the galactic plane made by Green (1974) using the Molonglo cross. Two sources were used for both flux and position calibration: PKS 1643-22 and PKS 1814-51. The data for these sources are shown in Table I. The fluxes were based on the Wyllie (1969a, 1969b) scale, and the above positions and fluxes are from Hunstead (1972).

A comparison with the microwave maps shows that with the exception of the source G0.1 – 0.0 all the sources listed at microwaves appear on the 408 MHz map. An additional source is seen at 408 MHz which appears only as a suggestion in the contours of the source G1.1 – 0.1 at microwaves. This new source, which has a flux comparable to G1.1 – 0.1, is thus clearly nonthermal in character and has been designated G1.05 – 0.1 after Downes and Maxwell (1966).

F. J. Kerr and S. C. Simonson, III (eds.), Galactic Radio Astronomy, 491–497. All Rights Reserved.

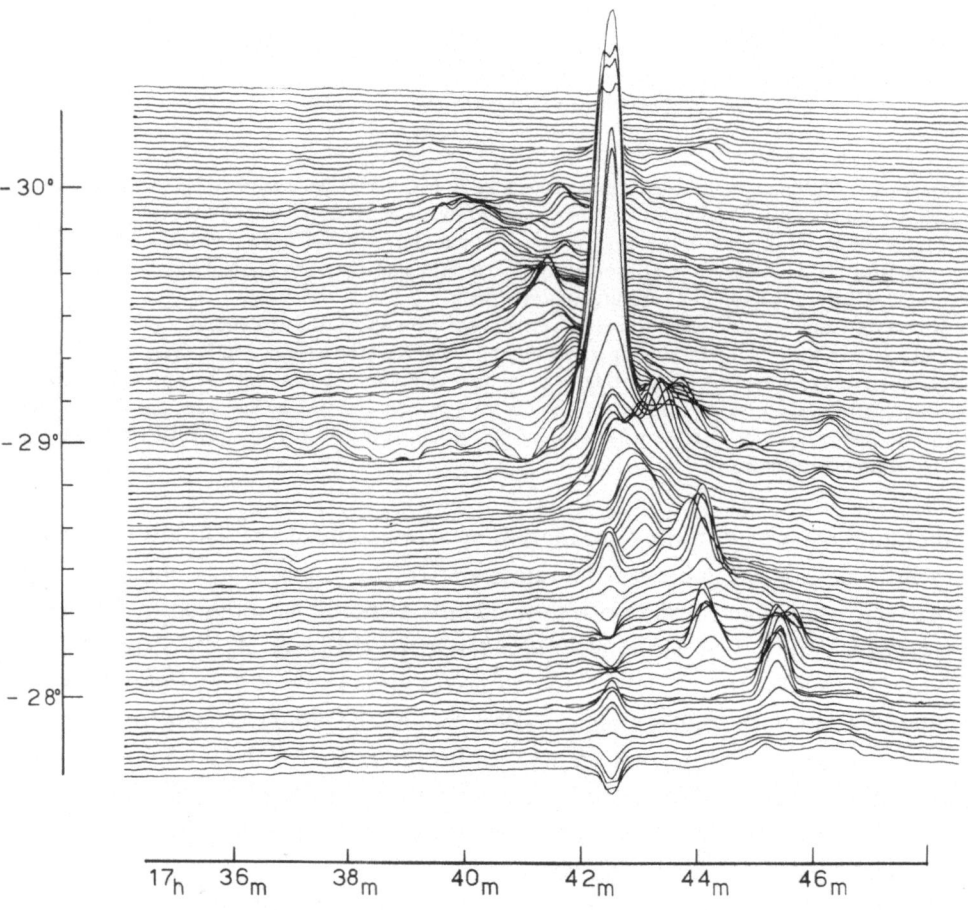

Fig. 1.   (1) Line scans of the galactic centre at 408 MHz (Sgr A is saturated at this sensitivity).

The integrated flux densities for all the sources are given in Table II together with values for the three microwave maps taken from Downes and Maxwell (1966).

It can be seen that the present observations support the view that apart from Sgr A and G1.0−0.1 all the sources are thermal in character. Moreover, the sources G0.2− −0.0, G0.5−0.0 and G0.7−0.0 appear to be optically thick at 408 MHz. The fact that G0.1−0.0 is not visible may be due to the source having a brightness temperature close to that of the hot background on which it is located.

In determining the flux densities there was the usual problem of separating the sources from the background. This has been done here by using the line scans, but it is by no means a precise operation. The temperatures of the background levels so determined are shown in the last column of Table II. The resulting uncertainty in the flux density values is of the order of 25%.

A value for the peak brightness temperature for each source has also been obtained and is given in Table III. However, because of the smoothing introduced by the beam,

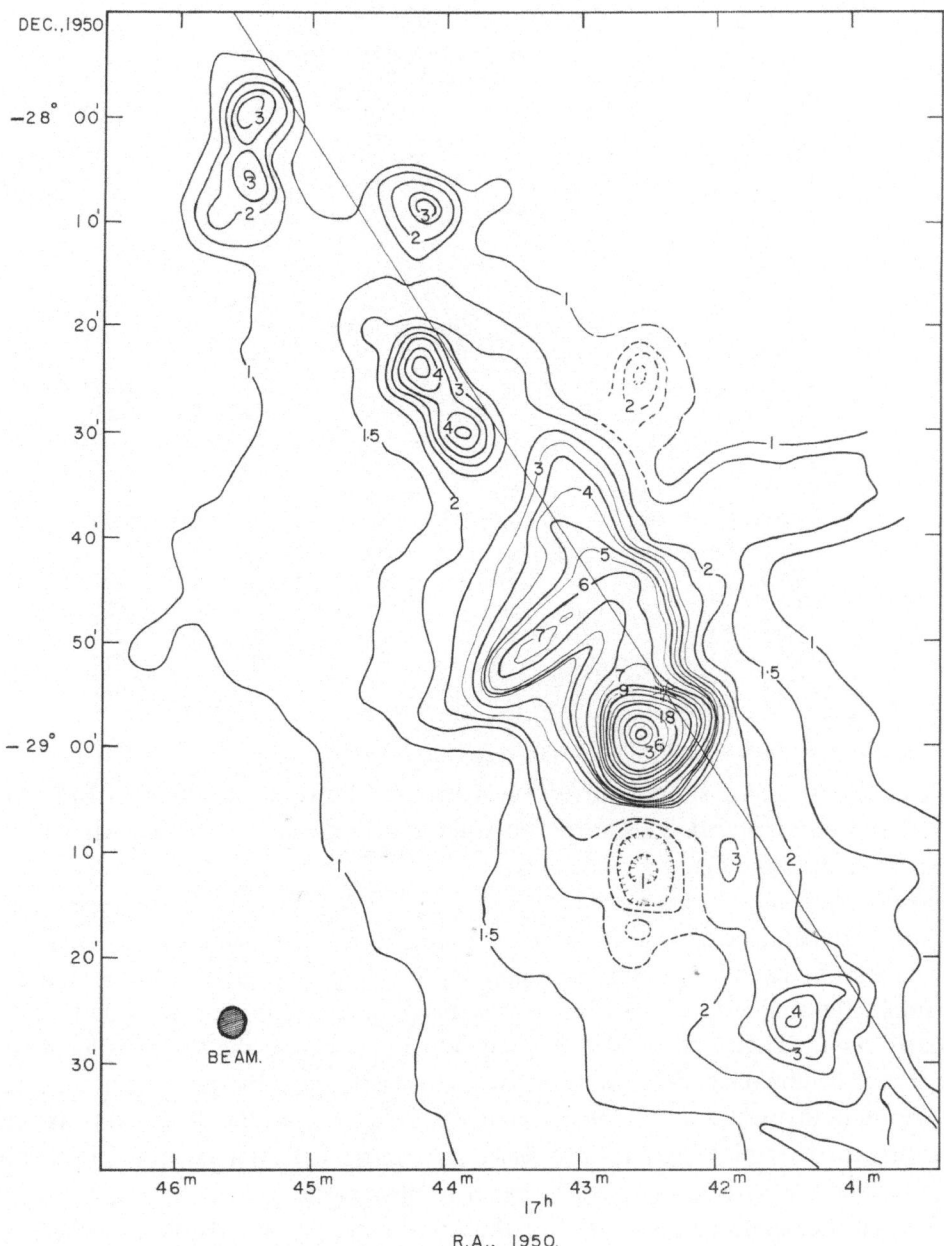

Fig. 2.   Contour map of the galactic centre region at 408 MHz. The lowest contour is 1075 K and the unit contour interval is 680 K.

TABLE I

Flux and position calibration sources

Source	Flux (Jy)	R.A. (1950)	Dec. (1950)
PKS 1643-22	5.46 ± 0.09	$16^h43^m04\overset{s}{.}8 \pm 0\overset{s}{.}09$	$-22°22'35\overset{''}{.}1 \pm 1\overset{''}{.}4$
PKS 1814-51	13.6 ± 0.2	$18^h14^m07\overset{s}{.}81 \pm 0\overset{s}{.}14$	$-51°59'22\overset{''}{.}1 \pm 1\overset{''}{.}6$

TABLE II

Source flux densities

Source	Flux (Jy)				Background temp. at 0.408 GHz (K)
	14.5 GHz	8.0 GHz	5.0 GHz	0.408 GHz	
G359.4−0.1	11	14	16	9.8	2095
G0.1−0.0	60	60	70	–	3115
G0.2−0.0	140	120	150	40	3115
G0.5−0.0	37	41	52	9.2	2095
G0.7−0.0	45	58	58	9.7	2095
G0.9+0.1	–	14	15	10	1211
G1.05−0.1	–	–	–	15.5	1075
G1.1−0.1	20	20	20	15.2	1075

these are low, and a simple correction using the chord construction of Bracewell (1955) has been applied. This corrected value is also given in Table III with the estimated error in the final temperatures.

For the optically thick sources it should be possible to make some estimate of the electron temperature. However, in order to do this, corrections have to be made for the background and foreground emission which are not negligible in this region. For example, the source G0.2−0.0 has a peak brightness temperature of 6400 K but is located on a background of 3115 K. Depending on whether the source is located in front of or behind the background and the relative amount of thermal and nonthermal radiation in the background, the correction to be applied to the observed peak temperatures to determine a value for the electron temperature could be quite substantial. It is beyond the scope of the present paper to discuss this.

The positions and angular sizes for the sources are given in Table IV, where the microwave values from Downes and Maxwell (1966) are indicated in parentheses. There are slight differences in position between the two, which may be simply related to the greater optical depth at 408 MHz since the sources are not symmetrical.

The measured value of flux density, position and angular size for Sgr A together with occultation values (Maxwell and Taylor, 1968) at 405 MHz are given in Table V. The present observations agree very well with the occultation results. For this source the background contour level was determined to be 7 units (5255 K).

Finally, the background survey (Green, 1974) mentioned earlier shows an extended source at the galactic centre which has a peak temperature of about 500 K and which is roughly 2°–3° in angular dimensions. This is illustrated in Figure 3, which shows the temperature variation (a) along the galactic plane at latitude zero and (b) transverse to the plane at longitude zero. The temperatures are average values over $0.5° \times 0.5°$ with all sources removed. This peak is almost certainly due to the nonthermal source observed at 85 MHz (Mills, 1956). At this frequency this source had two peaks of measured brightness temperature 34 000 K with an absorption dip at

TABLE III

Peak brightness temperatures for 408 MHz Sources

Source	Peak observed brightness temperature (K)	Corrected peak brightness temperature (K)
G359.4−0.1	3200	3500 ± 300
G0.1−0.0	5600	–
G0.2−0.0	5600	6400 ± 600
G0.5−0.0	3500	3900 ± 350
G0.7−0.0	3500	3800 ± 350
G0.9+0.1	2600	3200 ± 280
G1.05−0.1	2400	2700 ± 230
G1.1−0.1	2500	2800 ± 250

TABLE IV

Positions and angular sizes of 408 MHz sources

Source	Right ascension (1950)	Declination (1950)	Angular dimensions (arcmin)
G359.4−0.1	17^h41^m 25$\overset{s}{.}$8(22)	−29°26$\overset{'}{.}$0(26.6)	5.3 × 4.3 (5 × 3)
G0.1−0.0	17 − (42)− (34)	− (−28)−(51)	– (7 × 5)
G0.2−0.0	17 43 (42)19.7(59)	−28 50.4(47)	16.3 × 1.6 (17 × 5)
G0.5−0.0	17 43 52.3(49)	−28 30.2(29.1)	4.8 × 3.2 (8 × 4)
G0.7−0.0	17 44 11.2 (7)	−28 23.9(21.6)	4.9 × 2.9 (5 × 2)
G0.9−0.1	17 44 9.0 (7)	−28 8.7 (7.6)	3.8 × 3.4 (5 × 5)
G1.05−0.1	17 45 27.8	−28 5.8	6.8 × 3.9 –
G1.1−0.1	17 45 28.8(25)	−27 59.5(58.2)	5.7 × 4 (6 × 6)

TABLE V

Data on Sgr A

Frequency (MHz)	Flux (Jy)	R.A. (1950)	Dec.(1950)	Angular dimensions (arcmin)
408	229	$17^h42^m32\overset{s}{.}6 \pm 0\overset{s}{.}3$	−28°59′ ± 3″	2.6 × 3.4
Occultation measurements (Maxwell and Taylor, 1968)				
405	235	30.9	59′13″	3.5 × 4.4
405	225	32.2	58′51″	2.6 × 3.4

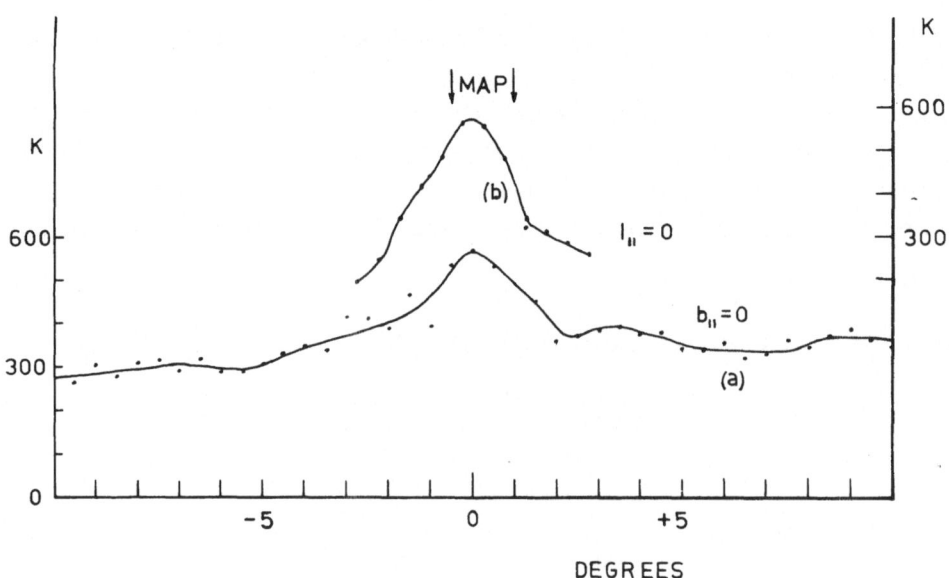

Fig. 3.   The brightness temperature measured (a) along the plane at $b=0°$ and (b) across the plane at $l=0°$ with all the sources removed. The points are average values over $0.5° \times 0.5°$ cells.

the centre. Thus its peak temperature is obviously higher – possibly up to 40–50 000 K. Scaling this to 408 MHz with a power law of $\nu^{-2.7}$ indicates a peak temperature of about 550–700 K. The present 408 MHz results are not too inconsistent with this, particularly if there is absorption at 408 MHz.

As Cooper and Price (1964) pointed out, there is a narrower source distribution approximately $1° \times 0.5°$ superimposed on that shown in Figure 3. This distribution, which surrounds the Sgr A source, is clearly seen on the present map and has a maximum brightness temperature at 408 MHz of 2100 K. From the microwave results quoted by Downes and Maxwell (1966) the brightness temperature of this region should be 940 K at 408 MHz, assuming a simple thermal spectrum. Allowing for the 500 K nonthermal source just discussed, there is about 1600 K to be accounted for, which is more than the expected 940 K from the microwave results. This source may then be not purely thermal.

Acknowledgements

The author would like to acknowledge the help of Dr David Clark for producing the diagram of the line scans of Figure 1 and Dr A. Green for making available previously unpublished data and Professor B. Y. Mills for many helpful discussions.

References

Bracewell, R. N.: 1955, *Australian J. Phys.* **8**, 200.
Broten, N. W., Cooper, B. F. C., Gardner, F. F., Minnett, H. C., Price, R. M., Tonking, F. G., and Yabsley, D. E.: 1965, *Australian J. Phys.* **18**, 85.
Cooper, B. F. C. and Price, R. M.: 1964, in F. J. Kerr and A. W. Rodgers (eds.), 'The Galaxy and the Magellanic Clouds', *IAU Symp.* **20**, 168.

Downes, D. and Maxwell, A.: 1966, *Astrophys. J.* **146**, 653.

Green, A.: 1974, *Astron. Astrophys. Suppl.* in press.

Hollinger, J. P.: 1965, *Astrophys. J.* **142**, 609.

Hunstead, R. W.: 1972, *Monthly Notices Roy. Astron. Soc.* **157**, 367.

Maxwell, A. and Taylor, J. H.: 1968, *Astrophys. Letters* **2**, 191.

Mills, B. Y.: 1956, *Observatory* **76**, 65.

Mills, B. Y., Aitchison, R. E., Little, A. G., and McAdam, W. B.: 1963, *Proc. Inst. Radio Engineers (Australia)* **24**, 156.

Wyllie, D. V.: 1969a, *Monthly Notices Roy. Astron. Soc.* **142**, 229.

Wyllie, D. V.: 1969b, *Proc. Astron. Soc. Australia* **1**, 235.

A. G. Little

Chatterton Astronomy Department,
School of Physics,
University of Sydney,
Sydney, N.S.W., Australia

(Discussion follows the paper by Swarup et al., p. 508.)

OCCULTATION OBSERVATIONS OF THE GALACTIC
CENTER REGION AT 327 MHz

G. SWARUP, GOPAL-KRISHNA, and N. V. G. SARMA

Tata Institute of Fundamental Research, Bombay, India

Abstract. Lunar occultation observations of the thermal sources Sgr B2, G0.9+0.1 and G1.1−0.1 at 327 MHz have been used to estimate their electron densities and temperatures. A new nonthermal source of size ∼ 10 × 5′ has been found about 7′ to the south of G1.1−0.1. A brightness contour diagram with a resolution of approximately 25 × 6′ is presented for the background radio emission near the sources Sgr A and Sgr B2.

I. Introduction

The region near the galactic centre consists of several thermal and nonthermal radio sources which have been studied at centimeter wavelengths with narrow pencil-beam antennas (Downes and Maxwell, 1966; Whiteoak and Gardner, 1973). We report here lunar occultation observations of the region made at 327 MHz using the Ooty radio telescope (Swarup *et al.*, 1971). Occultations of Sgr A, Sgr A-E, G0.1+0.0 and G0.2−−0.0 were observed on 1970, September 9 (hereafter referred to as OCC.I) and of sources G0.7−0.0, G0.9+0.1 and G1.1−0.1 on March 19, 1971 (OCC.II). Results for Sgr A and Sgr A-E have been reported elsewhere (Gopal-Krishna *et al.*, 1972). The data have provided information on spectrum and size for these sources. A new extended nonthermal source about 7′ to the south of G1.1−0.1 has been found whioh is perhaps a supernova remnant.

II. Observations

The observations were made at 327 MHz with the Ooty radio telescope, which provides twelve simultaneous beams separated by 4′/cos δ in declination. The radio telescope has a half-power beamwidth of 2° in the east-west direction. Its north-south beamwidth is 5.′6/cos δ and 4′/cos δ for the total-power and phase-switched modes respectively.

In Figure 1, lines *AB* and *CD* show the apparent path of the Moon for the two occultations. Positions of the 12 beams in declination are shown in the margin. During the observations the telescope tracked regions at these 12 declinations and at R.A. (1950) = $17^h42^m40^s$ for OCC.I and at R.A. (1950) = $17^h44^m55^s$ for OCC.II. The telescope was calibrated on PKS 1309-22 (for OCC.I) and PKS 0859-25 (for OCC.II), taking their flux densities at 327 MHz as 26.4 and 19.9 Jy respectively. These values are based on Wyllie's (1969) measurements at 408 MHz extrapolated to 327 MHz using spectral indices from the Parkes catalogue.

III. Results

Figure 2 shows the observed occultation records for OCC.I and OCC.II. The records

F. J. Kerr and S. C. Simonson, III (eds.), Galactic Radio Astronomy, 499–509. All Rights Reserved.

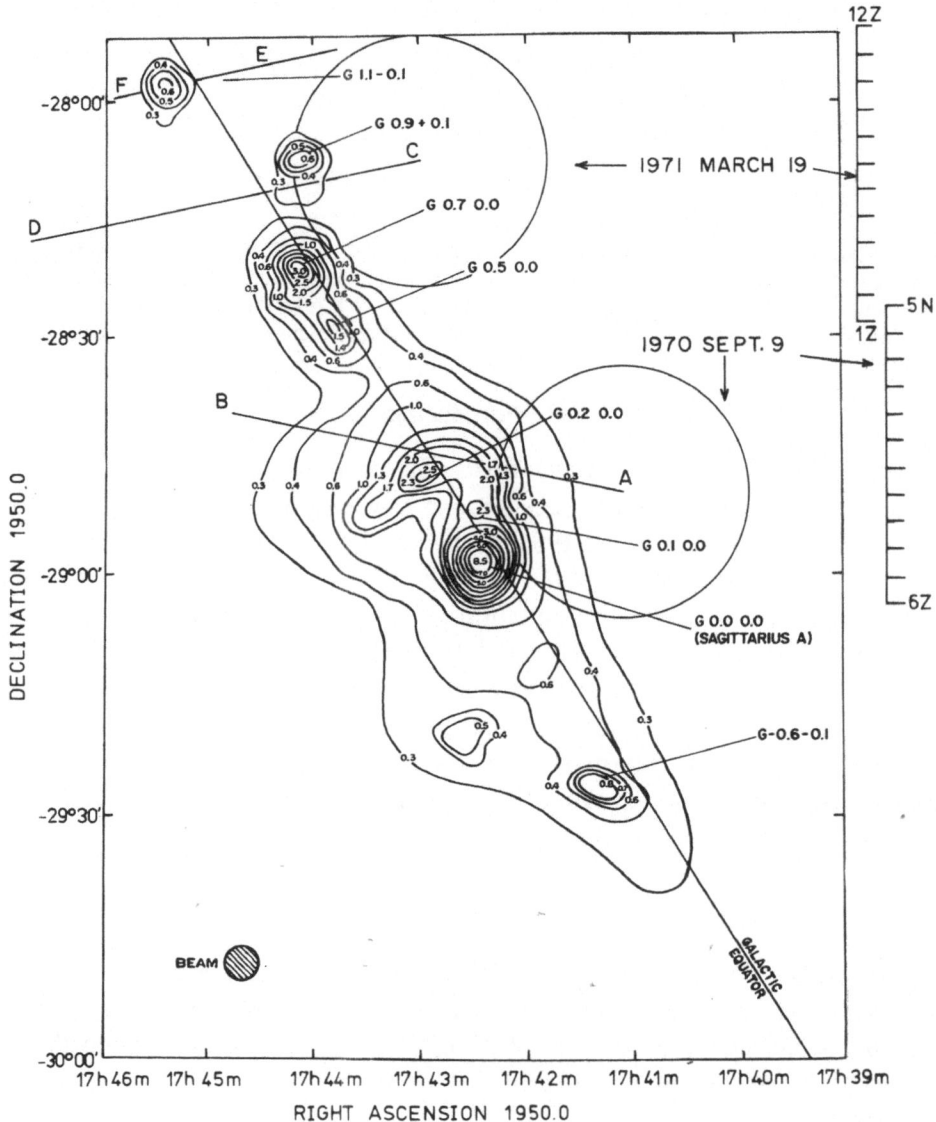

Fig. 1.   Lines *AB* and *CD* show the path of the Moon for the occultation observations on 1970, September 9, and 1971, March 19, and are superimposed on a map at 8 GHz taken from Downes and Maxwell (1966). Positions of the twelve beams of the Ooty radio telescope are shown by 6 *Z* to 5 *N* and 1 *Z* to 12 *Z* on the right hand side.

have been smoothed with a gaussian of 4 min half-power width in order to reduce the effects of reciever instabilities and of the ionospheric scintillations seen on both occasions. The curves for OCC.II have been differentiated and plotted in Figure 3, where the sources $G0.7-0.0$, $G0.9+0.1$, $G1.05-0.1$ and $G1.1-0.1$, marked as G0.7, G0.9, G1.0 and G1.1 respectively, can be seen superimposed over a background indicated by broken curves. The background was determined from two considerations: (a) by comparison with the occultation profiles for the neighbouring beams, particularly those in which discrete sources are not evident, and (b) by ensuring that the resultant immersion and emersion profiles for each source have equal area. The background curves of Figure 3 were integrated and the resultant profiles are shown in Figure 2 by broken curves. These curves represent the occultation profiles for the background without contribution from the discrete sources. Positions, flux-densities and angular widths for the observed sources were estimated from the known apparent motion of the Moon and from their occultation response in successive beams. The results are given in Table I. Position of Sgr B2 (G0.7 − 0.0) derived from the occultation

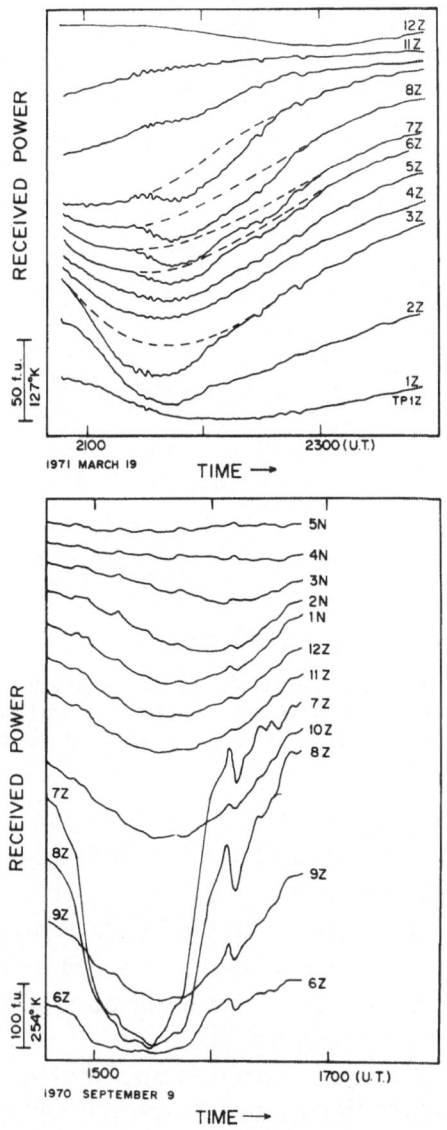

Fig. 2.   Records of occultation observations made on 1970, September 9, and 1971, March 19. Flux density and antenna temperature scales are given on the left.

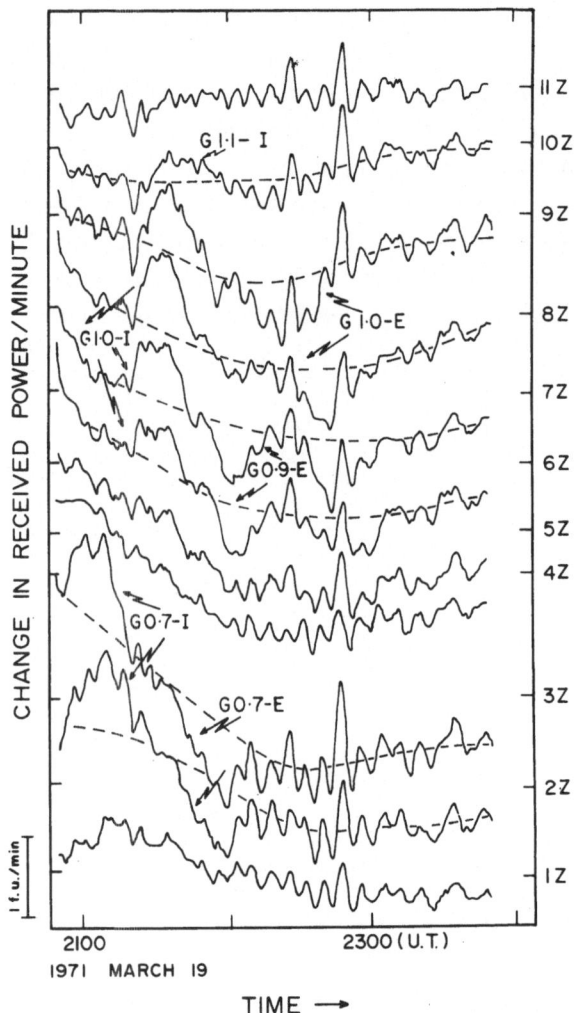

Fig. 3.   Differentiated outputs for occultation observations for beams 1 Z to 11 Z made on 1971, March 19.
Zero base levels for the curves of different beams are shown by dashes on both sides of the figure. I and E
after a source number refer to its immersion and emersion respectively.

observations at 327 MHz agrees with that of the continuum peak at 5 GHz (Gardner
et al., 1971; Whiteoak and Gardner, 1973) and also with the occultation positions for
the strong 1665 MHz OH emission features (Manchester *et al.*, 1969). The position
of G0.9+0.1 at 327 MHz also agrees with the 5 GHz continuum position given by
Whiteoak and Gardner (1973), but positions estimated by Downes and Maxwell
(1966) from high frequency observations are to the north of the 327 MHz positions
by about 1.5' for both Sgr B2 and G0.9+0.1.

The flux densities were determined from the response of the sources in the differentiated occulation curves after subtraction of the background as described above. For
the case of thermal sources, we added a correction ΔS for the estimated absorption of
that part of the background radiation which passes through the source. It is assumed

TABLE I

Measured parameters of the discrete sources at 327 MHz

Source	Position[a] (1950.0)	Flux-density (Jy)	Mean position angle (deg)	Angular half-power width (arcmin)
G0.1+0.0 (T) +G0.2−0.0	–	105±30	90 180	18 8
G0.7−0.0 (T) (Sgr B2)	$17^h44^m10\overset{s}{.}2\pm2^s$ $-28°23'06''\pm40''$	28±5	153 229	5.0±0.3 5.2±0.3
G0.9+0.1 (T)	$17^h44^m09^s\pm2^s$ $-28°09\overset{.}{'}2\pm1'$	11±3	180 286	4.2±1.5 4.4±0.7
G1.05−0.1 (NT)	$17^h45^m30^s\pm4^s$ $-28°05'\pm1\overset{.}{'}5$	27±6	80 350	5 ±2 10 ±3
G1.1−0.1 (T)	–	16±4	–	–

T = Thermal source;   NT = nonthermal source

[a] For G1.05−0.1, the given position refers to the peak of brightness which lies about 1' north of the centroid.

that the sources are located near the galactic center and half of the observed background emission T_b originates behind the sources. Therefore, the correction to the flux-density is $\Delta S = (2k\,\Delta T_c/\lambda^2)\,\Omega_s$, where Ω_s is the solid angle subtended by the source as found from the occultation observations and $\Delta T_c = \frac{1}{2}T_b\{1-\exp(-\tau)\}$, τ being the optical depth of the source at 327 MHz estimated iteratively. The flux densities given in Table I include a correction $\Delta S = 45$ Jy for (G0.1+0.0)+(G0.2−0.0) based on $T_b = 2400$ K (see Figure 6), and corrections of $\leqslant 4$ Jy for the other thermal sources for $T_b = 1500$ K. The flux density of Sgr B2 was reported as 9 Jy at 408 MHz by Little (1974), which is much lower than our estimate. We have re-examined our records and find it difficult to reconcile with Little's measurements, which might have been underestimated due to an overestimate of the background radiation and underestimate of the size of the source. There are similar difficulties in estimating the background radiation in the occultation observations, but the observed values of the size have small errors.

It may be seen from Figure 3 that there exists a broad structure extended in the north-south direction, marked by G1.1 and G1.0, for which the occultation profiles show response in six successive beams from 5 Z to 10 Z. The high frequency maps presented by Downes and Maxwell (1966) show a thermal source, G1.1−0.1, of ∼6 × 6′ in size. The beam 10 Z was pointed about 1′ to the north of this source. Occultation response in the output of beam 10 Z is expected to be mainly due to this source. Its response in beam 9 Z was expected to be about the same. Moreover, only about half the extent of this source was occulted on 1971, March 19, as may be seen from Figure 1 in which the path of the northern limb of the Moon is shown by line EF. It appears therefore that there exists another source ∼7′ to the south of the thermal source G1.1 −0.1, which is labelled by us as G1.05 −0.1. This source is also seen in a map of the region at 408 MHz presented by Little (1974) based on observations with the Mills Cross. The source has a nonthermal spectrum because if we assume its spectral index as -0.7, its expected brightness temperatures at 3 and 5 GHz are approximately 3 K and 1 K respectively, evidence for which can be found in the high frequency maps by Cooper and Price (1964) and Whiteoak and Gardner (1973). Figure 4 shows approximate brightness contour diagram for this source derived from the occultation observations after taking into account the curvature of the Moon's limb. The small extension at the northern end is due to the thermal source G1.1 $-$ 0.1. Most of the contribution is from the nonthermal source G1.05 $-$ 0.1, whose parameters are given in Table I. Considering its nonthermal spectrum, large angular extent and location near the galactic plane, we may consider the source to be a supernova remnant. However, it was not possible to identify any characteristics of a supernova remnant in its brightness distribution, probably due to insufficient signal-to-noise ratio of the occultation records. Using the surface brightness and the linear diameter relation for the galactic supernova remnants (Ilovaisky and Lequeux, 1972), we estimate that the distance of G1.05 $-$ 0.1 is about 9 kpc. It may be, therefore, that the source is located close to the galactic center.

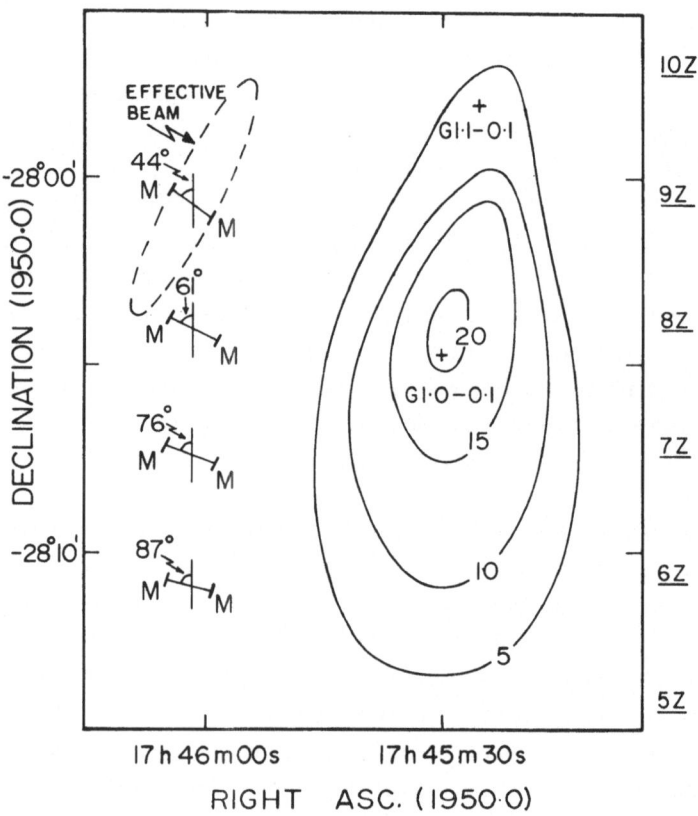

Fig. 4. Contour map of G1.05 $-$ 0.1 and G1.1 $-$ 0.1 at 327 MHz. Contour unit represents 110 K of brightness temperature above the background. Lines *MM* indicate resolutions and mean position angles of scan for different beams.

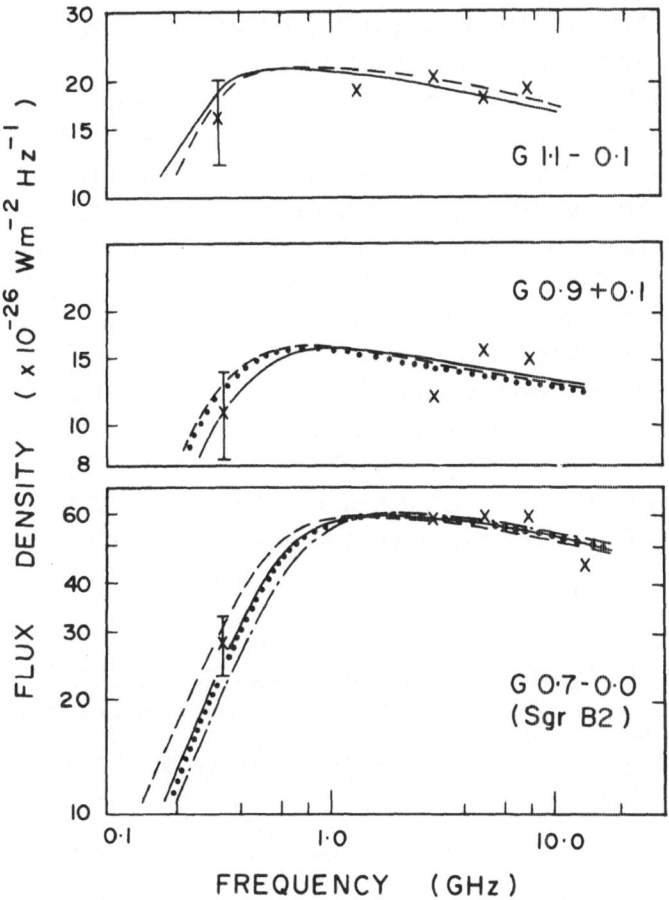

Fig. 5. Observed spectra of the thermal sources are compared with computed spectra for sets of T_e, N_e and θ_e listed in Table II.

The spectra of the sources G0.7−0.0, G0.9+0.1 and G1.1−0.1, shown in Figure 5, indicate that the radiation arises thermally. To derive their physical parameters, we assume that each source has a constant temperature T_e but a spherical gaussian distribution of electron-density with a peak value N_e and half-density angular width θ_e. The observed brightness distribution of such a source will vary with wavelength when the outer parts gradually become optically thick. For several different sets of the above parameters, we have computed spectra and half-power widths for the sources at 327 MHz. In Figure 5 and Table II we present the results of computations for a few of the cases which give a reasonable fit to the observations. The derived values of T_e, N_e, θ_e and the mass of ionized hydrogen (M_{HII}) for the sources are given in Table III. For G1.1−0.1, the size of the source could not be estimated reliably at 327 MHz, and we have taken $\theta_e = 6'$ to fit the observations by Downes and Maxwell (1966) and Whiteoak and Gardner (1973) at centimetre wavelengths. Considering the uncertainties in measured parameters, estimates of physical parameters for the sources G0.9+0.1 and G1.1−0.1 may be regarded as not well-determined.

From the observations made during OCC.I, we do not find evidence of any other source of small diameter except Sgr A and Sgr A-E, results for which have been reported earlier (Gopal-Krishna *et al.*, 1972). From the occultation records we estimate that at 327 MHz the thermal sources G0.1 + 0.0 and G0.2 − 0.0 are almost merged together with a size ∼ 18 × 8′. Their estimated optical depth at 327 MHz is ∼2.5.

We have also attempted to make a map of the galactic center region using the Moon as a screen (Stankevich *et al.*, 1970). It may be noted that the occultation outputs for the 12 total-power beams, when inverted, simply relate to the brightness temperature averaged over a strip equal to the Moon's area covered by the beam. The distribution of brightness temperature is given in Figure 6. The point $l = 1°1$ and $b = −0°8$ was used for zero calibration with estimated $T_b = 450$ K at 327 MHz (Altenhoff *et al.*, 1970). The north-south resolution is 6.4′. The east-west resolution for any point equals the angular width of the Moon's chord at the declination of the beam; the achieved east-west resolutions are shown in the margin of Figure 6 for all the beams for the start and the end of the observations. The background temperature seems higher for

TABLE II

Values of physical parameters used for computations of spectra shown in Figure 5.
The calculated and observed values of source widths are also shown.

Source	Physical parameters			Source width θ_a (327 MHz)		Spectral curve of Figure 5
	T_e (K)	N_e (cm^{-3})	θ_e (pc)	Calculated (pc)	Observed (pc)	
G0.7 − 0.0	4000	340	10	14.0	15 ± 1	—·—·—
	6000	360	10	12.7	15 ± 1	————
	4000	255	12	15.5	15 ± 1	· · · ·
	6000	270	12	14.0	15 ± 1	——————
G0.9 + 0.1	4000	170	10	10.1	13 ± 3	————
	5000	130	12	10.5	13 ± 3	· · · · ·
	4000	100	14	12.1	13 ± 3	——————
G1.1 − 0.1	3000	80	17	15.6	−	· · · · ·
	5000	85	17	14.3	−	————

TABLE III

Physical parameters of the observed thermal sources [a]

Source	T_e (K)	N_e (cm^{-3})	Diameter (pc)	M_{HII} ($M_\odot$)
G0.7 − 0.0 (Sgr B2)	5000 ± 1000	290 ± 40	11 ± 1	1.1×10^4
G0.9 + 0.1	4500 ± 1000	130 ± 40	13 ± 3	9.4×10^3
G1.1 − 0.1	4500 ± 1500	80 ± 20	17 ± 3	1.2×10^4

[a] Assumed distance of 10 kpc.

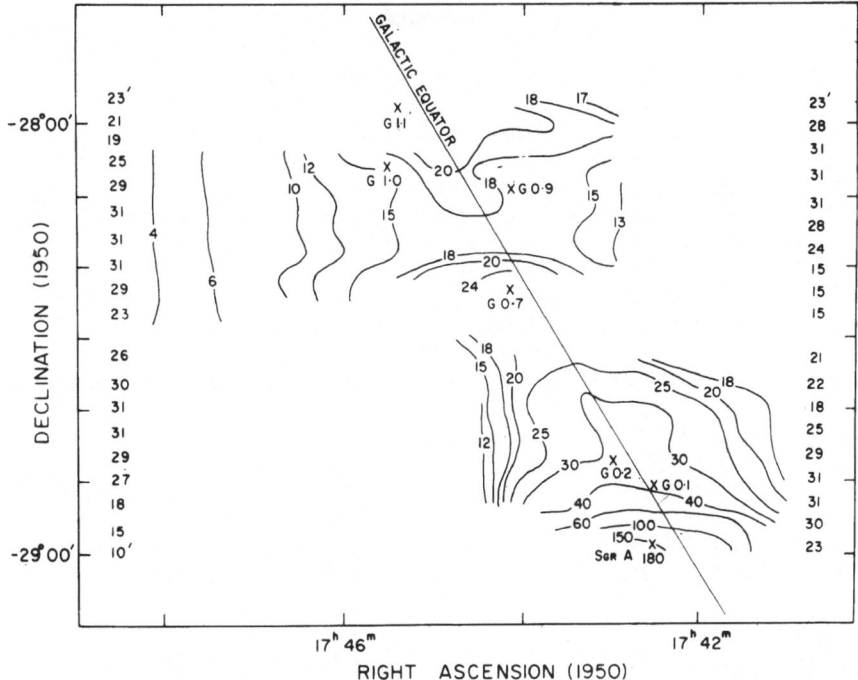

Fig. 6.   Contour map of the region near the galactic center made with resolution in declination of 6.4'
and in right ascension as given on either side of the map for different beams. Contour interval represents
110 K of brightness temperature.

positive compared to negative latitudes at $l \sim +1°$ (in the region around G0.9 + 0.1 and
G1.05 − 0.1). This can be seen also in the occultation records of Figure 2 for 1971,
March 19, where the power received is lower at the start of the occultation than at
its end.

Values of T_e found for Sgr B2 are of the same order as those estimated from recombination line observations (Mezger and Höglund, 1967; Reifenstein *et al.*, 1970),
indicating that departures from local thermodynamic equilibrium in this H II region
are not large, as also concluded by Shaver and Goss (1970) for several other H II
regions. Values of N_e and T_e determined from the continuum observations need not
be the same, however, as found from the recombination line observations, as the two
emission processes depend on different power factors of T_e and may not arise in the
same regions. Values of N_e and T_e determined from line emission refer to the entire
H II region while continuum observations at meter wavelengths give T_e only for its
outer part.

Acknowledgements

The authors are grateful to Professor T. K. Menon for valuable discussions and to
Mr V. K. Kapahi for reading the manuscript.

References

Altenhoff, W. J., Downes, D., Goad, L., Maxwell, A., and Rinehart, R.: 1970, *Astron. Astrophys. Suppl.* **1**, 319.

Cooper, B. F. C., and Price, R. M.: 1964, in F. J. Kerr and A. W. Rodgers (eds.), 'The Galaxy and the Magellanic Clouds', *IAU Symp.* **20**, 168.

Downes, D. and Maxwell, A.: 1966, *Astrophys. J.* **146**, 653.

Gardner, F. F., Ribes, J. C., and Sinclair, M. W.: 1971, *Astrophys. J. Letters* **169**, L109.

Gopal-Krishna, Swarup, G., Sarma, N. V. G., and Joshi, M. N.: 1972, *Nature* **239**, 91.

Ilovaisky, S. A. and Lequeux, J.: 1972, *Astron. Astrophys.* **18**, 169.

Little, A. G.: 1974, this volume, p. 491.

Manchester, R. N., Goss, W. M., and Robinson, B. J.: 1969, *Astrophys. Letters* **4**, 93.

Mezger, P. G. and Höglund, B.: 1967, *Astrophys. J.* **147**, 490.

Reifenstein, E. C., Wilson, T. L., Burke, B. F., Mezger, P. G., and Altenhoff, W. J.: 1970, *Astron. Astrophys.* **4**, 357.

Shaver, P. A. and Goss, W. M.: 1970, *Australian J. Phys. Astrophys. Suppl.*, No. 14, 133.

Stankevich, K. S., Wielebinski, R., and Wilson, W. E.: 1970, *Australian J. Phys.* **23**, 529.

Swarup, G., Sarma, N. V. G., Joshi, M. N., Kapahi, V. K., Bagri, D. S., Damle, S. H., Ananthakrishnan, S., Balasubramanian, V., Bhave, S. S., and Sinha, R. P.: 1971, *Nature Phys. Sci.* **230**, 185.

Whiteoak, J. B. and Gardner, F. F.: 1973, *Astrophys. Letters* **13**, 205.

Wyllie, D. V.: 1969, *Monthly Notices Roy. Astron. Soc.* **142**, 229.

G. Swarup
Gopal-Krishna
N. V. G. Sarma
Radio Astronomy Group,
Tata Institute of Fundamental Research,
Homi Bhabha Road,
Bombay 400 005, India

DISCUSSION

Hughes: We have extended our radio observations at 858 MHz in an attempt to confirm the existence of radio pulses from the direction of the galactic center. The previous results showed no correlation with the gravitational events as observed by Weber, though evidence was presented which showed that the radio pulses might have an origin in the direction of the galactic center. More recently, severe doubt has been cast on the evidence presented for the existence of gravitational pulses. Our most recent results with a sensitivity improved by a factor of about 4 show that the radio pulse rate is about the same as before, namely about 0.3 per hour. Though this could mean a real reduction in the pulse rate, at present we consider the origin of the radio pulses to be an open question.

Ekers: Do you know if the recombination line observations you have attributed to Sgr A comes from Sgr A itself, or could it be from the surrounding region?

Gordon: No. But if the background source stimulates the line emission, then the emitting region has the same dimensions as the background source.

Mezger: I want to draw attention to the source G0.9+0.0 which has an apparent thermal continuum spectrum even if the present low-frequency observations are considered. We searched for H109α and H85α lines in this source down to a very low level and within the velocity range $-300 \leqslant V_{LSR} \leqslant +300$ km s^{-1} without success.

Little and Swarup: The flux values from both the 327 MHz occultations and the 408 MHz pencil beam survey are lower than the microwave values for this source. A thermal spectrum fits the data reasonably well. The source is well separated from others in this region, and therefore the spectrum is well established.

Churchwell: Lowinger (1964) from his 15 GHz map has concluded that most of the extended emission seen at this frequency is thermal. T. Pauls, P. Mezger and I have detected recombination lines over this whole region, which would seem to confirm the thermal nature of this emission.

Gordon: There are conflicts between continuum maps, but I believe this map that Kapitsky and Dent made in 1973 with a magnificent instrument is highly reliable.

Davies: H166α recombination line observations have been made at Jodrell Bank using the Mark II radio telescope (beamwidth = 32') of the galactic center region ($l = 359°$ to $+1°$ in $0°5$ steps). Line emission comes from all these points and appears to originate in the extended thermal emission measured in the high-frequency surveys of Downs and Maxwell. An electron temperature of 10000 K was derived from these observations.

Gordon: The new 2 cm map indicates that the extended region is nonthermal. If so, the recombination lines cannot come from that component.

Parijskij: Now we have direct evidence of the nonthermal nature of at least some portion of the Sgr A complex. Polarization of the west part of Sgr A sources was found at 4 cm (with about 1' resolution) at Pulkovo. The angular size of the polarized region was $\sim 2'$, and the amount was $\sim 1\%$. The maximum of the polarized emission is displaced by $\sim 30''$ toward the west from the maximum of the unpolarized emission at that wavelength.

21-cm LINE STUDIES OF THE GALACTIC CENTER REGION

S. CHRISTIAN SIMONSON, III

University of Maryland, College Park, Md., U.S.A.

Abstract. Features of the 21-cm line radiation that have often been attributed to the galactic center or to activity there include, near the equator, the '3-kpc arm', the 'expanding arms' at $+70$ and $+135$ km s^{-1}, and the 'nuclear disk', and, away from the equator, features observed by Cugnon, by Shane, by van der Kruit, and by Sanders, Wrixon, and Penzias, among others. Models of the features in the equatorial plane are discussed on the basis of the form of the $T_b(l, V)$ contours that they generate. Both explosions in the nucleus and resonance with a spiral wave have been suggested as the cause of the 3-kpc arm and related features. Some of the models put limits on the mass distribution, particularly the amount of condensed mass at the center. Higher-latitude features have been less thoroughly modeled, but explosions in the nucleus and resonant orbits have also been suggested for their origins.

I. Introduction

In discussing the extensively observed 21-cm line radiation, it seems natural to define the 'galactic center' as including the region within the inner Lindblad resonance near the radius $R=4$ kpc, since features out to this distance have often been interpreted as originating near the galactic nucleus. The following subjects will be briefly reviewed here: (a) the observations and interpretations of features between $R=1$ and 4 kpc (mainly the '3-kpc arm'), (b) the same for the 'nuclear disk' at $R<1$ kpc, (c) features away from the plane of the galactic equator, and (d) suggested lines of further observation and theoretical development.

II. Features in the Galactic Plane

The important features of the 21-cm line profiles near the plane of the galactic equator that are thought to arise from the center region are shown in Figure 1b: (a) wide wings on the line profiles on both sides of the center, ending in an extreme-velocity ridge; (b) the well-known 3-kpc arm; (c) the 'expanding arm at $+70$ km s^{-1}', which is a rather weak and complex feature; (d) the 'expanding arm at $+135$ km s^{-1}', about which some dispute exists concerning its exact track in the $T_b(l, V)$ diagram; and (e) the 'nuclear disk'.

The first requirement of a theoretical interpretation of any of these features is that it reproduce the appearance of the observations. The model must therefore be put into the same form as the observations, namely, a $T_b(l, b, V)$ matrix giving the brightness temperature as a function of the three independent observed variables, galactic longitude and latitude and line-of-sight velocity with respect to the local standard of rest. Often some appropriate subset, such as a $T_b(l, V)$ contour map, is used. If the comparison with observation is satisfactory, a model can then be judged by such things as the plausibility of its symmetry properties, its energy and angular momentum requirements, or its originating mechanism.

F. J. Kerr and S. C. Simonson, III (eds.), Galactic Radio Astronomy, 511–519. All Rights Reserved.
Copyright © 1974 by the IAU.

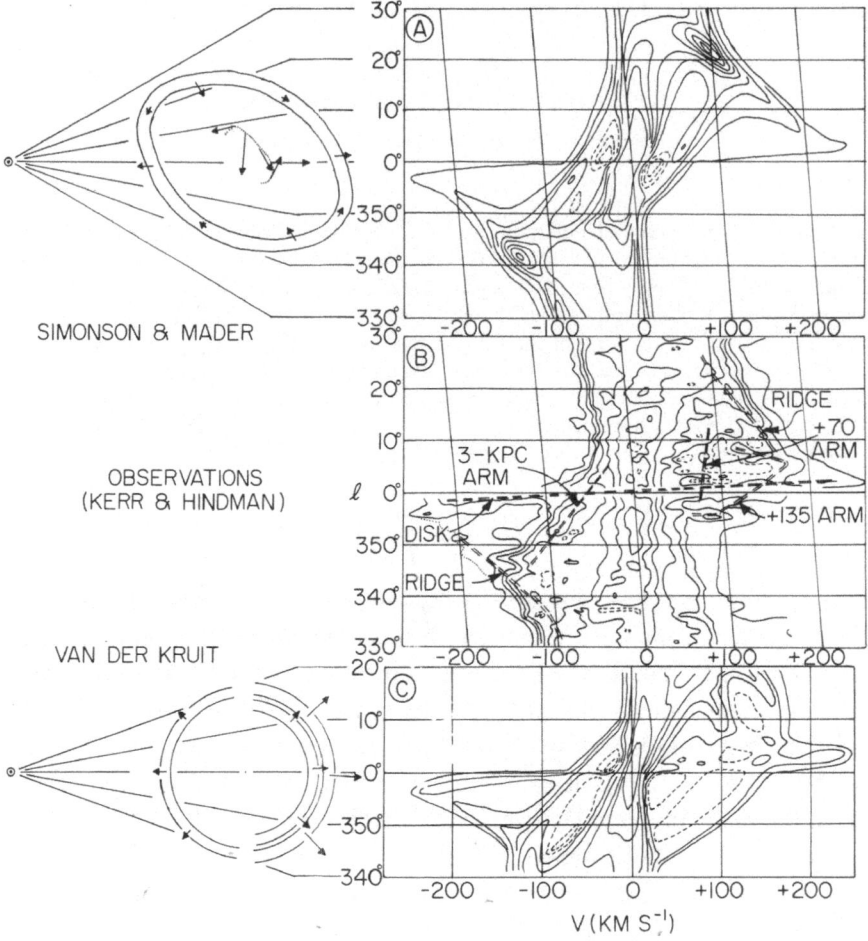

Fig. 1. $T_b(l, V)$ diagrams from (a) a model by Simonson and Mader (1973), illustrated at top left, (b) observations by Kerr and Hindman (1970), and (c) a model by van der Kruit (1971), illustrated at bottom left. Both models also contain differentially rotating disks. Arrows indicate the peculiar velocities. A schematic model for the $+135$ km s^{-1} feature is also shown in the center of Simonson and Mader's model. Features of the observations are labeled in (b).

For the 3-kpc arm and the $+70$ km s^{-1} expanding arm, two alternative interpretations have been offered in suitable form for comparison: an explosion model by van der Kruit (1971) and a dispersion-ring model by Simonson and Mader (1973). The models are shown in Figure 1 together with the $T_b(l, V)$ diagrams that they generate and the observed $T_b(l, V)$ diagram. A third model by Sanders and Wrixon (1973), which is similar to van der Kruit's, is shown in Figure 2. Within their restrictions of simplicity, both the explosion model and the dispersion-ring model give reasonable approximations to the observations, although further refinements are desirable. Common to both models is a differentially rotating disk extending all the way in to the center. One may conclude that the velocity field between $R = 1$ and 4 kpc is predominately that of rotation. It therefore seems possible to derive a rotation curve for this region, as has been done provisionally by Simonson and Mader, but more reliable observations are needed. The peculiar motions occur only in roughly circular structures of rather limited extent.

The choice between the two models is to be made mostly on the basis of the origin of the peculiar motions – an explosion in the galactic nucleus involving $10^7 \, M_\odot$ and 10^{55} erg for van der Kruit's model, or a resonance with a spiral wave whose existence is inferred from other evidence for the dispersion-ring model. One of the main difficulties with an explosion model is the short time scale of 10^7 yr for the presently observed phenomena. Sanders and Prendergast (1974), however, have calculated that a ring, once set in motion by an explosion, may continue to oscillate for 10^9 yr. The energy requirements and average mass flux are still extremely large in their model – 5×10^{59} erg and $1 \, M_\odot \, \mathrm{yr}^{-1}$.

The expanding arm at $+135$ km s^{-1} has so far been only roughly approximated in

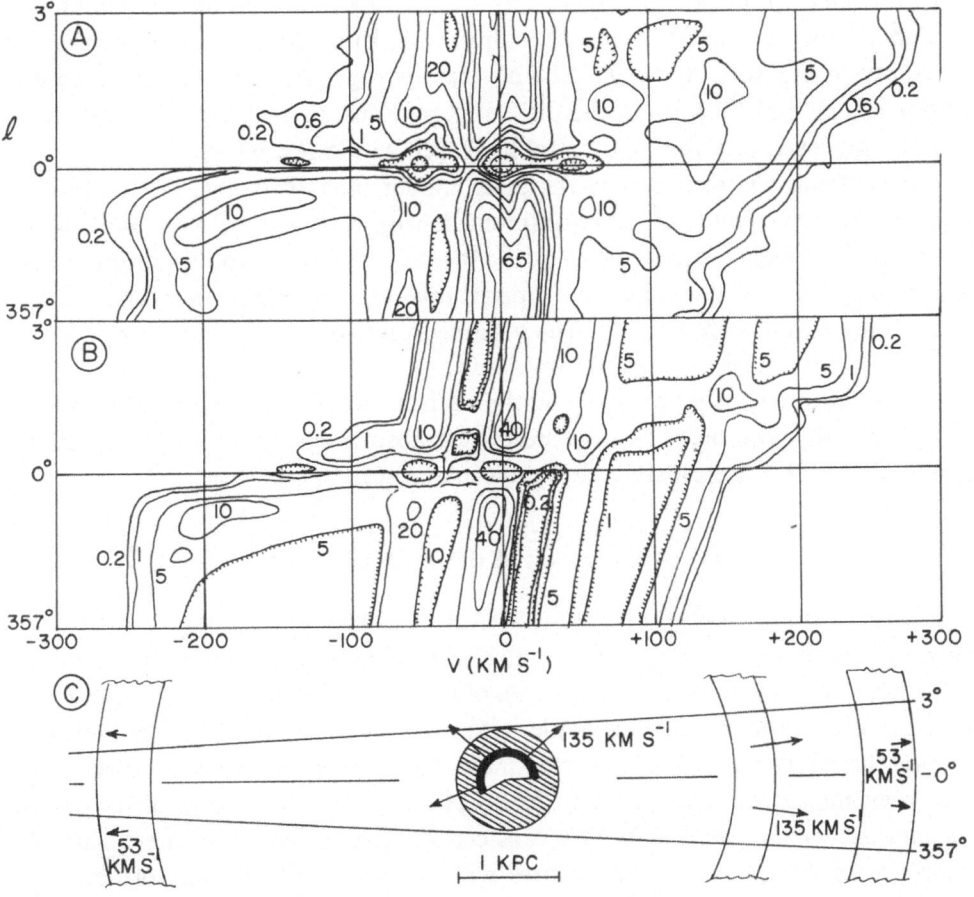

Fig. 2. $T_b(l, V)$ diagrams from (a) observations by Wrixon and Sanders (1973) and (b) a model by Sanders and Wrixon (1973), which is illustrated in (c). Contour levels are given in kelvins.

calculations, as illustrated in Figures 1c and 2. Simonson and Mader (1973) have pointed out inconsistencies between the $T_b(l, V)$ diagrams for expanding and rotating rings and the observations, suggesting instead a dispersion orbit interpretation as sketched in Figure 1a. Exact calculations of such an orbit await a better determination of the potential, which may include a central bar.

III. The Nuclear Disk

The feature in the observed $T_b(l, V)$ diagram attributed to the nuclear disk runs diagonally from $V = -230$ km s^{-1} at $l = 358°$ to $V = +250$ km s^{-1} at $l = 2°$. The region within 3° of the center has been newly observed by Wrixon and Sanders (1973) with the National Radio Astronomy Observatory 43-m telescope and low-noise receiver. From comparison of model $T_b(l, V)$ diagrams with their observations, Sanders and Wrixon (1973) find evidence for pure rotational motions for $l < 360°$. They are able to place a dynamical limit of $10^8 M_\odot$ on any condensed mass at the galactic center. The line radiation at $l > 0°$ is not symmetrical with that at $l < 360°$ but instead suggests a model with an expanding ring. Their model with its $T_b(l, V)$ diagram and the observations are shown in Figure 2. At $l < 360°$ the neutral hydrogen density increases smoothly from $n = 0$ cm^{-3} at the edge of the disk, at $R = 500$ pc, to $n = 5$ cm^{-3} at $R < 110$ pc. At $l > 0°$, however, they find no neutral hydrogen within $R < 300$ pc.

In considering the thickness of the disk, Sanders and Wrixon (1973) find a total thickness between points at half maximum density of 100 pc between $R = 300$ and 500 pc, in agreement with Rougoor's (1964) results. While it would require an rms velocity dispersion of 100 km s^{-1} to support this thickness, the maximum permitted by the observations is 24 km s^{-1}. The total neutral hydrogen mass within 500 pc of the galactic center is estimated at $2 \times 10^6 M_\odot$.

Another feature thought to be within the nuclear disk is a small cloud at $l = 359°97$, $b = -0°067$, that produces an extremely wide absorption dip at $V = +40$ km s^{-1} (Sandqvist, 1973). It is best studied in molecular lines.

IV. Features out of the Galactic Plane

The features of low intensity lying away from the plane of the galactic equator are shown schematically in Figure 3. Theoretical interpretations have not yet been brought to the point of calculating T_b as a function of l, b, and V for these features, nor indeed have the observations yet been made with sufficient sensitivity and resolution to warrant this effort in most cases. Furthermore, the surveys with the greatest sensitivity have been made from the Northern Hemisphere, where this region of the sky lies low on the horizon, and it is by no means certain that our census of these features is complete. In the following, the features will be considered in order of decreasing longitude as shown in Figure 3; some of the points are more extensively discussed by Simonson and Mader (1973).

(i) Shane's (unpublished) cloud at $l = 8°$, $b = -6°$, $V = -215$ km s^{-1}, while at a

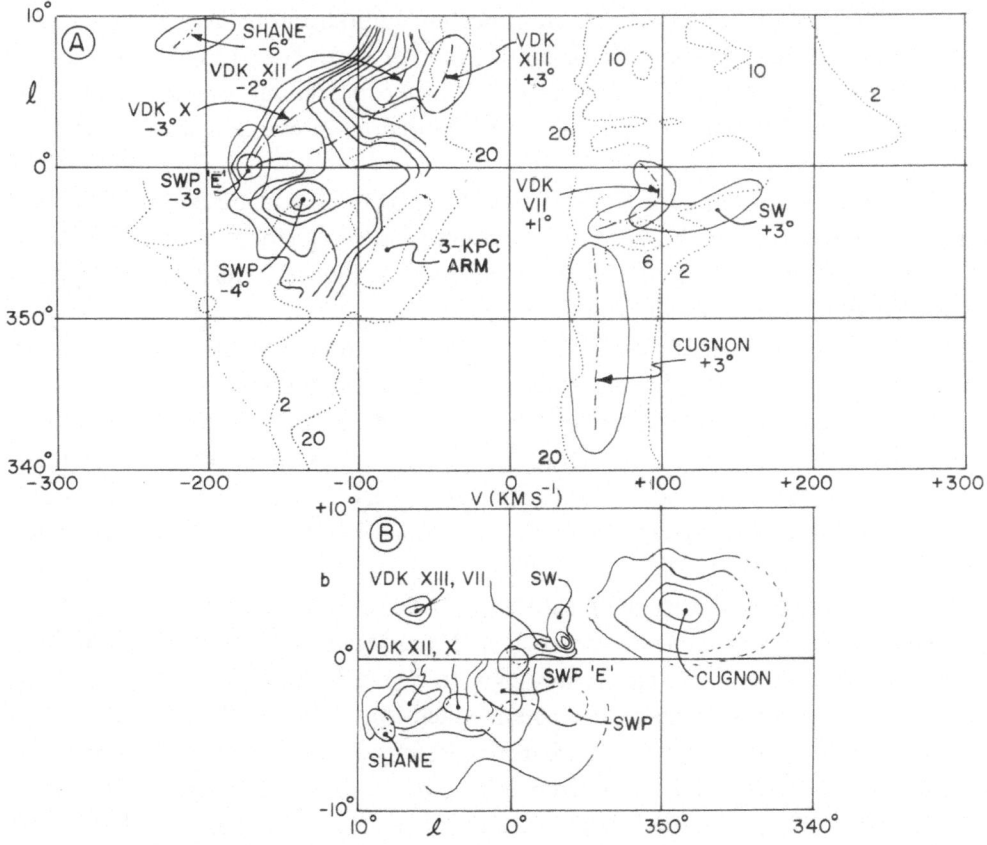

Fig. 3. (a) $T_b(l, V)$ diagram for features out of the galactic plane, superimposed on dotted contours taken from Figure 1b. Representative latitudes are given. (b) $T_b(l, b)$ diagram for the same features. Abbreviated notations: VDK = van der Kruit (1970), SWP = Sanders, Wrixon and Penzias (1972), and SW = Sanders and Wrixon (1972b).

position and velocity that might suggest expulsion from the center, has characteristics similar to 'high-velocity clouds' found at high latitudes. It may not be a typical feature of the center regions.

(ii) Van der Kruit's (1970) Feature XIII is the only positive-latitude feature found at negative velocities at $l > 0°$.

(iii) Van der Kruit's (1970) Feature XII and its extensions, denoted 'SWP' in Figure 3, were studied by Sanders et al. (1972) and Sanders and Wrixon (1972a). The features have been interpreted by these authors as an expanding and rotating ring. Sanders and Wrixon (1972a) suggest $v_{exp} = 128$ km s^{-1}, $v_{rot} = 151$ km s^{-1}, and $R = 2.4$ kpc as characteristic parameters. They propose a connection between the negative-velocity features and the $+135$ km s^{-1} expanding arm, the structure being viewed tangentially near $l = 14°$. Simonson and Mader (1973) suggest tracing the connection through the feature at $l = 9°$, $b = -0°.6$, $V = +135$ km s^{-1}, which has more of the appearance expected of a tangential point. Davies and Cohen (in preparation) find no connection of all the various individual components into a continuous ring, but they suggest

$v_{\mathrm{exp}} = 100$ km s^{-1}, $v_{\mathrm{rot}} = 60$ km s^{-1}, and $R = 2$ kpc for the features at negative velocities. The non-zero rotational velocity is an important limitation on theories to account for the origin of these features. For example, van der Kruit (1971) suggests expulsion from some distance out in the nuclear disk in order to account for the angular momentum, which the feature could not have acquired by interaction with material in the galactic plane.

(iv) Feature 'E' of Sanders, Wrixon and Penzias (1972) is denoted "SWP 'E'" in Figure 3. It is regarded by them as a separate feature, although there is some suggestion of a blend with van der Kruit's (1970) Feature X in Figure 3. Its position right at $l = 0°$ is suggestive of expulsion from the center.

(v) Van der Kruit's (1970) Feature VII is seen in emission in the direction of the continuum sources near the center, indicating outward radial motion.

(vi) The 'positive-velocity ridge' of Sanders and Wrixon (1972b) is denoted 'SW' in Figure 3. They suggest it is related to the 20-cm continuum jets observed by Kerr and Sinclair (1966). However, it has characteristics very similar to those of the $+135$ km s^{-1} expanding arm, which indicates that it may belong instead with the latter feature.

(vii) Cugnon's (1968) prominent feature may belong to the center region and have originated in an expulsion from the nucleus. More recent observations by Lindblad (1974), however, give somewhat more weight to one of Cugnon's alternative suggestions, namely, that it belongs to the distant, tilted-up, outer spiral structure.

Most of the explanations for the features described in this section have invoked explosions in the nucleus. If all of them so originated, it would have involved about 10^6 $M_{\odot}$ and 10^{54} erg (van der Kruit, 1971). If only feature 'E' of Sanders *et al.* (1972) had been ejected, it would have taken about 10^{52} erg, according to these authors, which is quite in keeping with the scale of other activity presently observed near the center (Lynden-Bell and Rees, 1971). Alternative explanations, such as resonant orbits in the field of a central bar, deserve to be explored.

V. Present Needs

The state of current observations is summarized in Table I. Much observational work remains to be done. The region is difficult to observe because geometrical effects produce extremely broad and weak line profiles here. We will ultimately want to survey the entire region with the highest resolution and sensitivity achievable. The main requirements at present are more modest. The most important task is to sort out the structures in the disk. This requires close sampling between $b = -2°$ and $+2°$ from $l = 345°$ to $15°$ with high angular resolution ($\sim 10'$), adequate sensitivity to measure profiles as weak as 1 K, and accurately known baselines covering 600 km s^{-1}. Only moderate velocity resolution is necessary, say 2 km s^{-1}. Above the disk, from $|b| = 2°$ to $10°$, the main thing is to detect all the hydrogen; its precise location is not so important as knowing how much there is. For this purpose high sensitivity and moderate angular resolution ($\sim 30'$) are needed, again with close sampling.

TABLE I

Current 21-cm line observations of the galactic center regions

Author (date of publication)	Telescope	Limits and sampling interval[a] or resolution		
		l(deg)	b(deg)	V^b(km s^{-1})
Burton (1970)	25 m	354(0.6)120	0	$-340(1.7)+340$
Cugnon (1968)	25 m	345(1)356	$+1(1)+7$	$-50(3.4)+180$
Davies and Cohen (1973)	30 m and 76 m	355(0.5)10	$-5(0.5)+5$	$-500(7)+500$
Kerr (1969)	64 m	296(0.2)63.5	0	$-300(7.0)+300$
		300(1)60	$-2(0.2)+2$	$-300(7.0)+300$
Kerr, Harten, and Ball (1973)	64 m	270(1)342	$-2(0.2)+2$	$-150(2.1)+60$
Kerr and Hindman (1970)	64 m	185(1)63	0	$-300(7.0)+300$
Kerr and Vallak (1967)	64 m	355(0.2)5	0	$-165(7.0)+165$
		359(0.1)1	$-1(0.2)+1$	$-165(7.0)+165$
van der Kruit (1970)	25 m	350(1)10	$-5(1)+5$	$-300(10.5)+300$
Lindblad (1974)	43 m	339(3)12	$-10(0.3)+10$	$-150(1)+150$
Menon and Ciotti (1970)	43 m	350(0.3)16	$-0.4(0.3)+0.3$	$-264(1.7)+264$
Rougoor (1964)	25 m	352(0.6)22	$-0.6(0.5)+0.9$	$-300(5)+300$
Sanders and Wrixon (1972a)	6 m horn	350(2)358	$-10(2)0$	$-340(16)-40$
Sanders and Wrixon (1972b)	6 m horn	355(2)5	$-5(2)+5$	$-300(16)+300$
Sanders, Wrixon, Penzias (1972)	6 m horn	358(2)12	$-10(2)0$	$-340(16)-40$
Sandqvist (1973)	43 m + moon {$l, b = 359.7, +0.46$ to 0.35, -0.94 by 0.004}			$-120(1.7)+120$
Simonson and Sancisi (1973)	25 m	356(0.6)10	0	$-120(3.4)+120$
		356(0.5)2	0(0.5)+5	$-120(3.4)+120$
		10(0.5)24	$-8(0.5)+5$	$-60(3.4)+180$
Weaver and Williams (1973)	26 m	10(0.5)250	$-10(0.5)+10$	$-140(2.1)+240$
Westerhout (1973)	91 m	11(0.2)233	$-2(0.2)+2$	$-60(2)+180$
Wrixon and Sanders (1973)	43 m	357(0.3)3	$-3(1)+3$ $-1(0.3)+1$	$-500(5.5)+500$

[a] Sampling interval given when it exceeds beamwidth.
[b] Velocity limits usually vary with longitude; the range near $l=0°$ is tabulated.

Theoretical developments to be hoped for center mostly about mechanisms. The models calculated until now have been largely kinematical. For dispersion rings, while individual resonant orbits have been calculated, collective effects have yet to be included. Hopefully, this would provide values for the peculiar velocity amplitudes and sizes of the structures. There is also the question of motions in the field of a possible central bar. Gas dynamics must be treated along with the stellar dynamics. Explosive theories also suffer from lack of definite values and must, in addition, account for the prominent asymmetries that are observed.

References

Burton, W. B.: 1970, *Astron. Astrophys. Suppl.* **2**, 261.
Cugnon, P.: 1968, *Bull. Astron. Inst. Neth.* **19**, 363.
Davies, R. D. and Cohen, R. J.: 1973, private communication.
Kerr, F. J.: 1969, *Australian J. Phys. Astrophys. Suppl.*, No. 9, 1.
Kerr, F. J. and Hindman, J. V.: 1970, *Australian J. Phys. Astrophys. Suppl.*, No. 18, 1.
Kerr, F. J. and Sinclair, M. W.: 1966, *Nature* **212**, 166.
Kerr, F. J. and Vallak, R.: 1967, *Australian J. Phys. Astrophys. Suppl.*, No. 3, 1.

Kerr, F. J., Harten, R. H., and Ball, D. L.: 1973, *Astron. Astrophys. Suppl.*, in preparation.
Kruit, P. C. van der: 1970, *Astron. Astrophys.* **4**, 262.
Kruit, P. C. van der: 1971, *Astron. Astrophys.* **13**, 405.
Lindblad, P. O.: 1974, *Astron. Astrophys. Suppl.* **16**, 207.
Lynden-Bell, D. and Rees, M.: 1971, *Monthly Notices Roy. Astron. Soc.* **152**, 461.
Menon, T. K. and Ciotti, J. E.: 1970, *Nature* **227**, 579.
Rougoor, G. W.: 1964, *Bull. Astron. Inst. Neth.* **17**, 381.
Sanders, R. H. and Prendergast, K. H.: 1974, *Astrophys. J.* **188**, 489.
Sanders, R. H. and Wrixon, G. T.: 1972a, *Astron. Astrophys.* **18**, 92.
Sanders, R. H. and Wrixon, G. T.: 1972b, *Astron. Astrophys.* **18**, 467.
Sanders, R. H. and Wrixon, G. T.: 1973, *Astron. Astrophys.* **26**, 365.
Sanders, R. H., Wrixon, G. T., and Penzias, A. A.: 1972, *Astron. Astrophys.* **16**, 322.
Sandqvist, Aa.: 1973, *Astron. Astrophys. Suppl.* **9**, 381.
Shane, W. W.: 1972, *Astron. Astrophys.* **16**, 118.
Simonson, S. C. and Mader, G. L.: 1973, *Astron. Astrophys.* **27**, 337.
Simonson, S. C. and Sancisi, R.: 1973, *Astron. Astrophys. Suppl.* **10**, 283.
Weaver, H. and Williams, D. R. W.: 1973, *Astron. Astrophys. Suppl.* **8**, 1.
Westerhout, G.: 1973, Maryland-Green Bank Galactic 21-cm Line Survey, 3rd edn., Univ. of Maryland, College Park, Maryland.
Wrixon, G. T. and Sanders, R. H.: 1973, *Astron. Astrophys. Suppl.* **11**, 339.

S. Christian Simonson, III
Astronomy Program,
University of Maryland,
College Park, Md. 20742, U.S.A.

DISCUSSION

Menon: There is a tendency in the literature to imply that all radial motions in the galactic plane cease with the '3-kpc expanding arm'. However, it has been known for some time that radial motions at slightly reduced velocities persist to at least 5 kpc. Hence in discussing the energies and ages of the phenomena involved we should discuss motions at least up to 5 kpc from the center.

Simonson: In our model, the motions extend up to 4.7 or even 5 kpc from the center.

Kahn: You mentioned the possibility that there might be a bar at the galactic center. If this bar makes an appreciable contribution to the gravitational field, the system would no longer have axial symmetry. Will this help with your angular momentum problem?

Simonson: It may help to organize the gas into resonant orbits, but orbits in a θ-dependent potential are hard to calculate.

Pishmish: I would like to comment on the rotating and expanding ring structure which is proposed to explain the kinematics of the molecular clouds in the central region of the Galaxy. It seems to me that the hypothesis of a ring is an oversimplified one. If one accepts that the expansion is a result of mass ejected from a rotating nucleus, the locus of the ejected matter will not be a circle but a spiral. The pitch, of course, depends on the velocity of expansion and the velocity of rotation. It appears that it is easier to produce a spiral than to produce a ring, as the latter would require an isotropic expulsion of matter, a phenomenon much more difficult to explain than a localized ejection. The formation of such a spiral locus was discussed by Huang and Pishmish back in 1960 (*Bol. Obs. Tonantzintla and Tacubaya*, No. 19).

Simonson: In fact, if you want to adopt an explosion model for the $+135$ km s^{-1} feature, some such thing as that would be probably all that is necessary, so that it is not a ring but has more of a spiral shape. Both models for the 3-kpc arm show expansion when one looks toward the center but circular velocities at an angle from the center, so both velocities are probably present, but it is hard to get the right match.

Oort: You mentioned the possibility of a bar in the central part. Don't you think this would be difficult to reconcile with this very strongly differentiately rotating nuclear disk that you find in the center? The gravitational field wouldn't seem suitable for a bar at all.

Simonson: In view of the fact that the nuclear disk is asymmetrical north and south, that material might not be exactly in the form of a nuclear disk but have properties resembling resonant orbits in the field of a bar. I'm not prepared to do any more than just raise the suggestion at this point. We need higher angular resolution there.

Baldwin: Do you imply that some changes are necessary in the inner parts of the galactic rotation curve or the distribution of mass?

Simonson: Schmidt took no observed points between $R=1$ and 4 kpc but used just a point mass at the center. We find his curve to be about 15% low, reaching a minimum of 185 km s^{-1}, at $R=2$ kpc, while observations suggest 210–215 km s^{-1} as the minimum at $R=3$–4 kpc. Adding a Gaussian disk of mass $4 \times 10^8 \, M_\odot$ and radial dispersion 1 kpc to Vandervoort's (*Astrophys. J.* **161** (1970), 67) model would about do it.

Burke: One way of avoiding the angular momentum problem inherent in explosion models of the galactic center is to assume that the explosion provides the energy (by a beam of relativistic particles, for example) but not much mass or momentum. The matter and the momentum is that of the gas in the disk, which would be partly sprayed up above the plane by the explosion, but preserving its angular momentum.

Churchwell: In the explosion models so far proposed which require energies of 10^{55}–10^{59} erg it seems to me that the gas would have to be ionized; however, what is observed is neutral hydrogen. What is the time scale for recombination for the expelled high velocity H I clouds?

Simonson: In van der Kruit's model, the gas starts off ionized but cools rapidly on a time scale of 10^6 yr; we see it now recombined after 10^7 yr.

Davies: Observations of H I in the galactic center ($l=5°$ to $+10°$, $b=-5°$ to $+5°$) have been made at Jodrell Bank with the Mark II (32′ beamwidth) and Mark IA (12′) telescopes. These observations were made at high sensitivity (rms noise ~ 0.1 K) to provide new information on the structure of the galactic center region. The full angular resolution and sensitivity were necessary to delineate many of the galactic center features.

KINEMATICS OF MOLECULES AT THE GALACTIC CENTRE

B. J. ROBINSON

Division of Radiophysics, CSIRO, Sydney, Australia

Abstract. Dense gas clouds containing OH, CO, NH_3 and H_2CO are found in the inner part of the H I nuclear disk. The molecular spectral lines allow direct observations of the kinematics of the gas near the galactic centre. Strong absorption of the thermal continuum sources by OH and H_2CO shows that much of the gas on the near side of the centre can be located in a massive 'ring' expanding at 130 km s^{-1}, which may have originated close to the nucleus about 10^6 yr ago. Observations of CO emission from beyond the centre show that the far side of the 'ring' is expanding at a lower velocity, less than 90 km s^{-1}. Observations of CO and NH_3 emission with positive velocities for $l < 360°$ are needed to establish whether the 'ring' is a continuous structure.

OH and H_2CO are also observed to be falling towards the centre. There is no agreement as to the location of this infalling matter.

The nuclear regions of the Galaxy are compared with those of NGC 253, particularly in regard to expansional velocities, IR and radio emission, and OH absorption.

I. Introduction

The most direct observations of the galactic centre can be made in the infrared, in the radio continuum and recombination line emission, and by means of molecular spectral lines. At visual wavelengths our view is obscured by 27 mag of absorption in the dust found near the centre and in the intervening spiral arms. High-velocity hydrogen near the centre can be observed at 21 cm wavelength, but for H I with radial velocities between ± 50 km s^{-1} the central regions are blanketed by the wings of the 21-cm emission from the spiral arms in front of and behind the centre.

The molecular spectral lines provide a particularly powerful probe of the central regions. The density of molecules there is several orders of magnitude greater than in the spiral arms, so that there is negligible confusion from gas in the intervening (or distant) arms. The high molecular densities are found in the inner parts of the 'nuclear disk' inferred from 21 cm observations. The kinematics of the region are defined by the radial velocities of the four molecules which are widely distributed: hydroxyl, formaldehyde, carbon monoxide and ammonia. Many exotic molecules have been observed in the dense clouds near Sgr B2 (G0.7 − 0.0) and Sgr A (G0.0 − 0.0), but they have not been seen at other longitudes and so provide little kinematic information. Hydroxyl and formaldehyde are seen in absorption against the continuum sources near the centre, while carbon monoxide and ammonia appear in emission and so can be detected on the far side of the continuum sources.

II. The Nuclear Disk

High-velocity 21 cm emission is seen between longitudes $356° < l < 0°$ with radial velocity $-250 < V < -70$ km s^{-1}. This gas has been associated with a rapidly rotating

F. J. Kerr and S. C. Simonson, III (eds.), Galactic Radio Astronomy, 521–535. All Rights Reserved.

nuclear disk extending to a radius R of about 800 pc from the galactic centre (Rougoor, 1964). For $l>0°$ the distribution of the high-velocity H I is confused by emission from H I beyond the centre, and the structure of the disk has been inferred from models. The basic parameters of the disk derived by Oort (1971) and by Sanders and Wrixon (1973) are shown in Table I.

TABLE I

Parameters of the H I nuclear disk

Parameter	Oort (1971)	Sanders and Wrixon (1973)
Radius (pc)	750	750
Thickness (pc)	100 for $R<300$ pc	100 for $300<R<500$ pc
	250 for $R>300$ pc	120 on average
Maximum rotational velocity (km s^{-1})	230	230
Velocity dispersion (km s^{-1})	–	13
H I density (atoms cm^{-3})	3 for $R<100$ pc	5 for $R<110$ pc
	0.3 on average	1.5 on average
Total H I mass ($M_\odot$)	4×10^6	2×10^6 for $R<500$ pc
Dust mass ($M_\odot$)	10^5	–
Mass of H$_2$ ($M_\odot$)	10^7	–

Oort's (1971) model is symmetric about the centre, with the gas increasing in density smoothly towards the centre and rotating in circular orbits. The model of Sanders and Wrixon (1973) is based on higher-resolution observations and has *no* neutral hydrogen in the inner 300 pc on the positive longitude side of the disk; at negative longitudes the H I density increases smoothly to within 100 pc of the centre.

We should bear in mind that the techniques which best probe the galactic centre *all* show a marked asymmetry about $l=0°$. The 100 μm IR observations (Hoffman *et al.*, 1971) and the radio continuum observations (reviewed by Gordon in this volume) show a much greater extent (and strength) of the emission at positive longitudes than at negative longitudes. The molecules show an even more pronounced asymmetry.

III. Distribution and Motions of OH Molecules

The most complete survey of the distribution of a molecular species near the galactic centre has been made for OH in absorption at 18 cm wavelength (Robinson and McGee 1970; McGee *et al.*, 1970). As shown in Figure 1, absorption extends in longitude from $359°<l<2°$ for $-180<V<+90$ km s^{-1}. Weaker absorption can be traced up to $l\approx4°$ at $b=0°$, but considerable searching has failed to reveal OH absorption in the range $357°<l<358°30'$. At a distance of 10 kpc, $l=2°$ corresponds to a radial distance of 350 pc from the centre.

We see that the OH is concentrated in the inner part of the H I nuclear disk on the positive longitude side. The OH can be traced nearly to the edge of the disk, to $R=700$ pc (corresponding to $l=4°$). Much of the OH has large radial motions towards

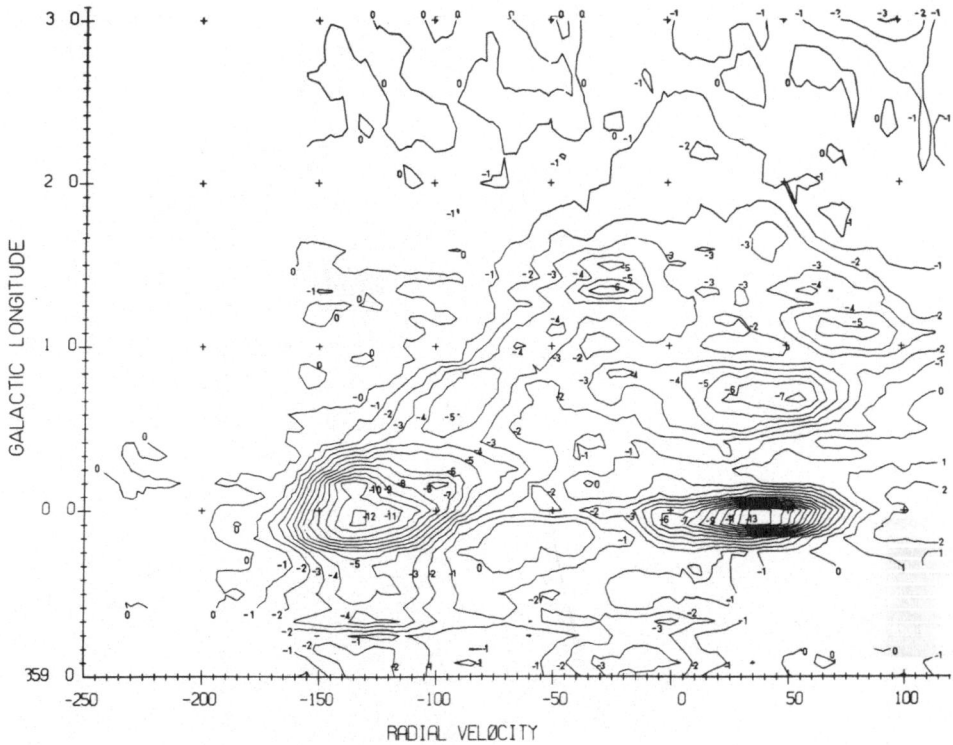

Fig. 1.   Galactic longitude – radial velocity $(l-V)$ plot of 1667 MHz OH absorption for $b = -0°10'$. Beamwidth is 12.2′, velocity resolution is 6.6 km s^{-1}. Observations were made every 5′ in longitude. Contour unit is 1.7 K in antenna temperature. (From Robinson and McGee, 1970).

and away from the centre, unlike the circular motions adopted in the H I nuclear disk model.

The thickness of the hydroxyl layer for $R \leqslant 350$ pc is about 50 to 70 pc (see McGee, 1970), compared to 100 pc for the H I disk. For $R > 400$ pc the hydroxyl layer becomes somewhat thinner while the H I layer expands.

For $R \leqslant 350$ pc the hydroxyl absorption reaches a maximum about 10′ south of the galactic plane, corresponding to $z = -30$ pc. No similar displacement is seen in the radio continuum or 100 μm IR emission.

It might be argued that the asymmetry in longitude of the OH absorption simply reflects the asymmetry in distribution of the radio continuum. But McGee (1970) has shown that the asymmetry is even more marked in the distribution of apparent OH opacity. If we compare the apparent opacity in Figure 2 with the directly observed absorption in Figure 1, we see that the influence of strong continuum sources like Sgr A, Sgr B2 (G0.7 − 0.0) and G1.1 − 0.1 has been almost eliminated, and the highest opacity is found near $l = 1°30'$ and $l = 3°$.

Kaifu *et al.* (1972) drew attention to the continuity of the OH absorption in the $l - V$ plane shown by Figure 2, and suggested that the locus of the absorption was an ellipse as shown in Figure 3. Such a locus could be produced by a ring of OH with

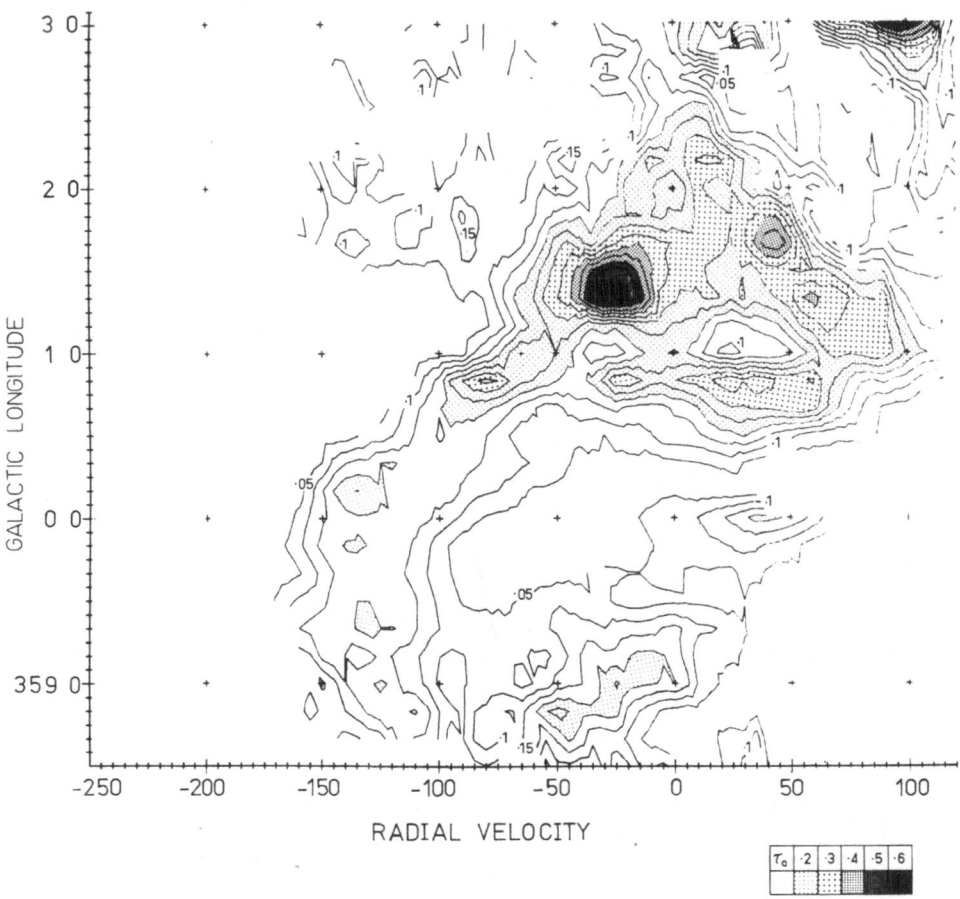

Fig. 2.   Contours of apparent OH opacity at 1667 MHz for $b = -0°10'$, derived from Figure 1.
(From McGee, 1970).

$R = 270$ pc, rotating at 50 km s^{-1} and expanding at 130 km s^{-1}. The positive velocity side of the ring is not observed, since the OH would be behind the continuum sources. However, Kaifu *et al.* showed that the velocities of some NH$_3$ emission lines observed by Knowles and Cheung (1971) at $l = 359°35'$ and $l = 1°12'$ (marked by squares in Figure 3) fell on the locus "... thus confirming the existence of the expanding ring...". Models for this 'ring' will be discussed in a later section.

IV. Distribution and Motions of H$_2$CO Molecules

Absorption by the $1_{10} \leftarrow 1_{11}$ transition of formaldehyde at 6 cm has been surveyed by Gardner and Whiteoak (1970), Scoville *et al.* (1972) and Scoville and Solomon (1973). Figure 4 shows an $l - V$ contour diagram of the apparent opacity at $b = -0°12'$. In general outline it shows marked similarities to the OH opacity in Figure 2, but more detail is apparent because of the four times higher resolution in velocity; the

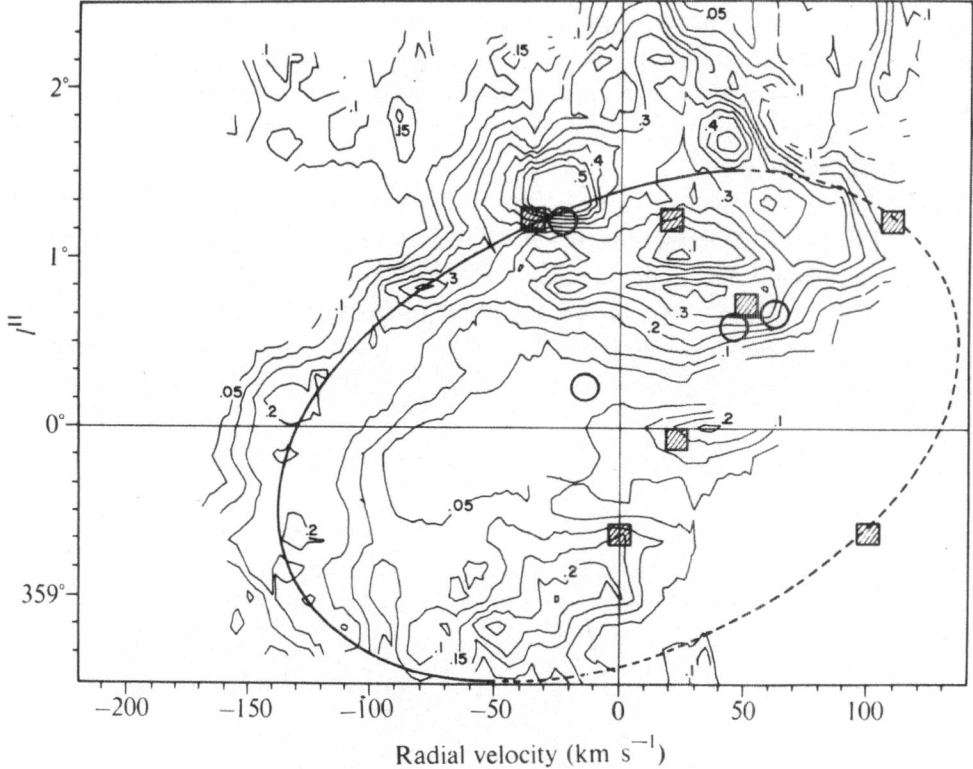

Fig. 3. Locus of 270 pc expanding molecular ring suggested by Kaifu *et al.* (1972) from OH absorption (Figure 2) and NH_3 emission velocities (shown by hatched rectangles). The circles indicate the hydrogen recombination line velocities of the H II regions near the galactic centre.

observations were undersampled in longitude. Narrow-band absorption by the 3 kpc expanding arm (near $V = -50$ km s^{-1}) and in the nearby spiral arms (near $V = 0$ km s^{-1}) is much more marked in the H_2CO observations than for OH. As was the case for OH, the H_2CO opacities are higher below the plane ($b = -0°12'$) than close to the plane ($b = -0°02'$); this is most marked for the negative velocity gas. H_2CO observations at other latitudes have not been published.

Scoville (1972) showed that the maximum H_2CO opacity also lies on part of an elliptical locus in the $l - V$ plane. This locus is shown by the continuous line in Figure 5, superimposed on contours of the average H_2CO opacity at $b = -0°02'$ and $b = -0°12'$ The elliptical locus would correspond to a ring of gas of radius about 220 pc rotating at 50 km s^{-1} and expanding at 145 km s^{-1}; the ring extends further at positive longitudes ($R = 305$ pc) than at negative longitudes ($R = 218$ pc). As with the OH observations in Figure 3, data on the H_2CO distribution at positive velocities are incomplete, presumably because the H_2CO then lies beyond the continuum sources near the centre.

V. Distribution and Motions of CO Molecules

A preliminary CO emission line survey at $\lambda = 2.6$ mm covering a strip at $b = -0°02'$

from $l = 359°7$ to $l = 2°8$ has been published by Solomon *et al.* (1972). As well as being restricted to positive longitudes the survey is also restricted to positive velocities $(+28 < V < +130$ km s$^{-1})$. The survey is also heavily undersampled in longitude, observations with a 1 arcmin beam being separated by 6' (identical with the spacing used for OH and H$_2$CO).

The $l - V$ contour map for CO at $b = -0°02'$ is shown in Figure 6. Between the centre and $l = 1°8$, CO emission is observed over almost the complete velocity range of the measurements. The high-velocity CO features near $l = 0°$ $(60 < V < 95$ km s$^{-1})$, $l = 0°7$ $(80 < V < 105$ km s$^{-1})$ and between these longitudes $(60 < V < 90$ km s$^{-1})$ are extremely weak or missing from the OH and H$_2$CO maps. The CO at these velocities is presumably beyond the centimetric continuum sources which are absorbed by OH and H$_2$CO.

At $l = 1°4$ scans perpendicular to the plane showed that the CO emission extended beyond $b = +14'$ and $b = -6'$.

The CO observations indicate high densities and low temperatures in the molecular

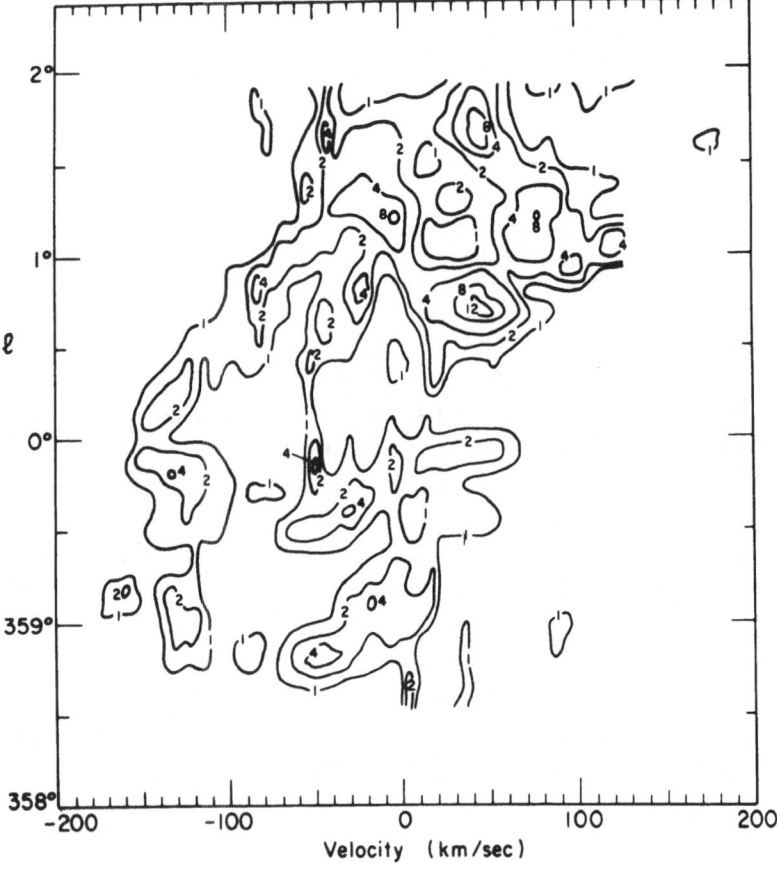

Fig. 4. Contours of apparent H$_2$CO opacity at 4830 MHz for $b = -0°12'$. Beamwidth is 6.6', velocity resolution is 1.63 km s^{-1}. Observations were made every 6' in longitude. (From Scoville and Solomon, 1973).

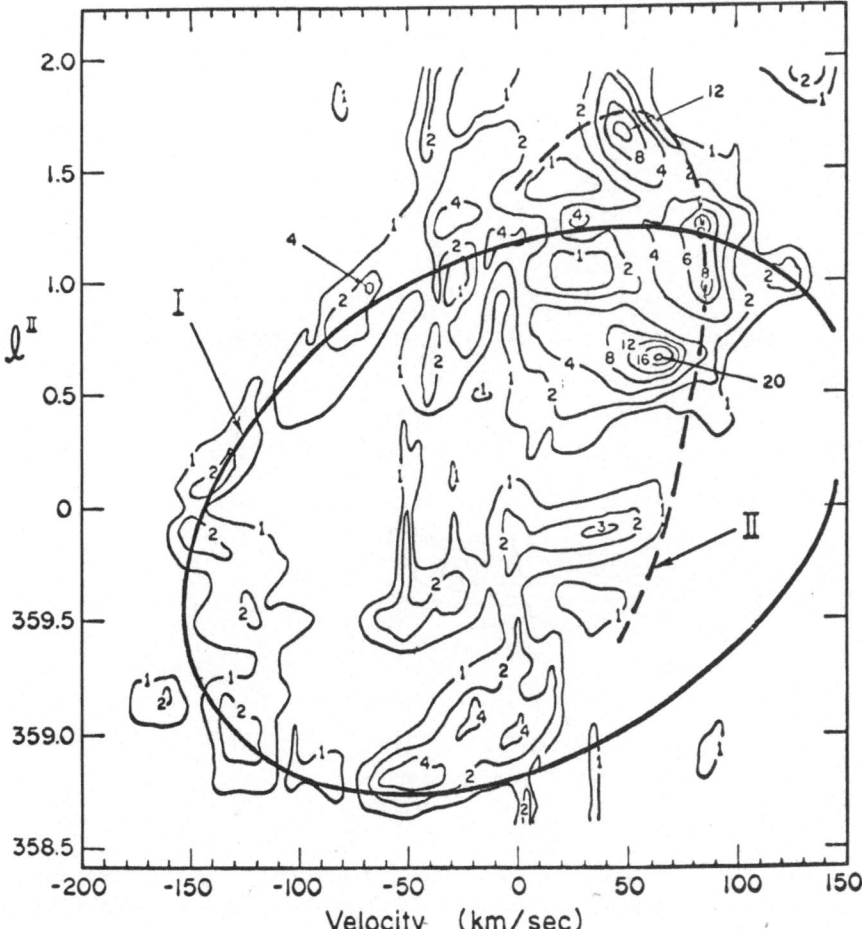

Fig. 5.   Locus of expanding molecular ring suggested by Scoville (1972) from H_2CO absorption.
The H_2CO opacity contours are the average of those at $b = -0°02'$ and $b = -0°12'$.

clouds. The upper state $(J = 1)$ of the 2.6 mm CO transition has a lifetime of 1.6×10^7 s and will be appreciably populated by collisions when the hydrogen density exceeds 10^3 atoms and/or molecules per cubic centimetre. The observation of CO emission against the isotropic background radiation implies that the molecular clouds near the centre have hydrogen densities $\geqslant 10^3$ cm^{-3}. A density exceeding 10^3 cm^{-3} is also required to excite the 1.3 cm ammonia emission lines. The 21 cm model for the nuclear disk has $0.3 < n_H < 5$ cm^{-3}, so the hydrogen in the clouds must be mainly in the molecular form.

Comparison of $^{12}C^{16}O$ and $^{13}C^{16}O$ lines shows that the ^{12}CO line is optically thick. Thus the observed brightness temperatures of 13 to 17 K are a measure of the kinetic temperature in the molecular clouds. Solomon *et al.* (1972) have pointed out that this temperature is close to the IR brightness temperature at 100 μm, suggesting

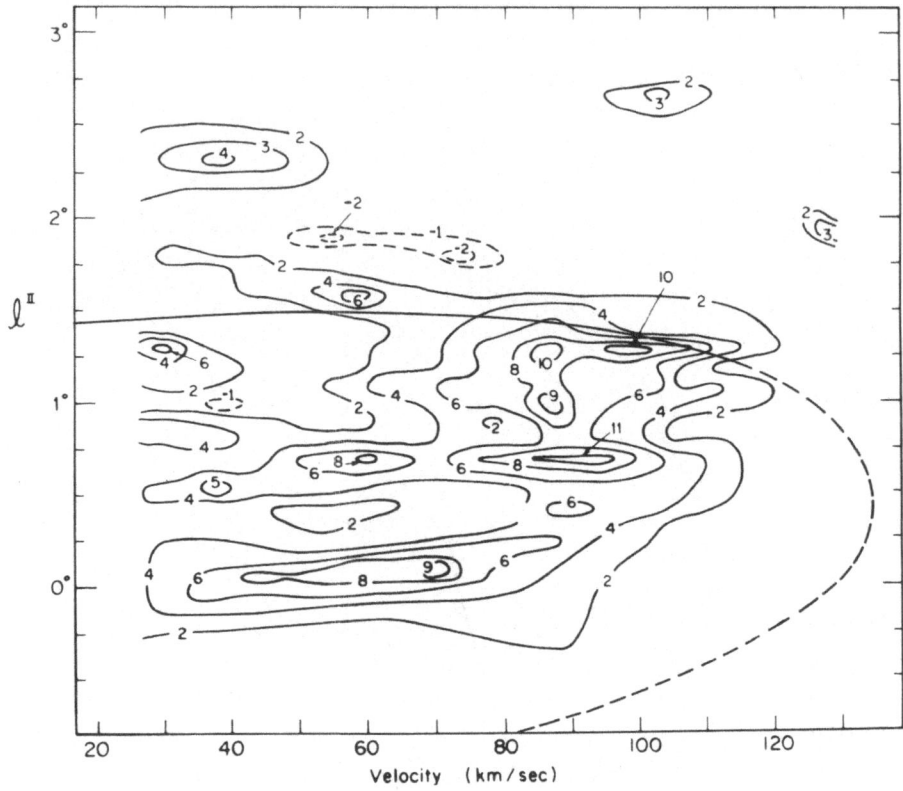

Fig. 6. Contours of 115 GHz CO emission for $b = -0°02'$. Beamwidth is 1', velocity resolution 5 km s^{-1}. Observations were made every 6' for $359°7 < l < 2°$, every 12' for $l > 2°$. The observations were restricted to $+28 < V < +130$ km s^{-1}. (From Solomon *et al.*, 1972). Part of the elliptical locus from Figure 3 is superimposed on the contours.

that the grains and gas are in thermal equilibrium and that the clouds are optically thick at 100 μm. This would imply 200 mag of extinction at optical wavelengths.

VI. Models for the Molecular Ring

The model for the expanding and rotating molecular ring derived by Kaifu *et al.* (1972) from OH absorption and NH$_3$ emission is shown in Figure 7, and the parameters describing it are listed in Table II. The model proposed by Scoville (1972) from the H$_2$CO absorption is shown in Figure 8, with the parameters also in Table II.

Kaifu *et al.* (1972) have noted that *positive* velocity OH and H$_2$CO absorption is seen against Sgr A, Sgr B2 and other continuum sources near the center, and have put the infalling material in a smaller rotating and *contracting* ring with radius 140 pc. The recombination line velocities of the H II regions enable them to be located on this contracting ring (except for G1.1 − 0.1, which is located on the expanding ring). Scoville's (1972) model accounts for the positive velocity molecular absorption by

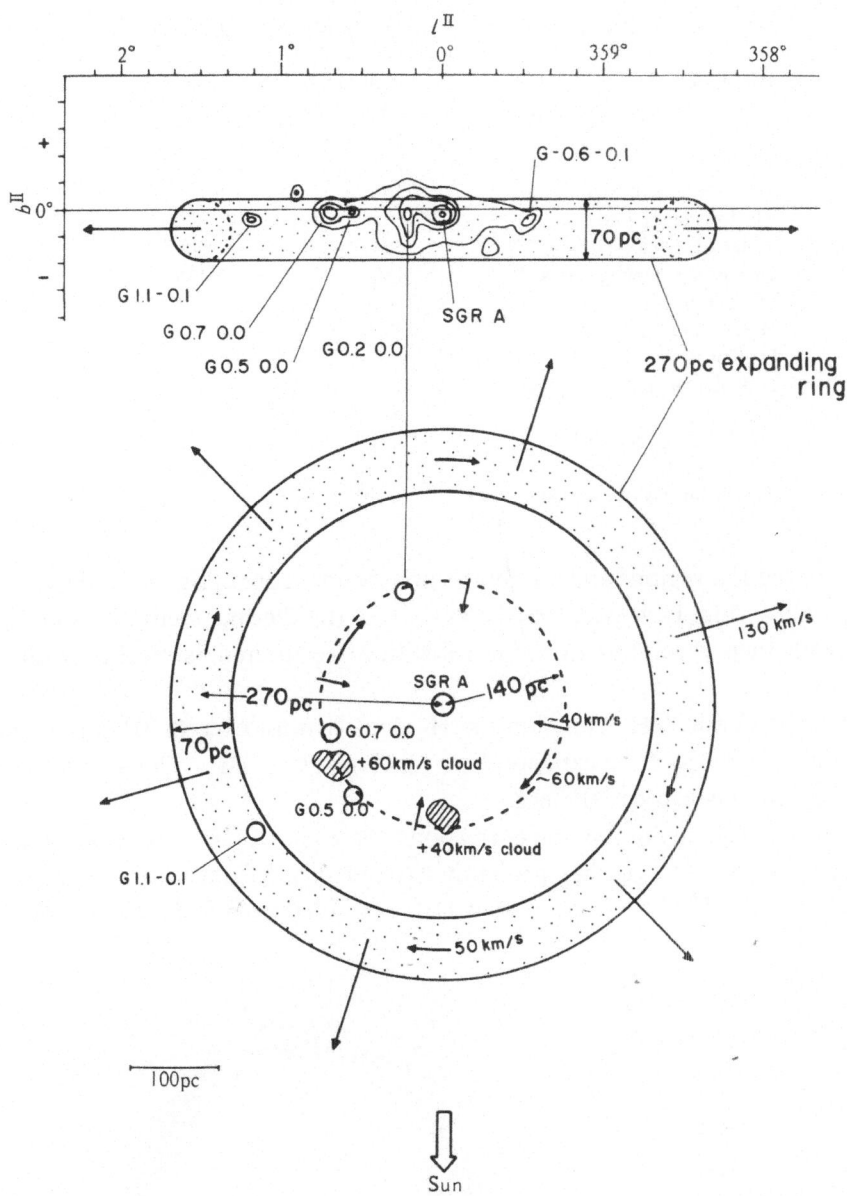

Fig. 7. Model for the 270 pc expanding molecular ring proposed by Kaifu *et al.* (1972) from OH absorption and NH$_3$ emission velocities. The broken circle shows the inner contracting ring, containing OH clouds (shaded patches) and H II regions (circles). The contours in the upper picture are 8 GHz continuum contours.

TABLE II

Parameters for the expanding molecular ring

Parameter	Kaifu *et al.* (1972)	Scoville (1972)
Radius (pc)	270	218
Thickness Δz (pc)	70	–
Tilt to galactic plane	$6°$	$3°$
Rotational velocity (km s^{-1})	50	50
Expansion velocity (km s^{-1})	130	145
OH column density (cm^{-2})	10^{16}	–
H$_2$CO column density (cm^{-2})	2×10^{13}	–
H$_2$ density n_{H_2} (cm^{-3})	$\approx 10^3$	$\leqslant 10^3$
H density n_H (cm^{-3})	2	–
n_H/n_{H_2}	$\approx 10^{-3}$	–
Total mass of ring ($M_\odot$)	$\approx 10^8$	$\geqslant 10^8$
Expansion time (years)	2×10^6	$\approx 10^6$
Kinetic energy of expansion (erg)	$\approx 10^{55}$	2×10^{55}

putting all the H II regions and all but the nonthermal core of Sgr A on the far side of the expanding ring. However, this disposition of the thermal sources is quite inconsistent with their recombination line velocities (see the points marked with circles on Figure 3).

To excite CO and NH$_3$ emission the H$_2$ density must exceed 10^3 cm^{-3}, and this makes the total mass of the expanding ring as high as 10^8 $M_\odot$. The kinetic energy of the expansion may exceed 10^{55} erg.

The angular velocity of the ring corresponds to that at a radius of about 50 pc from the centre, so that the gas has probably expanded out from near the centre. At a constant speed of 135 km s^{-1} the expansion would have taken 2×10^6 yr. We do not

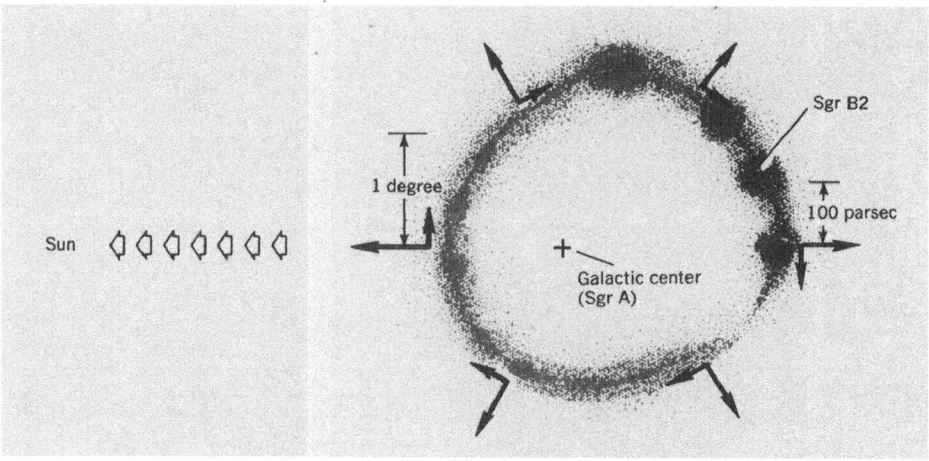

Fig. 8.   Model for the expanding molecular ring proposed by Scoville (1972) from H$_2$CO absorption velocities. The label Sgr A refers only to the nonthermal continuum component. The thermal continuum sources are located on the far side of the ring.

understand what force could drive such an expansion and resort to an explanation in terms of 'an explosive event' in the nulceus about 10^6 yr ago. Sanders and Lowinger (1972) have computed circular velocities V_c in the nuclear disk; at $R = 270$ pc. V_c is about 230 km s^{-1} and so the escape velocity is about 330 km s^{-1}. We see that the velocity of the expanding ring is far below the escape velocity. Thus unless there is some force to accelerate the ring the material will ultimately spiral back towards the nucleus.

Kaifu et al. (1972) note that the expansion velocity of 130 km s^{-1} far exceeds the sound velocity in the region (1 to 10 km s^{-1}), and regard the expanding ring as a shock front. Strong compression by the shock leads to the formation of the observed molecular clouds and of stars and H II regions. Their contracting ring with $R = 140$ pc may be a shock wave propagating towards the centre as the counter motion of gas after being swept by the outgoing shock wave (Kaifu et al., 1973).

We should take a critical look at the experimental data on which these models are based. Is the 'expanding ring' a continuous structure? Can we discriminate between a ring and an expanding spiral arm (or arms)? It is obvious from Figures 3, 5 and 6 that there is a serious lack of positive velocity data for OH, H_2CO and CO for $l < 360°$. Only one point appears in this quadrant, the NH_3 velocity of $+100$ km s^{-1} at $l = 359°35, b = -0°15$ (Figure 3). There is an urgent need for a survey of CO emission in this quadrant.

For $l > 0°$, the CO emission in Figure 6 for $80 < V < 120$ km s^{-1} should arise beyond the centre, as little absorption at these velocities is seen for OH and H_2CO. However, when we draw the locus of the expanding ring (from Figure 3) on the CO contours in Figure 6, we see that the CO data give no support for the symmetric ring models in the part of the locus shown by the dashed line. The expansion velocity of the far side of the ring cannot be more than 90 or 100 km s^{-1}.

VII. Similarity of the Nuclear Regions of the Galaxy and NGC 253

Because high obscuration prevents any visual observations of the galactic centre it is fruitful to draw analogies with optical data on external galaxies. The popular model has been the nucleus of M31 (see Oort, 1971).

In M31 Rubin and Ford (1970) have found a striking asymmetry of the radial velocities near the nucleus; on the NE side the gas out to $R = 800$ pc rotates at about the same velocity as the nuclear disk in the Galaxy; on the other side the motions are quite irregular. In the nucleus of M31 Sandage et al. (1969) found that the distribution of the 2.2 μm IR radiation is identical with that at optical wavelengths; the 2.2 μm distribution within $R = 50$ pc is also identical with that in the Galaxy (Becklin and Neugebauer, 1968). This suggests that the 2.2 μm distribution in the nucleus of the Galaxy can be used as a measure for the density distribution of the stars; within $R = 20$ pc the stellar mass must be about 2×10^8 $M_\odot$. Circular velocities computed for this stellar distribution are similar to those required for the H I nuclear disk.

An even better model for the nucleus of the Galaxy is provided by the nucleus of

NGC 253. It is strikingly similar to the nucleus of the Galaxy in the IR, contains a nonthermal radio source similar to Sgr A, and contains a high density of OH molecules. Optical and OH observations reveal expansion from the nucleus.

A full comparison of the nuclei of the Galaxy and of NGC 253 is made in Table III. Expansion at a velocity of 120 km s^{-1} has been measured optically by Demoulin and Burbidge (1970) within $R = 750$ pc of the nucleus of NGC 253. In addition, the first spiral arm at $R = 3.5$ kpc is expanding at $V = 40$ km s^{-1}, figures which are strikingly similar to those for our '3 kpc expanding arm'.

TABLE III

Comparison of the nuclei of the Galaxy and NGC 253

Property	Galactic centre	NGC 253
Expansion:		
At $R < 750$ pc	$V_E = 130$ km s^{-1}	$V_E = 120$ km s^{-1}
Of first spiral arm	$V_E = 53$ km s^{-1}	$V_E = 40$ km s^{-1}
(radius)	($R = 4$ kpc)	($R = 3.5$ kpc)
IR source:		
Size	300 pc × 600 pc	150 pc × 150 pc
Luminosity	$3.4 \times 10^8 \, L_\odot$	$3 \times 10^9 \, L_\odot$
Radio source:		
Nonthermal radio energy	3×10^{50} erg	2×10^{53} erg
Molecules:		
OH absorption $- V_{expansion}$	-130 km s^{-1}	-50 km s^{-1}[a]
OH masers (number)	1	2

[a] Extends to -150 km s^{-1}.

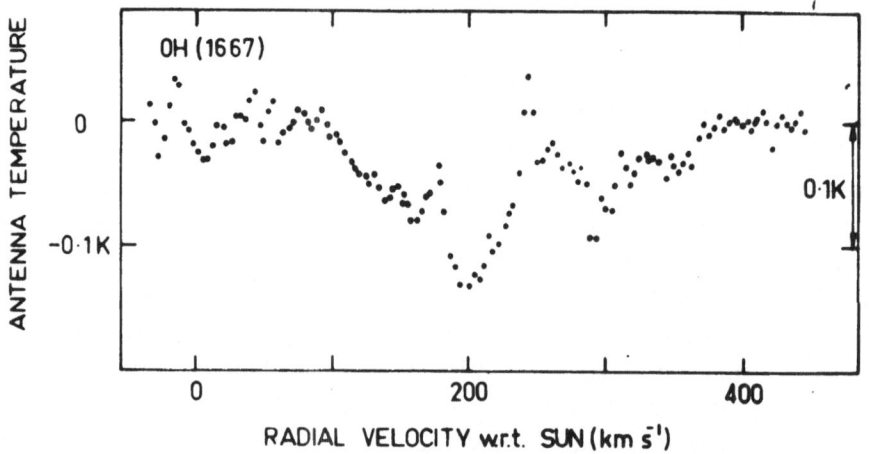

Fig. 9. OH absorption at 1667 MHz of the nuclear radio source in NGC 253 (Whiteoak and Gardner, 1974).

The far IR source in NGC 253 (Becklin *et al.*, 1973) is somewhat smaller than the one at the centre of the Galaxy, but about ten times as luminous. The IR observations show that both nuclei have Seyfert characteristics. The nonthermal radio source in NGC 253 is nearly three orders of magnitude more energetic than Sgr A is at the present epoch.

Wideband OH absorption of the nonthermal radio source in NGC 253 is shown in Figure 9 (Whiteoak and Gardner, 1974). The systemic velocity is 250 km s^{-1}, so that the strongest OH absorption corresponds to expansion at a velocity of 50 km s^{-1} there is less dense OH expanding at velocities up to 150 km s^{-1}. Weaker OH absorption at velocities higher than the systemic velocity shows that there must also be infalling OH with $V \approx 50$ km s^{-1}. Two narrow spikes on the profile may be OH maser sources. In all, the OH profile in Figure 9 is strikingly similar to the OH profile of Sgr B2 (see Robinson *et al.*, 1970, Figure 7).

VIII. Conclusions

Observations of hydroxyl, formaldehyde, carbon monoxide and ammonia lines near the galactic centre provide evidence for a section of a dense ring of matter about 270 pc in radius, expanding at about 130 km s^{-1} (see Table II). The ring must have a mass of about 10^8 $M_\odot$. It lies well within the rapidly rotating H I nuclear disk (see Table I); few molecules are seen in the outer parts of the nuclear disk, or inside the radius of the molecular ring.

The molecular ring has a rotational velocity of 50 km s^{-1} and so does not share the rotation of the H I nuclear disk – which would be about 230 km s^{-1} at $R = 270$ pc. The expansion velocity of the ring is well below the escape velocity at $R = 270$ pc, which would be about 340 km s^{-1}.

Kaifu *et al.* (1972) have suggested that the molecular ring has been produced by a shock wave travelling outwards from about 100 pc from the centre. The strong compression produced by the shock wave would lead to the formation of molecules and of young stars and H II regions. Since the expansion time of the ring would be about 2×10^6 yr, this model implies that the formation time for the molecules is about 10^5 to 10^6 yr. This requires high densities. High densities are also indicated by the excitation of CO and NH_3 to emit.

The observations of OH and H_2CO show that material is also falling in towards the galactic centre. Kaifu *et al.* (1972) locate these positive velocity molecular clouds on an inner ring of radius about 140 pc, contracting inwards at $V = 40$ km s^{-1} and rotating at 50 km s^{-1}. The recombination line velocities of many of the H II regions allow them to be located on this inner ring. The inner ring is perhaps a counter shock produced by particles spiralling back towards the nucleus under its gravitational attraction. Scoville (1972) places the positive velocity molecular clouds and the H II regions on the far side of the 270 pc expanding ring, but this location conflicts with the observed recombination line velocities.

Observational data about the far side of the 270 pc expanding ring are very limited.

The CO data for $l > 0°$ do not support the elegant symmetry of the ring proposed by Kaifu *et al.* (1972) and Scoville (1972). Observations of CO and NH_3 emission with positive velocities for $l < 360°$ are badly needed to define the structure of the ring, or to test whether it may be an expanding spiral arm (or arms).

Finally, attention is drawn to the strong similarities between the nuclei of the Galaxy and of NGC 253 in the infrared, radio continuum and OH absorption. Optical and OH velocities for NGC 253 show similar expanding and contracting motions to those found at the centre of the Galaxy.

References

Becklin, E. E. and Neugebauer, G.: 1968, *Astrophys. J.* **151**, 145.
Becklin, E. E., Fomalont, E. B., and Neugebauer, G.: 1973, *Astrophys. J. Letters* **181**, L27.
Demoulin, M. H. and Burbidge, E. M.: 1970, *Astrophys. J.* **159**, 799.
Gardner, F. F. and Whiteoak, J. B.: 1970, *Astrophys. Letters* **5**, 161.
Hoffmann, W. F., Frederick, C. L., and Emery, R. L.: 1971, *Astrophys. J. Letters* **164**, L23.
Kaifu, N., Iguchi, T., and Kato, T.: 1974, *Publ. Astron. Soc. Japan* **26**, 117.
Kaifu, N., Kato, T., and Iguchi, T.: 1972, *Nature* **238**, 105.
Knowles, S. H. and Cheung, A. C.: 1971, *Astrophys. J. Letters* **164**, L19.
McGee, R. X.: 1970, *Australian J. Phys.* **23**, 541.
McGee, R. X., Brooks, J. W., Sinclair, M. W., and Batchelor, R. A.: 1970, *Australian J. Phys.* **23**, 777.
Oort, J. H.: 1971, in D. J. K. O'Connell (ed.), *Nuclei of Galaxies*, Pontificiae Academiae Scientiarum Scripta Varia, North-Holland Publishing Co., Amsterdam, p. 321.
Robinson, B. J. and McGee, R. X.: 1970, *Australian J. Phys.* **23**, 405.
Robinson, B. J., Goss, W. M., and Manchester, R. N.: 1970, *Australian J. Phys.* **23**, 363.
Rougoor, G. W.: 1964, *Bull. Astron. Inst. Neth.* 17, 381.
Rubin, V. C. and Ford, W. K.: 1970, *Astrophys. J.* **159**, 379.
Sandage, A. R., Becklin, E. E., and Neugebauer, G.: 1969, *Astrophys. J.* **157**, 55.
Sanders, R. H. and Lowinger, T. L.: 1972, *Astron. J.* **77**, 292.
Sanders, R. H. and Wrixon, G. T.: 1973, *Astron. Astrophys.* **26**, 365.
Scoville, N. Z.: 1972, *Astrophys. J. Letters* **175**, L127.
Scoville, N. Z. and Solomon, P. M.: 1973, *Astrophys. J.* **180**, 55.
Scoville, N. Z., Solomon, P. M., and Thaddeus, P.: 1972, *Astrophys. J.* **172**, 335.
Solomon, P. M., Scoville, N. Z., Jefferts, K. B., Penzias, A. A., and Wilson, R. W.: 1972, *Astrophys. J.* **178**, 125.
Whiteoak, J. B. and Gardner, F. F.: 1974, *Astrophys. Letters*, in press; *Nature* **247**, 526.

B. J. Robinson
CSIRO Division of Radiophysics,
P.O. Box 76,
Epping, N.S.W. 2121, Australia

DISCUSSION

Morimoto: Shock wave calculations by Kaifu and others show that a secondary shock traveling outward with particles moving inward (thus appearing as a contracting ring) can develop as the particles behind the expanding ring fall back to the equilibrium position. This can explain the positive velocity features. The place where you left the symmetrical ellipse for an expanding ring in the l, v diagram is where confusion from the contracting ring is the heaviest.

Robinson: At the outer edges of the ring, the galactic rotation curve deduced from the infrared observations gives circular velocities between 200 and 250 km s^{-1}. The rotational velocity in the ring is on the order of 50 km s^{-1}. If it has been blown out from the center, it started much closer in and has angular momentum comparable with that distance much closer in. There is then a problem of how it is going to stay up there without spiraling back into the center.

Morimoto: Matter is just starting to go back in.

Caswell: If they are real, how strong are the maser emission sources in NGC 253 compared with the strongest ones in our own Galaxy?

Gardner: It is somewhat more than 10 times greater than W49.

Batchelor: To what extent is the southern velocity quadrant of the OH map influenced by the continuum background?

Robinson: We worked very hard to get absorption down there at 18 cm. The CO observations are not affected by the continuum, so you should see CO, but it certainly doesn't show up on the map.

Zuckerman: Those masers in NGC 253 must be really strong, since that was at 1667 MHz and the strongest galactic masers are at 1665 MHz.

Robinson: It is comparable with the 1667 MHz emission from G330.9 − 0.4.

Zuckerman: Those southern sources!

Robinson: You'll have to take our word for it that they exist.

OBSERVATIONS OF THE $J=6$ TO 5 LINE OF
OCS IN SGR B2

KENJI AKABANE, MASAKI MORIMOTO, and KIYOSHI NAGANE

Tokyo Astronomical Observatory, Tokyo, Japan

and

NORIO KAIFU

University of Tokyo, Tokyo, Japan

Abstract. The $J=6$ to 5 line of OCS at 72.9768 GHz was detected in Sgr B2 with antenna temperature 0.3 K, velocity $+62$ km s^{-1}, and velocity half width of ~ 25 km s^{-1}.

The $J=6$ to 5 line of OCS (72.9768 GHz) was detected in Sgr B2 with an antenna temperature of 0.3 K, centered at $+62$ km s^{-1} with a half width of about 25 km s^{-1}. This is the first detection of this line in celestial sources. The line strength gives a column density of OCS molecules around 10^{15} cm^{-2}, similar to that given by Jefferts *et al.* (1971) from their observations of the $J=9$ to 8 line, and the peak velocity also agrees well. However, our observations give twice as large a line width as given by

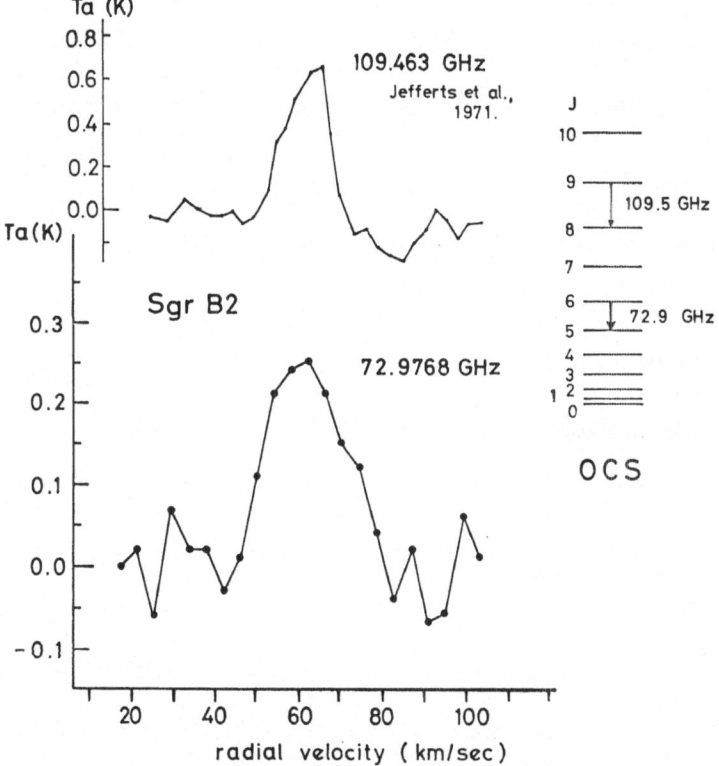

Fig. 1. Profiles of $J=9$ to 8 (from Jefferts *et al.*, 1971) and $J=6$ to 5 lines of OCS. Ordinates denote the line of sight velocity, and the abscissa the antenna temperatures.

F. J. Kerr and S. C. Simonson, III (eds.), Galactic Radio Astronomy, 537–538. All Rights Reserved.

Jefferts *et al.*, which may be due to the large beamwidth of our observations. A large beam tends to pick up emission from the outer part of the source where the velocity can be different from the inner part, giving rise to a larger line width.

In this search we detected $1_{01} - 0_{00}$ line of para-formaldehyde in Ori A and Sgr B2 but failed to detect the OCS line in Ori A, to a limit of 0.2 K peak to peak. Results of a search for other lines are summarized in Table I.

TABLE I

Negative results

Molecule	Frequency (GHz)	Object	Peak-to-peak noise (K)	τ (h)
H_2CO	72.8381	IRC + 10216	0.4	7.0
$1_{01} \rightarrow 0_{00}$		W51	0.4	8.2
		Heiles cloud 2	0.5	
		Heiles cloud 4	0.5	
OCS	72.9768	Ori A	0.2	15.3
$6 \rightarrow 5$		IRC + 10216	0.4	7.0
		W51	0.4	8.2
HCCCN	72.7851	W51	0.3	4.2
$8 \rightarrow 7$				
SO_2	72.7583	IRC + 10216	0.4	1.8
$6_{06} \rightarrow 5_{15}$				

A fuller account of this work will be published in the *Publications of the Astronomical Society of Japan.*

Reference

Jefferts, K. B., Penzias, A. A., Wilson, R. W., and Solomon, P. M.: 1971, *Astrophys. J. Letters* **168**, L111.

Kenji Akabane
Masaki Morimoto
Kiyoshi Nagane
Tokyo Astronomical Observatory,
Mitaka, Tokyo, Japan

Norio Kaifu
Department of Astronomy,
University of Tokyo,
Tokyo, Japan

THE GALACTIC CENTRE

J. H. OORT

Sterrewacht, Leiden, The Netherlands

Abstract. The phenomena displayed by the interstellar medium in the galactic centre are considered. The asymmetries shown by the features between 1 and 3 kpc from the centre together with the presence of material lying out of the galactic plane favour the expulsion hypothesis for their origin. The nuclear disk shows a perturbation which might have resulted from such expulsion. The dense molecular clouds in the disk may well be considered as the most direct evidence that matter is expelled from the nucleus and that this occurs at a high rate. The $+50$ km s^{-1} feature in the direction of Sgr A may be the most recently expelled body of molecular gas. New observations of the central radio source, Sgr A, have revealed details on a very small scale, and the infrared core also shows a complicated structure. Probably a number of individual concentrations of gas and dust are present. While the position of the actual nucleus seems now to have been defined to within a few arcseconds, no indication has yet been found concerning its nature nor concerning the mechanism that enables it to expel the vast expanding masses of gas observed in the central region.

The phenomena displayed by the interstellar medium in the central region of the Galaxy can be roughly divided in the following groups, arranged in order of decreasing distance from the centre.

§	Phenomenon	Gaseous mass ($M_\odot$)	R (pc)	Radial motion (km s^{-1})	Time scale (yr)
I	3-kpc arm	10^7			
	Expanding arm at $+135$ km s^{-1}	10^7	1000–3000	100	$(5-15) \times 10^6$
	High-velocity features outside galactic plane	10^5			
II	Nuclear disk	4×10^6	800	–	–
III	Expanding molecular clouds	10^6	100–700	200	$(1-2) \times 10^6$
IV	Radio sources		0–700		
V	$+50$ km s^{-1} absorption feature	10^5	5–10	50	10^5
VI	Infrared nuclear disk and radio fine structure	10^4	0.5	?	10^3

These data should be considered in conjunction with (1) those on the *stellar* mass density as found from the rotation of the nuclear disk and from the *near*-infrared radiation, which give roughly 10^9 $M_\odot$ within $R=100$ pc, 10^8 $M_\odot$ within 10 pc and 10^7 $M_\odot$ within 1 pc, and (2) the distribution of *far*-infrared radiation (around 100 μm), which shows a central concentration of $38 \times 15'$, or 110×45 pc, diameter (cf. Borgman, 1974).

I. Features Between 1 and 3 kpc from the Centre

The 3 kpc arm is the best-defined feature. It can be followed from l about 338° to 5° where it merges with low-velocity gas. The part we observe lies in front of the centre. At $l=0°$ its radial velocity is -53 km s^{-1}. In a sector of about 90° galactocentric longitude it has an HI mass of 2×10^7 $M_\odot$.

F. J. Kerr and S. C. Simonson, III (eds.), Galactic Radio Astronomy, 539–547. All Rights Reserved.

The $+135$ km s^{-1} expanding arm lies beyond the centre. It is seen from $l \sim 355°$, where it seems to break off rather suddenly (which suggests that it may at this point be situated within 1 kpc from the centre), to either 12° (according to Simonson and Mader [1973]) or 22° (according to Rougoor's [1964] early study). Its mass is of the same order as that of the 3 kpc arm. Figure 1, taken from a sketch prepared by Rougoor (1964). shows a possible situation of the two arms.

The most important problem in connection with these two features is whether they are caused by gas expelled from the galactic nucleus or whether they are parts of general dynamical phenomena in the Galaxy.

If they are due to explosions from the nucleus very large masses must have been involved. According to the computations by van der Kruit (1971) a mass of between 5 and $10 \times 10^6\ M_\odot$ would have had to be thrown out about 12×10^6 yr ago at velocities around 600 km s^{-1} in order to explain the outward motion of the 3 kpc arm. In his model the expulsion took place in two opposite directions, at angles of 25°–30° with the galactic plane. Because of this inclination it might not have completely destroyed

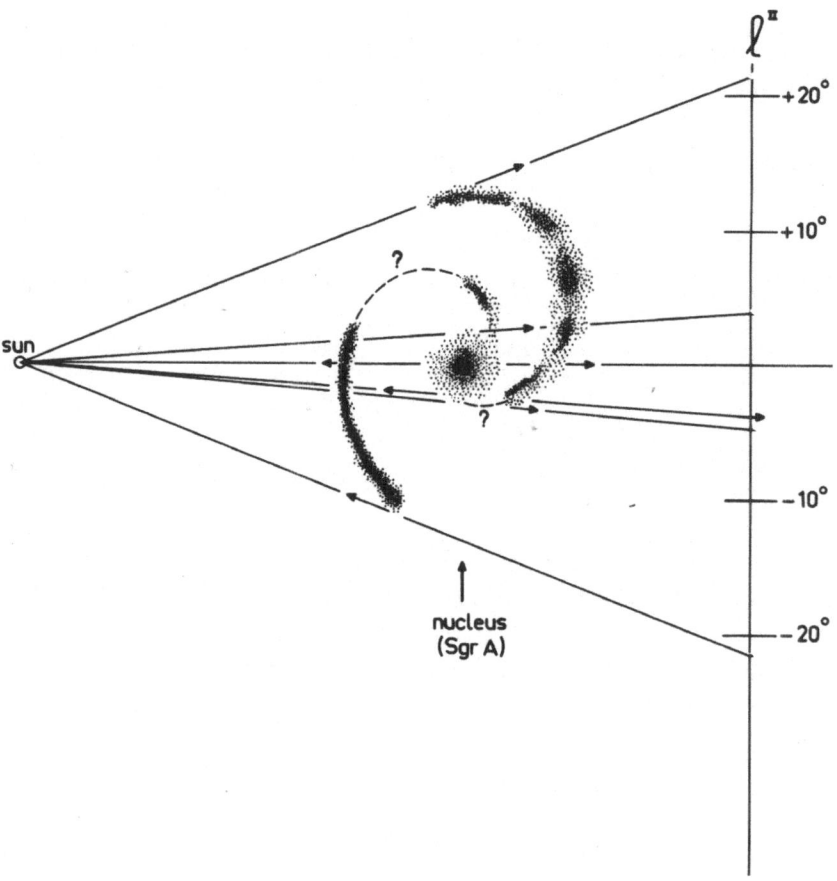

Fig. 1. Schematic sketch of the 3 kpc arm, the expanding arm at $+135$ km s^{-1} and the nuclear disk (from Rougoor, 1964).

the nuclear disk. Sanders and Prendergast (1974) have investigated a model with isotropic expulsion. They suggest that the 3 kpc arm could have gone through an entire epicycle; the expulsion would then have occurred longer ago, and sufficient time might have elapsed to form a new nuclear disk after the earlier disk had been blown away. This model, however, appears to me rather artificial.

Some evidence which appears to support an explosion model, and especially a non-isotropic one, is provided by the existence of large gas complexes in the central region *outside* the galactic plane. These features, originally found by W. W. Shane, were investigated in detail by van der Kruit (1970), and later also by Sanders *et al.* (1972). The observations by the latter authors show a concentration of hydrogen between $-2°$ and $-6°$ latitude and $356°$ and $8°$ longitude which at $l=0°$ has a velocity of about -135 km s^{-1}. The total mass is roughly $10^5 \, M_\odot$. This is often referred to as van der Kruit's Feature XII. Sanders, Wrixon and Penzias have suggested that it might form part of a 'ring', but the evidence for this is not at all convincing. They have also found some HI features closer to the centre and making large angles with the galactic plane (distances up to a few hundred parsecs, velocities around 200 km s^{-1} and masses of a few times $10^4 \, M_\odot$). These various phenomena suggest that gas is being expelled in some favoured directions making considerable angles with the plane. Recent observations at Jodrell Bank which show more detail have fully confirmed this.

EXPULSION OR DISPERSION ORBITS?

The expulsion hypothesis implies that gas masses of the order of $10^7 \, M_\odot$ have been thrown out by the galactic nucleus within the last 10 or 15×10^6 yr. Nothing is yet known about the mechanism that could provide the required kinetic energy nor about the origin of this vast quantity of interstellar gas.

It is therefore very important to explore exhaustively other possible explanations of the observed large radial motions.

The possibility that at least some of the expanding features might be parts of a dispersion ring has been studied by Shane (1972), and, in greater detail, by Simonson and Mader (1973). Simonson has reported on this investigation. He concluded that the observations can indeed be interpreted in this manner, and that such an interpretation might be preferred because the alternative requires an ad hoc hypothesis on the nature of the nucleus, while the dispersion orbit can in a plausible way be identified with the inner Lindblad resonance.

There are, however, a number of difficulties in this interpretation. The most serious one is the absence of a proper counterpart of the 3 kpc arm on the other side of the centre. The so-called $+70$ km s^{-1} arm, which has been invoked as a possible candidate, is a weak and uncertain feature, in no way comparable to the 3 kpc arm. It has been suggested that this lack of symmetry might be caused by the action of spiral arms; but then the dispersion ring model loses its simplicity and much of its attractiveness. There is, in the second place, the $+135$ km s^{-1} expanding arm to account for. This lies behind the centre, but can hardly form part of a dispersion ring containing the 3 kpc arm.

Simonson and Mader suggest that it might belong to a second, inner, dispersion ring. However, there is no indication that makes the existence of a second dispersion ring plausible. Moreover, this would again have to be strongly asymmetrical.

Asymmetries such as shown by these two major features seem, on the other hand, to be common in expulsion phenomena. Extreme cases of this are shown by NGC 1275 and M87.

A third difficulty is presented by the phenomena outside the galactic plane, such as van der Kruit's Feature XII. In the dispersion ring model this can only be explained in a rather artificial way.

Finally, it should be remembered that, though we do not understand how matter is expelled from the galactic nucleus, we must in any case accept the expulsion of *some* features, such as the Feature E found by Sanders, Wrixon and Penzias, and the molecular clouds near the centre. The latter contain probably at least a million solar masses. That expulsion *can* take place on a very large scale is moreover demonstrated convincingly by the phenomena in such galaxies as NGC 1275, NGC 4258 and M82.

II. The Nuclear Disk

I now want to discuss briefly the problems around the fast-rotating nuclear disk of about 4×10^6 $M_\odot$ introduced by Rougoor and Oort (1960). I shall not repeat the, in my opinion, very strong arguments that we are indeed observing a rotating disk, and that the velocities observed on the negative-longitude side probably correspond closely to circular velocities in the gravitational field of the Galaxy. There is, however, the problem how the disk could have survived such vast explosions from the nucleus as would be required to explain the radial motions of the expanding arms. A plausible way out of this difficulty is to suppose that the expulsion took place in a strongly anisotropic manner, and at considerable angles with the galactic plane (such as, for instance, in van der Kruit's model), the expelled gas falling back into the general galactic disk at a few kpc from the centre, where it transferred its radial momentum to the disk gas. The inner part of the nuclear disk would, however, still have been seriously perturbed. There are, indeed, clear signs of such a perturbation within about 2° on the positive longitude side, as indicated by Sanders and Wrixon (1972). Due to its rapid and strongly differential rotation the disk might largely recover from disturbances in a period of the order of 10^7 yr.

The distribution of the H I density in the disk has recently been studied by Sanders and Wrixon (1973). It varies from ~ 0.2 cm^{-3} at $R = 600$ pc to ~ 5 cm^{-3} around $R = 100$ pc.

III. The Molecular Clouds

These have been fully described by Robinson (1974). Most, if not all, are moving away from the centre and have therefore probably been expelled from the nucleus. They appear to be confined to a very thin disk, considerably thinner than the H I disk. This is remarkable, especially in view of the fact that much of the H I expulsion seems to occur at large angles with the galactic plane. It may be that the dense molecular clouds originate in some way through an interaction between outgoing streams and the disk.

Dense clouds of molecules are observed from about $358°5$ to nearly $+4°$ longitude. Because no dense clouds have been found below $l = 358°7$ it is probable that a large fraction lies within 250 pc from the centre. At $l = 0°$ the clouds in front of Sgr A are moving towards us at a velocity of 140 km s^{-1}, and should thus have been expelled during the past one or two million years. The observations of these molecular clouds may well be considered as the most direct evidence that matter is expelled from the galactic nucleus, and that this occurs at a high rate, for the total mass of these clouds must be of the order of 10^6 $M_\odot$. The negative-velocity clouds between $l = 358°8$ and $l = +0°8$ appear to form a more or less continuous structure which looks like a sector of an expanding ring, such as proposed by Scoville (1972) (cf. Figure 8 in the communication by Robinson). The evidence for a complete ring is, however, unconvincing. The molecular clouds behind the centre are mostly moving at a much lower velocity and are considerably more patchy. In addition, there are large concentrations of dense clouds with positive velocities extending to almost $l = 4°$ which certainly do not fit in with a single ring model. One gets rather the impression that there have been several periods of intensive expulsion, favouring different directions in the disk.

Most of the molecular clouds appear to lie embedded in the disk, However, they have not been stopped in their outward motion by the disk gas, nor carried along by its rotation. In fact, their present column densities are probably so high that they can move practically unimpeded through the disk. Many molecular clouds in the central region show molecular emission for which densities of at least 10^3, in some cases even up to 10^5, H$_2$ molecules per cubic centimetre are required; with diameters of several parsecs their column densities must be of the order of 10^{22} cm^{-2} or more.

IV. Radio Sources

An interesting phenomenon is the unexplained asymmetry in distribution: The sources north of the centre are mainly thermal, while those south of the centre are nonthermal.

The discrete sources are embedded in a general concentration of thermal as well as nonthermal radiation with a half-intensity diameter of about 300 pc.

The central source, Sgr A, has a very complicated structure, interesting details of which have recently come to light (cf. Figures 3 and 4 of the contribution by Gordon). These data have a very special importance, because they reveal details of nuclear structure which are two orders of magnitude smaller than what can be observed in any other galaxy except M31. The main structure is elongated in an approximately north–south direction, and has variously been called 'component A', or 'Sgr A West'. It contains the centre of the near-infrared concentration of stars as well as the centre of the farther-infrared disk discussed in Section VI. High-resolution observations at 4 cm show pronounced fine structure in the region enclosed by this infrared disk, to which I shall return below. Roughly $1''7$ east of the centre lies a nonthermal component, which has been called Sgr A East, and seems to be the central concentration of a broader distribution. The importance of this component is well brought out in a recent observation with the Fleurs interferometer at 1415 MHz.

V. The $+50$ km s^{-1} Absorption Feature

An intriguing phenomenon connected with Sgr A is the very broad absorption, observed particularly in OH and H_2CO, which is centred at a velocity of roughly $+50$ km s^{-1}. At first sight it looks as though this shows gas flowing into the galactic nucleus. Recent observations with the CalTech interferometer by Whiteoak, Rogstad, and Lockhart, for information on which I am indebted to Whiteoak, show, however, that the formaldehyde absorption at $+50$ km s^{-1} is not at all seen against component A, but mainly against the nonthermal source 1.7 east of the centre. This makes it possible to interpret the $+50$ km s^{-1} band as caused by gas ejected from the galactic nucleus and seen in absorption against Sgr A east, which may well be centred some 5 pc beyond the nucleus (see Figure 2). The proposed configuration receives

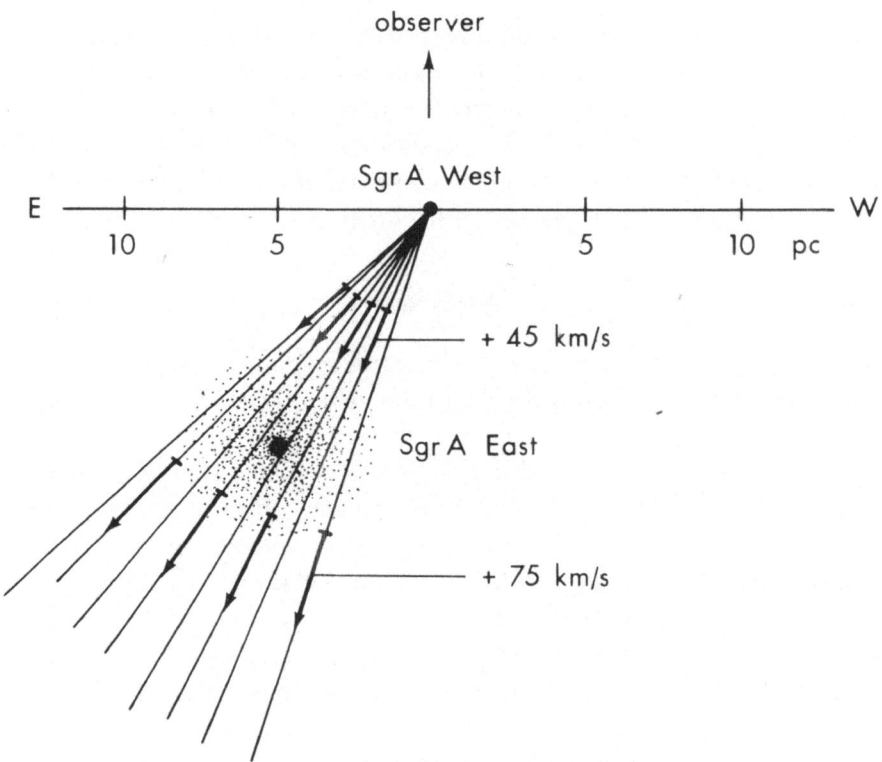

Fig. 2.   Possible situation of the gas showing CO emission between $+30$ and $+90$ km s^{-1}, and seen in absorption in the OH and H_2CO absorption lines between $+30$ and $+60$ km s^{-1}.

support from the observations of CO emission around Sgr A. This emission, which is concentrated within 3′ or 4′ (10 pc) from Sgr A, extends over a velocity interval from $+30$ to $+90$ km s^{-1}. The formaldehyde absorption, however, takes place only between $+30$ and $+60$ km s^{-1}. The higher-velocity gas therefore lies beyond the source against which the formaldehyde is seen in absorption. As shown in the diagrams by Solomon *et al.* (1972), this latter gas is centred some 2′ east of the lower-velocity material. The distribution in space and velocity may therefore be like that drawn in the sketch (Figure 2).

If this interpretation is correct the $+50$ km s^{-1} feature shows that expulsion of molecular gas has occurred quite recently and may well be continuing at present.

VI. The Infrared Nuclear Disk

Most interesting information on structure near the galactic nucleus has come from the infrared observations. After Becklin and Neugebauer (1968) and others had obtained rather convincing evidence from 2.2 μm observations for a concentration of *stars* at the centre of the Galaxy, and had obtained what appear to be reliable data on the density distribution and total star density from about 0.2 to 50 pc from the centre (cf. Figure 1 of Borgman's review), a remarkable new discovery was made by Rieke and Low (1973). They found that radiation between 4 and 20 μm is concentrated within a circle of about 16″, or 0.8 pc, diameter, which I shall denote by the term 'infrared disk,' or 'infrared core' (Figure 3). Within this disk at least five separate concentrations can be seen. One of these, roughly at the centre, coincides closely with the 'stellar' centre as observed at 2.2 μm and may well be identical with this stellar core. Of the other concentrations one may be an infrared star; the nature of the others is unknown. The close resemblance with the fine structure recently found in the same region at radio wavelengths indicates that they represent individual concentrations rather than fluctuations in the overlying absorption. If they are extended objects their free paths would probably be about equal to the dimension of the disk. The corresponding life times would then be of the order of 10^3 yr.

Radiation and absorption by solid particles must play an important role in the nuclear region. The bulk of the observed radiation is emitted between 75 and 125 μm. Hoffmann *et al.* (1971) have shown that this comes from a source of 38′ × 15′ half-width. Borgman states that it has a total luminosity of $7 \times 10^7 L_\odot$. The radiation can easily be provided by the central concentration of stars.

VII. The Actual Nucleus

The observations discussed in this session on the galactic centre indicate strongly that the true galactic nucleus lies within the 16″ infrared nuclear disk, probably close to its centre, and coinciding with the radio fine structure observed within Sgr A West. Its position thus seems to be defined to within a few arcseconds. This is an important advance.

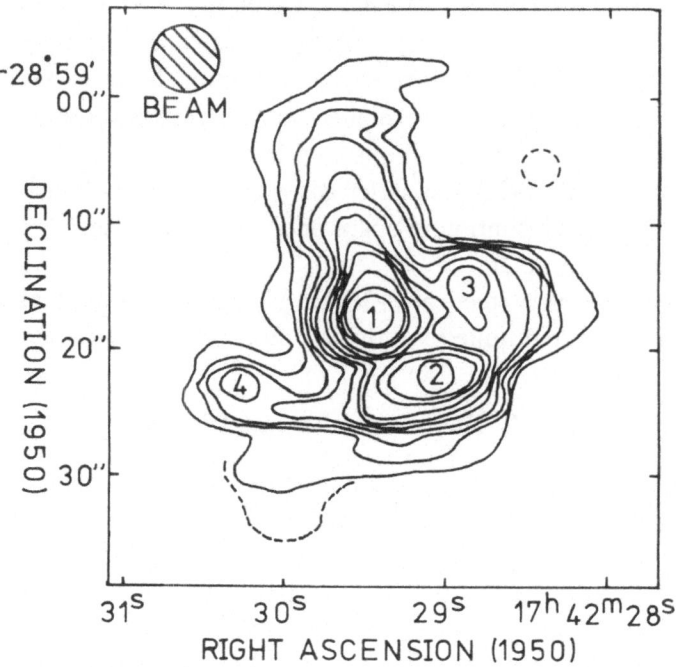

Fig. 3.   Distribution of 10.5 μm radiation in the infrared core of the Galaxy (Rieke and Low, 1973).

On the other hand no indication has yet been found concerning its nature, nor concerning the mechanism which enables it to expel the vast expanding masses of gas that are observed in the central region.

As regards the mass of this unknown nucleus the only information we have at present comes from the rotation of the nuclear H I disk. The smallest distance from the centre where its rotation could so far be observed is about 80 pc. From these observations Sanders and Wrixon (1973) have derived an upper limit of 10^8 $M_\odot$ for a point mass in the nucleus. The only way conceivable at present in which one might get dynamical information from regions much closer to the centre is through infrared observations of the distribution and velocities of planetary nebulae, but this appears to be still impracticable with existing instruments.

References

Becklin, E. E. and Neugebauer, G.: 1968, *Astrophys. J.* **151**, 145.
Borgman, J.: 1974, this volume, p. 465.
Hoffmann, W. F., Frederick, C. L., and Emery, R. J.: 1971, *Astrophys. J. Letters* **164**, L23.
Kruit, P. C. van der: 1970, *Astron. Astrophys.* **4**, 262.
Kruit, P. C. van der: 1971, *Astron. Astrophys.* **13**, 405.
Rieke, G. H. and Low, F. J.: 1973, *Astrophys. J.* **184**, 415.
Robinson, B. J.: 1974, this volume, p. 521.
Rougoor, G. W.: 1964, *Bull. Astron. Inst. Neth.* **17**, 381.
Rougoor, G. W. and Oort, J. H.: 1960, *Proc. Nat. Acad. Sci.* **46**, 1.

Sanders, R. H. and Prendergast, K. H.: 1974, *Astrophys. J.* **188**, 489.

Sanders, R. H. and Wrixon, G. T.: 1972, *Astron. Astrophys.* **18**, 467.

Sanders, R. H. and Wrixon, G. T.: 1973, *Astron. Astrophys.* **26**, 365.

Sanders, R. H., Wrixon, G. T., and Penzias, A. A.: 1972, *Astron. Astrophys.* **16**, 322.

Scoville, N. Z.: 1972, *Astrophys. J. Letters* **175**, L127.

Shane, W. W.: 1972, *Astron. Astrophys.* **16**, 118.

Simonson, S. C. and Mader, G. L.: 1973, *Astron. Astrophys.* **27**, 337.

Solomon, P. M., Scoville, N. Z., Jefferts, K. B., Penzias, A. A., and Wilson, R. W.: 1972, *Astrophys. J.* **178**, 125.

J. H. Oort
Sterrewacht,
Leiden,
The Netherlands

DISCUSSION

Robinson: The molecular cannonballs with $n_{H_2} \approx 10^3$ cm^{-3} can clearly travel with ease through the H I nuclear disk with $n_H \leqslant 10$ cm^{-3}. But it is strange that the molecular clouds have such low rotation velocities (≈ 50 km s^{-1}) about the nucleus in the gravitational field that makes the H I disk rotate at 250 km s^{-1}. The molecular clouds have insufficient velocity to escape from the nucleus and insufficient angular momentum to maintain their present radii. They must soon spiral back towards the nucleus.

In Oort's summary he has shown a total mass of only 10^6 $M_\odot$ for the molecular clouds near the galactic center. However, Kaifu has computed 10^8 to 10^9 $M_\odot$ for the ring from H_2CO/CO observations. Oort's mass of 10^6 $M_\odot$ looks much too small – just the molecular cloud associated with Sgr B2 is believed to have a mass of at least 5×10^5 $M_\odot$.

Palmer: While the estimates of densities and consequently masses of molecular clouds are the best we can make at the present time, Oort is certainly correct in being conservative in his mass estimates. Effects such as radiative transfer in the cloud and possible unresolved structure have usually not been properly taken into account to date. When these are worked out in detail, the mass estimates may be reduced.

Menon: If such molecular clouds exist so close to the nuclei of galaxies, should we not expect to find early type stars in the nuclei of galaxies?

Mezger: Observations of H II regions suggest the highest rate of O star formation encountered in our Galaxy is within 100 pc from the center.

Oort: As the time scale for the molecular clouds near the center is only of the order of 10^6 yr, stars may not have had time to form. But Mezger is, of course, right in pointing out that the H II regions like that found in Sgr B2 show that OB stars must have formed. This would be possible even in the short time indicated above if the gas density is quite high.

Menon: I would just like to ask, do we *see* OB stars in the nuclei of other galaxies?

Robinson: In this regard also, the nucleus of NGC 253 is similar to that of our Galaxy. The nucleus of NGC 253 is not like that of M31 but consists of a number of giant H II regions.

Oort: And there are other galaxies like this as well.

Swarup: In your model of expulsion of the $+50$ km s^{-1} cloud from the galactic nucleus, would there not exist serious difficulties of formation of molecules at the nucleus?

Oort: One does not know the mechanism by which the clouds are expelled. It might either be such that molecules would not be destroyed or else the densities might be so high that they would have been formed in a very short time after the expulsion.

Simonson: Leaving aside the nucleus itself, the largest source of kinetic energy in the Galaxy is rotation. We have merely tried to tap a small fraction of this energy to trap the gas into resonant orbits.

PART 6

LARGE-SCALE GALACTIC STRUCTURE

"The Magellanic Stream is coming down on the galactic plane, and it's going to get its own back on the Galaxy, you see, and give it a great boot."
D. S. Mathewson, in the discussion following the paper co-authored by M. N. Cleary and J. D. Murray

CURRENT PROBLEMS IN 21 cm LINE STUDIES OF THE LARGE-SCALE STRUCTURE OF THE GALAXY

W. B. BURTON

National Radio Astronomy Observatory* Charlottesville, Va., U.S.A.

Abstract. A number of current problems in 21 cm line studies of the Galaxy as a whole are discussed. Because of the difficulties involved with straightforward mapping, it is important to isolate integrated and other properties of the hydrogen profiles, the interpretation of which does not require accurate distance determinations. In addition, methods of analysis are necessary which either account for or exploit the sensitivity of hydrogen profiles to velocity irregularities and to geometrical configurations. The model-fitting approach to the interpretation of the hydrogen profiles is useful in this respect. Extragalactic hydrogen studies which show the relative ordering of the various components of spiral structure can inspire research in our own Galaxy. Such investigations are necessary for an understanding of the forces governing the spiral structure. It seems that the neutral hydrogen is primarily a tracer of locations where the overall distribution of stars is producing a gravitational sink. Other spiral tracers, in particular the molecules, are better considered as tracers of regions where the gas has been compressed, perhaps (at least on a large scale) by the shock front predicted by the density-wave theory.

I. Introduction

My charge is to report on 'the present status of 21 cm work on the spiral structure of the whole Galaxy'. Excluded from this report (but included in others in these proceedings) are the high-velocity clouds, the galactic center region, and the spur and loop features. It seems reasonable to restrict the report to phenomena showing some continuity over length scales typically 1 kpc or larger, at latitudes less than about 10° from the galactic plane. Insofar as my remarks pertain to continuing investigations, they will be rather general.

The emphasis in 21 cm line work on the large-scale galactic structure has changed somewhat since the previous IAU symposium on galactic radio astronomy held in Noordwijk (van Woerden, 1967). Relevant areas in which research activity has increased include investigations of the kinematic characteristics of the neutral hydrogen and the relationship of these characteristics to theoretical predictions, the motion and spatial distribution of hydrogen relative to other constituents of the galactic disk, and integrated and other properties of the hydrogen profiles which do not require very accurate distance determinations for their satisfactory interpretation. On the other hand, there have been fewer attempts at detailed spatial mapping of regions near the galactic plane.

II. Kinematic Aspects of the Neutral Hydrogen Distribution

Since the first observations of the 21 cm line in 1951, what was sought was the density and velocity distribution of the hydrogen. It has been clear that these problems are

---

* Operated by Associated Universities, Inc., under contract with the National Science Foundation.

F. J. Kerr and S. C. Simonson, III (eds.), Galactic Radio Astronomy, 551–572. All Rights Reserved.
Copyright © 1974 by the IAU.

intimately related, since distances are calculated from a measured velocity and thus require previous knowledge of the velocity field throughout the Galaxy. For the derivation of the well-known Leiden-Sydney map (see Oort *et al.*, 1958), which displayed contours of the volume-density of the hydrogen in spatial coordinates, it was reasonable to base the density derivation on the working hypothesis of circular symmetry. Even though it was clear that this hypothesis was not exactly true, it was not clear whether the deviations from circular symmetry involved mainly the density distribution or mainly the velocity one. If the latter case, it was thought that the deviations from circular rotation, which over most of the galactic disk are typically only a few per cent of the circular motions of differential rotation, would result in errors in the distances of this same order.

One way to demonstrate the lack of circular symmetry is to divide the Galaxy by a line through the Sun and the galactic center. If the density and velocity fields of the Galaxy were circular, then the borders of the 21 cm profiles taken at corresponding longitudes on either side of this line would occur at oppositely signed, but otherwise equal, velocities. Such a comparison is made in Figure 1. Dots in the figure

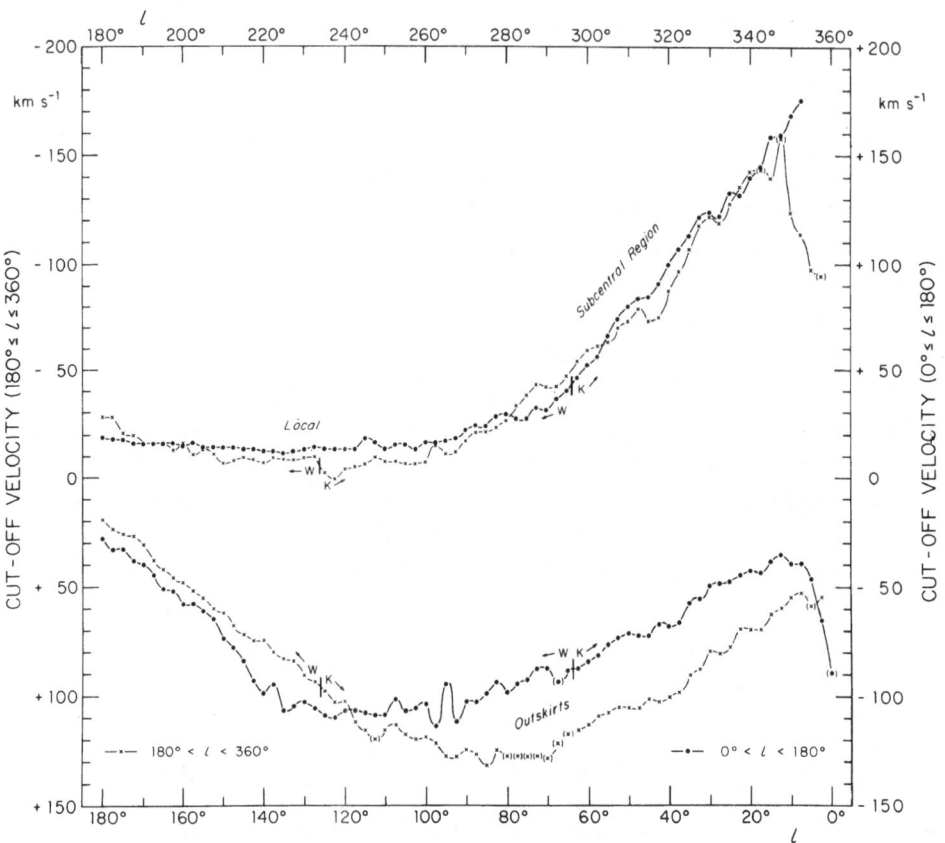

Fig. 1.   Non-circularity of the galactic hydrogen distribution as indicated by systematic differences in the cut-off velocities measured from the wings of profiles observed (by Westerhout, 1966; and by Kerr and Hindman, 1970) on either side of the Sun-center line (Burton, 1973).

represent the velocity cut-offs of profiles observed in the longitude range 0° to 180° and crosses represent the same thing taken from profiles observed on the other side of the Sun-center line, from 180° to 360°. The fact that the crosses and dots are non-coincident proves the existence of deviations from circular symmetry.

At least some of the deviations from circular symmetry require explanation in velocity terms. The cut-off velocities of profiles representing hydrogen near the sub-central ('tangential') points show irregularities which were originally (Kwee *et al.*, 1954) attributed to subcentral-point regions that did not contain enough gas to determine the profile cut-off. However, Shane and Bieger-Smith (1966) showed that this explanation in terms of the density distribution requires that there be essentially no hydrogen (about a factor 100 less than the average) along regions of 4 or 5 kpc in extent, and that these regions are furthermore preferentially oriented with respect to the observer. This is unacceptably artificial, and in addition is not consistent with other characteristics of the emission profiles. Consequently the irregularities in the subcentral point cut-off velocities, (which are the same irregularities which show up as bumps on the rotation curve), can better be attributed to deviations from circular motion with length scales of about 1 kpc and amplitudes of the order of 5 km s^{-1}. Amplitudes of 5 km s^{-1} amount to less than 3% of the circular motion over most of the Galaxy.

The opinion has been expressed that, because of these deviations from the circular rotation curve adopted as a working model, the resulting maps should be considered drawn on a rubber sheet which might be stretched by amounts corresponding to distance errors caused by the velocity irregularities. This opinion seems now to be somewhat too optimistic. It appears that the sort of velocity irregularities known to exist throughout the Galaxy probably determine the appearance of the line profiles in a way that dominates over density manifestations (Burton, 1971, 1972; Tuve and Lundsager, 1972, 1973). This sensitivity is demonstrated by Figure 2. In the upper part of the figure is the profile observed at $l = 90°$ in the galactic plane. In the lower part of the figure the velocity (with respect to the local standard of rest) which would be expected at this longitude in the case of circular Schmidt (1965) rotation, is plotted as the full-drawn line. Deviations from such circular rotation would result in perturbations to this smooth curve of the sort illustrated by the dashed line. These perturbations imply that at some velocities the velocity will vary slowly with distance relative to the smooth-curve situation; there will be more hydrogen contributing to the profile at such a velocity than at a velocity where the variation of velocity with distance is rapid. The sensitivity of line profiles to such velocity irregularities is suggested by a model profile calculated for the case of a thoroughly uniform density distribution but in the presence of the perturbed velocity-field. The resulting profile (shown by the dashed line in the figure) shows the same sort of structure as the observed profile, including the positive-velocity peak, which would be impossible to account for in terms of Schmidt-type rotation. It would require density contrasts of 50 or 100 to 1 to account for intensity differences equivalent to those obtained by systematic streaming motions of only a few km s^{-1}.

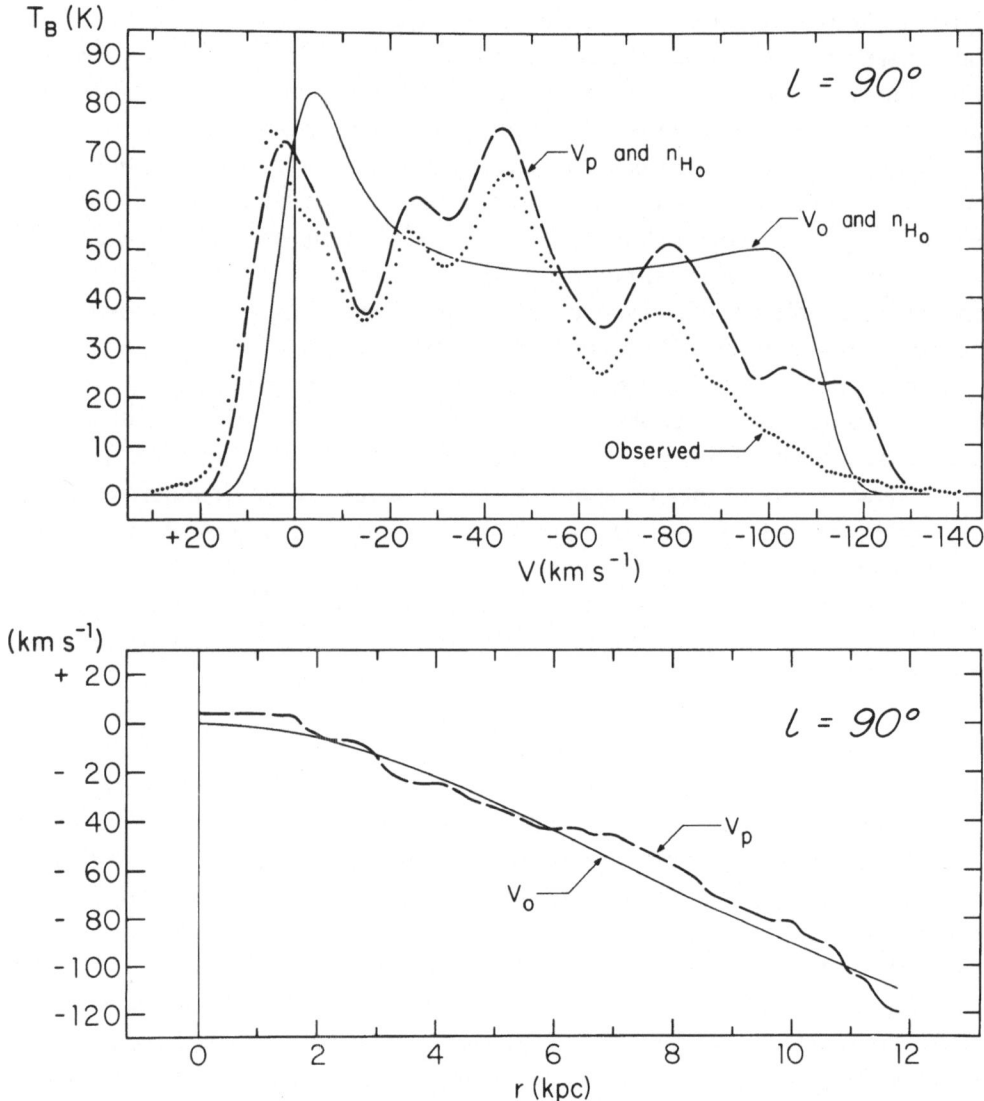

Fig. 2. Diagram illustrating both the sensitivity of hydrogen line profiles to velocity irregularities and
the profile-fitting method used in the derivation of Figure 4. The profile observed at $l = 90°$, $b = 0°$, is
indicated, as are model profiles calculated for the case of uniform density with a circular-rotation velocity-
distance relationship, V_0, and with a perturbed relationship, V_p.

Although it *is* clear on general grounds that irregularities in the velocity, density
and temperature will accompany one another, what is *not* clear is the extent to which
a particular peak in a hydrogen observation owes its characteristics to streaming
motions, to a density concentration, or to a variation in temperature. This means
that hydrogen density variations near the galactic plane cannot be determined with
any accuracy directly from observations; it is also generally not possible to determine
the true velocity dispersion from observations near the plane.

III. Model-Fitting Approach to Investigating the Galactic Distribution of Hydrogen

The opinion has been expressed, too, that because of these problems, the realistic end product of large-scale hydrogen studies is a complete velocity-longitude diagram, rather than a spatial map. This seems rather pessimistic. In my opinion the problems are primarily procedural. On the one hand, new methods of galactic structure analysis are needed, whereas at the same time efforts should be made to isolate problems which can be investigated without requiring very accurate distances or density determinations.

As far as the methods of analysis are concerned, it is worth mentioning one approach to the derivation of the large-scale galactic structure which has recently been applied to a variety of problems involving hydrogen observations. The model-fitting approach involves selecting a model's physical parameters according to the appearance of simulated observations. In the case of the hydrogen, it is required that the adopted velocity, density and temperature distribution reproduce the appearance of the observed profiles. Figure 3 illustrates the model-making approach for the interior portion of the Galaxy between longitude 40° and 90°. In the upper left of the figure is the observed velocity-longitude map of temperature contours. In the upper right is a contour map composed from model profiles; the model's parameters were chosen by comparing the observed with the model contour maps. The map in the lower part of the figure indicates the spatial distribution of the chosen parameters; in this case the regions of above-average density and the sense of the streaming motions are indicated. The classical approach involves a direct transformation from the observed velocity space to the spatial distribution (upper-left observations to lower map). The model-fitting approach requires agreement between the two upper maps, which are both in the observed space. In this way the effects inherent in the complicated geometry of the transformation from velocity to distance are accounted for. Once agreement between the two upper maps is established, the model's parameters can be further judged in other ways.

The model-fitting approach is also useful in investigating the compatibility of various theoretical predictions with the observations.

IV. Predictions of the Density-Wave Theory Relevant to the Overall Distribution of the Hydrogen

The density-wave theory provides for the maintenance of structures in the disk, against the shearing forces of the rapid differential rotation (e.g., Lin *et al.*, 1969; Lin, 1971). Stars and gas moving in approximately circular orbits experience the assumed spiral component of the gravitational potential in such a way that they tend to pile up in troughs of low potential. For the circumstances pertaining to our Galaxy, the gas motions involved with this piling-up are predicted to have amplitudes of about 6 km s^{-1}, with components in both the azimuthal and radial directions. The azi-

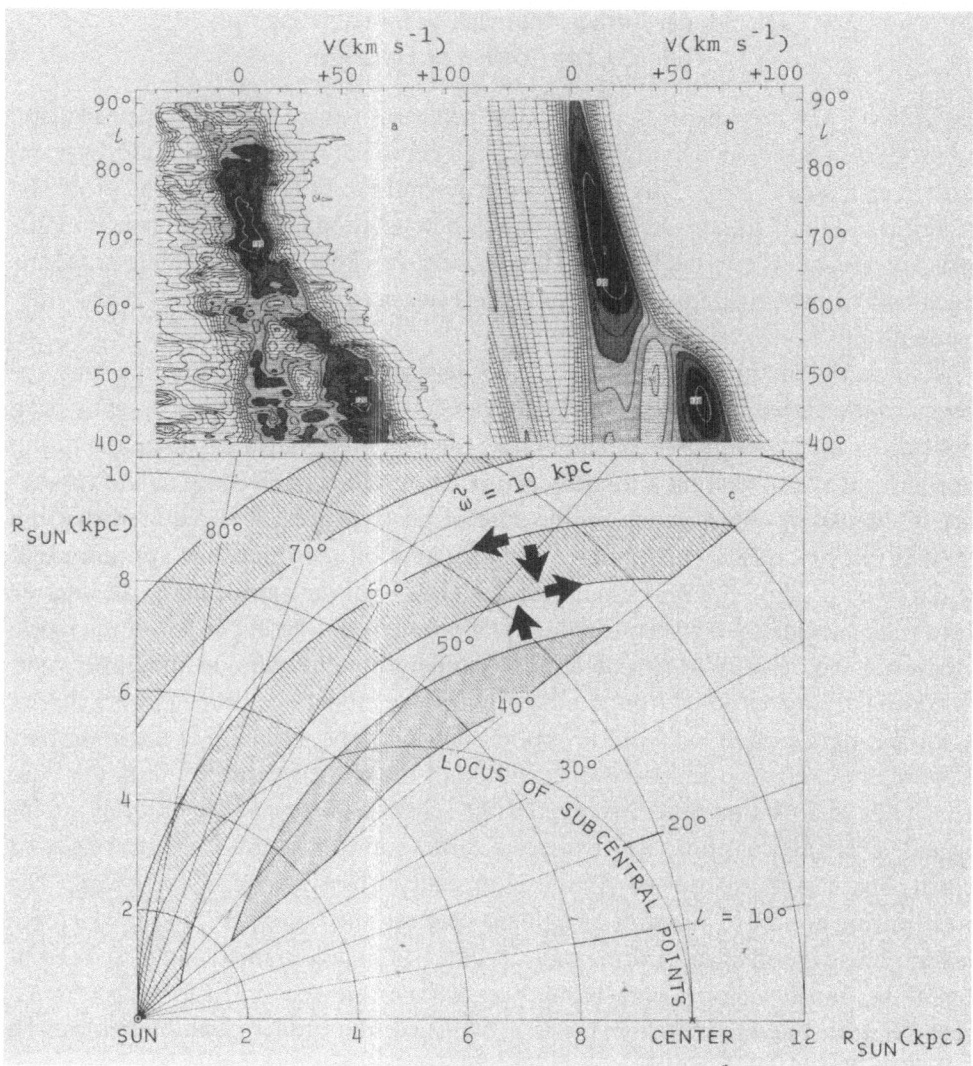

Fig. 3.   Diagram illustrating the model-making approach for a portion of the galactic plane (Burton, 1971). The classical approach involves a direct transformation from the observations to the spatial map. The model-fitting approach involves choosing the spatial distribution of the various parameters by examining simulated observations.

muthal streaming component gives an entirely satisfactory explanation of the bumps observed in the galactic rotation curve (Barbanis and Woltjer, 1967; Yuan, 1969; Burton and Shane, 1970; Burton, 1971). Evidence for the predicted radial component in the gas is less direct, but it does seem that a systematic flow pattern is necessary to produce the arm-interarm contrasts of the sort observed in the profiles (Burton, 1971).

Since the velocity field is the physical parameter to which the hydrogen profiles are most sensitive, and since knowledge of the velocity-field is crucial for a complete observational confrontation of any theory of spiral structure, it seems reasonable to take the consequences of the premise that the velocity dominates the hydrogen profiles' appearance. By assuming for the sake of argument that *all* profile structure has a kinematic origin, a model profile may be fit to each observed profile by per-

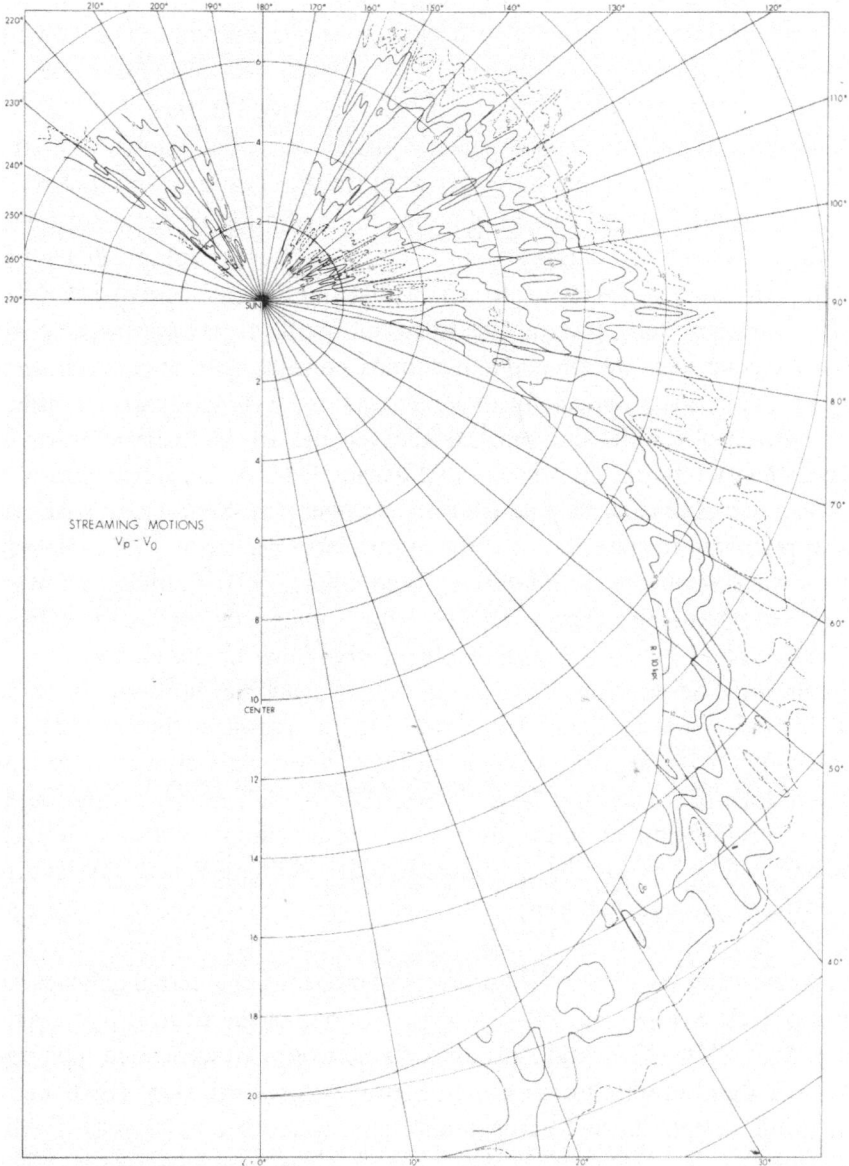

Fig. 4.   Line-of-sight streaming motions required, by themselves, to account for the structure of profiles observed in the galactic plane (Burton, 1973). Full-drawn contours indicate that the sense of the peculiar motion is away from the observer; the contour interval is 5 km s^{-1}.

turbing the line-of-sight velocity from the basic circular velocity situation, maintaining a completely uniform density distribution (Burton, 1972). The method is illustrated in Figure 2. The resulting line-of-sight streaming motions, which are the differences between the perturbed and basic velocity-fields, are shown in Figure 4 plotted in spatial coordinates. Along a particular line-of-sight the peculiar motions vary approximately sinusoidally. There is also a trend in the contours to be elongated along what one might (perhaps through analogy with other galaxies) allow to be called 'spiral arms'. The amplitude and distribution of the motions are consistent with those predicted by the density-wave theory. This approach also provides a velocity field which can be compared with observations of other spiral tracers.

Model-fitting applied to typical galactic plane regions has shown that streaming motions of about 5 km s^{-1} are necessary to fit the observations, and that reasonable fits to most profile features can be obtained if, in addition to these motions, an effective velocity dispersion of about 4 km s^{-1} and an average volume density of emitting hydrogen atoms of 0.4 cm^{-3} (for gas with a spin temperature of 135 K) also are adopted. A uniform density cannot be ruled out by the gross appearance of the profiles, but it can be ruled out on general grounds, considering that coherent structures in velocity and in density must, for dynamic reasons, accompany one another. However, discussion of the density contrast between the arm and the interarm regions requires a theory relating the velocity and density fields. A density contrast between 3 and 7 to 1 seems reasonable and is also consistent with theoretical predictions.

It is appropriate to mention here the controversy excited by the hydromagnetic theory for spiral structure put forward by Piddington (1973). Piddington's alternative to the gravitational-wave theory involves a hydromagnetic wave caused by a magnetic field exerting its force perpendicular to the plane of the Galaxy. A complete and quantitative description of the theory is not available; however, the most concrete prediction made by the hydromagnetic theory concerns the projected gas surface density of the spiral arms. As opposed to the approximately 5 to 1 arm-interarm density contrast of the density-wave theory, Piddington's theory firmly requires the hydrogen to be distributed uniformly or in a random manner. This requirement does not seem plausible. The continuity from longitude to longitude of profile features over angles corresponding to, in a number of cases, at least 5 kpc proves that some physical parameters are ordered over these lengths. The most unambiguous evidence in this respect comes from the recent Westerbork observations of external galaxies (see proceedings of IAU Symposium No. 58). Observations of the neutral hydrogen in, for example, M51, NGC 4258, and M81 all show hydrogen concentrated to spiral arms. In addition, these observations reveal an overall 'grand design'. (Evidence for an overall 'grand design' in the hydrogen distribution in our own Galaxy is still rather inconclusive.)

A number of spiral structure problems are more easily studied in external galaxies than in our own. Two related areas in which extragalactic hydrogen studies can inspire work in our own Galaxy involve the relative behavior of the various spiral arm constituents and the detection of evidence for the predicted galactic shock.

V. Distribution of the Hydrogen Relative to Other Constituents of the Galaxy

The problems of the relative ordering and of the general states of motion of the various components of spiral structure are fundamental to an understanding of the forces ordering this structure. In other galaxies, Lynds (1970) in particular has demonstrated the tendency of dust lanes to mark out narrow bands on the inside of luminous spiral arms. Observations of the continuum radio emission from M51 by Mathewson *et al.* (1972) showed that the maximum intensity of the radio emission also follows the dust bands on the inner edge of the bright arms. Hydrogen-line observations of M51 (Shane *et al.*, in preparation) show on the other hand that the neutral gas arms follow the stellar ones. The shock version of the density-wave theory formulated by Roberts (1969) has been successful as a framework in which these phenomena may be discussed. The theory provides for a compression which triggers the gravitational collapse of large concentrations of gas and dust and at the same time leads to synchrotron radiation enhancement caused by the compression of the magnetic fields which are frozen into the gas. The star formation process will take about 60 million years, and during this time the shock front will move to its presently observed position. It is clear that in this sort of situation the various spiral tracers will not all trace out the same physical parameters. It seems promising to pursue these problems also in our own Galaxy.

The neutral hydrogen seems to be primarily a tracer of the locations where the overall distribution of stars is producing a gravitational sink. It is an efficient tracer because of its ubiquity. An observational problem relevant to studies of our Galaxy, as well as of other galaxies, involves locating the spiral structure defined, say, by stars of type A. In general the problem is to have information on the distribution of the *total* mass of a spiral arm, so that we know the strength of the gravity in the gravitational wave. Composite photographs of the sort made by Zwicky (1955) of M51 show an underlying yellow-red spiral structure which is smoother than the knotty arms delineated by the H II regions. It seems plausible that the characteristics of the velocity fields of the hydrogen and other young tracers may show more regularity than shown by the density distributions.

It is expected that stellar populations with velocity dispersions less than about 35 km s^{-1} also will respond to the gravitational wave. Humphreys (1970, 1972) in particular has isolated systematic motions amongst the O and B supergiants in two regions which she interprets in these terms. In Figure 5 the locations of O and B stars are plotted in spatial coordinates. If one were to find structure in the spatial distribution of the stars in this figure, relevant problems include the question of the validity of the working hypothesis that the effects of the interstellar absorption are well enough known so that they do not dominate the analysis, and the question of the completeness of the data (some regions being studied in more detail than others). These problems become less important if the kinematic information is also used. Symbols in the figure are open or filled depending on the sign of the observed velocity deviations from circular rotation as defined by the Schmidt (1965) law. The additional information makes it easier to locate patterns of 'openedness' or 'filledness' amongst the sym-

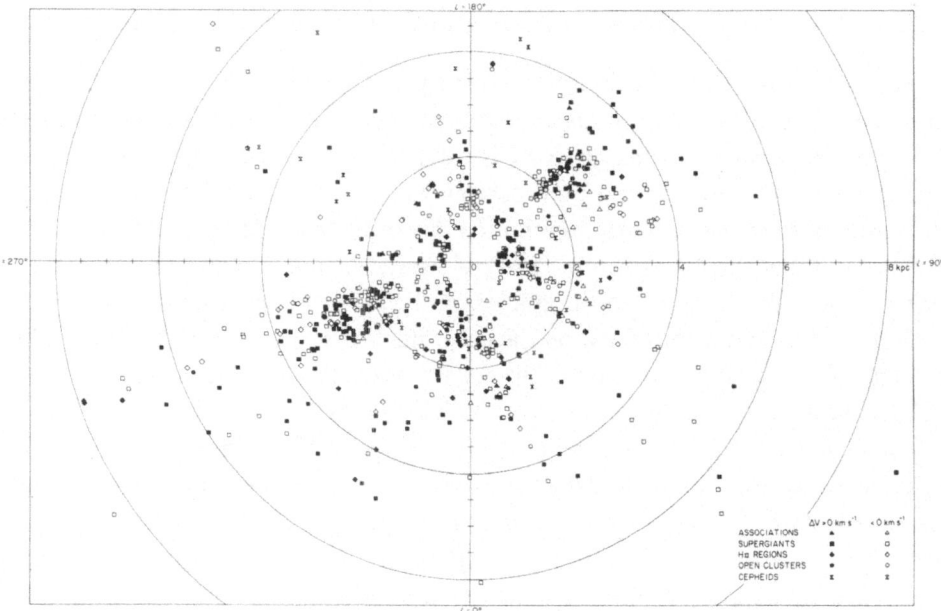

Fig. 5.   Spatial distribution of a compilation of young optical spiral tracers. The sense of the line-of-sight
peculiar motion of each object is indicated.

bols than it is to follow patterns in the density of symbols. Burton and Bania (1974)
have modeled the kinematic patterns using a velocity field of the sort predicted by the
linear density-wave formulation of Lin *et al.* (1969). At locations corresponding to the
spectroscopic distances of each of the observed O and B stars they calculated a model
velocity residual which was the difference between the LSR velocity predicted by the
theory and the Schmidt velocity at that location. The various parameters specifying
the model were varied in order to maximize the correlation between the observed
and model peculiar motions. Figure 6 shows the response of the correlation coefficient to changes in the parameters specifying the locations of two nearby arms. The
correlations show definite maxima at values indicating strong statistical significance,
suggesting that ordered large-scale streaming motions of the density-wave sort exist
in the young stars and that kinematic model fitting is a useful approach to isolating
patterns in these motions. The velocity field inherent in the model best-fit in this way
is shown in Figure 7. Contours represent streaming motions relative to the LSR.
The locations of the potential minima which in the framework of the linear theory
would govern these motions are also shown in the figure. Other best-fit parameter
values sought by maximizing the correlation coefficients in the same way are the
streaming motion amplitudes and the tilt angle. The resulting values are 5 km s^{-1}
and 8°. It is interesting that the kinematically derived tilt angle, 8°, is in good agreement with the value derived from hydrogen observations. These same stars' apparent
spatial distribution gives the conflicting value of ≈ 20°. There also seems to be no

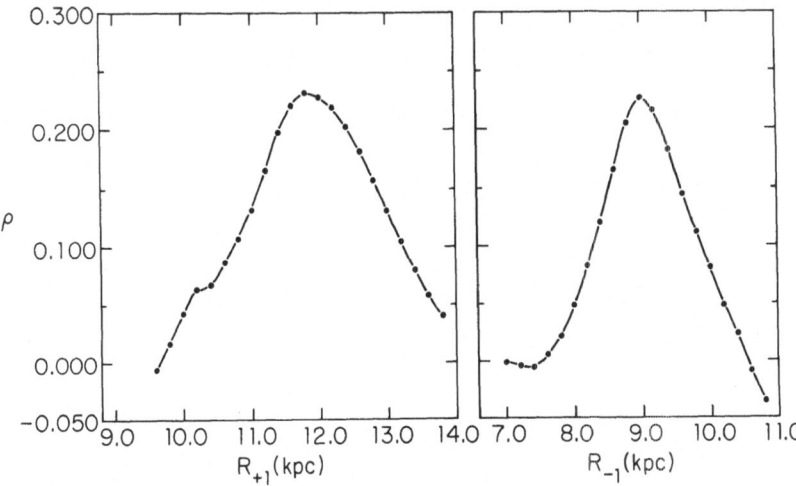

Fig. 6.   Response of the coefficient describing the degree of correlation between the observed and the modelled kinematics of the optical tracers to variations in the parameters R_{+1} and R_{-1} specifying the galactocentric distances at which two model spiral arms cross the Sun-center line. The responses peak at statistically significant correlation levels.

evidence in the kinematics for the interarm link a number of authors have identified as the 'Orion spur', or for a major local spiral arm.

The optical velocity residual is a quantity analogous to the velocity residual which was derived for the gas by exploiting the extreme sensitivity of the hydrogen line profiles to perturbations of the velocity field. A comparison of the optical residuals with the gas residuals found at the stars' location is shown in Figure 8. There is a correlation, and although it is weak, application of Student's t-test shows that it is certainly significant. This positive correlation between the stellar and gaseous motions supports the general prediction that these motions are ordered by gravitational forces.

Opposed to the relatively smooth underlying structure of the older spiral population is the extremely patchy appearance of the H II regions. In our Galaxy, observations of H II regions and of most of the molecules probably contain less direct information on the overall structure than the neutral hydrogen observations do. These spiral tracers are probably best considered to be tracers of regions of star formation, or, in general, of regions where the gas has been compressed, and in this respect their investigation may prove to be the best way to locate the shock front predicted by the density-wave theory. There is no very direct observational evidence yet for the existence of large-scale spiral shocks in this Galaxy.

There is a very close degree of correlation amongst the velocities and amongst the locations of clouds of OH, H_2CO, CO, and H II regions. It has been reported that the carbon monoxide molecule is spread ubiquitously throughout the Galaxy, and that its velocity and spatial coverage is similar to that of the neutral hydrogen. At this time, however, there have still been few observations of CO at directions not coincident with known H II regions; observations made at 'typical' regions by different groups do not appear to agree. Although the results are still very preliminary, ob-

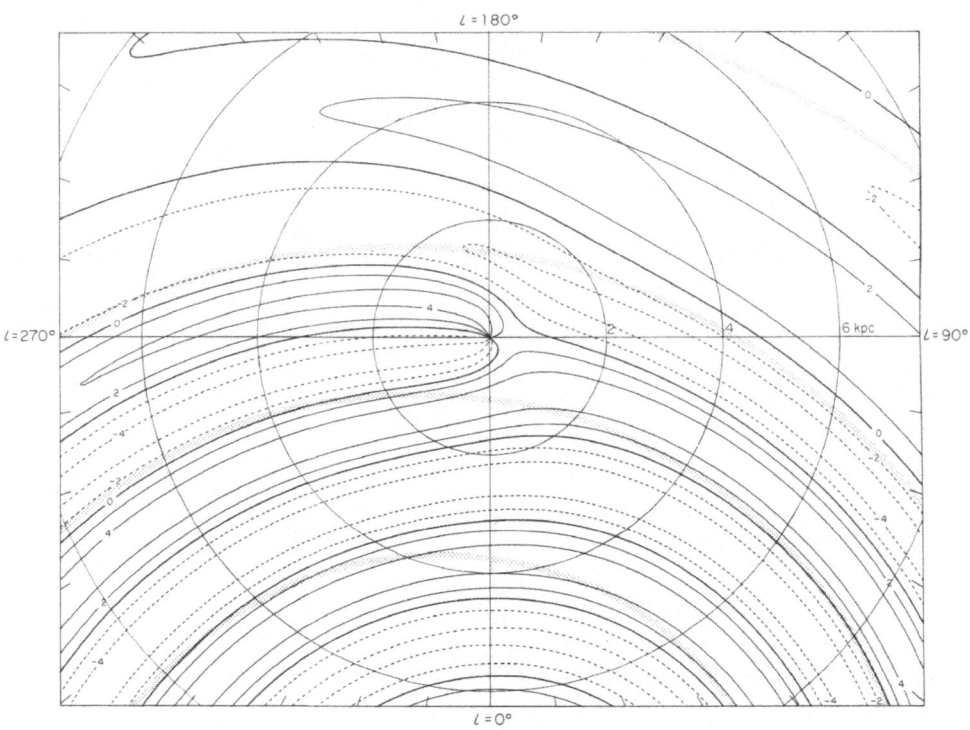

Fig. 7. Line-of-sight streaming-motion patterns derived as a best fit, in the framework of the linear density-wave theory, to the kinematics of the compiled observations of supergiants and associations. The stippled bands show the locations of gravitational potential minima. The most important parameters of the adopted best-fit model are the following: $R_{+1} = 11.4$ kpc, $R_{-1} = 8.9$ kpc, tilt angle $= 8°$, streaming amplitude $= 5$ km s^{-1}.

servations made at NRAO by Gordon, Lockman, Bania and myself at a limited number of regions, in the galactic plane, chosen to be away from H II regions do not show CO – to a peak-to-peak detection limit of about 0.3 K – at all the velocities covered by the neutral hydrogen. It is our group's suspicion that CO, as the other molecules, is concentrated to compressed regions and so is a tracer of such regions, and thus perhaps of the predicted shock front, but is not a tracer in the same sense as the neutral hydrogen.

The patchiness and the very complicated environment expected in star formation regions make comparison of the distribution of H I and of molecules difficult on a detailed scale. However, some aspects of the relative distribution do emerge if the comparison is carried out on a large scale. Figure 9 shows observations of neutral hydrogen taken at longitude 25° displayed as contours of antenna temperature in latitude-velocity coordinates. The lower panels show, for comparison also in latitude-velocity coordinates, some results from B. E. Turner's (private communication) extensive survey of OH emission and absorption made on the NRAO 140-ft telescope. To allow a qualitative comparison of the patchy OH distribution with the smoother hydrogen distribution, the hydroxyl data were added over a 15° range of longitude.

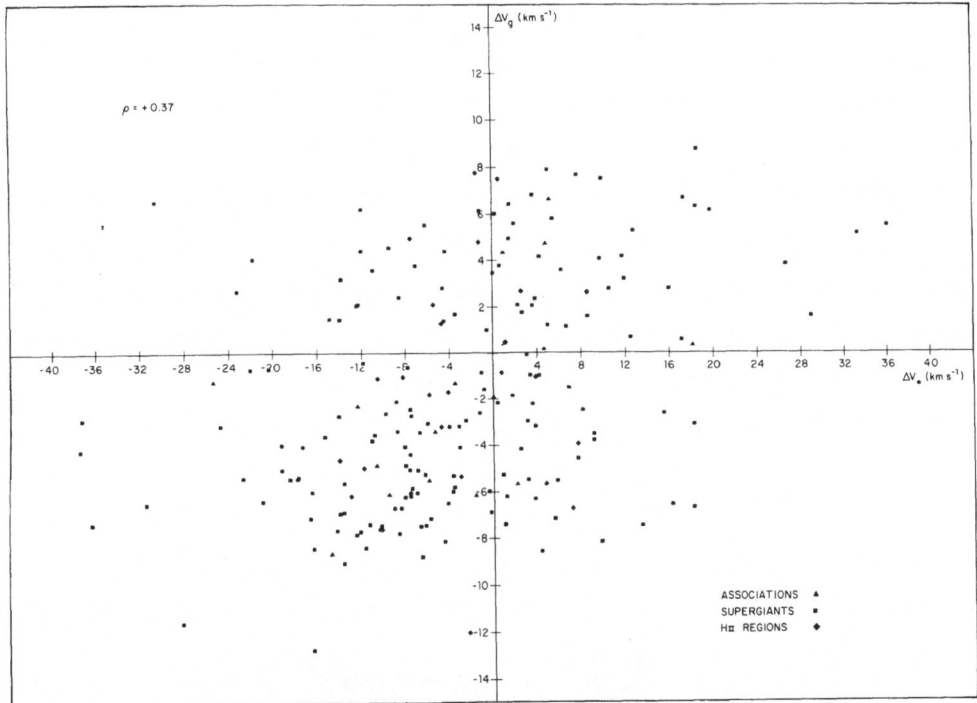

Fig. 8. Comparison of line-of-sight motions of the optical objects in Figure 5 with the motions of the gas found, at locations corresponding to the optical objects, from the data in Figure 4. Figures 5–8 are from Burton and Bania (1974).

Comparison of the hydrogen and hydroxyl situations in terms of velocity coverage indicates that there is agreement in coverage at the positive velocities, which at these longitudes correspond to the portion of the Galaxy interior to the solar orbit. On the other hand, little hydroxyl is observed at the negative velocities where the hydrogen emission is still strong. This comparison indicates the already familiar conclusion that regions associated with star formulation do not extend to such large galactic radii as does the neutral hydrogen.

Figure 9 also allows a comparison of the scale height of the neutral hydrogen with that of the OH. It is well known that interior to the solar orbit the hydrogen gas is confined to a thin and quite flat layer. The average full thickness of the layer to half-density points is about 220 pc between $R=4$ and 9 kpc. Over much of this region the deviation of the centroid of hydrogen emission from the galactic equator is less than 30 pc (Gum *et al.*, 1960). A scale height of 220 pc corresponds at $l=30°$ to an angle of $1°.5$. The scale height of the OH indicated in the figure is substantially less than that. Although Turner's observations extend beyond $|b|=1°$, very little OH emission or absorption is found more than 1° from the galactic plane. The OH radiation has essentially fallen to zero at the half-intensity level of the hydrogen. An estimate of the scale height characterizing the OH in this region is 100 pc, less than half of the hydrogen thickness. Molecular data are only now becoming complete enough to allow this sort of comparison, but it seems that this sort of investigation is desirable

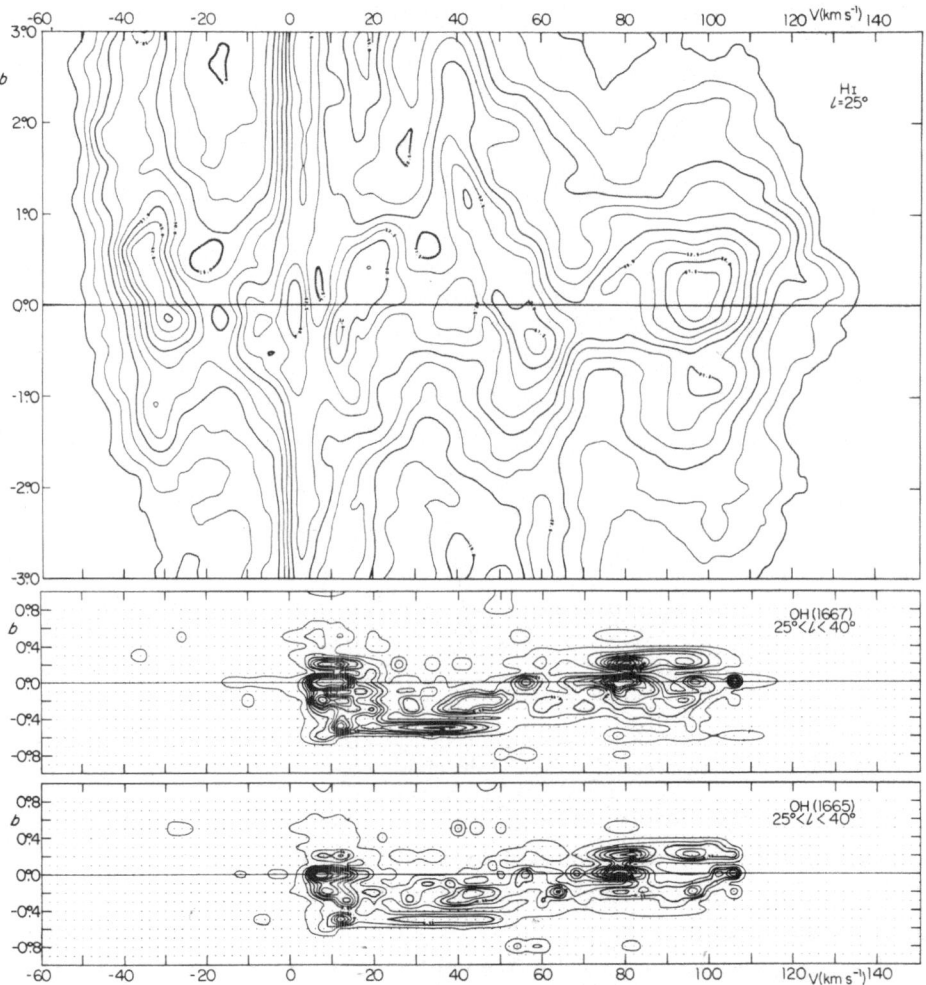

Fig. 9.   Comparison, on the same scale, of Hı and OH latitude-velocity contour maps showing that the
z-height of the neutral hydrogen's distribution is about twice that of the hydroxyl radical's distribution.
The H ı observations in this figure and in Figures 10 and 11 are from Burton and Verschuur (1973). The OH
data are from the survey of B. E. Turner; the amplitude of emission and absorption lines
observed at $25° < l < 40°$ are indicated.

in view of the increasing evidence that as galactic tracers the molecules are tracers of
regions of compression whereas the hydrogen is most efficient as a tracer of the galactic gravitational potential. If a galactic shock is responsible for the compression, then
the scale height of the molecules indicates the z-height to which this shock occurs.
This height will depend strongly on the hydrogen density because of the strong dependence of the Alfvén speed on density (see Wentzel, 1972). There will be no more
shock, thus no more compression and presumably no star formation, at heights above
the critical hydrogen density at which the Alfvén speed equals the shock speed.

In general, investigations of z-dependences in the hydrogen seem promising; however, both observations and theoretical predictions presently lag behind the state of

affairs obtaining in the galactic plane. Figure 10 shows hydrogen observations made in a cut through the galactic plane at constant longitude $l = 75°$. Evident here are a number of well-known phenomena. These include the deviation of the mean hydrogen layer from $b = 0°$ in the outer regions of the Galaxy (at this longitude negative velocities are attributed to galactocentric distances greater than 10 kpc). Distances are of

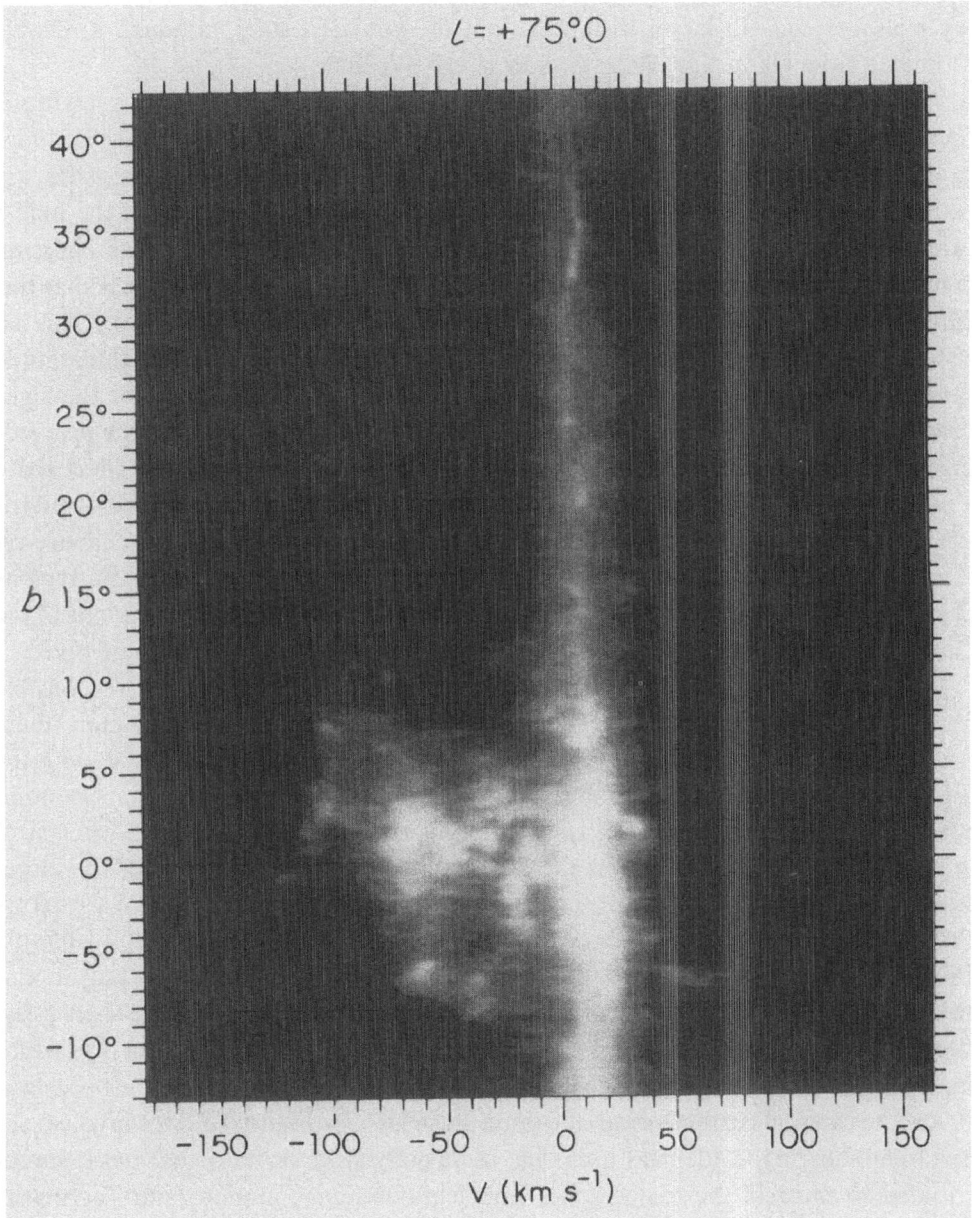

Fig. 10.   Grey-scale representation of H I intensities in a latitude-velocity scan through the galactic plane at $l = 75°$. These observations illustrate the hat-brim effect, the high-z extensions and the apparent shearing motions of hydrogen at large distances from the galactic center.

course harder to measure in the outer regions than in the inner ones: in the inner regions one is worried about deviations of a few per cent from a reasonably well-determined basic state of rotation, whereas in the outer regions this basic state itself remains an extrapolation based on a dynamic model. We do know that the hydrogen layer extends beyond the tracers found in the compression zone tentatively associated with the region interior to the density-wave's co-rotation distance. The hydrogen layer also extends to larger distances than the youngest stars, although a reliable estimate of the H I radius of the Galaxy is not available.

The well-known 'hat-brim' effect, in which the hydrogen in the outer region is found to bend up to positive latitudes in the first quadrant, but to negative latitudes in the fourth quadrant, is also apparent in Figure 10. The lower-level intensities are rather underexposed in this figure, although an example is apparent of the high-z extensions (Habing, 1966; Kepner, 1970; Verschuur, 1973) found as weak emission extending from the plane to z-distances of 3 or 4 kpc. What is intriguing is that this emission is rather clearly associated with prominent structures near $b=0°$. This association is evident in velocity and in terms of continuity over substantial longitude ranges. Similar continuity in longitude and in velocity has been found for the high-velocity clouds by Verschuur (1973), Davies (1972), and Mathewson (these proceedings). The high-z extensions, the high-velocity clouds, and the hat-brim effect share a number of systematic regularities, although of course these observational similarities do not require that only one dynamic process is at work. There is no evidence yet that any of the other spiral tracers, in particular the tracers associated with regions of star formation, partake in these non-planar phenomena in the outer regions of the Galaxy. More generally, it is not known to what extent the total mass is involved.

It has been suspected for some time that the Magellanic Clouds may be responsible for the hat-brim effect (e.g., Burke, 1957; Kerr, 1957); recently Mathewson (these proceedings) suggested that the relationship of the Large Magellanic Cloud with the high-velocity clouds also is causal. It is a pity that most of these non-planar phenomena have only been studied with any thoroughness in the parts of the sky accessible to northern observatories. From the standpoint of current problems of the large-scale galactic structures, more observations from the southern hemisphere are necessary. The new surveys by the Argentine group (Garzoli, 1972; Bajaja and Colomb, 1973) and especially the Parkes survey of Kerr *et al.* (1974) are important in this respect. Extensive southern hemisphere observations at latitudes more than a few degrees from the plane are still lacking. Table I lists the most recent surveys which have been made for studies of the large-scale galactic structure.

Another area of current research can be illustrated by Figure 10. This involves the apparent shearing, evident as a change of velocity with latitude, that has been attributed to most of the major spiral arm features (e.g., Fujimoto and Tanahashi, 1971a, 1971b; Harten, 1971). A very severe observational problem which plagues much work on the hydrogen distribution at low latitudes is the problem of blending whereby two or more features occur at approximately the same velocity in the profiles. Throughout the Galaxy structural features are expected to cross, blend and

TABLE I

Recent surveys of hydrogen emission from near the galactic equator

Authors	Publication date	Beam	Bandwidth (kHz)	Approximate region	Approximate interval	Form of publication
N. H. Dieter	1972	36'	10	$l = 10°$ to $250°$ $b = -15°$ to $15°$	$\Delta l = 2°$ $\Delta b = 2°$	$(b, V)\vert_l$ maps
S. L. Garzoli	1972	28'	10	$l = 270°$ to $310°$ $b = -7°$ to $2°$	$\Delta l = 1°$ $\Delta b = 2°$	$(l, V)\vert_b, (b, V)\vert_l,$ and $(l, b)\vert_V$ maps
E. Bajaja and F. R. Colomb	1973	28'	10	$l = 220°$ to $294°$ $b = -29°$ to $-11°$	$\Delta l = 2°$ $\Delta b = 2°$	$(l, V)\vert_b, (b, V)\vert_l,$ and $(l, b)\vert_V$ maps
W. B. Burton and G. L. Verschuur	1973	30'	10	$l = 20°$ to $230°$ $b = -20°$ to $20°$	scanned $\Delta b = 2°$	$(l, V)\vert_b$ maps
		10'	10	$l = 15°$ to $140°$ $b = -30°$ to $30°$	$\Delta l = 5°$	$(b, V)\vert_l$ maps
R. D. Davies and R. J. Cohen	1973	32'	33	$l = 355°$ to $10°$ $b = -5°$ to $5°$	$\Delta l = 1°0$ $\Delta b = 0°25$	$(b, V)\vert_l$ maps
P. O. Lindblad	1974	12'.5	10	$l = 12°$ to $72°$ $b = 1°$ to $10°$	$\Delta l = 3°$	$(b, V)\vert_l$ maps
		26'	5	$l = 339°$ to $12°$ $b = -15°$ to $15°$	scanned $\Delta l = 3°$	$(b, V)\vert_l$ maps
S. C. Simonson and R. Sancisi	1973	36'	16	$l = 354°$ to $24°$ $b = -5°$ to $5°$	scanned $\Delta l = 0°5$ $\Delta b = 0°5$	profiles, $(l, V)\vert_b$ and $(b, V)\vert_l$ maps
M. A. Tuve and S. Lundsager	1973	51'	10	$l = 336°$ to $270°$ $b = -16°$ to $16°$	$\Delta l = 4°$ $\Delta b = 2°$	profiles
H. Weaver and D. R. W. Williams	1973	36'	10	$l = 10°$ to $250°$ $b = -10°$ to $10°$	$\Delta l = 0°5$ $\Delta b = 0°25$	profiles
G. Westerhout	1973	12'.5	10	$l = 13°$ to $235°$ $b = -2°$ to $2°$	scanned $\Delta b = 0°2$	$(l, V)\vert_b$ maps
F. J. Kerr, R. H. Harten, and D. L. Ball	1974	15'	10	$l = 236°$ to $345°$ $b = -2°$ to $2°$	$\Delta l = 1°$ scanned	$(b, V)\vert_l$ maps

generally be rather tangled in the observed velocity space (see Knapp, 1972). It is extremely difficult to detect physically realistic velocity gradients unless the various features can be unambiguously separated. In some cases, observations are necessary with a larger velocity resolution than now available. All of the major large-scale surveys of low-latitude hydrogen have been carried out with a velocity resolution of about 2 km s^{-1}. It is in practice difficult to separate features closer together than about three bandwidths. The complexity is present even in the highest angular-resolution observations. Figure 11 shows observations made on the 300-ft telescope at $b = +2°0$ in the first quadrant of longitude. This figure also illustrates that separation of features is undoubtedly easier in the outer regions of the Galaxy where the basic velocity-distance relationship is single-valued, although the large-scale noncircular motions and the bending of the mean layer can contribute themselves to apparent velocity shearing motions (see Yuan, these proceedings). In the inner parts of the Galaxy separating features is especially difficult. In velocity-longitude maps of observations made near the galactic plane in the first quadrant there is an obvious tendency for intensities at positive velocities to be higher than at negative velocities. This is because of the double-valuedness of the velocity-distance relationship at positive velocities and offers a simple proof that the hydrogen gas seen in emission is optically thin, on the largest scale, at the longitudes where the drop-off of intensities occurs near zero velocity. Because of the doubled-valuedness at positive velocities there will on the average be twice as much hydrogen contributing there, if the gas is thin, than at negative velocities. A similar cut-off is also observed at the corresponding longitudes in the southern data.

Besides the current areas of large-scale 21 cm research already mentioned, a few additional problems are listed here:

– length scales of typical coherent patches of gas found near the galactic plane, as a function of galactocentric distance R;

– variation of the apparent velocity dispersion as a function of R, and, to the extent possible, of z;

– observational consequences of self-absorption features to the overall interpretation of the emission profiles;

– observational consequences of the two-component model of the neutral gas;

– isolation of saturation effects as a means of identifying spiral arms and for the investigation of the characteristic temperature and scale-height of the gas.

For these problems very accurate distances are not crucial; what is crucial is a careful identification in the profiles of the relevant effects.

Acknowledgements

I am indebted to B. E. Turner for providing access to his unpublished OH observations used in Figure 9, and to NRAO's Photo-Graphics department and to T. R. Cram for their special efforts to produce Figures 10 and 11 using a still-experimental CRT display system.

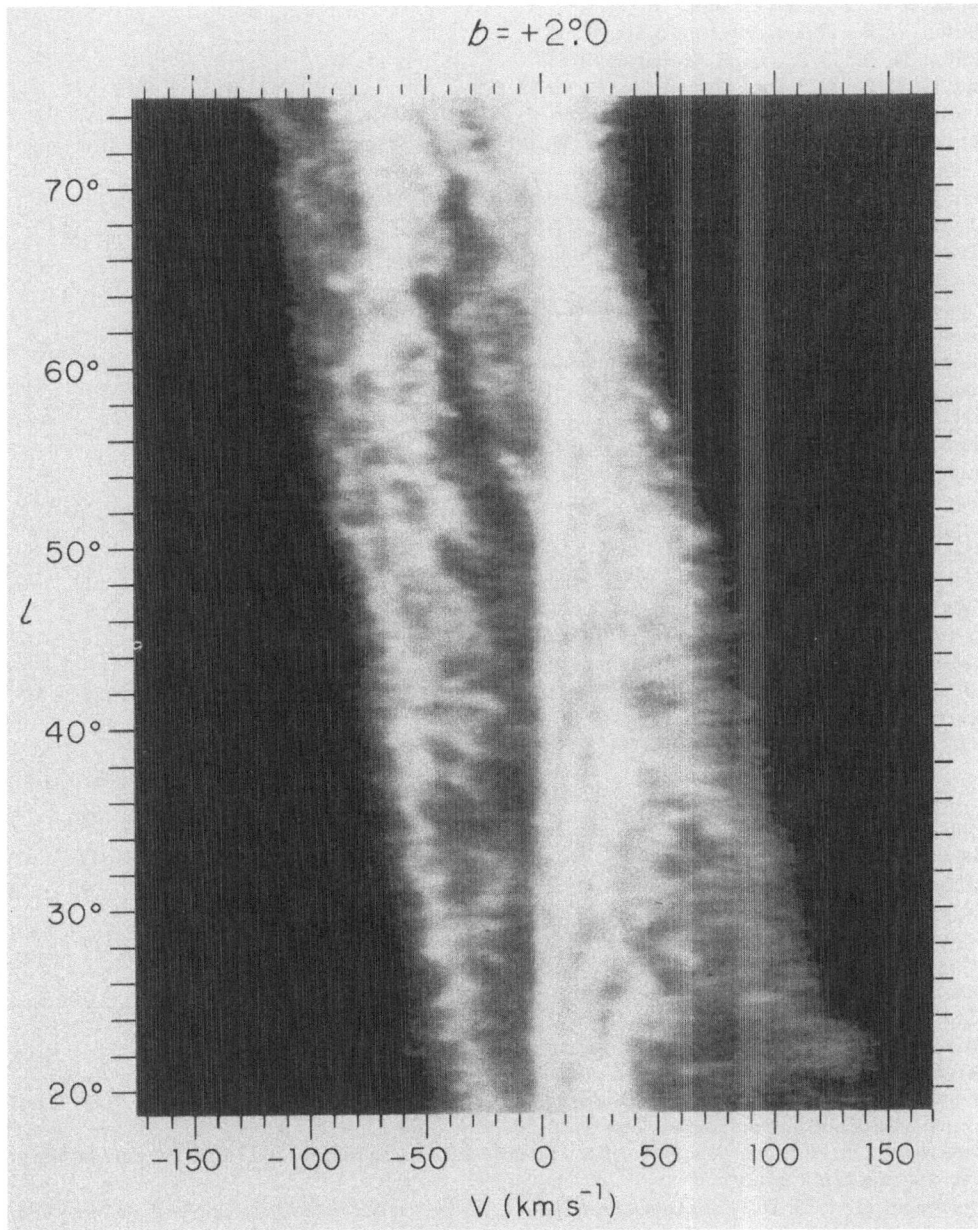

Fig. 11.   Grey-scale representation of H I intensities in a longitude-velocity scan parallel to the galactic plane at $b = +2°0$. The intensity drop-off near zero velocity, suggesting that the emitting gas is optically thin, is especially evident.

References

Bajaja, E. and Colomb, F. R.: 1973, *Carnegie Inst. Washington Publ. No. 632.*
Barbanis, B. and Woltjer, L.: 1967, *Astrophys. J.* **150**, 461.

Burke, B. F.: 1957, *Astron. J.* **62**, 90.

Burton, W. B.: 1971, *Astron. Astrophys.* **10**, 76.

Burton, W. B.: 1972, *Astron. Astrophys.* **19**, 51.

Burton, W. B.: 1973, *Publ. Astron. Soc. Pacific* **85**, 679.

Burton, W. B. and Bania, T. M.: 1974, *Astron. Astrophys*, in press.

Burton, W. B. and Shane, W. W.: 1970, in W. Becker and Contopoulos (eds.), 'The Spiral Structure of our Galaxy', *IAU Symp.* **38**, 397.

Burton, W. B. and Verschuur, G. L.: 1973, *Astron. Astrophys. Suppl.* **12**, 145.

Davies, R. D.: 1972, *Monthly Notices Roy. Astron. Soc.* **160**, 381.

Davies, R. D. and Cohen, R. J.: 1973, *Monthly Notices Roy. Astron. Soc.*, in preparation.

Dieter, N. H.: 1972, *Astron. Astrophys. Suppl.* **5**, 21.

Fujimoto, M. and Tanahashi, Y.: 1971, *Publ. Astron. Soc. Japan* **23**, 7.

Fujimoto, M. and Tanahashi, Y.: 1971, *Publ. Astron. Soc. Japan* **23**, 13.

Garzoli, S. L.: 1972, *Carnegie Inst. Washington Publ. No. 629.*

Gum, C. S., Kerr, F. J., and Westerhout, G.: 1960, *Monthly Notices Roy. Astron. Soc.* **121**, 132.

Habing, H. J.: 1966, *Bull. Astron. Inst. Neth.* **18**, 323.

Harten, R. H.: 1971, Ph.D. Thesis, University of Maryland.

Humphreys, R. M.: 1970, *Astron. J.* **75**, 602.

Humphreys, R. M.: 1972, *Astron. Astrophys.* **20**, 29.

Kepner, M.: 1970, *Astron. Astrophys.* **5**, 444.

Kerr, F. J.: 1957, *Astron. J.* **62**, 93.

Kerr, F. J. and Hindman, J. V.: 1970, *Australian J. Phys. Astrophys. Suppl.* No. 18, 1.

Kerr, F. J., Harten, R. H., and Ball, D. L.: 1974, *Astron. Astrophys. Suppl.*, in press.

Knapp, G. R.: 1972, *Astron. Astrophys.* **21**, 163.

Kwee, K. K., Muller, C. A., and Westerhout, G.: 1954, *Bull. Astron. Inst. Neth.* **12**, 117.

Lin, C. C.: 1971, in C. de Jager (ed.), *Highlights of Astronomy*, Vol. 2, Reidel Publ. Co., Dordrecht, p. 377.

Lin, C. C., Yuan, C., and Shu, F. H.: 1969, *Astrophys. J.* **155**, 721.

Lindblad, P. O.: 1974, *Astron. Astrophys. Suppl.* **16**, 207.

Lynds, B. T.: 1970, in W. Becker and G. Contopoulos (eds.), 'The Spiral Structure of Our Galaxy', *IAU Symp.* **38**, 26.

Mathewson, D. S., Kruit, P. C. van der, and Brouw, W. N.: 1972, *Astron. Astrophys.* **17**, 468.

Oort, J. H., Kerr, F. J., and Westerhout, G.: 1958, *Monthly Notices Roy. Astron. Soc.* **118**, 379.

Piddington, J. H.: 1973, *Monthly Notices Roy. Astron. Soc.* **162**, 73.

Roberts, W. W.: 1969, *Astrophys. J.* **158**, 123.

Schmidt, M.: 1965, *Stars and Stellar Systems* **5**, 513.

Shane, W. W. and Bieger-Smith, G. P.: 1966, *Bull. Astron. Inst. Neth.* **18**, 263.

Simonson, S. C. and Sancisi, R.: 1973, *Astron. Astrophys. Suppl.* **10**, 283.

Tuve, M. A. and Lundsager, S.: 1972, *Astron. J.* **77**, 652.

Tuve, M. A. and Lundsager, S.: 1973, *Carnegie Inst. Washington Publ. No. 630.*

Verschuur, G. L.: 1973, *Astron. Astrophys.* **22**, 139.

Weaver, H. and Williams, D. R.: 1973, *Astron. Astrophys. Suppl.* **8**, 1.

Wentzel, D. G.: 1972, *Publ. Astron. Soc. Pacific* **84**, 225.

Westerhout, G.: 1969, *Maryland-Green Bank Galactic 21-cm Line Survey*, 2nd ed., University of Maryland, College Park, Maryland.

Westerhout, G.: 1973, *Maryland-Green Bank Galactic 21-cm Line Survey*, 3rd ed., University of Maryland, College Park, Maryland.

Woerden, H. van: 1967, in H. van Woerden (ed.), 'Radio Astronomy and the Galactic System', *IAU Symp.* **31**.

Yuan, C.: 1969, *Astrophys. J.* **158**, 871.

Zwicky, F.: 1955, *Publ. Astron. Soc. Pacific* **67**, 232.

W. B. Burton
National Radio Astronomy Observatory,
Edgemont Road,
Charlottesville, Va. 22901, U.S.A.

DISCUSSION

Price: Can you derive an arm to interarm ratio for the H I density from your modeling studies?

Burton: The arm/interarm density contrast cannot be determined in a direct way, because of the general inaccessibility of density values. However, a ratio between arm and interarm densities greater than about 20 to 1 can be ruled out by the appearance of the observed profiles, considering especially the persistent ridge of high intensities near $b = 0°$ at velocities attributed to the locus of subcentral points. Along this locus the line of sight scans across several regions identified as arm ones and also across several interarm regions. A uniform density, on the other hand, is not realistic. A density contrast of 5 to 1 is consistent with model-fitting to the observations and with various theoretical predictions. So far no one has applied the model fitting approach using a two-component gas composition. The results might be different since equal densities of gas at different temperatures will of course make unequal contributions to the simulated observations.

Van Woerden: Burton's beautiful review clearly shows that research into galactic structure and dynamics makes progress, notwithstanding the grave problems of separating the velocity and density distributions. Important clues may come from the detailed analysis of H I distributions and motions now possible with synthesis telescopes. As examples, I might mention the work on M81 and M101 reported by Oort and by Allen *et al.* at Canberra (*IAU Symp.* **58**) and by myself at the IAU General Assembly.

Pishmish: It is well known that in reducing the 21 cm line data, it is assumed that neutral hydrogen is distributed in rings. But obviously this is only an approximation. If the H I concentration forms a spiral instead of a circle the maxima (or minima) of the radial velocities will again correspond to the tangential points. But the radius vector drawn to the tangent point will not be perpendicular to the line of sight. Hence the determination of the distance from the center will be erroneous if one ignores the spirality of the H I concentration. It is possible that if the reductions are performed properly assuming a spiral instead of a circle the north-south asymmetry of the rotation curve may be explained. Has anyone tried this?

Burton: Unless essentially all hydrogen were absent from the circles' tangential points, which seems unlikely, the argument of Shane and Bieger-Smith (1966) is relevant. If you adopted the tangent-to-spirals situation you would have to explain why no low-level emission is observed at the profile cut-offs used to determine the rotation curve. In addition, although I see that your approach would result in rotation-curve irregularities which would be located at different longitudes in the 'North' than in the 'South', these irregularities would be perturbations to the same basic rotation curve. Actually there seems to be a secular difference between the two observed curves.

Price: How do you derive your value of dispersion velocity of 4 km s^{-1} in our Galaxy, and isn't it a bit lower than previously used?

Burton: Model profiles calculated using a dispersion of 4 or 5 km s^{-1} show the same general shape as observed profiles. Results of Gaussian analysis of hydrogen profiles observed near the galactic plan show that typical profile features are broadened by 6 or 7 km s^{-1}, but this should be regarded as an upper limit primarily because of the inevitable blending which contaminates almost all features. The only part of low latitude profiles which probably give a direct indication of random cloud velocities is the wing of the profile contributed by gas near the subcentral points. This cut-off of the profiles is a kinematic one and yields dispersions in the range 4 to 6 km s^{-1}. I think dispersions larger than 6 or 7 km s^{-1} do not apply for typical features near the plane in our Galaxy. Of course, special features such as the high-velocity clouds have much larger dispersions.

Oort: You showed a diagram of the relation between the velocity deviations from the Schmidt rotation curve found in the optical spiral tracers and the source deviations obtained for the neutral hydrogen. There was some correlation between the two but not any strong one. Can you explain why the correlation is not more pronounced?

Burton: There are several sources of uncertainty, both in the observations and in the analysis, which would probably weaken what might be stronger correlations. A particular source of error which may enter the analysis is the possible lack of consistency between the stellar (spectroscopic) distances and the gas (kinematic) distances. On the other hand, the correlation may in part be weak due to the different dispersions of the gas and stars, different effects of magnetic fields, etc. There certainly remains a lot of work to be done in attempting comparisons among the various spiral tracers.

Yuan: I would like to comment on the motion of young stellar objects in your discussion. One should be aware of the presence of the strong selection effect, since most of young objects are situated in the galactic shock and their motions are best described by the non-linear theory. I believe that the unsatisfactory correction between gas and stars Oort pointed out will be resolved if the non-linear theory is used.

Burton: The correlation between the motions of the young stars and of the gas has been looked into, in more detail than I was able to present here, by Bania (1973, M.A. Thesis, University of Virginia) and in a paper by myself and Bania (in press). The correlation improves if it is taken into account that the stars, whose formation was triggered in a shock, will be kinematically decoupled from the interstellar medium because the viscous drag on such small objects as protostars is small compared to that acting on the gas.

Price: Can information on H I velocity dispersion in external galaxies be derived from synthesis observations yet?

Van Woerden: The Westerbork line receiver does not at present allow reliable determination of velocity dispersions; the filter half width is 27 km s^{-1}, equivalent to a dispersion of 12 km s^{-1}. In 1975 we shall have a new receiver with various options of bandwidth, and I expect the problem of velocity dispersions for H I in external galaxies will then be investigated.

SOME ASPECTS OF GALACTIC STRUCTURE DERIVED FROM THE BERKELEY LOW LATITUDE SURVEY OF NEUTRAL HYDROGEN

HAROLD WEAVER

University of California, Berkeley, Calif., U.S.A.

Abstract. The principal problem in the study of spiral structure is to reduce the data to a small enough number of pictures or numbers that can be easily comprehended. A new map has been produced, based on circular rotation, but the high probability of large radial motions makes it likely than any picture of spiral structure derived on current theories will be in error. New computer pictures have made it possible to see the structure of the outer arm, which has many cloud-like structures lying above it on one side, some reaching heights of more than 1 kpc. Some very weak extended features in the outer regions are described, and examples of holes and jets are discussed.

I. Some Statistics of the Berkeley Low Latitude Survey of Neutral Hydrogen

Table I displays the statistics of the Berkeley Low Latitude Survey of Neutral Hydrogen. The observational data have been published in the form of profiles (Weaver and Williams, 1973). Contour maps derived from these data in the form $T_A(v, b \mid l)$ are in press. The data are also available on magnetic tapes.

An extension of the latitude coverage of the original survey was made to permit investigation of spiral structure at large z distances. The statistics of this secondary program are also shown in Table I. In the form of profiles and contour maps the data of the extended program are in press.

TABLE I

Observational data relating to the Berkeley Low-Latitude Survey of Neutral Hydrogen

Sky coverage:	Primary program: Every $0°25$ in galactic latitude between limits $-10°$ to $+10°$; every $0°5$ in galactic longitude over range $10°$ to $250°$ (38 961 profiles). Extension program: Every $0°5$ in galactic latitude between limits $-10°$ to $-30°$, $+10°$ to $+30°$; every $2°5$ in galactic longitude starting at $l=12°$, extending to $249°5$.
Frequency coverage:	Each profile covers 250 km s^{-1}; 200 information points per profile, spaced every 5 kHz except at end points of profile.
Frequency resolution:	Information channels 10 kHz (2.11 km s^{-1}) wide between half-power points.
Integration time:	72 s per profile information point
Rms fluctuations:	0.38 K (system noise $\leqslant 160$ K)
Angular resolution:	H.P.B.W. 35.5'

F. J. Kerr and S. C. Simonson, III (eds.), Galactic Radio Astronomy, 573–586. All Rights Reserved.

II. Examples of Use of the Data for Studies of Galactic Structure

(a) SPIRAL STRUCTURE

The principal problem in the study of spiral structure is to compress the data sufficiently to produce a small number of pictures or descriptive numerical quantities that
can be readily comprehended. The basis of the compression technique employed here
is illustrated in Figure 1. If neutral hydrogen is distributed in space in spiral form as
illustrated in Figure 1a, and if the motion of the gas is governed by a rotation curve
monotonically decreasing with increasing radius, we will observe the longitude, velocity diagram shown in Figure 1b. Corresponding points on the loci in the spatial
(R, θ) and observational (l, v) coordinate systems are identified.

It has been pointed out that the picture of high-contrast gas arms is oversimplified.
Velocity perturbations arising from gravitational or other effects can produce the observed velocity-antenna temperature structures. Clearly, however, whatever the cause
of what is observed, the l, v map provides the fundamental basis for displaying the
observational data and delineating the large-scale structure of the system. Our principal task is to find a way to produce the most appropriate and useful l, v map or
maps from the observational data.

Various options have been built onto the computer program used to produce l, v
maps from the data of the Berkeley Low Latitude Survey. (a) We can specify the range
of antenna temperatures from which the map will be made. For example, we may
wish to map only that gas that produces an antenna temperature T_A greater than
some cutoff value $T_c = 50$ K, say. In this case we map only the most intense features
of the observational distribution of hydrogen in the Galaxy. (b) We can map at any
latitude or we can take integrals over any specified latitude range and use these integrals in the mapping process. Such integral provides an l, v map which represents a

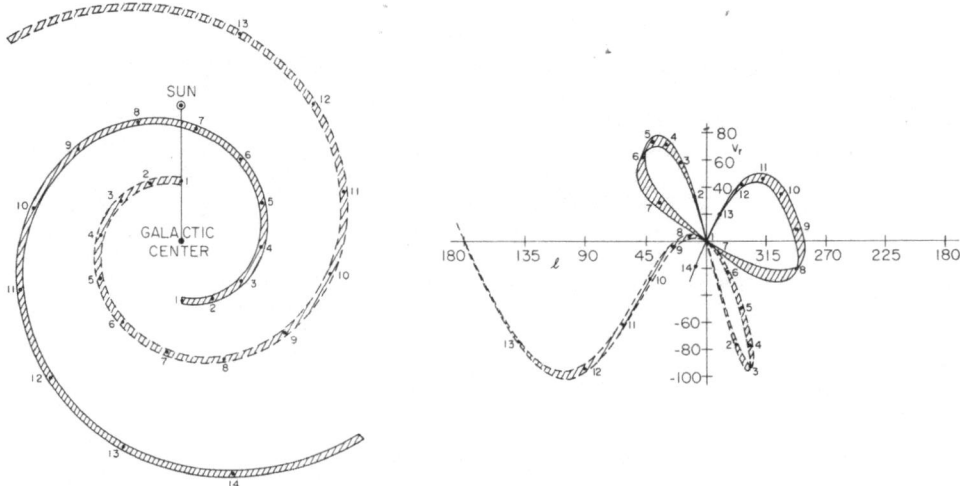

Fig. 1a.   Distribution of optically thin neutral hydrogen gas in a model spiral pattern.

Fig. 1b.   Observed velocity-longitude diagram for gas distributed as shown in Figure 1a.

quasi surface distribution of the gas projected onto the plane of the Galaxy. The basis
of this statement is as follows. The function

$$I(v \mid l, T_A \geqslant T_c) = \int_{-10°}^{+10°} T_A(v, b \mid l, T_A \geqslant T_c) \, db$$

measures the integrated hydrogen brightness derived by summation over latitude for
a given l, v column. If we take velocity as a measure of distance, r, in the line of sight,
an l, v column is analogous to an l, r column. Summation over b is analogous to summation over z, hence $I(v \mid l, T_A \geqslant T_c)$ represents quasi surface brightness of hydrogen
found by projecting onto the galactic plane (that is, summing over z) all that hydrogen
having a given l, r value.

An illustration of l, v map production follows. Figure 2a shows a characteristic
contour map made from the Berkeley data. We specify $T_c = 50$ K, eliminating all
lower antenna temperatures as shown in Figure 2b. The contour map in Figure 2b
is integrated over b to form $I(v, l \mid T_A \geqslant 50$ K) illustrated in Figure 2c. The distributions $I(v, l \mid T_A > 50$ K), one for each longitude, are combined to produce the l, v distribution map shown in Figure 3. In these l, v maps the value of $T(v, l \mid T_A > T_c)$ is
represented by blackness. Derived in a similar way, Figure 4 shows the l, v map for
$T_c = 25$ K, while Figure 5 shows the l, v map for $T_c = 10$ K.

Whatever the physical cause – velocity crowding or arm, interarm hydrogen gas
contrast – the general similarity of the observational l, v diagram to the model calculation, Figure 1b, is striking. The sinusoidal loci characteristic of outer spiral features are clearly shown, as is the loop structure of the next inner spiral arm. The observational picture is, of course, more complicated than the smooth theoretical one.
Observationally there are many bifurcated features and interarm spurs. The observed

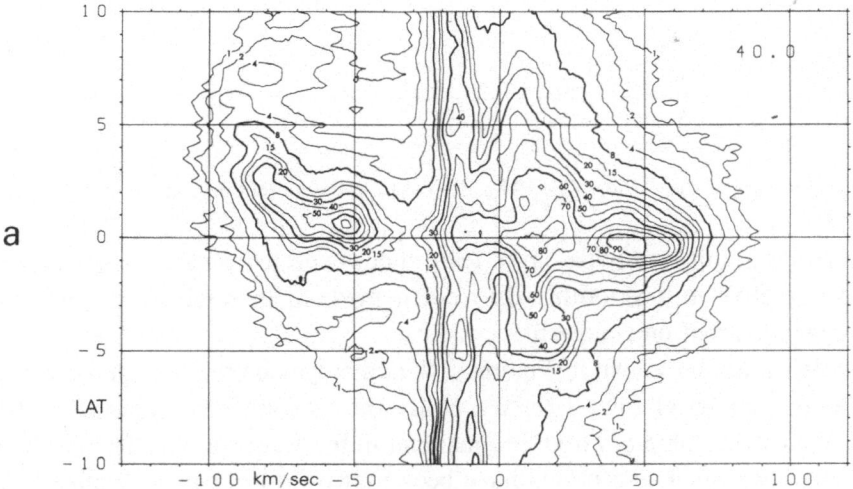

a

Fig. 2a.   Sample contour map derived from the Berkeley Low Latitude Survey of
Neutral Hydrogen; $l = 40°$.

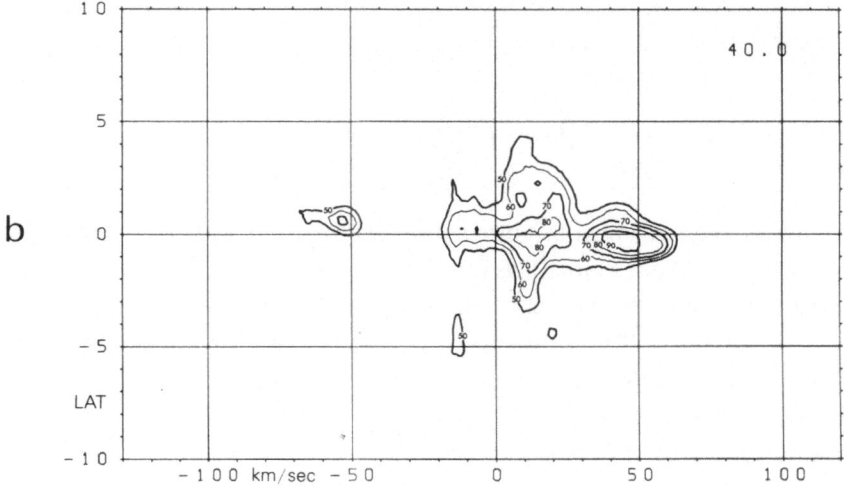

Fig. 2b.   Contour map for $T_A \geqslant T_c = 50$ K for $l = 40°$.

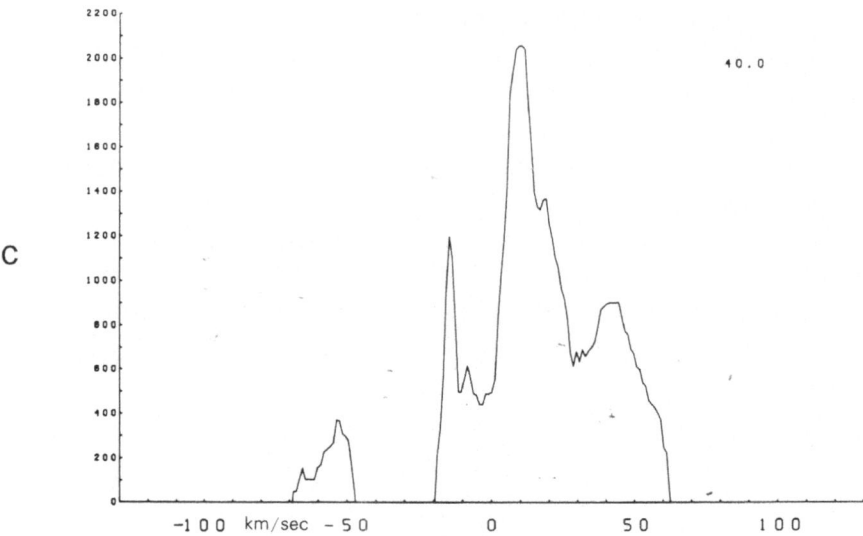

Fig. 2c.   The integral $I(v \mid l = 40°; T_A \geqslant 50$ K), derived from the contour map shown in Figure 2b.

'arm' structures are distinctly patchy, a feature not expected on the basis of velocity crowding operative in a smoothly distributed medium in which a smooth spiral gravitational perturbation is present.

The l, v distributions shown in Figures 3, 4, and 5 provide the basis for deriving a line graph, the analog of Figure 1b. We utilize the computer to find maxima from Figure 5. We fix the value of l and find maxima in the v-coordinate. The results are displayed in Figure 6. These points have been joined into curves in Figure 7. This last step required human intervention; the computer was instructed which points in Figure 6 to join together to form the observational analog of Figure 1b.

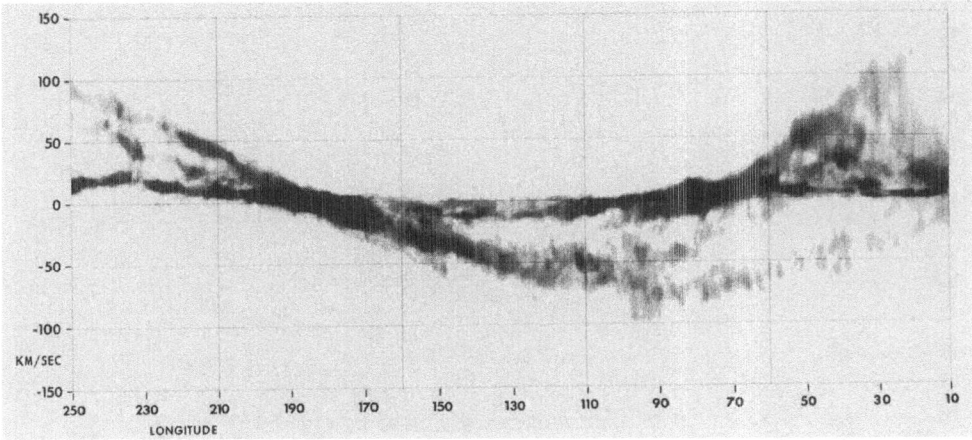

Fig. 3. The observed l, v distribution derived for $T_c = 50$ K.

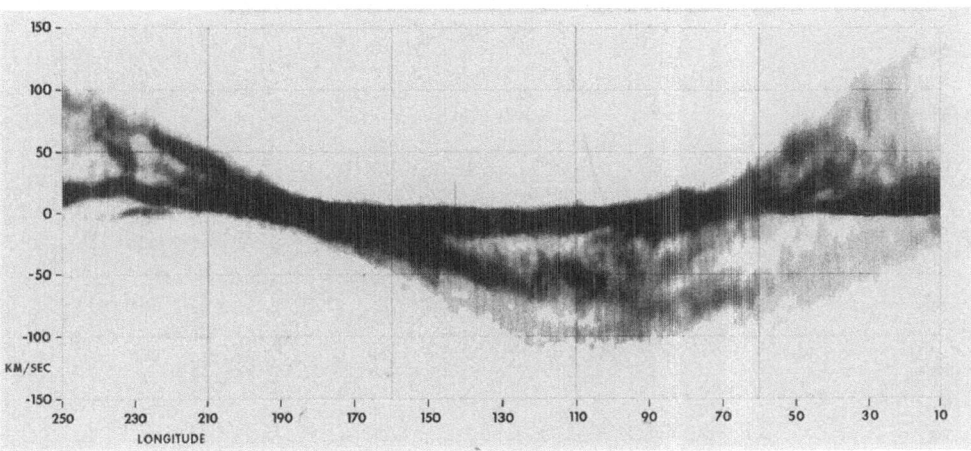

Fig. 4. The observed l, v distribution derived for $T_c = 25$ K.

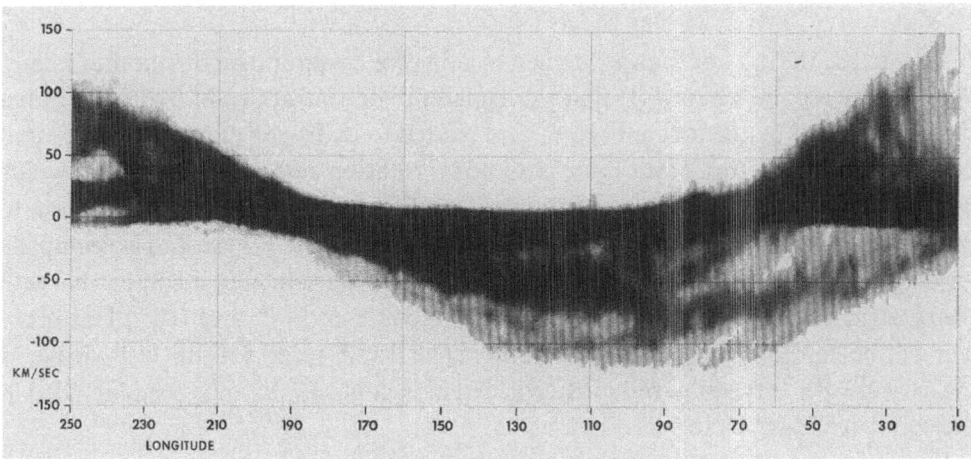

Fig. 5. The observed l, v distribution derived for $T_c = 10$ K.

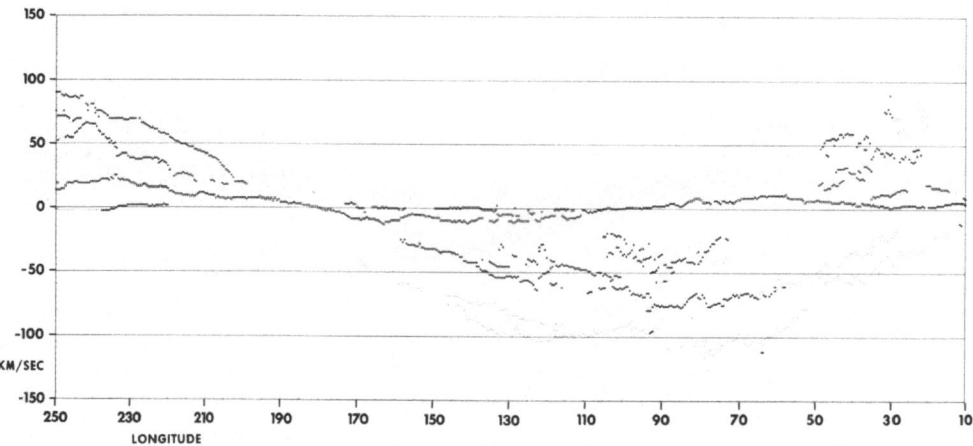

Fig. 6.   Maxima in the *l*, *v* plane derived from the data of Figure 5.

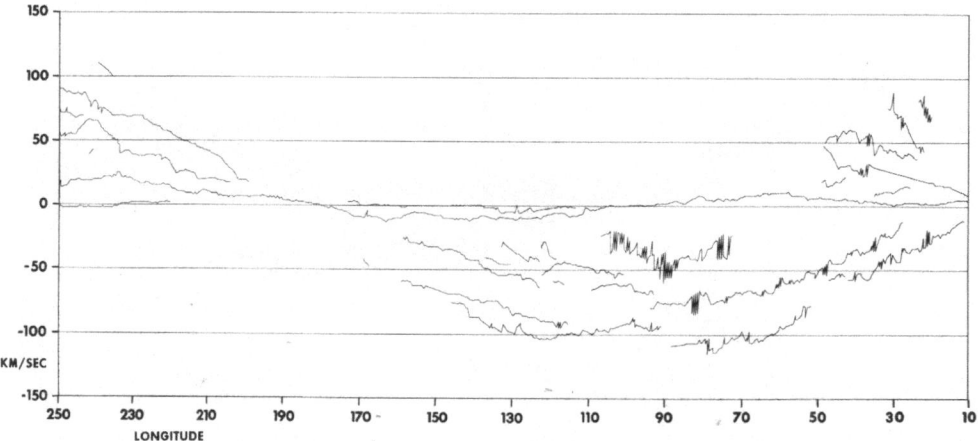

Fig. 7.   The *l*, *v* diagram formed by joining maxima pictured in Figure 6.

We can use the data of Figure 7 to derive a picture of spiral structure in the Galaxy. We adopt the point of view (1) that the apparent concentrations of hydrogen radiation represented by the loci in Figure 7 are related to hydrogen maxima in space that define spiral structure; (2) that there is a known rotation curve (Schmidt, 1965) which permits conversion of velocity to distance. Transformation of the *l*, *v* line diagram to spatial coordinates produced the spiral pattern shown in Figure 8. The complete *l*, *v* diagram in Figure 7 cannot be transformed; there are velocities incompatible with the rotation curve. (Nonzero observed radial velocities at *l*=0° and 180°.) This problem will be considered more fully presently. The 'local' spiral feature, well visible as the very low velocity gas of full longitude extent in Figure 7, has not been included in the spiral structure picture in Figure 8.

Except for the single step of instructing the computer as to what sequence of points is to be joined, the diagram in Figure 8 is completely computer produced. The near-far ambiguity of the next inner spiral arm is resolved by the existence of the loop in the l, v diagram, Figure 7. On the basis of latitude extent, the very high velocity structures in the longitude range 10° to 30° have been assigned to the 'far' solution.

Figure 8 is probably as good a picture of spiral structure as can now be produced, given the basis on which it was constructed: transformation of an l, v diagram to spatial coordinates by means of a rotation curve. (This is the process by which all spiral structure pictures of the character of Figure 8 have been produced in the past.)

The l, v diagram (Figures 3–5 and 7) is extremely well established; the precise physical significance of the spiral picture (Figure 8) is, however, questionable. This is the case not because of problems such as velocity pile-up, which has been thoroughly discussed by Burton, but because of what appears to me to be a much more fundamental problem: there are large radial velocities observed at $l=0°$ and $l=180°$. The well-known 'three kpc arm' shows a radial velocity of -53 km s^{-1} at $l=0$; at $l=180°$

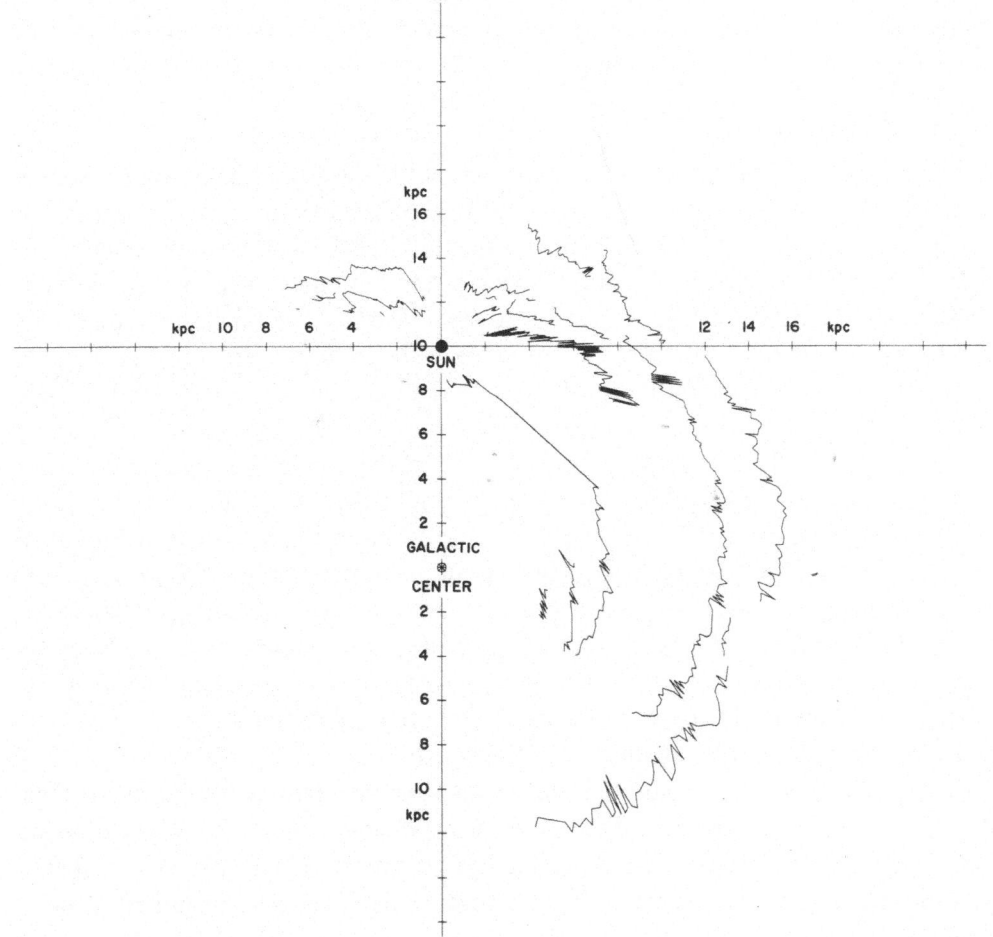

Fig. 8.   The spiral pattern derived from the data in Figure 7 with the aid of the Schmidt (1965) rotational model of the Galaxy.

the outermost spiral arm shown on Figures 5 and 7 exhibits a velocity of -30 km s^{-1} as found earlier by P. O. Lindblad. It is not possible within the present theoretical framework of galactic dynamics to account for such large velocities. Specifically, they lie outside the range of values encompassed by the current spiral density wave theory. They call into question the whole process currently used to produce pictures of spiral arms by kinematic effects.

Anomalous velocities have been pointed out at $l=0°$ and $180°$. It is not likely that a velocity anomaly exists only at those specific longitudes. It is probable that the highest velocity spiral feature shown in Figures 5 and 7 in the $l=10°$ to $30°$ range is directly related to the three kpc arm. The outermost arm extends over a long range of longitude. There is no velocity discontinuity in that arm near $l=180°$, hence there is no reason to expect that the identified velocity anomaly exists only at $180°$. The probability is high that radial motions of quite large size exist everywhere along these spiral arms, that distance estimates to these arms, made on the basis of any current dynamical model, are incorrect, and that consequently the picture of spiral structure, derived on the basis of velocity-distance transformation in any current theory, will be in error.

(b) PICTURES ON THE SKY

The high density of points in the sky coverage of the Berkeley Low Latitude Survey makes it possible to produce pictures of individual spiral features as they appear on the sky in hydrogen radiation. We employ the following technique of production. In Figures 5 and 7 the velocity-longitude locus of the outermost arm of the Galaxy is well shown. Over the longitude range $10°$ to $100°$ we draw, as a function of longitude, a velocity cut-off that separates this outer feature from adjacent velocity features (Figure 9). At each l, b point observed in the longitude range $10°$ to $100°$ and latitude range $-10°$ to $+10°$, we integrate the observed profile over velocity from the velocity cut off (Figure 9) to the greatest negative velocity of the profile. The resultant integral is proportional to the brightness of the outer arm at the l, b point. From the 14 661 data points available in the l, b-ranges of interest, the pictures shown in Figure 10 were produced. The three pictures, which show the outer arm over the l-range $10°$ to $100°$, b-range $-10°$ to $+10°$, are analogous to optical photographs taken on plates of different speeds and contrasts. The spiral arm pictures were produced by using in the computer program three different brightness, blackening relations, the analogs of different HD curves in the case of photographic plates.

Several features of these pictures are noteworthy.

(i) At the ends of the longitude range covered in the pictures we appear to look through one spiral feature to see the outer arm, which, throughout its length in the longitude range covered by this picture, lies high above the galactic plane (the galactic equator is indicated by the white line); it appears flattened on its southern (lower) side.

(ii) Many cloud-like high-z structures lie above the arm. Such features have been observed elsewhere in the galaxy by Kepner (1970). In the case of the outer arm the

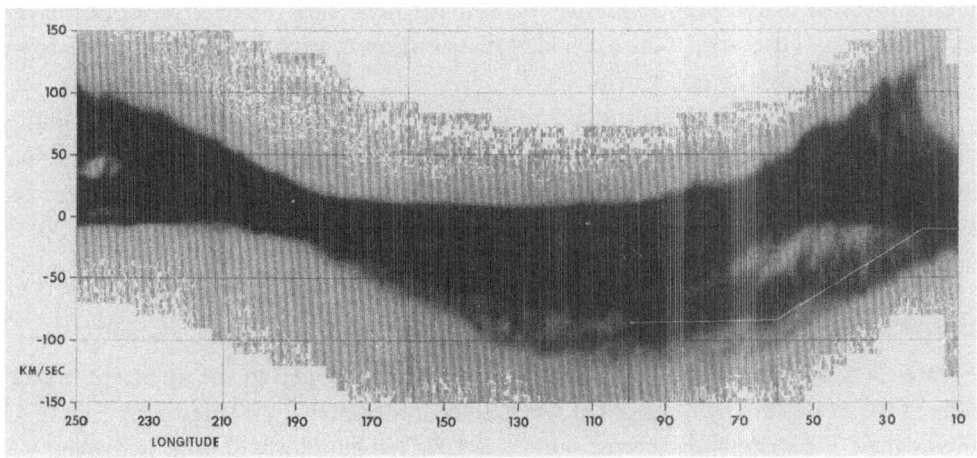

Fig. 9. The l, v distribution shown in this figure was derived with $T_c = 1$ K. On this distribution has been drawn as polygonal locus the velocity cut-off that separates the outer spiral arm from other velocity features.

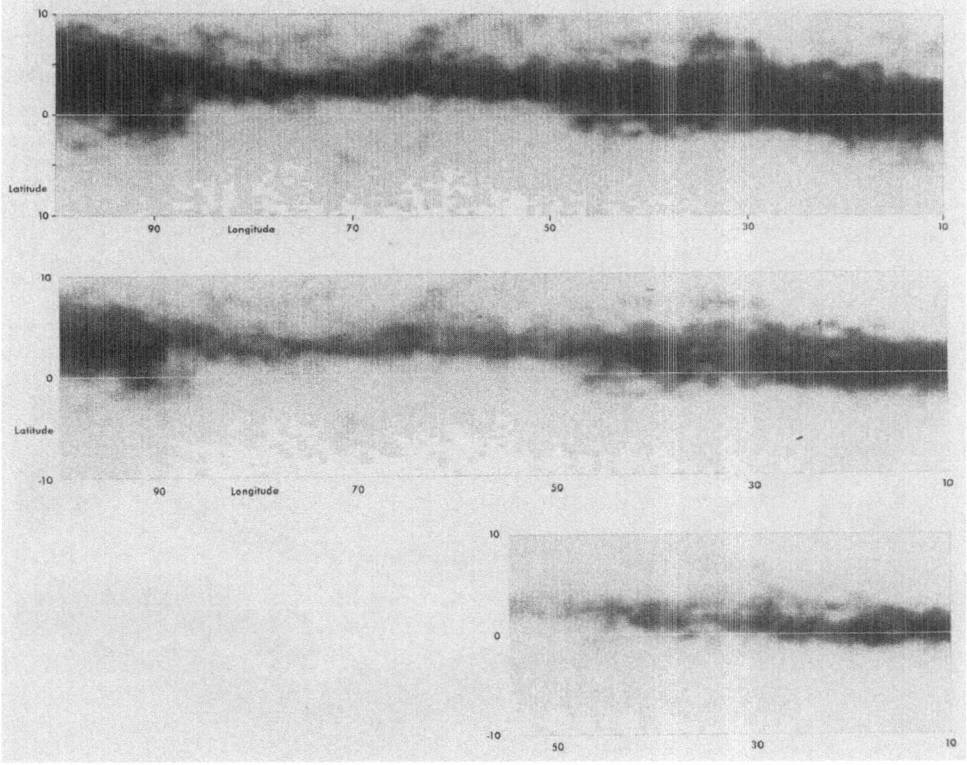

Fig. 10. Pictures of the outer arm of the Galaxy as seen on the sky in the 1420 MHz radiation of neutral hydrogen. The three pictures correspond to optical photographs taken with plates of different speed and contrast.

high-z features occur predominantly on the northern side of the arm, opposite the flattened side of the arm. This one-sided distribution of high-z objects appears to be a property of only the outer arm.

(iii) The cloud-like extensions rise to heights of a kpc and more above the plane; in some instances they are of 1–2 kpc length along the arm. Masses of individual features reach values $> 10^6 \, M_\odot$; many are in excess of $10^5 \, M_\odot$. The faint very extensive arc of gas in the negative latitude range at approximate longitude $70°$ is under detailed investigation.

(c) VERY FAINT FEATURES IN THE GALACTIC HYDROGEN

Figures 3, 4, and 5 exhibit the l, v distributions for hydrogen in the apparent brightness range extending from bright (Figure 3) to moderately faint (Figure 5). Figure 11 shows the l, v distribution derived with $T_c = 1$ K. We here look at faint hydrogen.

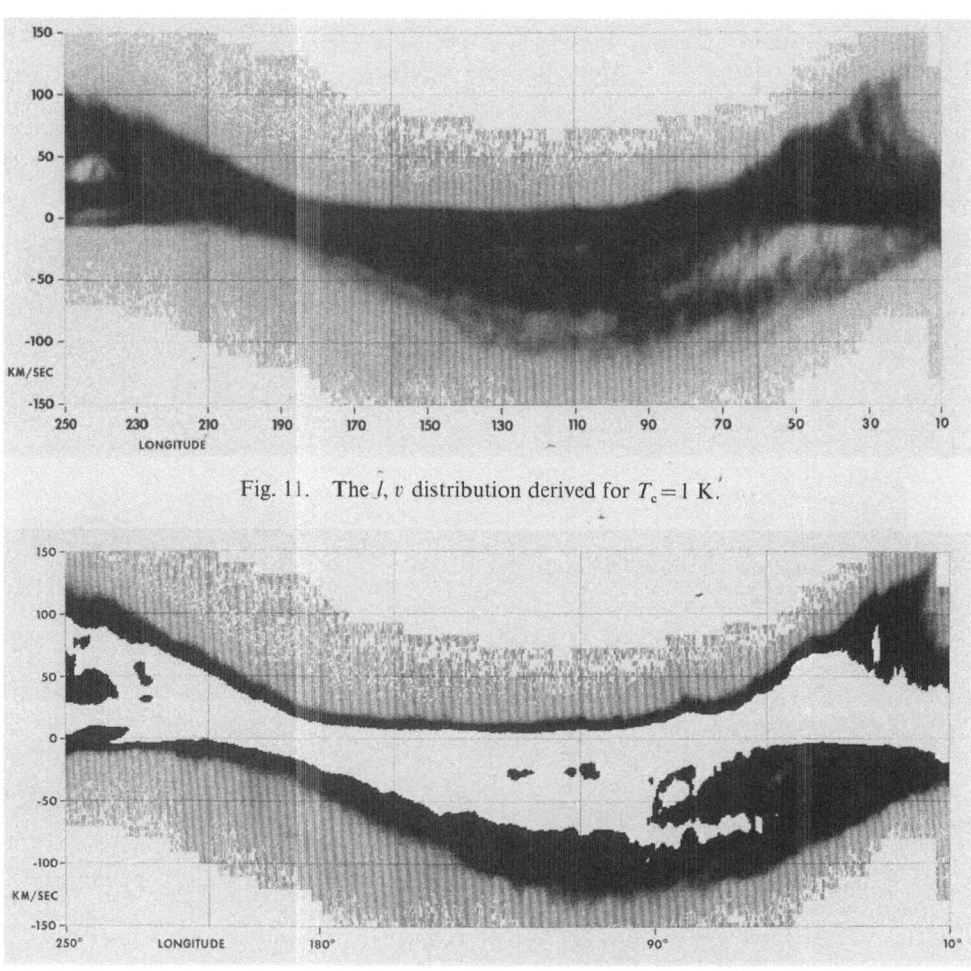

Fig. 11.   The l, v distribution derived for $T_c = 1$ K.

Fig. 12.   The l, v distribution derived for $T_c = 1$ K. The contrast has been increased over that shown in Figure 11 by computer techniques.

In Figure 11 there is a well defined negative velocity 'edge' to the distribution, which ends rather abruptly with the outer arm. In Figure 12 we show the results of increasing the picture contrast of the $T_c = 1$ K distribution by computer techniques. The negative velocity edge of the distribution remains quite sharp; only a few weak tassel-like extensions appear predominantly on the negative velocity side of the distribution. Many of these are of such longitude extent that they are composed of many different observations made at widely different times. They do not appear to be observational artifacts; they show strong correlation with negative velocity extensions found by Dieter (1971).

Of special interest is the long negative extension in the longitude range 190–200°,

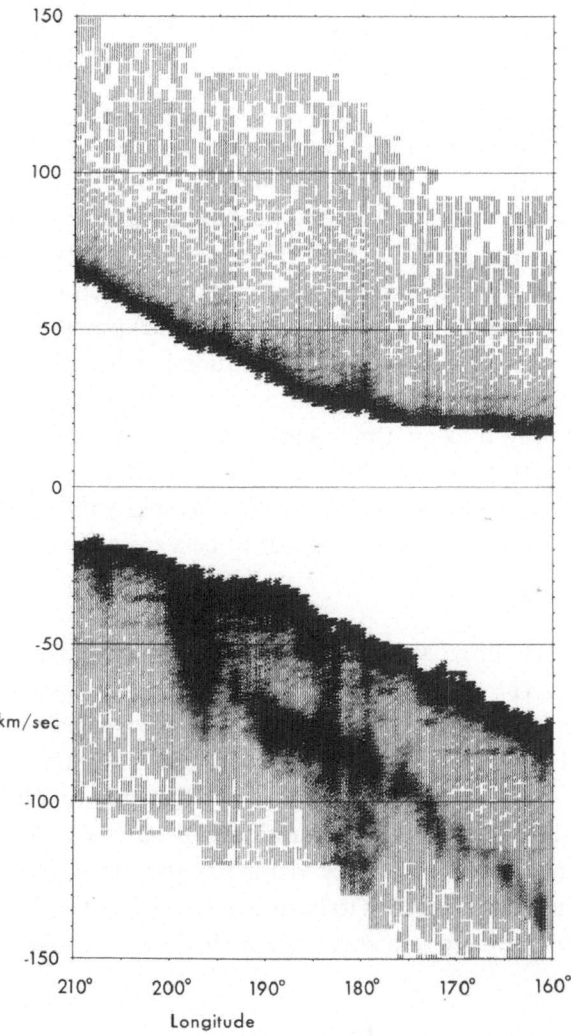

Fig. 13.   The l, v distribution derived for $T_c = 1$ K for a section of the galaxy. Contrast in this picture is very high, exceeding that of Figure 12.

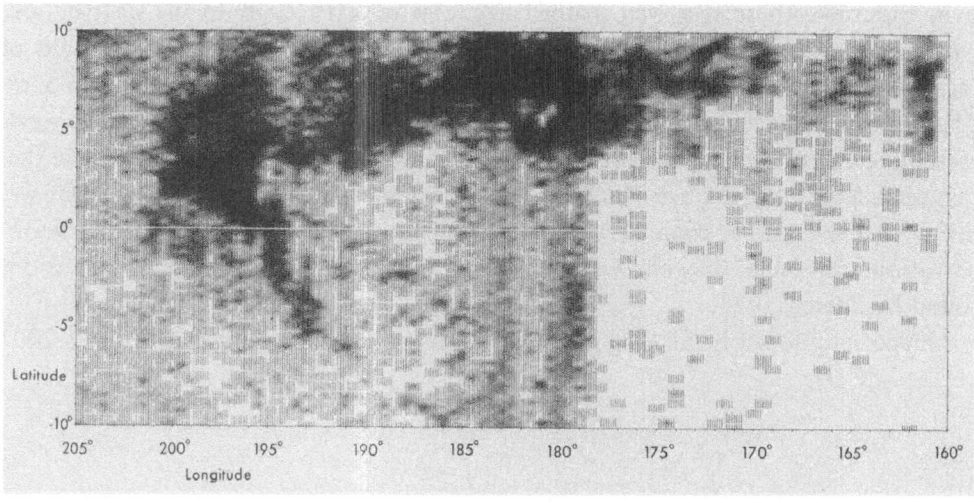

Fig. 14.   Computer-produced picture of the feature shown in Figure 13 as it would appear on the sky in
the 1420 MHz radiation of neutral hydrogen.

and the weak feature showing a velocity run extending from -50 km s^{-1} to -100 km s^{-1} in the longitude range 200° to 160°. These features are exhibited more clearly in Figure 13, where the contrast has once more been increased in the computer. The jet-like extension in the region $l = 200°$ has been discussed previously (Weaver, 1970). The feature showing the velocity run with longitude has been identified as a spiral arm by Verschuur (1973). A picture of the gas producing the feature pointed out above and identified by Verschuur as an arm is shown as a picture on the sky in Figure 14. The structure of the feature is very different in appearance from the arm shown in Figure 10; it is of large extent in latitude; it is bifurcated with a weak extension to negative latitudes. The feature does not appear to be a spiral arm of the Galaxy.

(d) INTERESTING OBJECTS

The Berkeley Survey has disclosed many jets, holes and other types of structures of astrophysical interest. The feature located at $l = 196°$ and discussed earlier provides an example of a large jet. An example of an extensive hole is visible in Figures 4 and 5 at longitude 80° and velocity -77 km s^{-1}. The diameter of this feature is approximately 1.5 kpc. The expansion velocity is 10 km s^{-1}. The mass involved is of the order several $\times 10^6\ M_\odot$, while the energy of expansion is $\sim 10^{52}$ erg. There are a number of supernova remnants in the immediate vicinity of the object, but these appear to possess rather small angular diameters to be directly associated with, or the immediate cause of, the hole, which we are now investigating in detail.

References

Dieter, N. H.: 1971, *Astron. Astrophys.* **12**, 59.
Kepner, M.: 1970, *Astron. Astrophys.* **5**, 444.

Schmidt, M.: 1965, *Stars and Stellar Systems* **5**, 528.
Verschuur, G.: 1973, *Astron. Astrophys.* **22**, 139.
Weaver, H.: 1970, *IAU Symp.* **39**, 22 (see p. 48 especially).
Weaver, H. and Williams, D. R. W.: 1973, *Astron. Astrophys. Suppl.* **8**, 1.

Harold Weaver
Radio Astronomy Laboratory,
University of California,
Berkeley, Calif. 94720, U.S.A.

DISCUSSION

Mathewson: What is your model for the perturbations in the velocity field of the local gas?

Weaver: This model is similar to the one suggested by Wesselius and Fejes. If there is some energetic event that interrupts the circular motion of the gas the material will start in an elliptical orbit. If it has less angular momentum than the Sun we run into it. From our position it is like a jet stream coming over our heads at a distance of around 70 pc. It shows negative velocities at $l \sim 110° - 120°$ and positive velocities $180°$ further around in longitude but at the same latitude. The mass is not large ($\sim 2 \times 10^3 \, M_\odot$).

Mezger: When you make the transition from velocity space to spiral structure a spiral arm emerges inside the solar circle. I was very glad to see this feature, since usually one sees on maps of spiral structure derived from 21-cm observations inside the solar circle features which reflect the author's best guess rather than hard observational facts. On the other hand, we know that giant H II regions disappear at distances of about 12 kpc from the galactic center. Therefore, if we limit interpretations of 21-cm observations to features outside the solar circle we may actually miss the major spiral structure of our Galaxy. Now, my question is, how did you manage to resolve the distance ambiguity for features located inside the solar circle?

Weaver: There are only two positions there where there is any distance ambiguity. One of them is resolved by the loop in the l, v diagram which automatically tells which is near and which is far. It also has a vastly different latitude extent on the two sides of the loop. It is very easy to resolve. If there were a lot of spiral features in the interior as seen in the hydrogen, I think there might be problems. But there is not a large amount of structure that appears in this organized way in the interior of the Galaxy.

Mezger: Do you tell us the spiral arms also are tilted?

Weaver: They are tilted only $8°$.

Simonson: It is important to make models that contain the velocity field for comparison with the observed $T_b(l, v)$ diagram. Mapping the peak intensities in the profiles may not really trace the density maxima.

Weaver: I am in complete agreement. If we wish to obtain a 'precise' picture of the spiral structure of the Galaxy, we certainly must have a 'precise' model of the velocity field and we must transform the $T(l, v \mid T_A \geqslant T_c)$ distribution. As I tried to emphasize in my remarks, however, the spiral picture I presented was not meant to be such a 'precise' picture. It was derived by the same techniques used to derive the spiral pictures published in the literature during the past several years and was meant to be comparable with those pictures. The main purpose of the part of my discussion dealing with the spiral picture was to emphasize that while we can now derive a very precise and beautiful $T(l, v \mid T_A \geqslant T_c)$ distribution, we cannot, I believe, make the transformation to a precise and beautiful distribution in space. I believe that we do not yet understand the dynamics of the Galaxy well enough to derive the velocity field required to make the transformation. The feature of the $T(l, v \mid T_A \geqslant T_c)$ distribution that convinces me of our inadequate knowledge is the very large radial motions we observe in the gas.

Simonson: Would you care to model the -70 km s^{-1} motion at $l = 180°$ with a dispersion ring as for the -53 km s^{-1} motion at $l = 0°$?

Weaver: No, I would not care to do so. In my view the -70 km s^{-1} feature at $l = 180°$ is not an arm of the Galaxy.

Van Woerden: I admire Weaver's computerized and photographic techniques for reduction and display of data. However, some of the computer-defined ridges of maxima of N_H in the (l, v) plane should not be interpreted as density maxima in the galactic (L, R) plane, but rather as effects of long line of sight and/or velocity crowding – as Burton has clearly demonstrated. But this should not make us unduly pessimistic. Burton's review shows that, with careful analysis and due regard for dynamical considerations, progress in understanding our Galaxy is possible.

Weaver: Ridges will occur if there is velocity pile-up, ridges will occur if there is contrast between arm and interarm densities. The computer technique will pick out ridges, there's no question about that; but that was not saying what makes the ridges. That I think is the part that I am more and more pessimistic about. I can see how to make excellent pictures in the observational space, but I do not see how at the present time to produce a convincing picture of the spatial distribution of gas. I lose my optimism because of these large velocities which I think are not present in the theory.

PROGRESS REPORT ON THE MARYLAND-GREEN BANK GALACTIC 21-CM LINE SURVEY

GART WESTERHOUT

University of Maryland, College Park, Md., U.S.A.

Abstract. The survey has been completely re-observed using the resurfaced NRAO 300-ft telescope. It covers the latitude ranges $b = -1.5°$ to $+1.6°$ from $l=11°$ to 60°, $b = -2.2°$ to $+2.4°$ from $l=60°$ to 132°, and $b = -1.9°$ to $+2.4°$ from $l=132°$ to 235°. Both the new survey and the old survey are available on magnetic tape and 70-mm film.

In 1971–72 the entire survey was re-observed using the resurfaced 300-ft NRAO telescope. The new survey therefore does not suffer from the effects of the error beam, which influenced the old survey in the sense that small-diameter features appeared less intense with respect to the broad background by a factor of about 1.5. The old survey is available on magnetic tape and 70-mm film (antenna temperature in $[R.A., v]$ contour diagrams at constant δ). It includes many observations made by other observers and in some places goes up to $b=5°$.

The new survey covers the region $b = +1.6°$ to $-1.5°$ between $l=11°$ and 60° and $b = +2.4°$ to $-1.9°$ between $l=60°$ and 235°. From $l=60°$ to 132° it is extended to $b = -2.2°$. The effective beamwidth is 0.22° (13′), and the effective bandwidth is 2 km s^{-1}. The magnetic tapes give lines profiles, i.e., brightness temperatures as a function of velocity, at intervals of 0.1° in l and b. They are sorted in two ways: profiles at constant b, with l varying, and profiles at constant l with b varying (i.e., parallel and perpendicular to the galactic plane). If a limited region in l is to be studied, the 'perpendicular tapes' are obviously to be preferred. The total survey is available at cost on eight tapes in binary or 12 tapes in BCD (approximately; twice this number if both parallel and perpendicular tapes are needed). Limited sections in l or b may also be obtained.

The data have also been reproduced in the form of computer-generated contour maps. These are likewise available both parallel and perpendicular to the galactic plane: contour maps giving brightness temperature in an $l-v$ coordinate system, every 0.2° in b, and in a $b-v$ coordinate system, every 0.2° in l. A sample of one page of perpendicular maps is given in Figure 1. Altogether, there are 535 pages containing 1400 contour maps. Currently, these are on 70-mm film and are available in that form. It is hoped that funds can be found to publish these maps as the final edition of the survey.

Acknowledgements

This investigation was partially supported by the U.S. National Science Foundation. NRAO is operated by Associated Universities, Inc., under contract with the National Science Foundation.

F. J. Kerr and S. C. Simonson, III (eds.), Galactic Radio Astronomy, 587–589. All Rights Reserved.

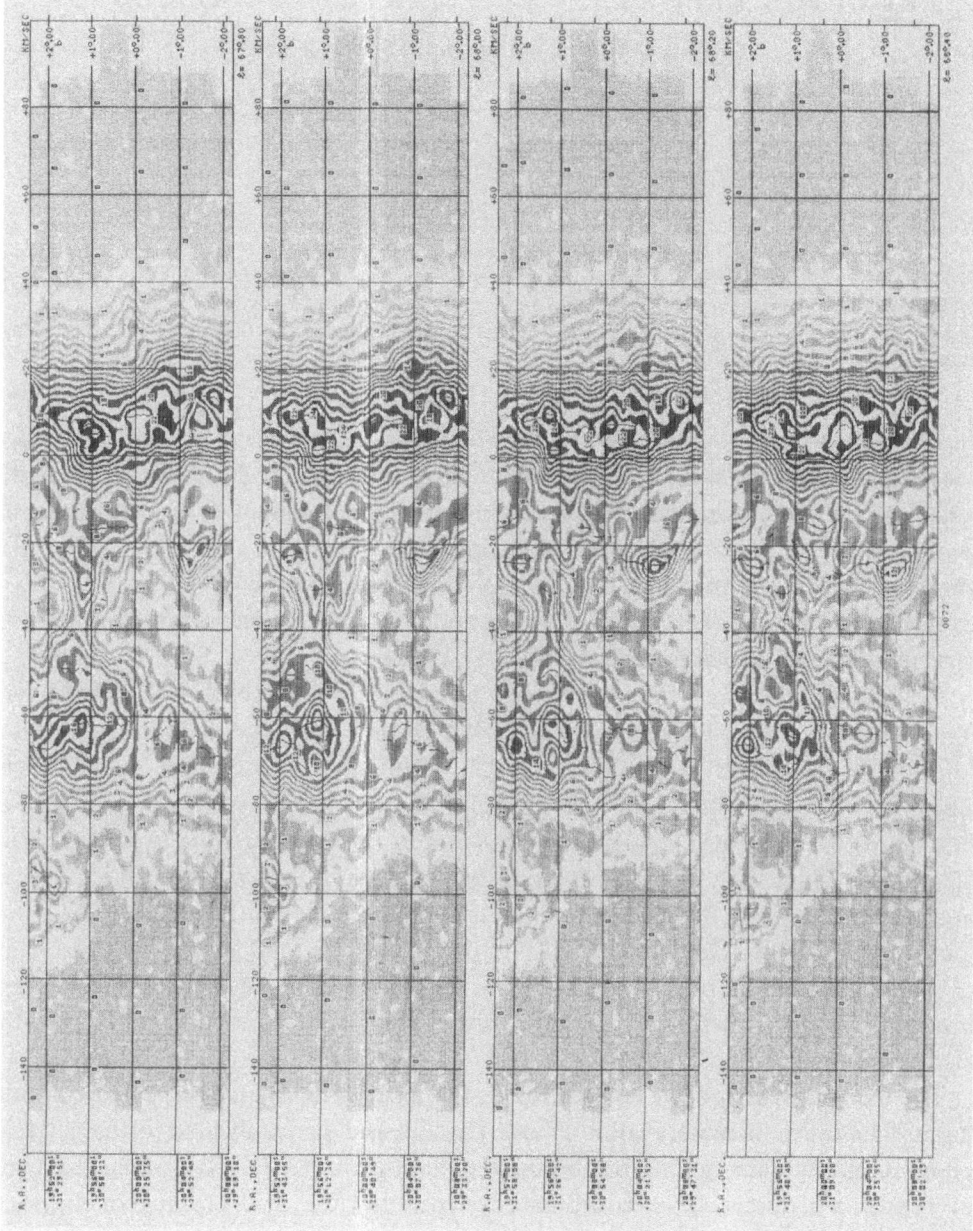

Fig. 1.   Sample page from the survey. Contour intervals are 5 K in brightness temperature from 5 to 20 K and 10 K from 30 K up (label units are 5 K). Contour edges are 1.25 K above and below their nominal value (2.5 K from 30 K up); thus, the 50 K dark area goes from 47.5 K to 52.5 K, the blank area from 52.5 K to 57.5 K, etc. Wiggly black lines give the positions of maxima in the line profiles. R.A. and Dec. are for 1950.

Gart Westerhout

Astronomy Program,

University of Maryland, College Park, Md. 20742, U.S.A.

DISCUSSION

Burton: One characteristic which almost all low-latitude hydrogen profiles share is the sharp cut-off in intensities at the edge of the profiles corresponding to the outskirts of the Galaxy (negative velocities at $0° < l < 180°$). The cut-off is about as sharp as the kinematic cut-off contributed by hydrogen at the sub-central points. P. L. Baker (NRAO) is presently collecting more complete observations on this. Do you have any explanation for the sharpness of this cut-off? One gets the impression that the hydrogen boundary of the Galaxy is very abrupt.

Simonson: The sharp cut-off in the profiles at negative velocities for $l < 180°$ is probably due to the hat-brim effect. We are seeing the bottom of the hydrogen layer rather than its edge.

Burton: This may be true at $b = 0°$, but the effect is also present at latitudes where the line-of-sight intercepts the 'edge' of the hydrogen layer.

Weaver: As Kerr remarked, the velocity cut-off for the 'outer edge' of the Galaxy appears to be very sharp. This is true even where an entire contour map showing very low intensities is considered.

A SIMPLE INTERPRETATION OF THE 'ROLLING MOTIONS'

CHI YUAN and LANCE WALLACE

City College of New York, New York City, N.Y., U.S.A.

Abstract. The rolling motion phenomenon is interpreted as an apparent motion in the observations. It is caused by (i) the combined effect of the differential rotation and the bending of the galactic plane for the spiral arms beyond the solar circle, or (ii) the displacement of the spiral arms from the central galactic plane for those within the solar circle.

I. Introduction

In their study of the galactic central regions, Oort and Rougoor first noticed that the motion of the 3 kpc arm "appears also to vary slightly with latitude: Relative to the center of the arm the part at average distance of 70 pc north of the plane is coming toward us with a velocity of 2.5 km s^{-1}, while the southern half of the arm is moving away with a similar velocity" (Oort, 1962, p. 15). Later, this phenomenon was interpreted explicitly by Rougoor (1964) as the gas motion rolling about the axis of the spiral arm. After this discovery, the 'rolling motions' have been found subsequently in almost all the major spiral arms in the Galaxy (Henderson, 1967; Burton, 1970; Shane, 1971; Harten, 1972). And Rougoor's interpretation was also tacitly accepted in general.

The first theoretical work in support of the concept of the actual rolling motions in spiral arms appeared accidentally in a paper by Fujimoto and Miyamoto (1969) which dealt only with the helical magnetic field in the Galaxy. After recognizing the implications of that paper, Fujimoto and Tanahashi (1971a) examined the observations of the rolling motions in the Perseus arm and found that they are compatible with their results. Furthermore, they proposed that the rolling motions in spiral arms are caused by the free precession of the Galaxy, a mechanism suggested by Lynden-Bell (1965) to account for the bending of the galactic plane (Fujimoto and Tanahashi, 1971b). Other suggestions on the cause of the rolling motions were given in Harten's thesis, but he did not develop a theoretical study to support his ideas.

All of these studies share one common view, namely, the spiral arms are actually rolling. We have taken an entirely different approach to this problem. The rolling motion phenomenon is interpreted as an apparent motion in the observations which is caused by the combined effect of the differential rotation and the bending of the galactic plane for the spiral arms beyond the solar circle or the displacement of the spiral arms from the central galactic plane for those within the solar circle. In this report, only a synoptic description of the observations, the method, and the results will be presented. A detailed exposition is published elsewhere (Yuan and Wallace, 1973).

F. J. Kerr and S. C. Simonson, III (eds.), Galactic Radio Astronomy, 591–597. All Rights Reserved.

II. Summary of the Observations

Almost all the published data of the 21 cm line surveys of the Galaxy contain information on the rolling motions of spiral arms. In his thesis, Harten has completed a very extensive review of the observations of the rolling motion phenomenon. Since the measurement of the rolling motions in the latitude-velocity diagram is very subjective, we also made independent measurements for all the major arms, directly from the observations of Kerr (1969), Hindman and Kerr (1970) and Henderson (1972). Most of Harten's general results are confirmed.

The observations may be summarized as follows:

(i) Despite a drastic change of distance to the Sun from the far branch to the near branch along any given inner arm (those within the solar circle), the gradient dv/db measured from the latitude-velocity diagrams undergoes no appreciable change. This, of course, immediately implies that the actual velocity gradient dv/dz is several factors greater in the near branch than in the far branch of the same spiral arm. For illustration, Harten's results on the Northern Sagittarius Arm are reproduced in Table I.

TABLE I

Rolling motions in the Northern Sagittarius Arm by Harten[a]

l	dv/db (km s^{-1} deg^{-1})		Distance (kpc)		dv/dz (km s^{-1} kpc^{-1})	
	Near	Far	Near	Far	Near	Far
50°	−0.3	−0.2	5.6		−3	
49.5	−0.2	+0.2	4.7	6.5	−2	+2
49	−0.8	+0.4	4.4	6.9	−10	+3
48.5	−1.5	−1.0	4.2	7.3	−20	−8
48	−2.4	−1.3	4.1	7.6	−33	−10
47.5	−2.1	−1.9	4.0	7.8	−29	−14
47	−1.8	−2.5	3.8	8.1	−26	−18
46.5	−2.4	≐1.9	3.7	8.3	−36	−13
46	−3.3	−2.5	3.6	8.5	−51	−17
45.5	−3.5	−2.0	3.5	8.7	−56	−13
45	−3.9	−3.0	3.4	8.9	−64	−19
44.5	−3.5	−2.1	3.3	9.1	−59	−13
44	−2.3	−1.6	3.3	9.3	−39	−10
43.5	−2.7	−1.7	3.2	9.5	−47	−10
43	−2.7	−1.7	3.2	9.7	−47	−10
42.3	−1.4	−0.5	3.1	9.8	−25	−3
41.3	−1.5	−0.5	3.0	10.2	−28	−3
40.3	−0.5	0.8	2.9	10.5	−10	−4
39.3	−0.4		2.8	10.8	−8	
38.3	−0.1	−0.2	2.8	11.1	−2	−1
37.3	−0.6	−0.1	2.7	11.4	−12	−5
36.3	−0.9	−0.8	2.6	11.7	−19	−4
35.3	−1.4	−0.8	2.6	11.9	−29	−4

[a] This table is reproduced for the purpose of demonstrating the great difference between dv/db and dv/dz. Harten's measurements in dv/db are somewhat larger than ours.

(ii) The gradients dv/db in the outer arms (those beyond the solar circle) and the '3 kpc arm' are predominantly negative for both the northern and the southern skies with an average of -2 km s^{-1} deg^{-1}. Most of the rolling motions can be followed continuously along a given spiral arm.

(iii) The values of the gradient dv/db in the inner arms (except the '3 kpc arm') are decisively smaller in magnitude and more evenly distributed in sign than those in the outer arms. Rapid variations are always found in the directions where the line-of-sight is tangent to a spiral arm.

III. A Simple Interpretation

If, for some unknown region, the gas in spiral arms were rolling in the way that Rougoor described, these rolling motions would likely be uniform along a spiral arm. In other words, the velocity gradient dv/dz must assume a fairly constant value for that arm. It is simply inconceivable that a portion of the spiral arm is turning at a speed two or three times faster than the other directly connected portion. Even more inconceivable is that the near portion of the spiral arm is always the faster turning one. Clearly, the first observational fact in the previous section constitutes strong evidence against the interpretation that the rolling motions in spiral arms actually exist.

As the second and the third observational facts indicate, the systematic and prominent 'rolling motions' are found mainly in the regions exterior to the position of the Sun. (Due to the presence of the Lindblad resonance, the '3 kpc arm' is not treated here.) This seems to suggest that the 'rolling motion' might have something to do with the bending of the galactic plane, which thus far stands as the most distinctive characteristic for the outer regions of the Galaxy. In fact, we shall show that the very fact of the bending of the galactic plane together with the differential rotation will be sufficient to cause the apparent 'rolling motions' in the observations.

The situation may be best illustrated in Figure 1, where an integral sign cross section of the Galaxy is shown schematically. The flattened oval shapes on the distorted portions of the galactic plane represent equal-density (gas) contours of two outer spiral arms. When these oval contours are mapped onto the latitude-velocity diagram, the corresponding contours for equal brightness temperature will appear tilted in the negative sense as shown in the lower part of Figure 1. Because of the bending of the galactic plane, the upper tangent point T_1 is situated farther from us than the lower tangent point T_2 on the same density contour: hence T_1 is more negative in velocity in the latitude velocity diagram. Similar results are obtained in the southern sky. It is remarkable that the negative sense of tilt is simultaneously achieved on both sides of the sky.

As pointed out by Henderson (1967) and Westerhout (1968), the inner spiral arms are not centered on the galactic plane. They are displaced up or down in a random manner, usually 10–30 pc from the central plane. If we take this into consideration, then we find that the same argument may be applied to the interpretation of the

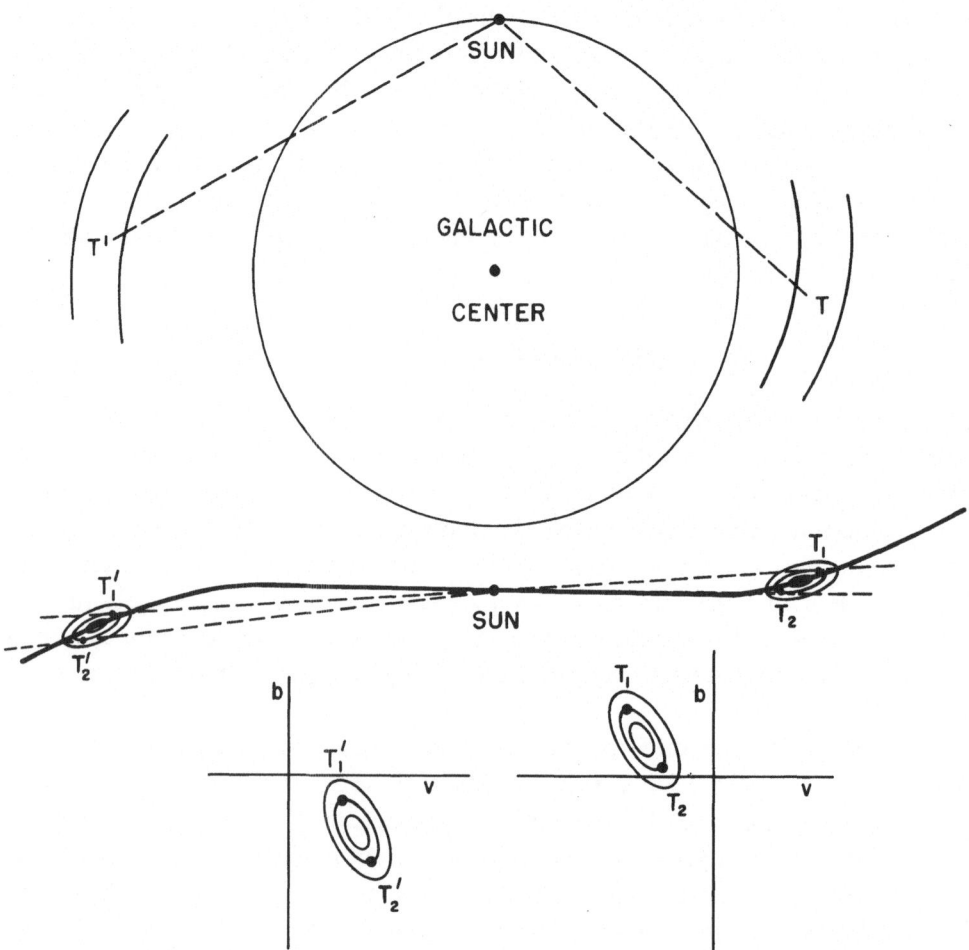

Fig. 1.   Geometric explanation of 'rolling motion' for the outer arms. (Top) Plan view showing outer arms. (Middle) Cross-section of the Galaxy showing the bending of the plane and the equal density contours of the two spiral arms. Points T_1 and T_2 are the two tangent points to the lines of sight on a given density contour. (Bottom) The corresponding contours mapped onto the $b-v$ diagram. Note that dv/db is negative in both cases.

'rolling motions' of the inner arms as apparent motions also. The situation, however, is more complicated, because it will depend on whether the arm is above or below the plane and whether the arm is the far branch or the near branch. The signs of the gradients, therefore, are expected to be more mixed, and the degree of the tilt, corresponding to displacements 10–20 pc, is expected to be less pronounced. It might be noted here that the rapid variation of the gradients near the tangent direction can be explained by the velocity crowding. Gas clouds distributed along a considerably lengthy interval centered on the tangent point are clustered into a small region of the velocity scale, hence producing rather impressive magnitudes of dv/db with mixed signs.

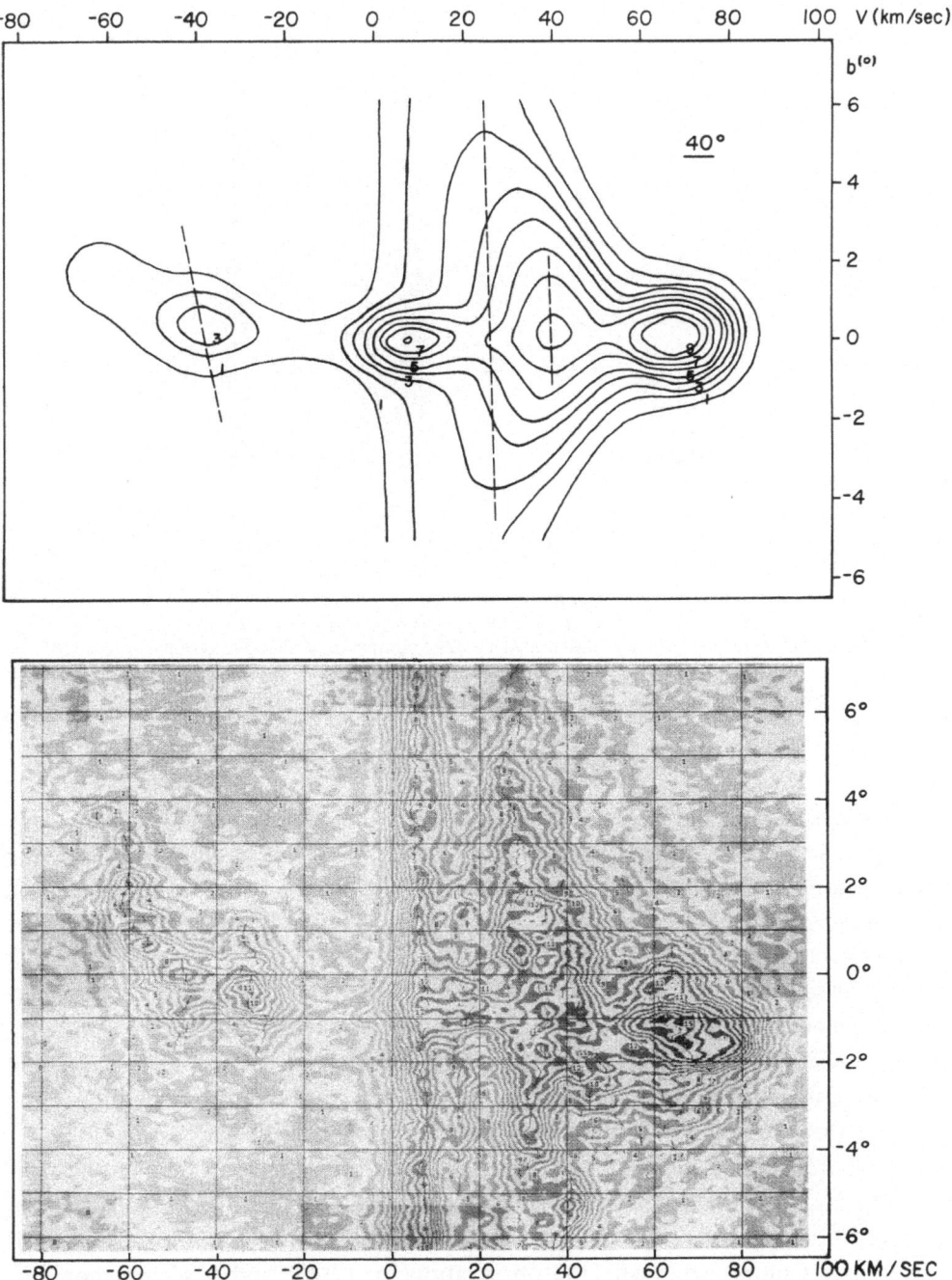

Fig. 2. Theoretical latitude-velocity diagram at $l = 40°$ compared with observations by Henderson at the same longitude.

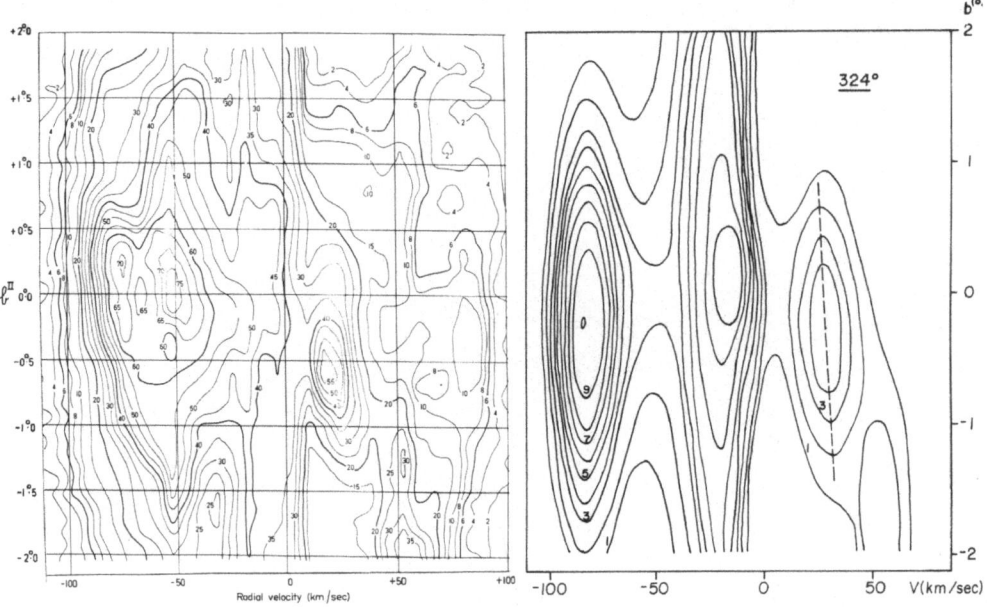

Fig. 3. Theoretical latitude-velocity diagram at $l=324°$ compared with observations by Kerr at the same longitude.

IV. Results of Numerical Calculations

In the above section, we have shown qualitatively that the rolling motions are only apparent. And these apparent motions are a natural result of the bending of the galactic plane or the displacement of an inner arm from the plane together with the differential rotation. In this section, we shall present quantitative results to support our interpretation.

Extensive numerical calculations have been carried out to simulate the observed latitude-velocity diagrams, based on the linear density-wave theory. The computation scheme is the same as that in constructing theoretical 21-cm profiles (e.g., Burton, 1970), except that the three dimensional motion of the gas and the bending of the plane (or the displacement of an inner arm) are included in the calculations.

The theoretical latitude-velocity diagrams were constructed for a number of longitudes for both the northern and the southern sky. It is demonstrated that the simple interpretation we have proposed also agrees well with the observations on a quantitative basis. Comparisons with the observation for $l=40°$ and $324°$ are shown in Figures 2 and 3.

V. Concluding Remarks

(i) Because of the extremely complicated nature of the '3 kpc arm', we have not tried to resolve the rolling motions in the '3 kpc arm' with our analysis.

(ii) With the present interpretation, we have avoided not only the difficulty of the origin of the rolling motions, if they are real, but also the problem of breaking the over-all gas flow in the Galaxy from the symmetry which is strongly established by the even distribution of the mass within the galactic plane.

(iii) The large-scale magnetic field with the strength and the topology observed in the Galaxy is not likely to be capable of introducing or maintaining an actual rolling motion. Fujimoto and Miyamoto's local analysis on the circular arm model is not free from the winding dilemma.

Acknowledgements

This work was supported in part by CUNY Faculty Research Award Program No. 1653 and by NASA Grant NGL-33-013-040. We would like to thank Dr R. H. Harten for allowing us to reproduce some data in his Ph.D. Thesis.

References

Burton, W. B.: 1970, *Astron. Astrophys. Suppl.* **2**, 291.
Fujimoto, M. and Miyamoto, Y.: 1969, *Publ. Astron. Soc. Japan* **21**, 194.
Fujimoto, M. and Tanahashi, Y.: 1971a, *Publ. Astron. Soc. Japan* **23**, 7.
Fujimoto, M. and Tanahashi, Y.: 1971b, *Publ. Astron. Soc. Japan* **23**, 13.
Harten, R. H.: 1972, Ph.D. Thesis, University of Maryland.
Henderson, A. P.: 1967, Ph.D. Thesis, University of Maryland.
Henderson, A. P.: 1972, *Latitude-Velocity Maps* $l = 16°$ to 230°, $\Delta l = 5°$, $b = 10°$ to $+10°$, Astronomy Program, University of Maryland.
Hindman, J. V. and Kerr, F. J.: 1970, *Australian J. Phys. Astrophys. Suppl.* No. 18, 43.
Kerr, F. J.: 1969, *Australian J. Phys. Astrophys. Suppl.* No. 9, 1.
Lynden-Bell, D.: 1965, *Monthly Notices Roy. Astron. Soc.* **129**, 299.
Oort, J. H.: 1962, in L. Woltjer (ed.), *Interstellar Matter in Galaxies*, Benjamin, New York, p. 3.
Rougoor, G. W.: 1964, *Bull. Astron. Inst. Neth.* **17**, 384.
Shane, W. W.: 1971, *Astron. Astrophys. Suppl.* **4**, 1 and 135.
Westerhout, G.: 1970, in H.-Y. Chiu and Muriel (ed.), *Galactic Astronomy*, Gordon and Breach, New York, Vol. 1, p. 167.
Yuan, C. and Wallace, L.: 1973, *Astrophys. J.* **185**, 453.

C. Yuan
L. Wallace
City College of New York,
New York, N.Y. 10031, U.S.A.

DISCUSSION

Baldwin: Are there any positions where the slope of dV/db is of the opposite sign, and how large are such regions?

Yuan: The positive slopes are seen rather frequently in the spiral arms interior to the position of the Sun. This is consistent with our mechanism. As for the spiral arms beyond the position of the Sun, positive slopes are rarely found, and in no case can they be followed more than a few subsequent longitudes.

HIGH VELOCITY NEUTRAL HYDROGEN CLOUDS

R. D. DAVIES

University of Manchester, Nuffield Radio Astronomy Laboratories, Jodrell Bank,
Cheshire, United Kingdom

Abstract. A review is given of the observations of neutral hydrogen high velocity clouds ($|v| > 80$ km s^{-1}) in and near the Galaxy. The positive and negative clouds are seen to have different distributions in the sky, following roughly the velocity pattern of galactic rotation. A characteristic of the majority of the clouds is their distribution in elongated bands or strings. The various theories of origin of HVCs are discussed; the possible role of the tidal interactions between the Magellanic Clouds and the Galaxy is emphasized. Tests are suggested to distinguish between the Oort theory of the infall of intergalactic material and theories which envisage the HVCs as originating in the outermost spiral structure.

I. Introduction

The major part of the neutral hydrogen in the Galaxy is confined to a disk-shaped region with a thickness at half intensity of a few hundred parsecs. Outside the Sun's distance ($R \gtrsim 10$ kpc) a significant departure from a plane occurs in the neutral hydrogen distribution in the sense that in the sector $l \sim 20°$ to $140°$ (galactocentric longitudes $L = 30°$ to $160°$) it lies above the galactic plane and in the sector $l = 240°$ to $340°$ ($L = 210°$ to $330°$) it lies below the plane. (See, for example, Kerr and Westerhout, 1965.) The centres of the arms at $R \simeq 15$ kpc lie at ~ 0.5 kpc from the plane in these regions. One has to look outside the Galaxy for the origin of this warp in the neutral hydrogen distribution. The cause is probably the tidal effect of the Magellanic Clouds (see, for example, Hunter and Toomre, 1969).

Within the Galaxy there is evidence for departures from non-circular motions of a few tens of km s^{-1}. Rickard (1968) has reported a systematic deviation of 20 to 30 km s^{-1} in a region $30°$ in extent on the galactic plane in Cassiopeia-Perseus. A number of investigators (Berkhuijsen *et al.*, 1971; Fejes and Verschuur, 1973) have also found velocities of 20–30 km s^{-1} associated with the continuum spurs.

Although the full width at half intensity of spiral arms near the Sun is in the range 200–300 pc there are significant amounts of material (a few per cent of the peak intensity) at the velocity of the arms, and therefore belonging to them, which reach 1–2 kpc from the galactic plane. This gas is seen for example in the Sagittarius and Perseus arms in the low latitude survey by Weaver and Williams (1973).

The present review will discuss the neutral hydrogen data which are not so obviously related to the narrow disk component. This gas is called the intermediate and high velocity gas. Much of it is in localized regions which have become known as intermediate velocity clouds (IVCs) and high velocity clouds (HVCs). Some of the gas is also in extended regions of lower brightness underlying the IVCs and HVCs. The division of clouds between IVCs and HVCs has somewhat arbitrarily been made at $|v| = 70$ km s^{-1} (Blaauw and Tolbert, 1966) or 80 km s^{-1} (Hulsbosch, 1973). Of course gas at velocities up to ~ 130 km s^{-1} relative to the LSR is well known in the plane of

F. J. Kerr and S. C. Simonson, III (eds.), Galactic Radio Astronomy, 599–616. All Rights Reserved.

the Galaxy and is compatible with circular motion. Indeed most IVCs are believed to be closely related to the spiral structure of the Galaxy (Blaauw and Tolbert, 1966). Our discussion will therefore be concerned with the HVCs, and it will be limited to gas with $|v| > 80$ km s^{-1}.

II. Surveys of HVCs

The Dutch groups first made surveys for HVCs with the Dwingeloo telescope (beam-width $= 0.6°$) and catalogued many HVCs (Muller et al., 1966; Hulsbosch and Rai-mond, 1966; Tolbert, 1971; Hulsbosch, 1968; van Kuilenberg, 1972a, 1972b). Another important survey was made at Ohio State University by Mathewson et al. (1966) and by Meng and Kraus (1970). This demonstrated that many of the brighter HVCs and IVCs are embedded in large elongated areas of low intensity emission at similar ve-locities. A series of high sensitivity observations of areas of the northern sky was made at a 2° angular resolution with the Bell Telephone Laboratories 20-ft horn reflector (Wannier and Wrixon, 1972; Wannier et al., 1972; Encrenaz et al., 1971). These showed extensive low intensity emission at both negative and positive velocities.

The available data on HVCs are summarized in Figure 1. This also includes the recent positive velocity data for extended emission features in the southern sky ob-served by Mathewson et al. (1973). They found relatively few negative velocity clouds. The dimensions plotted for the various clouds give an indication of their half intensity contours. The continuous bands of emission in which some of these concentrations lie are not shown. HVCs are plotted for $|v| > 80$ km s^{-1} with the addition of some positive velocity clouds in the range $v = +60$ to $+80$ km s^{-1} listed by van Kuilen-berg. A description of interesting features of the distribution will now be given.

(a) THE 'BAND' EXTENDING FROM $l = 50°$, $b = 10°$, TO $l = 130°$, $b = +55°$

Figure 1 shows a quasi-continuous band of HVCs beginning at $l = 50°$, $b = +10°$, passing through $l = 90°$, $b = +40°$, and continuing to $l = 130°$, $b = +55°$. HVCs with velocities in the range -80 to -165 km s^{-1} are seen; 80% are in the range -100 to -150 km s^{-1}. Further, there appears to be no systematic trend of velocity along the 'band'; the mean velocity at the high latitude end is -132 km s^{-1} compared with -114 km s^{-1} at the low latitude end.

The data available to the present are inconclusive on the question of whether this 'band' is a physical entity. They do, however, indicate that large areas of emission are continuous in the sky. Meng and Kraus describe the area from $l = 50°$ to $80°$ as a 'large quasi-continuous group'. Similarly, the section from $l = 90°$ to $138°$ has a com-mon envelope of detectable emission as shown for example in Figure 6b by van Kuilenberg (1972a). This latter region includes the brighter emission which was origi-nally designated complexes C I, C II and C III. The complexes M I and M II lie at the upper latitude end of this 'band'; observations so far do not suggest a connection between M I, M II and the 'band'.

The only break discernible in the published data for the 'band' is in the vicinity of $l = 80°$, $b = +30°$, and even then there is an isolated cloud at $l = 77°$, $b = +31°$ (Ohio

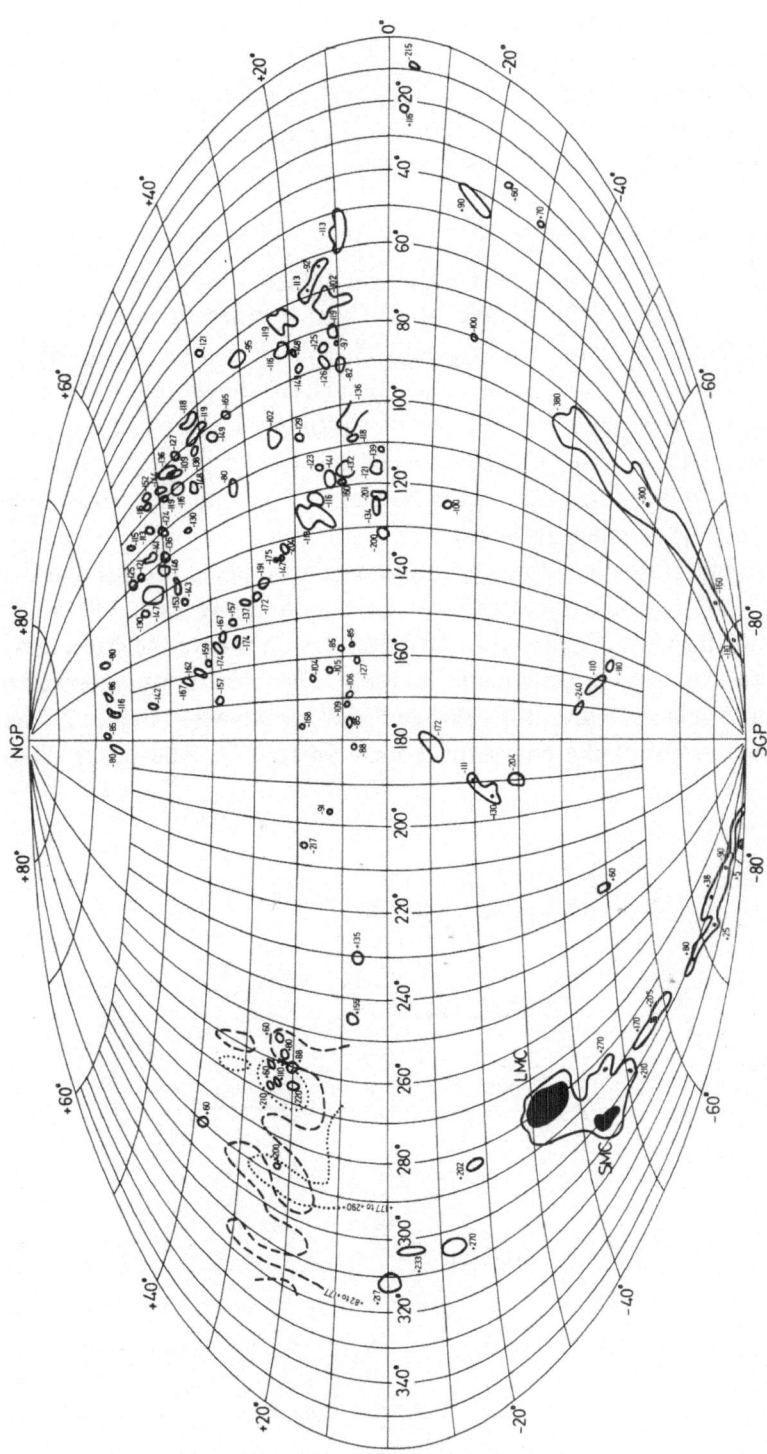

Fig. 1. The distribution of HVCs in galactic coordinates. The contours give an indication of the half-intensity dimensions of the clouds. Each cloud is labelled with its mean velocity relative to the LSR. In the case of the Magellanic stream (SGP feature) some velocities are indicated along its length. The Large and Small Magellanic Clouds are plotted to indicate their relationship to the Magellanic stream. An outline only is given of the HVC distribution at positive latitudes in the region $l = 250°$ to $320°$. Data have been taken from Hulsbosch (1968, 1973), Meng and Kraus (1970), van Kuilenberg (1972b), Mathewson et al. (1973) and Wannier and Wrixon (1972).

cloud OTH 468). High sensitivity observations will be required to detect any emission link through the region of the break.

Another interesting feature of the lower latitude section of this 'band' is the clear indication of a fork at $l=65°$, $b=+12°$. A lower arm of the fork runs at constant galactic latitude to $l=90°$, $b=+12°$. It has similar velocities to those seen in the upper arm.

(b) THE BAND EXTENDING FROM $l=100°$, $b=+7°$, TO $l=168°$, $b=+54°$

This band, which can first be distinguished at $l=100°$, $b=+7°$, and extends to $l=168°$, $b=+54°$, is the best defined band of HVCs at northern galactic latitudes. The high latitude end of this band was originally designated complex A. More sensitive studies (Hulsbosch, 1968; Giovanelli *et al.*, 1973) have shown that it is a quasi-continuous feature which extends for 30°, and its width is seldom more than 3°. It consists of bright small diameter features superposed on a low brightness background; these small-scale features will be discussed in Section V. Velocity differences in adjacent complexes may be as much as 50 km s^{-1}.

The low latitude end of this band merges into the galactic plane neutral hydrogen near $l=100°$, $b=+7°$, which has velocities at least as negative as -130 km s^{-1}. This situation is illustrated in Figure 6 of Hulsbosch (1973), where the band can be seen superposed on extensive galactic plane emission. High sensitivity observations in sections across the galactic plane at $l=100°$ and 120° are shown in Figure 2. These show the extended nature of the emission at $v<-96$ km s^{-1}. The peaks on the

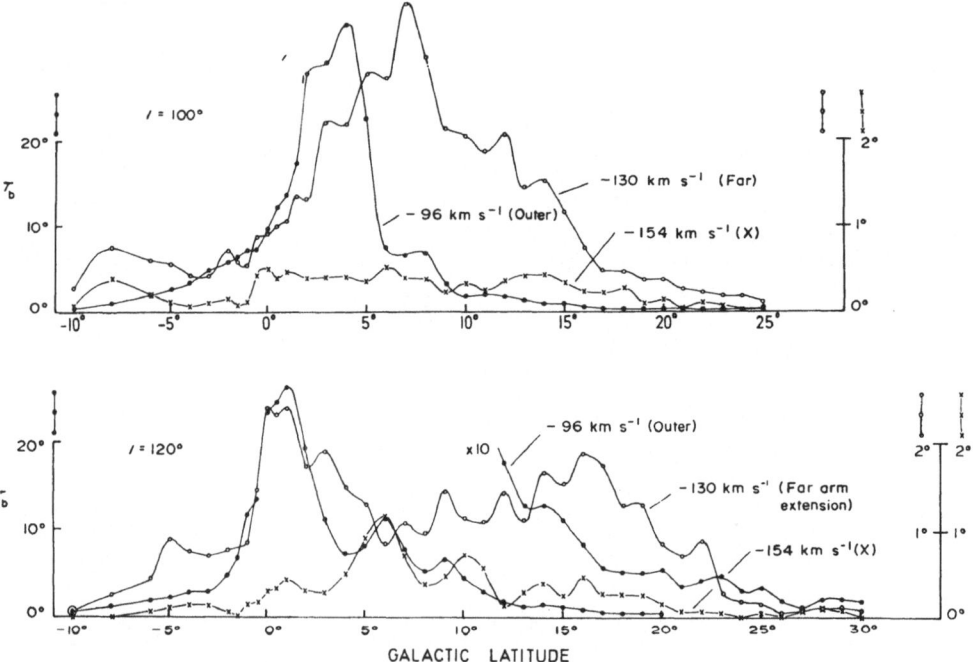

Fig. 2.   Sections across the galactic plane at $l=100°$ and 120° showing the broad distribution of high velocity gas at velocities in the range -96 to -154 km s^{-1} (from Davies, 1972).

$v = -130$ km s^{-1} plots in both sections were recognized as HVCs in early work and are plotted as individual clouds in Figure 1.

The existence of weak emission between the high latitude end of the band discussed in this section and the band discussed in Section IIa has been detected by Encrenaz *et al.* (1971). It is not clear whether it is physically associated with either of the bands; it could even be a connecting bridge between the two bands.

(c) THE BANDS OF EMISSION IN THE ANTICENTRE REGION

A number of bands of emission having substantial negative velocities are seen in the anticentre region. These are of particular interest since they cannot be understood in terms of any circular model of galactic rotation.

One of these bands runs approximately parallel to the galactic plane at $b = +8°$ between $l = 152°$ to $192°$. The principal emission peaks lie in the velocity range -70 to -130 km s^{-1}. This was labelled feature Q by Kepner (1970), arm U by Davies (1972) and arm E by Verschuur (1973).

Other elongated features are found along the line $l = 180°$, $b = -10°$, to $l = 195°$, $b = -25°$. These have been designated the anticentre complex (AC). Another string of HVCs is found along the line $l = 154°$, $b = -52°$, to $l = 171°$, $b = -44°$. These two lines contain clouds with a large negative velocity range (-110 to -240 km s^{-1}).

(d) POSITIVE VELOCITY CLOUDS IN THE REGION $l = 250°$ TO $320°$, $b = +10°$ TO $+30°$

A survey of positive velocity hydrogen above the plane in the range $l = 252°$ to $322°$ has been made by Wannier *et al.* (1972) with a high sensitivity system capable of detecting a neutral hydrogen column density of $\sim 4 \times 10^{18}$ cm^{-2}. A wealth of positive velocity emission was seen in this area with velocities extending to $+280$ km s^{-1}. Figure 1 indicates the areas of emission in the two velocity ranges $+82$ to $+177$ km s^{-1} and $+177$ to $+290$ km s^{-1}. The previously discovered positive velocity HVCs of Kerr (1964) and van Kuilenberg (1972a, b) lie within or adjacent to this area. The survey of Wannier *et al.* was taken on a $2° \times 2°$ grid with a $2°$ beamwidth, and accordingly there is no information on the fine structure which can be compared with the structure of the negative velocity gas in the northern sky.

(e) THE SOUTH GALACTIC POLE FEATURE AND THE MAGELLANIC STREAM

A high velocity complex was first observed near the South Galactic Pole (SGP) by Dieter (1965) and Hulsbosch and Raimond (1966). Subsequent observations by Hulsbosch (1968) and van Kuilenberg (1972a) showed a narrower region some $30°$ in extent extending to their southern horizon. It was unusual in having a velocity gradient along its length; more negative velocities are at the northern end. The very remarkable nature of this band was demonstrated by the high sensitivity survey of Wannier and Wrixon (1972) who found a lower intensity extension northwards to $l = 90°$, $b = -35°$. At this point the mean velocity relative to the LSR was -380 km s^{-1}.

The origin of this band was elucidated by the southern sky survey of HVCs by Mathewson *et al.* (1973) who showed that this feature emanated from the vicinity of the Magellanic clouds. This filament, which they named the Magellanic Stream, follows a great circle over an arc 150° long; they also describe an extension on the far side of the Magellanic Clouds which reaches to the galactic plane near $l=306°$. The velocity of the main Magellanic stream changes monotonically from a value of -380 km s^{-1} at $l=90°$, $b=-30°$, to $+250$ km s^{-1} in the vicinity of the Magellanic Clouds.

(f) A SUMMARY OF THE HVC PROPERTIES

Several dominant characteristics of HVCs emerge from their distribution as plotted in Figure 1. The first is the preponderance of negative velocities in the region at positive latitudes between $l=40°$ and 190°. This is within the region first surveyed at northern observatories and led to a belief that HVCs were principally negative velocity objects. The positional asymmetry about the galactic plane in this longitude range is in the sense of the well known tilt of the galactic plane; the possible significance of this fact will be discussed in later sections. The HVC velocities in the southern sky are mainly positive. Thus the sense of the velocities in both the northern and southern Galaxy are the same as produced by galactic rotation at the equator. The magnitude of the velocities, however, ranges up to values somewhat larger than normally expected for galactic rotation.

Another feature of the HVCs already commented upon in Section II and clearly indicated in Figure 1 is their confinement to elongated strings or bands. This is most vividly demonstrated in the band extending from $l=100°$, $b=+7°$, to $l=160°$, $b=+46°$. Other groups of HVCs in Figure 1 lie within elongated emission features. So far relatively little high angular resolution work has been done on the positive latitude regions between $l=250°$ to 320°, and it is not yet possible to say whether much elongated structure exists there.

Probably the most remarkable feature discovered so far is the Magellanic stream. It shares with some other HVCs the elongated band structure; it contrasts with them in possessing a large velocity gradient, at least when measured relative to the LSR, although this difference would disappear if the velocities were corrected for the effects of galactic rotation. The question of whether the Magellanic stream is a completely different feature from the majority of the HVCs described above will be discussed in later sections.

On the basis of the foregoing description of the main properties of the distribution of the HVCs we will now consider theories of their origin.

III. Extragalactic Interpretations of HVCs

The characteristic of HVCs shown by the early observations which appeared to distinguish them from the well-known galactic spiral arm gas was their greater range in velocity and greater spread in latitude. Furthermore, the clouds seemed, at least in the low sensitivity observations then available, to be isolated features. These facts led Oort (1966, 1969, 1970) to suggest that the HVCs were principally galactic gas swept up by the infall of intergalactic matter. Verschuur (1969) and Kerr and Sullivan (1969) considered the possibility that the HVCs were intergalactic clouds either at distances for a few hundreds or a few tens of kiloparsecs.

(a) THE HYPOTHESIS OF OORT

The preponderance of negative velocities found in the early observations of HVCs was interpreted by Oort (1966, 1969, 1970) to indicate that they were gas clouds moving into the Galaxy. He supposed that intergalactic material coming in at the free-fall velocity of ~ 500 km s^{-1} (as seen at the Sun) was slowed down to the observed velocities of ~ 120 km s^{-1} on sweeping up about three times their mass of galactic gas. This material was supposed to be in the halo of the Galaxy or in the upper reaches of the spiral arms. The corresponding rate of inflow of intergalactic gas was 1–2 $M_\odot$ per year, which would increase the mass of the Galaxy by 1% per 10^9 yr. The intergalactic density in the vicinity of our Galaxy would be $\sim 3 \times 10^{-4}$ atoms cm^{-3}. In order to maintain the galactic halo which supplies $\frac{3}{4}$ of the infalling mass in the HVCs, Oort proposed that superexplosions eject material from the galactic plane.

On this model the cloudy structure of the HVC gas is a consequence of irregularities in the halo. The problem of cooling the intergalactic gas after collision with the halo gas is overcome by the presence of coolants in the halo component which originated on the galactic plane (Savedoff et al., 1967).

The Oort model has difficulty in explaining the large body of positive velocity material in the southern sky. Also it has no immediate explanation of the elongated band-like structure of the HVC complexes making large angles with the galactic plane.

(b) HVCs AS LOCAL GROUP OBJECTS

Verschuur (1969) explored the possibility that the HVCs may be aggregates of matter (protogalaxies?) in the Local Group of galaxies which have not contracted to galactic dimensions or densities. He showed that the galactocentric velocities of HVCs and Local Group galaxies were similar. If the HVCs are required to be gravitationally stable their distances must be at least several hundred kpc. At a distance of ~ 400 kpc the H I masses of the complexes are $10^9 - 10^{10}$ $M_\odot$. These parameters Verschuur argued would be consistent with the HVCs being protogalaxies in the Local Group.

(c) HVCs AS SATELLITES OF THE GALAXY

Kerr and Sullivan (1969) also proposed that the HVCs were extragalactic but at distances such that they may be satellites of the Galaxy. They were able to derive orbits which fitted the galactocentric velocities and sky distribution of the HVCs with $|b| = 20°$ to 75°. The best-fitting orbits had (i) semimajor axes in the range 30 to 80 kpc, (ii) eccentricities of 0.5 to 0.8 and (iii) inclinations to the galactic plane of 40° to 70°. These distances are comparable with those of the Magellanic Clouds (55 kpc). The total mass of H I in HVCs would be $\sim 2 \times 10^8$ $M_\odot$, a value of the same magnitude as the Magellanic Clouds. Kerr and Sullivan then comment favourably on the suggestion by Toomre that galactic tidal forces may have broken this material away from the Magellanic Clouds during an earlier passage. Subsequent events have shown that such an explanation is applicable to the Magellanic stream – although it may not be applicable to all the northern HVCs which show close connections with the galactic plane H I.

(d) SHELL OF INFALLING MATERIAL

Dieter (1971) made a high sensitivity survey of the neutral hydrogen in the northern sky within $\pm 15°$ of the galactic plane. On the basis of these observations a model was constructed of the high velocity gas which included an infalling intergalactic component along the lines envisaged on the Oort model.

Dieter found extensive high negative velocity emission near the galactic plane which was intimately connected with the low velocity material. The lower latitude HVCs were situated in the extended emission. The maximum observed radial velocity varied with longitude but not exactly in the way expected for galactic rotation. These observations were explained in terms of a shell of high velocity gas that outlines the outermost spiral arms and extends ± 3.5 kpc from the galactic plane. This gas has a rotational velocity appropriate to its distance from the centre with an additional infall velocity of 125 km s^{-1} directed towards the galactic centre. The mass of infalling material situated at the outer edge of the Galaxy would be $\sim 2 \times 10^8 \, M_\odot$. Dieter follows Oort in ascribing this infall to intergalactic gas falling towards the centre and colliding with the outer regions of the Galaxy, thereby being slowed down to 125 km s^{-1}. The radial motion is a remnant from galaxy formation and is taken to be principally in the galactic plane. This model does not seek to explain the HVCs at higher latitudes ($|b| > 15°$) nor does it account for the extensive positive velocity emission in the southern sky.

IV. High Velocity Gas and Galactic Structure

The alternative approach to the understanding of the HVCs is to consider them as an integral part of the outer galactic spiral arm system. The well-observed outer spiral structure can be traced out to velocities of ± 130 km s^{-1}. Emission at these velocities, although within the range classified as high velocity clouds, is strongest on or near the galactic plane. Habing (1966) demonstrated that in the longitude range $l = 42°$ to 142° this negative velocity emission extended above the plane and was still strong at $b = +10°$ to $+15°$. A more extensive survey of the region $l = 48°$ to 228°, $b = +6°$ to $+20°$, by Kepner (1970) revealed a number of high velocity features extended in longitude and latitude which were not simple continuations of known spiral arm structures. Some of these are illustrated in Figure 3a, where velocities of up to -150 km s^{-1} are seen. However, following the Oort model, she chose to explain the five features O^*, OP^*, P^*, Q and R as 'high velocity features resulting from collisions with intergalactic gas'. Outer extensions of the well-documented spiral arms were traced to 1–2 kpc above the plane in many cases.

A more direct argument for the majority of the HVCs to be associated with outer

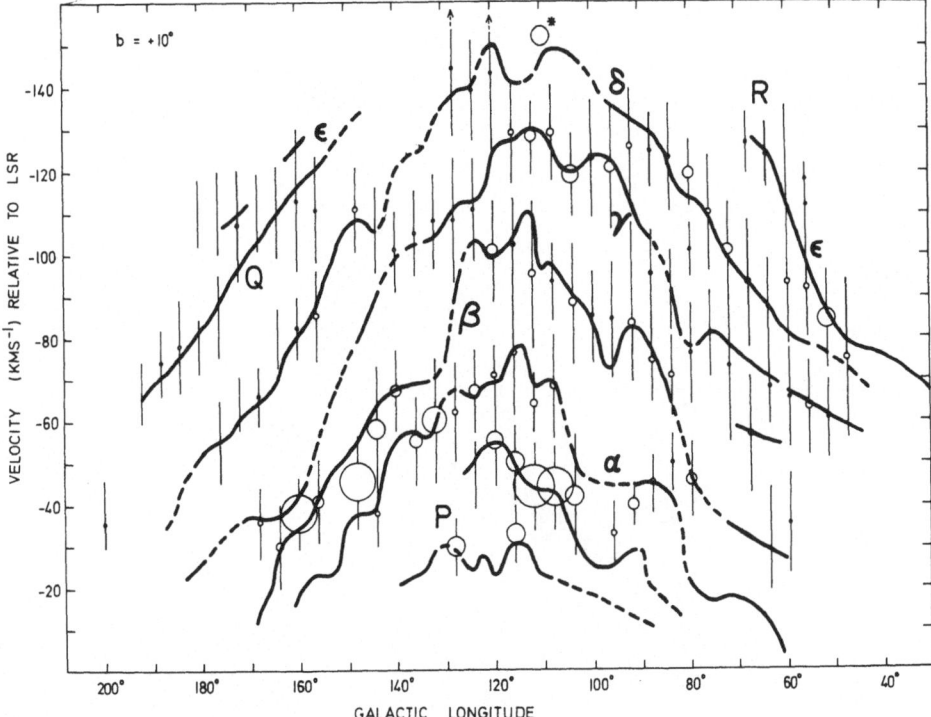

Fig. 3a. The longitude-velocity diagram for negative velocities at $b = +10°$ according to Verschuur (arms α, β, γ, δ, and ε shown as heavy lines) and to Kepner (as vertical lines through circles whose diameters indicate the intensity.

spiral arms was made on the basis of extensive new observations by Davies (1972, 1973) and Verschuur (1973) independently. Davies made latitude cuts at fixed longitudes around the plane which showed the intimate connection of high velocity material with the galactic plane. The latitude cuts at $l = 100°$ and $120°$ are shown in Figure 2, which illustrate the broad distribution of negative velocity gas about the galactic plane, with a dominant extension above the plane. The brighter peaks in these cuts were previously classified as HVCs. Verschuur's observations consisted of longitude cuts at $b = +5°$, $+10°$ and $+15°$. The velocity versus longitude plots from these two series of observations are shown in Figures 3a and 3b. They outline well-defined ridges or arm-like structures lying outside the well-known arms with velocities in the range -150 to $+150$ km s^{-1} and can be traced in both the northern and southern skies. It was argued that these features had the characteristics of spiral arms, viz. (i) they appear as separate velocity features on the spectra, (ii) they are restricted in latitude to near the galactic plane, (iii) they are extended in longitude by at least several times their extent in latitude and (iv) they show changes in velocity with longitude consistent with galactic rotation. Wannier et al. (1972) also showed that there was extensive emission in the region $l = 252°$ to $322°$ and $b = +10°$ to $+30°$ at velocities in the range $+82$ to $+290$ km s^{-1} which they suggested was most likely a high-latitude extension of the outer spiral arms.

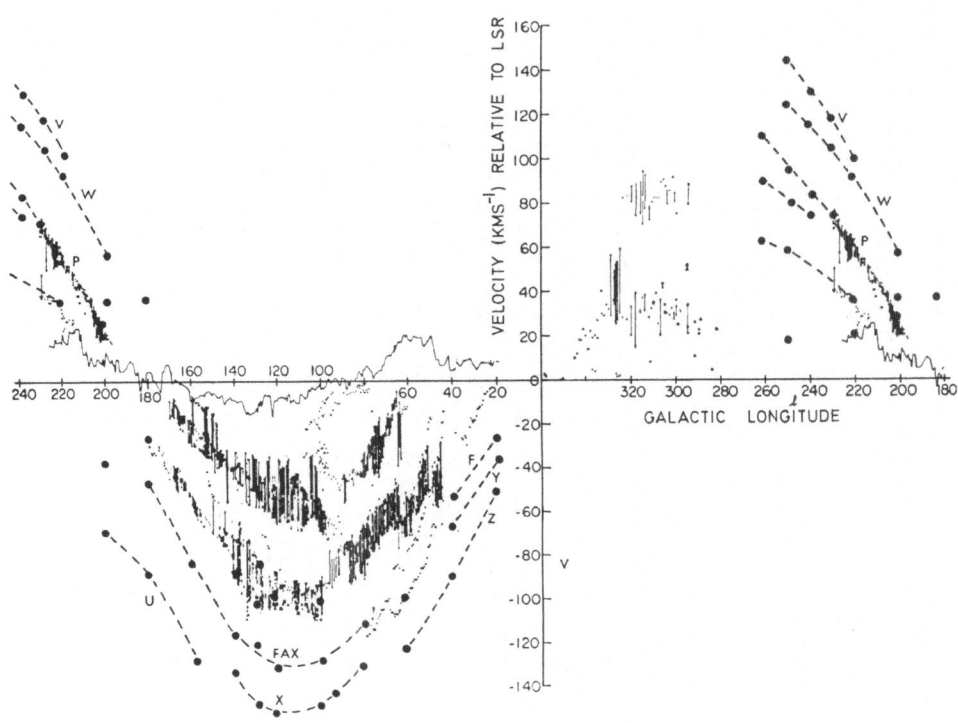

Fig. 3b.   The longitude-velocity diagram for features near the galactic plane in the longitude range 20°
to 260°. This shows the high velocity 'spiral arm' features studied by Davies (1972) superposed on the bright
features plotted by Weaver (1971).

There can be little doubt now that the majority of the H I at low latitudes (say
$|b| < 15°$ to 20°) is intimately connected with the Galaxy and its outer spiral structure.
Many of the features classified as HVCs and plotted in Figure 1 are a part of these
spiral arms. A characteristic of the high velocity gas distribution in the region $l = 40°$
to 180° is the preponderance of material at positive latitudes; this extends the pattern
already well known as the tilt of the plane of the stronger spiral arms outside the Sun's
distance. The actual location of these outer spiral arms is difficult to assess with any
precision because the galactic rotation curve is not well-determined in the outer parts
of the Galaxy, and moreover systematic velocities of a few tens of km s^{-1} which are
found elsewhere will produce further uncertainties in position. They are obviously
outside the well-determined outer arms ($R \sim 15$ kpc). Davies and Verschuur place
the arms in the range $R = 15$ kpc to perhaps 25 kpc or more.

Davies argued that those HVC in the bands emanating from near the galactic plane
(e.g., those beginning near $l = 50°$, $b = +8°$, $l = 100°$, $b = +6°$, and $l = 158°$, $b = +8°$)
were also part of the outermost spiral structure. These bands were envisaged as being
warped further out of the galactic plane than the nearer spiral arms. This greater displacement from the galactic plane is of the magnitude and sense that would be expected in theories where gravitational tides are produced in the Galaxy by the Magel-

lanic Clouds. The small fraction of HVCs with velocities grossly different from values expected for galactic rotation, for example those at high latitude and those in the anticentre at large negative velocities, were supposed to have been wrenched out of the Galaxy by the tidal forces. This is in accordance with the tidal theory and observations of external galaxies (Hunter and Toomre, 1969; Toomre and Toomre, 1972; Wright, 1972; Clutton-Brock, 1972; Eneev et al., 1973). No new phenomenon is required to explain any of the HVC data.

Verschuur's discussion of the relationship between HVCs and outer spiral arms essentially follows the discussion outlined above – emphasizing the physical connections between HVCs and spiral arms near the plane. His explanation of the noncircular motions includes the further possibilities that they may be due to a magnetohydrodynamic wave running along a spiral arm or to the winding-up of the outer spiral arms.

V. Fine Structure in HVCs

The early surveys of HVCs showed that both individual HVCs and the structures within high velocity complexes had sizes of a few degrees. The velocity width at half intensity points was measured to be 20 to 25 km s^{-1} (Hulsbosch, 1973). The structure in IVCs was found to be similar.

Observations with higher resolution in angle and velocity as well as with higher sensitivity showed that this is an oversimplified picture. Many of the HVCs reveal finer structure as may be seen in Figures 4a and 4b taken from Davies and Buhl (1974). These diagrams are RA-velocity cuts through the HVC complexes A and MII with the 10′ beam of the NRAO 300-ft telescope at Green Bank. Structure is evident on the scale of 0.7° in RA and 10 km s^{-1} in velocity (halfwidth). Verschuur et al. (1972) found bright cores in HVC $112+2-139$ and HVC $132+23-204$ with dimensions of 0.5° and velocity widths of 5 km s^{-1}. A similar situation was found by Giovanelli et al. (1973) in complexes A, M I, M II, C I and C III. Another characteristic of many of the complexes is the large range of velocities seen in adjacent positions, sometimes as great as 50 km s^{-1}. This suggests an intrinsic spread of velocities in the gas constituting the HVC bands. Given a distance, it sets a limit on the lifetime of the bands. For example, taking the main band through $l=140°$, $b=+28°$, to be at a distance of 15 kpc, a systematic velocity difference of 20 km s^{-1} would separate features by 8° in 10^8 yr – a value consistent with the observations.

The fine structure in the northern HVC bands can be compared with that of the Magellanic stream and the region between the Large and Small Magellanic Clouds. Figure 5, taken from Murray et al. (1974), shows the structure in a cut through the Magellanic stream at Dec$= -50°$. Two main branches of the stream are intersected in this cut – at RA$=00^h55^m$, $+90$ km s^{-1}, and RA$=01^h23^m$, $+140$ km s^{-1}. Weaker extended emission surrounds the main peak in each branch.

The average structural parameters of the following types of cloud are compared in Table I: (i) HVCs in the bands, (ii) the Magellanic stream, (iii) the region between the Magellanic clouds and (iv) spiral arm clouds in the galactic plane with $|v| \sim 100$ km s^{-1}.

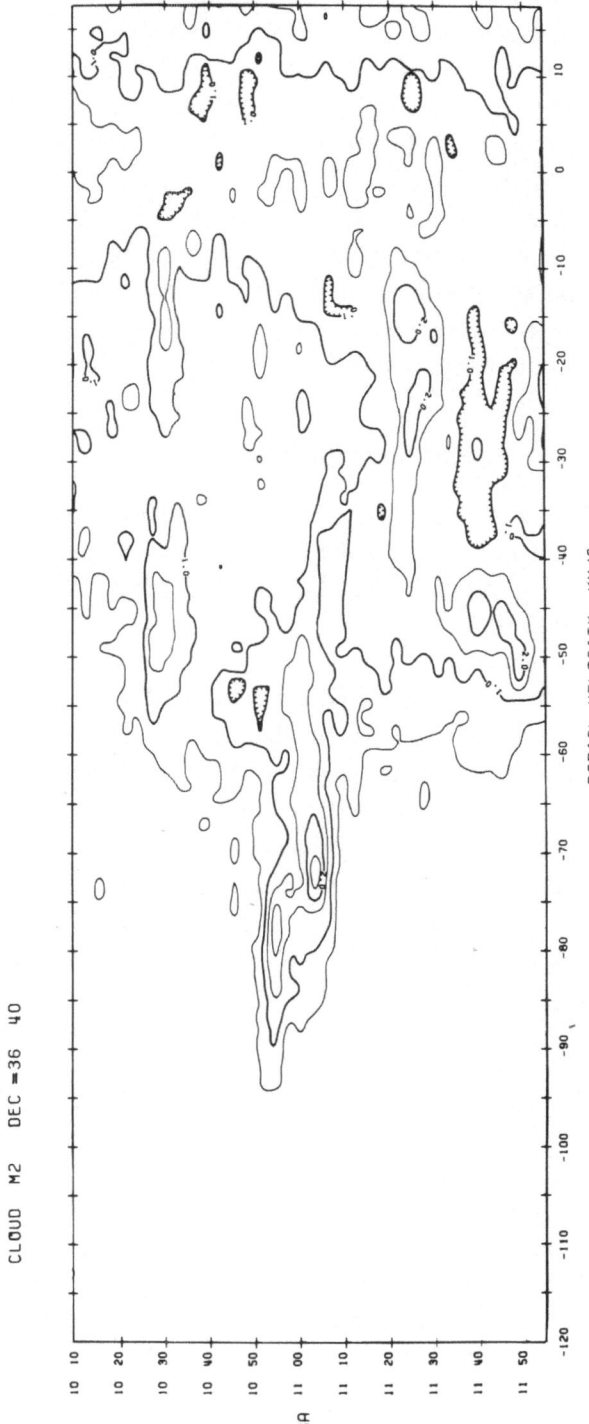

Fig. 4a, b. (a) Right ascension-velocity sections through cloud complexes M II (Dec = 36°40′) and (b) A (Dec = 62°30′) with a beamwidth of 10′ and velocity resolutions of 1.7 and 3.5 km s^{-1} showing fine structure in velocity and position (Davies and Buhl, 1974).

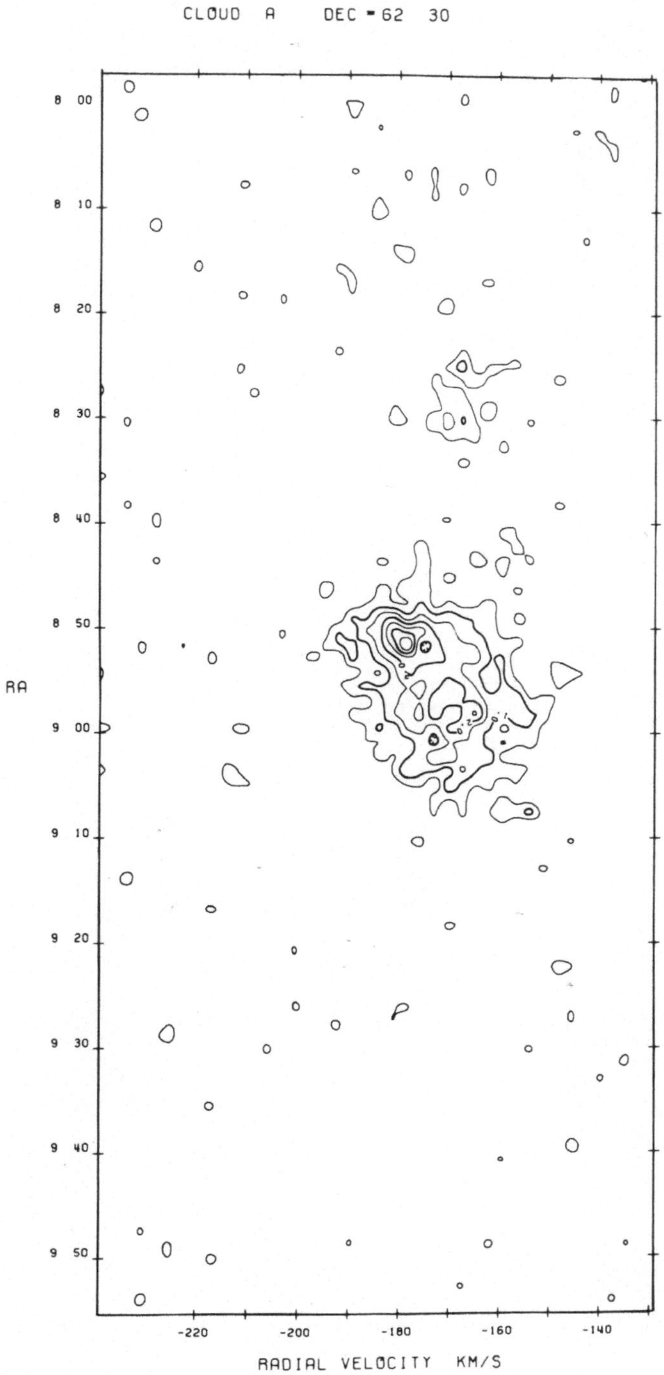

Fig. 4b.

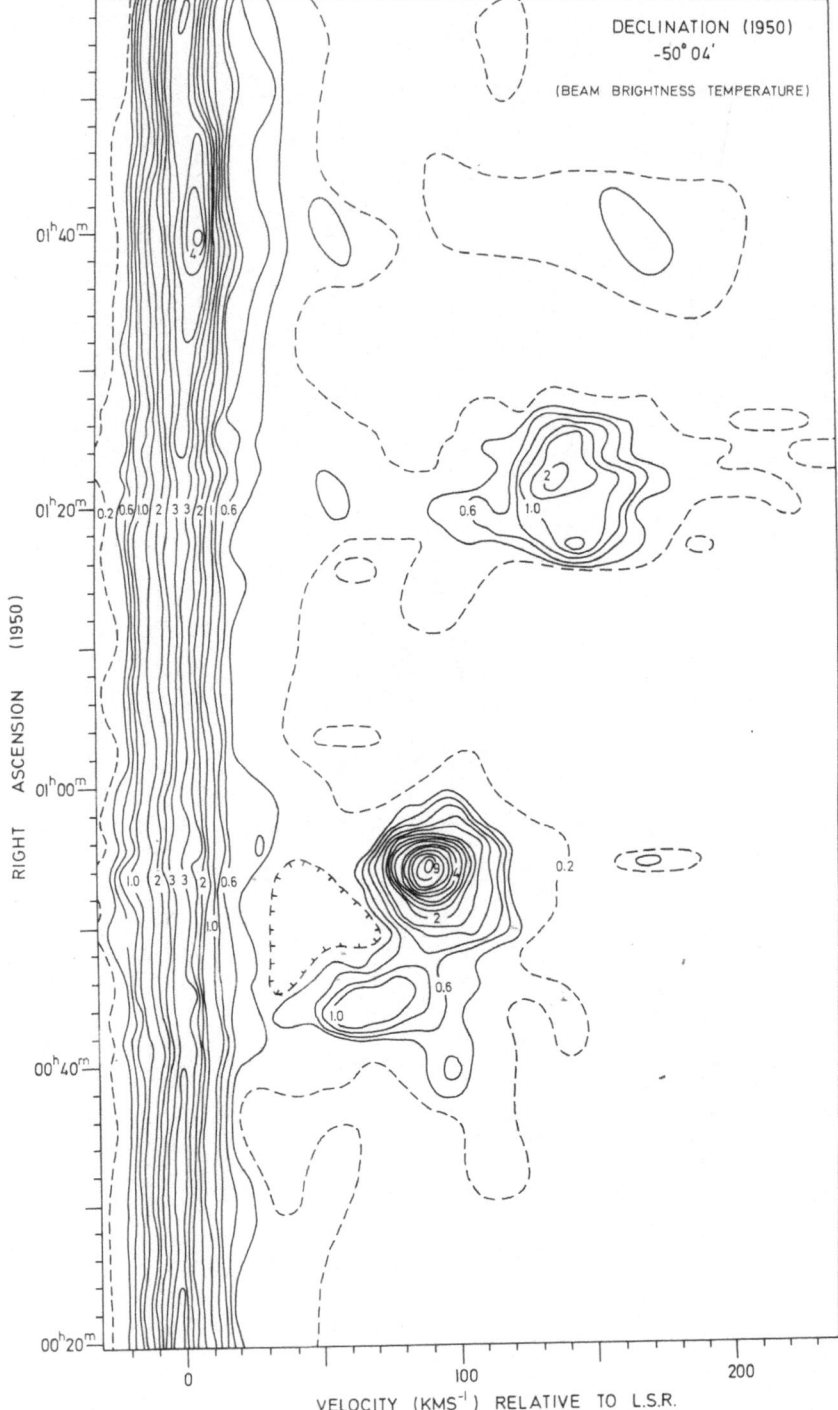

Fig. 5.   A right ascension-velocity section through the Magellanic stream at Dec = −50°04′ (1950) with
a resolution of 14 arcmin in angle and 7 km s⁻¹ in velocity (Murray *et al.*, 1974). Two branches of the stream
are shown in this section, one at 0ʰ55ᵐ and one at 01ʰ23ᵐ.

It can be seen that the HVCs have similar structural parameters to the Magellanic streams and the inter-Cloud region. The spiral arm clouds have about half the velocity spread of the other clouds.

A further important consideration is whether the clouds listed in Table I are gravitationally stable. If we take a typical HVC with a dispersion of 8 km s^{-1} (half width $= 20$ km s^{-1}) and a diameter of 0.5° at a distance of 20 kpc, the mass required to stabilize the cloud is $\sim 10^6$ $M_\odot$. However, in a cloud with $N_H = 10^{20}$ cm^{-2} the total hydrogen mass would be $\sim 2 \times 10^4$ $M_\odot$ at the assumed distance. We are accordingly left with a deficiency of a factor of ~ 30 when account is taken of the contribution due to helium (and possibly an ionized component). This calculation demonstrates, as has been known since the earliest observations, that the HVC complexes are not gravitationally bound at distances of a few tens of kpc. However, the narrow velocity components of the HVCs could well be gravitationally stable with velocity dispersion of ~ 3 km s^{-1} and a rather greater distance.

TABLE I

Structural parameters of HVCs

	Average half power width	
	Velocity (km s^{-1})	Angle
HVCs in bands	20 (10*)	0°5 (0°3 [a])
Magellanic stream	26	0.8
Region between Magellanic Clouds	21	0.74
Spiral arm clouds with		
$\quad |v| \sim 100$ km s^{-1}	10	0.9

[a] Velocity structure within a cloud.

Table I also shows that the parameters of the clouds in the Magellanic streams and in the region between the Clouds are similar. Taking a distance of 60 kpc for both, it is again found that the cloud masses are insufficient (by a factor of ~ 10 on the average in this case) to make them gravitationally stable.

There is no reason *a priori* why the HVCs should be gravitationally bound. Indeed, the argument can be inverted to give an age for this clouds in the various complexes. An expansion velocity of 10 km s^{-1} in a cloud complex 2.0° radius at a distance of 20 kpc leads to an upper limit to the age of $\sim 10^8$ yr. Alternatively, the small components within the HVCs with a velocity dispersion of 3 km s^{-1} and a radius of 0.25° would have a lifetime of $\sim 3 \times 10^7$ yr. At the distance of the Magellanic Clouds the ages would be ~ 3 times these values. These arguments suggest that we are dealing with clouds produced in the recent past ($\sim 10^8$ yr) if their distances are 20–60 kpc as the evidence suggests. We shall argue in Section VII that it is not a coincidence that this is of the same order as the time of the estimated perigalacticon passage of the Magellanic Clouds.

VI. HVCs and Other Phenomena

Several correlations between HVCs and optical phenomena have been investigated in order to throw light on the nature of HVCs. The observations of interstellar absorption lines in the spectra of high-z O and B stars by Münch and Zirin (1961) showed several intermediate velocity absorbing clouds in the velocity range -70 to $+78$ km s^{-1} relative to the LSR. Habing (1969) noted that these intermediate velocity optical components could be found in stars at sufficiently large distances from the galactic plane; the components were also seen in H I emission spectra taken in the direction of the stars. He concluded that the clouds were at distances of ~ 1 kpc from the galactic plane. Hulsbosch (1973) reports that Herbig searched unsuccessfully for optical absorption lines in the spectra of several stars lying in the direction of a few HVCs. An intermediate velocity cloud was seen, however, at a distance of between 400 and 1700 pc. All that can be concluded from these studies is that IVCs lie within a distance of ~ 1 kpc; they give no evidence about the distance of HVCs except that they are further than several kpc.

Minkowski *et al.* (1972) noted that there was a tendency for H II regions of Lynds' catalogue to cluster in the vicinity of high velocity ($|v| > 100$ km s^{-1}) gas. They speculate that the excitation may be causally related to the high-velocity gas – a 100 km s^{-1} shock would photoionize substantial amounts of interstellar gas. Until the association is confirmed by velocity measurements of the H II regions it is too early to pursue this further.

On the theoretical side Silk (1973) has made the proposal that the negative velocity HVCs may be propelled towards the Galaxy by the background soft X-rays. These will ionize the side of the cloud away from the Galaxy (the near side is protected by the Galaxy) and accelerate it inwards by the Oort-Spitzer rocket effect. There are several difficulties with this model: (i) it cannot explain the presence of positive velocity clouds, (ii) it is difficult to obtain a sufficient amount of gas in the neutral state at large distances from the Galaxy in the required X-ray flux and (iii) adequate soft X-rays have not been detected to effectively maintain this process.

VII. Conclusion

The main observational characteristics of the HVCs have been reviewed above. The salient properties which have to be taken into account in any theories of their origin are: (i) there are extensive areas of both negative and positive velocity gas, (ii) the major part of the HVC gas lies in bands or strings in which there are concentrations having small velocity spread and small angular diameter and (iii) much of the HVC material at low galactic latitudes ($|b| < 20°$) is part of the outer spiral structure of the Galaxy.

One new element in the understanding of the HVCs is the observation by Mathewson *et al.* of the streams emanating from the Magellanic Clouds. This establishes clearly that the Clouds are being tidally disrupted by the Galaxy during a close approach in which perigalacticon passage must have occurred several times 10^8 yr ago. There is now little doubt about this close approach between the Galaxy and the Magellanic Clouds; in recent years it has been the most widely accepted explanation of the well-known warp in the H I distribution in the Galaxy. The timescale and the distance of the perigalacticon passage required for the tidal effects in the Galaxy and the Magellanic Clouds are closely similar. The possible (and likely) contribution of this interaction to material in the vicinity of the Galaxy must be kept in mind in any discussion of HVCs.

More than one origin may have to be sought for the full range of HVC phenomena. Clearly the Magellanic streams have been drawn from the Magellanic Clouds by the tidal influence of the Galaxy. On the other hand, the bulk of the low latitude HVCs are generally agreed to be a part of the galactic spiral structure which also suffers a tidal warp in the outer reaches of the Galaxy. The discussion of the origin of the HVCs centres around the remainder, namely those HVCs at higher latitudes and those at lower latitudes with a large component in their velocities not attributable to galactic rotation. The line of argument favoured by the reviewer would be that this remaining category of HVCs would also be a result of the interaction between the Galaxy and the Magellanic Clouds. Those clouds at higher latitudes in the region above the galactic plane between $l = 40°$ and $200°$ and the positive velocity gas near the plane at $l = 250°$ to $320°$ would be gas originating in the Galaxy and pulled out of the plane by the tidal interaction in the manner discussed by Hunter and Toomre. This would leave the HVCs at the highest latitudes and in the anticentre region which could be either material that essentially has been pulled from the Galaxy *or* the Magellanic Clouds and is now falling towards the Galaxy. The velocities observed are all energetically possible. It is now necessary to calculate orbits for material drawn out of the Galaxy and the Magellanic Clouds by the encounter and see whether a satisfactory fit can be made to the observed orientation and velocity pattern in this latter group of HVC bands. This is one of the most important tasks to be done to elucidate the origin of thise HVCs.

One is then left to consider the role of the infalling intergalactic gas proposed by Oort. It is one possible mechanism to explain the velocity of those clouds which depart strongly (in the negative sense only) from galactic rotation velocities. The theory also requires an intergalactic medium with a density of $\sim 3 \times 10^{-4}$ cm^{-3} and a neutral hydrogen halo surrounding the galactic disk. The existence of an intergalactic medium of this density and a mechanism for ejecting gas from the plane can only be conjectured. Further confirmatory evidence for these phenomena would be of help in assessing the contribution of the intergalactic infall mechanism. One possible check for the existence of infalling intergalactic gas has been suggested by Oort at *IAU Symp.* **58** where he drew attention to the possible influence of this gas on the dynamics of the Magellanic streams.

References

Berkhuijsen, E. M., Haslam, C. G. T., and Salter, C. J.: 1971, *Astron. Astrophys.* **14**, 252.
Blaauw, A. and Tolbert, C. R.: 1966, *Bull. Astron. Inst. Neth.* **18**, 405.

Clutton-Brock, M.: 1972, *Astrophys. Space Sci.* **17**, 292.
Davies, R. D.: 1972, *Nature* **237**, 88.
Davies, R. D.: 1973, *Monthly Notices Roy. Astron. Soc.* **160**, 381.
Davies, R. D. and Buhl, D.: 1974, in preparation.
Dieter, N. H.: 1965, *Astron. J.* **70**, 552.
Dieter, N. H.: 1971, *Astron. Astrophys.* **12**, 59.
Encrenaz, P. J., Penzias, A. A., Gott, R., Wilson, R. W., and Wrixon, G. T.: 1971, *Astron. Astrophys.* **12**, 16.
Eneev, T. M., Kozlov, N. N., and Sunyaev, R. A.: 1973, *Astron. Astrophys.* **22**, 41.
Fejes, I. and Verschuur, G. L.: 1973, *Astron. Astrophys.* **25**, 85.
Habing, H. J.: 1966, *Bull. Astron. Inst. Neth.* **18**, 323.
Habing, H. J.: 1969, *Bull. Astron. Inst. Neth.* **20**, 177.
Hulsbosch, A. N. M.: 1968, *Bull. Astron. Inst. Neth.* **20**, 33.
Hulsbosch, A. N. M.: 1973, Thesis, Sterrewacht te Leiden.
Hulsbosch, A. N. M. and Raimond, E.: 1966, *Bull. Astron. Inst. Neth.* **18**, 413.
Hunter, C. and Toomre, A.: 1969, *Astrophys. J.* **155**, 747.
Kepner, M.: 1970, *Astron. Astrophys.* **5**, 444.
Kerr, F. J.: 1964, Lagonissi Lecture.
Kerr, F. J. and Sullivan, W. T.: 1969, *Astrophys. J.* **158**, 115.
Kerr, F. J. and Westerhout, G.: 1965, *Stars and Stellar Systems* **5**, 167.
Kuilenberg, J. van: 1972a, *Astron. Astrophys.* **16**, 276.
Kuilenberg, J. van: 1972b, *Astron. Astrophys. Suppl.* **5**, 1.
Mathewson, D. S., Clearly, M. N., and Murray, J. D.: 1973, *Astrophys. J.* **190**, 291.
Mathewson, D. S., Meng, S. Y., Brundage, W. D., and Kraus, J. D.: 1966, *Astron. J.* **71**, 863.
Meng, S. Y. and Kraus, J. D.: 1970, *Astron. J.* **75**, 535.
Minkowski, R., Silk, J., and Siluk, R. S.: 1972, *Astrophys. J. Letters* **175**, L123.
Muller, C. A., Raimond, E., Schwarz, U. J., and Tolbert, C. R.: 1966, *Bull. Astron. Inst. Neth. Suppl.* **1**, 213.
Münch, G. and Zirin, H.: 1961, *Astrophys. J.* **133**, 11.
Murray, J. D., Davies, R. D., Mathewson, D. S., and Cleary, M. N.: 1974, in preparation.
Oort, J. H.: 1966, *Bull. Astron. Inst. Neth.* **18**, 421.
Oort, J. H.: 1969, *Nature* **224**, 1158.
Oort, J. H.: 1970, *Astron. Astrophys.* **7**, 381.
Rickard, J. J.: 1968, *Astrophys. J.* **152**, 1019.
Savedoff, M. P., Hovenier, J. W., and van Leer, B.: 1967, *Bull. Astron. Inst. Neth.* **19**, 107.
Silk, J.: 1973, *Astron. Astrophys.*, in press.
Tolbert, C. R.: 1971, *Astron. Astrophys. Suppl.* **3**, 349.
Toomre, A. and Toomre, J.: 1972, *Astrophys. J.* **178**, 623.
Verschuur, G. L.: 1969, *Astrophys. J.* **156**, 771.
Verschuur, G. L.: 1973, *Astron. Astrophys.* **22**, 139.
Verschuur, G. L., Cram, T. R., and Giovanelli, R.: 1972, *Astrophys. Letters* **11**, 57.
Wannier, P. and Wrixon, G. T.: 1972, *Astrophys. J. Letters* **173**, L119.
Wannier, P., Wrixon, G. T., and Wilson, R. W.: 1972, *Astron. Astrophys.* **18**, 224.
Weaver, H.: 1971, in W. Becker and G. Contopoulos (eds.), 'The Spiral Structure of Our Galaxy', *IAU Symp.* **38**, 126.
Weaver, H. F. and Williams, D. R. W.: 1973, *Astron. Astrophys. Suppl.* **8**, 1.
Wright, A. E.: 1972, *Monthly Notices Roy. Astron. Soc.* **157**, 309.

R. D. Davies
University of Manchester,
Nuffield Radio Astronomy Laboratories,
Jodrell Bank, Cheshire SK11 9DL, United Kingdom

(Discussion follows the paper by D. S. Mathewson et al., p. 623.)

THE MAGELLANIC STREAM

D. S. MATHEWSON and M. N. CLEARY

Mount Stromlo and Siding Spring Observatory, Australian National University, Canberra, Australia

and

J. D. MURRAY

Division of Radiophysics, CSIRO, Sydney, Australia

Abstract. A southern sky survey of H<small>I</small> in the velocity range -340 km s^{-1} to $+380$ km s^{-1} has shown that a long filament of H<small>I</small> extends from the Small Magellanic Cloud (SMC) region down to the South Galactic Pole and connects with the long H<small>I</small> filament discovered recently by Wannier and Wrixon (1972) and van Kuilenberg (1972). There is also some evidence that this continues on the other side of the Magellanic Clouds and crosses the galactic plane at $l=306°$. This filament, which follows very closely a great circle over its entire 180° arc across the sky, is given the name 'The Magellanic Stream'. It may have been produced by gravitational interaction between the SMC and the Galaxy during a close passage (20 kpc) of the SMC some 5×10^8 yr ago, although it is impossible to account for the observed radial velocities along the Stream unless some force other than gravity is invoked to act on the Stream as well.

Recently a southern sky survey has been made for H<small>I</small> in the velocity range -340 km s^{-1} to 380 km s^{-1} using the 18-m reflector at Parkes.

The main discovery of this survey is shown in Figure 1a, and that is the long filament of H<small>I</small> which extends from the region of the Small Magellanic Cloud (SMC) to the South Galactic Polar cap and beyond. This is given the name 'The Magellanic Stream'. All H<small>I</small> is plotted in this diagram except the 'zero-velocity' (i.e., local) gas. The average velocity half-width of the H<small>I</small> in the Magellanic Stream is about 30 km s^{-1} which is much broader than the average half-width of 7 km s^{-1} of the local spiral arm gas.

Figure 1b shows the areas of high positive velocity H<small>I</small> (greater than 200 km s^{-1} with respect to the local standard of rest and clearly resolved from spiral arm emission) discovered on the other side of the SMC and extending through the galactic plane at $l=306°$ and up to $b=+30°$. This gas may also belong to the Magellanic Stream. The two clouds centered on $l=289°$, $b=20°$, and $l=268°$, $b=20°$, were discovered by Wannier *et al.* (1972); the latter cloud has probably been produced by the expanding shell of the Gum nebula and should not be included in the Magellanic Stream.

Figure 2 shows the full extent of the Magellanic Stream on an Aitoff projection of the sky in galactic coordinates. The section between $l=320°$, $b=-80°$, and $l=80°$, $b=-65°$, had been surveyed previously by van Kuilenburg (1972); also see Dieter (1965) and Hulsbosch and Raimond (1966). Wannier and Wrixon (1972), using a more sensitive receiver, increased the observed length of this section of the Stream up to $l=90°$, $b=-30°$. The crosshatched patches are the clouds of high velocity H<small>I</small> discovered much earlier by northern hemisphere observers mostly at positive latitudes in the longitude range 80° to 180°.

A striking feature of the Magellanic Stream is that it follows closely a great circle over its entire 180° across the sky. If all the H<small>I</small> in the Stream were at the distance of the

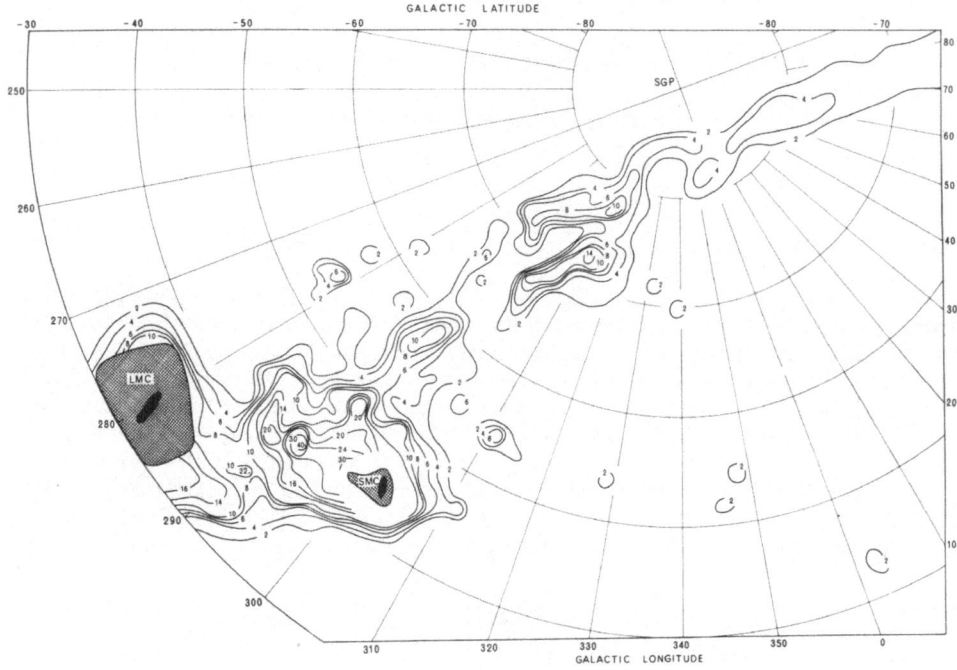

Fig. 1a. The section between the Magellanic Clouds and the South Galactic Polar region; all H I within the velocity range -340 km s^{-1} to $+380$ km s^{-1} is plotted except the 'zero-velocity' H I i.e., local spiral arm gas. The contours give the surface densities of the H I in the Magellanic Stream obtained using the 18-m reflector at Parkes. The contour unit is 2×10^{19} atoms cm^{-2}. The cross-hatched regions represent the approximate optical extent of the Large Magellanic Cloud (LMC) and Small Magellanic Cloud (SMC).

SMC (63 kpc), its mass would be about 10^9 $\mathscr{M}_\odot$, which is equal to the combined mass of H I in the LMC and SMC. The gas mass of the inter-Cloud region between the SMC and the LMC accounts for half of this total.

Figure 3 shows the variation of radial velocity (V_{GSR}) with angular distance (θ) along the Stream using the coordinate system of Wannier and Wrixon (1972) in their Figure 2. The radial velocities referred to the local standard of rest (LSR) have been corrected for the galactic rotation at the Sun. These corrected velocities ($V_{GSR} = = V_{LSR} + 225 \sin l \cos b$) are essentially velocities with respect to a non-rotating Galaxy. The systematic variation of radial velocity found by Wannier and Wrixon is seen to continue until the Magellanic Clouds are reached. However, on the other side of the Clouds there is no systematic variation of radial velocity, which produces a nagging doubt as to whether this H I is really part of the Stream.

Many interpretations of the Magellanic Stream have been considered, but the most favoured one is that gravitational interaction between the SMC and the Galaxy has pulled out the Magellanic Stream from the SMC. Toomre (1973) in unpublished work (referred to by Mirabel and Turner, 1973) has constructed an orbit for the SMC

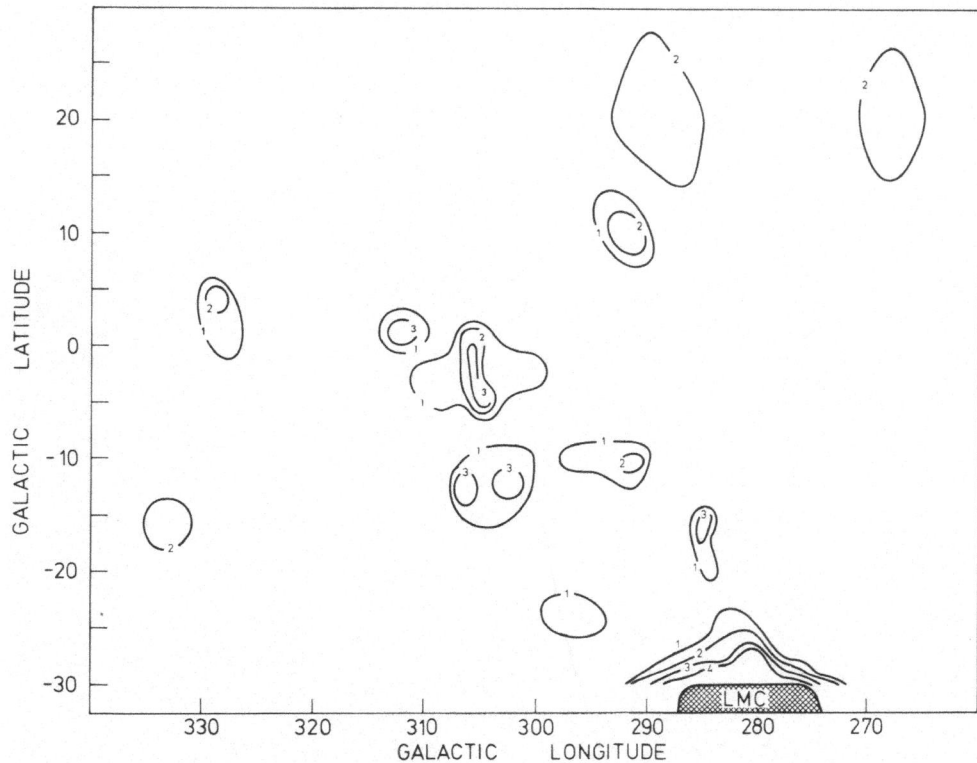

Fig. 1b.   The section between the Magellanic Clouds and $+30°$ galactic latitude; all H$_I$ greater than $+200$ km s^{-1} relative to the local standard of rest is plotted. The two clouds centered on $l=289°$, $b=20°$, and $l=268°$, $b=20°$, were taken from Wannier *et al.* (1972). The latter cloud has probably been produced by the expanding shell of the Gum nebula and should not be included in the Magellanic Stream.

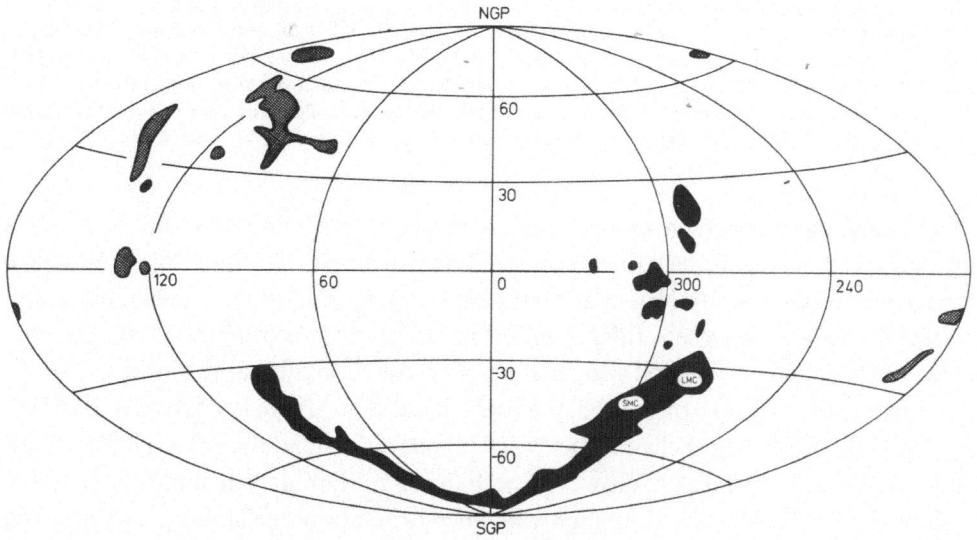

Fig. 2.   The Magellanic Stream is drawn on this Aitoff projection in galactic co-ordinates. The cross-hatched areas are the high velocity H$_I$ clouds discovered much earlier by northern hemisphere observers (cf. Hulsbosch, 1972).

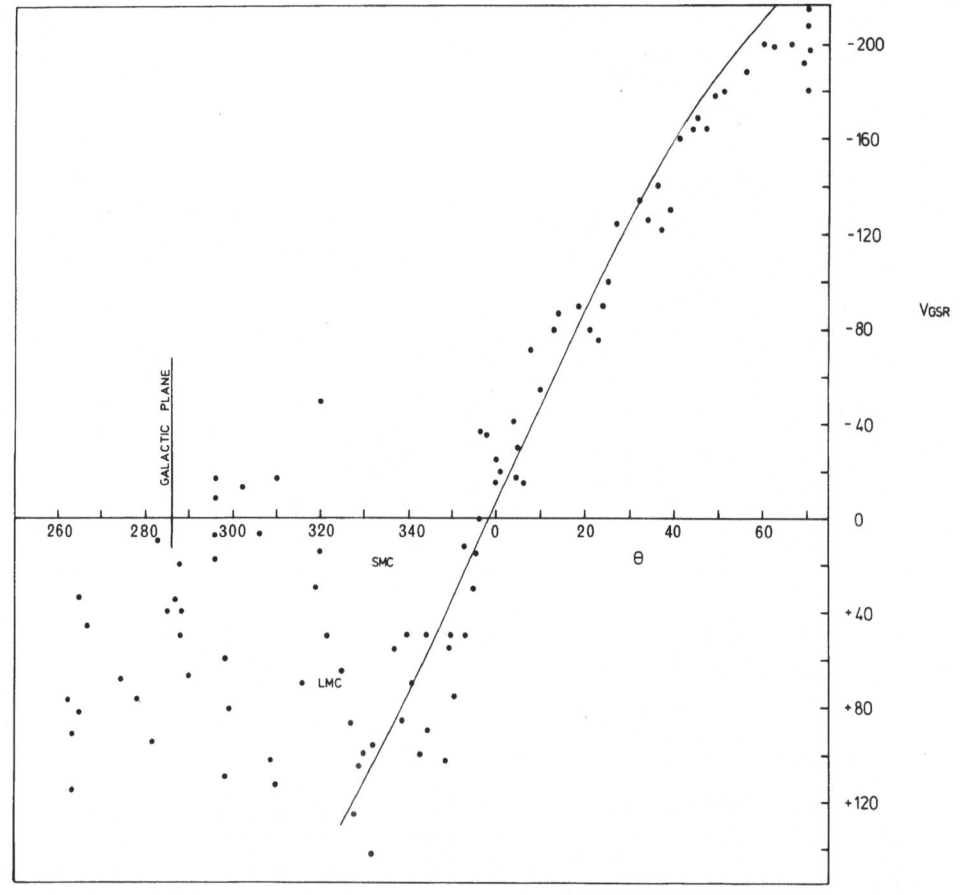

Fig. 3. Plot of radial velocity in km s^{-1} (V_{GSR}) versus the angular distance (θ) along the Magellanic Stream using the co-ordinate system of Wannier and Wrixon (1972) in their Figure 2. Their radial velocities have been used in the section from $\theta = 70°$ to $\theta = 30°$. The systemic radial velocities of the SMC (20 km^{-1} s) and the LMC (68 km s^{-1}) have also been plotted. The radial velocities (V_{GSR}) refer to a non-rotating galaxy and are obtained from the velocities relative to the local standard of rest by the relation $V_{GSR} = V_{LSR} + 225 \sin l \cos b$. The full line represents the relation $V_{GSR} = -240 \sin(\theta + 2°)$.

about the Galaxy in which perigalactic passage occurred some 5×10^8 yr ago at a distance of 20 kpc from the galactic center. Toomre found that the SMC would have undergone severe disruption, material being drawn out in a bridge and a tail characteristic of such violent tidal events. The tail lay well behind the LMC but projected onto the line joining the two galaxies. Toomre thought that the inter-Cloud H I (Hindman *et al.*, 1963; Turner, 1968) may be the tail. The bridge lay between the SMC and the Galaxy (see Mirabel and Turner, 1973) but unfortunately was angled at about 40° to the Magellanic Stream. However, perhaps on reexamination, it may be possible to allow the SMC to have a more highly inclined orbit about the Galaxy, which would shift the bridge closer to the South Galactic Pole and more nearly coincident with the Magellanic Stream.

Clutton-Brock (1972) has produced plots to show the tearing of a galaxy by the close passage of a very much larger galaxy. He finds that under these conditions, the gas and stars are drawn out from the smaller galaxy to form a very prominent bridge and tail. If the orbit of the SMC was highly inclined to the galactic plane then it may be seen from Clutton-Brock's Figures 4g and 4i that in 5×10^8 yr after perigalactic passage, a bridge and tail would be formed by the SMC giving a very similar spatial distribution to that of the Magellanic Stream. Wright (1973) has made similar computations to those of Clutton-Brock and has reached a similar conclusion.

However, whilst the existence and position of the Magellanic Stream may be explained in this way, it is impossible to explain the radial velocities observed along the Stream. In particular the high negative radial velocity ($V_{GSR} = -216$ km s^{-1}) at the tip of the Stream at $l = 90°$, $b = -30°$, is almost an order of magnitude greater than that predicted by the models of Clutton-Brock and Wright, which give a radial velocity in this region of only -30 km s^{-1}.

Oort (1973), who worked independently on the interpretation of the Magellanic Stream, overcomes this difficulty by suggesting that the Magellanic Stream 'snow-ploughs' through an intergalactic wind and is thereby braked and blown towards the galactic center. This increases the component of the velocity of the Stream along our line of sight and also increases the velocity of infall of the Stream. Oort calculates that an intergalactic wind of density 2×10^{-4} atoms cm^{-3} would be sufficient to produce the observed radial velocity at the tip of the Stream. This seems quite a reasonable speculation, as an intergalactic gas density of this order is necessary to explain the stability of the Local System (Oort, 1970). In addition Oort invokes an intergalactic wind of this density to explain the high velocity H I clouds found by northern hemisphere observers at positive latitudes between longitudes 80° and 180°.

However, there are difficulties with this concept because denser parts of the Stream such as that around $l = 300°$, $b = -73°$, would be relatively unaffected by snow-ploughing through the intergalactic wind and would continue along their original orbits so that they should have more positive radial velocities than the nearby less dense sections. However, the reverse is the case. In addition it is difficult to explain the positive radial velocities in the Stream near the Magellanic Clouds (e.g., 130 km s^{-1} at $l = 286°$, $b = -47°$, and 90 km s^{-1} at $l = 294°$, $b = -58°$) compared to the systemic radial velocities of 20 km s^{-1} for the SMC and 68 km s^{-1} for the LMC.

It is tempting to speculate that part of the Magellanic Stream may already have hit the galactic plane within 5 kpc of the Sun at $l \approx 100°$ and that its momentum pushed some disk gas to z distances of several kiloparsecs. As the width of the Stream is about 5 to 10 kpc, the disk gas would have been disturbed over quite a broad range of galactic longitudes around $l = 100°$. At this point the mechanism put forward by Oort (1970) to explain the origin of the high velocity H I clouds (cross-hatched in Figure 2) may operate. That is that the gas falls back into the galactic plane under the pressure of the intergalactic wind and the gravitational field of the Galaxy. This action of the Magellanic Stream on the H I in the plane replaces the need for super-explosions postulated by Oort (1970) to replenish the gas in the halo.

This interesting discovery has many follow-up observations, the most important of which is to search for optical emission. In this regard Bok (1966) commented that a striking feature of the distribution of globular clusters in the Clouds is that many of them are far from the two galaxies and indeed some of them cannot be assigned membership specifically to either Cloud because of this (Gascoigne and Lyngå, 1963).

This 'discovery' paper would not be complete without mentioning some earlier papers which, although not necessarily correct, were 'hot on the trail' of the Magellanic Stream. Nearly twenty years ago de Vaucouleurs (1954) discussed the possibility of a connection between the Magellanic Clouds and the Galaxy and coined the name 'The Magellanic Stream'; and Kerr and Sullivan (1969) considered that the high velocity H I clouds known at that time may be satellites of the Galaxy (perhaps debris scattered around the orbit of the LMC) at distances of the order of 50 kpc. It should also be noted that Hulsbosch (1972) when discussing the long South Polar filament found by van Kuilenburg and Wannier and Wrixon, suggested that "it may be a tidal arm expelled from the LMC by an encounter with the Galaxy".

Acknowledgements

The authors acknowledge valuable discussions about the interpretation of this feature with Prof. Toomre, Dr Kalnajs and Dr Hulsbosch, and they thank Prof. Oort for communicating the results of his work on the interpretation of the Magellanic Stream prior to their publication. The authors would also like to thank David Cooke, Frank Trett and Bob Phelps of the resident engineering staff at Parkes for keeping the receiver operational and Gary da Costa and Graham White from Mt. Stromlo for assistance with the observations.

References

Bok, B. J.: 1966, *Ann. Rev. Astron. Astrophys.* **4**, 95.
Clutton-Brock, M.: 1972, *Astrophys. Space Sci.* **17**, 292.
Dieter, N. H.: 1965, *Astron. J.* **70**, 552.
Gascoigne, S. C. B. and Lyngå, G.: 1963, *Observatory* **83**, 38.
Hindman, J. V., Kerr, F. J., and McGee, R. X.: 1963, *Australian J. Phys.* **16**, 570.
Hulsbosch, A. N. M.: 1972, *Studies on High-Velocity Clouds*, dissertation, Leiden University.
Hulsbosch, A. N. M. and Raimond, E.: 1966, *Bull. Astron. Inst. Neth.* **18**, 413.
Kerr, F. J. and Sullivan, W. T., III: 1969, *Astrophys. J.* **158**, 115.
Kuilenberg, J. van: 1972, *Astron. Astrophys.* **16**, 276.
Mirabel, I. F. and Turner, K. C.: 1973, *Astron. Astrophys.* **22**, 437.
Oort, J. H.: 1970, *Astron. Astrophys.* **7**, 381.
Oort, J. H.: 1973, private communication.
Toomre, A.: 1973, private communication.
Turner, K. C.: 1968, *Annual Report of the Director, Department of Terrestrial Magnetism 1967–68*, Carnegie Inst. of Washington, p. 291.
Vaucouleurs, G. de: 1954, *Observatory* **74**, 23.
Wannier, P. and Wrixon, G. T.: 1972, *Astrophys. J. Letters* **173**, L119.
Wannier, P., Wrixon, G. T., and Wilson, R. W.: 1972, *Astron. Astrophys.* **18**, 224.
Wright, A. E.: 1973, private communication.

D. S. Mathewson
M. N. Cleary
Mt. Stromlo & Siding Spring Observatory,
Private Bag,
Woden, A.C.T. 2606, Australia

J. D. Murray
CSIRO Division of Radiophysics,
P.O. Box 76,
Epping, N.S.W. 2121, Australia

DISCUSSION

Pishmish: How are the distances of the high velocity clouds determined?

Davies: Where the HVC is at lower latitudes and associated clearly with the outer spiral structure a dynamical distance may be estimated. Otherwise, the neutral hydrogen data by themselves do not give a distance (e.g., HVCs at high latitudes or in the anticenter).

Habing: Distances to high-velocity clouds can be obtained in principle by observing interstellar lines in distant stars. This method has so far not been successful although in a few cases distances of at least a kiloparsec been obtained.

Habing: What makes you think that those separate clouds belong to the Stream that lie on the extrapolation of the Stream beyond the Magellanic Clouds? Their velocities do not seem to indicate a relationship to the Stream, and their position beyond the Clouds makes, at first sight, their relation to the Stream doubtful.

Mathewson: The reasons why they have been included in the Stream are (1) they have very high positive velocities, which sets them apart from the spiral arm emission, (2) they lie close to the great circle followed by the Magellanic Stream on the other side of the Clouds, and (3) they are elongated in the direction of the Stream, which is at right angles to the galactic plane. However, the lack of systematic variations in the radial velocities is worrying and produces a doubt as to whether they do belong to the Stream, partiularly as the high positive velocity cloud at $l = 268°$, $b = 20°$, is almost certainly associated with the Gum nebula.

Oort: Davies has described the interpretation that Verschuur and he have suggested for the high-velocity clouds as parts of distant spiral arms bent away from the galactic plane. They have suggested also that the elongated structures shown by several of the high-velocity features connect up with gas close to the plane. The possibility of such a connection has recently been investigated in detail by Hulsbosch for the large high-velocity complex which has been called cloud A. He found no evidence for such a connection. In my opinion the real nature of high-velocity 'clouds' or 'streams', can best be seen in those at high and intermediate latitudes. A characteristic of these streams is an extreme unevenness in density as well as velocity. In cloud A one finds velocity jumps from one region to another of 40 km s^{-1} or more; in the so-called anticenter complex these differences run up to 100 km s^{-1}. These high internal motions suggest that what we observe in these streams is an interaction between two streams of gas. This is also indicated by the sharp edges found in some of these objects. An interaction of such a nature could occur as a consequence of the collision of an infalling wind of inter-galactic gas with the highly irregular gaseous halo of the Galaxy.

Davies: I agree that the higher latitude HVCs will give important clues about the origin of HVCs. Since the lower latitude HVCs are closely associated with the outer spiral arms, I would argue that we should first look for an origin involving material in some way associated with this outer spiral structure. On this basis the known tidal interaction between the Milky Way and the Magellanic Clouds seems to offer a ready explanation in terms of material pulled from the plane in the outer regions of the Milky Way or indeed material actually torn out of the Magellanic Clouds. Both negative and positive velocities can be produced in this way, and furthermore the material is likely to be in bands having a substantial velocity dispersion.

Mathewson: I think probably it's more unified than that. The Magellanic Stream may actually cause the high velocity clouds of northern latitudes. The Stream is coming down on the galactic plane, and it can quite easily eject the disk material. It is coming in at 250 km s^{-1}, and it has got a lot of mass. It can quite easily carry the gas in the disk afterwards. There is a bit of a gap here, but it only takes 10^7 yr to cover that gap. I think the clouds may have shot up 3–5 kpc. What the Northern Hemisphere observes is this gas falling back into the disk, probably pushed by the inter-galactic wind as Oort has suggested.

Van Woerden: The interpretation of most of the high-velocity clouds at high *b* as halo gas pushed down by infalling intergalactic matter appears to be supported by:

(i) the presence of a large intermediate – negative – velocity (INV) complex at very high positive latitudes: $b \sim +70°$ (Blaauw and Tolbert: 1966, *Bull. Astron. Inst. Neth.* **18**, 405);

(ii) the velocity field of this complex, indicating a flow at 70 km s^{-1} coming from $l \sim 120°$, $b \sim 40°$ (Wessalius and Fejes: 1973, *Astron. Astrophys.* **24**, 20, Fig. 3, and p. 33, Fig. 11);

(iii) the hole in the distribution of low-velocity hydrogen centered at $l \sim 160°$, $b \sim +70°$ (Wesselius and Fejes: 1973, *Astron. Astrophys.* **24**, 25, Fig. 6) and coinciding approximately with the INV complex.

Guélin: Evidence of intergalactic gas of tenuous density has been brought in evidence in the multiple system M81-M82-NGC 3077 (R. Davies, *IAU Symp.* **58**; M. Guélin, XVth General Assembly).

Pishmish: Has the search of the high velocity clouds been systematic? Are all galactic longitudes covered by the search?

Mathewson: Yes.

THE LARGE-SCALE STRUCTURE OF LOCAL Hi, DUST,
AND GALACTIC RADIO CONTINUUM

CARL E. HEILES

University of California, Berkeley, Calif., U.S.A.

Abstract. The correlation of Hi and dust at high latitudes gives different results depending on whether one uses galaxy or reddening data for the dust distribution. It is not certain to what extent this depends on differences in the gas to dust ratio or on difficulties in determining the amount of dust from the optical data. The correlation of Hi and continuum radiation shows Hi structures associated with the large-scale loops. There is also a coincidence of a hole in both the Hi and continuum.

I want to discuss two aspects of the interstellar medium that have been illuminated to some small degree by the Hat Creek survey of Hi at intermediate galactic latitudes (Heiles and Habing, 1974; Heiles and Jenkins, 1974). I discuss them here because they both have relevance to the structure of the Galaxy within a few hundred parsecs of the Sun.

I. Hi and Dust

(a) RADIO DATA

Many people have correlated the gas, as determined from 21-cm emission, and dust, as determined either from reddening measurements or from galaxy counts. Lilley (1955) formed the first correlation, which has been confirmed rather well, quantitatively speaking, by most of the later studies. However, a recent study by Wesselius and Sancisi (1971) using the Lick galaxy counts by Shane and Wirtanen (1967) and the 21-cm line data of Tolbert (1971), both averaged over 10° by 10° squares, seemed to show that the correlation was not so good as previously thought and perhaps even nonexistent. This is also in contrast to the recent study using UV line measurements and reddening by Savage and Jenkins (1972), who found a good correlation, albeit with large scatter.

Optical astronomers have in recent years gathered a wealth of data on interstellar reddening, and the Shane and Wirtanen (1967) galaxy counts yield data on total extinction. The Hat Creek intermediate latitude 21-cm line survey provides an essentially completely sampled data base for the Hi column density. Comparison of these sets of data provides new degrees of statistical confidence due to the large number of independent samples on the sky. These comparisons are currently being performed by Heiles and Jenkins (1974), and I would like to give a progress report on our present results and their possible implications.

Earlier in this symposium you have seen some of the 'photographs' of Hi column density in the sky made by Heiles and Jenkins (1973) from the Hat Creek survey data. We have made a similar photograph of the Shane and Wirtanen (1967) galaxy counts from a machine-readable copy of the counts kindly provided to us by Professor Peebles. Comparison of the two photographs is rather striking in that the number of

F. J. Kerr and S. C. Simonson, III (eds.), Galactic Radio Astronomy, 625–630. All Rights Reserved.
Copyright © 1974 by the IAU.

galaxies drops where the hydrogen is present in large quantities, and vice versa. Comparison of these two 'photographs' then shows the qualitative correlation of H I and dust, which simply reflects the old result of Lilley (1955). A quantitative comparison can be made by considering all of the some 30000 individual data points in the two sets of data. We can then make a point correlation diagram comparing the number of galaxies and the amount of H I. This diagram shows a very high degree of correlation with an average relation which differs somewhat from the correlation presented earlier in the symposium, derived from H I versus reddening data, by Kerr.

However, the distribution of points around the average relation looked suspicious and led us to break the data up into separate latitude ranges. This result is quite startling in that the slope of the average relation between galaxies and H I is a rather smooth and well-defined function of galactic latitude. The slope is largest at low latitudes and falls smoothly to zero at high latitudes. The qualitative behavior of the latitude dependence of slope was the same for positive and negative galactic latitudes, with some small but significant quantitative difference.

If we accept the galaxy counts as reliable indicators of extinction, this behavior leads to two possible conclusions concerning the correlation of extinction and hydrogen:

(i) There is no extinction at high latitudes

(ii) There is extinction at high latitudes, but it is not correlated with the presence of H I

There is one simple way to test the possible conclusion (i). If there is no extinction at high latitudes then the distribution of galaxy counts with latitude should not vary as expected for a plane-parallel distribution of dust. Shane and Wirtanen (1967) investigated the latitude dependence of galaxy counts and found it to be accurately consistent with a plane-parallel distribution of dust in the Galaxy; in this way they were able to derive the extinction towards the pole of the Galaxy, about 0.5 mag (photographic). This argues against interpretation (i).

There is another piece of evidence against interpretation (i). This is the dependence of H I on galactic latitude. Averaging over all longitudes we find that there is *less* hydrogen at high latitudes than would be expected from a plane-parallel distribution of gas. The pictures shown earlier in the symposium by van Woerden, made from the Dwingeloo survey at high latitudes (Wesselius and Fejes, 1973), also show this phenomenon. This argues that the reason the correlation between gas and dust breaks down at high latitudes is that the gas is absent at high latitudes (relatively speaking) while the dust is distributed in a fashion completely consistent with the plane-parallel distribution.

(b) OPTICAL DATA

This interpretation, however, is at variance with the reddening results of optical astronomers which have recently been used by Sandage (1972, 1973) to argue against the existence of any extinction at all above latitudes of 50°. In short, the galaxy counts imply the existence of substantial extinction at high latitudes whereas the reddening data imply the existence of no extinction.

We are thus left with an ambiguous situation in which we have the following two alternatives:

(i) *Favored by most optical astronomers: The galaxy counts do not reliably measure extinction, while the reddening measurements do.* In this case the dust/gas ratio goes to zero towards the galactic poles in the vicinity of the Sun. It is important to realize that this happens *only* in the vicinity of the Sun; if it happened everywhere in the Galaxy, we would be back to our plane-parallel distribution of dust/gas and hence both dust and gas separately.

In this case the universe is anisotropic with an excess of galaxies towards our own galactic poles. This excess, combined with the non-plane-parallel distribution of dust, is distributed in such a manner as to precisely mimic what we would observe if instead the galaxies were distributed isotropically and the dust in a plane-parallel layer.

(ii) *Apparently favored by galaxy counts and H I data: Galaxy counts do reliably measure extinction, while the reddening measurements don't.* In this case the ratio of total to selective extinction becomes larger than about 12 at the higher latitudes. This implies that dust grains towards the galactic poles in the vicinity of the Sun are bigger than normal. This can only occur in the vicinity of the Sun; otherwise the reddening would also exist in a plane-parallel distribution. In this case the dust and H I cannot be uniformly mixed and gas/dust ratio becomes smaller at the higher latitudes in the vicinity of the Sun. It is tempting to speculate that some of the H I has been converted to unobservable (in the 21-cm line) H II by the same process responsible for the existence of intermediate and high velocity gas. In this case the new distance scale of Sandage (1972) is incorrect in the sense that the Hubble constant is larger than his derived value by something like 20%.

At the moment I cannot tell you which of these alternatives is correct. We are currently analyzing the different sets of data and hope to have an answer within a few months.

II. H I and Radio Continuum Radiation

Association of low-velocity H I with radio continuum radiation of the North Polar Spur (Loop I) has been noticed by several authors, most recently by Berkhuijsen *et al.* (1971). There is a low velocity H I loop lying a few degrees outside the continuum loop. These authors also noticed correlations with continuum Loop III, using only a single H I scan at $b = +30°$ (from Grahl *et al.*, 1968). At the longitudes where Loop III crosses this latitude they found small-diameter features with high velocity dispersions, greater than 40 km s^{-1}, and high velocities as well. Fejes and Verschuur (1973) find what appears to be a significant correlation of H I density and velocity structure with Loop III, again using data restricted to a small segment of the circumference of the loop. Their correlations take the form of an absence of gas at one place and a split of low velocity gas into two velocity components at another place.

The H I color photographs of Heiles and Jenkins (1973), in which color indicates velocity, together with a similar 'photograph' of radio continuum radiation made from the data of Berkhuijsen (1971), enable these comparisons to be made over the whole circumference of the continuum loops. The H I loop just outside the North Polar Spur (Loop I) contains filamentary structure, some of which has been mapped by Dieter (1964) and Verschuur (1974). The photograph shows that the continuum loop also contains filamentary structure which is similar in appearance to the H I filaments.

The correlation of high velocity dispersion in the H I with radio continuum Loop III is confirmed and extended to include intermediate and high velocity gas. The color photograph shows that much of the intermediate velocity H I at positive latitudes exists in a loop which is coincident with continuum Loop III. In addition, the high velocity gas at positive latitudes (as summarized by Verschuur, 1973) also exists in a loop; this loop lies slightly outside the intermediate velocity H I loop. Both of these lie on the continuum loop, whose width is much bigger than that of the hydrogen loops. At negative latitudes, intermediate velocity H I also appears inside Loop II; the statistical significance of this association is questionable, however, because the H I data do not exist for the full range of langitude at the negative latitudes due to an absence of southern hemisphere data.

The photographs show two new correlations, previously unsuspected. As pointed out in this symposium by Sancisi, there is an absence of continuum radiation in a region about $10°$ in diameter centered near $l = 165°$, $b = -18°$; this is coincident with an absence of H I. Second, near the north celestial pole, there is a very striking elliptically-shaped H I filament, about $25°$ in diameter, centered near $l = 135°$, $b = +35°$. Inside this filament there is little hydrogen as compared with the amount in the filament. The inner boundary of this filament corresponds almost exactly to the boundary of a similarly-shaped hole in the continuum radiation. Most of this filament contains low velocity gas with considerable velocity structure. At one end of the filament the gas is moving at intermediate negative velocities.

It has been fashionable in some astronomical circles to associate the continuum loops with some sort of explosive phenomenon. It is easy to continue this analogy to the continuum holes mentioned in the above two paragraphs. In my own mind I have tended to do this, with little justification other than having been influenced in attitude by Berkhuijsen *et al.* (1971). In the discussion following this paper, however, we will hear some potent arguments against the explosion interpretation.

Note added in proof. More recent work shows that the relation between galaxy counts and extinction differs by a factor of about 2 from that usually assumed; that the extinction as inferred from galaxy counts has a dependence on galactic latitude, separate from any dependence on the intensity of 21-cm emission; and that the latitude-dependent slope mentioned above is probably more properly described as a threshold phenomenon. The effect of these developments is to reduce, but probably not completely eliminate, the discrepency between extinctions derived from galaxy counts and from reddening measurements. At present progress is rapidly being made and the reader is urged to consult either the author or the paper which will, hopefully, appear on this subject.

References

Berkhuijsen, E. M.: 1971, *Astron. Astrophys.* **14**, 359.

Berkhuijsen, E. M., Haslam, C. G. T., and Salter, C. J.: 1971, *Astron. Astrophys.* **14**, 252.

Dieter, N. H.: 1964, *Astron. J.* **69**, 288.

Fejes, I. and Verschuur, G. L.: 1973, *Astron. Astrophys.* **25**, 85.

Grahl, B. H., Hachenberg, D., and Mebold, U.: 1968, *Beitr. Radioastron.* **1**, 1.

Heiles, C. and Habing, H.: 1974, *Astron. Astrophys. Suppl.* **14**, 1.

Heiles, C. and Jenkins, E.: 1973, papers presented at AAS meetings 139 and 140 and IAU General Assembly 1973.

Heiles, C. and Jenkins, E.: 1974, in preparation.

Lilley, A. E.: 1955, *Astrophys. J.* **121**, 559.

Sandage, A.: 1972, *Astrophys. J.* **178**, 1.

Sandage, A.: 1973, *Astrophys. J.* **183**, 711.

Savage, B. and Jenkins, E.: 1972, *Astrophys. J.* **172**, 491.

Shane, C. D. and Wirtanen, C. A.: 1967, *Lick Publ.* **22**, part 1.

Tolbert, C. R.: 1971, *Astron. Astrophys. Suppl.* **3**, 349.

Verschuur, G. L.: 1973, *Astron. Astrophys.* **22**, 139.

Verschuur, G. L.: 1974, in preparation.

Wesselius, P. R. and Fejes, I.: 1973, *Astron. Astrophys.* **24**, 15.

Wesselius, P. R. and Sancisi, R.: 1971, *Astron. Astrophys.* **11**, 246.

Carl Heiles

Astronomy Department,
University of California,
Berkeley, Calif. 94720, U.S.A.

DISCUSSION

Menon: In your plots of the gas to dust correlations did you assume that all the gas is at the same temperature? If so, how can you say that every time there is a decrease in intensity, it is due to molecule formation?

Heiles: Decrease in H I dust could be due *either* to molecule formation *or* to low temperatures. In some areas, especially near the galactic center, we believe it is low temperatures. Others, like Orion and Perseus, are almost certainly molecular because other types of data are more consistent with that approach. In other areas we don't know.

Shakeshaft: If Sandage is wrong in his belief that there is no extinction above 50° latitude, what difference does it make to his current value of the Hubble constant?

Heiles: The difference would be equivalent to roughly 0.5 mag, I believe. This is about 25% in distance.

Shakeshaft: You attribute the correlation between 'holes' in the H I and 820 MHz continuum maps to 'explosions' which have blown material away. Does not one generally associate explosions with enhancement of continuum radiation rather than reduction?

Heiles: You are perfectly correct. As Mathewson has pointed out to me, another thing which argues against 'explosions' is the narrowness of the H I lines. Explosions should provide big line widths as well as large velocities, but large widths aren't seen.

Sancisi: E. Berkhuijsen and I have compared in detail the distributions of dust, neutral hydrogen and continuum radiation in the Taurus-Perseus region. We find that to the minimum of H I emission centered at about $l = 166°$, $b = -18°$, (diameter $\approx 12°$) there corresponds an equally striking minimum in the continuum. The regions of largest interstellar extinction as shown by star and galaxy counts do not lie in the direction of the 'hole' but at its low and high longitude side, and tend to coincide approximately with the H I concentrations making up the Perseus and the Taurus cloud complex.

Heiles: Yes, I agree. But the dust actually seems to be located partially inside the hole, on its edge and definitely inside the main hydrogen peak.

Mathewson: People talk about all these loops being supernova remnants. But there is just no widening of the velocity profile. I think your H I maps show the loops to be structural features of the magnetic field, not the general magnetic field but the field in our vicinity. This is particularly so where the neutral hydrogen filaments come right across the plane. If you put the whole loop as defined in the continuum on the H I, they just wouldn't coincide over the whole arc. This is an important point. It helps to establish that the optical polarization measurements are right in assuming a continuation across the plane. There is dust in the neutral hydrogen, because we get optical polarization, a lot of it.

Weaver: Holes in the gas can be made by processes other than explosions. The local arm, moving at circular velocity, can run into a mass of gas moving at less than circular velocity. The slow-moving gas will smash a hole right through the local gas. This appears to be what is happening, for example, in the region $l = 120°$, $b = +40°$, where the low velocity gas is very weak. A jet of gas here is moving over our heads at high velocity; the low velocity hydrogen is very weak.

Kerr: In the work I reported on observations in the directions of globular clusters, we found the gas-to-dust ratio to be independent of latitude. We estimated the amount of dust from reddening effects, whereas yours comes from extinction. There is apparently some important difference between reddening and extinction phenomena. Another possible interpretation additional to those you mentioned is that the clumpiness of the dust (which may vary with latitude) may be affecting different types of observations in different ways.

Van Woerden: Plots of hydrogen column density, N_H, versus latitude, b, averaged over all longitudes, may hide important information and, therefore, lead to erroneous conclusions. Fejes and Wesselius (1973, *Astron. Astrophys.* **24**, 10) have investigated the run of N_H with b separately for small longitude intervals. They show that much of the anomaly in this run can be attributed to the 'low-velocity hole' (a prominent minimum in the sky distribution of neutral hydrogen, centered at $l \sim 160°$, $b + 70°$, and extending over some 70°) and to the '*Scheve Schijf*' (a disk-like feature of low-velocity hydrogen, tilted 45° to the galactic plane).

LOW- AND INTERMEDIATE-VELOCITY HYDROGEN AT POSITIVE GALACTIC LATITUDES

K. TAKAKUBO

Astronomical Institute, Tohoku University, Sandai, Japan

Abstract. Low- and intermediate-velocity Gaussian components of hydrogen emission profiles observed at positive galactic latitudes are investigated. The low-velocity hole, as called by Wesselius and Fejes (1973), is a phenomenon which appears in the layer composed of the gas emitting narrow Gaussian components but does not in that of wide components. Both Models I and II suggested by Wesselius and Fejes contradict this result. A hypothesis is proposed that a stream motion in the layer emitting wide Gaussian components, which may be identified with the intercloud medium, swept away the interstellar clouds emitting narrow Gaussian components in the region of the hole.

A remarkable deficiency of low-velocity hydrogen gas at high positive latitudes has been shown by several investigators. Among them, Wesselius and Fejes (1973) called it the low-velocity hole and showed that there was a good correspondence in position and shape between the hole and a negative-velocity hydrogen complex.

In the present paper, we will consider low- and intermediate-velocity gas at positive latitudes and point out that the hole appears only in the layer of the gas emitting narrow Gaussian components but not in one emitting wide Gaussian components.

The data are 309 hydrogen emission profiles selected from observations by Tolbert (1971) at high galactic latitudes and by Takakubo and van Woerden (1966) at intermediate latitudes. These observed points more or less uniformly cover the northern hemisphere at latitudes higher than $10°$ ($\Delta b \sim 5°$, $\Delta l \cdot \cos b \sim 10°$).

Each profile observed is transformed into optical depth, $\tau(V)$ as a function of the radial velocity, V, with respect to the local standard of rest, assuming a uniform temperature of 125 K. Then it is expressed by a superposition of Gaussian components (τ_i, V_i, σ_i):

$$\tau(V) = \sum_i \tau_i \exp\{-(V-V_i)^2/2\sigma_i^2\}.$$

The column density, N_{Hi}, of a component is proportional to the product $\tau_i \sigma_i$ and does not depend on assumed temperature so much if the layer is optically thin.

Among these Gaussian components 1041 have radial velocities V_i between -60 km s^{-1} and $+60$ km s^{-1}. We will use them as low- and intermediate-velocity components.

As pointed out in a previous paper (Takakubo, 1967), Gaussian components have rather different natures according to the values of their radial-velocity dispersion, σ_i ($=0.85$ half half-width). We will, thus, divide them into two classes,

narrow components: $\sigma_i \lesseqgtr \sigma_{bound}$,

wide components: $\sigma_i > \sigma_{bound}$,

F. J. Kerr and S. C. Simonson, III, Galactic Radio Astronomy, 631–635. All Rights Reserved.

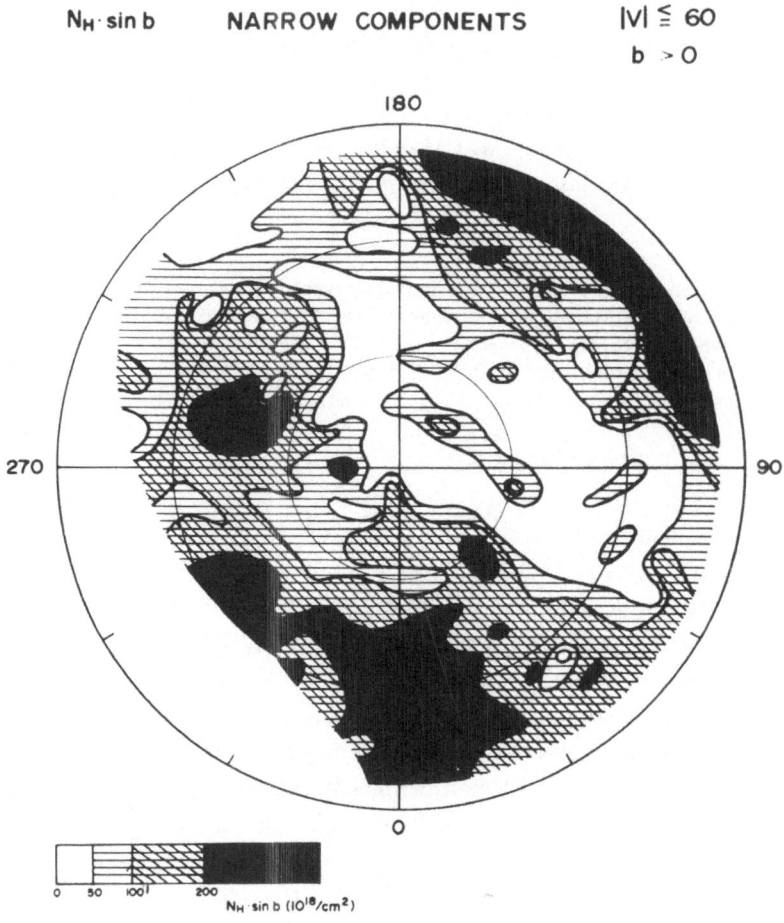

$N_H \cdot \sin b$ NARROW COMPONENTS $|V| \lesssim 60$

$b > 0$

180

270   90

0

$N_H \cdot \sin b \ (10^{18}/cm^2)$

Fig. 1.   The contour map of $\sum N_{Hi} \cdot \sin b$ for narrow components ($\sigma_i \leqq 7.0$ km s^{-1}). The north galactic pole is at the centre; the three circles indicate $b = 60°$, $30°$, and $0°$, respectively. Note that the low-velocity hole extends to considerably low latitude. The contour lines of this kind of figure cannot always be determined uniquely.

with

$$\sigma_{bound} = 7.0 \text{ km s}^{-1}.$$

The narrow components are of groups S and M in the previous paper, and the wide components are of group L. Although the boundary value $\sigma_{bound} = 7.0$ km s^{-1} was adopted more or less arbitrarily, it is a rather reasonable value (see Takakubo, 1967). The statistical investigation by Mebold (1972) also supports it, although he adopted a slightly smaller value of σ_{bound}. The classification in the present paper is essentially the same as his, and we may consider in a crude treatment that the narrow components are mainly due to the emission of the interstellar clouds and the wide ones to that of the intercloud medium.

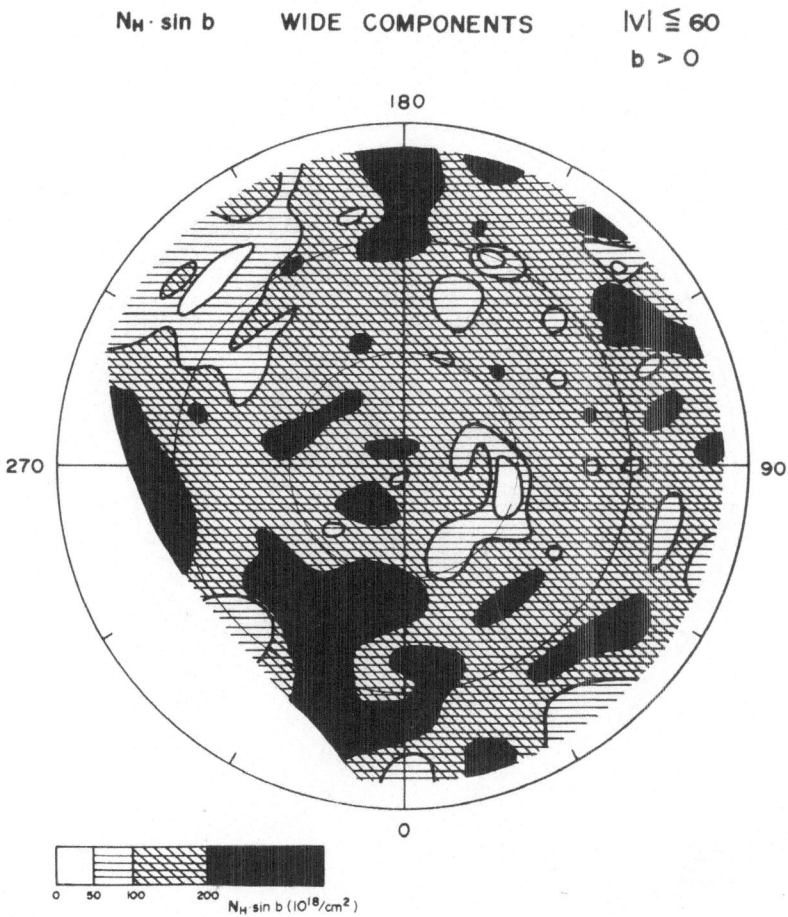

Fig. 2. The contour map of $\sum N_{\text{Hi}} \cdot \sin b$ for wide components ($\sigma_i > 7.0$ km s^{-1}).

At each point (l, b)

$$\sum_i N_{\text{Hi}} \cdot \sin b$$

is computed, where the summation extends over all the components in a profile. If the hydrogen gas is stratified parallel to the galactic plane, this quantity should be constant over the sky. But it is not the case, as is expected from previous investigations. Now we treat the narrow- and the wide-component classes separately and see the distribution of the quantity over the sky.

In the class of narrow components $\sum N_{\text{Hi}} \cdot \sin b$ is not constant at all and has marked deviations from the plane parallel stratification (Figure 1). Patterns due to the low velocity hole, the tilted disk ('*Scheve Schijf*') and the Perseus Arm are conspicuous. Particularly, the low-velocity hole extends to much lower latitudes than Wesselius and Fejes (1973) suggested.

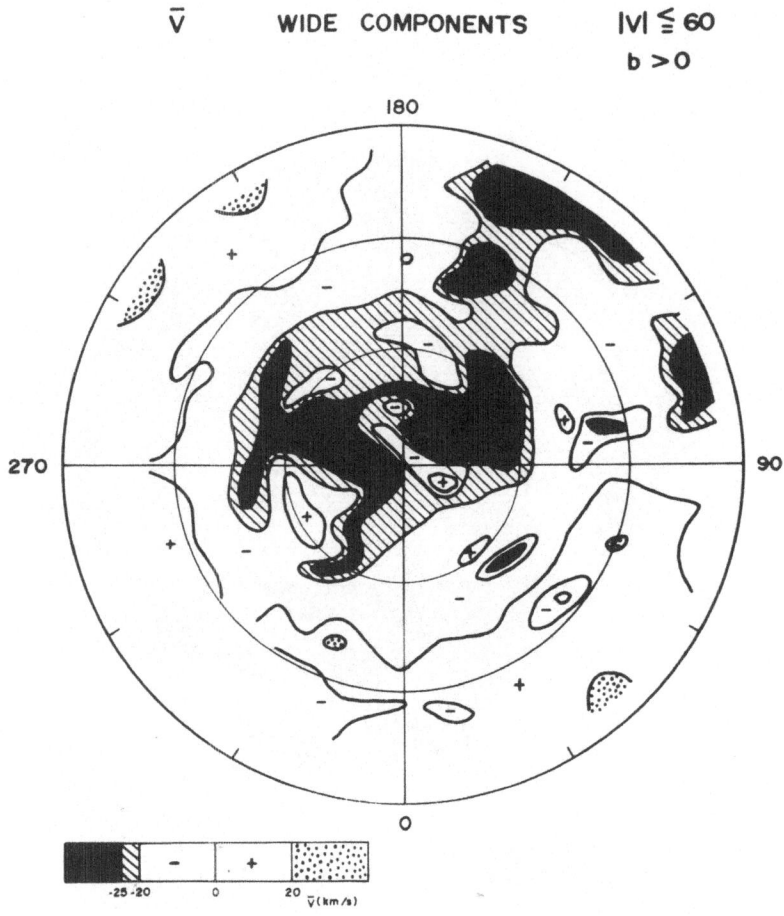

Fig. 3.   The contour map of the average radial velocity $\bar{V}_i$ for wide components.

On the other hand, in the class of wide components the distribution of $\sum N_{Hi} \cdot \sin b$ is nearly uniform (Figure 2). Only in a few small regions does it deviate from the average, and no outstanding patterns appear, except one due to the tilted disk.

The difference in the distribution of $\sum N_{Hi} \cdot \sin b$ of the two classes is also clear from the frequency distribution of $\sum N_{Hi} \cdot \sin b$. We may conclude that the gas emitting wide Gaussian components is stratified parallel to the galactic plane and its column density perpendicular to the plane is 1.4×10^{20} cm^{-2}.

It is concluded, therefore, that the hole is a phenomenon in the layer of narrow Gaussian components, but it is absent in that of wide components. If the hole is due to either a nearby supernova explosion (Model I of Wesselius and Fejes, 1973) or a colliding big gas complex (Model II of theirs), the absence of the hole in the inter-cloud medium is hardly understandable, since in these models the hole is expected also in the intercloud medium. A hypothesis that the internal motion of interstellar clouds in the region of the hole is increased and the emission becomes broader by some reason meets a difficulty, since, if this is the case, the value of $\sum N_{\mathrm{Hi}} \cdot \sin b$ of wide components should be increased in the hole region by a factor of 2 or so.

A possible explanation may be to assume a stream motion in the layer of wide components. Since the thickness of the layer of wide components is larger than that of narrow components by a factor of 1.5 (see Takakubo, 1967), a stream in the layer of wide components originating high above the galactic plane does not contain the interstellar clouds emitting narrow components. If the stream penetrates into lower layers, the clouds along its passage will be blown away. Indeed, the radial velocity of wide components (Figure 3) shows marked negative deviation in the region to lower latitudes at around a galactic longitude of 150°; it is natural to assume the stream comes from this longitude direction. It is noticed that we find a high density part, called the tilted disk, just in the opposite direction.

It is not yet clear, however, if the stream hypothesis is compatible with the continuity of mass in the layer of wide components.

Acknowledgement

The computations were carried out by using a NEAC-2200 at the Computer Center at Tohoku University.

References

Mebold, U.: 1972, *Astron. Astrophys.* **19**, 13.
Takakubo, K.: 1967, *Bull. Astron. Inst. Neth.* **19**, 125.
Takakubo, K. and van Woerden, H.: 1966, *Bull. Astron. Inst. Neth.* **18**, 488.
Tolbert, C. R.: 1971, *Astron. Astrophys. Suppl.* **3**, 349.
Wesselius, P. R. and Fejes, I.: 1973, *Astron. Astrophys.* **24**, 15.

K. Takakubo
Astronomical Institute,
Tohoku University,
Sendai, Japan

CONTINUUM RADIO EMISSION AND GALACTIC STRUCTURE

R. M. PRICE

Dept. of Physics and Research Laboratory of Electronics, Massachusetts Institute of Technology, Cambridge, Mass., U.S.A.

Abstract. Recent studies of high latitude brightness have shown that in several regions a significant proportion of it is due to loop features. The largest of these are now thought to be local (within a few hundred parsecs) and perhaps to be due to supernova events.

Aperture synthesis maps of 45 external spiral galaxies have shown large-scale continuum nonthermal features in the disks of some and, in general, an absence of spherical halos around the galaxies. This supports the view that the radio emission from our Galaxy can best be described as coming from a thick disk. Any large-scale spherical component could not have a volume emissivity greater than 1.5% of the average volume emissivity of the disk at meter wavelengths.

A comparison of the disk radiation with new disk models suggests that the galactic plane radiation has two components: a base disk and a spiral component. Each of these contributes $\sim 50\%$ of the total power output of the disk at 150 MHz.

I. Introduction

At meter wavelengths, most of the radio emission from within our Galaxy has its origin in synchrotron radiation from relativistic cosmic ray electrons radiating in the magnetic fields of the Galaxy. Only a small percentage of the brightness temperature observed at meter wavelengths is due to the thermal emission from H II regions. This paper deals only with the nonthermal (synchrotron) emission. For a discussion of the distribution of H II regions the reader is directed to Burbidge (1967), Burke (1968), and Reifenstein *et al.* (1970). (The nonthermal emission regions located near the center of our Galaxy are not discussed in this review.)

In order to estimate the locations of nonthermal emission regions in our Galaxy, we can utilize detailed studies of radio brightness distribution at both high and low galactic latitudes. We can also learn much about the expected distribution of nonthermal emission in 'normal' spirals from high-resolution aperture synthesis observations of external spiral galaxies.

Finally, we must combine data from these studies with model fitting to arrive at reasonable estimates of the distribution of the nonthermal radio emission regions (i.e., galactic magnetic field and cosmic ray electrons) within our Galaxy.

II. General Features of the Radio Sky

A map of the brightness distribution on the sky at meter wavelengths (e.g., Landecker and Wielebinski, 1970) is shown in Figure 1. On this map three notable characteristics are apparent.

(i) The radiation is strongly concentrated to the galactic equator.

(ii) Loops or spur features are visible in many portions of the sky, often appearing to 'come out of' the plane. These features are seen even at galactic latitudes higher than 60°.

F. J. Kerr and S. C. Simonson, III (eds.), Galactic Radio Astronomy, 637–648. All Rights Reserved.
Copyright © 1974 by the IAU.

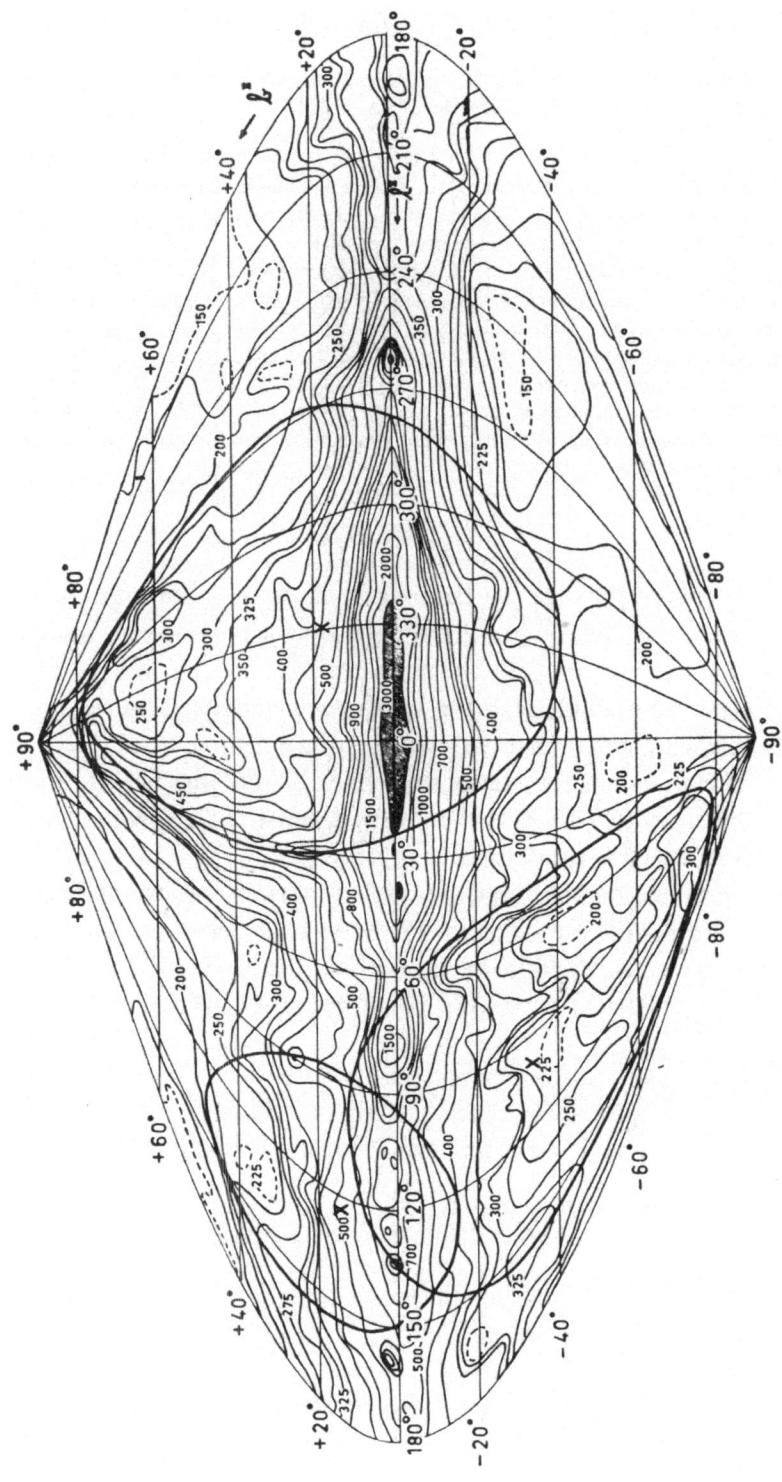

Fig. 1. Sky map at 150 MHz prepared by Landecker and Wielebinski (1970). Small circles of Loops I, II, and III are shown (after Berkhuijsen, 1971) with circle centers indicated by ×'s.

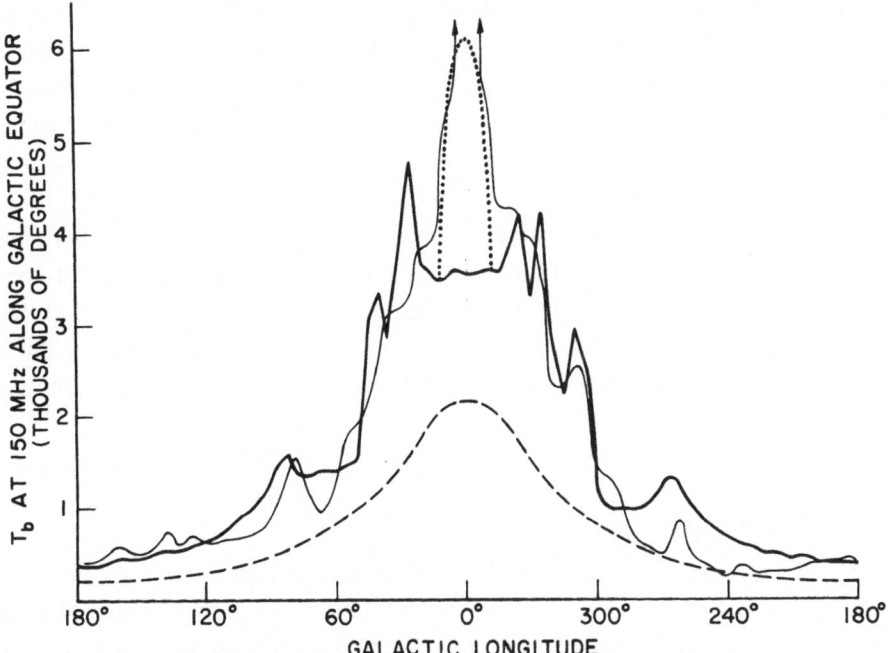

Fig. 2.   Profile of brightness temperature at 150 MHz along the galactic equator (Landecker and Wiele-binski, 1970) is shown by light solid line. Heavy solid line shows profile predicted by the model of base disk and spiral components (Price, 1974). Dashed curve: base disk component. Dotted line: expected non-thermal contribution from a region with constant volume emissivity and 2000 pc radius located at the galactic center.

(iii) The brightness temperatures at high latitudes are relatively high, as high as 30% of those found along the galactic equator at longitudes greater than 60° from the center.

Information about the distribution of nonthermal emission regions in the galactic disk can be derived from an examination of the profiles of the brightness temperatures along the galactic equator shown in Figure 2. This profile clearly shows that the brightness is highly concentrated within 60° of the galactic center in longitude, regions greater than 60° from the center do not differ greatly in brightness temperature except for two obvious features at $l \approx 80°$ and $l \approx 265°$, and bumps or steps are noticeable on the central concentration.

Each of these characteristics of the galactic brightness distribution gives information on a different aspect of the distribution of nonthermal emission regions in the Galaxy. These aspects will be discussed separately.

III. Loops

Loops or spurs are a well-known feature of the radio sky. The best-fit small circles to three of the loops are shown in Figure 2. Portions of some of the loop and spur features lie at high galactic latitudes (e.g., the North Polar Spur, which passes near the north galactic pole). They contribute significantly to the observed brightness temperatures away from the galactic equator. There are indications that they contribute not only along their main ridges but also 'inside' the ridges and beyond the outer boundaries of the bright ridges. A complete description of the radio properties of the loops can be found in Berkhuijsen (1971). Information and references concerning these features are listed in Table I.

On the basis of an 820 MHz survey, combined with polarization measurements, Berkhuijsen (1971) has concluded that the loop features have the following properties.

(i) They contribute significantly to the brightness temperature distribution in high latitude regions. Their radiation comes not only from the bright rims of the loop features but also from regions closer to the loop centers and outside of the bright rims.

(ii) They are nearby features, i.e., located within several hundred parsecs in many cases.

(iii) They are probably supernova remnants.

The question of whether some or all of the loops are supernova remnants is very important. The shapes of many of these features favor such a hypothesis. They show a large degree of circular symmetry (over the identifiable arc of the observed feature). In some cases the continuum radio ridges are associated with optically observed filaments or neutral hydrogen features. Spoelstra (1972, 1973) has shown that the observed brightness and polarization distributions are in rough agreement with results of models of supernova remnants in the local magnetic field. Also, Berkhuijsen (1973) has established that five of the loops lie near the $\Sigma - D$ relationship for known supernova remnants.

There are obstacles that raise uncertainty about the supernova origin of the loops.

(i) The low internal velocity of the clouds of neutral hydrogen associated with the North Polar Spur is not consistent with the idea of explosive origin.

(ii) The very large and regular outer envelopes of the loops (~ 100 pc) are hard to explain unless the interstellar medium is exceedingly homogeneous, of low density, and has a larger scale height above the galactic plane than is estimated at present.

(iii) Similar features have not been identified in other regions of the Galaxy. This can be a very difficult observational identification to make because of the low surface brightness of loop objects and the confusion at lower latitudes at which the more distant loops would be expected.

Other suggestions have been offered to explain the loop features. Brandt and Maran (1972) have suggested that they might be fossil Strömgren spheres produced by supernova outbursts. Kafatos and Morrison (1973) have estimated the radio and X-ray emission that might be expected from such an object and conclude from available observational information that such an origin is consistent.

Mathewson (1968) has suggested that the loops are 'radio tracers' of the helical component of the local magnetic field, but more recent studies of the magnetic fields in the vicinity of the sun do not appear to support this hypothesis (e.g., Manchester, 1972).

TABLE I

Galactic loop features

Feature	Center		Diameter	Estimated distance (pc)	Additional data		Discussion of objects, References
	l	b			HI	Optical	
Loop I North Polar Spur	329±1°.5	+17°.5±3°	116±4°	130±75 (1)	2		1, 3, 4, 5, 6, 7, 8, 9, 10, 11, 12
Loop II Cetus Arc	100±2°	−32°.5±3°	91±4°	110±40 (1)	13	14, 15, 16	1, 3, 4, 5, 7, 11, 12, 14, 16
Loop III	124±2°	+15°.5±3°	65±3°	150±50 (1)	17	18	1, 3, 4, 5, 7, 11, 12
Loop IV	315±3°	+48°.5±1°	39.5±2°	250±90 (1)	19		1, 3, 4, 7, 20
Origem Loop	194°.5	+0°.5	5°	1000±500 (1)			1, 20
Lupis Loop	330°.1	+15°.1	4°.5	400 (21)		20	20
Monoceros Loop	205°.7	−0°.1	3°.5	600 (21)	22		20, 23

1. Berkhuijsen (1973)
2. Berkhuijsen et al. (1970)
3. Berkhuijsen (1971)
4. Berkhuijsen et al. (1971)
5. Bingham (1967)
6. Bunner et al. (1972)
7. Haslam et al. (1971)
8. Holden (1969)
9. Large et al. (1966)
10. Merkelijn and Davis (1967)
11. Quigley and Haslam (1965)
12. Spoelstra (1972)
13. Hughes et al. (1971)
14. Elliott and Meaburn (1970)
15. Meaburn (1965)
16. Meaburn (1967)
17. Fejes and Verschuur (1973)
18. Elliott (1970)
19. Fejes (1971)
20. Spoelstra (1973)
21. Milne (1970)
22. Morgan et al. (1965)
23. Gebel and Shore (1972)

Loops or bubbles could also be produced in the galactic magnetic fields by cosmic ray pressure according to the theory of Parker (1965). This theory has not yet been examined in sufficient detail to determine whether the resulting structure would have the observed properties of the radio loops.

Thus the origin and the nature of the loop features is still uncertain. Regardless of their origin, there can be little doubt that the loops with the largest angular size are nearby features and that their contribution to local high latitude radiation must be considered in the interpretation of large-scale continuum radio galactic structure. If we calculate a space density for these objects based on those nearest the Sun, we may also conclude that a significant fraction of the nonthermal disk radiation might come from such objects distributed throughout the galactic disk.

IV. Galactic Plane

Surveys of the brightness distribution at low galactic latitudes (Table II) show the overall structure of the galactic disk.

Detailed studies of particular regions give additional information about the properties of individual sources. Many of the sources are H II regions. Some details of studies of these sources have been given by Goss and Shaver (1970) and by Felli and Churchwell (1972).

TABLE II

Surveys of brightness distribution at low galactic latitudes

Reference	Frequency (MHz)	Beam-width	Longitude range
Mathewson et al. (1962)	1440	~50'	256°–0°
Seeger et al. (1965)	400	~2°	0°–250°
Komesaroff (1965)	408	~48'	280°–355°
Wielebinski et al. (1968)	85	~3°7	290°–0–60°
	150	~2°2	
Hill (1968)	1410	~14'	280°–355°
Altenhoff (1968)	2700	~11'	345°–0–240°
Beard and Kerr (1969)	1410	~14'	27°–38°
	2650	~7'	27°–38°
Australian J. Phys.	2650	~7'	288°–307°
Astrophys. Suppl. Ser. No. 11,	2700	~7'	307°–330°
(1969)	2650	~7'	334°–345°
	2650	~7'	345°–0–5°
Altenhoff et al. (1970)	1414	~11'	335°–0–75°
	2650	~11'	335°–0–75°
	5000	~11'	335°–0–75°
Goss and Day (1970)	2700	~7'	6°–26°
Day et al. (1970)	2700	~7'	37°–47°
Sinclair and Kerr (1971)	1410	~14'	355°–0–5°
Berkhuijsen (1971, 1972)	820	~1°0	25°–220°
Green (1972)	408	~3'	240°–0–55°
Day et al. (1972)	2650	~7'	190°–290°
			46°–61°

As well as H II regions, 150 nonthermal galactic sources, which are presumed to be supernova remnants, have also been catalogued. These sources are discussed by Milne (1970), Downes (1971), Goss and Shaver (1970), and Clark, *et al.* (1973). Careful use of the $\Sigma - D$ relationships derived by using supernova remnents of known distances enables plotting the distribution of these objects in the Galaxy. Ilovaisky and Lequeux (1972a) have used this method to derive a surface density for supernova remnants in the disk.

There are too few catalogued nonthermal sources to obtain an accurate picture of their distribution in longitude. In latitude, they are concentrated to the galactic plane. More than half of the sources between $\pm 2°$ latitude lie within $\pm 0°.3$ of the galactic equator. Using the derived distances of all of the catalogued supernova remnants to obtain their distribution in z, we find that more than 75% lie within a disk of 300 pc total thickness.

The unresolved nonthermal disk radiation has a width in latitude of approximately 2° between half-brightness points (Altenhoff, 1968). Baldwin (1967) has pointed out that the equivalent thickness of the nonthermal radio disk is approximately 750 pc. A comparison of the distribution in latitude of the supernova remnants and the disk brightness temperature shows that the major portion of the nonthermal disk emission cannot be due to a superposition of a large number of nonthermal sources of the type now thought to be supernova remnants. From energy considerations Ilovaisky and Lequeux (1972b) also conclude that supernova remnants contribute less than 10% of the nonthermal galactic luminosity at 1 GHz.

There is a strong possibility, however, that supernovae could provide the energetic electrons responsible for the nonthermal disk emission. Lequeux (1971) has discussed this point in greater detail.

A map of the galactic plane by Green (1972) with a half-power beamwidth of approximately 3' has resolved the disk radiation. From this survey Green reached the following conclusions.

(i) There is only a slight clustering of disk sources (thermal and nonthermal) in longitude. The clustering is generally near the neutral hydrogen tangent points, as might be expected if the sources are weakly concentrated to spiral features.

(ii) The steps in the unresolved (i.e., 'smooth') radiation observed along the plane, first identified at 85 MHz by Mills (1959), are also clearly shown at 408 MHz.

(iii) The observed steps fit some model spirals, the best fits being 7° or 9° two-arm spirals and a 14° four-arm spiral.

V. External Normal Galaxies

The radio continuum structure of spiral galaxies has been studied by Cameron (1971) at 408 MHz with a 3' half-power beam. Studies have been carried out with the use of aperture synthesis techniques by Pooley (1969a, b), Mathewson, *et al.* (1972), and by van der Kruit (1971, 1973a, b) at 1400 MHz. The results of the studies can be summarized as follows.

(i) Nonthermal radiation exhibits large-scale structure in the disks of some spiral systems. It can be distributed in any or all of the following components: base disk (regular or irregular in distribution), spiral features, and sources or complexes of sources in the nuclei of the galaxies.

(ii) Large-scale halos are seldom associated with spiral galaxies. A possible exception is NGC 891, where some extensions of the nonthermal radiation are seen above the galactic disk. M31 also has a higher single-dish flux density than can be a accounted for by high-resolution measurements. This might be due to a large-scale halo around M31.

(iii) Observations show concentration of neutral hydrogen gas to spiral features in some systems, e.g., M101. This indicates the possibility of a corresponding compression of magnetic field and relativistic particles which would result in nonthermal emission from the spiral features.

The best examples of nonthermal radio spiral features that we now have are those observed in M51 (Mathewson *et al.*, 1972). It should be noted that the emission does not follow a regular spiral pattern; rather, it wanders somewhat, generally on the inner edge of the bright optical arms, also forming spurs or cross links between spiral features. This observation suggests that we should not expect a high degree of regularity in nonthermal spiral features in our Galaxy.

At the present time, there is no simple basis on which to predict the type of large-scale nonthermal emission that might be associated with a given spiral galaxy. This is a strong argument against close direct comparison of the structure of our Galaxy with any particular external system.

VI. Radio Disk and Spiral Features

At the time Mills (1959) first suggested that the steps and bumps observed in the brightness temperature profile along the galactic equator at meter wavelengths might be due to nonthermal radiation from spiral features there was no widely accepted model for the form of the spiral features. The Lin-Shu (1967) density wave theory now provides one method for calculating the form of the spiral patterns in galaxies. On the basis of the nonlinear density wave theory (Roberts, 1969) the relative locations of gas, dust, and young stars can be predicted. The location of nonthermal radio emission can also be predicted, under the assumption of the existence of a large-scale magnetic field and energetic electrons in a galaxy. The theory provides a satisfactory explanation of the observations of the nonthermal spiral structure in M51 (Mathewson *et al.*, 1972).

For our Galaxy, we can compare the observed radio brightness distribution at meter wavelengths with models to help determine the distribution of non-thermal emission regions. The approach suggested by Price (1974) is as follows.

(i) Estimate a base disk component by using a volume emissivity function that accounts for the bulk of the observed brightness temperature along the galactic plane, especially in longitudes toward the galactic center.

(ii) Compare the residuals (observations minus estimated base disk) with the contribution expected from nonthermal emission regions distributed in the form of spiral features.

Such an analysis suggests the following properties of the distribution of nonthermal emission in the disk of our Galaxy.

(i) The volume emissivity of the base disk falls off from the galactic center toward the outer regions of the Galaxy. It has been suggested (Price, 1974) that the volume emissivity follows a function of the form $e^{-\alpha R}$. This base disk component, as shown in Figure 2, contributes approximately one-half of the total power observed from the galactic disk at meter wavelengths.

(ii) Emission from spiral features provides most of the remaining total power emitted from the galactic disk. These spiral features are irregular both in volume emissivity (i.e., compression) and in shape. That is, they do not follow a precise spiral pattern but wander somewhat about some mean best-fit spiral. A comparison of observations with the profile predicted by a model incorporating a base disk and spiral features is shown in Figure 2. It can be seen that there is general agreement in the overall shape of the profile, but not in smaller details.

(iii) From the observed width of the concentration of brightness temperatures toward the galactic center, and from the brightness of the emission associated with the local arm (at $l = 80°$ and $265°$), it is inferred that the Sun is located near the inner edge of the local nonthermal radio spiral feature.

(iv) If the Sun is, in fact, located just within the local nonthermal feature, the nonthermal radiation produced locally will account for the relatively high brightnesses observed at high galactic latitudes. Yates (1968) has suggested a similar origin for the high latitude brightness. His analysis, however, requires that the Sun be at the center of the local spiral arm.

In any case, observations at meter wavelengths at high galactic latitudes ($|b| > 30°$) indicate that any large-scale spherical component around our Galaxy could not have a volume emissivity greater than 1.5% of the average volume emissivity of the disk (Price, unpublished).

VII. Observational Needs

The observational requirements listed by Baldwin (1967) have been carried out in part. The information derived from those studies suggests the need for further observational programs.

(i) Observation of more external normal galaxies to determine the frequency of occurrence of nonthermal radio features in their nuclei, disks, and spiral features.

(ii) Search for more distant loop features in the disk radiation of our Galaxy.

(iii) Search for additional nonthermal sources in the galactic plane, and a careful study of the characteristics and distribution of these sources.

(iv) Studies of the distribution of the nonthermal component of the disk at low and medium (to $b \approx 20°$) latitudes.

Acknowledgements

I would like to thank Dr Green and Dr Berkhuijsen for helpful discussions about their investigations, and also Dr Clark, Dr Caswell, and Dr Green for providing me with data in advance of publication. This work was supported by the U.S. National Science Foundation (Grant GP-21348A #2).

References

Altenhoff, W. J.: 1968, in Y. Terzian (ed.), *Interstellar Ionized Hydrogen*, Benjamin, New York, p. 519.
Altenhoff, W. J., Downes, D., Goad, L., Maxwell, A., and Rinehart, R.: 1970, *Astron. Astrophys. Suppl.* **1**, 319.
Baldwin, J. E.: 1967, in H. van Woerden (ed.), 'Radio Astronomy and the Galactic System', *IAU Symp*, **31**, 337.
Beard, M. and Kerr, F. J.: 1969, *Australian J. Phys.* **22**, 121.
Berkhuijsen, E. M.: 1971, *Astron. Astrophys.* **14**, 359.
Berkhuijsen, E. M.: 1972, *Astron. Astrophys. Suppl.* **5**, 263.
Berkhuijsen, E. M.: 1973, *Astron. Astrophys.* **24**, 143.
Berkhuijsen, E. M., Haslam, C. G. T., and Salter, C. J.: 1970, *Nature* **225**, 364.
Berkhuijsen, E. M., Haslam, C. G. T., and Salter, C. J.: 1971, *Astron. Astrophys.* **14**, 252.
Bingham, R. G.: 1967, *Monthly Notices Roy. Astron. Soc.* **137**, 157.
Brandt, J. C. and Maran, S. P.: 1972, *Nature* **235**, 38.
Bunner, A. N., Coleman, P. L., Kraushaar, W. L., and McCammon, D.: 1972, *Astrophys. J. Letters* **172**, L67.
Burbidge, E. M.: 1967, in H. van Woerden (ed.), 'Radio Astronomy and the Galactic System', *IAU Symp.* **31**, 209.
Burke, B. F.: 1968, in Y. Terzian (ed.), *Interstellar Ionized Hydrogen*, Benjamin, New York, p. 541.
Cameron, M. J.: 1971, *Monthly Notices Roy. Astron. Soc.* **152**, 439.
Clark, D. H., Caswell, J. L., and Green, A. J.: 1973, *Nature* **246**, 28.
Day, G. A., Warne, W. G., and Cooke, D. J.: 1970, *Australian J. Phys. Astrophys. Suppl.*, No. 13, 11.
Day, G. A., Caswell, J. L., and Cooke, D. J.: 1972, *Australian J. Phys. Astrophys. Suppl.*, No. 25.
Downes, D.: 1971, *Astronom. J.* **76**, 305.
Elliott, K. H.: 1970, *Nature* **226**, 1236.
Elliott, K. H. and Meaburn, J.: 1970, *Astrophys. Space Sci.* **7**, 252.
Fejes, I.: 1971, *Astron. Astrophys.* **15**, 419.
Fejes, I. and Verschuur, G. L.: 1973, *Astron. Astrophys.* **25**, 85.
Felli, M. and Churchwell, E.: 1972, *Astron. Astrophys. Suppl.* **5**, 309.
Gebel, W. L. and Shore, S. N.: 1972, *Astrophys. J. Letters* **172**, L9.
Goss, W. M. and Day, G. A.: 1970, *Australian J. Phys. Astrophys. Suppl.* No. 13, 3.
Goss, W. M. and Shaver, P. A.: 1970, *Australian J. Phys. Astrophys. Suppl.* No. 14, 1.
Green, A. J.: 1972, Ph.D. Thesis, University of Sydney, Australia.
Haslam, C. G. T., Kahn, F. D., and Meaburn, J.: 1971, *Astron. Astrophys.* **12**, 388.
Hill, E. R.: 1968, *Australian J. Phys.* **21**, 735.
Holden, D. J.: 1969, *Monthly Notices Roy. Astron. Soc.* **145**, 67.
Hughes, M. P., Thompson, A. R., and Colvin, R. S.: 1971, *Astrophys. J. Suppl.* **23**, 323.
Iloviasky, S. A. and Lequeux, J.: 1972a, *Astron. Astrophys.* **19**, 169.
Iloviasky, S. A. and Lequeux, J.: 1972b, *Astron. Astrophys.* **20**, 347.
Kafatos, M. C. and Morrison, P.: 1973, *Astron. Astrophys.* **26**, 71.
Komesaroff, M. M.: 1965, *Australian J. Phys.* **19**, 75.
Kruit, P. C. van der: 1971, *Astron. Astrophys.* **15**, 110.
Kruit, P. C. van der: 1973a, *Bull. Am. Astron. Soc.* **5**, 30.
Kruit, P. C. van der: 1973b, *Nature Phys. Sci.* **243**, 127.
Kruit, P. C. van der, Oort, J. H., and Mathewson, D. S.: 1972, *Astron. Astrophys.* **21**, 169.
Landecker, T. L. and Wielebinski, R.: 1970, *Australian J. Phys. Astrophys. Suppl.* No. 16.

Large, M. I., Quigley, M. J. S., and Haslam, C. G. T.: 1966, *Monthly Notices Roy. Astron. Soc.* **131**, 335.

Lequeux, J.: 1971, *Astron. Astrophys.* **15**, 42.

Lin, C. C. and Shu, F. G.: 1967, in H. van Woerden (ed.), 'Radio Astronomy and the Galactic System', *IAU Symp.* **31**, 313.

Manchester, R. N.: 1972, *Astrophys. J.* **172**, 43.

Mathewson, D. S.: 1968, *Astrophys. J. Letters* **153**, L47.

Mathewson, D. S., Healey, J. R., and Rome, J. M.: 1962, *Australian J. Phys.* **15**, 369.

Mathewson, D. S., Kruit, P. C. van der, and Brouw, W. N.: 1972, *Astron. Astrophys.* **17**, 468.

Meaburn, J.: 1965, *Nature* **208**, 575.

Meaburn, J.: 1967, *Z. Astrophys.* **65**, 93.

Merkelijn, J. K. and Davis, M. M.: 1967, *Bull. Astron. Inst. Neth.* **19**, 246.

Mills, B. Y.: 1959, in R. N. Bracewell (ed.), 'Paris Symposium on Radio Astronomy', *IAU Symp.* **9**, 431.

Milne, D. K.: 1970, *Australian J. Phys.* **23**, 425.

Morgan, W., Hiltner, W., Neff, J., Garrison, R., and Osterbrock, D.: 1965, *Astrophys. J.* **142**, 974.

Parker, E. N.: 1965, *Astrophys. J.* **142**, 584.

Pooley, G. G.: 1969a, *Monthly Notices Roy. Astron. Soc.* **144**, 101.

Pooley, G. G.: 1969b, *Monthly Notices Roy. Astron. Soc.* **144**, 143.

Price, R. M.: 1974, *Astron. Astrophys.* **33**, 33.

Quigley, M. J. S. and Haslam, C. G. T.: 1965, *Nature* **208**, 741.

Reifenstein, E. C., III, Wilson, T. L., Burke, B. F., Mezger, P. G., and Altenhoff, W. J.: 1970, *Astron. Astrophys.* **4**, 357.

Roberts, W. W.: 1969, *Astrophys. J.* **158**, 123.

Seeger, C. L., Westerhout, G., Conway, R. G., and Hoekema, T.: 1965, *Bull. Astron. Inst. Neth.* **18**, 11.

Sinclair, M. W. and Kerr, F. J.: 1971, *Australian J. Phys.* **24**, 769.

Spoelstra, T. A. Th.: 1972, *Astron. Astrophys.* **21**, 61.

Spoelstra, T. A. Th.: 1973, *Astron. Astrophys.* **24**, 149.

Wielebinski, R., Smith, D. H., and Garzon Cardenas, X.: 1968, *Australian J. Phys.* **21**, 185.

Yates, K. W.: 1968, *Australian J. Phys.* **21**, 147.

R. M. Price
Research Laboratory of Electronics and
Department of Physics,
Massachusetts Institute of Technology,
Cambridge, Mass. 02139, U.S.A.

DISCUSSION

Yuan: From Davies and Whiteoak we learn that there is a large-scale magnetic field in the direction $l = 90°$ and from Mathewson that there is a local magnetic field in the direction $l = 50°-60°$. Would you estimate the contribution from the local field in your observations? I feel somewhat uncomfortable with your placing the Sun in different locations with respect to the spiral arms without considering the local field.

Price: It is not possible to determine very much about the structure of the local magnetic fields by continuum brightness measurements alone. Some information can be obtained from polarization of local emission. However, any small rms component of field on top of a uniform field will preclude detection of large-scale synchrotron beaming and polarization effects. Continuum measurements give the integrated line-of-sight brightness, and we must in many cases resort to simple geometrical models to estimate the location of regions where the emission originates.

Mills: Green in her 408 MHz survey investigated different spiral patterns and found that many fitted her results reasonably well, the original 7° spiral being best. Thus the radio continuum results cannot provide a unique model but can be most effective for deciding between competing models.

Shakeshaft: In any dismissal of radio haloes, it is important to make clear the frequency range under consideration since it may be that, as Ginzburg emphasizes, any halo to our Galaxy may have a steep spectrum and be detectable only at low frequencies. As radio maps of other galaxies (such as M31 and M51) show, interpretation of the continuum features in terms of simple disks and spiral arms is likely to be misleading. One hopes, of course, to relate the nonthermal emission to the distribution of cosmic-ray electrons, which may in turn be related to the distribution of supernova remnants, or perhaps to whether there has been an explosion in the galactic center. Do you have any views on this?

Price: There is some evidence from the low-frequency work done by Hamilton and Ellis in Tasmania that there could be a halo at low frequencies. This should be carefully looked at. Certainly I would urge caution in the direct comparison of our Galaxy to any specific external galaxy. However, radio observations of external galaxies do provide us with information on what types of large-scale radio continuum emission features are possible and can thereby help us in the interpretation of the observed brightness distribution in our own Galaxy. I agree, no simple model will be comprehensive. We have no direct evidence for the source of the energetic electrons causing the synchrotron radiation. All we know for certain is that there is a definite fall-off in radio volume emissivity in going from the galactic center towards the outer parts of the Galaxy.

Westerhout: I distinctly remember that at the Noordwijk Symposium, Ginzburg started his discussion with the statement that there might not be a galactic halo, but that the galactic plane material must extend several kiloparsecs up. I agree with Price that there is at present no evidence for a halo, even at very low frequencies.

Price: The last phrase is not quite in agreement with what I said. I said there is no evidence for a large-scale radio halo – except there might be some evidence for such a halo at low frequencies, tens of megaherz.

Wielebinski: Maps of M31 at 408 MHz (Pooley) and at 2695 MHz (Berkhuijsen and Wielebinski) with similar resolutions of 4.5' are remarkably similar. The assertion that a halo is seen at lower frequencies in M31 needs careful reexamination.

THE BONN 408 MHz SURVEY

C. G. T. HASLAM, W. E. WILSON, D. A. GRAHAM, and G. C. HUNT

Max-Planck-Institut für Radioastronomie, Bonn, F.R.G.

Abstract. The 100 m telescope was used to complete the northern sky survey started at Jodrell Bank. The survey aimed at reaching the confusion limit of the telescope. Zero levels are consistent within ± 3 K, and scale errors are less than 10%. The survey will be published in the form of maps; a machine readable version is also available.

The Bonn 408 MHz survey was designed to fulfil two goals. The first was to provide a simple first experiment for the new 100 m telescope at Efflesberg. The second purpose was to complete the northern sky survey that was started in 1966 at Jodrell Bank (Haslam *et al.*, 1970) and which was designed to provide data for the study of large scale galactic features such as the loops and subsequently to be used for studies of background spectral distribution.

The Bonn experiment was begun in March 1971 with the construction of a correlation receiver which was specifically designed to have a wide dynamical linear response (> 40 dB) and to have continuous gain measurement. This allowed the survey to be made at high speed without having to reduce the sensitivity as the beam crossed the galactic plane and without having to stop to calibrate the gain. The antenna was designed to accept circular polarization to minimize the errors introduced by the ionosphere in the measurements of the background polarization. The experiment aimed to record a survey down to the confusion limit of the 100 m telescope. It is essential for survey work that the telescope beam have a low sidelobe level. In this experiment the beam was 37′ with first side lobe ≈ 25 dB down on the forward response.

As in the Jodrell Bank survey the maps were scanned by nodding scans along the meridian at a speed of 4 deg min^{-1}, the operational upper speed limit of the telescope at the time of the survey. This method has many advantages. The scans intersect many times under repeatable hour angle conditions. It is possible to analyse this data later to work out the receiver baselevel drifts by an iterative process. This method has been used successfully, each hour of right ascension having ~ 2500 intersections and requiring about 10 min of CDC 3300 time to converge to the survey noise level.

Two maps for each region were made using up and down scans separately. These were compared and found to agree to 1.7 K in the colder regions. The final maps were made by combining these two surveys. The overall temperature calibration was made by convolving the survey down to the resolution of the 404 MHz survey of Pauliny-Toth and Shakeshaft (1962). The comparison showed the zero levels to be consistent to ± 3 K and the scale errors to be less than 10%.

The survey is to be published in *Astronomy and Astrophysics Supplement Series* in the form of maps. However, we also have the data in machine readable form and have developed a library of computer processing routines which will enable us to

F. J. Kerr and S. C. Simonson, III (eds.), Galactic Radio Astronomy, 649–650. All Rights Reserved.

make maps of selected areas in any coordinate system and beam resolution less than or equal to the original maps. This library will enable us to make contour diagrams as direct overlays for the Palomar Sky Atlas prints. It also allows us to standardize other surveys at other frequencies and resolutions so that direct detailed spectral comparison can be made. We are currently compiling a library of continuum maps for this purpose. An additional use for the survey is to provide maps of the low order terms for high resolution surveys made by the large cross and synthesis telescopes.

At the present time work is in progress on the construction of a new 408 MHz receiver to be used with the Parkes 64 m telescope to survey the southern sky. The equipment has been extended to allow simultaneous measurements of the total power and linear polarised components of the 408 MHz continuum radiation.

References

Haslam, C. G. T., Quigley, M. J. S., and Salter, C. J.: 1970, *Monthly Notices Roy. Astron. Soc.* **147**, 405.
Pauliny-Toth, I. I. K. and Shakeshaft, J. R.: 1962, *Monthly Notices Roy. Astron. Soc.* **124**, 61.

C. G. T. Haslam
W. E. Wilson
D. A. Graham
G. C. Hunt
Max-Planck-Institut
für Radioastronomie,
Bonn, F.R.G.

AUTHOR INDEX